RECENT TRENDS IN PROCESSING AND DEGRADATION OF ALUMINIUM ALLOYS

Edited by **Zaki Ahmad**

Recent Trends in Processing and Degradation of Aluminium Alloys
Edited by Zaki Ahmad

Published by InTech
Janeza Trdine 9, 51000 Rijeka, Croatia

Publishing Process Manager Petra Nenadic

Technical Editor Teodora Smiljanic

Cover Designer InTech Design Team

Additional hard copies can be obtained from orders@intechopen.com

Recent Trends in Processing and Degradation of Aluminium Alloys, Edited by Zaki Ahmad
p. cm.
ISBN 978-953-307-734-5

For EU product safety concerns:
IN TECH d.o.o., Prolaz Marije Krucifikse Kozulić 3, 51000 Rijeka, Croatia,
info@intechopen.com or visit our website at intechopen.com.

Meet the editor

Professor Dr. Zaki Ahmad worked at King Fahd University of Petroleum and Minerals for thirty years in rendered distinguished services in teaching and research. He obtained his PhD from LEEDS University, UK. He is a chartered metallurgical engineer (C.Eng) from engineering council UK. He is a fellow of the institute of Materials, Minerals and Mining(FIMMM). He is a member European federation of corrosion and a fellow of institute of Metal Finishing. He substantially contributed to the founding activities in material science, corrosion engineering and nanotechnology at KFUPM and in Iran. He worked on international projects on aluminum with Aluminum, Ranshofen, Austria and Forschungzentrum, Geethscht, Germany and with Metallgesselscheft, Germany. He worked on international projects with Ministry of Technology, Germany. He is a founder contributor of center of excellence in corrosion at KFUPM, Dhahran, Saudi Arabia. He worked on the foundation and development of nanotechnology in Saudi Arabia in 2004. He is the author of "Principles of Corrosion Engineering and Corrosion Control" published by Elsevier in 2006. He has written over 95 research papers and international journals and over forty papers in international research conferences. His research activities include development of Al/SC alloys, Nanostructured superhydrophrobic surfaces, Nanocoatings and self-healing techniques. He was nominated for best researcher award in the Middle East by Energy Exchange in 2011. He is a consultant of several research organizations.

Contents

Preface

Aluminum is the second most plentiful element on the earth and it became a competition in 19th century. The huge demand of aluminum is projected to get 70 million tons, over 30 millions being obtained from recycled scrap. The scope of aluminum ranges from household to space vehicles. Dramatic advances in casting,welding,forming and eco friendly methods of production have made aluminum and aluminum alloys highly attractive candidate for automotive and aerospace industry, because of their wide range of attributes such as high strength, resistance to corrosion, low density, high reflectively, high ductility and proven reliability. The developments in the last two decades have been revolutionary and well documented both for wrought, heat treatable and non-heat treatable alloys. New developments in methods of casting, forming, welding, environmental degradation, grain refinement and particle size at nano and micro scale have made big impact on their demand for space and automotive industry.

Despite serious competition from composites, aluminum alloys are still the king in these industries, as exemplified by 777 Airbus. Aluminium is a very versatile metal and can be cast in any form, stamped, forged, machine, brazed and resin bonded.

Substantial evidence has been gathered on formability which affects the structural integrity of the components. The demand on quality and integrity of welding is increasing on military and commercial aircraft. This includes improved toughness, lower weight and increased resistance to corrosion fatigue. The progress made in welding, analysis of different techniques and their impact on micro structural characteristics has been discussed in several chapters in the hinder the section "welding, casting and forming". Ambriz Richardo has discussed this topic whereas Saravana Kumar has provided valued information on tailored blanks and deep drawing behavior of aluminum alloys. Because of high plastic strain, levels, rigid strength requirements and high quality controls are required for forming processes ranging from single to multiple stage because of increasingly dynamic and competitive market demands which includes outstanding toughness and a high resistance to corrosion fatigue. Various casting process such as direct chill casting, rotary die equal channel angular processing, counter gravity casting and centrifugal casting is described by Aremo Bolaji, Watazu Akira and others have shown new dimensions in the field of casting. Dramatic changes in surface treatment have proved highly effective in the life cycle of aluminum alloys as shown by Shoja Razavi Reza, Wierzbinska Malgorzata, Lopez-Cellajas Regulo, Metikas Theodore and Morwka-Nowotnik Grazyna have added new data to new information. Improvements in surface topography and stability are only achieved successfully by combining electrochemistry, microstructures and the role of micro/nano particles new methods of surface modifications are described by Akbarzadeh Abbas, Timelli Giulio and Kornev Vladimir. Corrosion is a serious thread to aluminum alloys and aluminum based metal matrix composites reinforced by silicon carbide. A brief mechanism of corrosion of composited is given by Zaki Ahmad. Environmental damages such as corrosion fatigue in aluminum alloys and their mechanism have been described by Ghazali Mariam and Eichlseder Wilfried. The principle behind the chapters in book was an analysis of the procedures such as casting, forming, welding and environmental degradation which have a strong bearing on the integrity of aerospace structures and automotive. The authors have addressed the problems of grain refinement, micro segragation, casting defects, crack growth, weld defects to show to what levels the aluminum alloys have been technically elevated. The chapters were selected on a rigid criteria of which novelty and new approaches were the main pillars.

I hope the book would be very useful for practicing engineers, technicians, senior students and all those interested in aluminum alloys in particular the technical staff of aerospace, automotive and defense industry. Chapters on casting, welding and others could be used to support their textbook at a graduate level. This book is profusely illustrated to make the concepts clear to the readers.

While editing the book I had the problems of shifting from one country other which prolonged my editing work for which I apologize.

I thank Mr. Mishaal Ahmed (my grandson), Manzar Ahmed, Intesar Ahmed, Shamsujehan, Huma Sabir and Abida Sultana, they provided me with the mental support for the work not the least, the spirit of my beloved dead son Intekhab Ahmed drove me through very hard times while reviewing the chapters.

I am very grateful to KFUPM who gave me the moral support. I am specially indebted to Dr Faleh Al Sulaiman vice rector of technology at KFUPM, Dr Nasir Aqeeli, Mr Faheemuddin and Mr Abdul Aleem of M.E department of KFUPM. I also thanks Dr M. Budair rector of Al-Jouf University for his moral support.

In the end I thank the Al Mighty to give me the moral courage to undertake the responsibility. I am grateful to InTech publisher for giving me the honor of being the chief editor of this book.

I would like to finish this preface by the famous saying of Albert Einstein "Any fool can make things bigger, but it takes a genius to make the things smaller". I hope this small book would prove an asset in pursuit of knowledge on aluminum alloys.

Dr. Zaki Ahmad (Professor Emeritus)
C. Eng, UK, FIMMM, UK
King Fahd University of Petroleum and Minerals, Dhahran
Saudi Arabia

Part 1

Casting and Forming of Aluminium Alloys

1

Aluminium Countergravity Casting – Potentials and Challenges

Bolaji Aremo[1] and Mosobalaje O. Adeoye[2]
[1]*Centre for Energy Research & Development, Obafemi Awolowo University, Ile-Ife,*
[2]*Department of Materials Science and Engineering, Obafemi Awolowo University, Ile-Ife, Nigeria*

1. Introduction

Counter-gravity casting, also called vacuum casting, is a mould filling technique in which low pressure created inside a mould cavity, causes prevailing atmospheric pressure on the melt surface to bring about an upward or counter-gravity movement of the melt into the mould cavity. The process was patented in 1972 by Hitchiner Manufacturing (Lessiter & Kotzin, 2002) and different variants of the process had evolved over the years. Greanias & Mercer (1989) reported a novel valve system that could potentially increase throughput by allowing mould disengagement prior to solidification while Li *et al* (2007) have developed a multifunctional system aimed at aggregating different variations of the technology into a single equipment.

The unique mould filling approach of the countergravity casting technique confers on it a set of unique advantages related to casting economics, defects elimination and attainment of net-shape in cast products. Such desirous attributes has ensured the growing importance of the technology, especially in power and automotive applications. A testament to the rising profile of this casting technique is its adoption in the production of a range of parts such as compressor wheels for turbo-chargers (TurboTech, 2011), automotive exhaust manifolds (Chandley, 1999) and a high-volume production (130,000 units/day) automotive engine Rocker Arm (Lessiter, 2000).

The growing importance of this casting technique in some metal casting sectors notwithstanding, there is scant awareness and interest in many mainstream casting spheres. This chapter thus seeks to present a technology overview of the countergravity casting technique. The shortcomings of conventional processes are highlighted alongside the unique advantages of the countergravity technique. Challenges of the countergravity technique are also presented with discussion of efforts and prospects for their redemption.

2. Description of the countergravity casting process

The basic process steps for the vacuum casting process are presented as follows. In the diagram in figure 1, a preheated investment mould with an integrated down-sprue (fill pipe) is positioned in the moulding flask.

The sprue, with a conical-shaped intersection point with the rest of the mould, pokes through and sits in the conical depression of the lock-nut. The "square" fit of the two, depicted in figure 2, ensures a sealing of the flask interior from the external environment.

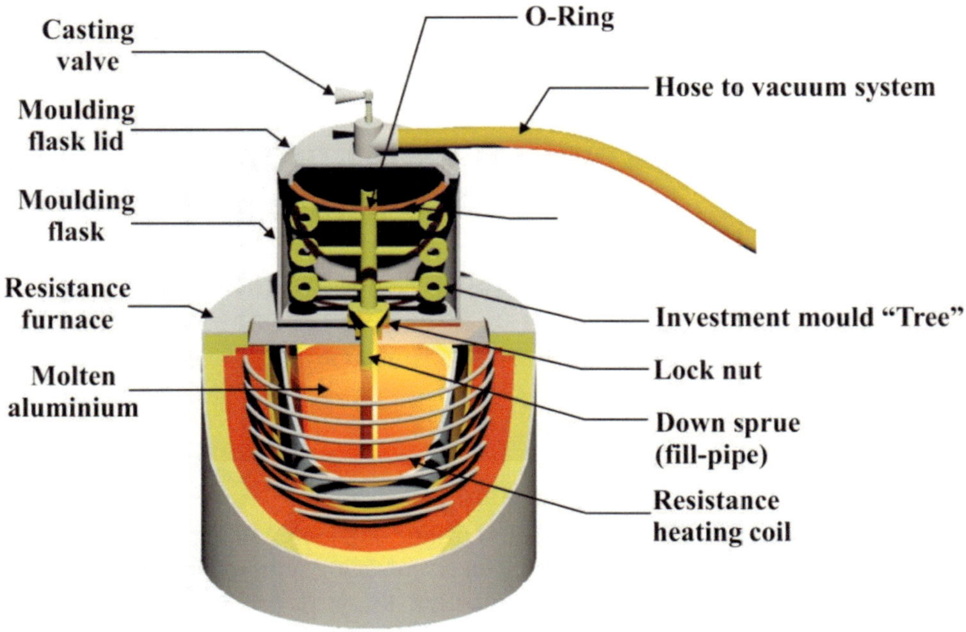

Fig. 1. Typical setup of the countergravity casting process

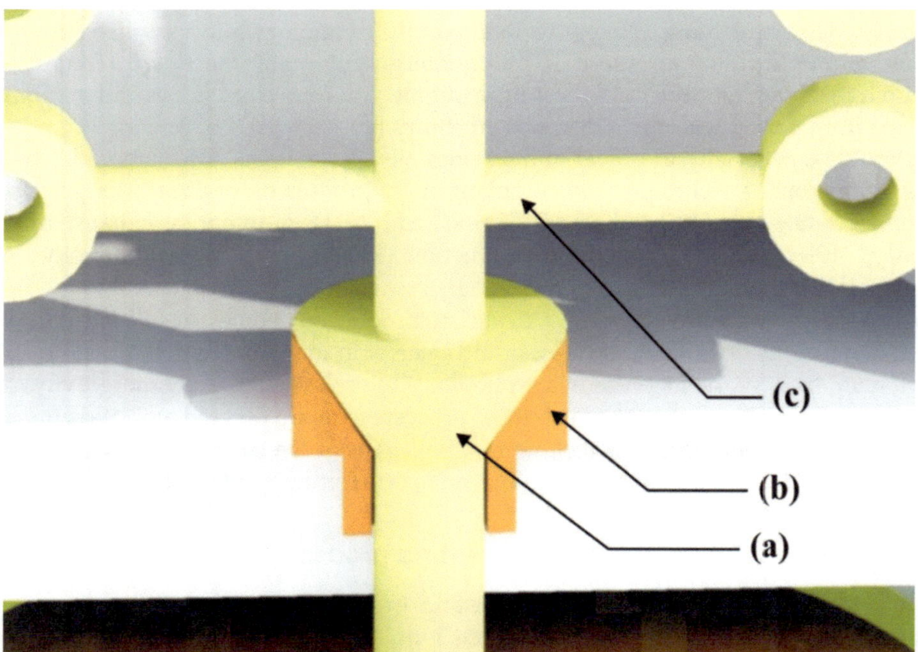

Fig. 2. Down sprue, with conical base (a) is integrated with the rest of the investment mould "tree" (c). The assemble rests inside the conical depression of the lock-nut (b)

The otherwise solid investment mould is made permeable by a single opening at its apex. This opening effectively connects the mould cavity with the interior space of the moulding flask, making it an extension of the moulding flask and enabling its evacuation along with the rest of the flask. The flask lid hosts the casting valve, a connecting hose to the vacuum system and lid locking mechanism. The electrical resistance furnace melts the aluminium charge, usually by a superheat of about 40 °C above the melting temperature (660 °C) of aluminium to reduce melt viscosity and ease melt up-flow into the mould. During countergravity casting, the moulding flask with the mould assembly inside, is placed on the furnace lid with the down-sprue poking through a hole in the furnace lid.
The vacuum system evacuates the moulding flask and the ensuing low pressure thus created causes ambient atmospheric pressure on the melt to push up the molten metal, up inside the mould. See figure 3.

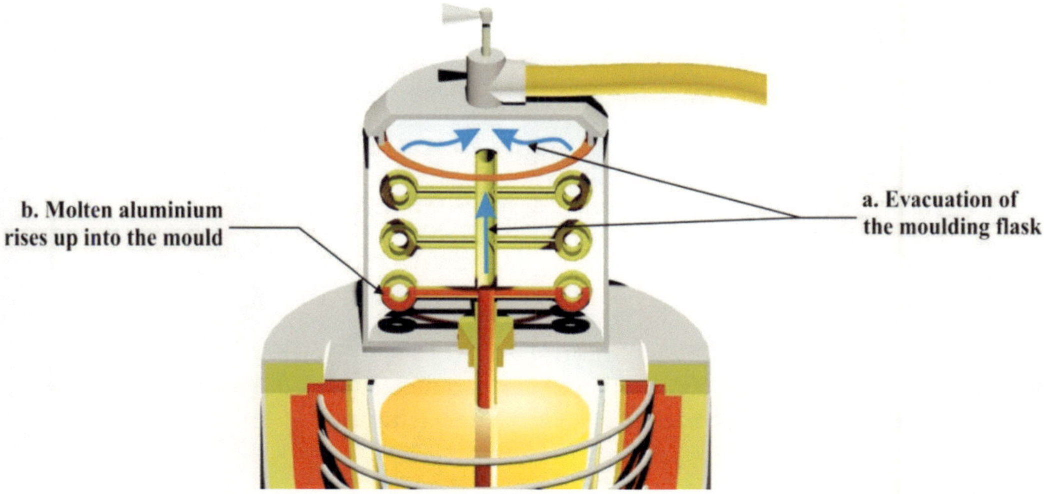

Fig. 3. The evacuation of the moulding flask (a) also evacuates the investment mould cavity. This causes molten aluminium to rise up into the mould cavity (b)

Apart from investment material, the mould could be a metal mould or a ceramic mould. The vacuum system is calibrated so that just the right volume of melt flows inside the mould for a period long enough for the melt to solidify. The vacuum is released after allowing enough time for melt solidification in the mould cavity. This allows un-solidified melt along the sprue length to be flow back into the furnace. The illustration in figure 4 shows the vacuum being maintained until the cavity is completely filled. Vacuum pressure is then released causing un-solidified melt in the sprue to flow back into the furnace

3. Conventional techniques and casting defects

Conventional gravity- or pressure-assisted aluminium metal casting techniques like sand casting, investment casting and die casting are fraught with problems. These include gas defects, melt oxidation, shrinkage defects and pouring defects. Defects are naturally undesirable because they can result in low strength, poor surface finish and high number of rejects in a batch of cast products.

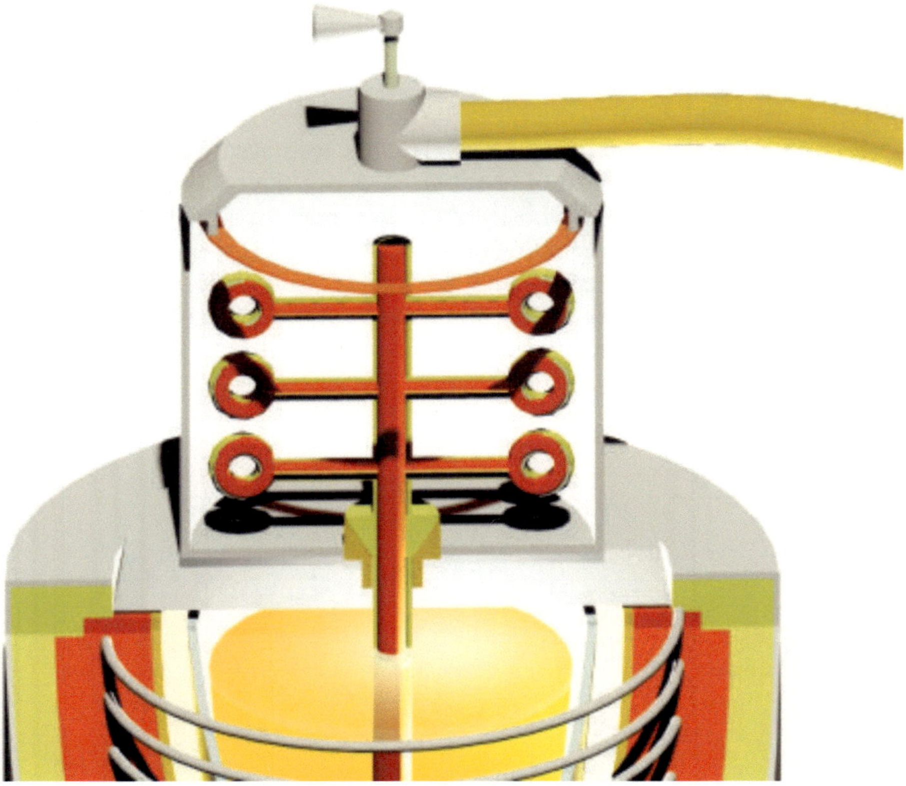

Fig. 4. The vacuum is maintained until the cavity is completely filled. Vacuum pressure is released causing un-solidified melt to flow back into the furnace

3.1 Gas defects

Molten aluminium is particularly susceptible to adsorbing significant quantities of hydrogen gas from atmospheric moisture, which leads to a high concentration of dissolved hydrogen in the melt. This may be further exacerbated by alloying element like magnesium which may form oxidation reaction products that offer reduced resistance to hydrogen diffusion into the melt (Key to Metals, 2010). This causes blow holes and gas porosity which combine to reduce strength of the cast part. The micrograph in figure 5 shows a blow hole defect, it can appear at any region of the cast microstructure and is exacerbated by damp mould materials which give off steam during casting. Figure 6 shows gas porosity defects in an aluminium casting, these are much smaller than blow holes and tend to form in clusters around the region of the grain boundaries.

3.2 Melt oxidation

Oxidation of the melt is another severe defect suffered by aluminium alloy castings. The elevated melt temperature promotes easy oxidation of the aluminium by ambient oxygen. The aluminium oxide thus formed is an undesirable non-metallic inclusion. Considerable efforts, through the use of in-mould filters, protected atmosphere, or alloying additions are often needed to reduce oxide formation and entrainment in the mould.

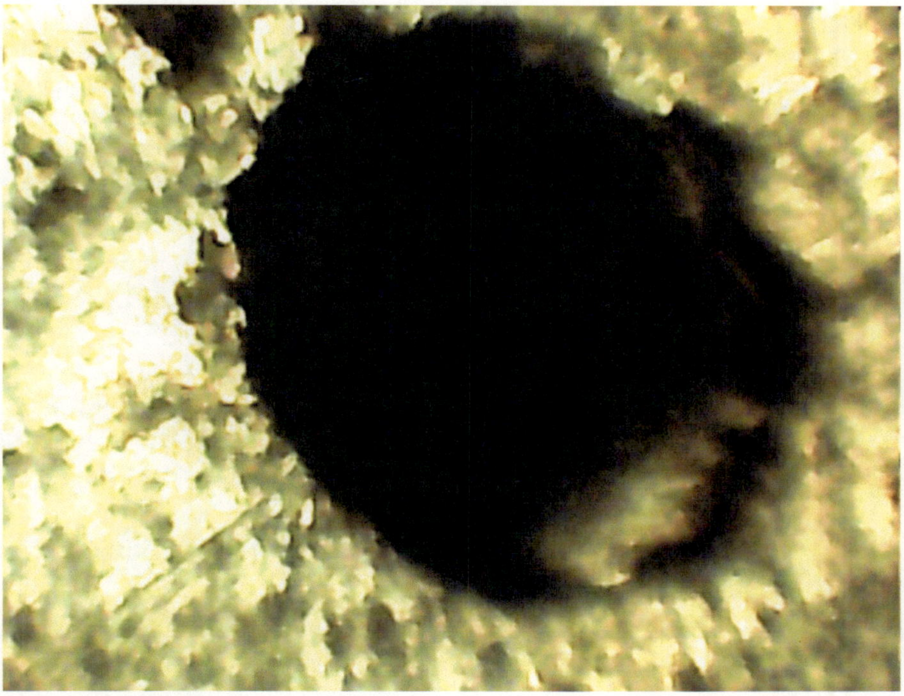

Fig. 5. A Blow hole defect in an aluminum casting at 100× magnification

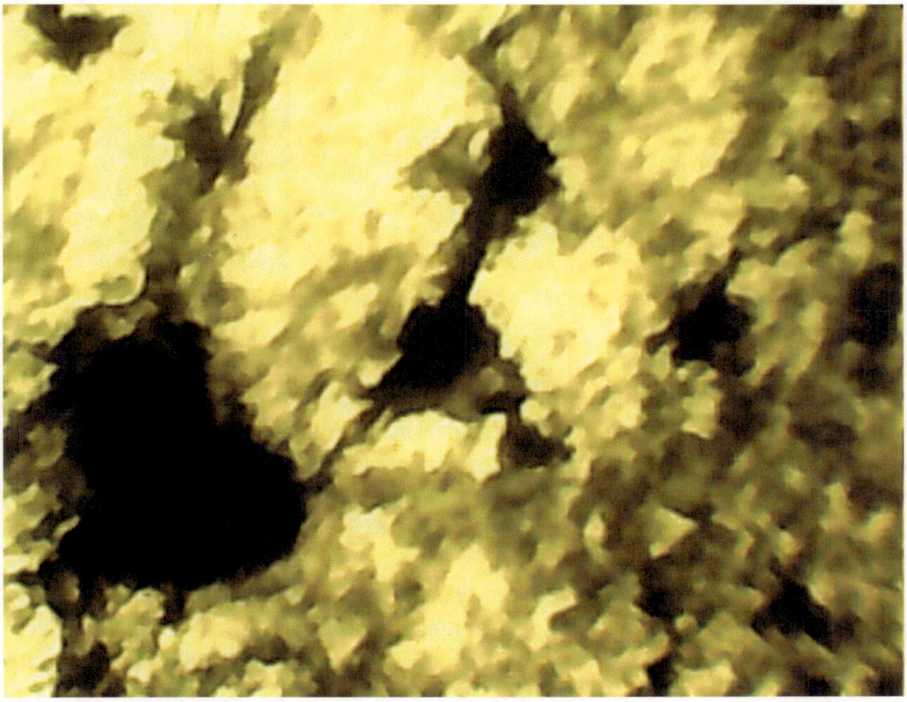

Fig. 6. Gas porosity in aluminium casting at 1000× magnification

3.3 Shrinkage

Shrinkage is the natural consequence of liquid to solid transformation of the melt during cooling and is common in most metals. Shrinkage is particularly severe in aluminium alloys. In aluminium alloys, the volumetric shrinkage ranges from 3.5% to 8% (Kaufman and Rooy, 2004). This manifests as shrinkage cavities in larger portions of the casting.

This is often counteracted by strategic placement of risers. Figure 7 shows the typical appearance of volumetric shrinkage defect in an aluminium section.

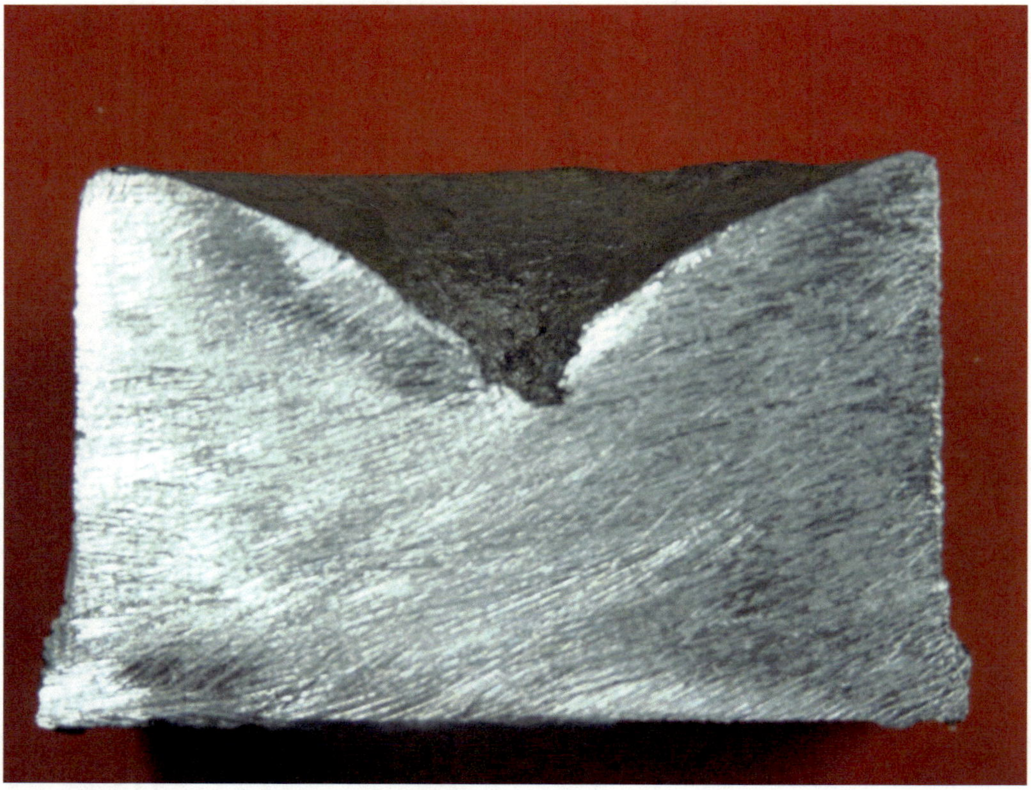

Fig. 7. Typical appearance of volumetric shrinkage defect in an aluminium section

3.4 Pouring defects

During pouring of the melt, there is considerable splashing and sloshing about of the melt. This entrains significant quantities of air and non-metallic inclusions in the mould. Such entrained material degrades casting quality. This problem is often mitigated by incorporation of complex gating systems designed using advanced Computational Fluid Dynamics (CFD) modules. Such casting simulation software is able to predict and avoid bubble streams in metals castings (Waterman, 2010).

Some of the problems outlined above have been resolved by advancements in pressure die casting, improved investment casting techniques and centrifugal casting. These techniques individually solve some, but often not all of the problems with gravity-assisted pour of an air-melt. For instance, in conventional die casting, melt is sprayed at high velocity into the die and cavity-atmosphere tends to be admixed and entrapped in castings during the turbulent cavity-fill (Jorstad, 2003). The process of air melting and pouring also inevitably introduces oxides, formed during melting, into the cast product. Significant inclusions segregation at grain boundaries are thus very common with gravity assisted sand casting.

4. Advantages of the countergravity casting technique

Numerous advantages for metal casters are endemic to the countergravity casting technique. These may be broadly categorized into defect reduction and elimination and casting economics.

4.1 Cleaner melt

For aluminium alloys, metal oxides formed and aggregated on the melt surface can be by-passed by taking clean melt from below the surface. The practice of de-slagging using a hand ladle or metal rod to scoop the slag layer off the melt surface unavoidably leaves pieces of slag in the melt which ultimately flows into the mould during casting. Countergravity casting also results in improved melt cleanliness, due to reduced turbulence during mould filling (Druschitz and Fitzgerald, 2000).

4.2 Elimination of shrinkage defect

Shrinkage is virtually eliminated in the countergravity casting technique. This is because a constant supply of fresh melt is maintained in the mould during casting. Hence, as portions of the mould begin to solidify, the down-sprue is the last to start solidifying. The reservoir of molten melt in the crucible acts as a riser, ensuring a steady supply of melt into the mould during solidification. This effectively eliminates the need for risering. Figure 8 shows the cross-section of a countergravity cast rod. The absence of volumetric shrinkage defect is evident from the convex meniscus at the top of the rod section.

4.3 Simplified gating system

In the countergravity technique, the gating system is considerably simplified as is depicted in figure 9. It consists merely of branches of flow channels emanating from the central sprue. This simplicity is possible because the interior of the mould is actually an extension of the vacuum system. The high pressure differential between the mould interior and the atmospheric pressure ensures that the molten metal will completely permeate every cavity in the mould. Complex in-gates, depending on gravity flow of melt are thus not needed. This considerably simplifies the mould design.

4.4 Economical

Countergravity technique significantly decreases the amount of gates that must be re-melted (Flemings et al, 1997). This was actually one of the original goals of the countergravity technique at its inception. Fettling time and costs are reduced while high quality melt is judiciously used.

5. Potentials and applications of the countergravity casting technique

The countergravity technique has numerous potentials, derivable from its advanatges over the conventional metal casting techniques. As such it is gradually making in-roads into traditional investment casting applications and also in novel materials production.

Fig. 8. Cross-section of a countergravity cast rod showing the absence of volumetric shrinkage defect as evident from the convex meniscus at the top of the rod

Fig. 9. An investment mould "tree", simplified structure is characteristic of the countergravity technique

Fig. 10. Ceramic mould at 400× magnification shows heavy segregation of impurities at the grain boundaries

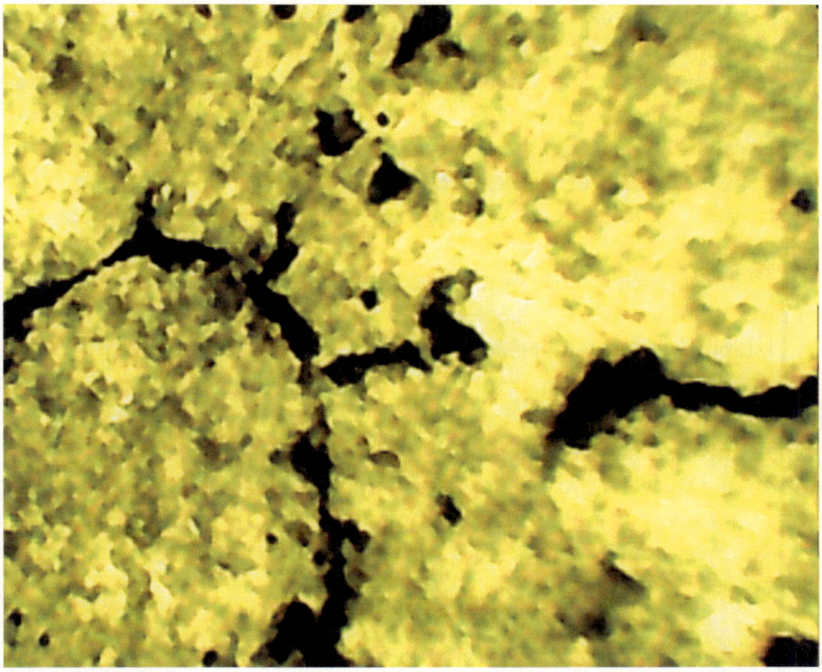

Fig. 11. Countergravity cast specimen at 400× magnification. Significant reduction of impurities at the gain boundaries indicates lesser intake of impurities from the melt

5.1 Scrap reduction and scrap usage

Due to the intrinsic ability of the casting technique to produce cleaner castings, it is more adaptable to the use of scraps and foundry returns. These types of foundry feedstock contain significant admixed impurities like moulding sand and oxide inclusions. Such scrap metals produce significant slag which float on the melt surface.

The process of taking the melt can actually be used to pump clean metal below the melt surface. Figures 10 and 11 respectively show the micrographs of gravity-pour ceramic mould cast samples and countergravity cast samples of scraps of aluminium foundry returns. The microstructure shows more segregation of melt impurities in the gravity-pour ceramic mould, while the vacuum cast specimen shows significant reduction in impurities.

5.2 Net-shape casting

The countergravity technique is well suited for producing net-shape cast products. It is especially suited for thin-walled sections and intricate details due to its excellent mould filling. This is possible due to the virtual elimination of shrinkage defects in the countergravity casting technique. Near net-shape castings of even higher temperature alloys, such as steels are possible. Such has been reported by Chandely *et al* (1997) in the production of thin-walled steel exhaust manifolds.

5.3 Improved strength

Countergravity cast products have improved strength over green sand and ceramic mould specimens. The technique may be thus deployed in the production of high strength parts hitherto produced by forging. High Counter-Pressure Moulding, a proprietary variant, has been reported to exhibit the same strength characteristics as forging in alloy wheel production, at little more than the price of cast wheel (Alexander, 2002). Countergravity techniques are increasingly becoming the preferred choice for the production of alloy wheels because of the added advantage of design flexibility over forging processes. Furthermore, the Cosworth process, which achieves countergravity melt flow by means of an electromagnetic pump, has been successfully used for high strength structural components for air frames, gun cradles, and air tanker re-fuelling manifolds (Bray, 1989).

Griffiths *et al* (2007) observed that countergravity filling method produced higher values of the Weibull modulus than conventional gravity mould filling methods. This is a pointer to the reduced variability of strength achievable in the countergravity technique.

5.4 Economical use of melt

There are often considerable wastages of melt in more conventional casting techniques due to provisions made for risering and complicated in-gates.

This also results in considerable fettling time and costs. Such wastages are virtually eliminated in countergravity casting since there is no need for risers and complex in-gates are not necessary.

5.5 Production of metal matrix composites

Use of the countergravity casting technique is gradually branching into novel materials production. An emerging field of application is the production cast Metal Matrix

Composites (MMC) which can be cast into complex, intricate geometries. These materials have found applications in diverse fields, from high quality reflective mirrors to optical and laser equipment (O'Fallon Casting, 2009). There has been increased interest in the use of cast aluminium/silicon carbide MMC for optoelectronics packaging due to its compatible coefficient of thermal expansion, high thermal conductivity, and potential to produce parts at low cost (Berenberg, 2003). In ring laser gyros, these MMCs are displacing traditional favourites like beryllium and stainless steel in the production of dimensionally stable mirrors that can withstand extreme thermal cycling (Mohn and Vukobratovich, 1988).

6. Challenges and limitations of countergravity casting

The afore-mentioned advantages notwithstanding, the process has some challenges militating against its wide-spread deployment.

6.1 Equipment cost

Spada (1998) reported the cost of countergravity mould and handling equipment to be typically between $50,000 to $1.25 million depending on complexity. Present day prices would naturally be much higher. This is so because the proper utilisation of a countergravity casting equipment requires an ecosystem of support facilities. These include high-temperature mould pre-heating ovens, mould and moulding flask positioning units, and sophisticated vacuum control systems. These added facilities add to the cost of setting up and operation of the technique. In some instances, licensing fees may also apply, further raising up the cost.

6.2 Size restriction of products

Countergravity casting is typically restricted to smaller sized components, usually less than 50 kg. This is because the moulding flask tend to be small, to allow for proper operation of the vacuum system. Larger flasks are more difficult to evacuate and maintain at desired partial vacuum.

6.3 Mould and sprue pre-heat temperature

It is essential for the mould and the sprue to be adequately heated prior to carrying out countergravity casting. The pre-heat prevents chilling of the melt as it flows up from the crucible. Improperly pre-heated sprue and mould will cause increased melt viscosity and a tendency for the melt to get stuck in the sprue or incomplete mould filling. Figure 12 shows premature solidification of melt inside the sprue due to inadequate pre-heat of the mould and sprue assembly.

6.4 Vacuum control

Proper control of vacuum pressure is paramount in countergravity casting. Too much vacuum will result in splatter of melt inside the moulding flask due to over-filling of the mould cavity. Loss of vacuum during casting is also a real problem for countergravity technique. This may be caused by improperly closed lid, damage to or cracks in the moulding flask, or a poor seal between the recess of the lock-nut and the conical connection point of the sprue on the mould. These non-ideal, but very real instances may require a more interactive vacuum system, wherein pressure feedback is used to constantly adjust the flask vacuum pressure.

6.5 Melt contamination by reusable sprue

An effort to bring down overall system costs have led to the use of re-usable sprues. These are usually in the form of metallic pipes. Re-usable sprues must however be used with caution because of the tendency of accumulated impurities in the sprue channel to contaminate the melt.

Fig. 12. Premature solidification of melt inside the sprue due to inadequate pre-heat

7. Benefits of countergravity casting

Some of the advantages highlighted for the countergravity casting technique may be achievable in other, more conventional processes. However, the countergravity technique provides a more complete solution. The process easily lends itself to automation for large scale production; while at the same time can be scaled down for small-scale and jobbing applications.

The possibility of more economical use of the melt is good for the bottom line of foundry operation and was actually the original goal of the countergravity technique. This has motivated a growing list of companies and industrial sector to adopt the technology.

The combination of precision near net shape and strength has resulted in countergravity die casting being used to produce parts formerly made of steel that required a significant amount of secondary machining (Aurora Metals LLC, 2009).

Net shape casting, particularly for thin sections is easily achievable in countergravity casting. Countergravity cast part may have walls as thin as 0.5 mm (National Institute of Industrial Research, 2005).

In order to make the benefits of this casting technique more accessible, low-cost countergravity equipment have been developed. A low-cost design developed by the authors is presented in figure 13.

The design utilizes a simplified vacuum control system and manual positioning of mould and moulding flask. Such low cost alternatives would be invaluable for small scale operations.

Fig. 13. A low-cost machine for countergravity casting

Size restrictions have been tackled by many recent designs. Jie *et al* (2009) reported a system using compressed air to assist the up-flow of melt for large-sized castings. The Check Valve (CV) process is has been developed Hitchiner for larger sized casting. This allows for portions of the melt in the down sprue to be returned to the furnace whilst keeping the portion delimited by the valve in the moulding flask.

Vacuum control in countergravity casting has benefited significantly from advances in control technology and instrumentation. Li et al (2008) demonstrated a pressure control system based on fuzzy-PID control and a digital valve system and achieved pressure error of less than 0.3 KPa. Other workers such as Khader et al (2008) have carried out extensive system modelling of the countergravity casting machine with the goal of developing an automatic controller for control of machine operation.

8. Conclusion

Metal casting is several millennia old, and yet it continues to evolve both in areas of applications and in the technologies of implementation. The increasing relevance of aluminium alloys in modern technology, from power applications to consumer products, makes it imperative to seek better, more cost-effective production routes.

The countergravity casting technique is an ingenious method for production of aluminium parts. The numerous permutations and mutations of this technique over the last four decades is a testament to its feasibility and flexibility; and a recognition of its inherent advantages. Aluminium alloy castings stand to benefit immensely from the unique attributes of the countergravity technique because the goals of net-shape casting and superior mechanical properties are truly achievable via this method.

9. References

Alexander, D. (2002). High-Performance Handling Handbook. MotorBooks International, ISBN 978-0760309483, Osceola, Wisconsin

Aurora Metals. (2009). Vacuum Cast Impellers. March 12th 2011, Available from: < http://www.aurorametals.com/vc.htm>

Berenberg, B. (2003). Metal Matrix Composites Advance Optoelectronics Package Design. In: *High Performance Composites*, 20th May, 2011, Available from: < http://www.compositesworld.com/articles/metal-matrix-composites-advance-optoelectronics-package-design>

Bray, R. (1989). Aluminium casting process finds new applications; developments in the Cosworth Process have spread its application to such fields as the aerospace and defence industries. *Modern Casting*, 20th May 2011, Available from; < http://www.highbeam.com/doc/1G1-8230613.html>

Chandley, G.D.; Redemske, J.A.; Johnson, J.N.; Shah, R.C. & Mikkola, P.H.(1997). Counter-Gravity Casting Process for Making Thinwall Steel Exhaust Manifolds *Society of Automotive Engineers (SAE), Technical Papers,* March 3rd 2011., Available from: < http://papers.sae.org/970920/>

Chandley, G.D. (1999). Use of vacuum for counter-gravity casting of metals. *Materials Research Innovations*, Vol. 1999, No. 3, pp 14-23.

Druschitz, A.P & Fitzgerald, D.C. (2000). Lightweight Iron and Steel Castings for Automotive Applications. *Proceedings of Society of Automotive Engineers (SAE) 2000 World Congress*, ISSN 0148-7191, Detroit Michigan, March, 2000

Flemings, M.; Apelian, D.; Bertram, D.; Hayden, W.; Mikkola, P. & Piwonka, T.S. (1997). *World Technology Evaluation Centre Report on Advanced Casting Technologies in Japan and Europe*. American Foundrymen's Society (AFS), ISBN 1-883712-45-9

Greanias, A.C. & Mercer, J.B. (1989). Vacuum countergravity casting apparatus and method with backflow valve, In: *United States Patent 4862945*, March 3rd 2011, Available from :< http://www.freepatentsonline.com/4862945.html>

Griffiths, W.D.; Cox, M.; Campbell, J & Scholl, G. (2007). Influence of counter gravity mould filling on the reproducibility of mechanical properties of a low alloy steel. *Materials Science and Technology*, Vol. 23, No. 2, pp 137-144, ISSN 0267-0836

Jie, W.Q.; Li, X.L. & Hao, Q.T. (2009). Counter-Gravity Casting Equipment and Technologies for Thin-Walled Al-Alloy Parts in Resin Sand Moulds. *Materials Science Forum*. Vols. 618-619 (2009), pp 585-589

Jorstad, J.L. (2003). High Integrity Die casting Process Variations, *Proceedings of International Conference on Structural Aluminium Casting,* Orlando Florida, November, 2003

Kaufman, G.J. & Rooy, E.L. (2004). *Aluminium Alloy Castings Properties, Processes, and Applications*. ASM International, ISBN 0-87170-803-5, Ohio.

Key to Metals (2010). Aluminium and Aluminium Alloys Casting Problems. March 6th 2011, Available from: < http://www.keytometals.com/Article83.htm>

Khader, A. M.; Abdelrahman, M. A.; Carnal, C.C. & Deabes, W. A. (2008). Modelling and Control of a Counter-Gravity Casting Machine. *Proceedings of 2008 American Control Conference*, Seattle Washington, June 2008

Lessiter, M.J. & Kotzin, E.L. (2002). Timeline of Casting Technology. *Engineered Casting Solutions*, Summer 2002, pp 76-80.

Lessiter, M.J. (2000). Engineered Cast Solutions for the Automotive Industry. *Engineered Casting Solutions*, Fall 2000, pp 37-40.

Li, X.; Hao, Q.; Jie, W. & Zhou, Y. (2008). Development of pressure control system in counter gravity casting for large thin-walled A357 aluminium alloy components. *Transactions of Nonferrous Metals Society of China*, Vol. 18, No. 4, pp 847-851

Li, X.; Hao, Q.; Li, Q. & Jie, W. (2007). Study on Technology of Multifunction Counter Gravity Casting Equipment. *Foundry Technology*, Vol. 07, No. 014,

Mohn, W.R. & Vukobratovich, D. 1988. Recent applications of metal matrix composites in precision instruments and optical systems. *Optical Engineering*, Vol. 27, No. 2, pp 90-98

National Institute of Industrial Research (NIIR) Board of Consultants & Engineers. (2005). The Complete Book on Ferrous, Non-Ferrous Metals with Casting and Forging Technology. National Institute of Industrial Research, ISBN: 8186623949, New Delhi

O'Fallon Casting. (2009). Silicon Carbide Metal Matrix Composite Alloys (SiC MMC), March 12th 2011, Available from: < http://www.ofalloncasting.com/MetalMatrix.html>

Spada, A.T. (1998). Hitchiner Manufacturing Co. – Turning the Casting World Upside Down. *Modern Casting*, Vol. 88, No. 7, pp 39-43

TurboTech Precision Products Ltd. (2011). Aluminium Compressor Wheel Castings for Turbochargers, March 3rd 2011, Available from:
< http://www.turbotech.co.uk/manufacturing-process.htm>

Waterman, P.J. (2010). Understanding Core-Gas Defects Flow-3D software helps trace, predict and avoid bubble streams in metal castings. In: *Desktop Engineering, Design Engineering and Technology Magazine*, March 6th 2011, Available from:
< http://www.deskeng.com/articles/aaaype.htm>

2

Intermetallic Phases Examination in Cast AlSi5Cu1Mg and AlCu4Ni2Mg2 Aluminium Alloys in As-Cast and T6 Condition

Grażyna Mrówka-Nowotnik
Rzeszów University of Technology, Department of Materials Science
Poland

1. Introduction

Cast Al-Si-Cu-Mg and Al-Cu-Ni-Mg alloys have a widespread application, especially in the marine structures, automotive and aircraft industry due to their excellent properties. The main alloying elements – Si, Cu, Mg and Ni, partly dissolve in the primary α-Al matrix, and to some extent present in the form of intermetallic phases. A range of different intermetallic phases may form during solidification, depending upon the overall alloy composition and crystallization condition. Their relative volume fraction, chemical composition and morphology exert significant influence on a technological properties of the alloys (Mrówka-Nowotnik G., at al., 2005; Zajac S., at al., 2002; Warmuzek M., at al. 2003). Therefore the examination of microstructure of aluminium and its alloys is one of the principal means to evaluate the evolution of phases in the materials and final products in order to determine the effect of chemical composition, fabrication, heat treatments and deformation process on the final mechanical properties, and last but not least, to evaluate the effects of new procedures of their fabrication and analyze the cause of failures (Christian, 1995; Hatch, 1984; Karabay et al., 2004). Development of morphological structures that become apparent with the examination of aluminium alloys microstructure arise simultaneously with the freezing, homogenization, preheat, hot or cold reduction, anneling, solution and precipitation heat treatment of the aluminium alloys. Therefore, the identification of intermetallic phases in aluminium alloys is very important part of complex investigation. These phases are the consequence of equilibrium and nonequilibrium reactions occurred during casting af aluminium alloy. It worth to mention that good interpretation of microstructure relies on heaving a complete history of the samples for analysis.

Commercial aluminium alloys contains a number of second-phase particles, some of which are present because of deliberate alloying additions and others arising from common impurity elements and their interactions. Coarse intermetallic particles are formed during solidification - in the interdendric regions, or whilst the alloy is at a relatively high temperature in the solid state, for example, during homogenization, solution treatment or recrystallization (Cabibbo at al., 2003; Gupta at al., 2001; Gustafsson at al., 1998; Griger at al., 1996; Polmear, 1995; Zhen at al., 1998). They usually contain Fe and other alloying elements

and/or impurities. In the aluminium alloys besides the alloying elements, transition metals such as Fe, Mn and Cr are always present. Even small amount of these impurities causes the formation of a new phase component. The exact composition of an alloy and the casting condition will directly influence the amount and type of intermetallic phases (Dobrzański at al., 2007; Warmuzek at al. 2004, Zając at al., 2002). Depending on the composition, a material may contain $CuAl_2$, Mg_2Si, $CuMgAl_2$, and Si as well as Al(Fe,M)Si particles, where M denotes such elements as Mn, V, Cr, Mo, W or Cu. During homogenization or annealing, most of the as-cast soluble particles from the major alloying additions such as Mg, Si and Cu dissolve in the matrix and they form intermediate-sized 0.1 to 1 μm dispersoids of the AlCuMgSi type. Dispersoids can also result from the precipitation of Mn-, Cr-, or Zr-containing phases. A size and distribution of these various dispersoids depend on the time and temperature of the homogenization and/or annealing processes. Fine intermetallic particles (<l μm) form during artificial aging of alloys and they are more uniformly distributed than constituent particles or dispersoids. Dimensions, shape and distribution of these particles may have also important influence on the ductility of the alloys. Therefore, a systematic research is necessary regarding their formation, structure and composition. For example, the coarse particles can have a significant influence on a recrystallization process, fracture, surface and corrosion, while the dispersoids control grain size and provide stability to the metallurgical structure. Dispersoids can also have a large affect on the fracture performance and may limit strain localization during deformation. The formation of particles drains solute from the matrix and, consequently, changes the mechanical properties of the material. This is particularly relevant to the heat-treatable alloys, where depletion in Cu, Mg, and Si can significantly change the metastable precipitation processes and age hardenability of the material (Garcia-Hinojosa at al., 2003; Gupta at al., 2001; Sato at al., 1985). Therefore, the particle characterization is essential not only for choosing the best processing routes, but also for designing the optimized alloy composition (Mrówka-Nowotnik at al., 2007; Wierzbińska at al., 208, Zajac at al., 2002; Zhen at al., 1998).
The main objective of this study was to analyze a morphology and composition of the complex microstructure of intermetallic phases in AlSi5Cu1Mg and AlCu4Ni2Mg2 aluminium alloys in as-cast and T6 condition and recommend accordingly, the best experimental techniques for analysis of the intermetallic phases occurring in the aluminium alloys.

2. Material and methodology

The investigation was carried out on the AlCu4Ni2Mg2 and AlSi5Cu1Mg casting aluminium alloys. The chemical composition of the alloys is indicated in Table 1.

Alloy	Cu	Mg	Si	Fe	Ni	Zn	Ti
AlSi5Cu1Mg	1.3	0.5	5.2	0.2	-	<0.3	0.18
AlCu4Ni2Mg2	4.3	1.5	0.1	0.1	2.1	0.3	-

Table 1. Chemical composition of investigated AlCu4Ni2Mg2 and AlSi5Cu1Mg aluminium alloys, Al bal (wt%)

Microstructure analysis was carried out on the as-cast and in T6 condition aluminium alloys. The alloys were subjected to T6 heat treatment: solution heat treated at 520°C for 5 h followed by water cooling and aging at 250°C for 5 h followed by air cooling. The

microstructure of examined alloy was observed using an optical microscope on the polished sections etched in Keller solution (0.5 % HF in 50ml H_2O). The observation of specimens morphology was performed on a scanning electron microscope (SEM), operating at 6-10 kV in a conventional back-scattered electron mode and a transmission electron microscopes (TEM) operated at 120, 180 and 200kV. The thin foils were prepared by the electrochemical polishing in: 260 ml CH_3OH + 35 ml glycerol + 5 ml $HClO_4$. The chemical composition of the intermetallics was made by energy dispersive spectroscopy (EDS) attached to the SEM.
The intermetallic particles from investigated AlCu4Ni2Mg2 and AlSi5Cu1Mg alloys in T6 condition were extracted chemically in phenol. The samples in the form of disc were cut out from the rods of ∅12 mm diameter. Then ~0.8 mm thick discs were prepared by two-sided grinding to a final thickness of approximately 0.35 mm. The isolation of phases was performed according to following procedure: 1.625 g of the sample to be dissolved was placed in a 300 ml flask containing 120 mm of boiling phenol (182°C). The process continued until the complete dissolution of the sample occurred ~10 min. The phenolic solution containing the residue was treated with 100 ml benzyl alcohol and cooled to the room temperature. The residue was separated by centrifuging a couple of times in benzyle alcohol and then twice more in the methanol. The dried residue was refined in the mortar. After sieving of residue ~0.2 g isolate was obtained. The intermetallic particles from the powder extract were identified by using X-ray diffraction analysis. The X-ray diffraction analysis of the powder was performed using a diffractometer - Cu Kα radiation at 40kV.
DSC measurements were performed using a calorimeter with a sample weight of approximately 80-90 mg. Temperature scans were made from room temperature ~25°C to 800°C with constants heating rates of 5°C in a dynamic argon atmosphere. The heat effects associated with the transformation (dissolution/precipitation) reactions were obtained by subtracting a super purity Al baseline run and recorded.

3. Results and discussion

DSC curves obtained by heating (Fig. 1a) and cooling (Fig. 1b) as-cast specimens of the examined AlSi5Cu1Mg alloy are shown in Fig. 1. DSC curves demonstrate precisely each reactions during heating and solidification process of as-cast AlSi5Cu1Mg alloy. One can see from the figures that during cooling the reactions occurred at lower temperatures (Fig. 1b) compared to the values recorded during heating of the same alloy (Fig. 1a). Solidification process of this alloy is quite complex (Fig. 1) and starts from formation of aluminum reach (α-Al) dendrites. Additional alloying elements such as: Mg, Cu, as well as impurities: Mn, Fe, leads to more complex solidification reaction. Therefore, as-cast microstructure of AlSi5Cu1Mg alloy presents a mixture of intermetallic phases (Fig. 2). The solidification reactions (the exact value of temperature) obtained during DSC investigation were compared with the literature data (Bäckerud at al., 1992; Li, et al., 2004) and presented in Table 2. Results obtained in this work very well corresponding to the (Bäckerud at al., 1992; Li, et al., 2004; Dobrzański at al., 2007).
Fig. 2 shows as-cast microstructure of AlSi5Cu1Mg alloy. The analyzed microstructure contains of primary aluminium dendrites and substantial amount of different intermetallic phases constituents varied in shape, (i.e.: needle, plate-like, block or "Chinese script"), size and distribution. They are located at the grain boundaries of α-Al and form dendritic network structure (Fig. 2).

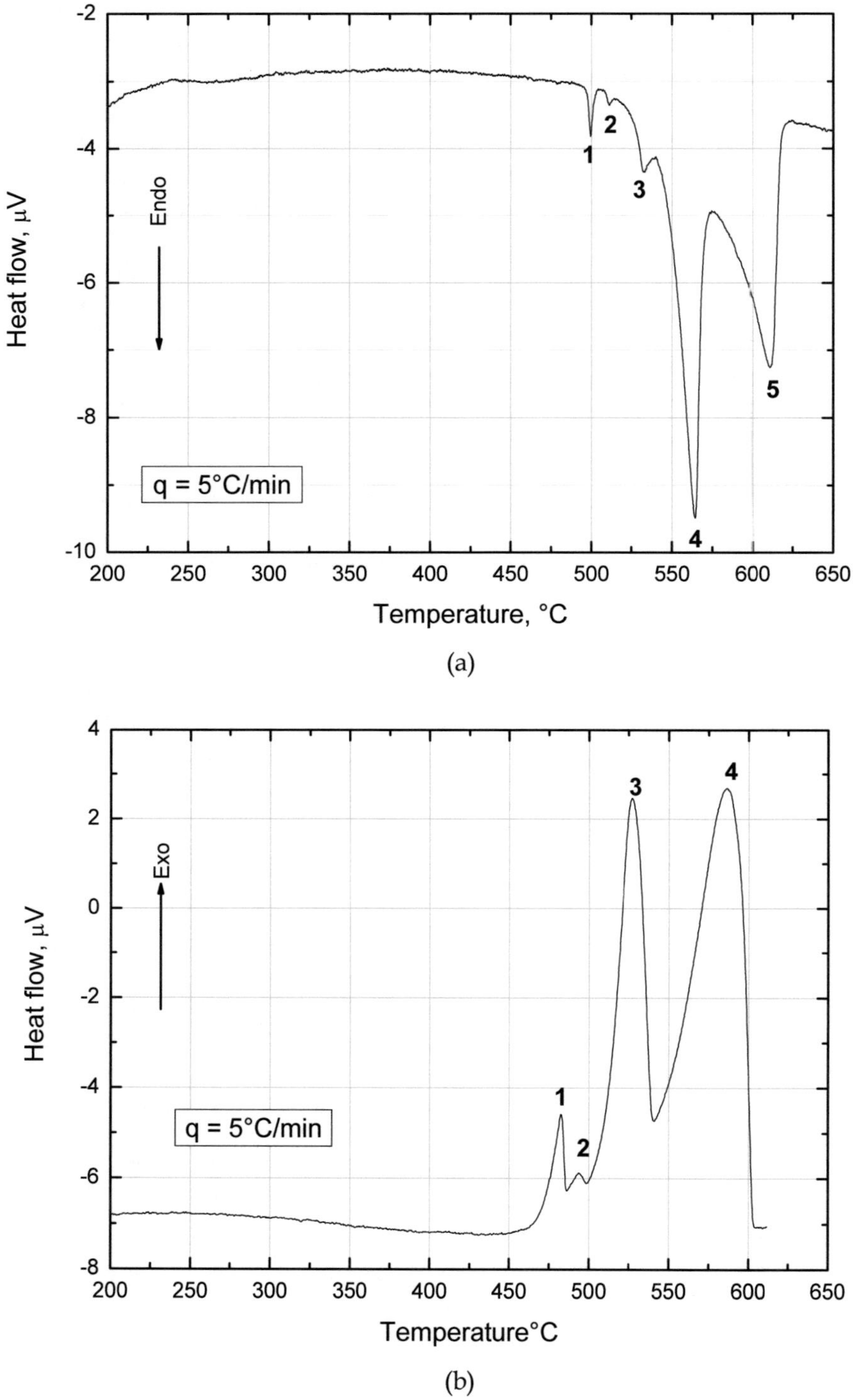

Fig. 1. DSC thermograms of as-cast specimens of AlSi5Cu1Mg alloy, obtained during a) heating and b) cooling at rate of 5°C/min

Bäckerud et al.	**Temp., °C**	**Li, Samuel et al.**	**Temp., °C**	**This work**
L→ (Al) dendrite network	609	(Al) dendrite network	610	610
L→(Al)+$Al_{15}Mn_3Si_2$+(Al_5FeSi)	590			
L→(Al)+Si+Al_5FeSi	575	Precipitation of eutectic Si	562	564
L→(Al)+Si+AlMnFeSi	558	Precipitation of $Al_6Mg_3FeSi_6$+Mg_2Si	554	532
L→(Al)+Al_2Cu+Al_5FeSi	525	Precipitation of Al_2Cu	510	510
L→(Al)+Al_2Cu+Si+ $Al_5Mg_8Cu_2Si_6$	507	Precipitation of $Al_5Mg_8Cu_2Si_6$	490	499

Table 2. Reactions occurring during the solidification of the AlSi5Cu1Mg alloy according to (Bäckerud at al., 1992; Li, Samuel et al., 2004)

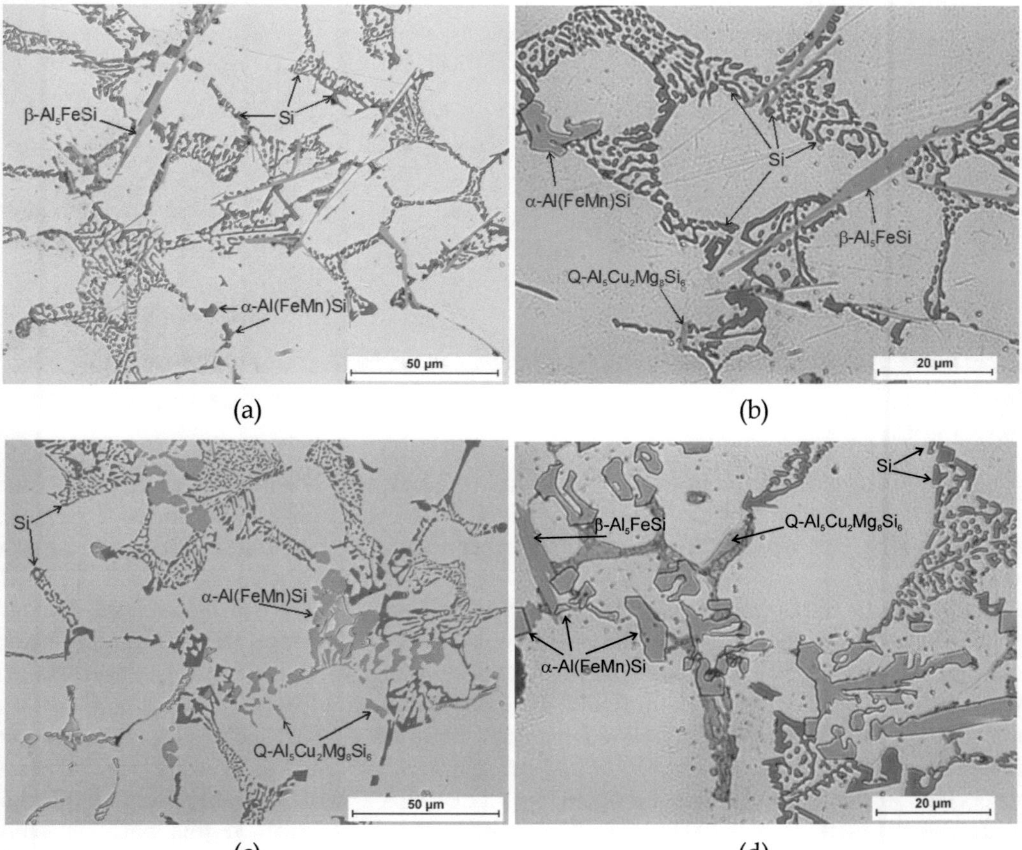

Fig. 2. Morphology of AlSi5Cu1Mg alloy in the as-cast state: (a,c) unetched and (b,d) etched

In order to identify the intermetallic phases in the examined alloy, series of elemental maps were performed for the elements line Al-K, Mg-K, Fe-K, Si-K, Cu-K and Mn-K (Fig. 3 and 4). The maximum pixel spectrum clearly shows the presence of Al, Mg, Fe, Si, Cu and Mn in the scanned microstructure. In order to identify the presence of the elements in the observed phases, characteristic regions of the mapped phase with high Mg, Fe, Si, Cu and Mn concentration were marked and their spectra evaluated (Fig. 5).

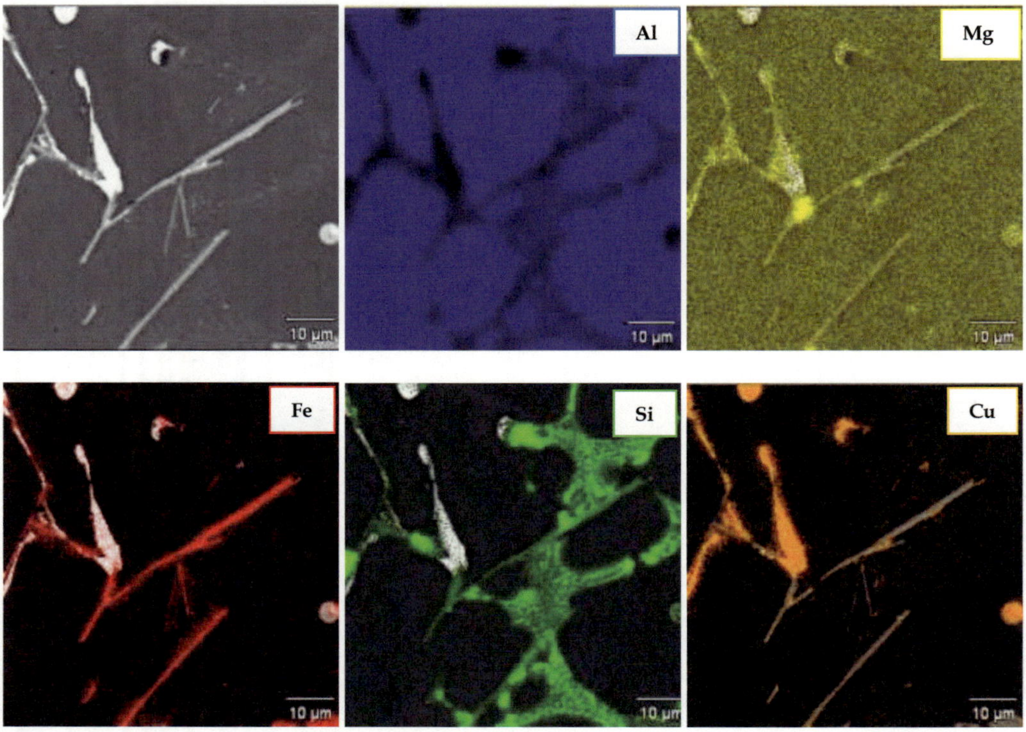

Fig. 3. SEM image of the AlSi5Cu1Mg alloy and corresponding elemental maps of: Al, Mg, Fe, Si and Cu

Fig. 5 shows the SEM micrographs with corresponding EDS-spectra of intermetallics observed in the as-cast AlSi5Cu1Mg alloy. The EDS analysis indicate that the oval particles are Al_2Cu (Fig. 5a). Besides Al_2Cu phase, another Cu containing phase $Al_5Mg_8Cu_2Si_6$ was observed (Fig. 4,5). In addition the Cu-containing intermetallics nucleating as dark grey rod, primary eutectic Si particles with "Chinese script" morphology were also observed. Fe has a very low solid solubility in Al alloy (maximum 0.05% at equilibrium) (Mondolfo, 1976), and most of Fe in aluminium alloys form a wide variety of Fe-containing intermetallics depending on the alloy composition and its solidification conditions (Ji et al., 2008). In the investigated as-cast AlSi5Cu1Mg alloy Fe-containing intermetallics such as light grey needle like β-Al_5FeSi (Fig. 5a) and blockly phase consisting of Al, Si, Mn and Fe (Fig. 5a) were observed. On the basic of literature date (Liu Y.L. et al., 1999; Mrówka-Nowotnik et al., a,b, 2007; Wierzbińska et al., 2008) and EDS results (Fig. 5 and Tab. 3) this particles were identified as α-Al(FeMn)Si phase.
Fig. 5 shows SEM micrographs with corresponding EDS-spectra of intermetallics observed in as-cast AlSi5Cu1Mg alloy. The EDS spectra indicate that the oval particles are Al_2Cu (Fig. 5a). Besides Al_2Cu phase, another Cu containing phase AlCuMgSi is observed (Fig 5b). The results of EDS analysis are summarized in Tab. 3 versus the results obtained by earlier investigators.

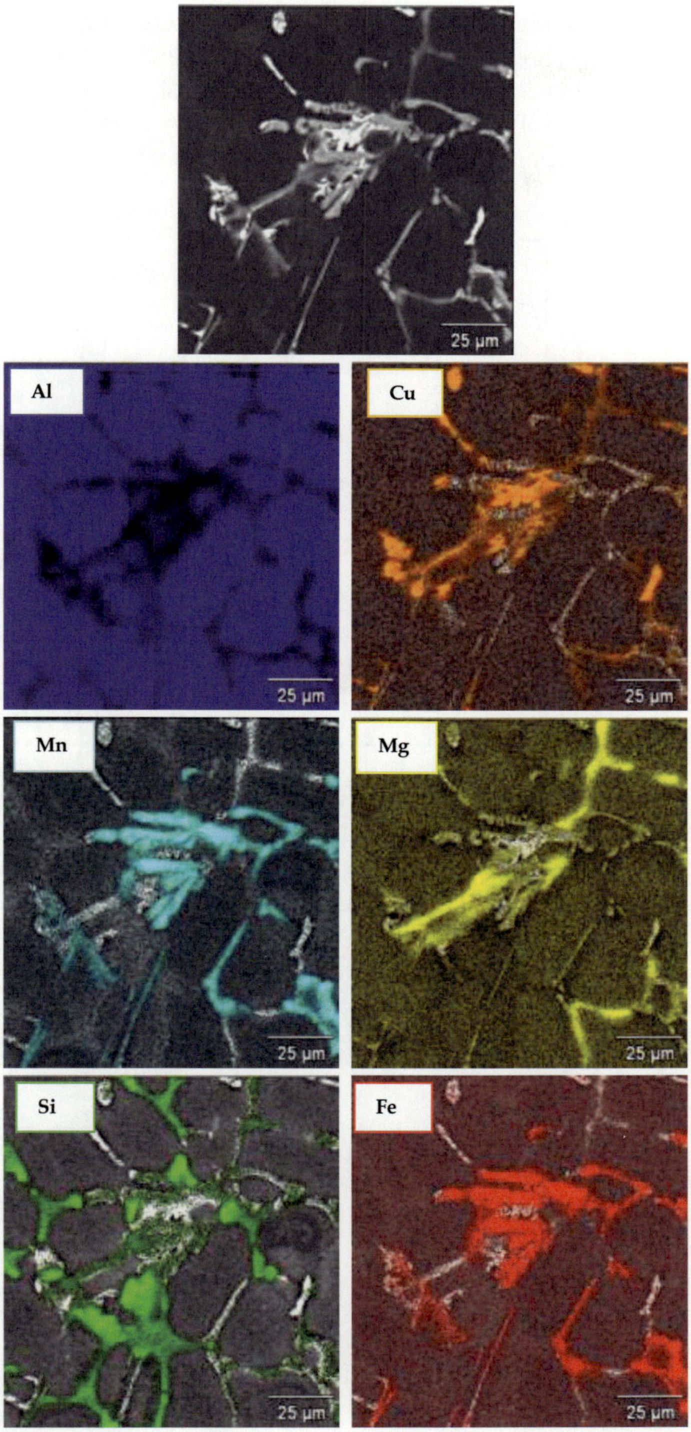

Fig. 4. SEM image of the AlSi5Cu1Mg alloy and corresponding elemental maps of: Al, Mn, Mg, Fe, Si and Cu

The following phases were identified in the as-cast AlSi5Cu1Mg alloy based on DSC results and microstructure - LM and SEM observations (Tab. 2 and 3, Fig. 1-5): Si, β-Al_5FeSi, $Al_5Cu_2Mg_8Si_6$, Al_2Cu, α-Al(FeMn)Si. These results can suggest, that in this alloys occur five solidification reactions (Tab. 4). The data presented in Tab. 4 shows that the solidification sequence of AlSi5Cu1Mg alloy differ only slightly from this obtained by Backerud and Li (Tab. 2).

No. of analyzed particles	Suggested type of phases	Chemical composition of determined intermetallic phases, (% wt)					References
		Si	Cu	Mg	Fe	Mn	
20	$Al_5Cu_2Mg_8Si_6$	19.2 15.2 17.97	31.1 26.9 27.48	33 29.22 28.49			Ji, 2008 Lodgaard, (2000) This work
25	β-Al_5FeSi	12-15 12.2 14.59 13-16			25-30 25 27.75 23-26		Mondolfo, 1976 Warmuzek, 2005 Liu, 1999 This work
12	α-$Al_{12}(FeMn)_3Si$	10-12 5.5-6.5 5-7 8-12			10-15 5.1-28 10-13 11-13	15-20 14-24 19-23 14-20	Mondolfo, 1976 Warmuzek, 2006 Liu, 1999 This work
10	Al_2Cu		52.5 49.51				Belov, 2005 This work
25	Si	85-95					This work

Table 3. The chemical composition of the intermetallic phases in AlSi5Cu1Mg alloy in the as-cast state

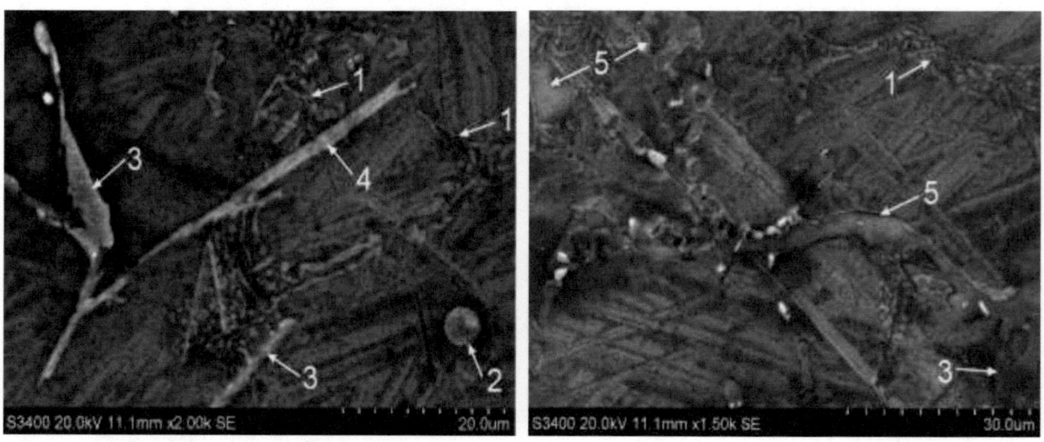

Fig. 5. a) SEM micrographs of the AlSi5Cu1Mg alloy in the as-cast state

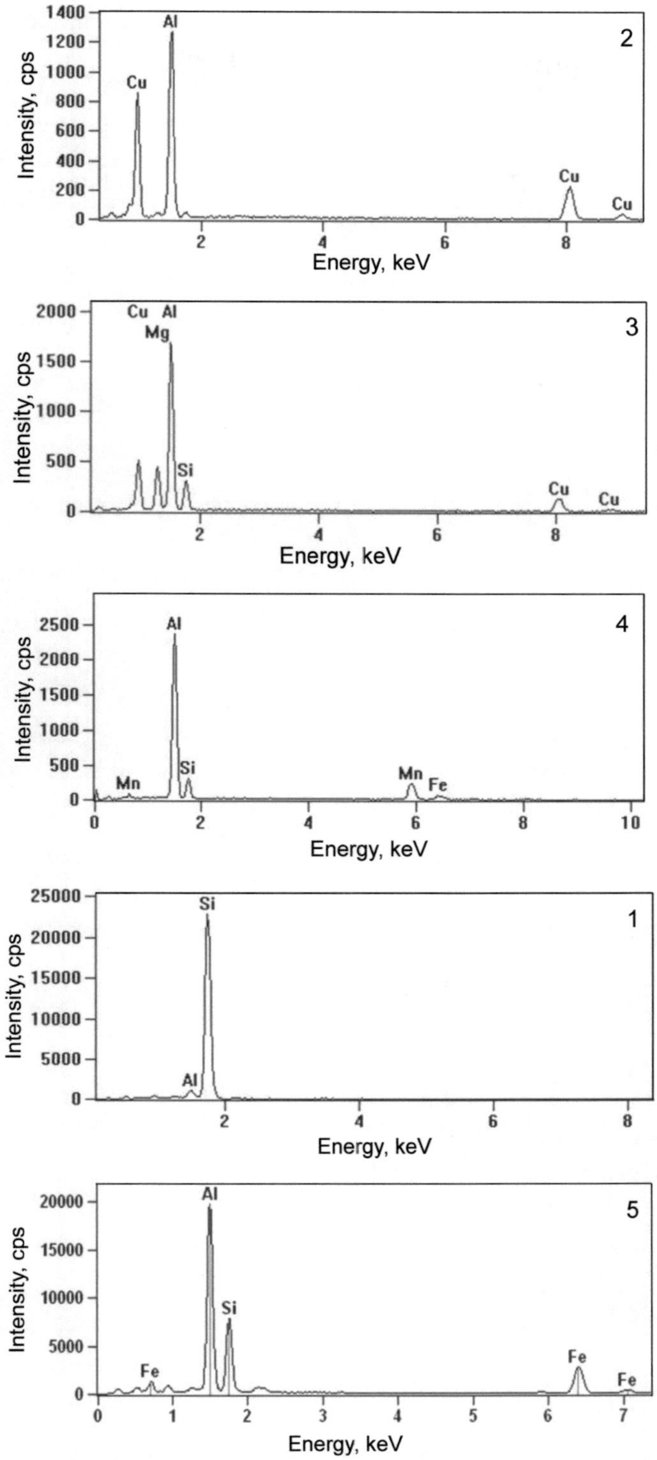

Fig. 5. b) The corresponding EDS-spectra acquired in positions indicated by the number 1-5

Reactions	Temperature, °C
L→ (Al) dendrite network	610
L→(Al)+Si+Al_5FeSi	564
L→(Al)+Si+AlMnFeSi	532
L→(Al)+ Al_2Cu+ Al_5FeSi	510
L→(Al)+ Al_2Cu+Si+$Al_5Cu_2Mg_8Si_6$	499

Table 4. Solidification reactions during nonequilibrium conditions in the investigated AlSi5Cu1Mg alloy, heating rate was 5°C/min

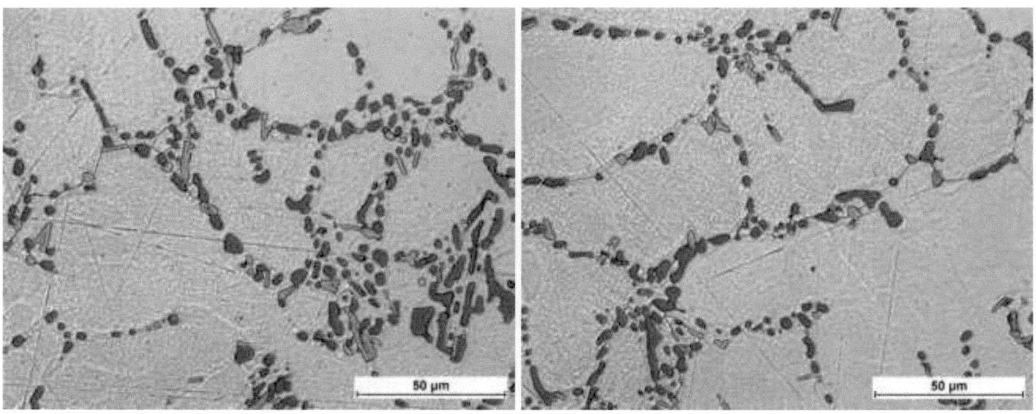

(a) (b)

Fig. 6. The microstructure of AlSi5Cu1Mg alloy in the T6 condition (a,b)

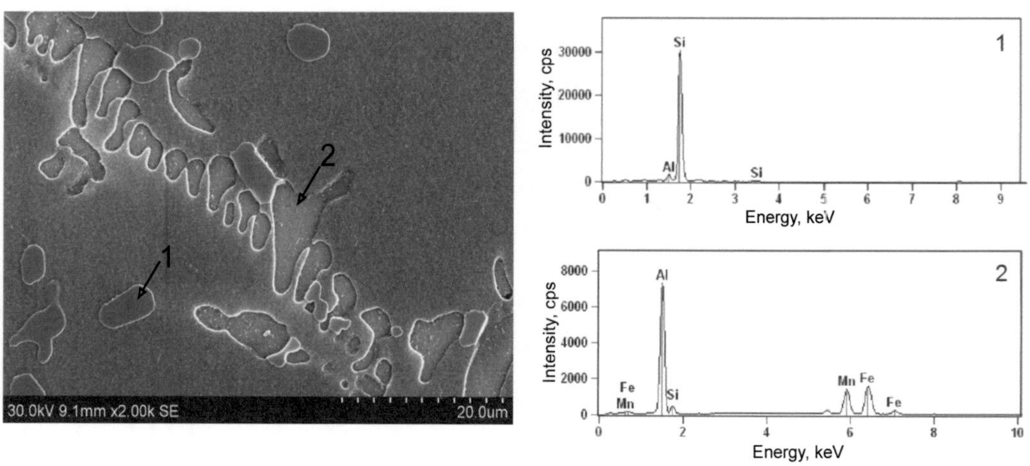

(a) (b)

Fig. 7. a) SEM micrographs of the AlSi5Cu1Mg alloy in the T6 condition; b) The corresponding EDS-spectra acquired in the positions indicated by the number 1 and 2

Microstructure of AlSi5Cu1Mg alloy in T6 condition is presented in Fig. 6. Analyzing the micrographs of the alloy after heat treatment at 520°C for 5h it had been found that during solution heat treatment the morphology of primary eutectic Si changes from relatively large needle like structure to the more refined "Chinese script" and spherical in shape particles. Most of the needle like particles of β-Al_5FeSi phase transform into spherical-like α-Al(FeMn)Si (Kuijpers at al, 2002; Liu at al., 1999; Christian, 1995) as shown in Figure 6 and 7. It has been found that Al_2Cu and $Al_5Cu_2Mg_8Si_6$ phases dissolve in the α-Al matrix during solution heat treatment. The subsequent aging heat treatment at 250°C for 5 leads to formation form the supersaturated solid solution fine intermetallic strengthening particles of Al_2Cu (<1μm).
Fig. 7 shows scanning electron micrographs and EDS analysis of particles in the investigated AlSi5Cu1Mg alloy in T6 condition. The EDS analysis performed on the phases presented in microstructure of the alloy revealed, that spherical in shape inclusions are the eutectic

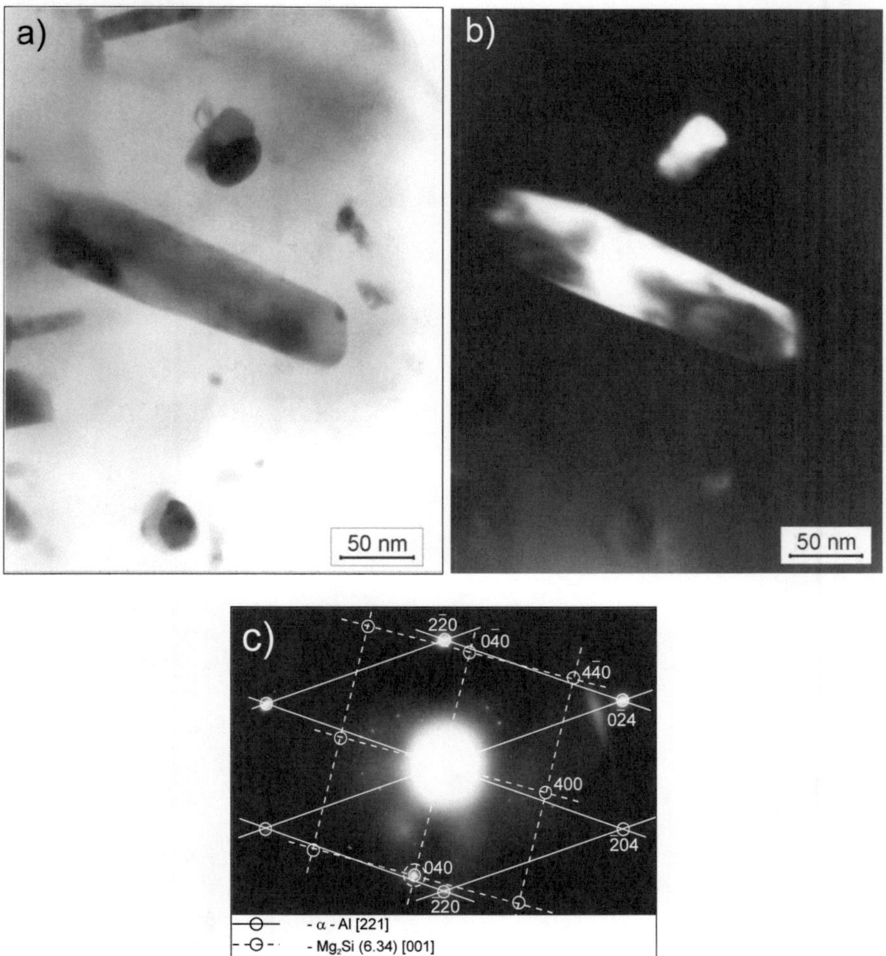

Fig. 8. TEM micrograph of AlSi5Cu1Mg alloy in T6 conditions showing the precipitate of the β-Mg_2Si phase (a,b), and corresponding electron diffraction pattern (c)

silicon ones, whereas the rod-like and "Chinese script" shaped, are inclusions of the phase consisting of Al, Si, Mn and Fe (Fig. 2,7 and Tab. 3).

Since it is rather difficult to produce detailed identification of intermetallic using only one method (e.g. microscopic examination) therefore XRD and TEM techniques was utilized to provide confidence in the results of phase classification based on metallographic study. The microstructure of the examined alloy AlSi5Cu1Mg in T6 state consists of the primary precipitates of intermetallic phases combined with the highly dispersed particles of hardening phases. The TEM micrographs and the selected area electron diffraction patterns analysis proved that the dispersed precipitates shown in Figure 8 and 9 were the precipitates of hardening phase β-Mg_2Si (Fig. 8) and θ'-Al_2Cu (Fig. 9).

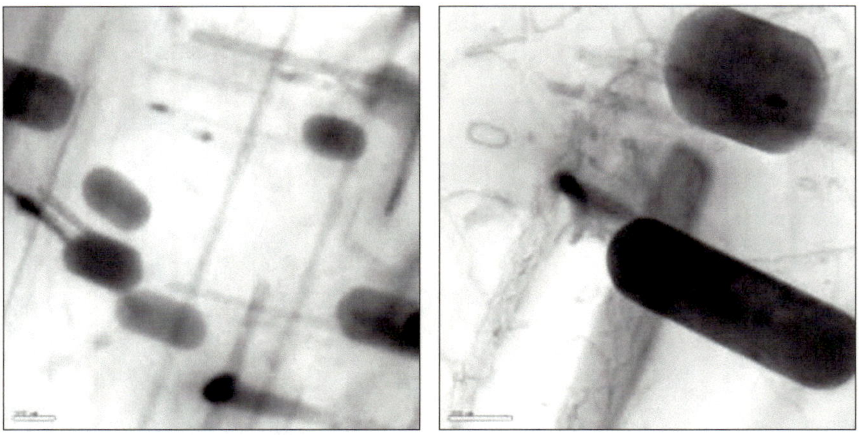

(a) (b)

Fig. 9. Precipitation of strengthening β-Mg_2Si i θ'-Al_2Cu phases in AlSi5Cu1Mg (a,b) – TEM

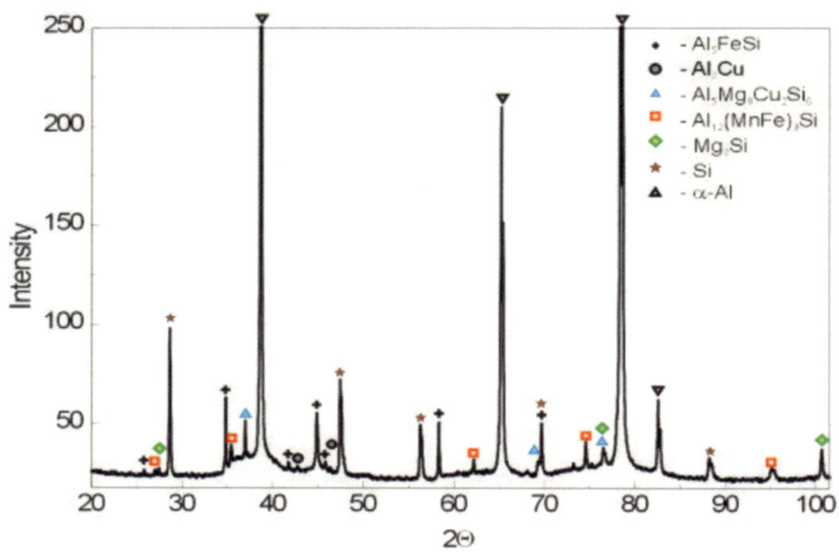

Fig. 10. X-diffraction pattern of AlSi5Cu1Mg alloy

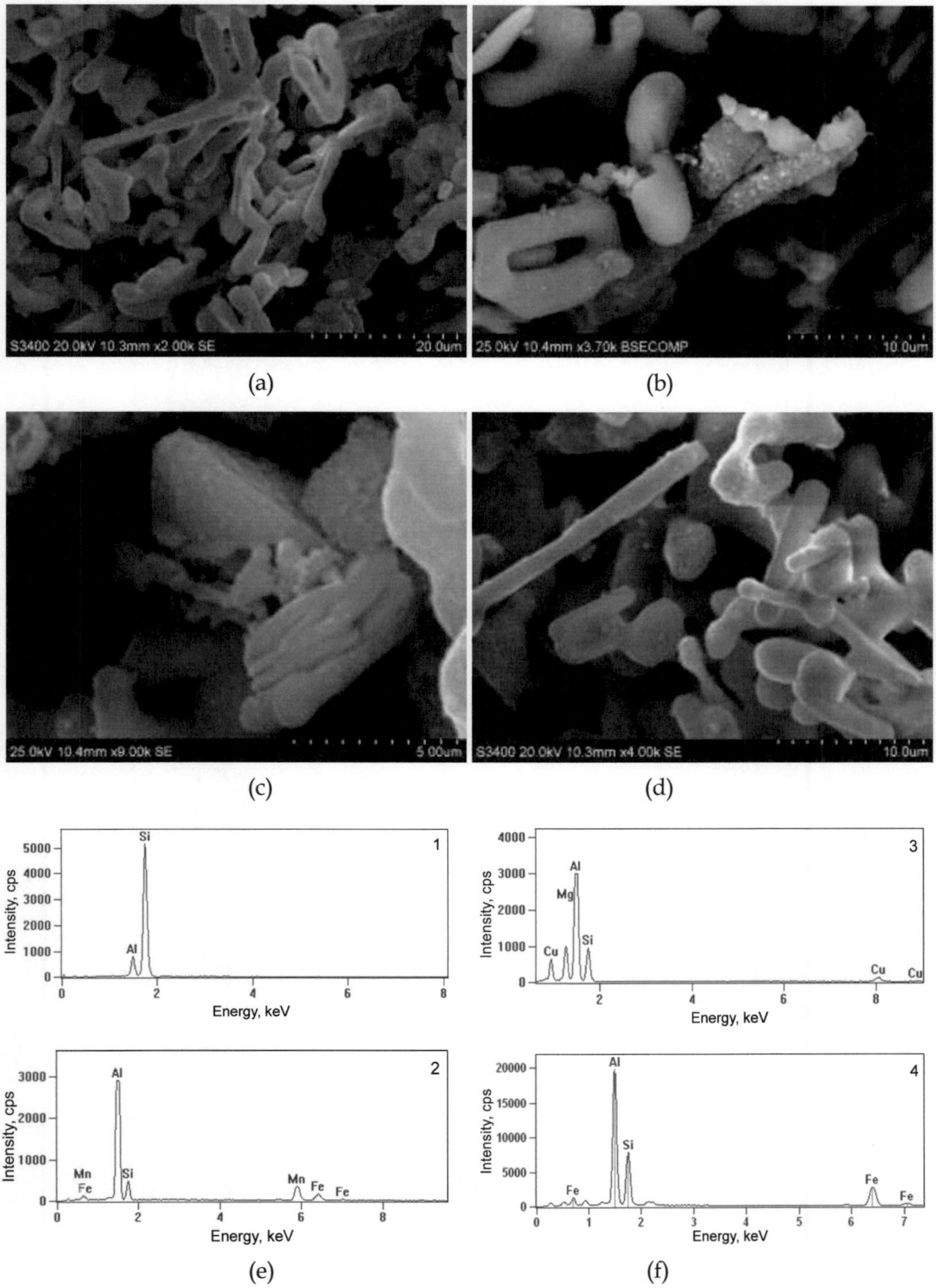

Fig. 11. SEM micrographs (a-d) of the particles extracted from the AlSi5Cu1Mg alloy and EDS spectra (e,f)

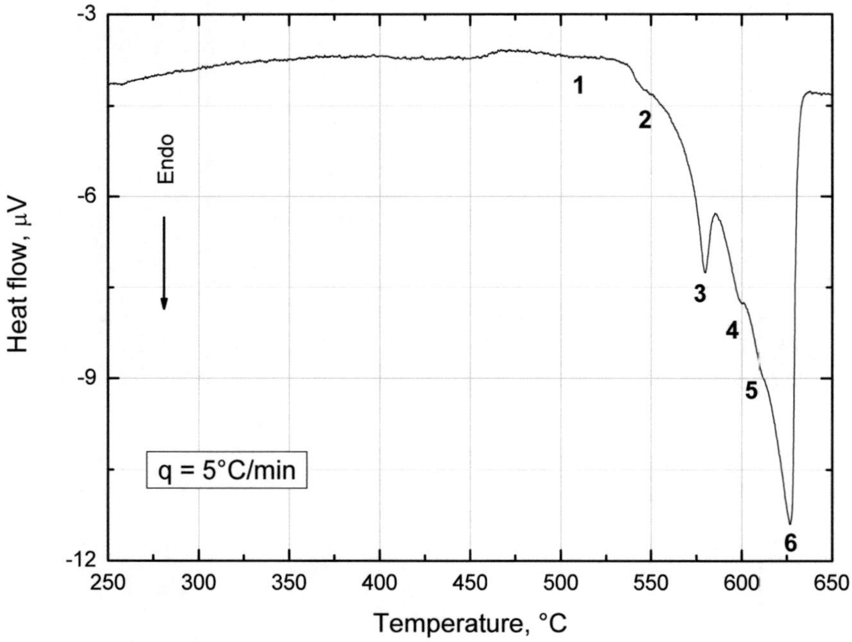

(a)

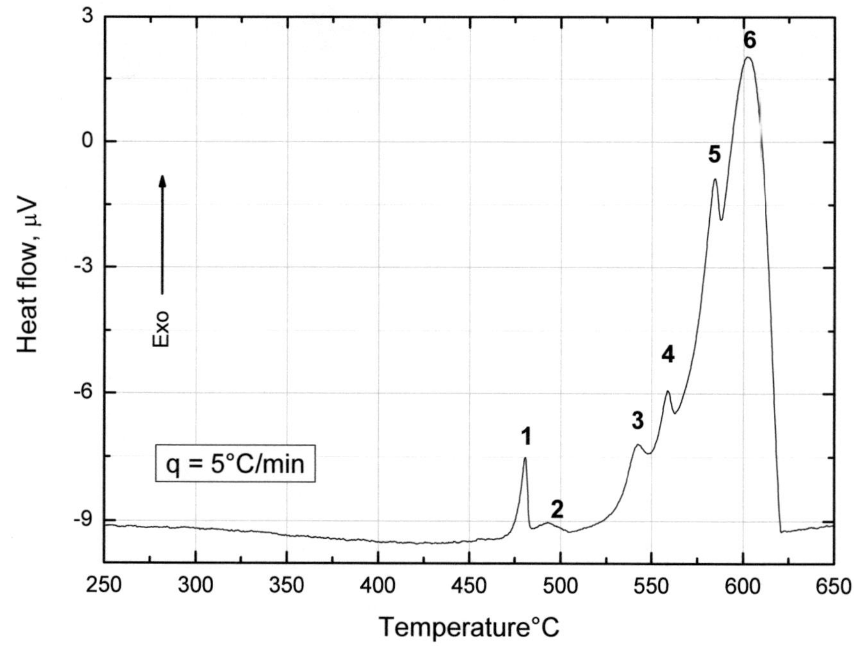

(b)

Fig. 12. DSC thermograms of as-cast specimens of AlCu4Ni2Mg2 alloy, obtained during a) heating and b) cooling at a rate of 5°C/min

The results of XRD investigation are shown in Fig. 8. X-ray diffraction analysis of AlSi1MgMn alloy confirmed metalograffic observation. Additionaly the presented above results were compared to the analysis of the particles extracted from the AlSi5Cu1Mg alloy using phenolic dissolution technique (Fig. 11). The EDS spectra revealed the presence of Al, Mg, Mn, Si, Fe and Cu - bearing particles in the extracted powder (Fig. 11). The EDS analysis results proof that analyzed particles extracted from the AlSi5Cu1Mg alloy were: Si, AlMnFeSi, Al_5FeSi, $Al_5Mg_8Cu_2Si_6$ phases.

DSC curves obtained by heating (Fig. 12a) and cooling (Fig. 12b) of as-cast specimens of $AlCu_4Ni_2Mg_2$ alloy are shown in Fig. 12. DSC curves demonstrate reactions which occurred during heating and solidification process of the alloy. The obtained results were similar to the peaks observed during cooling of the samples of AlSi5Cu1Mg alloy – the recorded peaks were shifted to the lower values (Fig. 12b).

The solidification sequence of this alloy can be quite complex and dependent upon the cooling rate (Fig. 12). Possible reactions which occurred during solidification of AlCu4Ni2Mg2 alloy are presented in Tab. 5. Aluminum reach (α-Al) dendrites are formed at the beginning of solidification process. Additional alloying elements into the alloys (Ni, Cu, Mg) as well as impurities (eg. Fe) change the solidification path and reaction products. Therefore, as-cast microstructure of the tested alloy exhibit the appearance of mixture of intermetallic phases (Fig. 13a). The solidification reactions (the exact value of temperature) obtained during DSC investigation presented in Tab. 5.

Possible reactions	Temperature, °C
$L \rightarrow (Al) + Al_6Fe$	612
$L \rightarrow (Al) + Al_4CuMg$	584
$L \rightarrow (Al)+Al_2Cu+Al_2CuMg$	558
$L \rightarrow (Al)+Al_2Cu+Al_7Cu_4Ni$	542
$L \rightarrow (Al)+Al_2Cu+ Al_2CuMg +Al_3(CuFeNi)_2$	493
Solidus	480

Table 5. Possible solidification reactions during nonequilibrium conditions in investigated AlCu4Ni2Mg2 alloy, at a heating rate 5°C/min

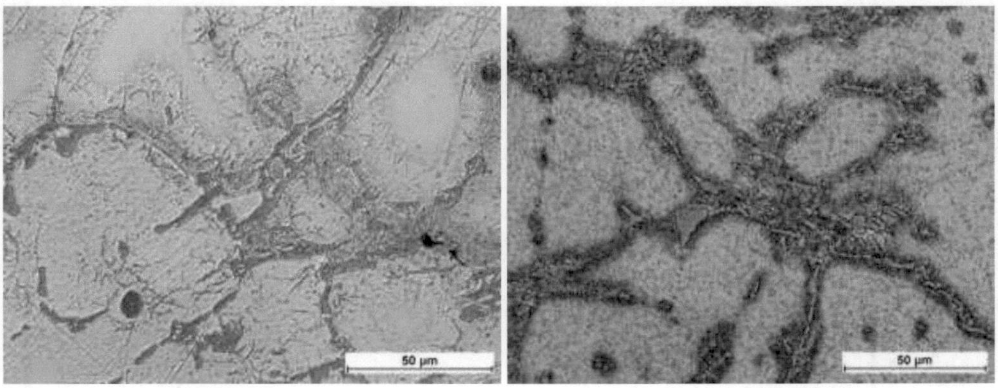

(a) (b)

Fig. 13. The microstructure of AlCu4Ni2Mg2 alloy in as-cast state (a) and the T6 condition (b)

The analyzed microstructure in as- cast state (Fig. 13a) contains of primary aluminium dendrites and substantial amount of different intermetallic phases constituents varied in shape, size and distribution. They are located at the grain boundaries of α-Al and form dendrites network structure (Fig. 13a).

The analyzed microstructure of investigated AlCu4Ni2Mg2 alloy in T6 condition (Fig. 13b) consists different precipitates varied in shape, i.e.: fine sphere-like, complex rod-like and ellipse-like distributed within interdendritic areas of the α-Al alloy. Large number of fine sphere-like strengthening phase are located in the boundary zone. However, small volume of this phase is also present homogenously throughout the sample (Fig. 13b). In order to identify the intermetallic phases in the examined alloy, series of distribution maps were performed for the elements line Mg-K, Al-K, Fe-K, Ni-K, Cu-K (Fig. 14). The maximum pixel spectrum clearly shows the presence of Ni and Cu in the scanned microstructure. In order to identify the presence of the elements in the observed phases, two regions of the mapped phase with high nickel and copper concentration were marked and their spectra evaluated.

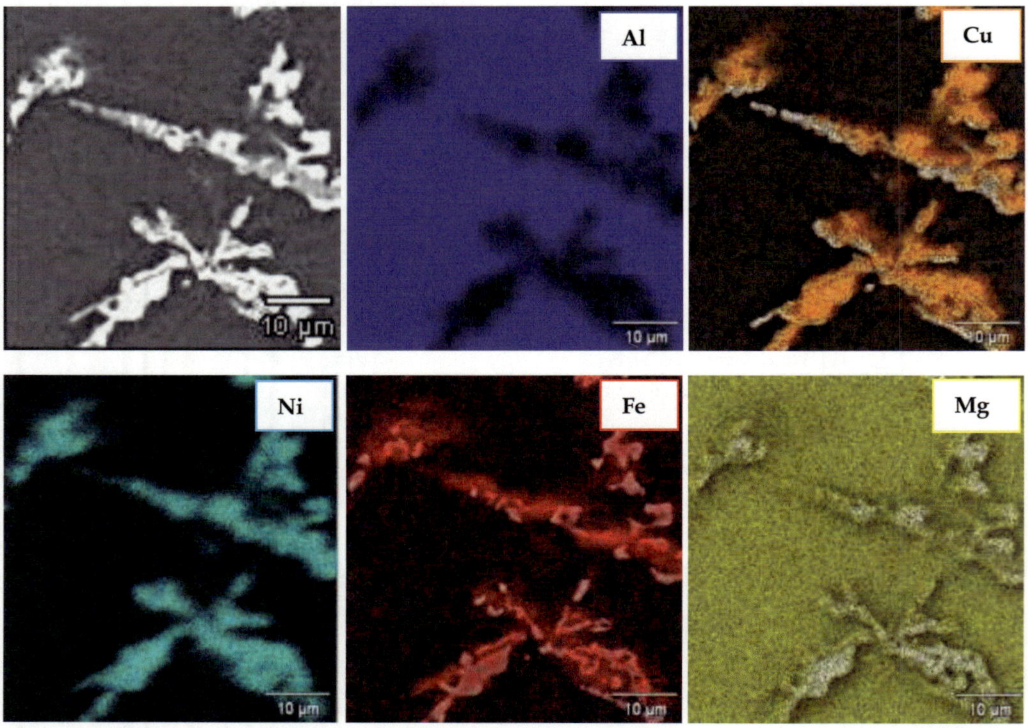

Fig. 14. SEM image of the AlCu4Ni2Mg2 alloy and corresponding elemental maps of: Al, Mg, Fe, Ni and Cu

As seen in the elemental maps in Fig. 14, the regions enriched in Ni and Cu correspond to the formation of type precipitates (complex rod-like) and ellipse-like precipitates observed in Fig. 13. Fig. 15 shows the scanning electron micrographs and EDS analysis of particles in the AlCu4Ni2Mg2 alloy.

The EDS analysis performed on the phases present in microstructure of the alloy revealed, that complex rod-like phase is the Al_7Cu_4Ni one, whereas the ellipse-like is $Al_3(CuFeNi)_2$ (Fig. 15 and Tab. 6)

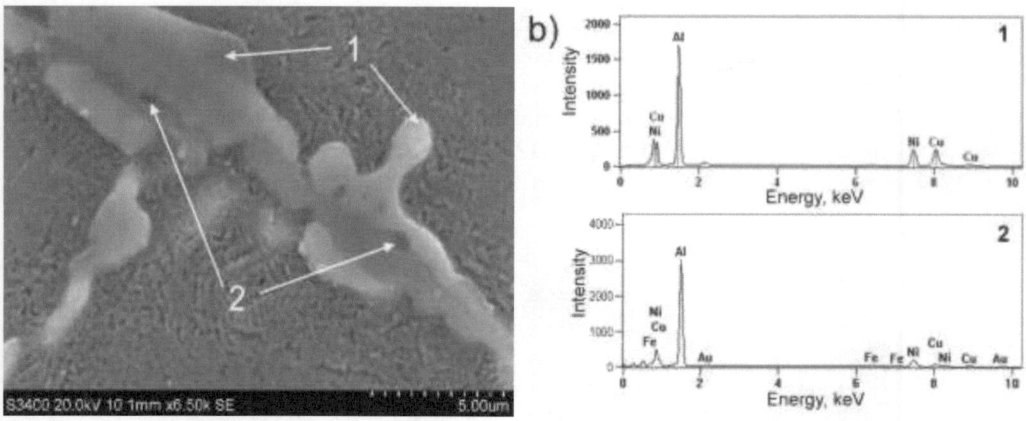

Fig. 15. a) SEM micrographs of the AlCu4Ni2Mg2 alloy in the T6 condition; b) The corresponding EDS-spectra acquired in positions indicated by the number 1 and 2

No. of analyzed particles	Suggested type of phases	Chemical composition of determined intermetallic phases, (%at)			Reference
		Ni	Cu	Fe	
20	Al_7Cu_4Ni	11.8÷22.2 18.08 14.2÷22.6	38.7÷50.7 34.33 29.7÷45.2		Belov, 2005 Chen, 2010 This work
25	$Al_3(CuFeNi)_2$	18÷22 17.1÷20.5	9÷15 10.5÷19.3	8÷10 7.2÷9.5	Belov, 2002 This work
12	Al_2Cu		52.5 47.7÷51.9		Belov, 2005 This work

Table 6. The chemical composition and volume fraction of the intermetallic phases in the AlCu4Ni2Mg2 alloy

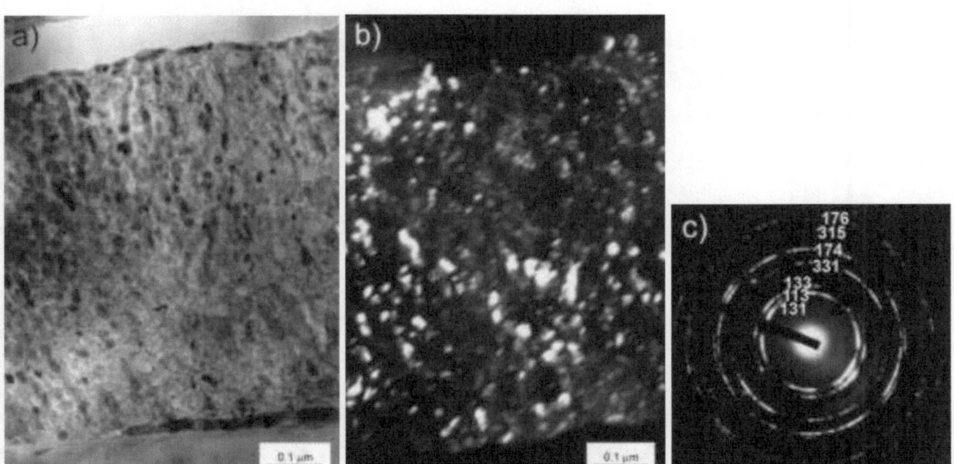

Fig. 16. TEM micrograph of AlCu4Ni2Mg2 alloy in T6 conditions showing the precipitate of the S-Al_2CuMg phase (a,b), and corresponding electron diffraction pattern (c)

The microstructure of the examined alloy AlCu4Ni2Mg2 in T6 state consists of the primary precipitates of intermetallic phases combined with the highly dispersed particles of hardening phases. The TEM micrographs and the selected area electron diffraction patterns analysis proved that the dispersed precipitates shown in Fig. 13b are the intermetallic phases S-Al_2CuMg (Fig. 16) and Al_6Fe (Fig. 17) besides the precipitates of hardening phase θ′-Al_2Cu were present in AlCu4Ni2Mg2 alloy (Fig. 18). The approximate size of the S phase was 0,5 μm.

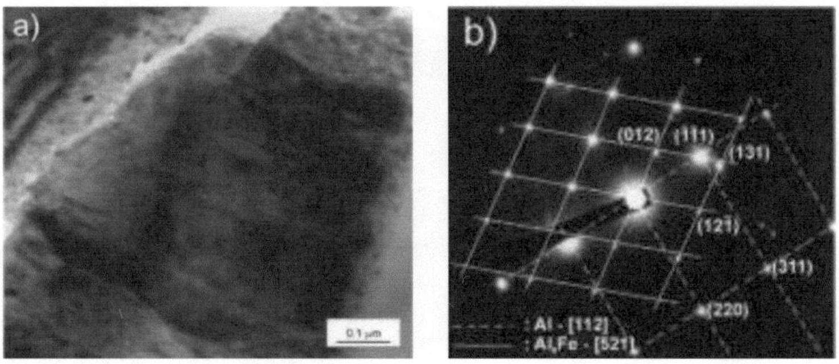

Fig. 17. TEM micrograph of AlCu4Ni2Mg alloy in T6 condition showing the precipitate of the Al_6Fe phase (a), and corresponding electron diffraction pattern (b)

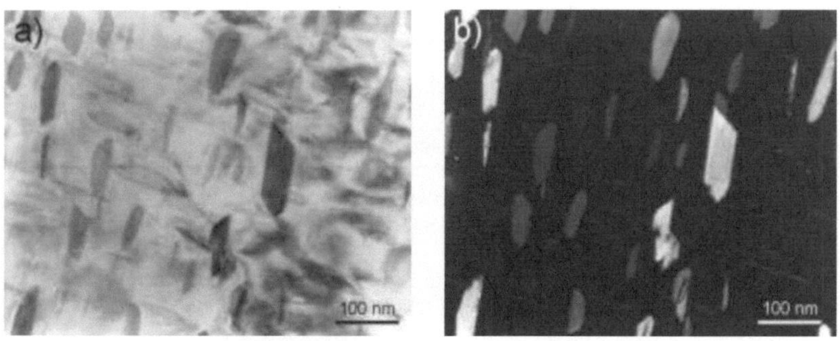

Fig. 18. TEM micrograph of AlCu4Ni2Mg alloy in T6 condition showing the precipitates of hardening phase θ′-Al_2Cu, bright field (a) and dark field (b)

The results of the SEM/EDS analysis of the particles extracted with boiling phenol from AlCu4Ni2Mg2 alloy (Fig. 19) were compared with X-ray diffraction pattern (Fig. 20). The observed peaks confirmed SEM and TEM results. The majority of the peaks were from Al_7Cu_4Ni, Al_6Fe, S-Al_2CuMg, and $Al_3(CuFeNi)_2$.

On the other hand, it is nearly impossible to make unambiguous identification of the all intermetallics present in an aluminium alloy which are rather complex, even applying all well-known experimental techniques. X-ray diffraction analysis is one of the most powerful and appropriate technique giving the possibility to determine most of verified intermetallics based on their crystallographic parameters. Our analysis shows that the difficulties of having reliable results of all the possible existing phases in a microstructure of the alloy is related to the procedure of phase isolation. The residue is separated by centrifuging and since some of the particles are very fine and available sieves are having too big outlet holes there is no chance prevents them from being flowing out from a solution.

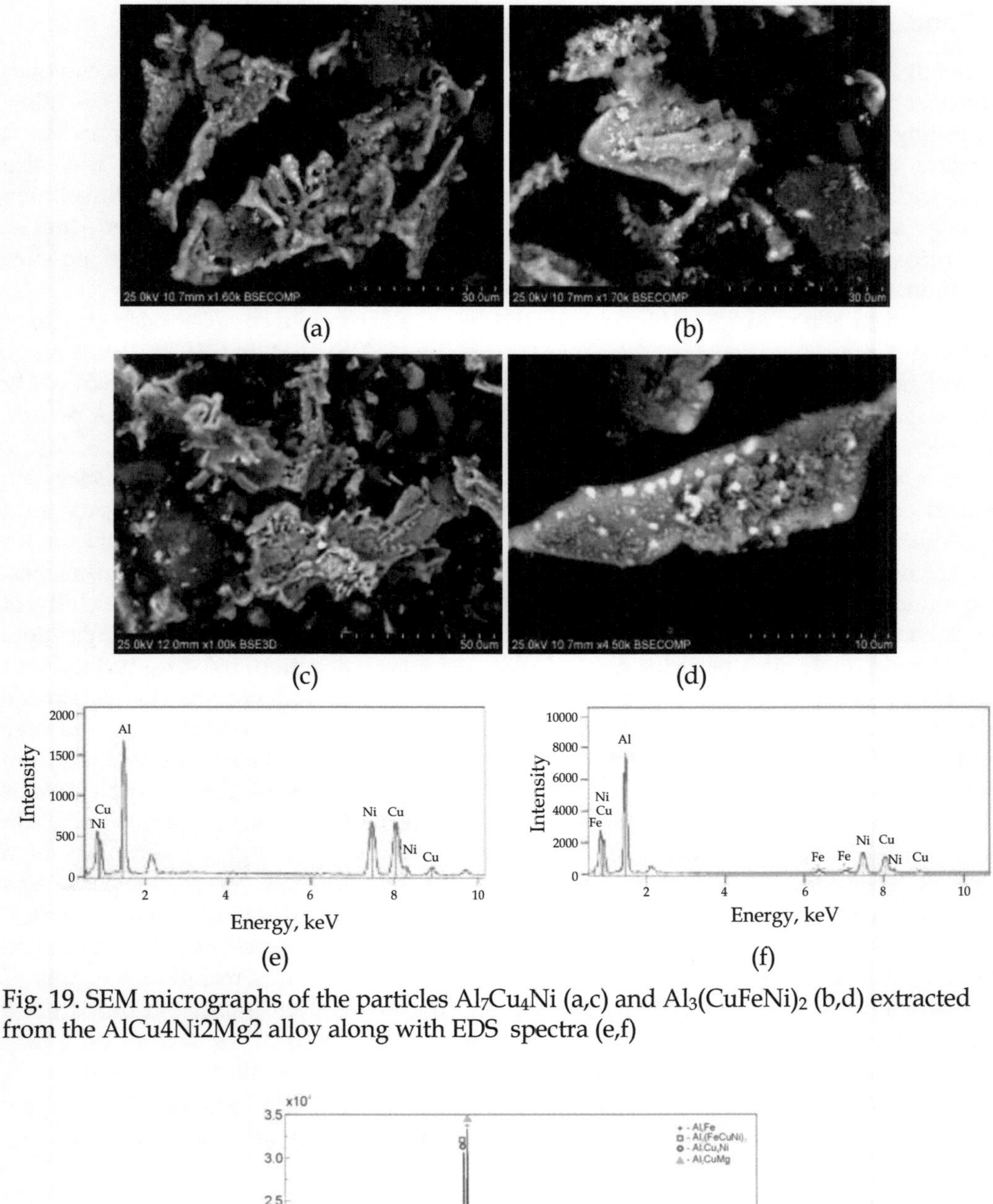

Fig. 19. SEM micrographs of the particles Al_7Cu_4Ni (a,c) and $Al_3(CuFeNi)_2$ (b,d) extracted from the AlCu4Ni2Mg2 alloy along with EDS spectra (e,f)

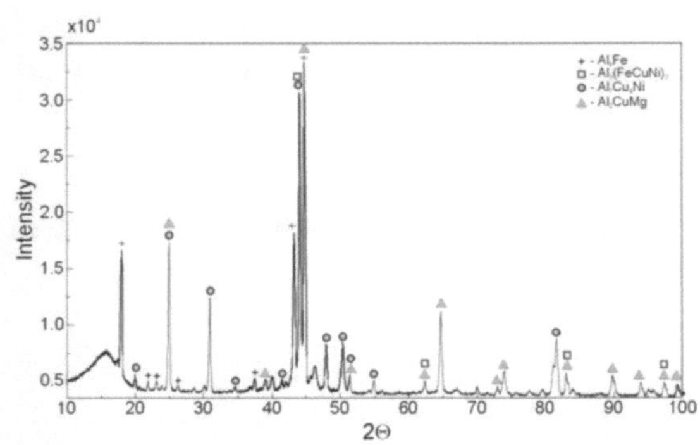

Fig. 20. The X-ray diffraction from the particles extracted from AlCu4Ni2Mg2 alloy

4. Conclusion

Currently, efforts are being directed towards the development of analytical techniques which rapidly achieve an accurate determination of phase components in an alloy. According to the obtained results, the applicability of the proposed methods provides a practical alternative to other techniques. The phenol extraction procedure was also successfully applied to the examined aluminium alloys. The main advantages of dissolution techniques are its reliability - when used properly you will always get pure residue - and its low price. The major disadvantageous of phenol extraction method are the possible contamination of the residue and the time needed.

The examined alloys AlSi5Cu1Mg and AlCu4Ni2Mg2 possessed a complex microstructure. By using various instruments and techniques (LM, SEM-EDS, TEM and XRD) a wide range of intermetallics phases were identified. The microstructure of investigated AlSi5Cu1Mg alloy included: β-Al_5FeSi, α-$Al_{12}(FeMn)_3Si$, Al_2Cu, Q-$Al_5Cu_2Mg_8Si_6$, Si and Mg_2Si phases. The microstructure of AlCu4Ni2Mg2 alloy included five phases, namely: Al_7Cu_4Ni, θ'-Al_2Cu, Al_6Fe, S-Al_2CuMg, and $Al_3(CuFeNi)_2$. A size and distribution of these various dispersoids depend on the time and temperature of the homogenization and/or annealing processes. Fine intermetallic particles (<1μm) are formed during artificial aging of heat-treatable alloys and are more uniformly distributed than constituent particles or dispersoids. Dimensions, shape and distribution of these particles may have important effects on the ductility of alloys and more needs to be known regarding their formation, structure and composition. For example, the coarse particles can influence the recrystallization, fracture, surface, and corrosion behavior, while the dispersoids control grain size and provide stability to the metallurgical structure. The dispersoids can also affect the fracture performance and may limit strain localization during deformation. The formation of particles drains solute from the matrix and, consequently, changes the strength properties of the material. This is specially relevant in the heat-treatable alloys, where depletion in Cu, Mg, and Si can significantly change the metastable precipitation processes and age hardenability of a material. Therefore, particle characterization is essential not only for choosing the best processing routes, but also for designing optimized alloy composition. Thus, particle characterization is important not only to decide what sort of processing courses should be applied, but also for designing optimized chemical composition of a material. A variety of microscopic techniques are well appropriate to characterize intermetallics but only from a small section of an analyzed sample. From commercial point of view it is extremely advantageous to provide use quick, reliable and economical examination technique capable of providing data of particles from different locations of a full scale-sized ingot. One of these methods is dissolving the matrix of an aluminium alloy chemically or electrochemically.

5. Acknowledgment

This work was carried out with the financial support of the Ministry of Science and Higher Education under grant No. N N507 247940

6. References

Bäckerud, L. & Chai, G. (1992). Solidification Characteristics of Aluminum Alloys 3, American Foundry Society, Des Plaines, Illinois

Belov, N.A., Aksenov, A.A. & Eskin, D.G. (2002). Iron in aluminium alloys, Taylor & Francis Inc, New York, ISBN 0-415-27352-8

Belov, N.A., Eskin, D.G. & Avxentieva, N.N. (2005). Constituent phase diagrams of the Al-Cu-Fe-Mg-Ni-Si system and their application to the analysis of aluminium piston alloys, *Acta Materialia*, No. 53 pp.4709-4722

Cabibbo, M., Spigarelli, S. & Evangelista, E. (2003). A TEM investigation on the effect of semisolid forming on precipitation processes in an Al-Mg-Si alloy, *Materials Characerisation*, No. 49, pp. 193-202

Chen, C.L. & Thomson, R.C. (2010). Study of thermal expansion of intermetallics in multicomponent Al-Si alloys by high temperature X-ray diffraction, *Intermetallics*, No. 18, pp 1750-1757

Christian, J.W. (1995). The theory of transformations in metals and alloys. Pergamon Press, Oxford, Anglia

Dobrzański, L.A., Maniara, R. & Sokolooki, J.H. (2007). Microstructure and mechanical properties of AC AlSi9CuX alloys, *Journal of Achievements in Materials and Manufacturing Engineering*, Vol. 24, No.2, pp. 51-54

Garcia-Hinojosa, J.A., González, C.R., González, G.M. & Houbaert, Y. (2003). Structure and properties of Al-7Si-Ni and Al-7Si-Cu cast alloys nonmodified and modified with Sr, *Journal of Materials Processing Technology*, No. 143-144, pp. 306-310

Gupta, A.K., Lloyd, D.J. & Court S.A. (2001). Precipitation hardening in Al-Mg-Si alloys with and without excess Si. *Materials Science and Engineering A*, No. 316, pp. 11-17

Gustafsson, G., Thorvaldsson, T. & Dunlop, G.L. (1986). The influence of Fe and Cr on the microstructure of cast Al-Mg-Si alloys, *Metallurgical and Materials Transactions. A*, No. 17A, pp. 45-52

Griger, A. & Stefaniay V. (1996). Equilibrium and non-equilibrium intermetallic phases in Al-Fe and Al-Fe-Si alloys, *Journal of Materials Science*, No. 31, pp. 6645-6652

Hatch, J.E (1984). Aluminium. Properties and Physical Metallurgy. Ed.., ASM Metals Park, Ohio, ISBN 0-87170-176-6

Ji, Y. Guo, F. & Pan, Y. (2008). Microstructural characteristics and paint-bake response of Al-Mg-Si-Cu alloy, *Transactions of Nonferrous Metals Society of China*, No.18, pp. 126-129

Karabay, S., Yilmaz, M. & Zeren, M. (2004). Investigation of extrusion ratio effect on mechanical behaviour of extruded alloy AA-6101 from the billets homogenised-rapid quenched and as-cast conditions. *Journal of Materials Processing Technology*, No. 160, pp. 138-147

King, F. (1987). Aluminium and its alloys. John Willey and Sons, New York, Chichester, Brisbane, Toronto

Kuijpers, N.C.W., Kool, W.H. & van der Zwaag, S. (2002). DSC study on Mg-Si phases in as cast AA6xxx. *Mater. Sci. Forum*, No. 396-402, pp. 675-680

Li, Z., Samuel, A.M., Samuel, F.H., Ravindran, C., Valtierra, S. & Doty, H.W. (2004). Parameters controlling the performance of AA319-type alloys Part I. Tensile properties, *Materials Science and Engineering*, No. 367, pp. 96-110

Liu, Y.L. Kang, S.B. &. Kim, H.W. (1999). The complex microstructures in as-cast Al-Mg-Si alloy, *Materials Letters*, No. 41, pp. 267-272

Lodgaard, L. & Ryum, N. (2000). Precipitation of dispersoids containing Mn and/or Cr in Al-Mg-Si alloys, *Materials Since and Engineering A*, No.283, pp. 144-152

Mondolfo, L.F, (1976). Aluminium Alloys: Structure and Properties. London-Boston, Butterworths

Mrówka-Nowotnik, G. & Sieniawski, J. (2005). Influence of heat treatment on the micrustructure and mechanical properties of 6005 and 6082 aluminium alloys, *Journal of Materials Processing Technology*, Vol. 162-163, No. 20, pp. 367-372

Mrówka-Nowotnik, G., Sieniawski, J. & Wierzbińska, M. (2007). Analysis of intermetallic particles in AlSi1MgMn aluminium alloys, *Journal of Achievements in Materials and Manufacturing Engineering*, No. 20 pp. 155-158.

Mrówka-Nowotnik, G., Sieniawski, J. & Wierzbińska, M. (2007). Intermetallic phase particles in 6082 aluminium alloy, *Archives of Materials Science and Engineering*, No. 282, pp. 69-76

Polmear, I.J. (1995). Light alloys. Metallurgy of light metals. Arnold, London-New York-Sydney-Auckland, ISBN 0-7506-6371-5

Sato, K. & Izumi I. (1985). Application of the technique for isolating and analysis intermetallic compounds to commercial aluminium alloys, *Journal of Japan Institute of Light Metals*, No. 35, pp. 647-649

Warmuzek, M., Sieniawski, J., Wicher, K. & Mrówka-Nowotnik, G. (2003). Analiza procesu powstawania składników fazowych stopu AlFeMnSi w warunakch zmiennej zawartości metali przejściowych Fe i Mn. *Inżynieria Materiałowa*, No. 137, pp. 821-824

Warmuzek, M., Mrówka, G. & Sieniawski, J. (2004). Influence of the heat treatment on the precipitation of the intermetallic phasesin commercial AlMn1FeSi alloy. *Journal of Materials Processing Technology*, 157-158, No. 20, (December 2004) pp. 624-632, ISSN 0924-0136

Warmuzek, M., Rabczak, K. & Sieniawski, J. (2005). The course of the peritectic transformation in the Al-rich Al–Fe–Mn–Si alloys, *Journal of Materials Processing Technology*, No. 162-163, pp. 422-428

Warmuzek, M., Sieniawski, J., Wicher, K. & Mrówka-Nowotnik, G. (2006). The study of distribution of the transition metals and Si during primary precipitation of the intermetallic phases in Al-Mn-Si alloys, *Journal of Materials Processing Technology* Vol. 1-3, No. 175, pp. 421-426

Wierzbińska, M. & Mrówka-Nowotnik, G. (2008). Identification of phase composition of AlSi5Cu2Mg aluminium alloy in T6 condition, *Archives of Materials Science and Engineering*, Vol. 30, No. 2, pp. 85-88

Zajac, B., Bengtsson, Ch. & Jönsson, (2002). Influence of cooling after homogenization and reheating to extrusion on extrudability and final properties of AA 6063 and AA 6082 alloys, *Materials Science Forum*, No. 396-402, pp. 675-680

Zhen, L. & Kang, S.B. (1998). DSC analyses of the precipitation behavior of two Al-Mg-Si alloys naturally aged for different times, *Materials Letters*. No. 37, pp. 349-353

3

Rotary-Die Equal Channel Angular Pressing Method

Akira Watazu
National Institute of Advanced Industrial Science and Technology (AIST)
Japan

1. Introduction

Light metals such as aluminum, magnesium, titanium and their alloys are useful for a wide range of applications such as in the automotive, railway, and aerospace industries. Engineering of fine-grained light metal materials is an indispensable technology that is expected to improve material properties such as tensile strength, elongation, corrosion resistance, fracture toughness, strain-rate plasticity, low-temperature plasticity, etc. The production of fine-grained light metals with excellent properties using severe plastic deformation methods, especially rolling and extrusion, has been intensively studied. With such processes, the size of the metal grain generally decreases because plastic deformation causes a decrease of grain size, by the principle shown schematically in Fig. 1.

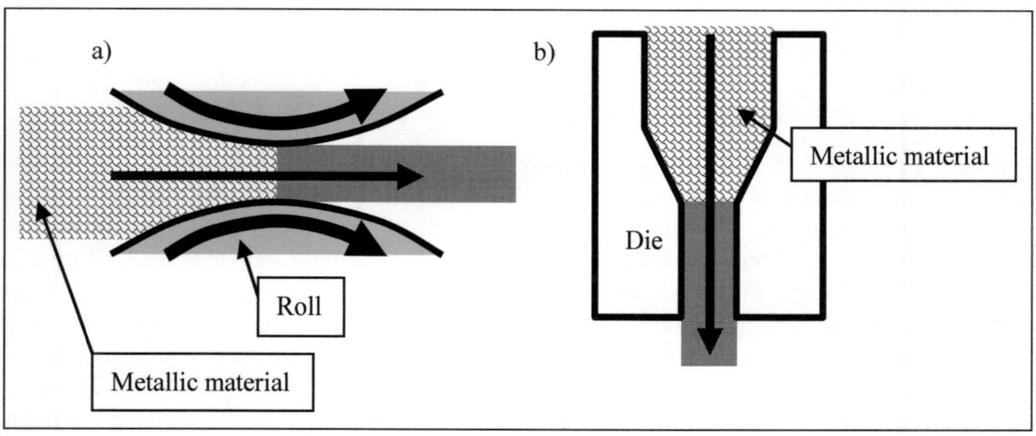

Fig. 1. Schematic diagram of a) rolling method and b) extrusion method

On the other hand, the equal channel angular pressing (ECAP) method invented by Segal et al. in 1981 has proven successful for fabricating fine-grained bulk metals. A schematic diagram of the ECAP method is shown in Fig. 2. In the ECAP method, a large strain can be introduced into a billet sample by simple shear deformation without changes in the cross-sectional area. In the ECAP process, the billet is extruded through a die consisting of two channels intersecting at an angle of 2Φ. The sample is set in the vertical channel and pressed into the second channel. The greatest advantage of the ECAP method is that the initial size and shape of the sample processed by the ECAP process are maintained. The sample is enhanced with the shear stress of the angle Φ to the extrusion direction in the ECAP process and its structure is fine-grained.

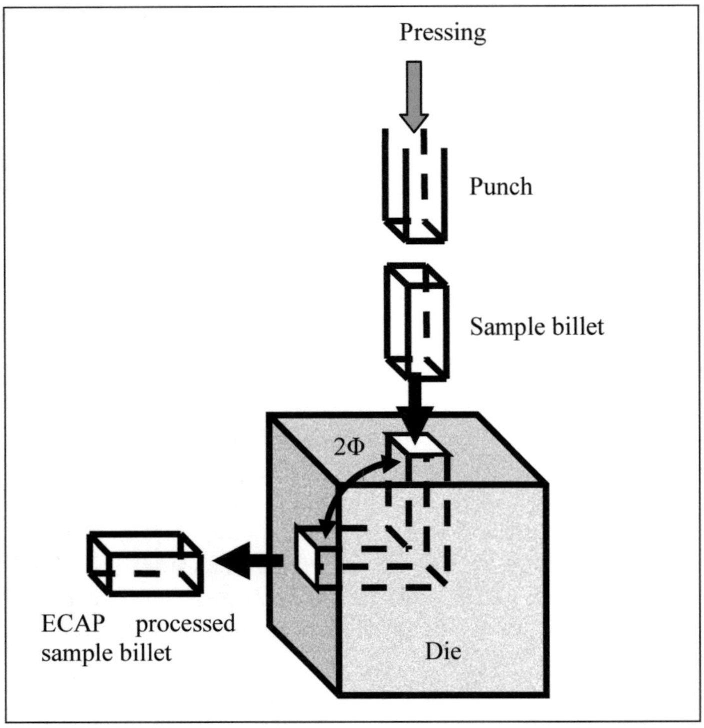

Fig. 2. Schematic diagram of equal channel angular pressing method. (V.M. Segal, V.I. Rexnikov, A.E. Drobysevsky and V.I. Kopylov: Metally Vol. 1 (1981), p. 115.)

The rolling method, the extrusion method and the ECAP method are severe plastic deformation methods, and are useful for grain refinement of light metals. In all of these methods, excellent grain refinement is generally expected with many passes. In the case of the rolling method, as shown in Fig. 3-a, the grain is continuously refined with each pass as the material's thickness decreases. In order to process the material with ECAP many times, either a continuous cycling method (Fig. 3-b) or a method of expulsion and reinsertion (Fig. 3-c) are possible. In the case shown in Fig. 3-b, twice or more pressure is necessary when continuing the second time, and there is a limit in the number of ECAP passes depending on the maximum pressure and the die strength. In conventional ECAP, as shown in Fig. 3-c, the pressed sample must be removed from the die and reinserted back for the next pressing, making the process inefficient. Not only does this process take a long time, the temperature of the sample is difficult to control.

A new ECAP process method called the rotary-die equal channel angular pressing (RD-ECAP) method was developed at Japan's National Institute of Advanced Industrial Science and Technology (AIST, formerly the National Industrial Research Institute of Nagoya (NIRIN)) to form fine-grained bulk materials such as aluminum alloys, aluminum composites, magnesium alloys, and titanium. Using the RD-ECAP method, ECAP processing of up to 2 passes can be done without sample removal, and samples processed over 30 cycles were obtained. One-pass RD-ECAP could be processed in 30 s. In this paper, the RD-ECAP process is explained and its use in the processing of light metals (aluminum alloys) is reported.

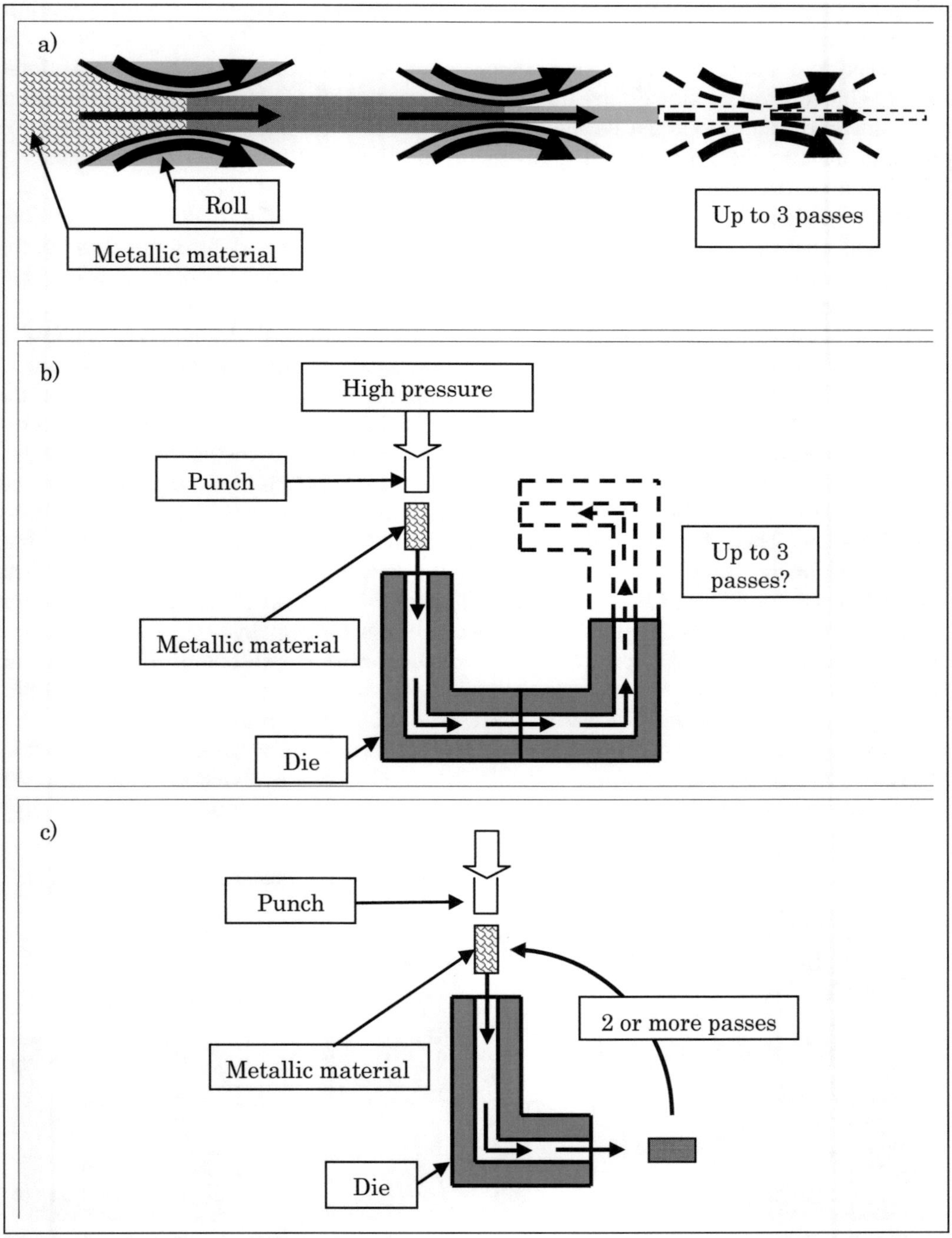

Fig. 3. Schematic diagram of a) rolling method and b, c) the equal channel angular pressing method in the case of two or more passes

2. Principle of rotary-die equal channel angular pressing method

The rotary-die equal channel angular pressing (RD-ECAP) method was developed for structuring fine grains in light metal, such as aluminum alloys, magnesium alloys, titanium and so on. In general, using equal channel angular pressing (ECAP), a large strain can be introduced into a billet by simple shear deformation without changes in the cross-sectional area; the billet develops fine grains after several passes of ECAP. However, in conventional ECAP method, the billet must be removed from the die and reinserted back for the next pressing, making the process inefficient. Using the RD-ECAP method, up to 4 passes of ECAP-style severe plastic deformation is possible without billet removal.

Schematic diagrams of the RD-ECAP method are shown in Fig. 4. It consists of four cylindrical channels meeting at the center of the rotary die and four punches in the corresponding channels. The sample is set into the center of the hole. Then, the four punches are placed into the holes from the four directions and the die is set on a die holder. The die is heated to about 500–700 K and a plunger presses the punch at the top. The sample is extruded to the left direction because the right punch and the bottom punch are locked in place due to contact with the die holder. The remaining two channels are used for the conventional ECAP extrusion process. The punch at the top is pushed completely into the die to complete one extrusion or RD-ECAP process. After this extrusion, the die is rotated clockwise 90° to the initial configuration with the exposed punch at the top, and a second pressing is performed. The process continues until the die returns to its former position, after 4 passes. Then, because the sample is not reduced and the die is enough big, temperature of the sample is able to control with control of temperature of the die. The informal name for RD-ECAP is the Japanese term "Mochitsuki", which is the common process of making rice cake by pressing steamed rice again and again.

In the present work, the samples can be processed under conditions of 573–773K at an approximately 0.9–2.4 mm/s punch speed at 300MPa or lower. By the RD-ECAP method, ECAP processing could be repeatedly done without sample removal. In addition, the temperature of the sample could be easily controlled. In our study, samples processed over 30 cycles (one cycle=one extrusion and 90° die rotation) were obtained. One RD-ECAP cycle could be processed in 30 s. Therefore, RD-ECAP has the advantage of being energy efficient.

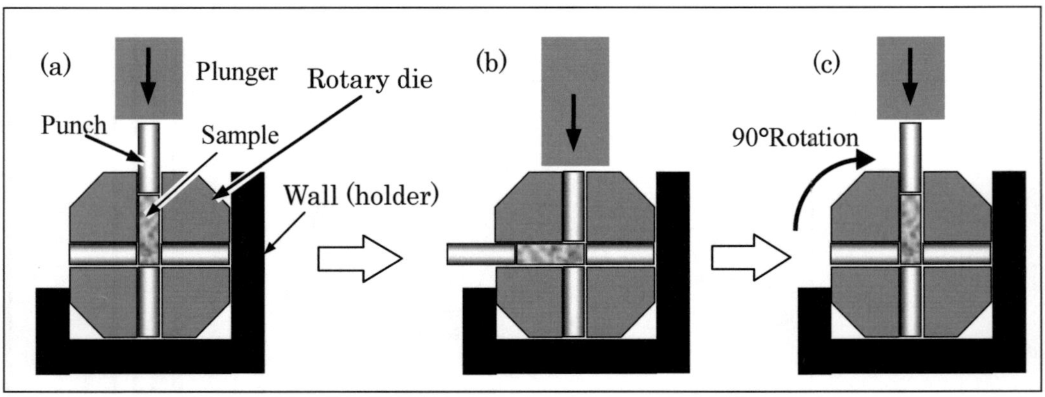

Fig. 4. Schematic diagram of rotary-die equal channel angular pressing. (a) initial state, (b) after one pass, and (c) after 90° die rotation

Compared with the conventional ECAP die consisting of two channels intersecting at an angle, the RD-ECAP die is easy to make because the channels in the RD-ECAP die are formed with two straight holes. Though there are many channels in the RD-ECAP die, the sample is always pressed from the same direction and general press equipment can be used.

3. RD-ECAP processed aluminum

3.1 AC4C aluminum alloy

AC4C (JIS, ISO; Al-Si7Mg(Fe)) casting aluminum alloy (Cu<0.20, Si 6.5-7.5, Mg 0.20-0.4, Zn <0.3, Fe<0.5, Mn<0.6, Ni<0.05, Ti<0.20, Pb<0.20, Sn<0.05) is an excellent material for observation of the RD-ECAP effect, such as breaking of the precipitated phase, because the alloy has primary crystal dendrite and a coarse Al-Si microstructure. An AC4C casting aluminum alloy material 20 mm in diameter and 50 mm in length was used. Cylindrical samples 19.5 mm in diameter and 40 mm in length long were prepared by lathing.
The RD-ECAP die had a two cylindrical holes 20 mm in diameter that intersect at 90° to form four channels. Three punches are pushed completely into the side and bottom channels, the sample is placed in the top hole, and the die is set onto a die holder, as shown in Fig. 4-a. Samples were processed under conditions of 543 K, 603K, 673 K at an approximately 0.9 mm/s punch speed from one pass (= one extrusion) to 20 passes.
Photographs of AC4C aluminum alloy samples processed by the RD-ECAP are shown in Fig. 5. The surfaces of the samples were dirty with lubricants but had no cracks or contamination after the RD-ECAP process.
An experimentally obtained load-displacement curve of the plunger for the rotary-die equal channel angular pressing at 603 K is shown in Fig. 6. The load increased with pressing, reached a maximum load, and then decreased with further sample deformation.
Change in the maximum stress with the number of rotary-die equal-channel angular pressing passes is shown in Fig. 7. The maximum load was lower at higher temperatures. At 673K, the first maximum load was about 150 MPa, and the fourth maximum load was about 100 MPa. At 603 K and 543 K, the maximum load decreased as RD-ECAP pass increased from the 1st to 6th pass. The decrease of the maximum load at 603 K was the highest.

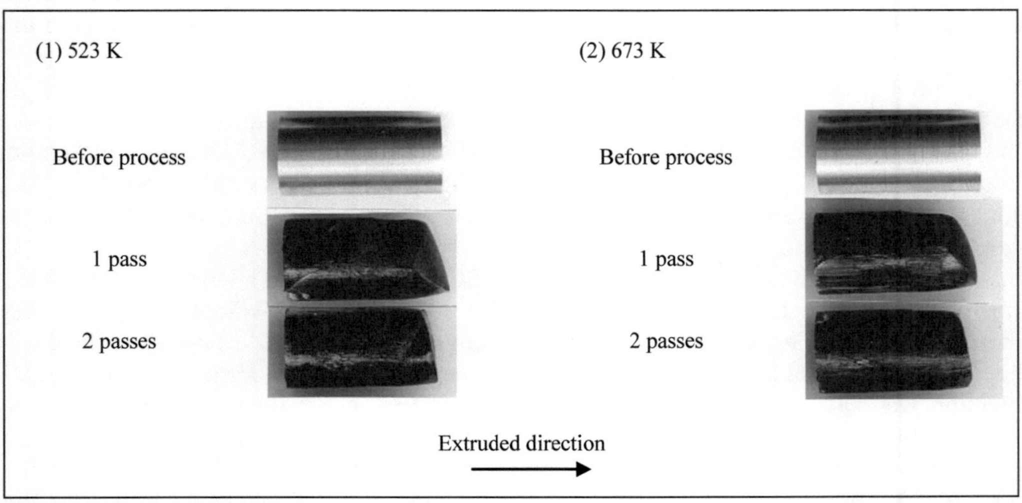

Fig. 5. Photograph of samples processed by rotary-die equal channel angular pressing

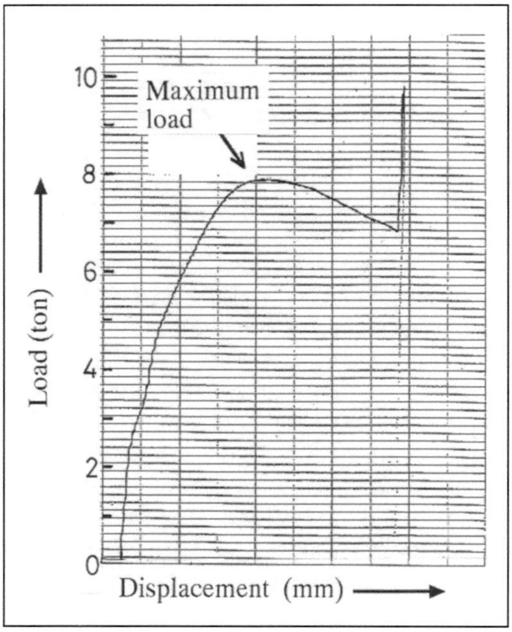

Fig. 6. Experimentally obtained load-displacement curve of the plunger for the rotary-die equal channel angular pressing at 603 K. (Y. Nishida, H. Arima, J.C. Kim and T. Ando: J. Japan Inst. Light Metals. Vol. 50-12 (2000), p. 655-659 in Jp.)

The microstructures of AC4C aluminum alloys processed by RD-ECAP with 1-20 passes at 603 K are shown in Fig. 8. The as-cast sample with 0 passes had a typical aluminium eutectic structure with dendrites. The dendrites were deformed after one pass. After 6 passes, the shape of the primary crystal dendrite disappeared and most eutectic structures were also broken. After 10 passes, the cast structure disappeared. After 20 passes, a uniform microstructure with fine primary-crystal aluminium and fine eutectic structure was observed. The microstructure became fine with increasing of RD-ECAP pass number. In addition, the distribution of the silicon particles appeared to have become more homogeneous with the rising number of RD-ECAP passes.

A TEM photograph of an AC4C aluminium alloy processed by 10 passes of rotary-die equal channel angular pressing at 603 K is shown in Fig. 9. The crystal grains were about 2–3 μm.

The relationship between the total elongation and strain rate of the AC4C aluminum alloy processed by the RD-ECAP at 603 K is shown in Fig. 10. The 6-pass sample had about 90 % elongation. The 10- and 20-pass samples had over 100 % elongation, and the maximum elongation was 126 %.

The appearance of the samples after 10-pass RD-ECAP at 603 K and a tensile test at 723 K is shown in Fig. 11. The samples processed by RD-ECAP had smooth surfaces. SEM photographs of the tensile test sample surfaces are shown in Fig. 12. The sample shown in Fig. 12-a had a detailed surface and 111 % elongation. Narrow structure along tensile direction was also shown in Fig. 12-b. By contrast, the as-cast 0-pass sample had many cracks on the 90° direction to the axis of tension and had a rough surface.

By RD-ECAP process, AC4C aluminium alloy hardly had any crack and had the elongation in the tensile test because the microstructure became fine and homogeneous with increasing of RD-ECAP pass number.

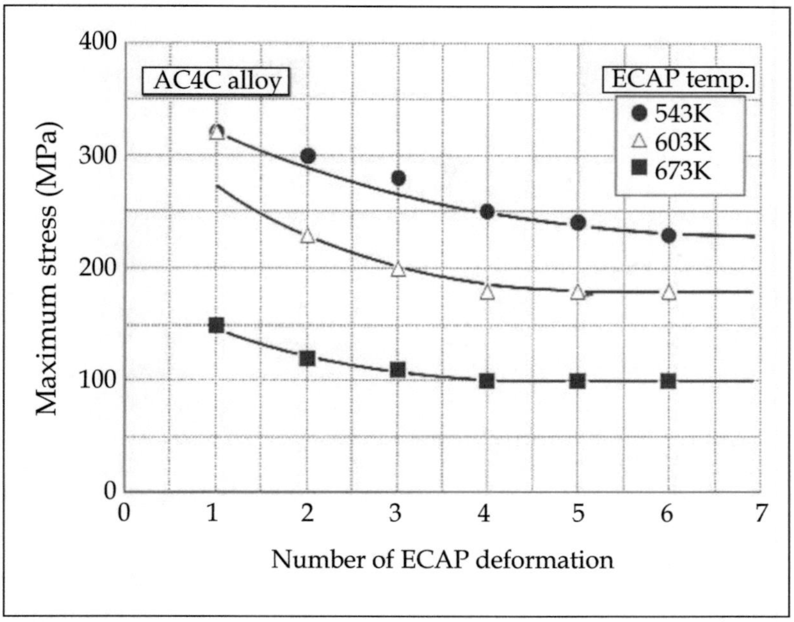

Fig. 7. Change of maximum stress with number of the rotary-die equal channel angular pressing passes. (Y. Nishida, H. Arima, J.C. Kim and T. Ando: J. Japan Inst. Light Metals. Vol. 50-12 (2000), p. 655-659 in Jp.)

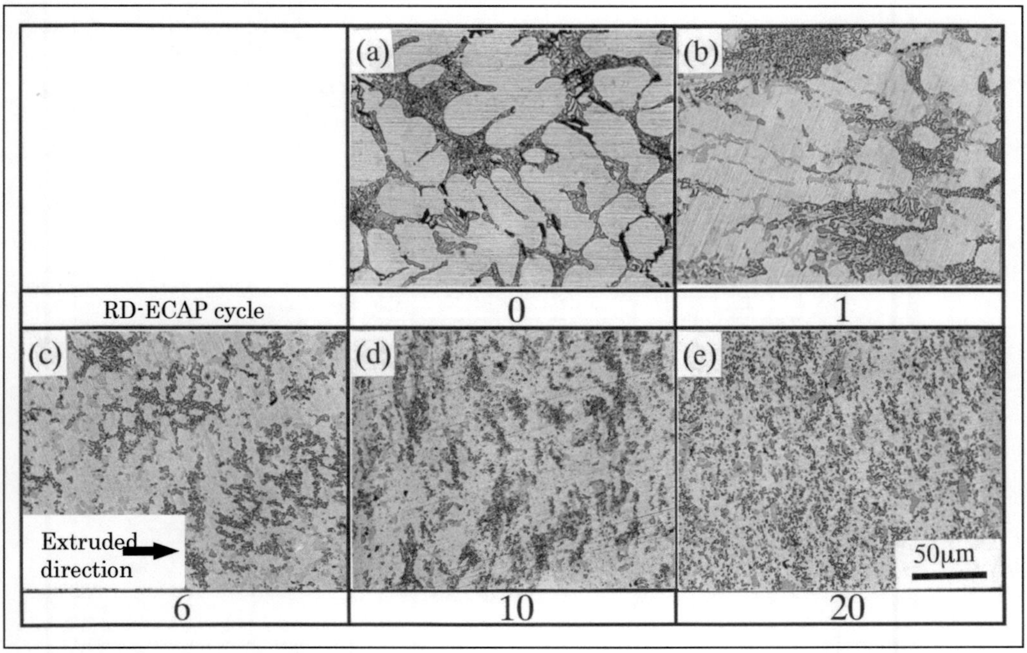

Fig. 8. Microstructures of AC4C aluminum alloys processed by the rotary-die equal channel angular pressing at 603 K. (a) initial state, (b) – (d), after 1–4 passes of RD-ECAP. (Y. Nishida, H. Arima, J.C. Kim and T. Ando: J. Japan Inst. Light Metals. Vol. 50-12 (2000), p. 655-659 in Jp.)

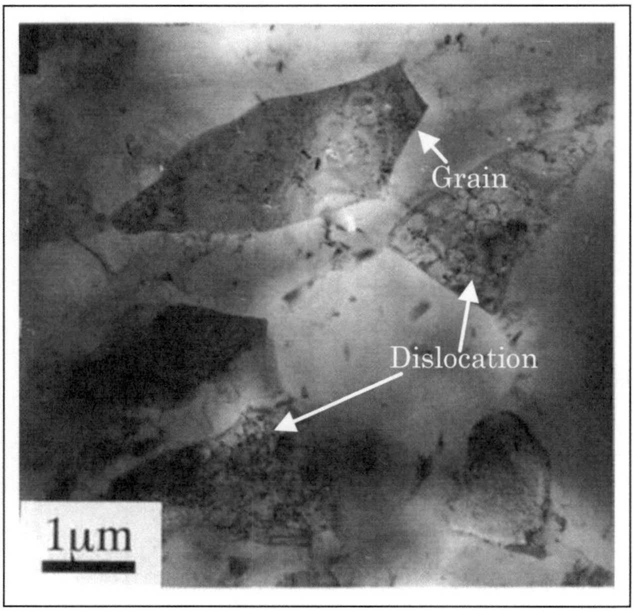

Fig. 9. TEM photograph of AC4C aluminium alloy processed by 10 passes of rotary-die equal channel angular pressing at 603 K. (Y. Nishida, H. Arima, J.C. Kim and T. Ando: J. Japan Inst. Light Metals. Vol. 50-12 (2000), p. 655-659 in Jp.)

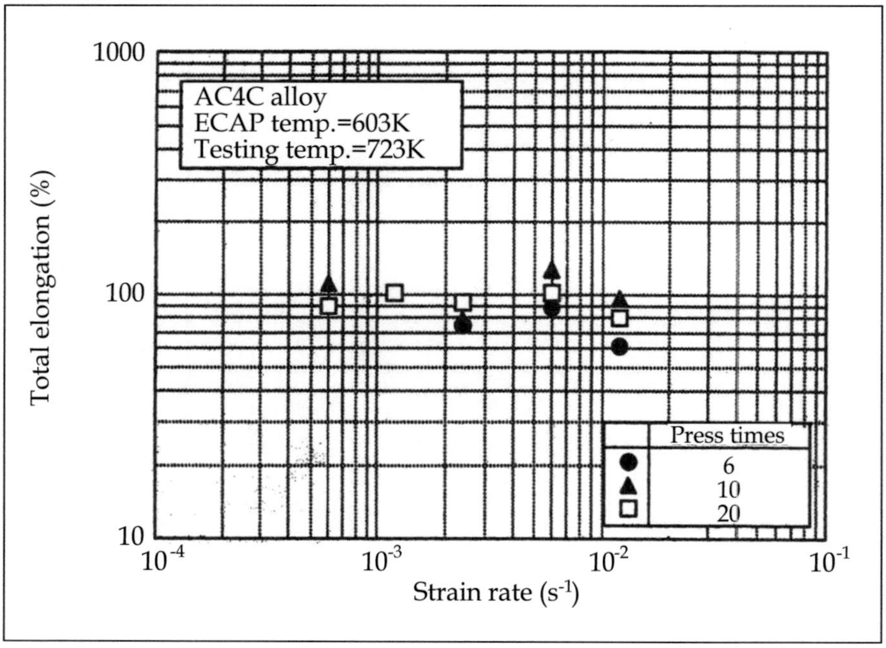

Fig. 10. Relationship between total elongation and strain rate of AC4C aluminum alloy processed by rotary-die equal channel angular pressing at 603 K. (Y. Nishida, H. Arima, J.C. Kim and T. Ando: J. Japan Inst. Light Metals. Vol. 50-12 (2000), p. 655-659 in Jp.)

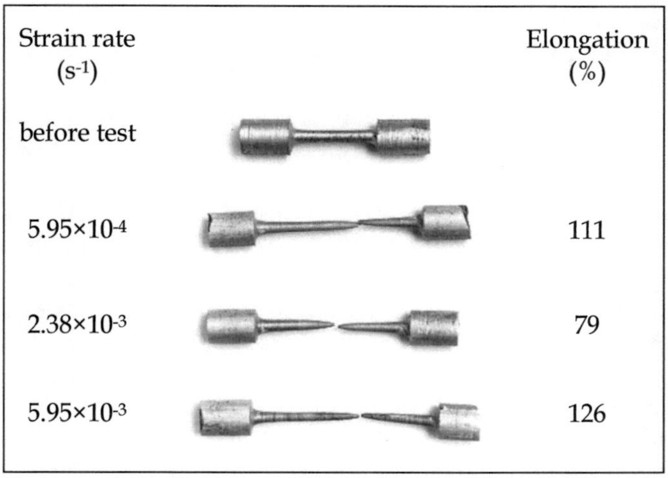

Fig. 11. Appearance of samples after 10 passes of rotary-die equal channel angular pressing at 603 K and tensile test at 723 K. (Y. Nishida, H. Arima, J.C. Kim and T. Ando: J. Japan Inst. Light Metals. Vol. 50-12 (2000), p. 655-659 in Jp.)

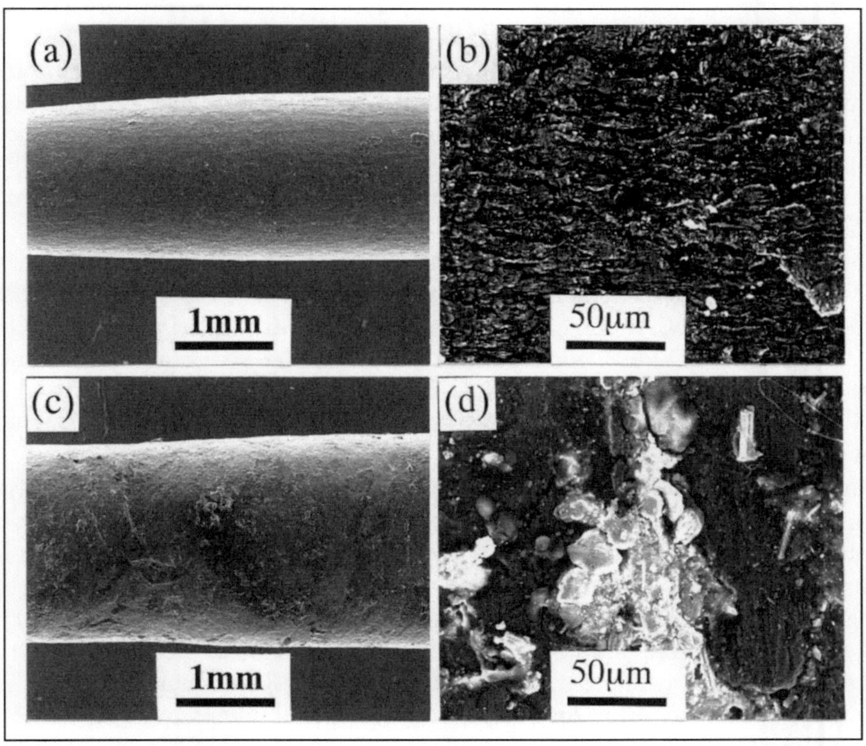

Fig. 12. Tensile test sample surfaces of AC4C alloy observed by SEM after tensile test at 723 K at 5.95×10^{-4} S^{-1}. (a) and (b) are processed by 10 passes of rotary-die equal channel angular pressing at 603 K; (c) and (d) are as-cast samples. (Y. Nishida, H. Arima, J.C. Kim and T. Ando: J. Japan Inst. Light Metals. Vol. 50-12 (2000), p. 655-659 in Jp.)

3.2 Al-11mass%Si alloy and impact toughness

Al–Si eutectic alloys are in wide use in industry, especially in the automobile industry, due to their good wear resistance, high tensile strength at elevated temperatures and amenability to casting. However, their low fracture toughness impedes their broader application. Their microstructure consists of eutectic silicon crystals and an aluminum alloy matrix. The silicon crystals, which have three-dimensionally complex shapes and are very brittle, congregate at the grain boundaries of the aluminum matrix. The low fracture toughness of these alloys originates in their microstructure, and is influenced by aluminum dendrite arm spacing and cell size, eutectic silicon characteristics (size and morphology) and eutectic silicon distribution. To improve the microstructure, several techniques are in use industrially: for example, the addition of elements like sodium and strontium. However, this treatment results in little improvement in toughness, since brittleness is thought to be inherent in these alloys.

There are several routes to improving the microstructure of alloys. These include rapid solidification, stirring during solidification, heat treatment, and plastic deformation, with the last being the most energy-efficient.

The Al–11mass%Si eutectic alloy used for the present research contains, by mass, 11.3% Si, 1.00% Cu, 1.13% Mg, 1.10% Ni and 0.277% Fe. The copper, magnesium and nickel are used to improve the mechanical properties of this alloy at elevated temperatures. These elements are present in the alloy as intermetallic compounds including Mg_2Si, Al_4CuNi, Al_9FeNi, Al_6Cu_3Ni and Al_3Ni. For RD-ECAP processing, the material of Al–11mass%Si alloy was machined to be a cylindrical billet 19.5 mm in diameter (the channel is 20 mm in diameter) and 40 mm in length. Due to the low billet aspect ratio (=2), the billet is subjected to non-uniform deformation, since there is minimal deformation of the billet end zone regions.

The RD-ECAP die had two cylindrical holes 20 mm in diameter that intersect at 90° to form four channels. Three punches are pushed completely into the side and bottom channels, the sample is placed in the top hole, and the die is set onto a die holder, as shown in Fig. 4-a. Then, The die is heated. The effect of RD-ECAP temperature on the impact toughness of the Al–11mass%Si alloy was examined at three temperatures: 573, 623 and 673 K. The billets were processed by RD-ECAP for 4, 8, 12, 16 and 32 passes at each temperature. In addition, four special routes of RD-ECAP were used in this work: (a) 8 passes at 673 K followed by 8 passes at 623 K; (b) 4 passes at 673K followed by 12 passes at 623 K; (c) 4 passes at 573 K followed by 4 passes at 673 K and 8 passes at 623 K; (d) 4 passes at 673 K followed by 4 passes at 623 K and 8 passes at 573 K.

Impact toughness test pieces were made from the RD-ECAP-processed (RD-ECAPed) billet by machining along the longitudinal direction. The size of the rectangular prism test pieces was 3 mm × 4 mm in cross-section and 34 mm in length, with a U-notch 1.5 mm in width and 1.5 mm in depth. A computer-aided instrumented Charpy impact test machine including software for tougher materials was used for measuring the absorbed energy of the samples as impact toughness during impact testing. The plot in the figure is the average value of four test pieces made from one billet (six pieces in all were made from one billet).

An as-cast alloy was also tested for comparison. An optical microscope and a transmission electron microscope (TEM) were used to observe the microstructures of the RD-ECAPed samples that had been cut from the longitudinal sections of the billets. Proven Solution for Image Analysis was used for investigation of the particle size distribution in the alloy. The maximum diameter of each particle was used as the particle size. A scanning electron microscope (SEM) was employed for observation of the fractured surface.

3.2.1 Microstructures

The effect of the number of RD-ECAP passes on the alloy microstructure is shown in Fig. 13, where (a) shows the microstructure of the as-cast alloy, and (b), (c) and (d) illustrate, respectively, the microstructures of samples processed with 8, 16 and 32 passes at 623 K via RD-ECAP. The as-cast (0 pass) sample consists of large grains, including the dendrites of the aluminum matrix, interdendritic networks of eutectic silicon plates and particles of other large intermetallic compounds present between the aluminum dendrite arms or grain boundaries, as shown in this Figure. After pressing by RD-ECAP for 8 passes, the large grains observed in 0 pass sample did not exist and no dendrite structure was found in the alloy. Though >20 μm eutectic silicon plates and its interdendritic networks were observed in 0 pass sample, the plates became fine in the samples pressed for multipasses and <6 μm plates or particles were observed after pressing by RD-ECAP for 32 passes. In addition, the distribution of the silicon particles appeared to have become more homogeneous with the rising number of RD-ECAP passes. The results indicate that stirring and deformation occurred in the sample by RD-ECAP.

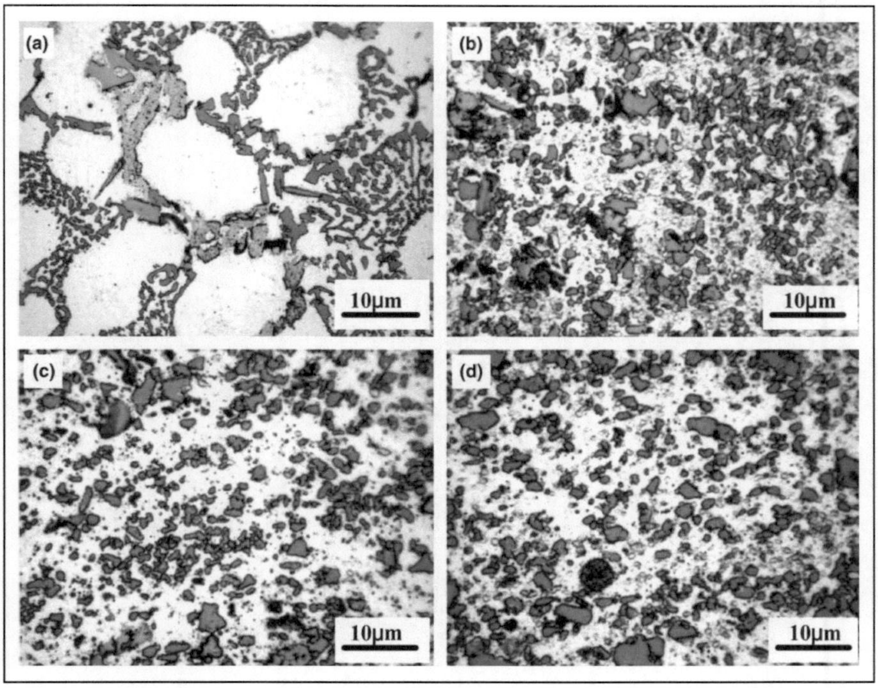

Fig. 13. Microstructures of the Al–11mass%Si samples processed by RD-ECAP at 623 K for (a) - (d) 0, 8, 16 and 32 passes, respectively. (A. Ma, K. Suzuki, Y. Nishida, N. Saito, I. Shigematsu, M. Takagi, H. Iwata, A. Watazu, T. Imura: Acta Materialia 53 (2005) 211–220.)

Fig. 14 illustrates the particle size distribution in the alloy after RD-ECAP. Over 60% of the particles are smaller than 1 μm in the samples processed with 8 and 16 passes at 623 K. After 32 RD-ECAP passes, over 70% of the particles were smaller than 1 μm. However, the large particle (over 2 μm in diameter) contents in the samples processed with 8, 16 and 32 passes were not significantly different. It is evident that particles smaller than 1 μm in the alloy increased with increasing number of RD-ECAP passes.

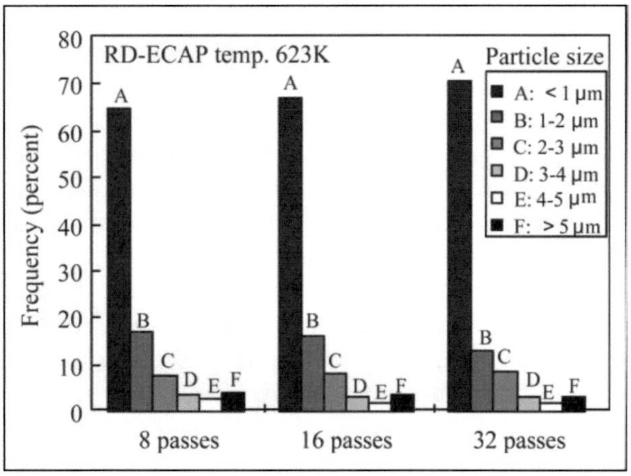

Fig. 14. Effect of the number of RD-ECAP passes on the particle size distribution in the alloy. (A. Ma, K. Suzuki, Y. Nishida, N. Saito, I. Shigematsu, M. Takagi, H. Iwata, A. Watazu, T. Imura: Acta Materialia 53 (2005) 211–220.)

Fig. 15 shows the microstructures of the alloy processed with 16 passes by RD-ECAP at three different temperatures: (a) 573 K, (b) 623 K, and (c) 673 K. The particle distribution, including eutectic silicon and intermetallic compounds, seems to have become more homogeneous when the processing temperature increased from 573 to 673 K.

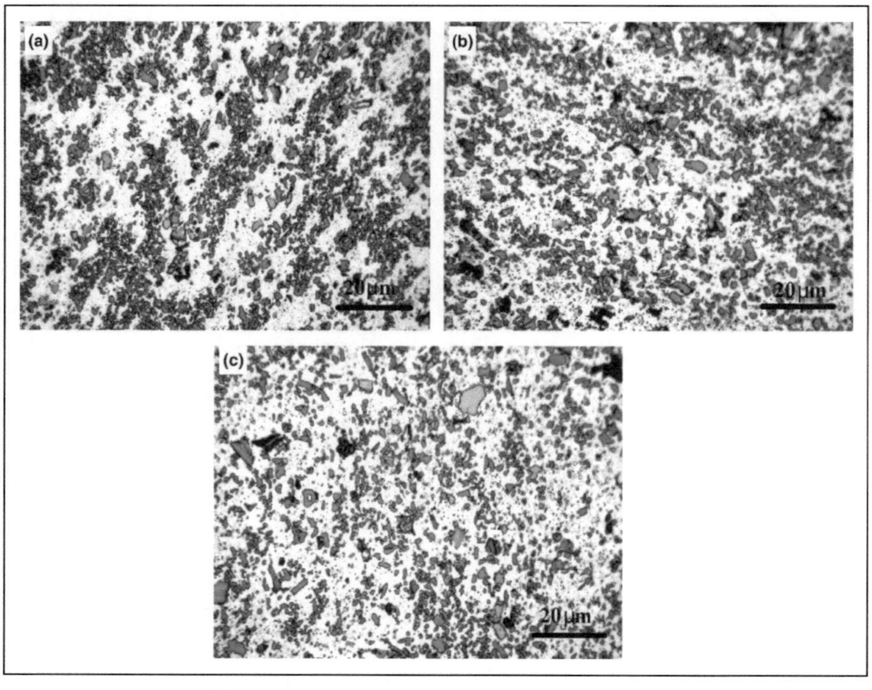

Fig. 15. Microstructures of the Al–11mass%Si alloy processed by RD-ECAP for 16 passes at: (a) 573 K, (b) 623 K, and (c) 673 K. (A. Ma, K. Suzuki, Y. Nishida, N. Saito, I. Shigematsu, M. Takagi, H. Iwata, A. Watazu, T. Imura: Acta Materialia 53 (2005) 211–220.)

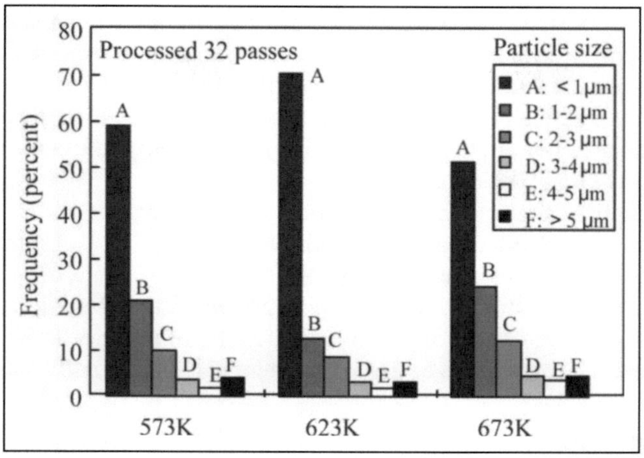

Fig. 16. Effect of the RD-ECAP processing temperature on the particle size distribution in the alloy. (A. Ma, K. Suzuki, Y. Nishida, N. Saito, I. Shigematsu, M. Takagi, H. Iwata, A. Watazu, T. Imura: Acta Materialia 53 (2005) 211–220.)

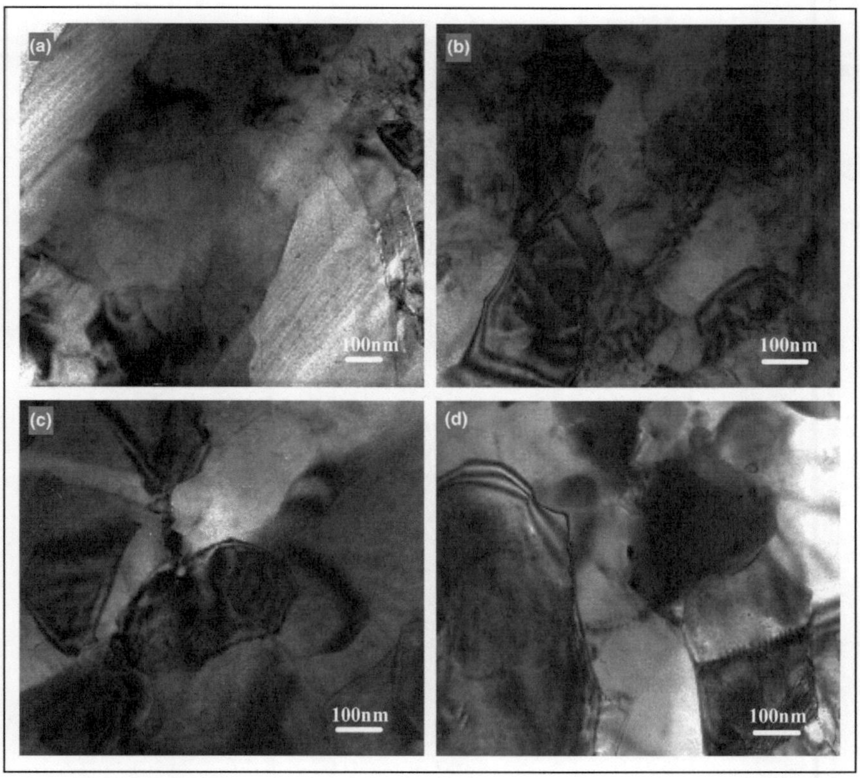

Fig. 17. Transmission electron micrographs of matrix of aluminium in the Al-11mass%Si alloy. (a) as-cast alloy, (b)processed 4 passes, (c) 16 passes, (d) 32 passes by RD-ECAP at 573 K. (A. Ma, K. Suzuki, Y. Nishida, N. Saito, I. Shigematsu, M. Takagi, H. Iwata, A. Watazu, T. Imura: Acta Materialia 53 (2005) 211–220.)

Fig. 16 illustrates the particle size distribution after 32 RD-ECAP passes at three different processing temperatures. Among the three samples, the sample processed at 623 K had the highest content of particles smaller than 1 μm. However, no difference in the large particle content (> 2 μm in diameter) was not clear among the samples processed at the three different temperatures. This result indicates that the processing temperature had little effect on the distribution of large particles.

Fig. 17 shows transmission electron micrographs of aluminum matrix in the Al–11mass%Si alloy, with (a) showing the microstructure of the as-cast alloy, and (b), (c) and (d) showing the microstructures of samples processed by RD-ECAP at 573 K with 4, 16 and 32 passes respectively. It is clear that the grain or grain fragment size of the aluminum was refined after only 4 passes. In spite of the further increase in the number of RD-ECAP passes to 32, the alloy maintained the same grain or grain fragment size of about 200–400 nm.

3.2.2 Impact toughness

Fig. 18 shows typical load–displacement curves for the Al–11mass%Si alloy, where curve (a) is the as-cast sample, and (b), (c) and (d) are samples processed by RD-ECAP at 623 K for 4, 16 and 32 passes, respectively. The area below the load–displacement curve of (a) shows the absorbed energy of the as-cast alloy, which is very small in comparison with the results from the other samples.

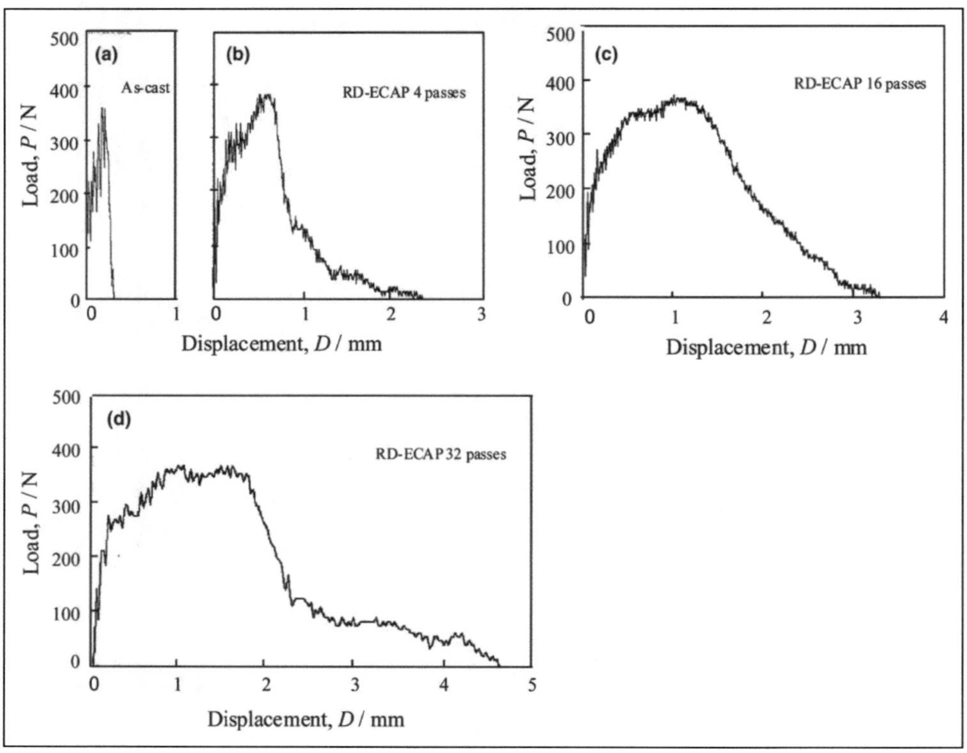

Fig. 18. Typical load–displacement curves of the Al–11mass%Si alloys: (a) as-cast state, (b) (c) and (d) processed by RD-ECAP at 623 K for 4, 16 and 32 passes, respectively. (A. Ma, K. Suzuki, Y. Nishida, N. Saito, I. Shigematsu, M. Takagi, H. Iwata, A. Watazu, T. Imura: Acta Materialia 53 (2005) 211–220.)

Fig. 19 shows the relationship between the absorbed energy of the sample during impact testing and the number of RD-ECAP passes. The absorbed energy of the as-cast Al–11mass%Si alloy was 0.9 J/cm2. After RD-ECAP, the absorbed energy increased markedly with the increasing number of RD-ECAP passes at all three processing temperatures, ultimately reaching 10 J/cm^2 after 32 passes at 623 K. This value is 10 times that of the as-cast Al–11mass%Si alloy. The relation of RD-ECAP temperature to impact toughness is also shown in Fig. 19, indicating little effect of temperature when the number of RD-ECAP passes is fewer than 12. However, when the number of RD-ECAP passes exceeds 12, a marked effect of RD-ECAP processing temperature on impact toughness is readily observed. This result indicates the existence of a better temperature for RD-ECAP when the number of RD-ECAP passes exceeds 12. For the alloy used in this study, the optimal temperature for RD-ECAP is around 623 K. The effect of the processing route of RD-ECAP on impact toughness is also illustrated in Fig. 19. Using the same number of pressing passes, 16, the additional four routes described in the above section and marked A, B, C and D in Fig. 19 were attempted to achieve high impact toughness. It is evident that the impact toughness of the Al–11mass%Si alloy samples processed by routes A and B were significantly higher than those by other routes; i.e., the samples processed by RD-ECAP for 8 or 4 passes at 673 K followed by 8 or 12 passes at 623 K exhibited relatively high impact toughness.

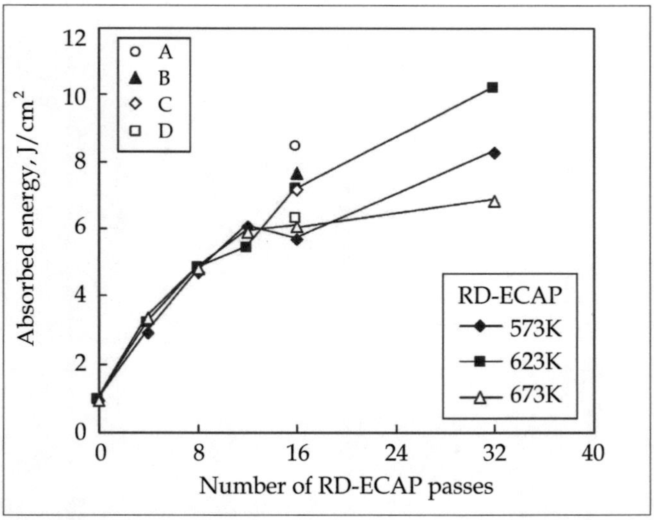

Fig. 19. The absorbed energy of the samples as a function of the number of RD-ECAP passes at 573, 623, 673 K and other routes: (A) at 673 K for 8 passes followed 8 passes at 623 K; (B) at 673 K for 4 passes followed by 12 passes at 623 K; (C) at 573 K for 4 passes followed by 4 passes at 673 K and 8 passes at 623 K; (D) at 673 K for 4 passes followed by 4 passes at 623 K and 8 passes at 573 K. (A. Ma, K. Suzuki, Y. Nishida, N. Saito, I. Shigematsu, M. Takagi, H. Iwata, A. Watazu, T. Imura: Acta Materialia 53 (2005) 211–220.)

Fig. 20 shows Charpy impact test pieces without notches after impact tests, in which (a) represents that processed by RDECAP at 623 K for 32 passes and (b) the as-cast alloy. Using the test piece without the notch, the absorbed energy of the sample processed by RD-ECAP at 32 passes could not be obtained because the test piece bent considerably, as shown in Fig. 20(a). However, the test pieces of the unpressed samples fractured in a brittle manner.

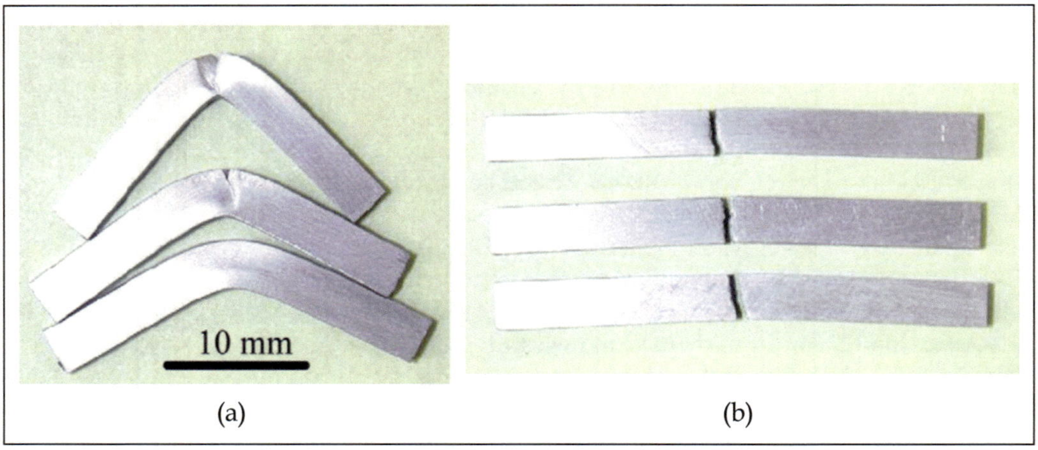

Fig. 20. Impact toughness test pieces without notches of the Al–11mass%Si alloy after testing: (a) processed by RD-ECAP at 623 K for 32 passes; (b) as-cast. (A. Ma, K. Suzuki, Y. Nishida, N. Saito, I. Shigematsu, M. Takagi, H. Iwata, A. Watazu, T. Imura: Acta Materialia 53 (2005) 211–220.)

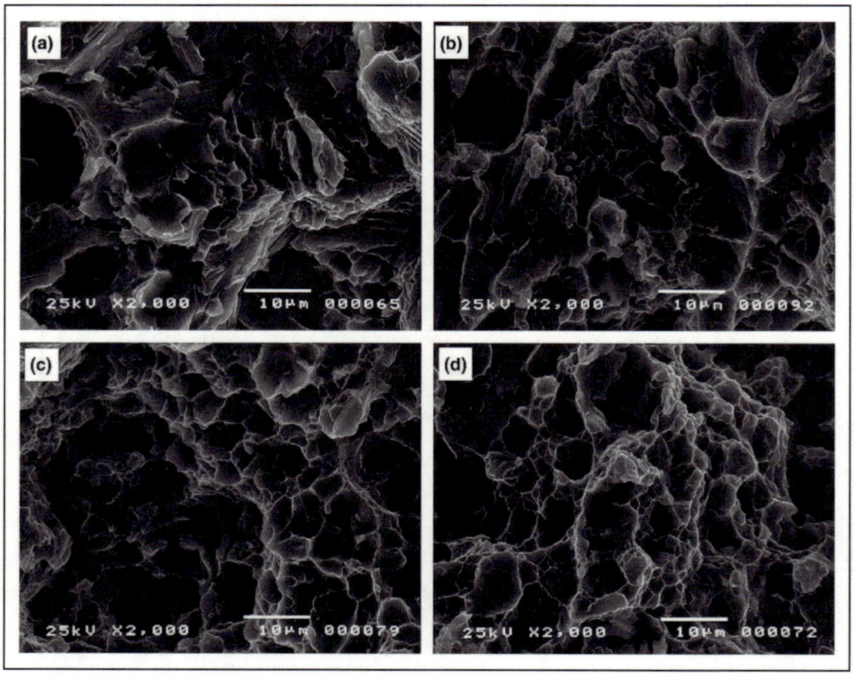

Fig. 21. SEM observations for the fractured surfaces of the Al–11mass%Si alloy: (a) as-cast, (b), (c) and (d) processed by RD-ECAP at 623 K for 4, 16 and 32 passes, respectively. (A. Ma, K. Suzuki, Y. Nishida, N. Saito, I. Shigematsu, M. Takagi, H. Iwata, A. Watazu, T. Imura: Acta Materialia 53 (2005) 211–220.)

Fig. 21 shows the fractured surfaces of the test pieces after impact testing, observed by SEM, with (a) representing the as-cast alloy, and (b), (c) and (d) the samples processed by

RD-ECAP at 623 K for 4, 16 and 32 passes, respectively. The as-cast sample shows a rough surface due to the large grains in the alloy and the particles of eutectic silicon and intermetallic compounds at the grain boundaries. The sample processed by RDECAP at 623 K for 4 passes, (b) shows a finer fracture surface, including a couple of dimples, compared to the as-cast sample. The sample RD-ECAPed for 16 passes, (c) shows a fine and homogeneous fracture surface with many dimples. However, little difference can be observed between (c) and (d). A side view of the ductile fracture surface of the Al–11mass%Si alloy processed by RDECAP at 623 K for 32 passes is shown in Fig. 22. The high magnification reveals a typical ductile fracture surface.

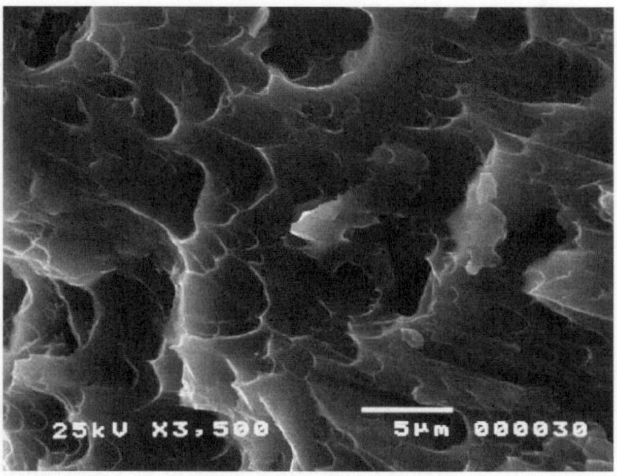

Fig. 22. Side view of a typical ductile fracture surface of the Al–11mass%Si alloy processed by RD-ECAP at 623 K 32 passes. (A. Ma, K. Suzuki, Y. Nishida, N. Saito, I. Shigematsu, M. Takagi, H. Iwata, A. Watazu, T. Imura: Acta Materialia 53 (2005) 211–220.)

3.2.3 Discussion

As illustrated in Fig. 19, the absorbed energy of the Al–11mass%Si alloy increases with increasing number of RD-ECAP passes. However, after an abrupt increase in the first few passes, generally 4, the increment of absorbed energy gradually levels off with increased RDECAP passes, indicating that the first four passes have the greatest effect on impact toughness. This result is related to the microstructure of the as-cast Al–11mass%Si alloy. As shown in Fig. 13, the microstructure of the as-cast Al–11mass%Si alloy consists of large aluminum grains, including large dendrites and interdendritic networks of eutectic silicon plates, which are the primary reason for the low impact toughness of this alloy. We therefore conclude that breaking up this microstructure and dispersing the eutectic silicon results in improved impact toughness. It appears that the first four RD-ECAP passes do most of the work of breaking the microstructure of the large aluminum dendrites and interdendritic networks of eutectic silicon in the alloy. In fact, during the first 4 RD-ECAP passes, the grain or grain fragment sizes of this alloy are also significantly refined, as shown in Fig. 17(b).

The signal effect of ECAP, as reported in several recent works, is the modification of the grain boundaries. Misorientation angles of grain boundaries are clearly modified during

ECAP. In the present study, RD-ECAP had a similar effect on the grain or grain fragment boundaries. Electron backscatter diffraction (EBSD) can be used to analyze the distribution of misorientation angles (h) for the aluminum matrix in the RD-ECAPed samples since the grains or grain fragments are very small; i.e., the second particle phase can be ignored. The results for EBSD show that the fraction of high angle boundaries with h > 15 are 65% and 73% for the samples processed by RD-ECAP at 623 K for 8 and 16 passes, respectively. Therefore, the modified grain boundaries produced during RD-ECAP are likely to be an important factor in enhancing impact toughness.

In fact, the experimental results suggest that modified boundaries affect impact toughness. As shown in Figs. 17(b)–(d) and 3, although the diameter of grains or grain fragments did not clearly decrease and the distribution of the larger particles (over 2 μm in diameter) in the sample did not greatly change with increasing the number of RD-ECAP passes, the absorbed energy steadily increased with increased number of RD-ECAP passes, as shown in Figs. 18(b)–(d) and 8. This means that, except for the grain or grain fragment size and the aspect of the particles, other factors such as modified grain or grain fragment boundaries appear to be having an effect during impact testing for improving toughness. We therefore think that if the modified boundaries were eliminated, the impact toughness of this alloy would greatly decrease. To investigate the effect of the modified grain boundaries and the larger silicon particles on the impact toughness, we made two kinds of samples and measured the absorbed energy using Charpy impact tests:

- Sample 1 was processed with 16 passes at 623 K by RD-ECAP, then heat-treated at 793 K for 2 h followed by water quenching (solution treatment).
- Sample 2 was heat-treated under T6 conditions (after solution treatment, aged at 443 K for 10 h) followed by RD-ECAP for 16 passes at 573 K.

Impact testing of sample 1 was carried out immediately after the solution treatment. In this case, the recrystallization of the aluminum alloy would take place during the solution treatment. However, since the precipitation of fine particles before impact testing may have been negligible, the effect of particle precipitation at the boundaries could be ignored. As shown in Fig. 12, large particles, including eutectic silicon and intermetallic compounds, other than the small amounts dissolved in the aluminum matrix during solution treatment, were still evenly distributed in the alloy due to the fact that they were evenly distributed during RD-ECAP. The grain or grain fragment size increased to around 8 μm, meaning the modified grain or grain fragment boundaries were eliminated; however, the average silicon particle size also increased with the disappearance of small particles, compared with Fig. 15(b). Impact testing results show that the absorbed energy of sample 1 fell markedly from 7.2 to 4.0 J/cm^2 after the solution treatment.

As shown in Fig. 13, the large particle size in sample 2 is as large as that in sample 1 but clearly larger than that in the sample shown in Fig. 4(a). However, sample 2 exhibited a higher impact toughness (absorbed energy is 7.1 J/cm2) than the sample shown in Fig. 4(a). This result indicates that increased silicon particle size is not the reason for the impact toughness reduction of sample 1. Therefore, three factors are likely to be the chief reasons for the loss of impact toughness after the solution treatment: (a) elimination of the modified boundaries, (b) increased grain or grain fragment size, and (c) disappearance of small particles. On the other hand, although the grain or grain fragment size did not clearly decrease with the increased number of RD-ECAP passes over 4 passes, the impact toughness still markedly increased on increasing the number of RD-ECAP passes from 4 to 32, as shown in Fig. 8. This result means that the incremental value of impact toughness may be related to grain or grain fragment boundary modification and the increase in the proportion of fine particles (smaller than 1 μm) because, as stated above, the degree of boundary modification and the fine particle content increased with increased numbers of RD-ECAP passes.

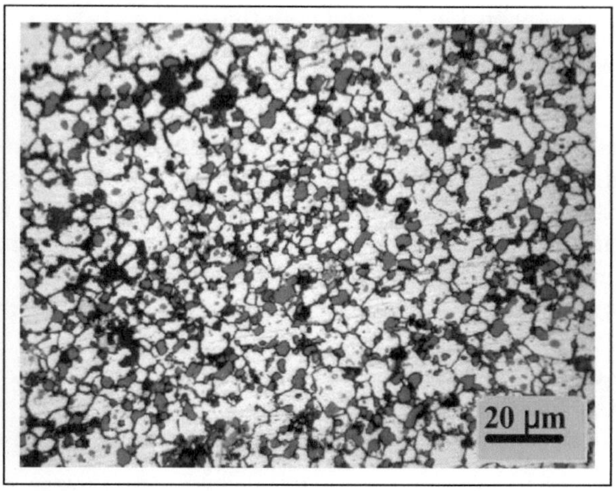

Fig. 23. Microstructure of the Al–11mass%Si alloy processed 16 passes at 623 K by RD-ECAP followed solution treatment at 793 K for 2 h. (A. Ma, K. Suzuki, Y. Nishida, N. Saito, I. Shigematsu, M. Takagi, H. Iwata, A. Watazu, T. Imura: Acta Materialia 53 (2005) 211–220.)

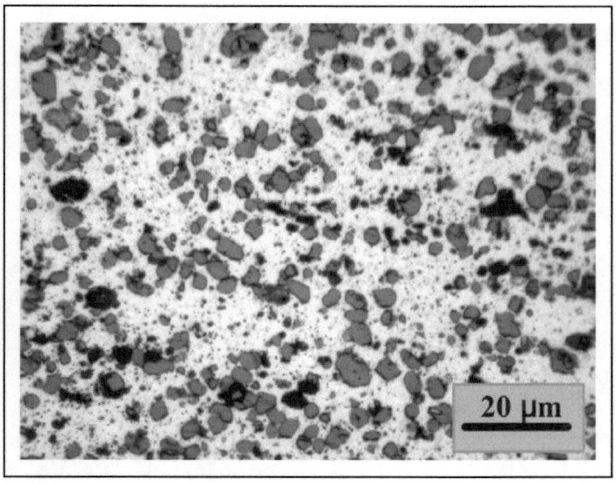

Fig. 24. Microstructure of the Al–11mass%Si alloy heat-treated under T6 conditions (793 K for 2 h then at 443 K for 10 h) followed RD-ECAP at 573 K for 16 passes. (A. Ma, K. Suzuki, Y. Nishida, N. Saito, I. Shigematsu, M. Takagi, H. Iwata, A. Watazu, T. Imura: Acta Materialia 53 (2005) 211–220.)

3.3 Other aluminum alloy

In the present work, various other aluminum alloys such as Al-23 mass% Si alloy, Si-whisker/extra super duralumin composite, etc., were studied for processing by

RD-ECAP. In the case of Al-23 mass% Si alloy, absorbed energy of a sample processed 32 passes was about 18 times higher than that of a sample processed 0 pass. In the case of Si-whisker/extra super duralumin composite, after pressing by RD-ECAP for 10 passes, grain size was 1.5-2 µm and Si-whisker distribution became homogeneous. Also, Mg alloys and titanium processed by RD-ECAP were studied and the results confirmed that RD-ECAP is useful for forming fine-grained light metal materials.

4. Conclusion

A new ECAP processing method called rotary-die equal channel angular pressing (RD-ECAP) was developed at Japan's National Institute of Advanced Industrial Science and Technology (AIST, formerly the National Industrial Research Institute of Nagoya (NIRIN)), to form fine-grained bulk materials such as aluminum alloys, aluminum composites, magnesium alloy, and titanium. RD-ECAP has the following features:

1. ECAP processing of up to 32 passes (one pass=one extrusion) can be done without sample removal.
2. RD-ECAP saves energy because there is no cooling and re-heating.
3. One-pass RD-ECAP can be processed in 30 s.
4. Over 30 cycles (one cycle=one extrusion and 90° die rotation) of processing were possible in a short time.
5. Aluminium material with fine grain and high impact toughness can be formed.

Researches on aluminium processing by ECAP deformation of more than 20 passes are still very few in the world. Therefore, other excellent effects or features are expected to be discovered in the future.

5. Acknowledgment

The author thanks Dr. Yoshinori Nishida and Dr. Aibin Ma for his advice and material offer.

6. References

Y.H. Zhao, Y.T. Zhu, X.Z. Liao, Z. Horita and T.G.Langdon: *Mater. Sci. Eng.* Vol. A463 (2007), p. 22-26.

R.Z. Valiev, A.V. Korznikove and R.R. Mulyukov: *Mater. Sci. Eng.* Vol. 168 (1993), p. 141.

V.M. Segal, V.I. Rexnikov, A.E. Drobysevsky and V.I. Kopylov: *Metally* Vol. 1 (1981), p. 115.

S.L. Semiatin, V.M. Segal, R.E. Goforth, N.D. Frey and D.P. Delo: *Metall. Mater. Trans. A,* Vol. 30A (1999), p. 1425.

Y. Nishida, H. Arima, J.C. Kim and T. Ando: *J. Japan Inst. Light Metals.* Vol. 50-12 (2000), p. 655-659.in Jp.

Y. Nishida, H. Arima, J.C. Kim and T. Ando: *J. Japan Inst. Metals.* Vol. 64-12 (2000), p. 1224-1229.in Jp.

A. Ma, K. Suzuki, N. Saito, Y. Nishida, M. Takagi, I. Shigematsu, H. Iwata: *Mater. Sci. Eng. A,* Vol. 399 (2005), p. 181-189.

A. Ma , K. Suzuki, Y. Nishida, N. Saito, I. Shigematsu, M. Takagi, H. Iwata, A. Watazu, T. Imura: *Acta Materialia* Vol. 53 (2005) , p. 211.

A. Watazu, I. Shigematsu, A. Ma, K. Suzuki, T. Imai, N. Saito: *Mater. Trans.* Vol. 46 (2005), p. 2098.

A. Watazu, I. Shigematsu, X. Huang, K. Suzuki and N. Saito: *Mater. Sci. Forum* Vol. 544 (2007), p. 419.

Part 2

Welding of Aluminium Alloys

4

Welding of Aluminum Alloys

R.R. Ambriz and V. Mayagoitia
Instituto Politécnico Nacional CIITEC-IPN,
Cerrada de Cecati S/N Col. Sta. Catarina C.P. 02250, Azcapotzalco, DF,
México

1. Introduction

Welding processes are essential for the manufacture of a wide variety of products, such as: frames, pressure vessels, automotive components and any product which have to be produced by welding. However, welding operations are generally expensive, require a considerable investment of time and they have to establish the appropriate welding conditions, in order to obtain an appropriate performance of the welded joint. There are a lot of welding processes, which are employed as a function of the material, the geometric characteristics of the materials, the grade of sanity desired and the application type (manual, semi-automatic or automatic). The following describes some of the most widely used welding process for aluminum alloys.

1.1 Shielded metal arc welding (SMAW)

This is a welding process that melts and joins metals by means of heat. The heat is produced by an electric arc generated by the electrode and the materials. The stability of the arc is obtained by means of a distance between the electrode and the material, named *stick welding*. Figure 1 shows a schematic representation of the process. The electrode-holder is connected to one terminal of the power source by a welding cable. A second cable is connected to the other terminal, as is presented in Figure 1a. Depending on the connection, is possible to obtain a direct polarity (Direct Current Electrode Negative, DCEN) or reverse polarity (Direct Current Electrode Positive, DCEP). The core wire of the coated electrode conducts the electric current and it provides filler metal to perform the weld.
The heat of the arc melts the wire core and the coating (flux) at the tip of the electrode. The melt material is transferring to the base metal in a drop shape, as is showed in Figure 1b. The molten metal is stored in a weld pool and it solidifies in the base metal. The flux due to its low density floats to the surface of the weld pool and solidifies as a layer of slag in the surface of the weld metal.
The electrode covering contents some chemical compounds, which are intended to protect, deoxidize, stabilize the arc and add alloy elements. There are basically four types of electrode coating types: (i) *Cellulosic* (20-60% rutile, 10-20% cellulose, 15-30% quartz, 0-15% carbonates, 5-10% ferromanganese), which promotes gas shielding protection in the arc region, a deep penetration and fast cooling weld. (ii) *Rutile* (40-60% rutile, 15-20% quartz, 0-15% carbonates, 10-14% ferromanganese, 0-5% organics), this is employed to form slags mainly for slag shielding, it presents high inclusion content in weld deposit. (iii) *Acid* (iron

ore-manganese ore, quartz, complex silicates, carbonates ferromanganese), which provides fairly high hydrogen content and high slag content in the weld. (iv) *Basic* (20-50% calcium carbonate, 20-40% fluorspar, 0-5% quartz, 0-10% rutile, 5-10% ferro-alloys), this coating brings low high hydrogen levels (≤ 10 ppm) and electrodes can be kept dry (Easterling, 1992). Because the temperature is high, the covering of the electrode produces a shielding gas for the molten metal. During the welding process, the covering of the electrode reacts to eliminate the oxides produced in the fusion process and it cleans the weld metal. Also, the slag formed in the solidification process protects the weld metal, especially when the temperature is too high. The electric arc is produced by the ionization of the gases (plasma), which conduct the electric current. Arc stabilizers are compounds that decompose into ions arc in the form of oxalates of potassium and lithium carbonate. They increase the electrical conductivity and improve to conduct the current in softer form. Additionally, the electrode covering also provides alloying elements and/or metallic powders to the weld metal. Alloying elements tend to control the chemical composition of the weld metal; metallic powders tend to increase the deposit rate.

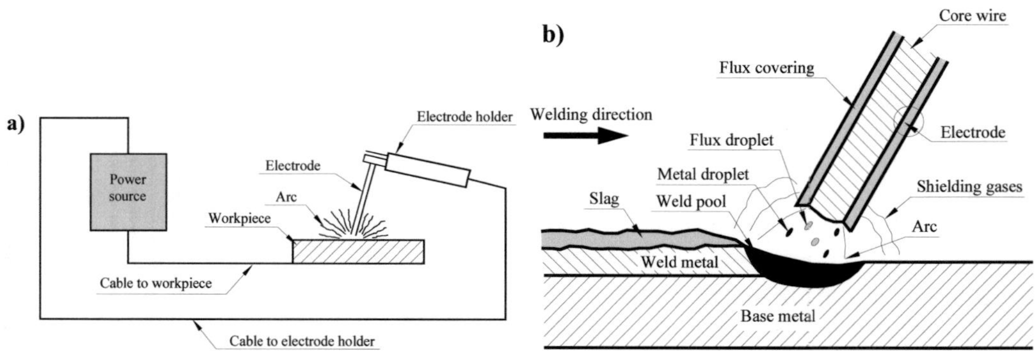

Fig. 1. Shielded metal arc welding process

Some advantages of the SMAW welding process is that it is portable and not expensive compared with others. These features allow that the SMAW process can be employed in maintenance, repair operations, production of structures or pressure vessels. However, in welding of aluminum alloys and titanium, the welding process does not provides a sufficient degree of cleaning because the gas produced by the coating is not enough to obtained welds free of defects and discontinuities. On the other hand, the deposit rate is limited because the electrodes must be changed continuously due to its length and the operator must be stop.

1.2 Gas metal arc welding (GMAW)

GMAW is a welding process that melts and joins metals by heating employing an electric arc established between a continuous wire (electrode) and metals to be welded, as is shown in Figure 2. Shielding protection of the arc and molten metal is carried out by means of a gas, which can be active or inert. In the case of aluminum alloys, non ferrous alloys and stainless steel Ar gas or mixtures of Ar and He are employed, whereas for steels the base of the shielding gases is CO_2. When using an inert gas, it is kwon as MIG process (Metal Inert Gas) and MAG when Metal Active Gas is used. GMAW process is one of the most employed to weld aluminum alloys.

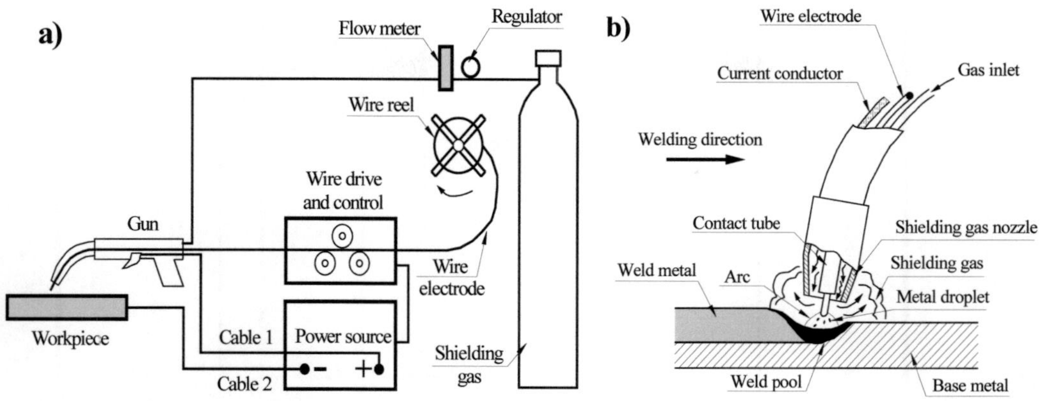

Fig. 2. Gas metal arc welding process

When using an Ar gas arc, the arc energy has a smaller spread than an arc of He, due to the low thermal conductivity of Ar. This aspect helps to obtain a metal transfer more stable and an axial Ar plasma arc. The shielding gas effect on aluminum welding is presented schematically in Figure 3. The penetration pattern is similar to a bottle nipple when using Ar, whereas when using He, the cross-sectional area has a parabolic penetration.

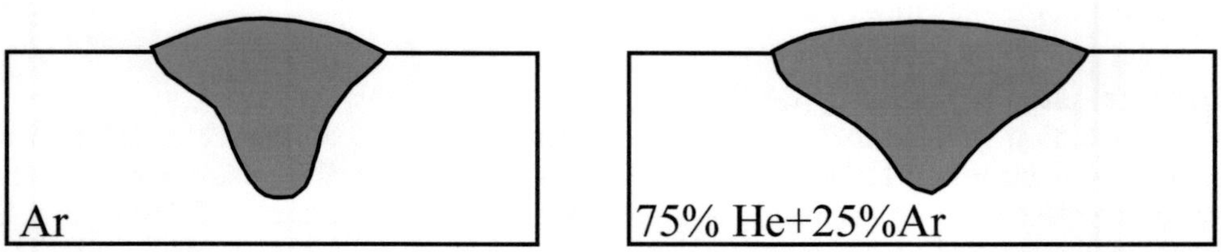

Fig. 3. Schematic representation in aluminum welds using different shielding gases

There are three basics metal transfer in GMAW process: *globular transfer*, *spray transfer* and *short-circuiting transfer* (Kou, 2003).

In the *globular transfer*, metal drops are larger than the diameter of the electrode, they travel through the plasma gas and are highly influenced by the gravity force. One characteristic of the globular transfer is that this tends to present, spatter and an erratic arc. This type of metal transfer is present at low level currents, independently of the shielding gas. However, when using CO_2 and He, globular transfer can be obtained at all current levels.

On the other hand, *spray transfer* occurs at higher current levels, the metal droplets travel through the arc under the influence of an electromagnetic force at a higher frequency than in the globular transfer mode. In this transfer mode, the metal is fed in stable manner and the spatter tends to be eliminated. The critical current level depends of the material, the diameter of the electrode and the type of shielding gas.

In *short-circuiting transfer*, the molten metal at the electrode tip is transferred from the electrode to the weld pool when it touches the pool surface, that is, when short-circuiting occurs. The short-circuiting is associated with lower levels of current and small electrode diameters. This transfer mode produces a small and fast-freezing weld pool that is desirable for welding thin sections, out-of-position welding and bridging large root openings. Figure 4, shows the typical range of current for some wire diameters.

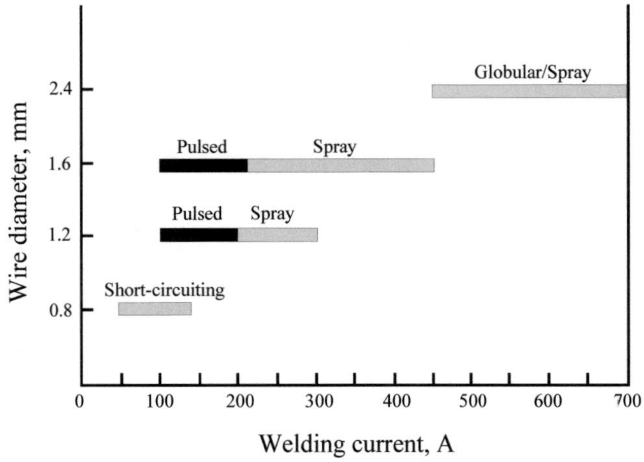

Fig. 4. Typical welding current ranges for wire diameter and welding current

The principal advantages of GMAW process with respect to SMAW process are: (i) There is not production of slag. (ii) Is possible to perform welds in all welding positions. (iii) Rate deposition is roughly two times than SMAW. (iv) Quality of the welds is very good. (v) Is possible to weld materials with a short-circuiting transfer mode, which tends to improve the reparation and maintenance operations.

1.3 Gas tungsten arc welding (GTAW)

This is a welding process that melts metal by heat employing an electric arc with a non consumable electrode. GTAW process employs an inert or active shielding gas, which protects the electrode and the weld metal. A schematic representation of GTAW process is showed in Figure 5. The arc functions as a heat source, which can be directly used for welding, with or without the use of filler materials. This process produces high quality welds, but the principal disadvantage is that the rate of deposition is slow and it limits the range of application in terms of thickness. For instance, in welding of aluminum alloys it is convenient to use this welding process in thickness no greater than 6 mm, since greater thicknesses require a large number of passes and the welding operation tends to be expansive and slow.

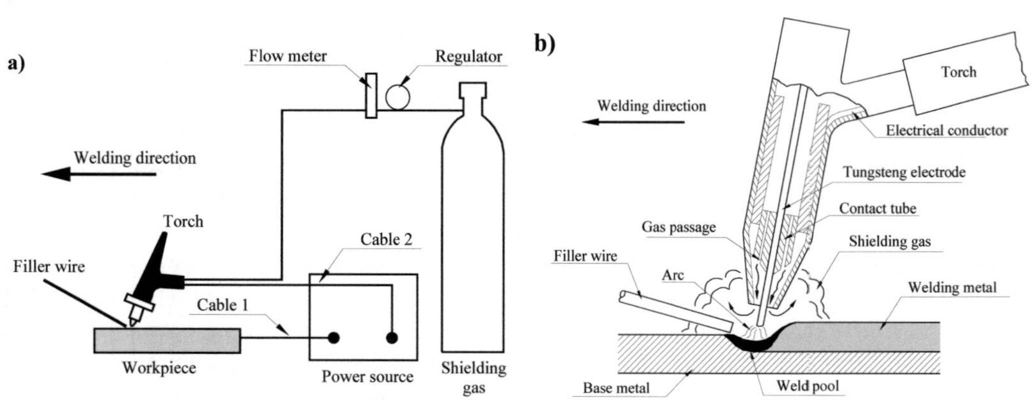

Fig. 5. Schematic representation of GTAW process

It is possible to use *Direct Current* (DC) or *Alternating Current* (AC) to weld by GTAW. In the case of DC, we can use direct polarity (electrode negative, DCEN) or reverse polarity (positive electrode, DCEP), Figure 6, shows the polarity effect on the weld.
Direct polarity is the most commonly employed in GTAW. In this case, the electrode is connected to the negative pole of the heat source and the electrons are emitted from the electrode and they are accelerated as they travel through the arc (plasma). This effect produces a high heat in the workpiece and therefore gives a good penetration and a relatively narrow weld shape.

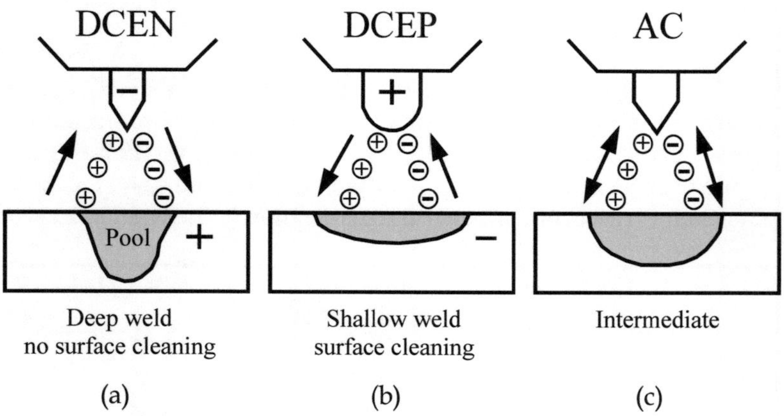

Fig. 6. Polarity in GTAW

On the other hand, when reverse polarity is used, the electrode is connected to the positive pole of the heat source. Now the effect of the heating due to the bombardment of the electrons is higher in the electrode than that of workpiece, which results in a wide weld bead and shallower than that generated by direct polarity. In this case, due to high energy concentration in the electrode it is necessary to employ a thicker diameter and a cooling system to eliminate the electrode tip melting possibility. The bombardment effect by positive ions of the inert gas removes the oxide film and produces a cleaning effect on the welding surface. Therefore, reverse polarity can be used to weld materials that are resistant to oxides such as aluminum and magnesium, if it is not required a high penetration.
When alternating current is used, is possible to obtain a good combination of oxides elimination (cleanliness) and penetration, as is presented in Figure 6. This polarity is the most employed to weld aluminum alloys.
There are several types of electrodes to weld by GTAW. These include pure tungsten and tungsten alloyed with thorium oxide (ThO_2) or zirconium oxide (ZrO_2), which are added to improve the arc ignition and to increase the life of electrode. In the last years some alloy elements have been incorporated, such as cerium and lanthanum, which also increase the life of the electrode and tend to decrease the risk of radiation that is produced when electrodes of high thorium content are employed. Zirconium electrodes are preferred for AC, because they present a higher melting point than pure tungsten or tungsten-thorium.
During the welding process, it is assumed that the electrode tip is hemispherical type. This is a very important aspect, because the arc stability depends in a greater manner of tip geometry. There are electrodes of varies diameters, which can range from 0.3 to 6.4 mm. Table 1 presents the recommended current ratings for different diameters of electrodes using Ar shielding gas.

Electrode diameter, mm	Current, A
1.0	20-50
1.6	50-80
2.4	80-160
3.2	160-225
4.0	225-330
5.0	330-400
6.4	400-550

Table 1. Recommended electrode diameters and current range employed with Ar shielding gas

Table 2, presents the gases used as a protection, in GMAW and GTAW. A shielding gas is selected according to their ionization potential, density, degree of protection and the effect of oxides removal. For example, it is easier to ionize an Ar gas (15.7 eV) than a He gas (24.5 eV), and due to this effect the arc ignition tends to be more easy. Furthermore, the Ar density is higher than He and consequently the penetration of the weld bead is better.

Gas	Chemical symbol	Molecular weigth, g/mol	Density, g/L	Ionization potential, eV
Argon	Ar	39.95	1.784	15.7
Carbon dioxide	CO_2	44.01	1.978	14.4
Helium	He	4.00	0.178	24.5
Hydrogen	H_2	2.016	0.090	13.5
Nitrogen	N_2	28.01	1.25	14.5
Oxygen	O_2	32.00	1.43	13.2

Table 2. Gas shielding properties employed in GMAW and GTAW (Kou, 2003)

1.4 Friction stir welding (FSW)

Friction-Stir Welding (FSW) is a solid-state, hot-shear joining process (Thomas et al.; 1991, Thomas & Dolby, 2003, Dawes & Thomas, 1996, Mishra & Ma, 2005). The process utilizes a bar-like tool in a wear-resistant material (generally tool steel for aluminum) with a shoulder and terminating in a threaded pin. This tool moves along the butting surfaces of two rigidly clamped plates placed on a backing plate. The shoulder makes a contact with the top surface of the plates to be welded. The heat generated by friction at the shoulder and to a lesser extent at the pin surface and it softens the material being welded. Severe plastic deformation and flow of this plasticised metal occurs as the tool is translated along the welding direction. The material is transported from the front of the tool to the trailing edge where it is forged into a joint. Figure 7 shows a schematic representation of FSW.

There are two principal parameters in FSW: tool rotation rate (ω, rpm) in clockwise or counterclockwise direction and the tool traverse speed (v, mm/min) along the line of joint. The rotation of the tool results in stirring and mixing of material around the rotating pin and the translation of the tool moves the stirred material from the front to the back of the pin and finishes welding process. Additionally, the angle of spindle or tool tilt and pressure are other important process parameters. A suitable tilt of the spindle towards trailing direction ensures that the shoulder of the tool holds the stirred material by threaded pin and move material efficiently from the front to the back of the pin. The heat generation rate,

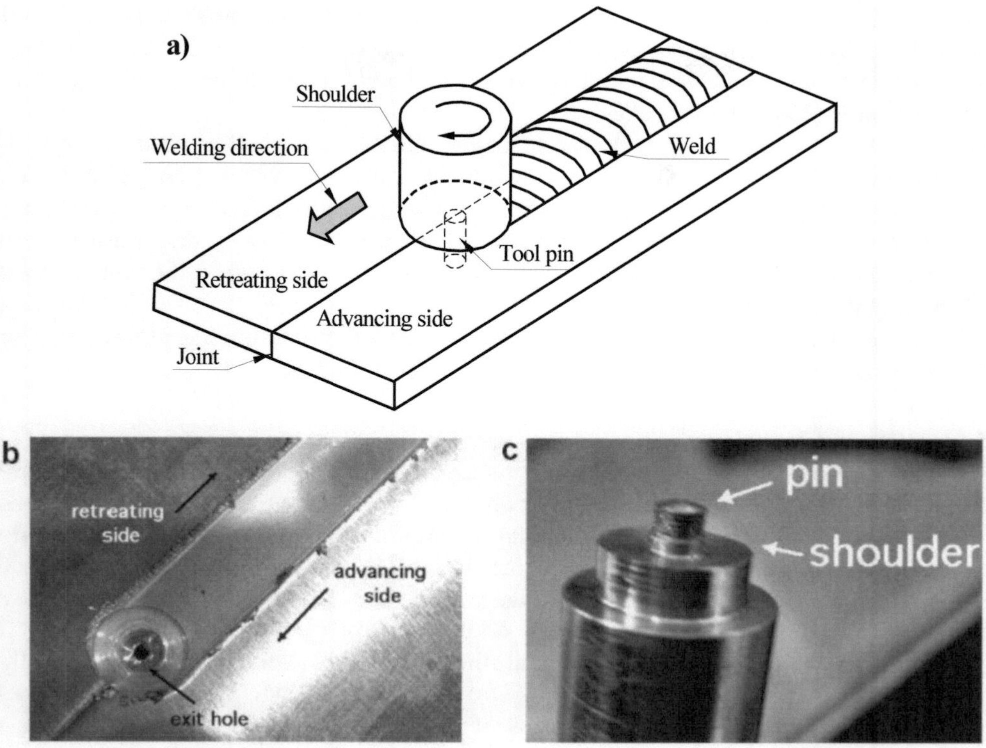

Fig. 7. Illustration of the friction-stir welding process, b) weld between aluminum sheets and c) An actual tool with a threaded-pin

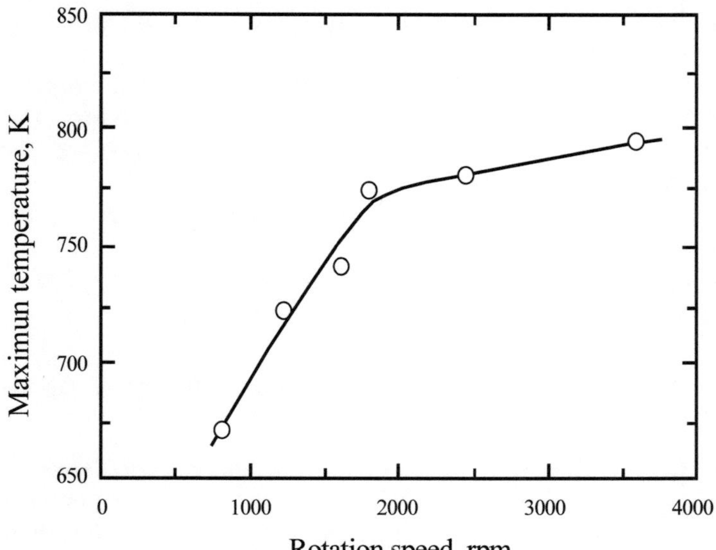

Fig. 8. Relationship between rotational speed and peak temperature in FSW of AA6063 (Sato et al., 2002)

temperature field, cooling rate, *x*-direction force, torque and power are totally depended of the welding speed, the tool rotation speed, the vertical pressure on the tool, etc. Figure 8, shows a relationship between rotational speed and peak temperature in FSW of a 6063 aluminum alloy (Nandan et al., 2008).

FSW enables long lengths of weld to be made without any melting taking place. This provides some important metallurgical advantages compared with fusion welding, i.e. no melting means that solidification and liquation cracking are eliminated; dissimilar alloys can be successfully joined; the stirring and forcing action produces a fine-grain structure. However, one disadvantage is that the keyhole (exit hole) remains when the tool is retracted at the end of the joint (Figure 7b).

Several alloys have been welded by FSW, they included the following aluminum alloys: 5083, 5454, 6061, 6082, 2014, 2219 and 7075 (Nandan et al., 2008).

1.5 Modified indirect electric arc welding technique (MIEA)

Although, welding of aluminum alloys is relatively easy employing friction stir welding, when the thickness is thick a fusion welding process is usually required to join these materials. In the case of a fusion welding process, a large amount of heat input can be dissipated via heat conduction throughout the base material close to the welded zone. Typically, this thermal dissipation induces localized isothermal sections where the thermal gradient can have important and detrimental effects on the microstructure and therefore on the mechanical properties of the material constituting the heat affected zone (HAZ), specially in aluminum alloys hardening by artificial ageing (Myhr et al., 2004). In order to improve the mechanical and microstructural conditions of the welded joint in aluminum alloys, the Modified Indirect Electric Arc (MIEA), has been developed (Ambriz at al., 2006, Ambriz et al. 2008). This welding technique is based on a simple joint modification which provided several advantages with respect to the traditional arc fusion welding process, for instance:

i. The high thermal efficiency that allows welding plates by using a single welding pass. As a result, the thermal effect is reduced and the mechanical properties of the HAZ are improved as compared to a multi-pass welding procedure,
ii. The dilution percent of the weld pool is higher; which tends to improve the hardening effect after performing a post weld heat treatment (PWHT) (Ambriz et al., 2008),
iii. The solidification mode promotes an heterogeneous nucleation and jointly diminishes the micro-porosity formation,
iv. The geometry of the welding profile improves the fatigue performance of the welded joint (Ambriz et al., 2010a).

MIEA welding technique employs the same equipment that is required to weld by GMAW. A schematic representation of the MIEA joint is present in Figure 9.

2. Importance of microstructure and mechanical properties on aluminum welds

After welding, the microstructure and mechanical properties conditions are the principal aspects that determine the appropriate perform in structures and components of aluminum alloys. It means that it is necessary to know exactly the mechanical behavior of the welded joint, including the global and local mechanical properties. This is necessary because the temperature susceptibility of some aluminum alloys tends to change in a great manner the microstructure conditions in the fusion zone and in the HAZ. Here are some results on the microstructural and mechanical conditions in welding of aluminum alloys, especially for FSW and MIEA.

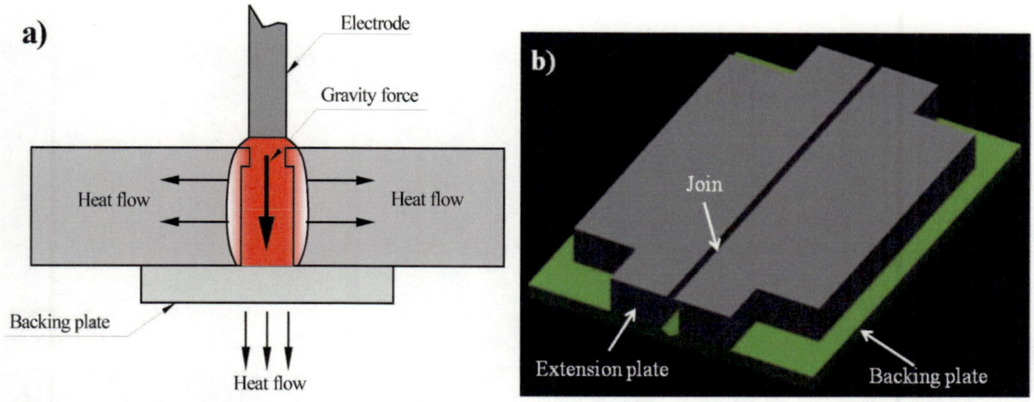

Fig. 9. Schematic representation of MIEA welding technique

2.1 Microstructure

After a fusion welding process two principal zones are identified in the welded joints named: Fusion Zone (FZ) and Heat Affected Zone (HAZ) whereas in the case of FSW three different zones are formed: stirred zone (nugget), Thermo-Mechanical Affected Zone (TMAZ) and the HAZ (Mishra & Ma, 2005). These zones are showed in the macrographs of Figure 10.

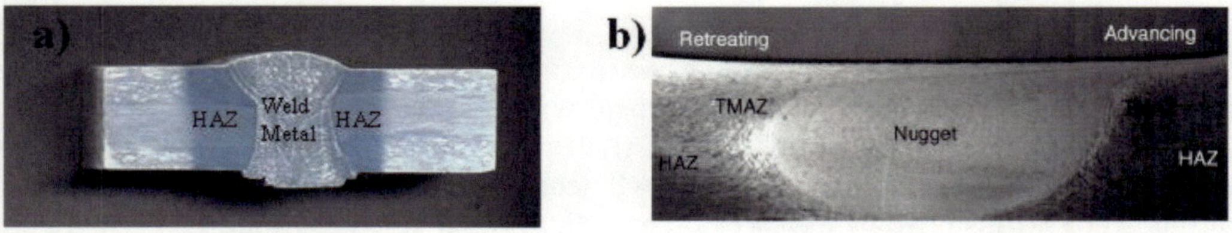

Fig. 10. Principal zones in welding of aluminum, a) MIEA welding technique (Al-6061-T6) and b) FSW (Al-7075-T651)

2.1.1 Weld metal microstructure in MIEA

In a fusion welding process, the heat input produces a fusion-solidification phenomenon, which is different to that obtained in the solidification of an ingot. (i) In an ingot, solidification begins with heterogeneous nucleation at the chill zone meanwhile in a weld pool the liquid metal partially wets the grains of the parent metal and epitaxial growth takes place from the partially melted grains of the parent metal (Davies et al., 1975). (ii) The rate of solidification in a weld pool, which depends on the traveling speed as well as the welding process, is by far faster than in an ingot. (iii) The macroscopic profile of the solid/liquid interface in welds progressively changes as a function of the traveling speed of the heat source whereas it exclusively depends on the time for an ingot. (iv) The movement of the liquid metal in a weld pool is greater than in an ingot due to the Lorentz forces which create turbulence within the molten metal (Grong, 1997). Figure 11 shows longitudinal views, which depict the direction of solidification of the welds, for a multi-pass welding and MIEA with different preheating conditions. The arrows indicate the displacing direction of the electric arc.

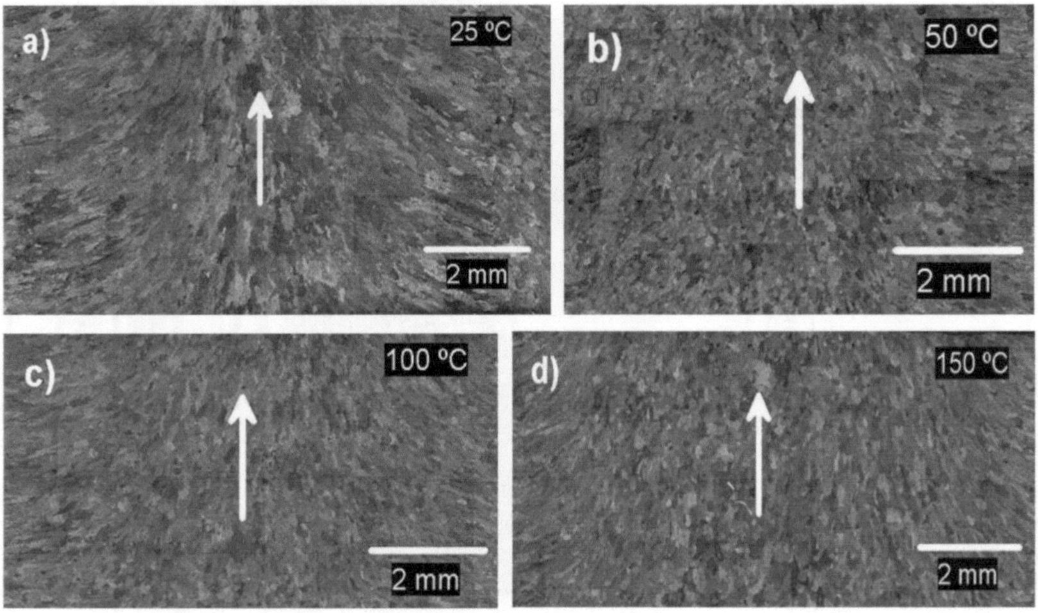

Fig. 11. Longitudinal top view of the weld metal grain structure at the mid plane for:
a) single V groove, b)-d) MIEA

Figure 11a corresponds to the single V groove joint and it shows the crystalline growth of a columnar-dendritic structure at a given angle with respect to the direction of the heat source. This feature is determined by the traveling speed of the welding torch. In this instance, the rate of the local crystalline growth tends to be that of the welding process. This phenomenon is illustrated by means of a schematic representation in Figure 12.

It is possible to observe that the local crystalline growth, R_L, is always larger than the nominal crystalline growth, R_N, since there are directions in which growth occurs preferentially. Thus, the rate of crystalline growth tends to be the traveling speed of the heat source, v, when the angles are less pronounced (when $\alpha \to 0$ and $\phi \to 0$), according to the following equation (Grong, 1997).

$$R_L = \frac{R_N}{\cos\phi} = \frac{v\cos\alpha}{\cos\phi} \tag{1}$$

The changes in direction are readily appreciated for the longitudinal view shown in Figure 11a. Competitive growth toward the heat source is evident, giving rise to columnar grains; this characteristic is typical of arc fusion welding processes. Analyzing equation (1) along with Figure 12, it is apparent the increase of the rate of crystalline growth as a function of the changes in orientation of the crystalline growth with regard to the largest thermal gradient of the weld pool. It is clear thus that local crystalline growth is favored due to the prevailing high temperature conditions.

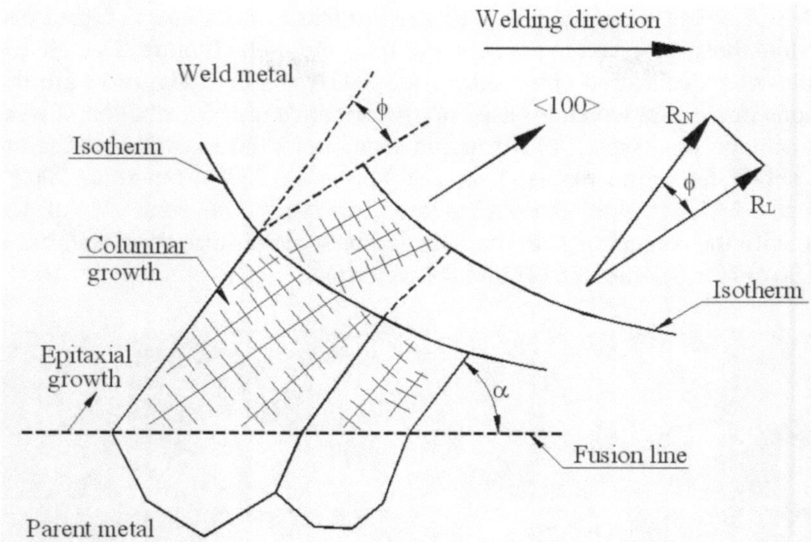

Fig. 12. Schematic representation of the local and nominal crystal growth rate (Ambriz et al., 2010b)

The longitudinal macrostructures for the MIEA joint, Figures 11b-d, exhibit significant differences with respect to the multi-passes single V groove joint. Irrespective of the preheating condition, the local crystalline growth maintains an angle nearly constant in relation to the moving heat source. The virtual non existence of changes in growth direction means that the local and nominal rates of crystalline growth tend to be equal. This phenomenon yields a significantly different grain structure in the weld metal for the MIEA joint as compared to the structure observed for the single V groove joint. It leads, in fact, to a grain refining effect which is obviously affected by the initial preheating temperature of the joint.

Figure 13 shows the grain structure at the bottom, mid height and top of the welds. These micrographs correspond to equivalent positions between welds and were captured at the same magnification. A dramatic change in the size and morphology of the grains is observed for the single V groove joint. Besides, some levels of porosity, as indicated by the arrows, are visible. The fine grain size present in the root pass is ascribed to the rapid cooling and/or to recrystallisation effects owing to subsequent welding passes which increased the heat input and caused grain growth toward the top of the weld. Microstructural examination of the fusion line revealed epitaxial growth from partially melted grains and columnar-dendritic grains. The micrograph in Figure 13 at the top of the single V weld shows that this solidification mechanism prevails between welding passes. On the other hand, the microstructures obtained for the MIEA joints do not exhibit major changes in morphology and size with regard to position. Thus, while the microstructures for the single V groove joint show that competitive growth occurs during solidification, the MIEA joint exhibit signs of heterogeneous nucleation which promotes grain refining. Figure 14 shows a micrograph obtained in the Scanning Electron Microscope showing heterogeneous nucleation in MIEA. Also, the levels of porosity in the MIEA joints decrease with preheating temperature (50, 100 and 150 °C) and are comparatively lower than that obtained in the single V groove joint. Epitaxial solidification is also observed at the fusion line of the MIEA welds, however, competitive columnar growth was restricted instead grain structures alike those observed in the centre of the weld metal (Figure 13) were present. The characteristics of solidification observed for the MIEA welds in Figure 13 are the result of heterogeneous nucleation which is based on the principle of the formation of a critical radii needed to achieve the energy of formation from potential sites for nucleation such as inclusions, substrates or inoculants (Ti or Zr) (Rao et al., 2008; Ram et al., 2000; Lin et al., 2003). For the MIEA welds, these sites are principally the sidewalls of the joint in conjunction with the content of Ti in the filler and base metal since the significant dilution of base metal favors incorporation of Ti into the weld pool.

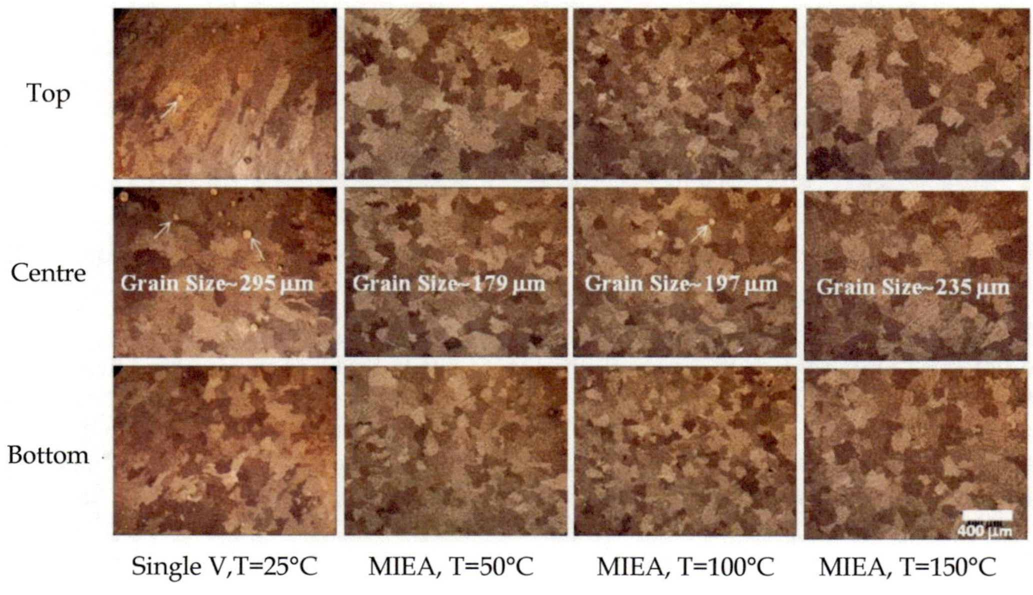

Fig. 13. Optical micrographs in welding of 6061-T6 for multipass welding process and MIEA with three different preheating (50, 100 and 150 °C)

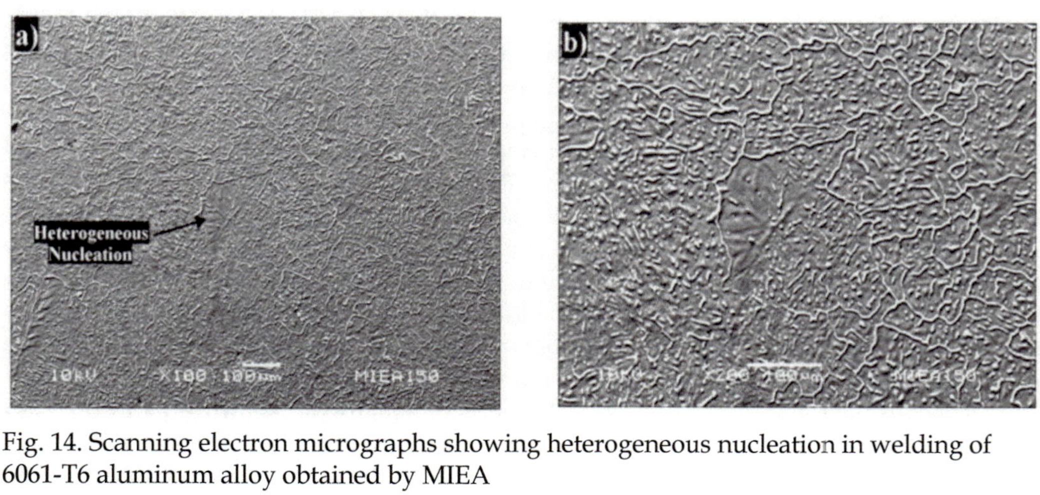

Fig. 14. Scanning electron micrographs showing heterogeneous nucleation in welding of 6061-T6 aluminum alloy obtained by MIEA

Generally speaking MIEA joint yields homogeneous grain refined microstructures in the weld metal, having the average grain size well below than that obtained with the conventional single V groove joint. The differences in grain size and morphology between single-V groove and MIEA joints are expected to have a significant impact on the mechanical performance of the welds. Before dealing with this aspect, it is worth to elucidate about the possible mechanism that gives rise to a self-refining effect when welding with the MIEA joint technique.

2.1.2 Nugget zone in FSW

The intense plastic deformation and frictional heating during FSW results in the generation of a recrystallized fine-grained microstructure within the stirred zone (Mahoney et al., 1998). This is usually referred to as a weld nugget (or nugget zone) or dynamically recrystallized zone. Also, under the same FSW conditions, onion ring structure is observed in the nugget zone, as is presented in Figure 15.

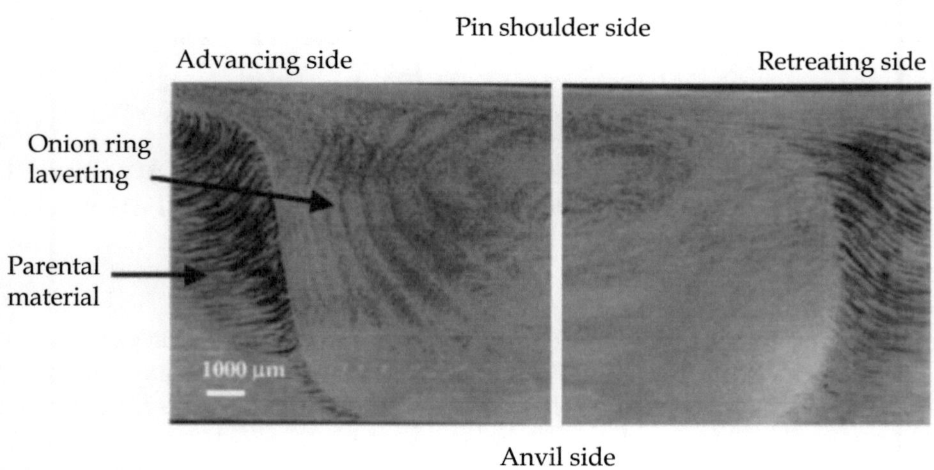

Fig. 15. Optical image showing the macroscopic features (Nandan et al., 2008) in a transverse section of FSW of 2195-T81 Al-Li-Cu alloy. Note the onion-ring and the adjacent large upward movement of material

Depending on the processing parameter, tool geometry, temperature of the workpiece, and thermal conductivity of the material, various shapes of nugget zone have been observed. Basically, nugget zone can be classified into two types, basin-shaped nugget that widens near the upper surface and elliptical nugget. Sato et al. (Sato et al., 1999) reported the formation of basin-shaped nugget on friction FSW of 6063-T5 aluminum alloy plate. They suggested that the upper surface experiences extreme deformation and frictional heating by contact with a cylindrical-tool shoulder during FSW, thereby resulting in generation of basin-shaped nugget zone. On the other hand, Rhodes et al. (Rhodes et al., 1997) and Mahoney et al. (Mahoney et al., 1998) reported elliptical nugget zone in the weld of 7075-T651 aluminum alloy.

In terms of grain size it is well know that FSW produces a fine structure, which is a direct function of the welding parameters like: tool geometry, chemical composition of the workpiece, temperature of the workpiece, vertical pressure and active cooling. For example, in FSW of 6061-T6 aluminum alloy is possible to obtain a grain size near to 10 μm (Liu et al., 1997). Figure 16 illustrates the characteristic microstructures in 2024 and 6061 aluminum alloys welds obtained by FSW. One of the principal parameters which affect the grain size in FSW is the tool rotation, as was reported previously (Sato et al., 2002).

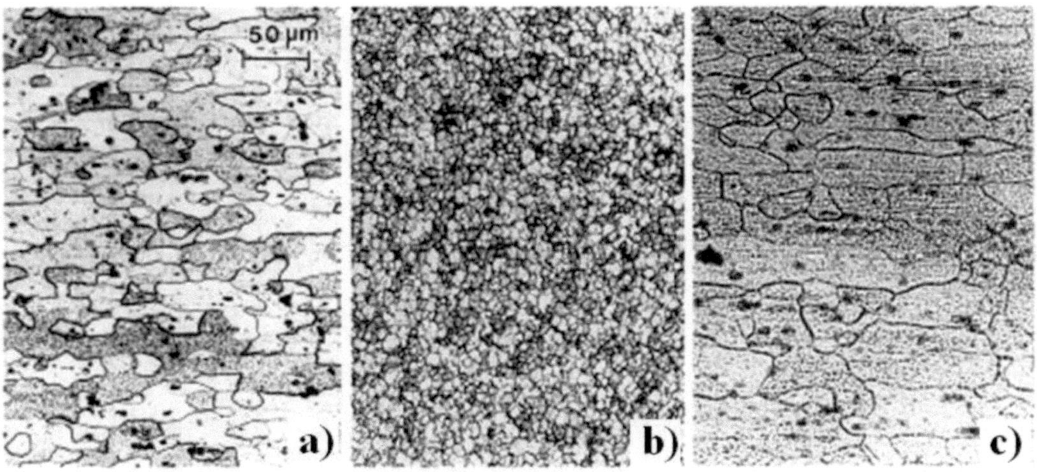

Fig. 16. Representative 2024Al/6061 Al FSW microstructure comparison, a) 2024 Al base plate grain structure, b) 2024 Al lamellar weld zone grain structure and c) 6061 Al base plate grain structure (Li et al., 1999)

2.1.3 Thermo-mechanically affected zone

In FSW a Thermo-Mechanically Affected Zone (TMAZ) is formed between the parent metal and the nugget zone, as shown in Figure 10b. The TMAZ experiences both temperature and deformation during welding process. The TMAZ is characterized by a highly deformed structure. Although the TMAZ underwent plastic deformation, recrystallization does not occur in this zone due to insufficient deformation strain. However, dissolution of some precipitates is observed, due to the high-temperature. The extent of dissolution depends on the thermal cycle experienced by TMAZ.

2.1.4 Heat affected zone

Heat Affected Zone (HAZ) is present in fusion welding as well as in FSW processes. The wide of this zone is a direct function of the heat input and the thermal conductivity of the materials to be welded. Obviously the HAZ in FSW tends to be lower than that obtained in a fusion welding process. The HAZ is very important in welding of aluminum alloys, especially in alloys hardened by precipitation (artificial ageing), for instance 2024-T6, 2014-T6, 6061-T6 and 7075-T6. During artificial ageing in Al-Mg-Si alloys (6000 series), a high density of fine, needle-shaped β'' particles are formed uniformly in the matrix (aluminum, α). This precipitate is the dominating hardening phase, which is produced according to the following precipitation sequence (Dutta & Allen, 1991).

$$\text{Super-Saturated Solid Solution (SSS)} \rightarrow \text{Solutes clustering} \rightarrow \text{GP zones (spherical)} \rightarrow \beta'' \text{ (needle)} \rightarrow \beta' \text{ (bar)} \rightarrow \beta$$

However, since these precipitates are thermodynamically unstable in a welding process, the smallest ones will start to dissolve in parts of the HAZ where the peak temperature has been above the ageing temperature (> 160 °C), while the larger ones will continue grow (Dutta & Allen, 1991). Close to the weld fusion line full reversion of the β'' particles is achieved. At the same time, coarse rod-shaped β' precipitates may form in the intermediate peak temperature range. This microstructural transformation is showed in Figure 17.

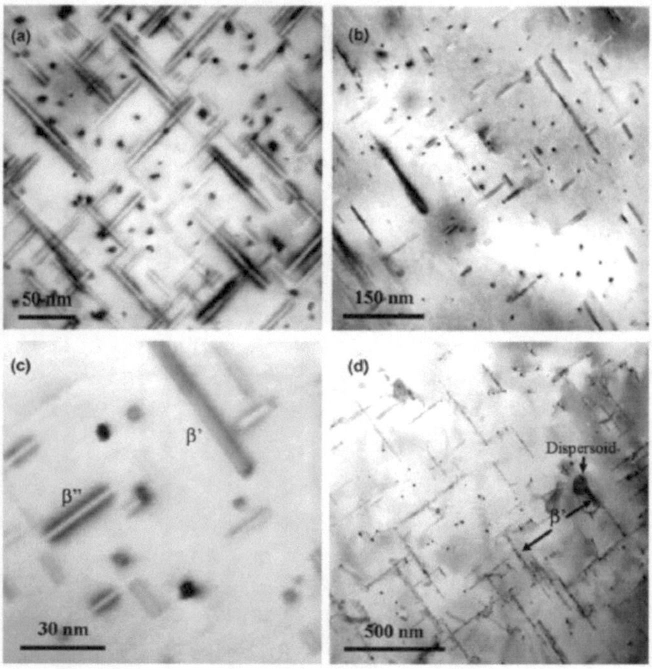

Fig. 17. TEM bright field images of microstructures observed in the ‹100› Al zone axis orientation after artificial ageing and Gleeble simulation (Series 1), a) Needle-shaped β'' precipitates which form after artificial ageing, b) Mixture of coarse rod-shaped β' particles and fine needle-shaped β'' precipitates which form after subsequent thermal cycling to T_P = 315 °C (10 s holding time), c) Close up of the same precipitates shown in b) above, d) Coarse rod-shaped β' particles which form after thermal cycling to T_P = 390 °C (10 s holding time) (Myhr et al., 2004)

2.2 Mechanical properties

2.2.1 Microhardness

In order to determine the effect of the welding process in aluminum alloys, a common practice is to perform a microhardness profile in a perpendicular direction to the weld bead, as is showed in Figure 18. Standard Vickers measurements are conducted with an appropriate penetration force and time, i.e. 1 N and 15 s. The indentation is measured and the hardness is calculated applying equation 2:

$$HV = 1.8544\frac{P}{d^2} \quad (2)$$

where HV is expressed in MPa if P is given in N and d, the indent diagonal, in mm.

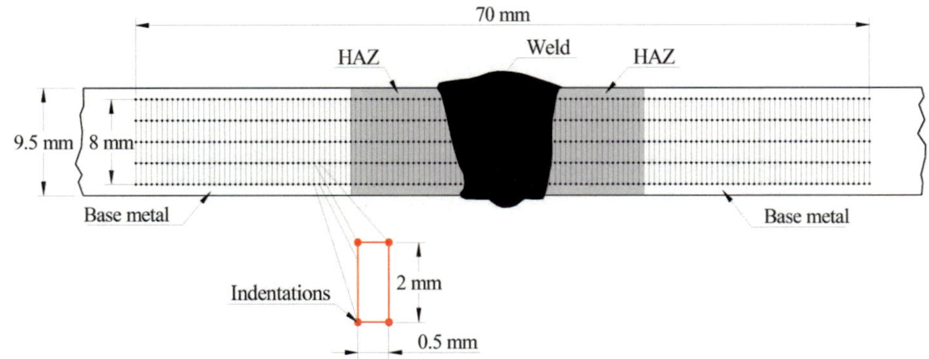

Fig. 18. Mesh definition for classical Vickers indentation measurements

Microhardness measurements give a general idea of the microstructural transformations and the variation of the local mechanical properties (Ambriz et al. 2011) produced after a welding process in aluminum alloys.

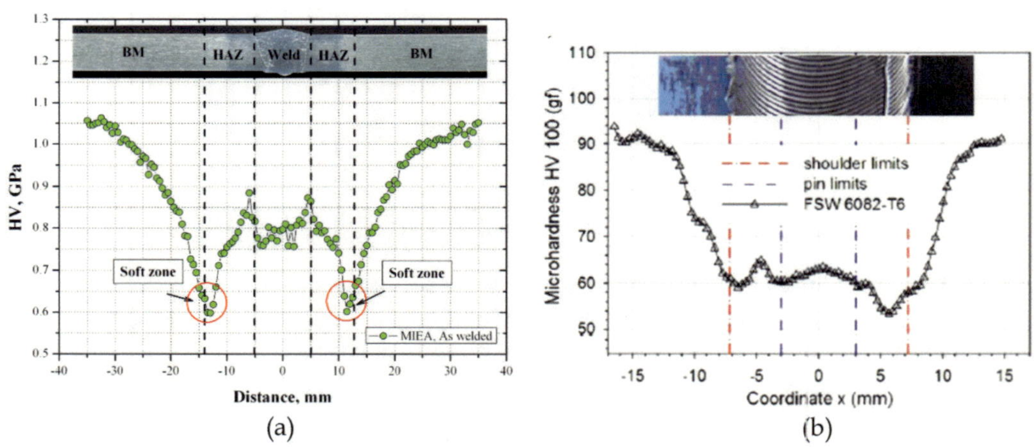

Fig. 19. Microhardness profile in aluminum alloys welds, a) 6061-T6 alloy welded by MIEA and b) 6082-T6 alloy welded by FSW (Moreira et al., 2007), note that $1HV=9.8\times10^{-3}$ GPa

Figure 19 presents the Vickers hardness number profile in two different aluminum alloys welds obtained by MIEA and FSW. In both cases a significant difference for the hardness number of the weld material and HAZ with respect to the base material (~ 1.05 GPa or 107.1 $HV_{0.1}$) is observed. Also, at the limit between the HAZ and the base metal, we note the presence of a soft zone which is formed nearly symmetrically in both sides of the welded joints. It should be note that the hardness obtained in this zone represents roughly 57 % of the hardness number of the base material. This seems to indicate that the tensile mechanical properties after welding process will be greatly different. Figure 20 visualizes the location of the soft zone highlighted by the Vickers hardness profile represented in Figure 19a, by means of a hardness mapping. In this figure, the hardness values for each zone of the welded joint are well-defined. It is clear that in the soft zone (HAZ) the hardness number range is between 0.55 to 0.7 GPa. This soft zone results from the thermodynamic instability of the β″ needle-shaped precipitates (hard and fine precipitates) promoted by the high temperatures reached during a fusion welding process. Indeed the temperatures reached during the welding process are favorable to transform the β′ phase, rod-shaped, according to the transformation diagram for the 6061 alloy.

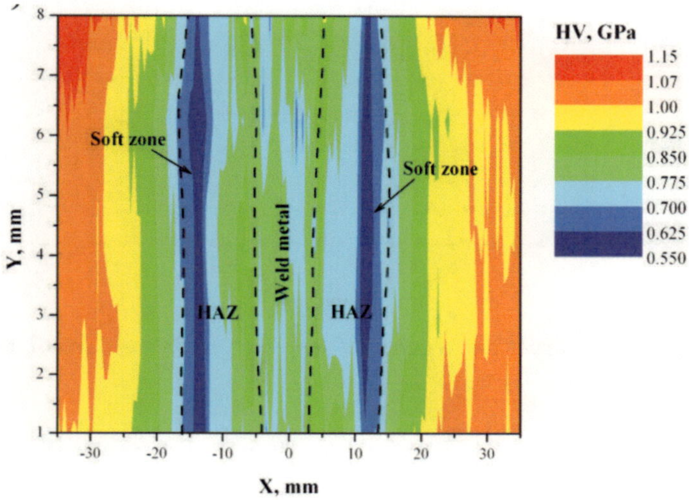

Fig. 20. Vickers hardness mapping over the welded zone

2.2.2 Tensile properties

The individual mechanical behaviour of the base metal, weld metal, HAZ and welded samples in as welded condition for 6061-T6 aluminum welds by MIEA is shown in Figure 21 as stress as function of strain graph.

From Figure 21, it can be observed that the experimental results for the base metal are in agreement with nominal values found in the literature for 6061-T6 alloy (American Society for Metals Fatigue and Fracture, 1996). Also, the base metal exhibits the best mechanical properties and well defined proportional limit. The tensile properties of the sample obtained from the HAZ presents a 41% and a 19 % reduction of the ultimate strength with respect to the base metal and weld metal respectively. The loss of mechanical strength, commonly referred to as over-aging, when welding a 6061-T6 alloy is a fairly well understood phenomenon and it is explained in terms of the precipitation sequence. During welding, however, the base metal adjacent to the fusion line is subjected to a gradient of temperature imposed by the welding thermal cycle. At certain distance from the fusion line, the cooling curve crosses the interval of temperatures between 383 to 250 °C in which the β′ phase, rod-shaped, is stable. It is thus the transformation of β″ into β′ the responsible of the decrease in hardening of the α matrix due to the incoherence of the β′ phase caused by the thermodynamic instability of β″ in a welding process.

On the other hand, in the case of FSW for 6061-T6, the same effect (over-ageing) is observed, although in this case the welded specimens represents an ultimate strength of 70% of the base material (Moreira et al., 2007). The conventional stress-strain curves are presented in Figure 22.

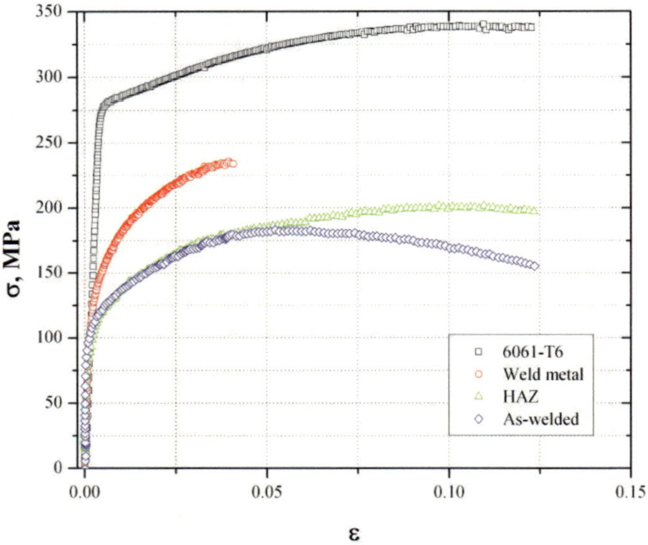

Fig. 21. True stress-strain curves for as-received 6061-T6 plates, weld metal, HAZ, and welds in the as-welded condition

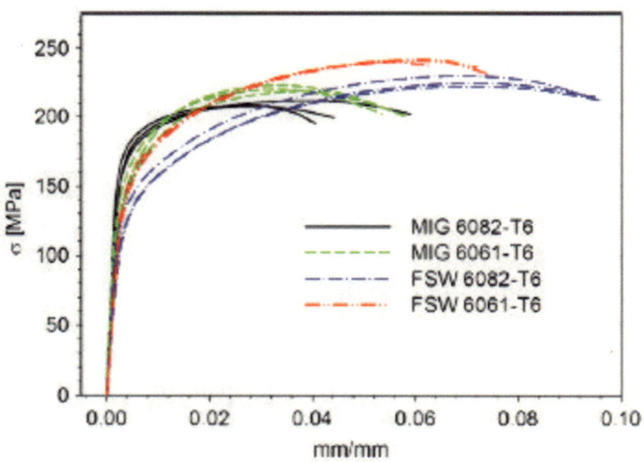

Fig. 22. Tensile tests of MIG and FS welded specimens (Moreira et al., 2007)

2.2.3 Fatigue crack growth

Fatigue behavior of aluminum alloys welded by conventional process has been investigated by some authors (Ambriz et al. 2010b; Branza et al., 2009; Seto et al., 2004). In terms of fatigue behavior considering FSW, some interesting studies have been published (Matic & Domazet, 2005, Chiarelli et al., 1999, James & Paterson, 1998). This part presents the experimental results in terms of Fatigue Crack Growth (FCG) in the weld metal, heat affected zone and base material of 6061-T6 aluminum alloy welded by MIEA. These results were compared in terms of FCG with those reported previously (Moreira et al., 2008) for FSW of the same alloy.

Figure 23, presents the crack length as a function of number of cycles for base metal, weld metal and HAZ in 6061-T6 welds by MIEA, for ΔP equal to 2.5 and 3.0 kN. In general, the *a-N* curves showed in Figure 23 reveal a notable difference in terms of crack length for each material as a function of the number of cycles, nevertheless the small change in ΔP (Ambriz et al., 2010b).

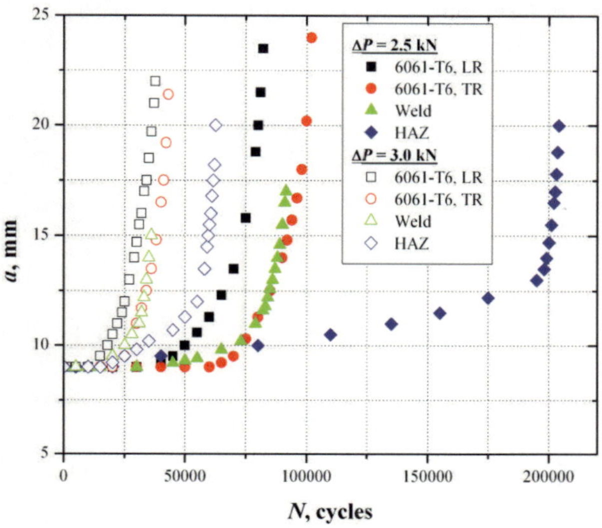

Fig. 23. Graph of crack length as function of number cycles, load ratio, R=0.1

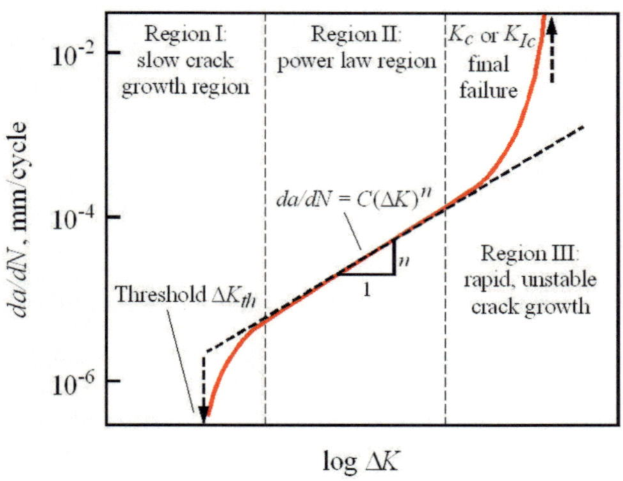

Fig. 24. Fatigue crack growth regimes versus ΔK

Taking into account the power-law region showed in Figure 24, the experimental results for *a*, were plotted in *da/dN* versus ΔK graphs according to Paris law:

$$\frac{da}{dN} = C(\Delta K)^n \tag{3}$$

where C and n are constants obtained directly from the fitting curve. Figure 25, presents the FCG data obtained for the base metal, weld metal and HAZ in MIEA, as well as the comparison with FSW data, found in the literature (Moreira et al., 2008). In general terms, the experimental results for MIEA welds adjust very well with equation 3.

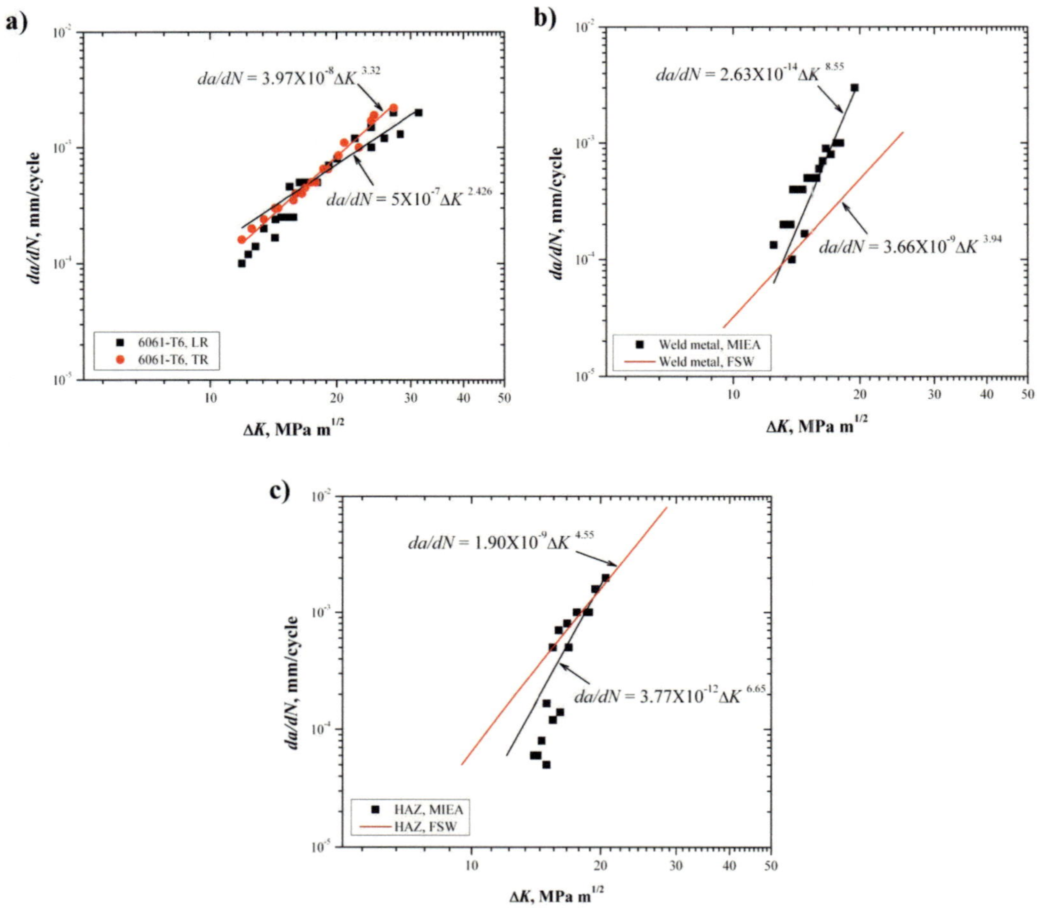

Fig. 25. Fatigue crack growth rate as function of stress intensity factor range

Figure 25a, shows the FCG for base metal in both directions. This graph shows that the microstructure aspect (anisotropy) does not have an important influence in terms of FCG as could be expected, taking into consideration that yield strength in the base metal parallel to rolling direction is higher than transverse direction. However this is not the case for the weld metal and HAZ (Figures 25b-c), in which the crack tends to propagate faster than base metal. Under this scenario, the FCG behavior for base metal (L-T) was taken as a basis to perform a comparative table between the weld metal and HAZ of MIEA and FSW. Table III, presents the crack growth rate, *da/dN* and the stress intensity factor range ΔK, for base metal, weld metal and HAZ corresponding to a critical crack length in MIEA welds. For comparison effects, values for *da/dN* in MIEA were taken to compute the ΔK in FSW.

	Base metal	MIEA		FSW	
		Weld metal	HAZ	Weld metal	HAZ
da/dN (mm/cycle)	1.981×10^{-3}	1.0×10^{-3}	1.413×10^{-3}	1.0×10^{-3}	1.413×10^{-3}
ΔK (MPa $m^{1/2}$)	30.41	17.27	19.46	23.98	19.51
$(da/dN)_i/(da/dN)_{BM}$	1.0	0.50	0.71	0.50	0.71
$(\Delta K)_i/(\Delta K)_{BM}$	1.0	0.57	0.64	0.79	0.64

BM = base metal, *i* corresponds to weld metal or HAZ for MIEA or FSW.

Table 3. Comparative table between MIEA and FSW based on a critical crack length

The results presented in table 3 indicate that, there is an important difference in ΔK for weld metal and HAZ, independently of the welding process. In this way, it should note that ΔK for weld metal in MIEA represents only 57% of the base metal, unlike the ΔK for weld metal in FSW, which reach a 79% with respect to base metal. This means that FCG rate are higher in MIEA weld metal than FSW, as can be seen in Figure 25b. This behavior is totally related to the joining processes; it means that MIEA is a welding technique based on a fusion welding process that employs a high silicon content filler metal, which produces a self grain refining, but a brittle microstructure in the weld metal (Ambriz et al., 2010c). On the other hand, FSW is a solid-state joining process that does not use a filler metal (Nandan et al., 2008). Thus, chemical composition in weld metal is similar to the base metal and microstructural characteristics related to dynamic recrystalization tends to be better than MIEA.

In contrast, Figure 25c, shows that FCG rate in MIEA and FSW is similar in the HAZ. The stress intensity factor relation was 64% with respect to base metal. It is noted that thermal effect produced by the microstructural transformation of very fine precipitates needle shape β'', to coarse bar shape β' precipitates, has a profound impact in the HAZ crack growth rate. It confirms that, independently of the welding process (MIEA and FSW), the crack growth conditions are directly influenced by the temperature within the HAZ, which is normally above of the aging temperature of the alloy, causing a hardening lost and important decrease in mechanical properties.

3. Conclusion

Some welding process can be employed to weld aluminum alloys. In this chapter the fundamental characteristics of the most common welding processes have been presented, such as: shielded metal arc welding (SMAW), gas metal arc welding (GMAW), gas tungsten arc welding (GTAW), friction stir welding (FSW), and a new welding technique named modified indirect electric arc (MIEA). Special attention has presented on welding of 6061-T6 aluminum alloy welded by MIEA and FSW. In the case of MIEA welds important microstructural characteristics in terms of morphology and grain size has been observed with respect to those obtained by a multi-pass welding process (GMAW). It means that when

MIEA is used, the solidification process tends to promote a heterogeneous nucleation, thus an auto-refinement of the grain size is promoted. However, when multi-pass welding process is employed (GMAW) columnar-epitaxial solidification prevails causing an increase in terms of grain size. On the other hand, the grain structure in the fusion zone produced by FSW has the better characteristics in terms of grain size (~10 μm).

A few mechanical properties after a welding process of 6061-T6 aluminum alloy have been presented. It is observed that quasi-static mechanical properties decrease in a dramatic manner in MIEA as well as in FSW, this aspect is totally related to the micro-structural transformation in the heat affected zone of very fine needle shape β'' precipitates to coarse bar shape β' precipitates produced by the thermal effect during the welding process (thermodynamic instability). This micro-structural transformation has been quantified by means of a micro-hardness map from which is possible to observe the soft zone formation where the failures are presented after a monotonic load (tension load).

Fatigue crack growth behaviors in weld metal, HAZ and base metal of 6061-T6 welded joints obtained by MIEA were quantified. It was observed that the worst crack growth conditions are presented in the fusion zone (weld metal), which are related to brittle micro-structure characteristics due to abundant presence of eutectic Si. A comparison between weld metal for FSW and the MIEA indicates that fatigue crack growth rate in the MIEA is higher than that in FSW; it means that for a critical crack length, the ΔK represents a 57% of the base material, whereas in the case of FSW it reaches a 79%. In addition, it was observed that the fatigue crack growth rate in the HAZ tends to be similar in both welding processes.

4. References

Ambriz, R.R., Barrera, G., & García, R. (2006). Aluminum 6061-T6 welding by means of the modified indirect electric arc process. *Soldagem and Inspecao*, Vol. 11, No. 1, pp. 10-17.

Ambriz, R.R., Barrera, G., García, R., & López V.H. (2008). Microstructure and heat treatment response of 2014-T6 GMAW welds obtained with a novel modified indirect electric arc joint. *Soldagem and Inspecao*, Vol. 13, No. 3, pp. 255-263.

Ambriz, R.R., Chicot, D., Benseddiq, N., Mesmacque, G. & de la Torre, S. (2011). Local mechanical properties of the 6061-T6 aluminium weld using micro-traction and instrumented indentation. *European Journal of Mechanics A/Solids*, Vol. 30, pp. 307-315.

Ambriz, R.R., Mesmacque, G., Ruiz, A., Amrouche, A., López, V.H. (2010). Effect of the welding profile generated by the modified indirect electric arc technique on the fatigue behavior of 6061-T6 aluminum alloy. *Materials Science and Engineering A*, Vol. 527, pp. 2057-2064.

Ambriz, R.R., Barrera, G., García, R. & López, V.H. (2010). The microstructure and mechanical strength of Al-6061-T6 GMA welds obtained with the modified indirect electric arc joint. *Materials and Design*, Vol. 31, No. 6, pp. 2978-2986.

Ambriz, R.R., Mesmacque, G., Ruiz, A., Amrouche, A., López, V.H. & Benseddiq, N. (2010). Fatigue crack growth under a constant amplitude loading of Al-6061-T6 welds obtained by modified indirect electric arc technique. *Science and Technology of Welding and Joining*, Vol. 15, No. 6, pp. 514-521.

Branza, T., Deschaux-Beaume, F., Velay, V. & Lours, P. (2009). A microstructural and low-cycle fatigue investigation of weld-repaired heat-resistant cast steels. *Journal of Materials Processing Technology*, Vol. 209, No. 2, pp. 944-953.

Chiarelli, M., Lanciotti, A. & Sacchi, M. (1999). Fatigue resistance of MAG welded steel elements. *International Journal of Fatigue*, Vol. 21, No. 10, pp. 1099-1110.

Davies, G.J. & Garland J.G. (1975). Solidification structures and properties of fusion welds. *International Metals Review*, Vol. 20, pp. 83-106.

Dawes, C.J. & Thomas, W.M. (1996). Friction stir process welds aluminum alloys. *Welding Journal*, Vol. 75, 3, pp. 41-45.

Dutta, I. & Allen, S.M. (1991). Calorimetric study of precipitation in commercial Al alloys. *Journal of Materials Science Letters*, Vol. 10, pp. 323-326.

Easterling, K. (1992). *Introduction to the Physical Metallurgy of Welding* (Second Edition), Butterworth Heinemann, ISBN 0750603941, Oxford.

Grong, O. (1997). *Metallurgical Modeling of Welding*, (Second Edition), The Institute of Materials, ISBN 1861250363, London.

James, M.N. & Paterson, A.E. (1997). Fatigue performance of 6261-T6 aluminium alloy – constant and variable amplitude loading of parent plate and welded specimens. *International Journal of Fatigue*, Vol. 19, No. 93, pp. 109-118.

Kou, S. (2003). *Welding Metallurgy* (Second Edition), John Wiley and Sons, ISBN 0-471-43491-4, United States of America.

Li, Y., Murr, L.E. & McClure, J.C. (1999). Flow visualization and residual microstructures associated with the friction-stir welding of 2024 aluminum to 6061 aluminum. *Materials Science and Engineering A*, Vol. 271, pp. 213-223.

Lin, D.C., Wang, G.X. & Srivatsan T.S. (2003). A mechanism for the formation of equiaxed grains in welds of aluminum-lithium alloy 2090. *Materials Science and Engineering A*, Vol. 351, pp. 304-309.

Liu, G., Murr, L.E., Niou, C.S., McClure, J.C. & Vega, F.R. (1997). Microstructural aspects of the friction-stir welding of 6061-T6 aluminum. *Scripta Materialia*, Vol. 37, No. 3, pp. 355-361.

Mahoney, M.W., Rhodes, C.G., Flintoff, J.G., Spurling, R.A. & Bingel, W.H. (1998). Properties of friction-stir-welded 7075 T651 aluminum. *Metallurgical Materials Transactions*, Vol. 29, pp. 1955.

Matic, T. & Domazet, Z., (2005). Determination of structural stress for fatigue analysis of welded aluminium components subjected to bending. *Fatigue and Fracture Engineering Materials and Structures*, Vol.28, No. 9, pp. 835-844.

Mishra, R.S. & Ma, Z.Y. (2005). Friction stir welding and processing. *Materials Science and Enginnering R*, Vol. 50, pp. 1-78.

Moreira, P.M.G.P, de Jesus, A.M.P., Ribeiro, A.S., & Castro, P.M.S.T. (2008). Fatigue crack growth in friction stir welds of 6082-T6 and 6061-T6 aluminium alloys: as a comparison. *Theoretical and Applied Fracture Mechanics*, Vol. 50, pp. 81-91.

Moreira, P.M.G.P, Figueiredo, de M.A.V. & Castro, P.M.S.T. (2007). Fatigue behaviour of FSW and MIG weldments for two aluminium alloys. *Theoretical and Applied Fracture Mechanics*, Vol. 48, pp. 169-177.

Myhr, O.R., Grong, O., Fjaer, H.G., & Marioara, C.D. (2004). Modelling of the microstructure and strength evolution in Al-Mg-Si alloys during multistage thermal processing. *Acta Materialia*, Vol. 52, pp. 4997-5008.

Nandan, R., DebRoy, T. & Bhadeshia H.K.D.H. (2008). Recent advances in friction-stir welding-process, weldment structure and properties. *Progress in Materials Science*, Vol. 53, pp. 980-1023.

Ram, G.D.J, Mitra, T.K., Raju, M.K. & Sundaresan, S. (2000). Use of inoculants to refine weld solidification structure and improve weldability in type 2090 Al-Li alloy. *Materials Science and Engineering A*, Vol. 276, pp. 48-57.

Rao, K.P., Ramanaiah, N. & Viswanathan, N. (2008). Partially melted zone cracking in AA6061 welds. *Materials Design*, Vol. 29, pp 179-186.

Rhodes, C.G., Mahoney, M.W., Bingel, W.H., Spurling, R.A. & Bampton C.C. (1997). Effects of friction stir welding on microstructure of 7075 aluminum. *Scripta Materialia*, Vol. 36, No. 1, pp. 69-75.

Sato, Y., Urata, M. & Kokawa H. (2002). Parameters controlling microstructure and hardness during friction-stir welding of precipitation-hardenable aluminum alloy 6063. *Metallurgical Materials Transactions A*, Vol. 33, No. 3, pp. 625-635.

Sato, Y.S. Kokawa, H., Enmoto, M. & Jogan, S. (1999). Precipitation sequence in friction stir weld of 6063 aluminum during aging. *Metallurgical and Materials Transactions A*, Vol. 30, No. 12, pp. 3125-3130.

Seto, A., Yoshida, Y. & Galtier, A. (2004). Fatigue properties of arc-welded lap joints with weld start and end points. *Fatigue and Fracture Engineering Materials and Structures*, Vol. 22, No. 12, pp. 1147-115.

Thomas, W.M. & Dolby R.E. (2003). Friction Stir Welding Developments, Proceedings of the Sixth International Trends in Welding Research, Materials Park, OH, USA.

Thomas, W.M., Nicholas, E.D., Needham, J.C., Murch M.G., Temple-Smith P. & Dawes C.J. (1991). Friction Stir Butt Welding. *International Patent Application No. PCT/GB92/02203*.

5

Prediction of Tensile and Deep Drawing Behaviour of Aluminium Tailor-Welded Blanks

R. Ganesh Narayanan[1] and G. Saravana Kumar[2]
[1]Department of Mechanical Engineering, IIT Guwahati, Guwahati
[2]Department of Engineering Design, IIT Madras, Chennai
India

1. Introduction

Tailor-welded blanks (TWB) are blanks with sheets of similar or dissimilar thicknesses, materials, coatings welded in a single plane before forming. Applications of TWB include car door inner panel, deck lids, bumper, side frame rails etc. in automotive sector (Kusuda et al., 1997; Pallet & Lark, 2001). Aluminium TWBs are widely used in automotive industries because of their great benefits in reducing weight and manufacturing costs of automotive components leading to decreased vehicle weight, and reduction in fuel consumption. The general benefits of using TWBs in the automotive sector are: (1) weight reduction and hence savings in fuel consumption, (2) distribution of material thickness and properties resulting in part consolidation which results in cost reduction and better quality, stiffness and tolerances, (3) greater flexibility in component design, (4) re-usage of scrap materials to have new stamped products and, (5) improved corrosion resistance and product quality[1]. The forming behaviour of TWBs is affected by weld conditions viz., weld properties, weld orientation, weld location, thickness difference and strength difference between the sheets (Bhagwan, Kridli, & Friedman, 2003; Chan, Chan, & Lee, 2003). The weld region in a TWB causes serious concerns in formability because of material discontinuity and additional inhomogeneous property distribution. Above said TWB parameters affect the forming behaviour in a synergistic manner and hence it is difficult to design the TWB conditions that can deliver a good stamped product with similar formability as that of un-welded blank. Designers will be greatly benefited if an expert system is available that can deliver forming behaviour of TWB for varied weld and blank conditions. Artificial neural network (ANN) modelling technique is found to show better prediction of any response variable that is influenced by large number of input parameters. Artificial Neural Networks are relatively crude electronic models based on the neural structure of the brain. The building blocks of the neural networks is the neuron, which are highly interconnected. In the artificial neural networks, the neurons are arranged in layers: an input layer, an output layer, and several hidden layers. The nodes of the input layer receive information as input patterns, and then transform the information through the links to other connected nodes layer by layer to the output nodes. The transformation behavior of the network depends on the structure of the

[1] http://www.ulsab.org

network and the weights of the links. A neural network has to go through two phases: training and application. During the learning, the training of a network is done by exposing the network to a group of input and output pairs. In the recognition phase, after it is trained, the network will recognize an untrained pattern. Application of ANN modelling technique in predicting the formability of TWB will definitely be helpful in understanding and designing the TWB conditions that can deliver a better stamped product.

This chapter describes the tensile and deep drawing forming behaviour of aluminium TWBs and prediction of the same using expert system based on ANN models. Standard tensile testing and square cup deep drawing set up are used to simulate the tensile and deep drawing processes respectively using elastic-plastic finite element (FE) method. The sheet base materials considered for the present work is a formable aluminium alloy. Global TWB tensile behaviour like yield strength, ultimate tensile strength, uniform elongation, strain-hardening exponent, strength coefficient, limit strain, failure location, minimum thickness, and strain path, and deep drawing behaviour viz., maximum weld line movement, draw depth, maximum punch force, draw-in profile are simulated for a wide range of thickness and strength combinations, weld properties, orientation, and location. Later, ANN models are developed to predict these tensile and deep drawing behaviour of TWBs. ANN models are developed using data set obtained from simulation trails that can predict the tensile and drawing behaviour of TWB within a chosen range of weld and blank conditions. To optimize the training data and thus the number of FE simulation, techniques from design of experiments (DOE) have been used for systematic analyses. The accuracy of ANN prediction was validated with simulation results for chosen intermediate levels. The results obtained are encouraging with acceptable prediction errors. An 'expert system framework' has been proposed by the authors (Veerababu et al., 2009) for the design of TWBs and the study described in this chapter is part of this framework to predict the formability of aluminium TWBs (Abhishek et al., 2011; Veerababu et al., 2009, 2010).

2. Formability studies on aluminium TWBs

The forming behaviour of aluminium TWBs is critically influenced by thickness and material combinations of the blanks welded; weld conditions like weld orientation, weld location, and weld properties in a synergistic manner. The impact of above said parameters on the tensile and forming behaviour of TWB in general viz., stress-strain curve, forming limit strain, dome height, deep drawability, and weld line movement can be understood from the existing work (Bhagwan et al., 2003; Chan et al., 2003, 2005). The variation of the experimental formability results found in the literature for aluminium TWBs appears to be large (Davies et al., 1999). Aluminium TWBs for automotive applications are particularly problematic because of the low formability of aluminium weld metal. Friction stir welding (FSW) is a process recently applied to aluminium TWBs that has the potential to produce a higher quality weld. Friction stir welding utilizes frictional heating combined with forging pressure to produce high-strength bonds virtually free of defects. Friction stir welding transforms the metals from a solid state into a plastic-like state, and then mechanically stirs the materials together under pressure to form a welded joint. In this process the tool is a dowel which is rotated at speeds depending on the thickness of the material. The pin tip of the dowel is forced into the material under high pressure and the pin continues rotating and moves forward. As the pin rotates, friction heats the surrounding material and rapidly produces a softened plasticized area around the pin. As the pin travels forward, the material behind the pin is forged under pressure from the dowel and consolidates to form a bond. In a study (Miles et al., 2004), three aluminium alloys: 5182-O, 5754-O, and 6022-T4 were considered and TWBs were made using gas tungsten arc welding process. All three of these alloys are being used to fabricate stamped automotive parts. The gas tungsten arc welding process has been used to make aluminium TWBs industrially, so results using this process were compared to FSW results. The results of tensile and formability tests suggest that the 5xxx series alloys had similar tensile ductility and formability regardless of the welding process. However, the 6022-T4 sheets joined using FSW had better formability than those joined using gas tungsten arc welding because FSW caused less softening in the heat-affected zone. Other welding processes like non-vacuum electron beam (NVEB) and Nd:YAG laser techniques have also been studied for welding aluminium TWBs (Shakeri et al., 2002). In that study, a limiting dome height (LDH) test is used to evaluate formability of the AA5754 sheet TWBs with gauge combinations 2 to 1 mm, 1.6 to 1 mm and 2 to 1.6 mm. Different weld orientations were considered and the failures of TWBs were studied. In general the failure occurs in welds and the thinner gauge. Weld orientation has a predominant effect on the formability of the TWBs. In a study (Stasik & Wagoner, 1998), laser welded 6111-T4 and 5754-O blanks were tensile tested, and longitudinal weld TWB's were formability tested using the OSU formability test. In the study, press formability was found to be much greater than the inherent weld ductility. Both materials had satisfactory TWB formability under longitudinal deformation, but 6111-T4 was severely limited under transverse loading, because of a softer heat-affected zone in the heat-treatable alloy. The effects of welds with transverse and longitudinal orientations on the formability of aluminium alloy 5754-O laser welded blanks using the swift cup test has been reported (Cheng et al., 2005). The results showed that longitudinal TWBs underwent considerable reduction in the forming limit when compared with transverse TWBs, and an un-welded blank. Transverse welded blanks exhibit approximately the same forming limit as that of an un-welded blank. However, the effect may change if the weld is placed at critical locations, say, at some offset from the centre-line.

Few research groups have aimed at predicting the formability of aluminium welded blanks by using different necking theories and FE simulations. For example, Jie et al. (2007) studied the forming behaviour of 5754-O Al alloy sheets, where in the forming limit curve (FLC) of welded blanks with thickness ratio of 1:1.3 was experimentally evaluated and predicted using localized necking criterion based on vertex theory. It is found from the analysis that the forming limit of the TWB is more closer to thinner material FLC and the experimental and predicted FLCs correlate well with each other. Davies et al. (2000) investigated the limit strains of aluminium alloy TWB (1:2 mm thickness), where in the FLCs predicted by Marciniak-Kuczynski (M-K) analysis are compared with the experimental results. Here the geometrical heterogeneity, i.e., the initial imperfection level, involved in the welded blank is modelled by using the strain-hardening exponent determined from miniature tensile testing together with the Hosford yield criterion, to determine a level of imperfection that exactly fits an failure limit diagram (FLD) to each experimentally evaluated failure strains. These empirically determined initial imperfection levels were statistically analyzed to determine the probability density functions for the level of imperfection that exists in un-welded and welded blanks. Since the imperfection level is not a single value and follows a statistical distribution, a means of selecting a single value for imperfection is formulated. Two different FLCs - namely, failure FLC and safe FLC - were defined. The first FLC was based upon an imperfection that represents a 50 per cent predicted failure rate and was designated the average or failure FLC. The second FLC was based upon an imperfection level that represents a 0.1 per cent predicted failure rate. This second FLC represents a failure rate of 1 part in 1000 and was defined as the safe FLC. The safe FLCs thus predicted are found to have good agreement with the experimental safe FLCs, except in the bi-axial stretching region. The influence of considering different constitutive behaviour of weld region on the prediction levels and strength imperfections across the weld region in the model are also discussed in the work. Recently, Ganesh & Narasimhan (2008) predicted the forming limit strains of laser welded blanks by using thickness gradient based necking theory incorporated into a FE simulation code PAMSTAMP 2G®. It is found that the predictions are good in drawing region of FLD, with deviation in stretching region.

Studies on deep drawability and other forming behaviour of welded blanks involving both experimental and simulation are available for some aluminium base materials. In a study (Buste et al., 2000) numerical prediction of strain distribution in multi-gauge, aluminium alloy sheet TWBs welded by NVEB and Nd:YAG process is performed by modelling LDH. The study indicates that the Nd:YAG process is superior to the NVEB process considered. Nd:YAG welded blanks generally fail in the thinner parent metal away from the fusion zone. In general, the model agrees with measured strain distributions relatively well, particularly in cases when weld failure dominates as in the NVEB welds. Heo et al. (2001) investigated the characteristics of weld line movement during rectangular cup deep drawing where in draw bead has been used in the thinner blank side to restrict the movement of weld zone. Finite element simulations were also performed and compared with experimental results. An analytical model has been developed by Bravar et al. (2007) and Kinsey & Cao (2003) to predict the weld line movement and dome height for a typical application. This has been compared with numerical simulations and results were quite satisfactory. An interesting work was done by Lee et al. (2009) in predicting the forming limit and load-stroke behaviour of FSW blanks. In this investigation, wide variety of automotive sheet materials including 5083-O aluminium alloys sheets were experimentally tested and their forming limit were predicted using M-K model. The predictions are in good agreement with the experimental FLCs. From the above discussion, it is clear that one has to follow a limit strain theory in conjunction with numerical or analytical methods to predict the forming limit strains of welded blanks for different base material and weld conditions.

Designing TWB for a typical application will be successful only by knowing the appropriate thickness, strength combinations, weld line location and profile, number of welds, weld orientation and weld zone properties. Predicting these TWB parameters in advance will be helpful in determining the formability of TWB part in comparison to that of un-welded base materials. In order to fulfil this requirement, one has to perform lot of simulation trials separately for each of the cases which is time consuming and resource intensive. Automotive sheet forming designers will be greatly benefited if an 'expert system' is available for TWBs that can capture the wealth of knowledge created by the simulations and experiments and deliver the forming behaviour for varied weld and blank conditions. Experts system like "TENSALUM" (Emri & Kovacic, 1997) have shown the significant advantage of using them for computer-assisted testing of aluminium and aluminium alloys according to various standards. The authors are presently working on an research scheme to develop an 'expert system' for welded blanks that can predict their forming behaviour including tensile, deep drawing behaviour under varied base material and weld conditions using different formability tests, material models, and formability criteria. The knowledge base is constructed using learn by analogy engines based on ANNs that can predict the tensile behaviour and other forming characteristics of TWBs for a wide range of thickness, strength combinations and weld properties. The knowledge is acquired using simulations that simulate the tensile or other forming process using computer aided analysis of material behaviour. The expert system framework created the knowledge acquisition and inference methods are described further in the following sections.

3. The expert system framework

Developing artificial intelligent system like expert system, especially in fields like material forming and deformation behaviour, die design, casting design, machining processes, energy engineering, metallurgy, condition monitoring etc. is of interest to manufacturing, design engineers and scientists for long time (Asgari et al., 2008; Cakir & Cavdar, 2006; Dominczuk & Kuczmaszewskim, 2008; Ebersbach & Peng, 2008; Palani et al., 1994; Stein et al., 2003; Yazdipour et al., 2008). There has been a sustained interest in the sheet metal industry to create and use expert systems. Computer aided blanking process planning for aluminium extruded material using an expert system has been reported (Ohashi et al., 2002). The system identifies blanking features and sequences them to prepare the final product from the raw material. Specific to sheet metal forming and deep drawing studies, Manabe et al. (1998) have created an expert system framework for predicting and controlling blank holding force in a drawing process as the friction changes during the process. In summary, one can see the significance of the application of expert system in sheet metal process industry.

An expert system is domain specific system which emphasizes the knowledge used by an expert for solving problems in that domain (Wang et al., 1991). Typical expert system has a knowledge acquisition facility, a knowledge base and an inference subsystem that helps the end user as well as in continuous updating of knowledge base. Expert systems incorporate three basic types of knowledge: factual or data-oriented knowledge, rule-based or judgmental knowledge, and procedural or control knowledge embodied within a model base. An important trend in knowledge bases is the convergence of these three kinds of knowledge within a single system. There are several expert system frameworks reported in literature for application in computer aided engineering and one of the earlier work (Dym, 1985) provides a comprehensive discussion. The present expert system is data driven system with the knowledge acquisition enabled by modelling and simulation and knowledge base using artificial neural networks (Veerababu et al., 2009). The proposed expert system design is shown in Figure 1. This expert system is expected to involve three different phases viz., Phase 1 where in input base materials, TWB conditions and material model selection will be done, Phase 2 where in different forming behaviour can be selected for prediction, and Phase 3 involves use of the results as well as updating of the expert system if the prediction errors with simulation results are not acceptable. All the three phases have a design mode of operation where an initial expert system is created and put in place. The created expert system is then operated in use and update mode.

In Phase 1, while the expert system is designed, a range of material properties and TWB conditions are defined within which ANN models are developed to predict the results as will be discussed in the later sections. The same phase while operated in the usage mode, the user selects base material properties and TWB conditions within the chosen range for application and prediction of formability. In this phase, user can select different material models viz., strain-hardening laws and yield theories to predict the forming behaviour.

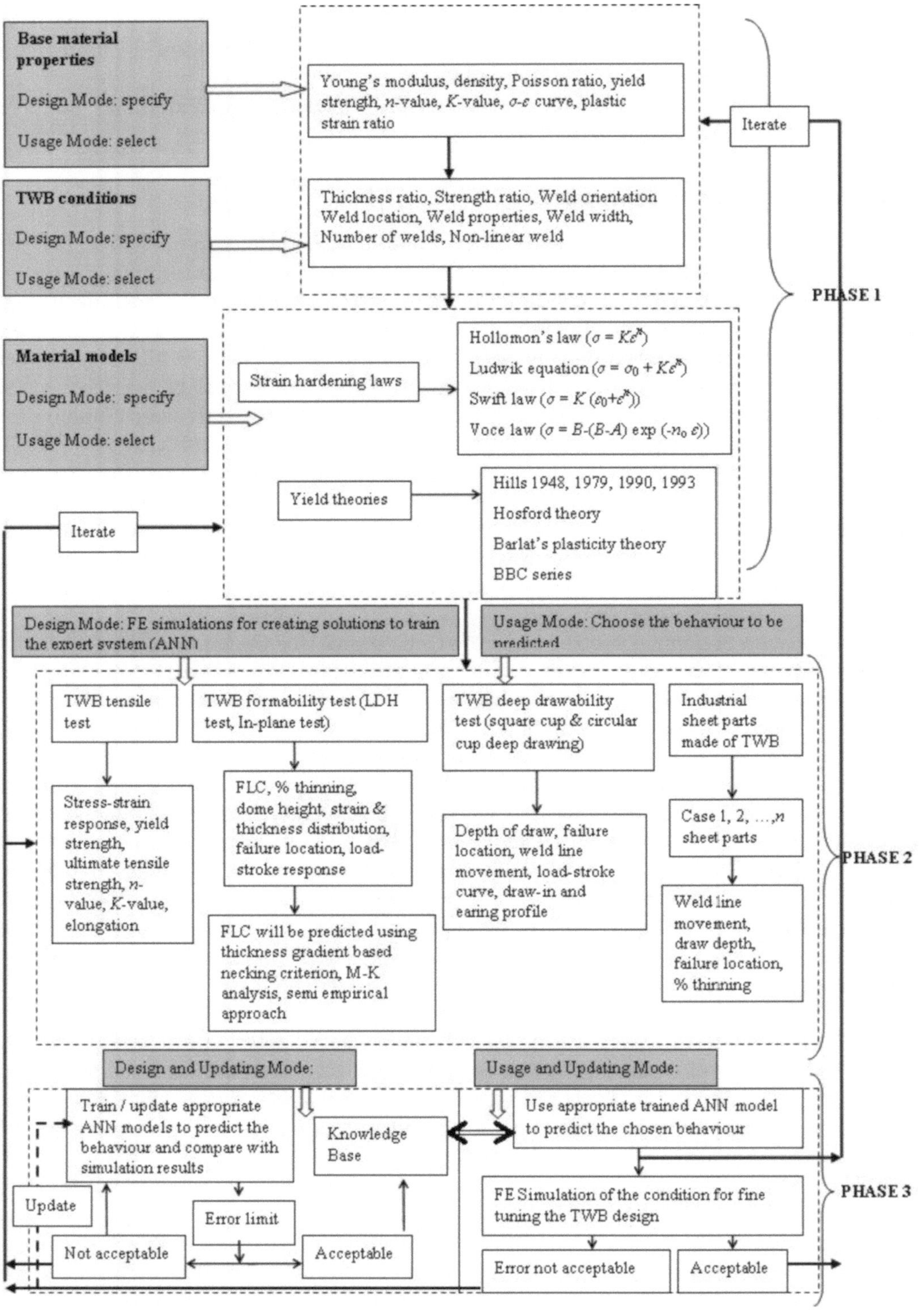

Fig. 1. Expert system framework (Veerababu et al., 2009)

There is no single strain-hardening law and yield theory that can predict the forming behaviour of TWBs made of varied sheet materials accurately. Hence in the design mode, ANN models will be developed to predict the forming behaviour using different material models. As a result, in the usage mode of the expert system, the user can opt for desired material models to predict the forming characteristics. The Phase 2 involves selecting the forming behaviour to be predicted for chosen base material and weld conditions. In the design mode, tensile behaviour, formability characteristics, deep drawability of welded blanks will be simulated by standard formability tests. Different category of industrial sheet parts will be simulated and expert system will be developed to predict their forming behaviour. For example, the global tensile behaviour of TWB viz., stress–strain curve, yield strength, ultimate tensile strength, elongation, strain-hardening exponent and strength coefficient will be monitored. Formability properties like forming limit curve, thinning percentage, maximum thinning, dome height at failure, failure location will be predicted by LDH test and in-plane stretching tests using different limit strain theories (say M–K analysis, thickness gradient based necking criterion, semi empirical approach). Cup deep drawing response like draw depth, weld line movement, drawing force, failure location, earring and draw-in profile can be predicted. Also it is planned to develop ANN model and expert system for predicting the formability of application specific sheet parts made of welded blanks. In the usage mode, the user selects the type of test results he is interested in predicting. In Phase 3 the training, testing, usage and updating the ANN predictions with simulation results will be performed. In the design mode operation, various ANNs are created and validated for predicting the forming behaviour (enumerated in Phase 2) for various combination of material properties and TWB conditions and constitutive behaviour (enumerated in Phase 1). In the usage mode, the user uses to predict the required forming behaviour for an initially chosen material, TWB condition and constitutive behaviour. If the forming behaviour predicted is not indicative of a good stamped product, the user changes the above said conditions till he gets satisfactory results.

In the absence of this expert system, the user will have to run time consuming and resource intensive simulation for this iterative stage. In the usage mode, if the results are not with in the expected error limit, user will have the choice of selecting different material models for predicting the required forming behaviour as described earlier and/or the expert system is updated with the specific case by updating the ANN models to predict the case within acceptable error limits. In this way, the expert system also learns form the application cases, enhancing the range, success rate of predictions. The three main functions of the expert system viz., knowledge base and its acquisition and inference are done in all the three phases as can be inferred from the design and usage mode of operation. The methods devised for these three functions for predicting formability of aluminium TWBs are discussed in the forthcoming section.

4. Simulation of formability behaviour of aluminium TWBs

The study presented here comprises of tensile and deep drawing behaviour of aluminium TWBs. The methodology followed for the study is described in Fig. 2. The first part of methodology involves FE simulation design and deals with the design of experiments to generate required data for expert system development.In order to conduct the exercise with optimum simulations, DOE using the Taguchi's statistical design (Taguchi, 1990) is followed. Simulation models for predicting the tensile as well as deep drawing behaviour of

TWBs are constructed as per the DOE parameter tables. The second part of the methodology is the ANN modelling and validation. The post processed results of FE simulations are used to train the ANN. Finally the ANN models for the expert system is validated with simulation results for chosen intermediate levels and other test data available. The methodology is discussed in the following subsections.

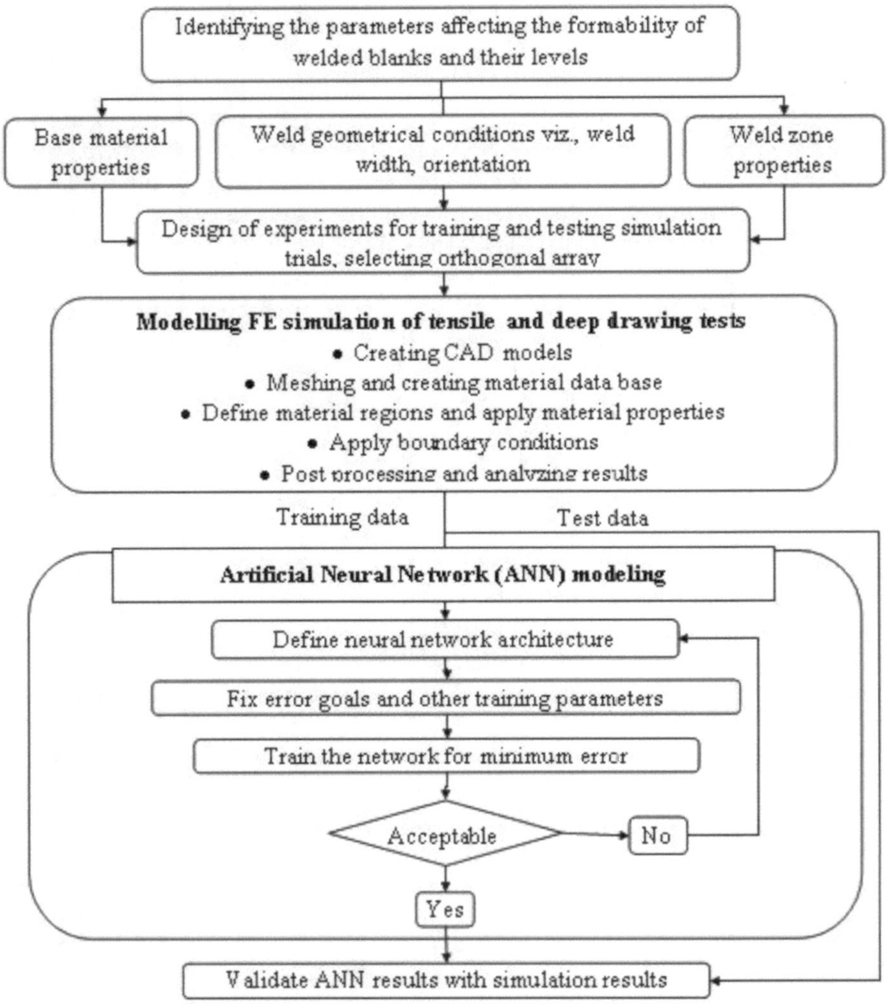

Fig. 2. Methodology of TWB simulation and developing expert system

4.1 Base material properties and TWB parameters

Initially for conducting simulation trials the material and process parameters that affect the TWB tensile and deep drawing behaviour are identified from available literature. The mechanical and forming properties of aluminium base metal and weld region used in FE simulations are shown in Table 1. The plastic strain ratios of weld zone are assumed to be 'one' in all the rolling directions as it is assumed isotropic. In order to generate the required data for expert system with optimum simulations, the Taguchi's statistical design is followed.

In this work, L_{27} orthogonal array with linear graph indicating the allocation of individual factors in orthogonal array is followed. Here L_{27} orthogonal array corresponds to three levels with six factors. However, this design fundamentally does not account for all the interaction among the processing parameters. In view of cost saving and time restriction higher order interactions are neglected. The six factors considered at three levels are shown in Table 2 for the tensile testing and deep drawing simulation of aluminium alloy TWBs. The five common TWB parameters considered for the analysis are thickness ratio, yield strength ratio, weld orientation, weld yield strength and weld width. For tensile testing weld '*n*' value and for deep drawing simulation weld location are considered as the sixth parameter. The schematic representation of these parameters for tensile testing and deep drawing are depicted in Fig. 3a & b. Each parameter has three levels (1, 2 and 3). The levels of parameters are chosen in such a way that the range covers practically all the combinations in typical experiments and industrial parts (Raymond et al., 2004; Saunders & Wagoner, 1996; Stasik & Wagoner 1998). In case of tensile test simulation the weld orientations that are significant i.e. longitudinal, transverse and 45° weld orientation are considered. In case of deep drawing simulation since both 0° and 90° orientations will be similar, an orientation of 60° was chosen as the third level. The weld zone yield strength was chosen such that it is higher or lower when compared to that of base materials as seen in most of the steel and aluminium alloy TWBs (Ganesh & Narasimhan, 2008; Miles et al., 2004; Stasik & Wagoner, 1998). Generally weld zone exhibits lesser ductility when compared to that of base materials (Ganesh & Narasimhan, 2008; Stasik & Wagoner, 1998) and hence strain-hardening exponent (n) of weld zone was selected such that it is lower than that of base materials. The average thickness of thinner and thicker sheets is assumed as weld zone thickness in simulation trials.

Since L_{27} orthogonal array is followed, 27 simulations are performed to generate data for ANN modelling. In case of tensile testing simulation, the tensile behaviour, viz., yield strength, ultimate tensile strength, uniform elongation, strain-hardening exponent (n), strength coefficient (K), limit strain, failure location, minimum thickness, and strain path are predicted for each test simulation. The important deep drawing behaviour predicted are maximum weld line movement, draw depth, maximum punch force, and draw-in profile for varied TWB conditions. These parameters are sensitive to the input conditions and are suitable representatives of the deep drawing behaviour of welded blanks (Ganesh & Narasimhan, 2006), specifically weld line movement and draw-in profile has industrial importance too.

Material properties	Aluminium alloy sheet	
	Base metal	Weld zone
Young's modulus (E), GPa	77	77
Density (ρ), kg/m^3	2700	2700
Poisson's ratio (ν)	0.3	0.3
r_0	0.7	1
r_{45}	0.6	1
r_{90}	0.8	1
Strain-hardening exponent (n)	0.172	See Table 2

Table 1. Material properties of aluminium alloy base material

Parameters	**Level 1**	**Level 2**	**Level 3**
Thickness ratio (T_1/T_2), mm/mm	0.5 (0.75/1.5)	0.75 (1.125/1.5)	1 (1.5/1.5)
Strength ratio (YS_1/YS_2), MPa/MPa	0.5 (190/380)	0.75 (285/380)	1 380/380)
Weld yield strength, (YS_w), MPa	150	300	400
Weld width (W), mm	2	5	10
Tensile testing			
Weld orientation (°)	0	45	90
Weld 'n' value (n_w)	0.1	0.13	0.15
Deep drawing			
Weld orientation (°)	0	45	60
Weld location, mm	0	10	20

Table 2. TWB parameters for aluminium alloy TWB and their levels

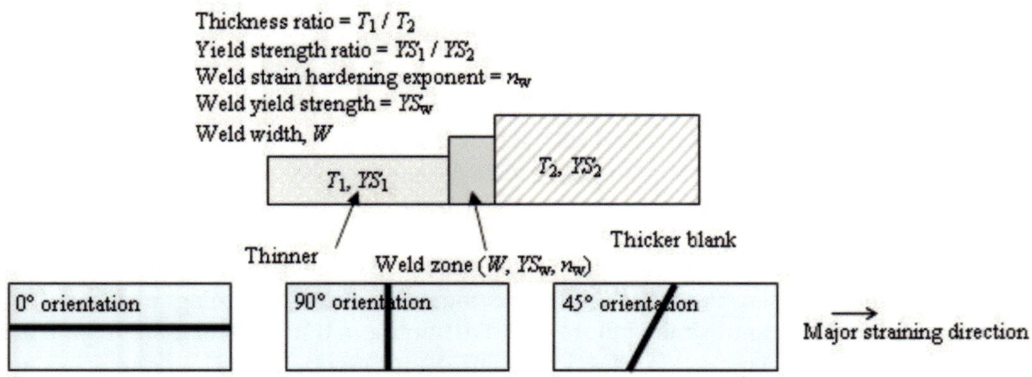

a) TWB parameters for tensile testing

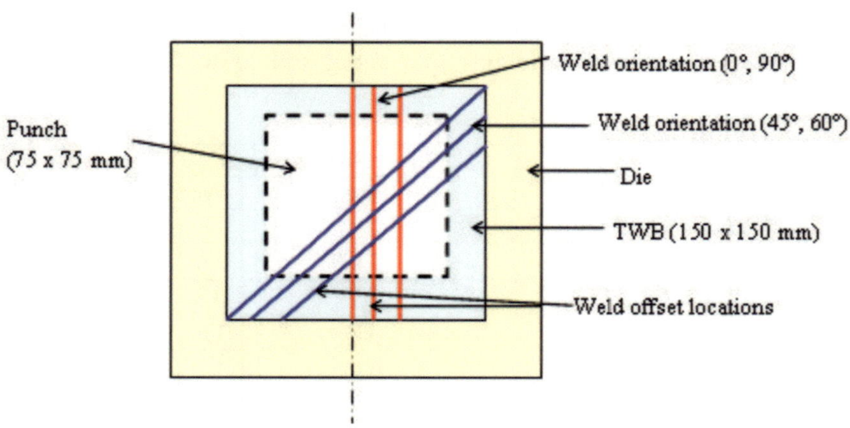

b) TWB parameters for deep drawing

Fig. 3. Schematic representation of control factors in tensile and deep drawing simulation

4.2 Modelling simulation of tensile test without pre-existing notch

Two sets of simulations were done to analyse the tensile behaviour. The first set of simulations consisted of observing the engineering stress-strain behaviour of TWB by monitoring the effective strain evolution at safe region of tensile sample for every small progression and later used to evaluate the tensile behaviours like yield strength, ultimate tensile strength, uniform elongation, strain-hardening exponent (*n*) and strength coefficient (*K*). The second set of simulations involved a notched specimen and observing the limit strains by allowing the failure to occur. For the first set of simulations CAD models of tensile specimen were generated as per the ASTM E 646-98 specifications (ASTM, 2000) (Fig. 4.) in Pro-E® a solid modelling software and imported into PAM STAMP 2G® an elastic plastic FE code for pre-processing, performing simulations and post processing. These CAD models were meshed using 'Deltamesh' facility in PAM STAMP 2G®. The meshing was done with quadrilateral shell elements of the Belytschko-Tsay formulation, with five through-thickness integration points. The meshed blank thus obtained was divided into three different regions viz., weld region (without HAZ), base material 1 and base material 2 (Ganesh & Narasimhan, 2006, 2007) to construct meshed models of TWB for varied weld orientations. A constant mesh size of 1 mm was kept in the weld region and base metal (Fig. 5.) as this has been reported to validly predict the forming limit of TWBs acceptably in Ganesh & Narasimhan (2008). The material properties were assigned to weld zone and base metals according to the different parameter levels (Tables 1 and 2) in the orthogonal array. Displacement boundary conditions (Fig. 6.) are applied to the tensile sample such that one end of the specimen is fixed and the other end is given some finite displacement with a velocity of 0.5 mm/min.

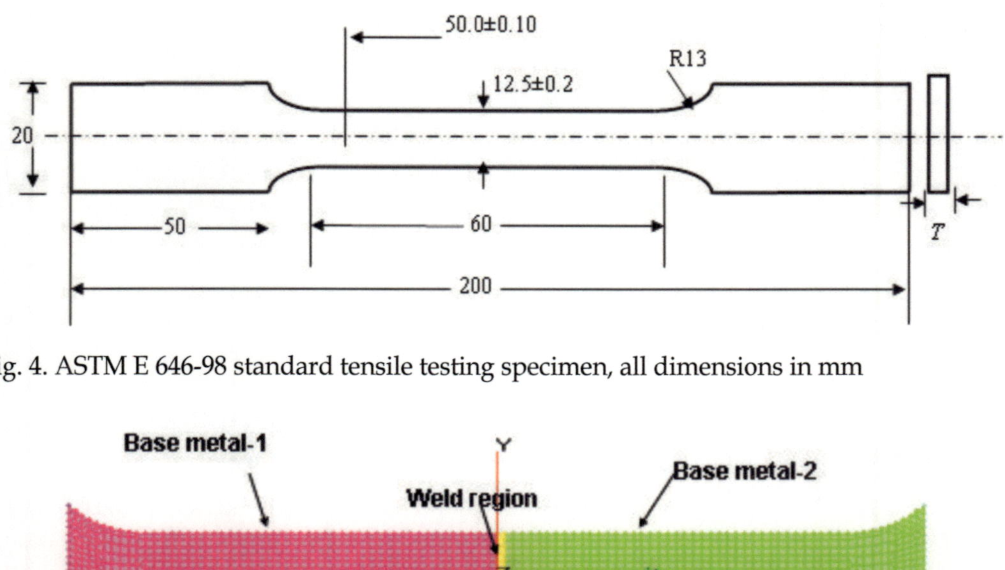

Fig. 4. ASTM E 646-98 standard tensile testing specimen, all dimensions in mm

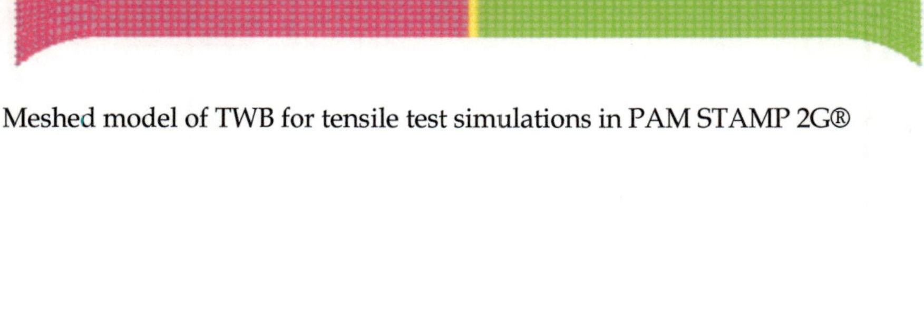

Fig. 5. Meshed model of TWB for tensile test simulations in PAM STAMP 2G®

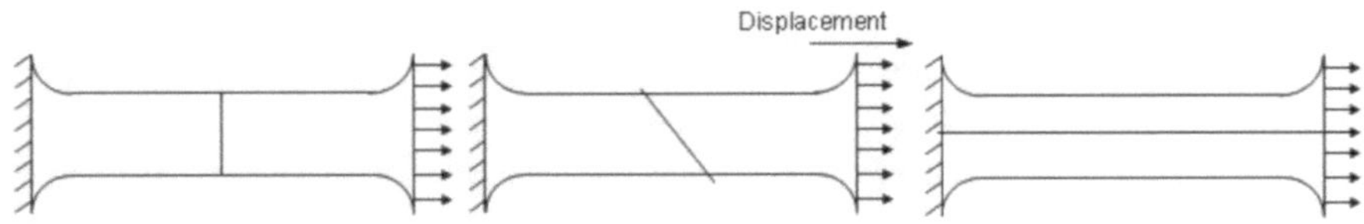

Fig. 6. TWB sample with different weld orientations and boundary conditions

For this set of simulations, Hollomon's power law ($\sigma = K\,\varepsilon^n$; where, K - strength coefficient and n - strain-hardening exponent) was used to describe the strain-hardening behaviour of base material and weld region. Hill's 1948 isotropic hardening yield criterion (Banabic, 2000) was used as the plasticity model for the aluminium alloy base material. This quadratic yield criterion has the form,

$$F(\sigma_{22}-\sigma_{33})^2+G(\sigma_{33}-\sigma_{11})^2+H(\sigma_{11}-\sigma_{22})^2+2L(\sigma_{23})^2+M(\sigma_{31})^2+N(\sigma_{12})^2=1 \quad (1)$$

where F, G, H, L, M, N are constants defining the degree of anisotropy and σ_{ij} are the normal and shear stresses. The tensile response i.e., stress-strain curve of TWB was obtained by monitoring effective stress and corresponding strain values in safe regions of TWB tensile sample for each unit of progression. From this engineering stress-strain data and required global mechanical properties of TWBs viz., yield strength, ultimate tensile strength, uniform elongation, strain-hardening exponent (n) and strength coefficient (K) were evaluated. The methodology for evaluating these properties is schematically described in Fig. 7 a-b. Similar procedure was followed for all the 27 tensile simulation of first set.

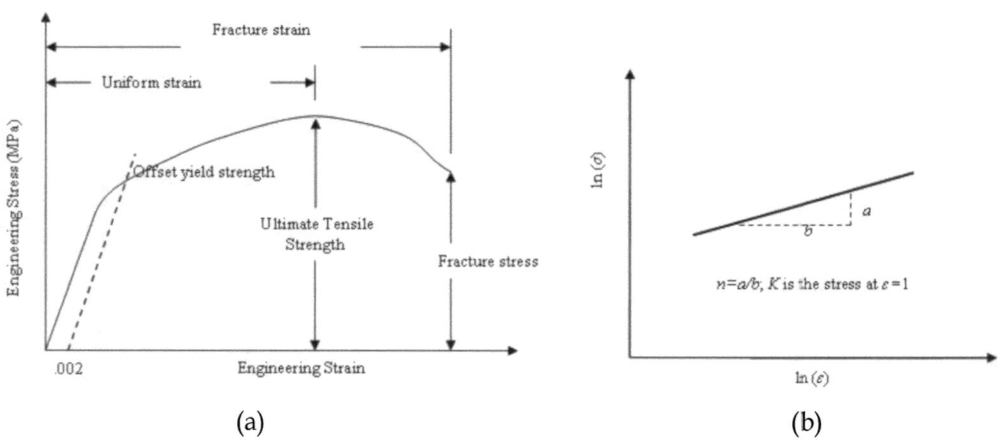

Fig. 7. a) Evaluating tensile properties from stress-strain curve, b) evaluating n, K values from true stress, strain data for TWBs

The results of the tensile test simulation of aluminium TWBs within the safe progression are discussed. Three modes of failure are generally seen in TWBs. They are, (i) Failure occurs perpendicular to weld region in the case of TWB with longitudinal weld, (ii) Failure occurs in the thinner or weaker base material in the case of TWB with transverse, stronger weld ($YS_{weld} > YS_{basematerials}$), (iii) Failure occurs in the weld region in the case of TWB with transverse, weaker weld ($YS_{weld} < YS_{basematerials}$). In TWB with longitudinal weld, higher load requirements are seen in the case of stronger weld zone when compared to weaker or softer weld zone. This can be understood from load sharing principle between the weld zone and base materials. In the case of transverse weld, TWB with stronger weld zone exhibits better tensile behaviour than TWB with softer weld zone. This is mainly because of the gauge effect, and TWB tensile behaviour is found to deteriorate with increase in thickness or strength ratio. Fig. 8. shows the failure location of TWB tensile sample for varied weld orientations. It is observed that (i) TWB with longitudinal weld witness failure normal to the weld region (Fig. 8a), (ii) Failure occurs only in the base material (Fig. 8b- c) in transverse and 45° weld zone because of stronger weld zone ($YS_{weld} > YS_{basematerials}$), and (iii) Weld failure is seen in the case of softer weld zone (Fig. 8d; $YS_{weld} < YS_{basematerials}$). This is consistent with results obtained in many literature including Ganesh & Narasimhan (2006). Fig. 9. depicts the engineering stress-strain data generated by simulations for varied TWB conditions for TWB with aluminium alloy base material. In this, curve numbers 1, 2,….27 represent stress-strain curves corresponding to 27 simulation trials in the orthogonal array. The TWB tensile behaviour viz., yield strength, ultimate tensile strength, uniform elongation, strain-hardening exponent, strength coefficient were evaluated from these curves. It is seen from Fig. 9. that (i) Longitudinal weld with stronger weld zone (curve 7) exhibit higher load requirements when compared to TWB with softer weld zone (curves 4, 1) for same strain values, (ii) Transverse weld with stronger weld zone exhibit base metal failure and hence shows better stress-strain behaviour (curves 3, 9, 13, 23, 26) when compared to the case with softer weld (weld failure is witnessed in this case; curves 6, 10, 16, 20), and (iii) In the case of transverse weld, with increase in thickness and strength ratio, the tensile behaviour is found to deteriorate.

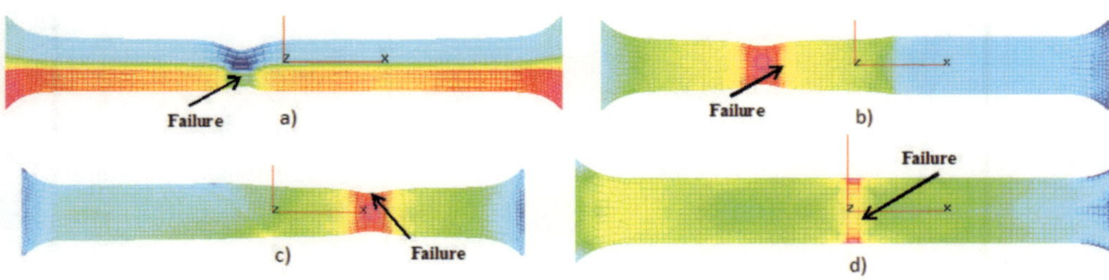

Fig. 8. a) Failure normal to weld region in TWB with longitudinal weld, b) base metal failure of TWB with transverse, stronger weld, c) base metal failure of TWB with 45°, stronger weld, d) Weld region failure of TWB with transverse, weaker weld

4.3 Modelling simulation of tensile test with pre-existing notch

The second set of tensile simulations were done with CAD models of tensile specimen as per geometry shown in Fig. 10. (Holmberg et al., 2004) in Pro-E® and imported into ABAQUS 6.7® for pre-processing, performing simulations and post processing. Since the aim of this set of simulations was to induce failure in the TWBs during simulation, a geometrical notch of 10 mm width is provided. The limit strains are predicted by thickness gradient based necking criterion. This notch geometry is decided based on trial simulations such that the entire deformation is concentrated only in that region and finally necking occurs, without much deformation happening in the shoulder region. For this, varied notch widths 14 mm, 10 mm, and 8 mm were simulated and compared with each other. Finally the notch of 10mm width is selected, wherein the effect of different TWB factors is not suppressed because of the notch effect and lesser deformation is observed in the shoulder region during simulations. The meshing, material assignment and boundary conditions are similar to the first set of simulations.

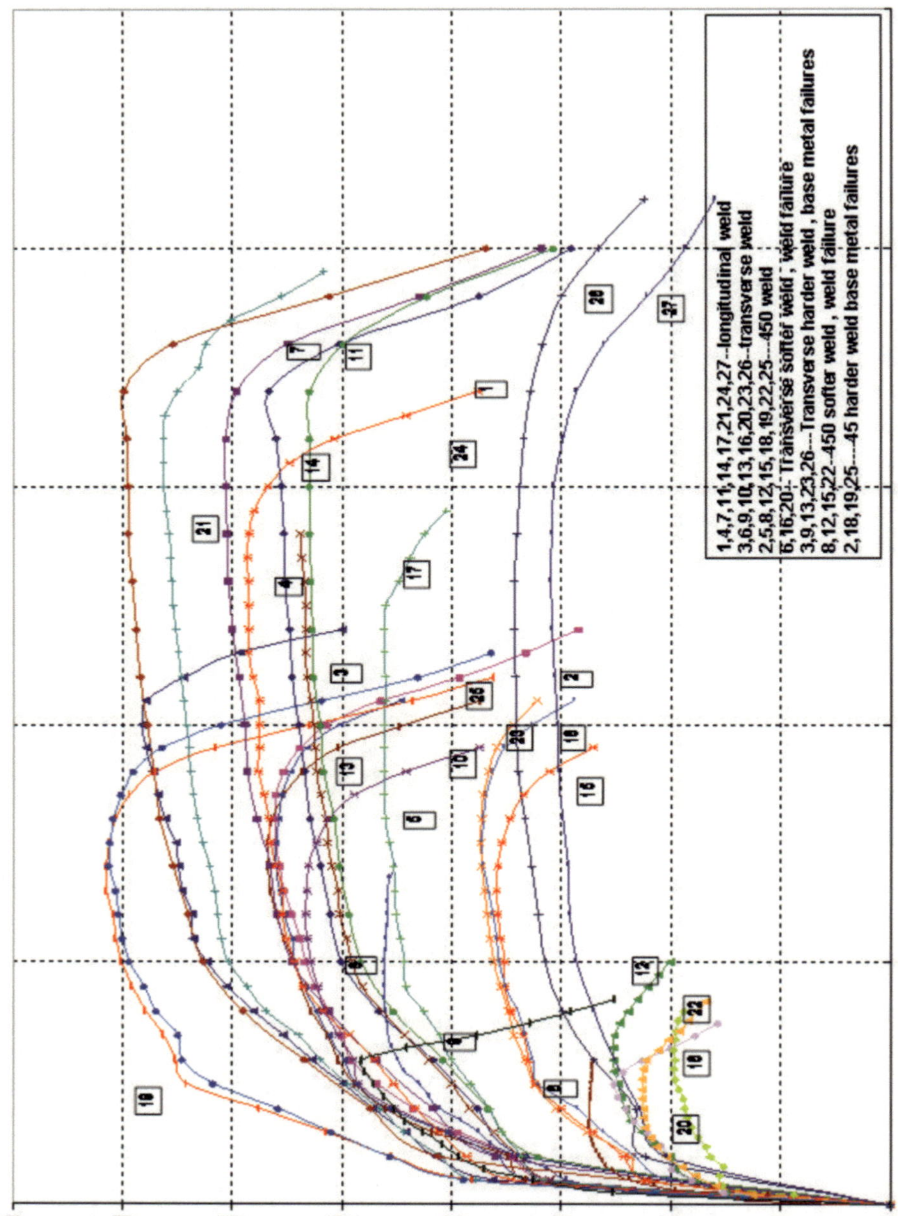

Fig. 9. Stress-strain behaviour of aluminium TWB (27 simulation data) (Veerababu et al., 2009)

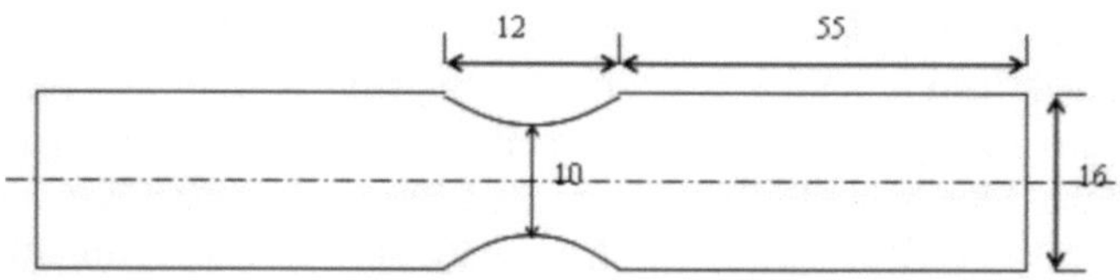

Fig. 10. Schematic of representation of notched tensile sample modelled in ABAQUS 6.7®, all dimensions in mm

For the second set of simulations, Swift law ($\sigma = K(\varepsilon_0 + \varepsilon_p)^n$, K – strength coefficient, n – strain-hardening exponent, ε_0 – pre-strain value of 0.003) is used as strain-hardening law describing the stress-strain relationship of weld and base material with Hill's 1948 isotropic hardening yield criterion. After simulations, five output parameters as described below are predicted for the TWBs.

Limit strain (major and minor strain): as per thickness gradient criterion, necking occurs when the thickness ratio between thinner and thicker element reaches 0.92 (Kumar et al., 1994). The major strain and minor strain of the thicker element, when the criterion is satisfied, is quantified as limit strain of that TWB condition. The thinner element has already failed and hence can not be referred for limit strain prediction. This means that the strain in the thinner element is above actual limit strain value. So the thicker element which is closer to thinner element is referred for the prediction work. This procedure is followed for all the 27 tensile simulations trials of this set. The limit strains are found to be in negative minor strain region of FLD (Fig. 11.), because of the presence of notch and tensile, plane-strain strain paths. Failure location is the distance from the fixed end to the thicker element in the progression where necking has occurred or criterion is satisfied. Minimum thickness is the minimum thickness of the element of specimen in the progression where necking has occurred. Strain path is the plot between major and minor strain from the starting progression to the progression where necking has occurred. This is quantified by the slope of the strain path curve (Fig. 11.).

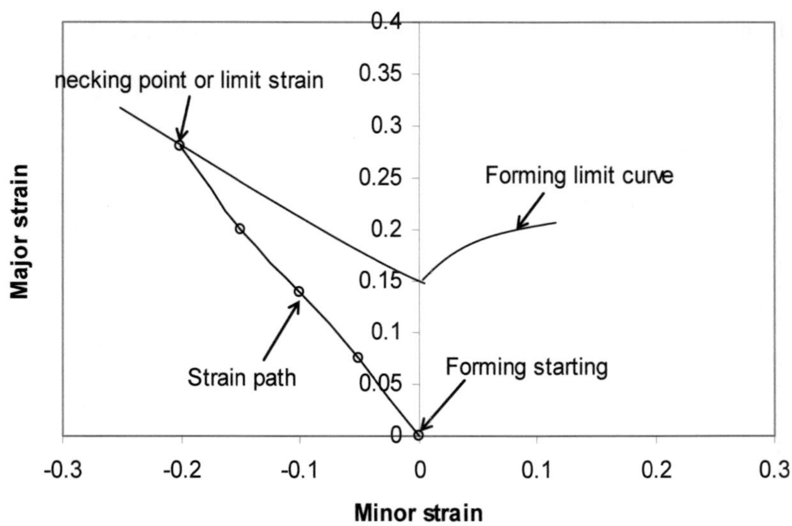

Fig. 11. Schematic representation of strain path with forming limit curve

The results of simulation showing the limit strain data for the 27 simulations are shown in Fig. 12. Though tensile test is simulated, most of the limit strains are close to plane-strain strain path (i.e., major strain axis) and not in tensile strain path. This is mainly because of the notch present in the tensile sample that is used for simulation for failure occurrence. Another important observation is that the strain path slope varies from 200.33 to 1.938. The lower slope values correspond to limit strain values that are away from plane-strain condition. The maximum limit strain is characterized by thickness ratio and strength ratio equal and occurs for a value of 1 and for longitudinal weld orientation. The failure location values for different experiments are shown in Fig. 13. The failure is expected to occur within the notch region, either in the base material or in the weld region depending on the weld width. It is clear from Fig. 13. that in almost all cases failure has occurred within the span, except in few cases wherein failure is seen just outside the notch region. The failure location is found to show significant effect on the minimum thickness achieved during TWB forming. Fig. 14. shows the variation of minimum thickness achieved for different experiments. The minimum thickness of 0.17 mm occurs in experiment 5, for which the failure location is at 65.93 mm, which is close to the notch edge AA in Fig. 13. (see inset).

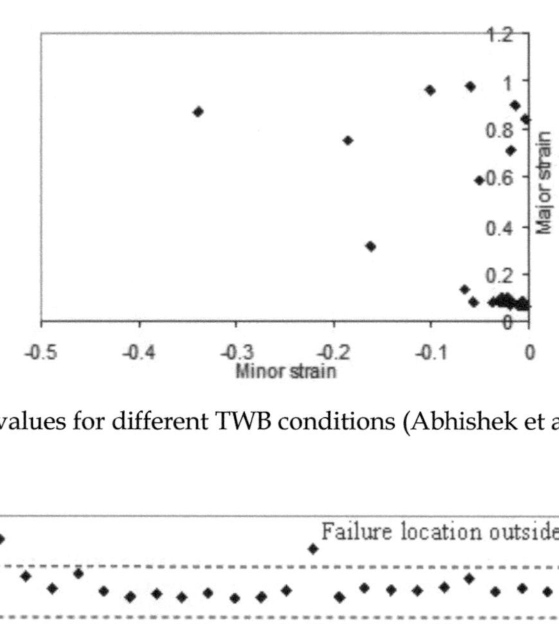

Fig. 12. Limit strain values for different TWB conditions (Abhishek et al., 2011)

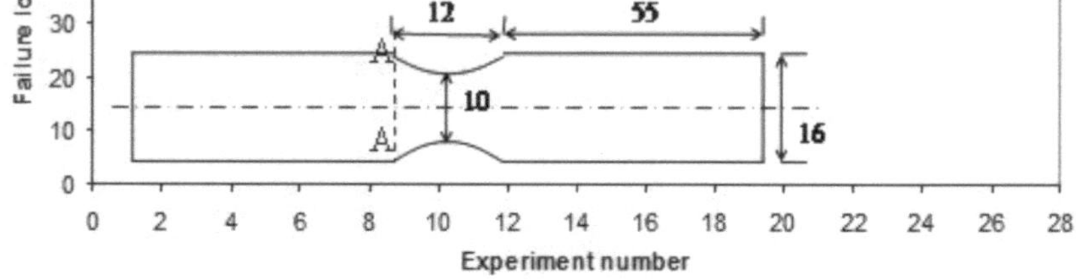

Fig. 13. Failure location in aluminium TWB (Abhishek et al., 2011)

4.4 Modelling simulation of deep drawing test for welded blanks

For the simulation of deep drawing a square cup deep drawing simulation set up constructed as per the NUMISHEET '93 benchmark specifications (Makinouchi et al., 1993) is used (Fig. 15.). CAD models of the tools (like die, punch, blank holder) in deep drawing are generated in Pro-E® and imported into PAM-STAMP 2G® for pre-processing, performing simulations and post processing. The meshing and material assignment are followed as discussed for the tensile simulations. Two shims are used to compensate the thickness difference in TWB, and these shims are exactly positioned above and below the thinner sheet. The shims are compressible with properties same as the stronger base metal. The friction coefficient between contact surfaces is taken as 0.12 as this approximates all forming conditions. The blank holding force is optimized during simulation to avoid wrinkling and extra thinning. Downward stroke is given to the punch with a velocity of 0.5 mm/min. The solution is mapped in such a way that the punch force is monitored for each unit of progression of the punch.

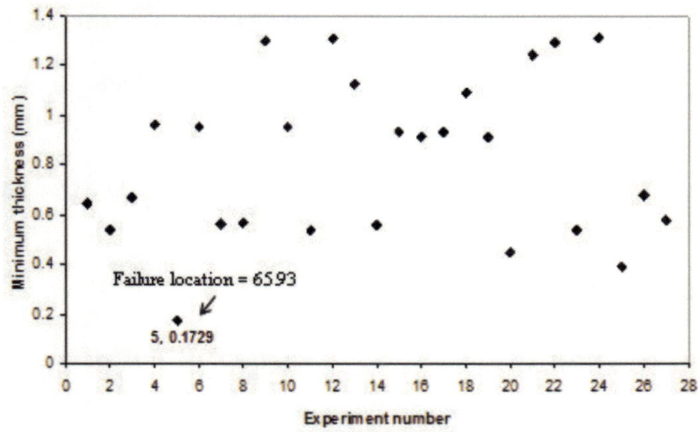

Fig. 14. Minimum thickness achieved for different simulation trials (Abhishek et al., 2011)

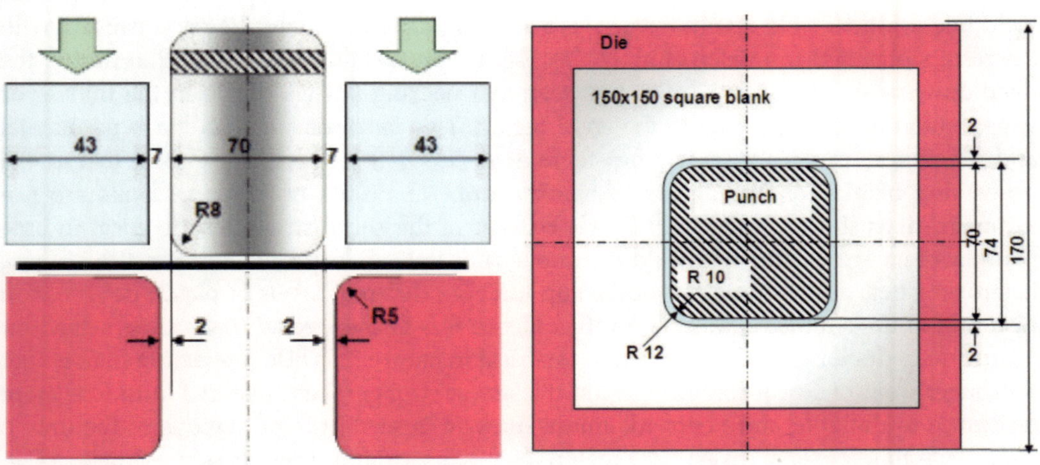

Fig. 15. Square cup deep drawing set up used for simulations (Makinouchi et al., 1993)

Hollomon's power law is used to describe the strain-hardening behaviour of base material and weld region, and Hill's 1948 isotropic hardening yield criterion is used as the plasticity model as before. The output deep drawing parameters monitored are maximum punch force, maximum weld line movement, draw depth and draw-in profile. The maximum punch force was obtained from force-progression data during deep drawing simulation. Draw depth was obtained after cup failure is witnessed. Fig. 16. shows that the draw-in profile of deep drawn TWB cup and was quantified by the dimensions DX, DY and DD. The draw-in profile is important and can be related to anisotropic sheet properties and earing behaviour of sheet metal. In order to include the impact of weld and base material conditions only, plastic strain ratios of base metal was kept constant throughout the work. Maximum weld line movement as observed is also represented in Fig. 16. This weld line movement is of practical importance as weld region should ideally be located in the safe region of the drawn cup. Similar procedure was followed for all the 27 deep drawing simulation trials for TWBs.

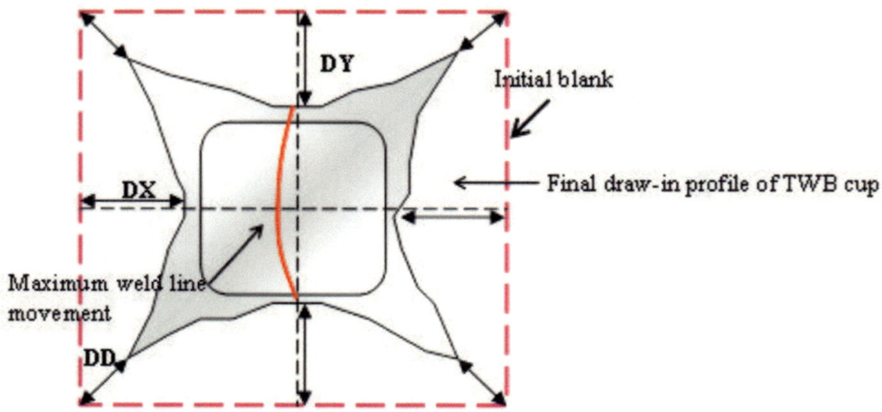

Fig. 16. Weld line movement and draw-in profile during deep drawing of TWB

The simulation of deep drawing test of aluminium TWBs show that with increase in initial weld line position, weld line movement is found to increase. The results are consistent with the results obtained from Heo et al. (2001). Fig. 17. presents the failure location during the deep drawing of square cup TWB. It is seen that necking always occurs in the thinner or weaker base material parallel to the weld region. This is consistent with the experimental and simulation results shown in Ahmetoglu et al. (1995). It is interesting to note that unlike un-welded blanks (or homogeneous blanks), draw-in profile of welded blanks are un-symmetrical as shown in Fig. 18a. This is because of thickness, strength differences in base materials that are welded and weld line movement during deep drawing. Because of these heterogeneities, different regions of the cup undergo different levels of plastic deformation resulting in un-symmetric draw-in profile. Hence it is also expected that earing behaviour during deep drawing will also be un-symmetrical in nature. It is also interesting to note that a stronger weld region at some angle and thicker (or stronger) base material introduce more resistance to drawing and hence minimum draw-in is seen in these regions of the drawn cup, while thinner (or weaker) base material show maximum draw-in as presented in Fig. 18b.

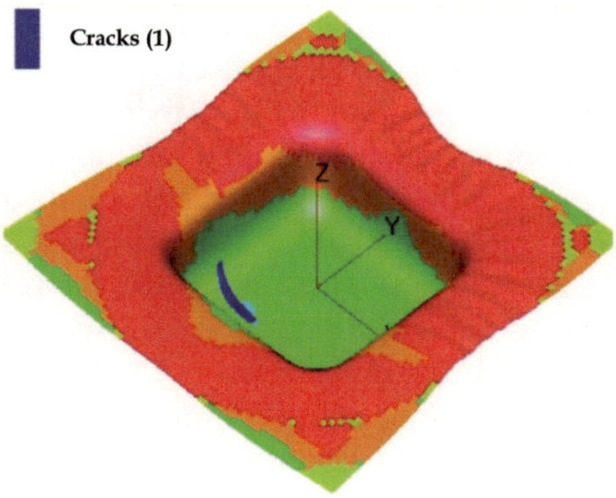

Fig. 17. Failure location seen in thinner sheet near weld line during deep drawing

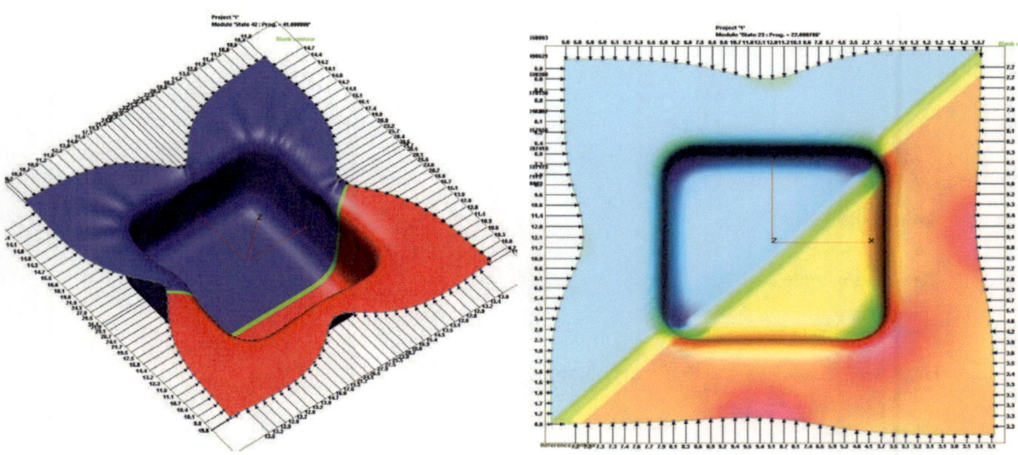

a) b)

Fig. 18. Simulation of deep drawn cup showing un-symmetric draw-in for different weld orientations; (a) 90° weld region, (b) Angular weld region

5. TWB formability prediction using ANNs

The set of tensile and deep drawing characteristics of TWB from the simulation trials are used for ANN modelling and expert system development. The ANN is trained to learn arbitrary nonlinear relationships between input and output parameters of TWB. The ANNs are inspired by biological neurons and have shown credible results in learning the arbitrary and complex relationships between the inputs that govern the outputs of an system. ANN consists of several layers of highly interconnected neurons which are the basic computing. The various architectural parameters of an ANN are number of hidden layers, neurons, and transfer functions which are optimized based on many trials to predict the outputs within certain errors that can be tolerated in an given application. In the present study a normalized error limit of 10^{-4} is taken. Using the simulation data obtained ANN with various network structures with one and two hidden layers with varying number of neurons in each layer and different transfer functions were examined. Optimized ANN architecture are found to model the tensile behaviour from the two sets of tensile simulation as well as the deep drawing behaviour. In all these cases, the ANN architecture consists of input layer with 6 input neurons (corresponding to 6 factors), and one / two hidden layers and output neurons corresponding to the number of outputs to be predicted. A feed forward back propagation algorithm is selected to train the network in Matlab® programming environment (Mathworks Inc., 2008). Here the scaled conjugate gradient algorithm (Mathworks Inc., 2008) is used to minimize the error. For each of the simulation trials based on the L_{27} orthogonal design of experiments, 27 data sets were used to train and two intermediate data sets were utilized for testing. The TWB tensile behaviour or deep drawing behaviour from FE simulations and ANN modelling for chosen two intermediate test sets or trials are compared to validate the accuracy of ANN predictions in each case. As an example, the ANN architecture used to predict the tensile behaviour without pre-existing defect based on the first set of tensile simulation data is shown in Fig. 19. Similar ANNs were trained for the other tensile simulation test for limit strain prediction as well as deep drawing simulation. The salient observations on the prediction of TWB tensile and deep drawing behaviour are described further.

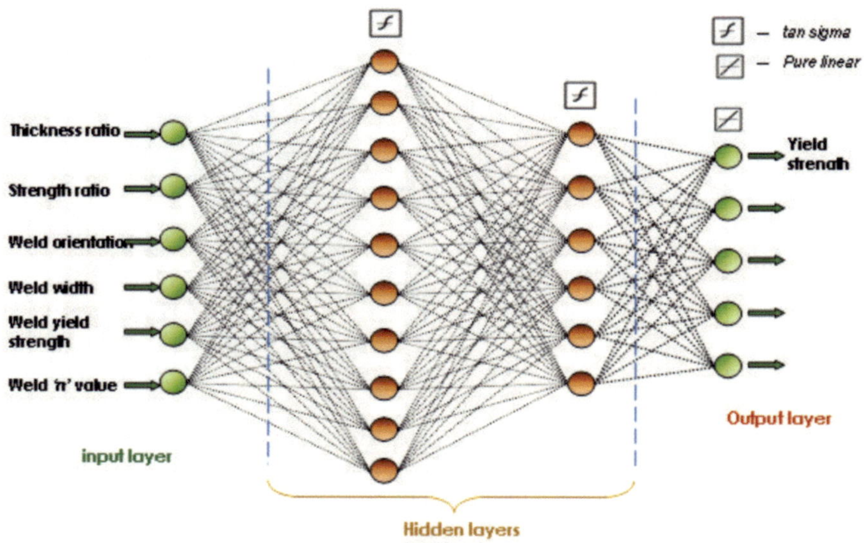

Fig. 19. Neural network architecture for TWB tensile behaviour prediction in safe region

The first set of 27 tensile simulation data (for the safe region of progression) was used to train an ANN and true stress-strain response, yield strength, ultimate tensile strength, uniform elongation, strain-hardening exponent and strength coefficient of welded blanks were predicted and validated with FE simulation results for two intermediate input levels. The comparison between ANN predicted true stress-strain behaviour and simulation results are shown in Fig. 20. The strain-hardening exponent (n) and strength coefficient (K) values obtained from ANN models were incorporated into Hollomon's equation ($\sigma = K\,\varepsilon^n$) for TWB made aluminium alloy base materials and true stress-strain curves were obtained. It should be noted that even though Hollomon's strain-hardening law is not accurate to predict the tensile behaviour of aluminium alloy base material, ANN predictions are quite accurate in predicting the same.

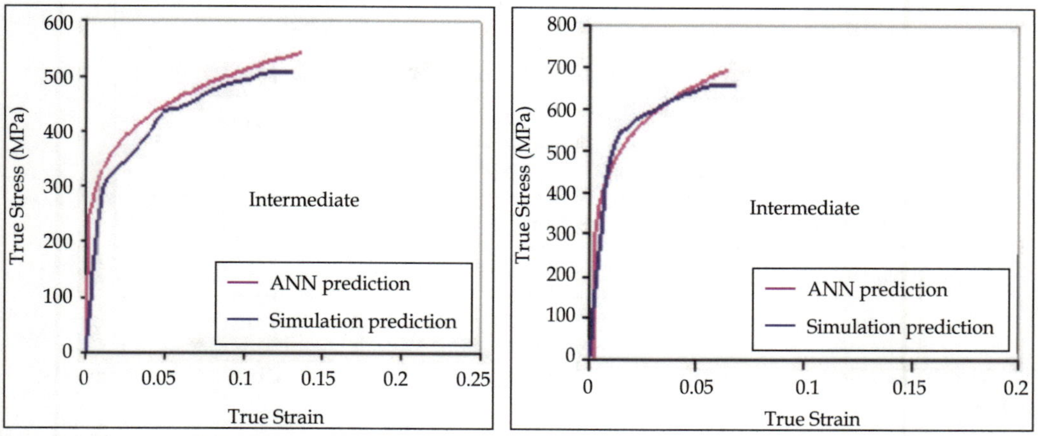

Fig. 20. Validating the true stress - strain behaviour predicted by ANN with FE simulation for two test data (Veerababu et al., 2009)

The TWB tensile properties from FE simulations and ANN modelling for chosen two intermediate trials were compared to validate the accuracy of ANN prediction. Table 3 summarizes the average error statistics pertaining to ANN prediction for training and testing with two intermediate test data not used for training. In the industrial application error range of 10-12% is considered acceptable and the same has been taken as a bench mark. It can be seen that almost all the output parameters are predicted within acceptable error limits. It is seen that only strain-hardening exponent (*n*) value of aluminium alloy TWB shows unacceptable error percentage (14.35%). This is possibly due to the smaller values of strain-hardening exponent which gives large percentage difference even if varied within small range (Veerababu et al., 2009). Similarly, the second set of tensile simulation with notched sample was used to train an ANN in a similar way. The limit strain (major and minor strain), failure location, minimum thickness, strain path were predicted using the trained ANN and validated with FE simulation results for two intermediate input levels. The prediction of failure location showed a higher level of prediction error (6.52 %) (Table 4). All other parameters show better prediction level with acceptable error range (Abhishek et al., 2011).

Output	**Training**		**Testing**	
	% error	**SD in error**	**% error**	**SD in error**
Yield strength (MPa)	0.18	7.08	11.91	50.04
Ultimate tensile strength (MPa)	0.05	13.71	5.03	35.63
Uniform elongation (mm)	0.09	0.10	4.45	1.41
Strain-hardening exponent '*n*'	0.01	0.01	*14.35*	0.01
Strength coefficient '*K*' (MPa)	0.01	13.01	10.49	36.04

Table 3. Validation of ANN model for tensile test simulation within safe progression limits

Output	Training		Testing	
	% error	SD in error	% error	SD in error
Major strain	0.007	0.23	5.23	3.51
Minor strain	0.067	0.92	2.79	0.87
Failure location	0.071	0.76	6.52	1.82
Minimum thickness	0.003	0.03	4.28	2.64
Strain path slope	0.052	0.42	3.91	0.36

Table 4. Validation of prediction by ANN for tensile simulation with necking induced failure

The deep drawing simulation data was used to train ANN to predict global TWB deep drawing behaviour viz., maximum weld line movement, draw depth, maximum punch force, draw-in profile for the chosen range of thickness and strength combinations, weld properties, orientation, and location. Two intermediate level data were taken for testing and validating the results as shown in Table 5. Fig. 21 presents the comparison between ANN and simulation results of draw-in profile of deep drawn cup. At different TWB conditions, the draw-in profile predicted by ANN model is well matched with the simulation results. All output parameters are predicted within acceptable error limits, except maximum weld line movement. Average error in this case is approximately 15% which is unacceptable. This possibly can be improved by using different strain-hardening laws and yield theories more suitable for aluminium alloy base materials. It is observed from Fig. 21a that the draw-in profiles are un-symmetric in shape. Minimum draw-in is seen along the angular weld region and in thicker material side, while thinner material shows maximum draw-in.

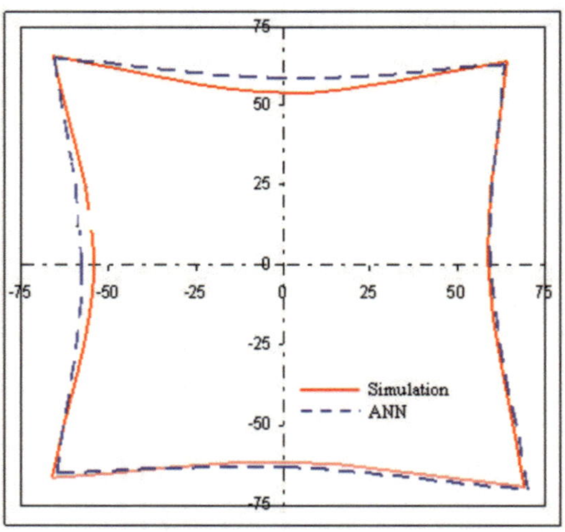

a) Test sample-1

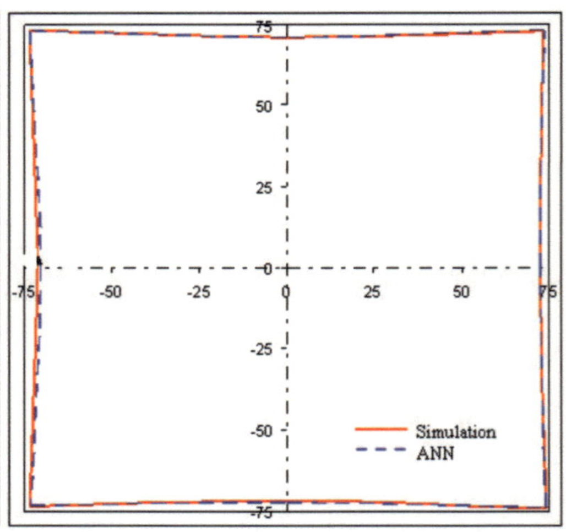

b) Test sample-2

Fig. 21. Comparison of draw-in profile between ANN prediction and FE simulation for two deep drawing test simulation of aluminium alloy TWB

Parameters	Test Data 1	Test Data 2
Thickness ratio (T_1/T_2), T_1 mm	0.6, 0.9	0.7, 1.05
Strength ratio (YS_1/YS_2), YS_1 MPa	0.7, 210	0.6, 180
Weld orientation (°)	35	55
Weld location, mm	14	7
Weld yield strength (YS_W), MPa,	250	325

Table 5. Input properties for validating the ANN deep drawing behaviour prediction of TWB

6. Conclusion

This chapter presented some studies on tensile and deep drawing behaviour of aluminium tailor-welded blanks. A finite element based numerical simulation method is used to understand the behaviour. The presence of thickness, strength heterogeneities and weld region deteriorates the formability of aluminium welded blanks in most of the cases. Designing TWB for a typical application will be successful only by knowing the appropriate thickness, strength combinations, weld line location and profile, number of welds, weld orientation and weld zone properties. Predicting these TWB parameters in advance will be helpful in determining the formability of TWB part in comparison to that of un-welded base materials. In order to fulfil this requirement, one has to perform lot of simulation trials separately for each of the cases which is time consuming and resource intensive. Automotive sheet forming designers will be greatly benefited if an 'expert system' is available for TWBs that can deliver its forming behaviour for varied weld and blank conditions. A artificial neural network based expert system is described which is being developed by the authors. The expert system is envisaged to be expanded with industrial applications also. For example, a sheet forming engineer who wants to develop expert system for some industrial TWB sheet part can just make it as part of existing system framework in the same line of thought, without introducing new rules and conditions. The relations between TWB inputs and outputs are non-linear in nature and hence it is complex to explicitly state rules for making expert system. But these complex relationships can be captured by artificial neural networks. The expert system proposed is a continuous learning system as the field problems solved by the system can also become a part of training sample. Though the expert system can not reason out the decisions/results unlike rule based systems, one can interpret the results by comparing the outputs of two different input conditions quantitatively with minimum knowledge in TWB forming behaviour.

7. References

Ahmetoglu, M. A., Brouwers, D., Shulkin, L., Taupin, L., Kinzel, G. L. & Altan, T. (1995). Deep Drawing of Round Cups from Tailor-welded Blanks, *Journal of Material Processing Technology*, Vol. 53, No. 3-4 , (September 1995), pp. 684-694

ASTM, (2000). Test Methods for Tensile Strain-hardening Exponents (*n*-values) of Metallic Sheet Materials. *Annual book of ASTM standards 2000 (E 646-98)*, Section 3, Vol. 03.01.

Asgari, S. A., Pereira, M., Rolfe, B. F., Dingle, M. & Hodgson, P. D. (2008). Statistical Analysis of Finite Element Modelling in Sheet Metal Forming and Springback Analysis. *Journal of Materials Processing Technology*, Vol. 203, No. 1-3, (July 2008), pp. 29–136

Banabic, D. (Ed.), (2000). *Formability of Metallic Materials*. Springer, Berlin, Heidelberg

Bhagwan, A. V., Kridli, G. H. & Friedman, P. A. (2003). Formability improvement in Aluminium Tailor-welded Blanks via Material Combinations. *Proceedings of NAMRC XXXI*, MF03-155, Hamilton, Ontario, Canada

Bravar, M., Krishnan, N. & Kinsey, B. (2007). Comparison of Analytical Model to Experimental and Numerical Simulations Results for Tailor-welded Blank Forming. *Journal of Manufacturing Science and Engineering*, Vol. 129, No. 1, (February 2007), pp. 211–215

Buste, A., Lalbin, X., Worswick, M.J., Clarke, J.A., Altshuller, B., Finn, M. & Jain, M. (2000). Prediction of Strain Distribution in Aluminum Tailor-welded Blanks for Different Welding Techniques. *Canadian Metallurgical Quarterly*, Vol. 39, No. 4, (October 2000), pp. 493-502

Cakir, M. C. & Cavdar, K. (2006). Development of a Knowledge-based Expert System for Solving Metal Cutting Problems. *Materials and Design*, Vol. 27, No.10, pp. 1027–1034

Chan, M., Chan, L. C. & Lee, T. C. (2003). Tailor-welded Blanks of Different Thickness Ratios Effects on Forming Limit Diagrams. *Journal of Materials Processing Technology*, Vol. 132, No. 1-3, (January 2003), pp. 95-101

Chan, L. C., Chang, C. H., Chan, S. M., Lee, T. C. & Chow, C. L. (2005). Formability Analysis of Tailor-welded Blanks of Different Thickness Ratios. *Journal of Manufacturing Science and Engineering*, Vol. 127, No. 4, (November 2005), pp. 743-751

Cheng, C. H., Chan, L. C., Chow, C. L. & Lee, T. C. (2005). Experimental Investigation on the Weldability and Forming Behaviour of Aluminum Alloy Tailor-welded Blanks. *Journal of Laser Applications*, Vol. 17, No. 2, (May 2005), pp. 81–88

Davies, R.W., Oliver, H.E., Smith, M.T. & Grant, G.J. (1999). Characterizing Al Tailor-welded Blanks for Automotive Applications. *Journal of Metals*, Vol. 51, No. 11, (November 1999), pp. 46-50

Davies, R.W., Smith, M.T., Oliver, H.E., Khaleel, M.A. & Pitman S.G. (2000). Weld Metal Ductility in Aluminium Tailor-welded Blanks. *Metallurgical and Materials Transactions A*, Vol. 31, No. 11, (November 2000), pp. 2755-2763

Dhumal, A. T., Ganesh Narayanan, R. & Saravana Kumar, G. (2011). Simulation based Expert System to Predict the Tensile Behaviour of Tailor-welded Blanks, *International Journal of Advanced Manufacturing Systems*, Vol. 13 (1), 2011, pp. 159-171

Dominczuk, J. & Kuczmaszewskim, J. (2008). Modelling of Adhesive Joints and Predicting their Strength with the Use of Neural Networks. *Computational Materials Science*, Vol. 43, No. 1, (July 2008), pp. 165–170

Dym, C. L. (1985). Expert Systems: New Approaches to Computer-aided Engineering, *Engineering with Computers*, Vol. 1, No. 1, pp. 9-25

Ebersbach, S. & Peng, Z. (2008). Expert System Development for Vibration Analysis in Machine Condition Monitoring. *Expert Systems with Applications*, Vol. 34, No. 1, (January 2008), pp. 291–299

Emri, I. & Kovacic, D. (1997). Expert System for Testing Mechanical Properties of Aluminum and Aluminum Alloys. *Expert Systems With Applications*, Vol. 12, No. 4, (May 1997), pp. 473-482

Ganesh Narayanan, R. & Narasimhan, K. (2006). Weld Region Representation During the Simulation of TWB Forming Behaviour, *International Journal of Forming Processes*, Vol. 9, No. 4, (2006), pp. 491-518

Ganesh Narayanan, R. & Narasimhan, K. (2007). Relative Effect of Material and Geometric Parameters on the Forming Behaviour of Tailor-welded Blanks. *International Journal of Forming Processes*, Vol. 10, No. 2, pp. 145-178

Ganesh Narayanan, R. & Narasimhan, K. (2008). Predicting the Forming Limit Strains of Tailor-welded Blanks. *Journal of Strain Analysis for Engineering Design*, Vol. 43, No. 7, (June 2008), pp. 551–563

Heo, Y., Choi, Y., Kim, H. Y. & Seo, D. (2001). Characteristics of Weld Line Movements for the Deep Drawing with Drawbeads of Tailor-welded Blanks. *Journal of Materials Processing Technology*, Vol. 111, No. 1-3, (April 2001), pp. 164–169

Holmberg S., Enquist B. & Thilderkvist P. (2004). Evaluation of Sheet Metal Formability by Tensile Tests. *Journal of Materials Processing Technology*, Vol. 145, No. 1, (January 2004), pp. 72-83

Jie, M., Cheng, C. H., Chan L. C., Chow C. L. & Tang C. Y. (2007). Experimental and Theoretical Analysis on Formability of Aluminium Tailor-welded Blanks. *Journal of Engineering Materials and Technology*, Vol. 129, No. 1, (January 2007), pp. 151-158

Kinsey, B. L. & Cao, J. (2003). An Analytical Model for Tailor-welded Blank Forming. *Journal of Manufacturing Science and Engineering*, Vol. 125, No. 2, (May 2003), pp. 344–351

Kumar, S., Date P. P. & Narasimhan K. (1994). A New Criterion to Predict Necking Failure under Biaxial Stretching, *Journal of Materials Processing Technology*, 1994, Vol. 45, No. 1-4, (September 1994), pp. 583-588

Kusuda, H., Takasago, T., & Natsumi F. (1997). Formability of Tailored Blanks. *Journal of Materials Processing Technology*, Vol. 71, No. 1, (November 1997), pp. 134–140

Lee, W., Chung, K-H., Kim, D., Kim, J., Kim, C., Okamoto, K., Wagoner, R. H. & Chung, K. (2009). Experimental and Numerical Study on Formability of Friction Stir Welded TWB Sheets based on Hemispherical Dome Stretch Tests. *International Journal of Plasticity*, Vol. 25, No. 9, (September 2009), pp. 1626-1654

Makinouchi, A., Nakamachi, E., Onate, E. & Wagoner, R. H. (Eds.). (1993). Benchmark Problems, Square Cup Deep Drawing, *Proceedings of NUMISHEET 1993, 2nd International conference on Numerical simulation of 3D sheet forming process – Verification of simulation with experiments*, Isehara, Japan, 1993, pp. 377-380

Manabe, K., Yang, M. & Yoshihara, S. (1998). Artificial Intelligence Identification of Process Parameters and Adaptive Control System for Deep Drawing Process, *Journal of Materials Processing Technology*, Vol. 80–81, (August 1998), pp. 421–426

Mathworks Inc. (2008). *Matlab Neural Network Toolbox User's Guide*, Version 7.0, The Mathworks Inc., MA

Miles, M.P., Decker, B.J. & Nelson, T.W. (2004). Formability and Strength of Friction-Stir-Welded Aluminum Sheets. *Metallurgical and Materials Transactions A*, Vol. 35, No. 11, (November 2004), pp. 3461-68

Ohashi, T., Saeki, Y., Motomura, M. & Oki, Y. (2002). Computer-aided Blanking Sequence Design of Extruded Aluminium Materials. *Journal of Material Processing Technology*, Vol. 123, No. 2, (April 2002), pp. 277-284

Palani, R., Wagoner, R. H. & Narasimhan, K. (1994). Intelligent Design Environment: A Knowledge Based Simulations Approach for Sheet Metal Forming. *Journal of Materials Processing Technology*, Vol. 45, No. 1-4, (September 1994), pp. 703–708

Pallet, R. J. & Lark, R. J. (2001). The use of Tailored Blanks in the Manufacture of Construction Components. *Journal of Materials Processing Technology*, Vol. 117, No. 1-2 , (November 2001), pp. 249-254

Raymond, S. C., Wild, P. M. & Bayley, C. J. (2004). On Modelling of the Weld Line in Finite Element Analyses of Tailor-welded Blank Forming Operations. *Journal of Materials Processing Technology*, Vol. 147, No. 1, (March 2004), pp 28–37

Saunders, F. I. & Wagoner, R. H. (1996). Forming of Tailor-welded Blanks, *Metallurgical and Materials transactions A*, Vol. 27, No. 9, (September 1996), pp. 2605-2616

Shakeri, H.R., Buste, A., Worswick, M.J., Clarke, J.A., Feng, F., Jain, M. & Finn, M. (2002). Study of Damage Initiation and Fracture in Aluminum Tailor-welded Blanks made via Different Welding Techniques. *Journal of Light Metals*, Vol. 2, No. 2, (May 2002), pp. 95–110

Stasik, M. C. & Wagoner R. H. (1998). Forming of Tailor-welded Aluminium Blanks, *International Journal of Forming Processes*, Vol. 1, No. 1, pp. 9-33

Stein, E. W., Pauster, M. C. & May, D. (2003). A knowledge-based System to Improve the Quality and Efficiency of Titanium Melting. *Expert Systems with Applications*, Vol. 24, No. 2, (February 2003), pp. 239–246

Taguchi, G. (1990). *Introduction to Quality Engineering*, Asian Productivity Organization, Tokyo

Veera Babu, K., Ganesh Narayanan, R. & Saravana Kumar, G. (2009). An Expert System based on Artificial Neural Network for Predicting the Tensile Behaviour of Tailor-welded Blanks. *Expert Systems with Applications*, Vol. 36, No.7, (September 2009), pp. 10683–10695

Veera Babu, K., Ganesh Narayanan, R. & Saravana Kumar, G. (2010). An Expert System for Predicting the Deep Drawing Behaviour of Tailor-welded Blanks. *Expert Systems with Applications*, Vol. 37, No. 12 , (December 2010). pp. 7802-7812

Wang, L., Porter, A.L. & Cunninghame, S. (1991). Expert Systems: Present and Future. *Expert Systems with Applications*, Vol. 3, No. 4, pp. 383-396

Yazdipour, N., Davies, C. H. J. & Hodgson, P. D. (2008). Microstructural Modelling of Dynamic Recrystallization using Irregular Cellular Automata. *Computational Materials Science*, Vol. 44, No. 2, (December 2008), pp. 566–576

Part 3

Surface Treatment of Aluminium Alloys

6

Laser Surface Treatments of Aluminum Alloys

Reza Shoja Razavi and Gholam Reza Gordani
Materials Science and Engineering Department, Malek Ashtar University of Technology, Shahin Shahr, Iran

1. Introduction

Advanced industrial applications require materials with special surface properties such as high corrosion and wear resistance and hardness. Alloys possessing these properties are usually very expensive and their utilization drastically increases the cost of the parts. On the other hand, failure or degradation of engineering components due to mechanical and chemical/electrochemical interaction with the surrounding environment is most likely to initiate at the surface because the intensity of external stress and environmental attack are often highest at the surface.

The engineering solution to prevent or minimize such surface region of a component through a procedure known as surface engineering. Conventionally practiced surface engineering techniques like carburizing, nitriding, etc. are often material specific, time/energy/manpower intensive and lacking in precision.

Among the surface engineering techniques, a relatively new and attractive method is laser surface treatment. In other words, laser surface treatment offers an excellent scope for tailoring the surface microstructure and/or composition of a component and proves superior to conventionally surface engineering.

For most engineering application, the laser, in simple terms, can be regarded as a device for producing a finely controllable energy beam, which, in contact with a material, generates considerable heat. The basic physics of laser surface treatment is simply heat generation by laser interaction with the surface of an absorbing material and subsequent cooling either by heat conduction into the interior, or by thermal reradiation at high temperatures from the surface of the material. Various laser surface treatment methods that are currently available are shown in figure 1.

2. Laser – assisted materials surface treatment requirements

Figure 2 shows general regimes of various laser surface treating parameters for both pulsed and continuous wave lasers. Short pulses (ns to fs) with high peak power densities are desirable for laser shock processing and ablation applications. In general, longer pulses (μ_s to ms) or continuous wave lasers are preferred for melting and heating processes (Nagarathnam & Taminger, 2001).

Laser chemical vapor deposition and laser surface transformation hardening require lower densities and interaction times as compared to processes involving meting and vaporization.

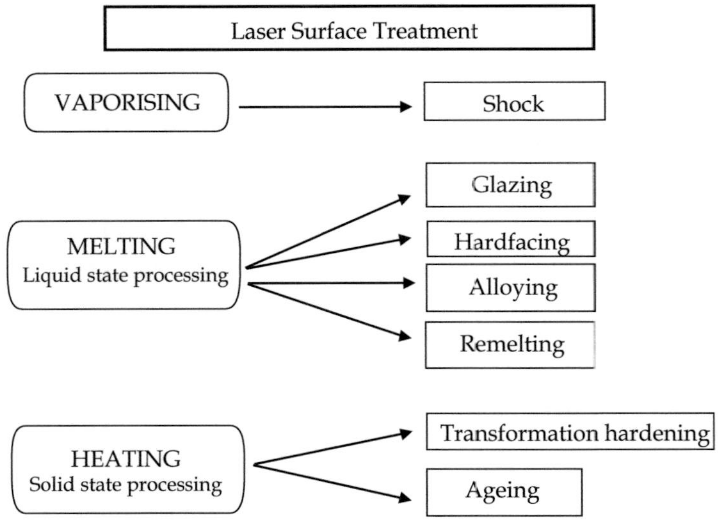

Fig. 1. Various laser surface treatment methods

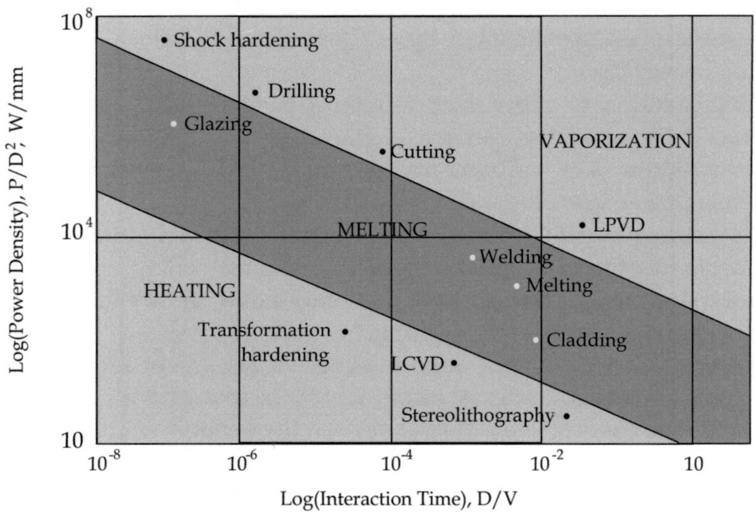

Fig. 2. Laser power density, specific energy and interaction times for various laser processing regimes (Nagarathnam & Taminger, 2001)

3. Laser surface alloying of aluminium alloys

Laser surface alloying (LSA) involves tailoring the surface microstructure and composition by rapid melting, intermixing and solidification of a pre/co deposited surface layer with apart of the underlying substrate (Majumdar & Manna, 2002). Also in this treatment, a shallow layer at the surface of the material is melted by the laser beam which becomes efficiently coupled to the surface, while alloying elements are added simultaneously to give a local composition having the desired surface properties on solidification (Renk et al., 1998). When alloying elements are added to the melted pool then they will start to

interdiffuse into the substrate. As soon as the laser pulse is finished the resolidification process begins from the liguid/solid interface towards the surface.
Laser surface alloying may induce an extreme heating/cooling rate of 10^4-10^{11}k/s, thermal gradient of 10^5-10^8 k/m and solidification velocity as high as 30m/s (Draper & Poate, 1985). Due to the high cooling rates, solid state diffusion can be neglected and homogeneous and fine solidification microstructures can be achieved with a wide variety of surface compositions without the limitations of conventional processes, for instance, to extend solid solutions and to obtain metastable structures or even metallic glasses (Damborenea, 1998).
Laser surface alloying of Al- alloys by different alloying elements and different techniques was investigated by several researchers. In most of their reports, it was shown that the structure of the zones of laser alloyed depends on the properties of the treated and alloying materials and on the dispersity of the alloying particles, the power of the laser radiation, and the duration of the irradiation (Aleksandrov, 2002). Figure 3 shows the structure of Al-alloy D16 saturated with NiO_2 and $NbSi_2$ particles. The saturation with NiO_2 is provided by convective mass transfer, which is confirmed by the vortex-like appearance of the structure of the molten zone (figure 3a). The well-manifested heterophase (figure 3b) in the surface layer is provided by the mechanism of penetration of particles of $NbSi_2$ into the molten pool.

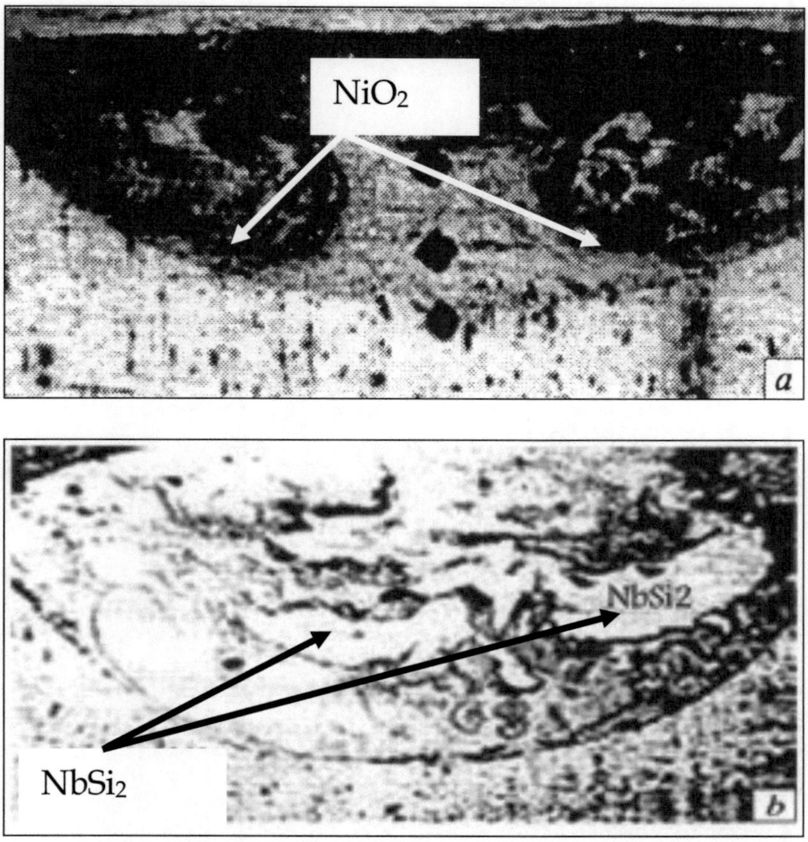

Fig. 3. Structure of the molten zone after laser alloying of Al-alloy D16 with NiO_2 (a) and $NbSi_2$ (b). x200. (Aleksandrov, 2002)

The structure of the laser surface alloyed of Al with Nb is shown in figure 4 (Almeida et al., 2001). A strong segregation of Nb in structure leading to the formation of a zone of resolidified Al solid solution and a zone with a high Nb concentration, consists of dendrites of Nb-free α-Al solid solution and undissolved particles of Nb (figure 4a), that some of these particles can be surrounded by a layer consisting of Al_3Nb dendrites in an α- Al matrix (figure 4b) showing incipient dissolution and partial redistribution of Nb due to convective flow. It is necessary to mentioned that the temperature and convective mass transport must be sufficient to allow for the complete homogenization of the alloyed layers. This, it can be seen in figure 4 that the temperature and convective mass transport were not sufficient to allow for the complete homogenization of the material (Almeida et al., 2001).
Mazumder (Majumdar & Manna, 2002) studied mass transport in melt pools using a numerical model and concluded that the extremely fast homogenization frequently observed in laser surface alloying can only be explained by the intense Marangoni convection caused by the high temperature gradients within the melt pool (Almeida et al., 2001), with diffusion having only a minor role.

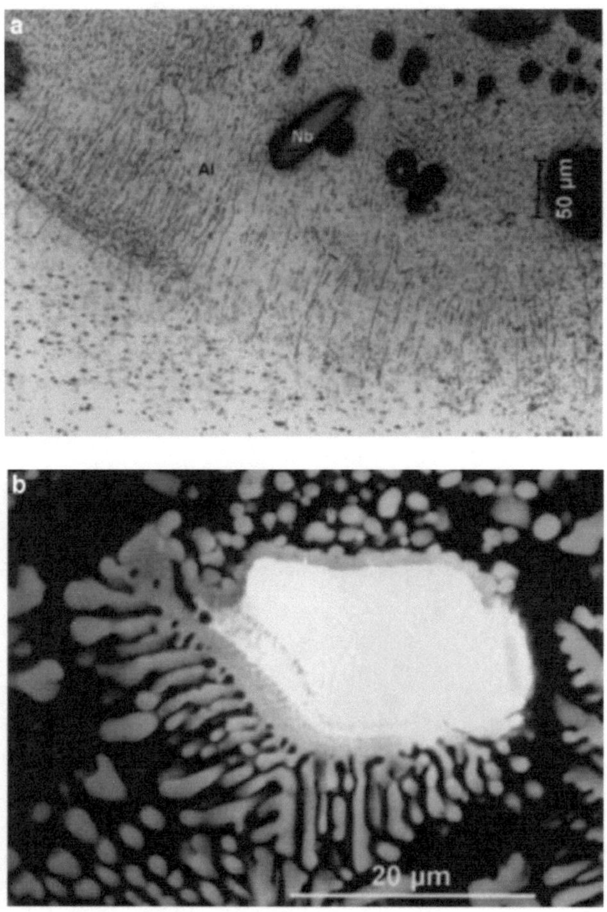

Fig. 4. (a) Structure of the bottom layer (A). (b) Undissolved Nb particle surrounded by a layer of Al_3Nb dendrites and α-Al (Almeida et al., 2001)

They suggested that the influence of convection in liquid homogenization could be characterized by the surface tension number, S, which relates thermocapillarity-induced convection velocity and laser beam scanning speed, and is give by:

$$S = \frac{\left(\frac{d\sigma}{dT}\right)qd}{\mu u_0 k} \tag{1}$$

Where (dσ/dT) is the temperature coefficient of the surface tension, q is the net heat flow from the laser beam, d is diameter of the laser beam, μ is the viscosity, u_0 is the scanning speed of the laser beam, and k is the thermal conductivity. When S is low ($S \leq 45000$), convection is negligible. Due to the short lifetime of the melt pool, mass transport will be insufficient for melt homogenization. When S is high, convection plays a dominant role in transport phenomena in the melt pool. In general for metals, the convection speed is several orders of magnitude higher than the scanning speed, leading to rapid homogenization. However, for liquid Al the temperature coefficient of the surface tension (dσ/dT) is relatively low (-0.155×10^{-3} $Nm^{-1}k^{-1}$), and therefore S will be low. Consequently, in some cases insufficient homogenization of the melt pool is to be expected for laser surface alloying of Al-alloys. For example, this was happened for recently mentioned Almeida's research that is shown in figure 4 (Almeida et al., 2001). A further difficulty arises when the alloying elements react with the melt pool material to form insoluble high melting temperature phases, such as intermetallic compounds. In the matter the diffusion phenomena is responsible to control of dissolution kinetics.
The dissolution kinetics was theoretically analysed by Costa and Vilar (Costa & Vilar, 1996) using a spherical geometry and dropping the quasi-steady–state approximation. Figure 5 shows the results of evaluation of intermetallic layer thickness of Al_3Nb as a function of time. This result is reported by Almeida and co-workers (Almeida et al., 2001). They calculated the dissolution time of Nb particles with a diameter of 100μm. These particles takes about 22s to transform to Al_3Nb, a time much longer than the interaction time used (0.24s) in their research.

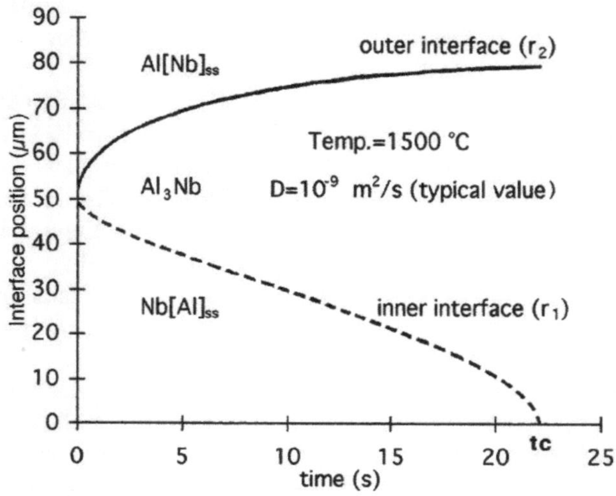

Fig. 5. Thickness of intermetallic layer and size of particle for any laser interaction time (Almeida et al., 2001)

In order to obtain significant particle dissolution, the temperature of the melt pool must be higher than the melting point of the intermetallic compounds.
Since convection-driven homogenization is negligible, and the melting point of alloying elements is higher than the melting temperature of Al, the latter starts to solidify before the alloying particles dissolve, and a layer consists of the starting material will be formed. In general, this happened in the bottom layers. In the upper layers, due to the higher temperature of the melt, the alloying particles dissolve in the liquid Al.
The microstructure of alloyed layer depends on solidification rate (R) and the temperature gradient at the solid-liquid interface (G) which in turn depend on heat and mass transfer in the system (Almeida et al., 2001). A simple relation exists between the local solidification rate (R) and the scanning speed gives by fallow equation:

$$R=V_S \cos \theta \qquad (2)$$

where θ is the angle between the normal to the solid-liquid interface and the scanning direction. The solidification rate increases with decreasing depth from 0 at the bottom of the melt pool ($\cos\theta=0$ at this point) to a value that remains lower that the scanning speed, because $\cos \theta<1$. Conversely, the thermal gradient G is higher at the bottom of the melt pool and decreases as depth decreases. Both solidification parameters vary rapidly during the first stages of solidification (near the bottom of the melt pool) to reach a value that remains approximately constant during most of the solidification process. Consequently, the microstructure in most of the re-solidified layer can be characterized by a single set of solidification parameters and should not change significantly.
Sometimes, in laser surface alloying the microstructure of alloyed layer appears as a texture. The texture effect increases with increasing solidification speed. The origin of this texture can be understood by considering the solidification mechanism in laser surface alloying and the variation of the shape of the melt pool as a function of scanning speed.
Gingo and et. al (Gingu et al., 1999) produced Al/SiC_P composite by laser surface alloying. Figure 6 presents the microstructural aspect of the alloyed layer produced at the surface of an AA413 alloy. There is an obvious difference between the base microstructure of the Al alloy, which is the classic eutectic AlSi12, characterised by dendrites grains dispose randomly in the eutectic mass (zone 1), and the very fine granulated microstructure of the alloyed layer (zone2).
In this process, depending on the processing parameters, it is possible to use or not use an adhesive layer at the material support. As can be seen in figure 7, in this case there is a perfect adherence of the alloyed layer at the AA413 support; this phenomenon can be explained by the perfect compatibility of the matrix of Al- alloy (Gingu et al., 1999).
In laser surface alloying of Al with Nb as an alloying element, the dendritic structure was observed by Almeida et. al (Almeida et al., 2001), showing that Al_3Nb grows with a dendritic solid/liquid interface. In this type of growth, there is a preferential growth direction usually a low index crystallographic direction. During the initial stages of solidification competition between neighboring dendrites with different orientations occurs, and those with the preferential growth direction nearest to the heat flow direction will be favored, leading to preferential orientation, and eventually to a strong texture. When the scanning speed of surface is increased the shape of melt pool increasingly elongated from semi-hemispherical. Also, when the scanning speed is low the heat flow direction changes progressively from the fusion line to the surface, leading to a variety of preferential growth directions of columnar grains (Almeida et al., 2001).

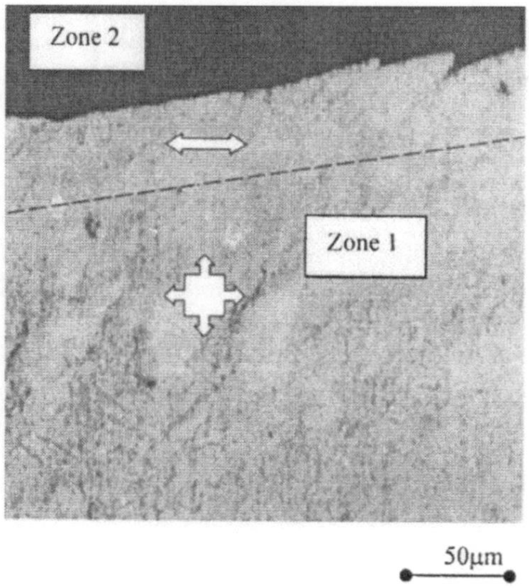

Fig. 6. The micro structural aspects of the laser alloyed layer (Gingu et al., 1999)

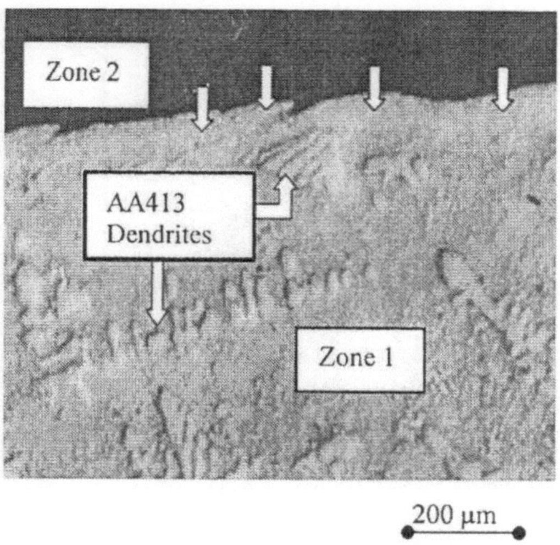

Fig. 7. The adherence aspect of the layer at the material support (Gingu et al., 1999)

A mathematical modeling of laser surface alloying with solid particles is established by Aleksandrov (Aleksandrov, 2002). According to this model, the presence of a transverse temperature gradient $\frac{\partial T}{\partial X}$ makes the particles move to the peripheral part of the molten pool. In the case of low gradients $\frac{\partial T}{\partial X}$ and $\frac{\partial T}{\partial Z}$ the particles simply drown in the field of the effective force of gravity (with allowance for the buoyancy force). The higher the difference in the densities of the alloying and Al alloys, the more effective the immersion of the particles. The equations of the motion of a single particle in Cartesian coordinates X, Z have the form (Aleksandrov, 2002):

$$\frac{4}{3}\pi p R^2 \ddot{x} = 2\alpha_0 \frac{\partial T}{\partial x} R - 6\pi\eta R \dot{x} \tag{3}$$

$$\frac{4}{3}\pi p R^3 \ddot{z} = 2\alpha_0 \frac{\partial T}{\partial z} R + \frac{4}{3}\pi\left(P - P_1\right) R^3 g - 6\pi\eta R \dot{z} \tag{4}$$

where ρ is the density of the particle, R is the radius of the particle, ρ_1 is the density of the Al alloy, η is the viscosity of the Al alloy, $\dot{x},\ddot{x}$ are the transverse velocity and acceleration of the particle respectively, and $\dot{z},\ddot{z}$ are the vertical velocity and acceleration of the particles, respectively.

The mechanism of the infusion and velocity of particles in melt pool affects on the formation of heterogeneous or homogenous structure in alloyed layer. The corresponding qualitative dependence $v_0(R)$ is plotted in figure 8.

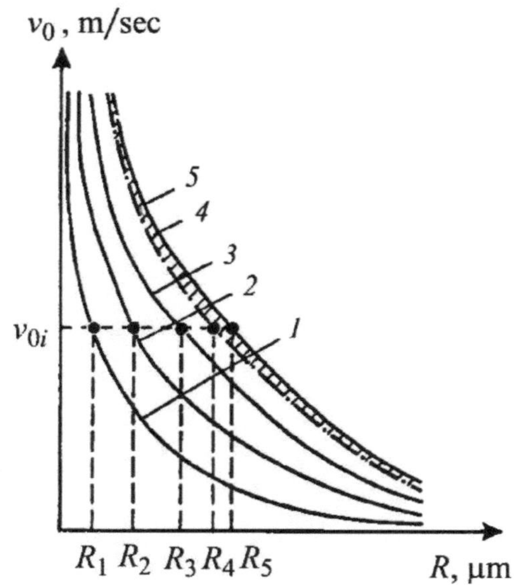

Fig. 8. Dependence of the initial velocity of the particles v_0 on their radius R at a fixed surface density (ρ_i) of the energy of laser radiation (P_1<P_2<P_3<P_4<P_5) (Aleksandrov, 2002)

The order of magnitude of the initial velocity of the particles v_o needed for the alloying is determined by the time τ of the action of laser radiation and the depth of the molten region H ($v_o = \frac{H}{\tau}$). Particles with a size ranging between R_2 and R_3 may acquire the requisite values v_o. Under actual conditions and fixed energy and time of the laser action, the dependences $v_o(R)$ corresponds to a certain domain (hatched in figure 8).

Aleksandrove (Aleksandrov, 2002) also studied the wear resistance of laser surface alloyed layer of Al alloy with Ni, $NbSi_2$ and Cr. The results of wear tests are presented in figure 9.

It can be seen that wear resistance of the hardened surface of aluminum alloy layer is much higher (by factor of 4-5) that the initial one or the one provided by LHT. Also, the friction coefficient tests show that laser surface alloying of Al alloys with Cr, $NbSi_2$ and Ni decreases the friction coefficient of the friction surface by about a factor of 3-4, which makes it possible to vary it by changing the filling factor of the surface and the filling of the alloyed zone with conglomerates of high-hardness particles. The use of lubricating materials improves the service properties of various friction pairs (Aleksandrov, 2002). Similar results are reported by Tomlinson and co- workers (Tomlinson & Bransden, 1996). They studied the effect of laser alloying of metallic elements such as Si, Ni, Fe, Cu, Mn, Cr, Co, Mo and Ti on hypoeutectic cast Al-Si alloys using a pre-placed coating method, and found an improvement in the wear resistance of aluminum. Senthile selvan and co-workers (Senthil Selvan et al., 2000), reported that when laser alloying of Al with Ni was carried out at the highest scan speed of 1.1 m min^{-1}, the hardness increased to 800-900 Hv with negligible fluctuations in the hardness behavior. This may be attributed to a uniform LAC with well distributed intermetallic phases. While, at a slightly increased scan speed, the hardness increased from 300 to 800 Hv, but with large fluctuations, which can be attributed to the homogeneous alloyed layer (figure 10).

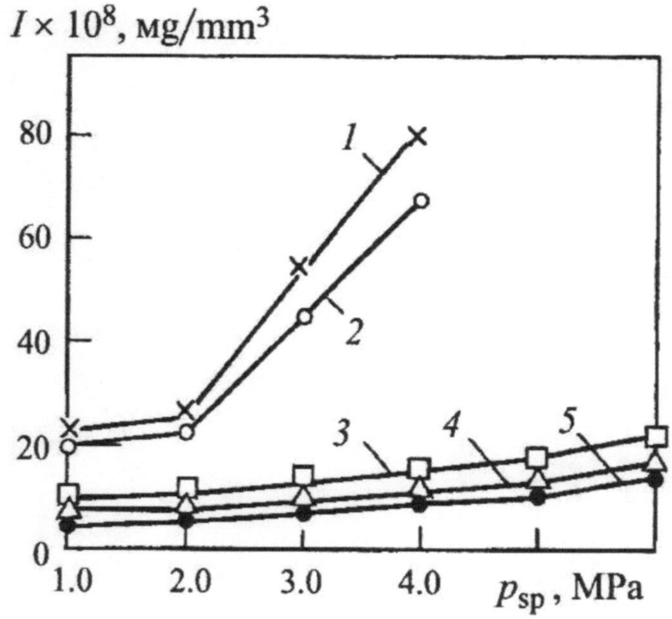

Fig. 9. Dependence of the wear intensity of aluminium alloy on the specific load in a wear test laser treatment at P=1kw, v=12.5 mm/sec, 1)Initial state, 2) after LHT 3,4,5) after alloying with Ni, $NbSi_2$, Cr, respectively (Aleksandrov, 2002)

The homogeneous distribution of hard intermetallic phases in Al matrix can prevent adhesion and abrasive wear during fretting. Yongqing Fu (Yongqing et al., 1998), reported that after a large number of fretting cycles, the rate of fretting wear depth decreases, which means that the wear volume loss is probably caused by an increase in fretting area rather by wear along the depth. This phenomenon is probably caused by the formation and compaction of fretting oxide debris, which can reduce the wear along the fretting depth. Laser surface alloying can decreases the fretting wear volume by a factor of three and decreases the coefficient of friction, probably due to the hardening effect of oxide debris which can prevent adhesion and abrasive wear during fretting, therefore, it can offers an effective means of preventing fretting wear (figure 11). The (Yongqing et al., 1998).

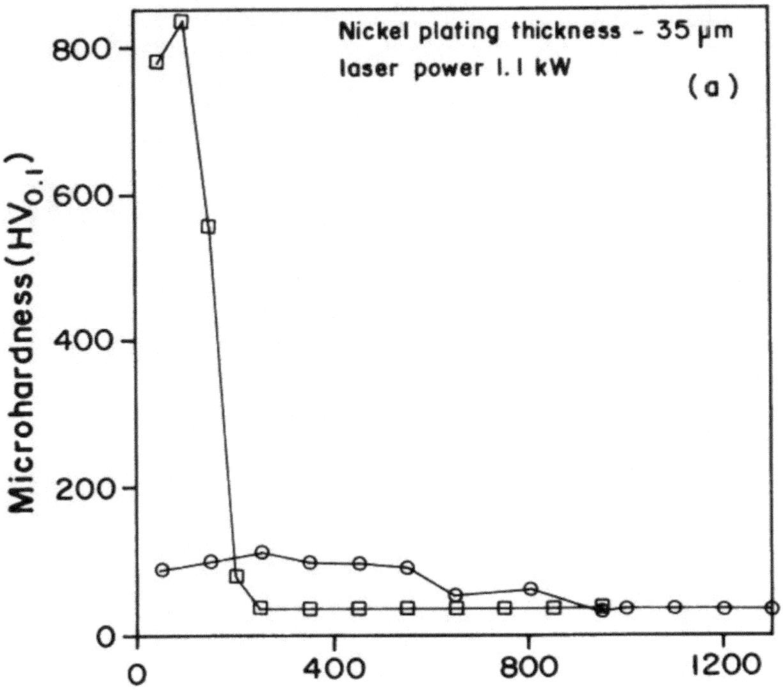
Nickel plating thickness - 35 μm
laser power 1.1 kW
(a)
Microhardness ($HV_{0.1}$)
800
600
400
200
0
0
400
800
1200

(a)

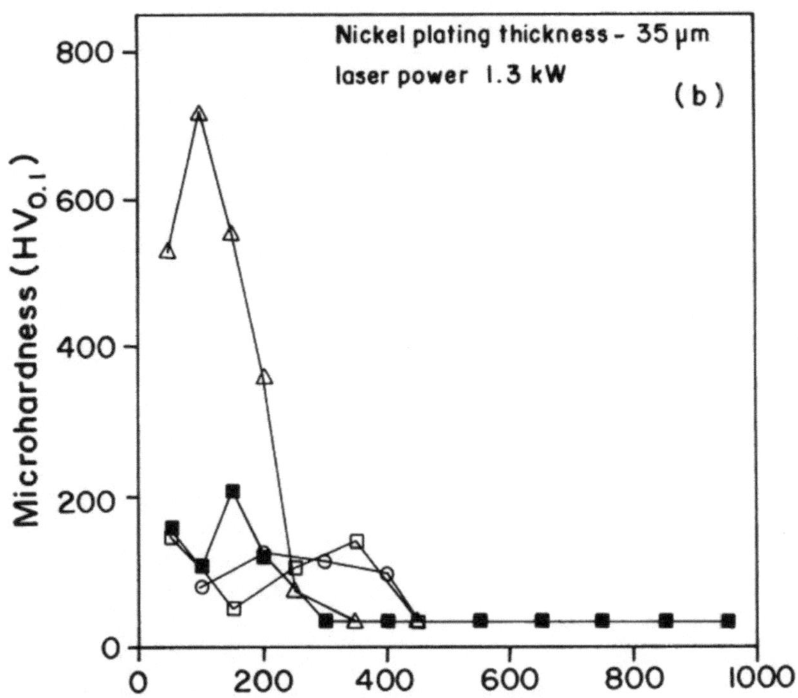
Nickel plating thickness - 35 μm
laser power 1.3 kW
(b)
Microhardness ($HV_{0.1}$)
800
600
400
200
0
0
200
400
600
800
1000

(b)

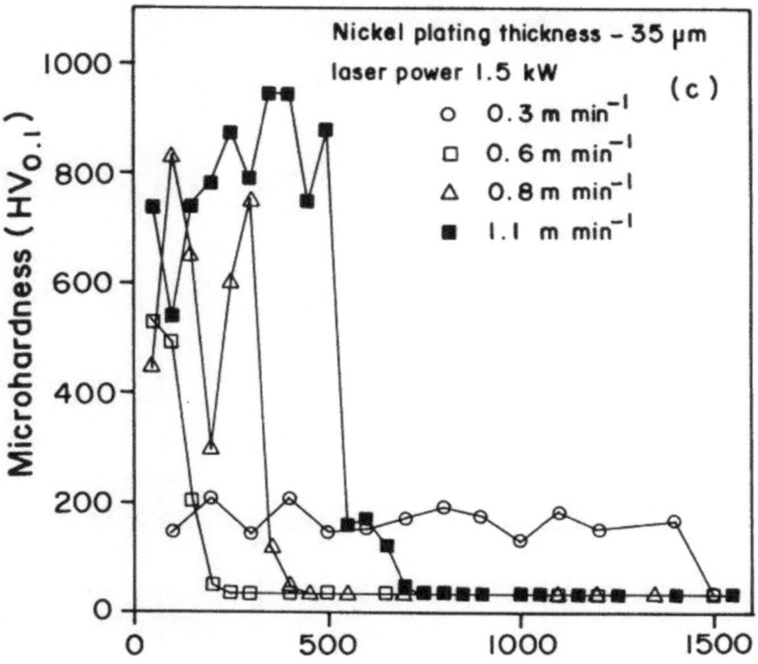

(c)

Fig. 10. Microhardness profiles of laser- alloyed Cp- Al with Ni at powers of a) 1.1, b) 1.3 and c) 1.5 kw at different scan speeds (Senthil Selvan et al., 2001)

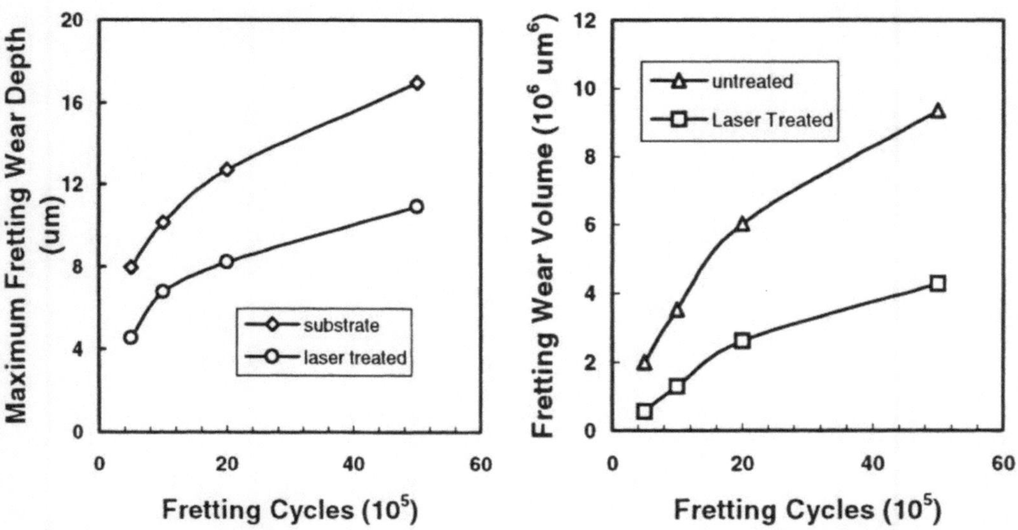

(a) The maximum fretting wear depth (b) Fretting wear volume

Fig. 11. The maximum fretting wear depth and retting volume for the Al6061 and laser-treated Al- alloy (normal load of 2N, amplitude of 50μm, frequency of 50Hz under unlubricated conditions) (Yongqing et al., 1998)

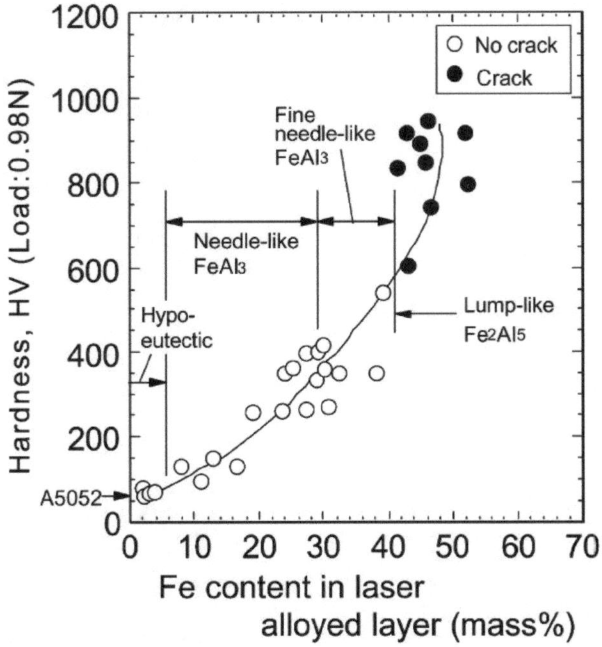

Fig. 12. Relation between Fe content and hardness of laser alloyed layer on aluminum (Tomida & Nakata, 2003)

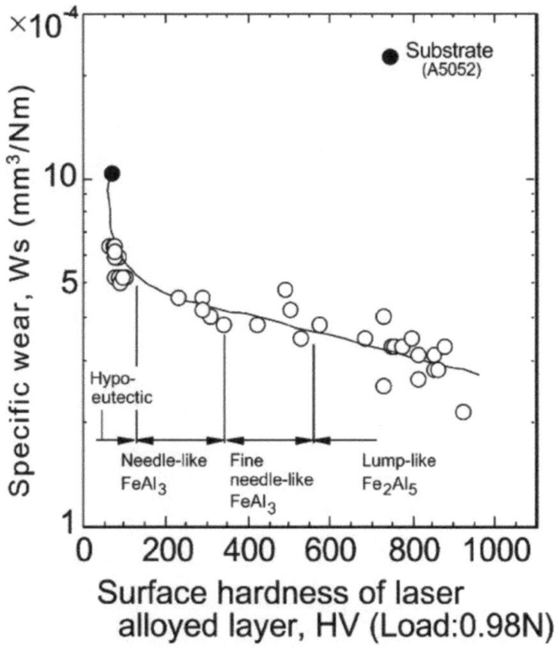

Fig. 13. Relation between surface hardness of laser alloyed layer and abrasive wear behavior. (Tomida & Nakata, 2003)

Laser surface alloying of Al with Fe is studied by Tomida and Nakata (Tomida & Nakata, 2003). They reported that the hardness of laser alloyed layer increases with increasing Fe content as shown in (Figure 12). However, cracking occurred in the alloyed layer with higher hardness than Hv600, because the brittle 1ump-like Fe_2Al_5 compound was produced in these layers. The wear resistance of the alloyed layer improved with increasing the hardness due to the formation of the fine Fe rich intermetallic compounds. This result is shown in figure 13.

4. Laser surface remelting of aluminium alloy

Laser surface melting (LSM) is a well established technology applied to many materials for hardening, reducing porosity and increasing wear and corrosion resistance.

LSM is a versatile and promising technique that can be used to modify the surface properties of a material without affecting its bulk property (Yue et al., 2004; Rams et al., 2007). The modifying in the surface properties of the material is due to rapid melting followed by rapid solidification. The intimate contact between the melt and the solid substrate causes a very fast heat extraction during solidification resulting in very high cooling rates of the order of 10^5 to 10^8 k/s. The high cooling rates to which this surface layer is submitted result in the formation of different microstructures from bulk metal leading to improved surface properties (Aparecida Pinto et al., 2003). Materials processed via rapid solidification tend to show advantages of refined microstructure, reduced microsegregation, extensive solid solubility and formation of metastable phases (Munitz, 1985; Zimmermann et al., 1989). It is generally accepted that the improvement in corrosion performance is due to refinement/homogenisation of microstructure and dissolution/redistribution of precipitates or inclusions, which result from rapid solidification (Chong et al., 2003). This was considered to be due to the presence of the compact oxide layers on top of the laser-melted zone. The layers mainly consisted of structures α-Al_2O_3, which is a homogeneous and chemically stable phase and serves as an effective barrier to protect the matrix against corrosion attacks. In untreated surfaces of Al alloys the microsegregation in relatively thin surface layer plays an important role in initiating pitting in the inhomogeneous structures. The schematic of the laser surface melting process is shown in figure 14 (Aparecida Pinto et al., 2003).

Some industrial laser sources such as CO_2, Nd:YAG, excimer and high power diode lasers were applied to surface melting of aluminium alloys. Since aluminum alloy have no solid phase transformation, if the surface of aluminium alloys should not be melted, the surface cannot be strengthened. In view of the basic physical properties of aluminium alloy, such as large specific heat, high heat conductivity and high reflectivity to laser power density than that for ferrous alloy (Wong et al., 1997). The controlling of laser parameters is very important factor for laser surface melting process.

Because the properties of a material depend largely on its microstructure, controlled formation of such microstructures is essential to develop new materials with desired properties (Aparecida Pinto et al., 2003). Laser parameters such as laser power density, interaction time and scan speed affect on solidification behaviour and thus the microstructure of melted zone can be changed.

The diagram shown in figure 15 associated the microstructural evolution with the solid/liquid front velocity (Aparecida Pinto et al., 2003).

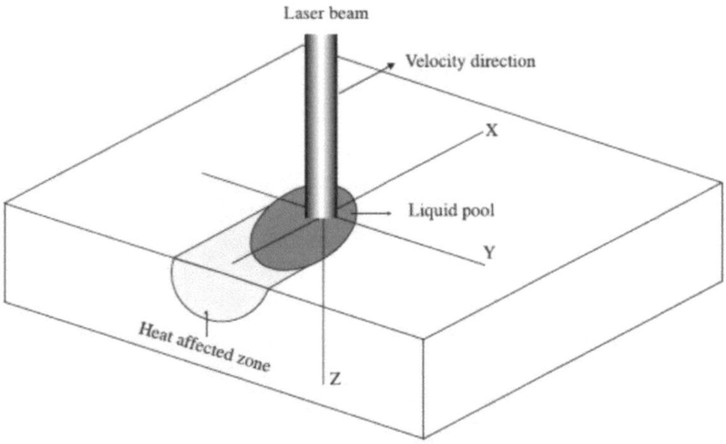

Fig. 14. Schematic illustration of the laser surface melting process (Aparecida Pinto et al., 2003)

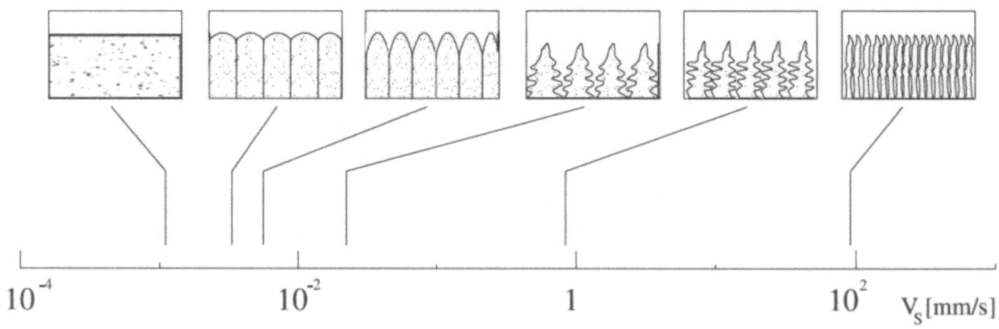

Fig. 15. Microstructure variation according to the solid/liquid front velocity (Aparecida Pinto et al., 2003)

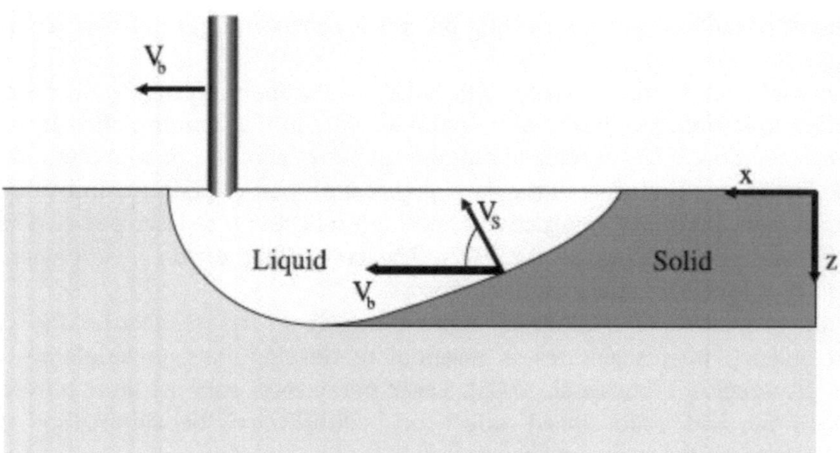

Fig. 16. Schematic representation of the relationship between solidification speed and laser beam speed (Aparecida Pinto et al., 2003)

Taking a longitudinal section through the centerline of the laser track, the speed of the solid/liquid front (V_s) is correlated to the beam speed (V_b) by (Aparecida Pinto et al., 2003):

$$V_s = V_b \cos \phi \tag{5}$$

where ϕ is the angle between V_s and V_b vectors that shown in figure 16.
This equation describes that V_s varies from zero at the bottom of the moltem pool to a approaching the value of V_b at the top of the molten pool.
Pinto et. al. (Aparecida Pinto et al., 2003) investigated the microstructure of Al-Cu alloy after laser surface melting. The influence of V_b on the microstructure is shown in figure 17. The lower beam speed of 500 mm/min has permitted a more extensive cellular zone to be formed and a later transition from a cellular to a dendritic structure when compared with the structure developed under a speed of 800 mm/min.

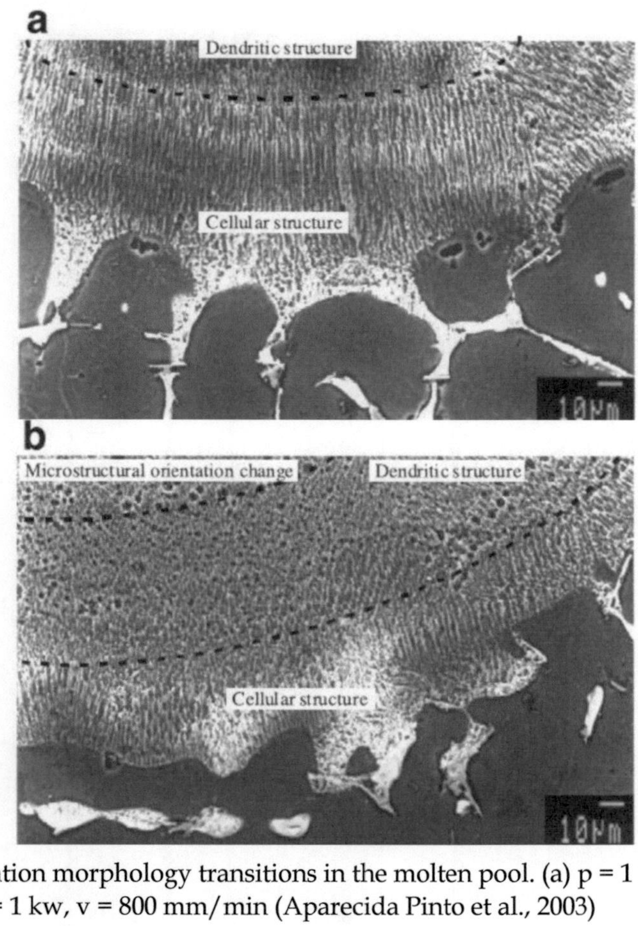

Fig. 17. Solidification morphology transitions in the molten pool. (a) p = 1 kw, v = 500 mm/min; (b) p = 1 kw, v = 800 mm/min (Aparecida Pinto et al., 2003)

From the result of hardness tests, Pinto reported that the mean hardness increase from 75 Hv in the unmelted zone to 160 Hv in the cellular structure. In contrast, a higher value of 210 Hv was measured in the dendritic structure due to the fineness of the microstructure (Aparecida Pinto et al., 2003).

Leech (Leech, 1989) studied the laser surface melting of Al-Si alloys as a function of the beam interaction time τ, that determined by following equation:

$$\tau = \frac{l}{V} \tag{6}$$

Where l is the beam diameter and V is the scanning velocity. The microstructure features in the laser-melted zone consisted of a highly refined dendritic growth at beam traverse speed of 100 mm/min. Within the structure there is a progressive change in dendrite morphology from a planar melt-substrate interface (figure 18) at the maximum melt depth, through a region of oriented columnar dendrite growth (figure 19), to a central region which at more rapid scan rates comprised a fine, filamentary eutectic (figure 20).

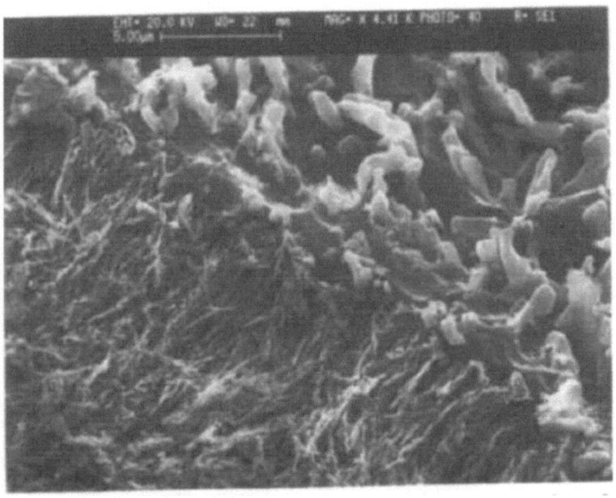

Fig. 18. SEM micrograph showing the melt-substrate interface in the Al-Si alloy (beam traverse speed, 100 mm/min (Leech, 1989)

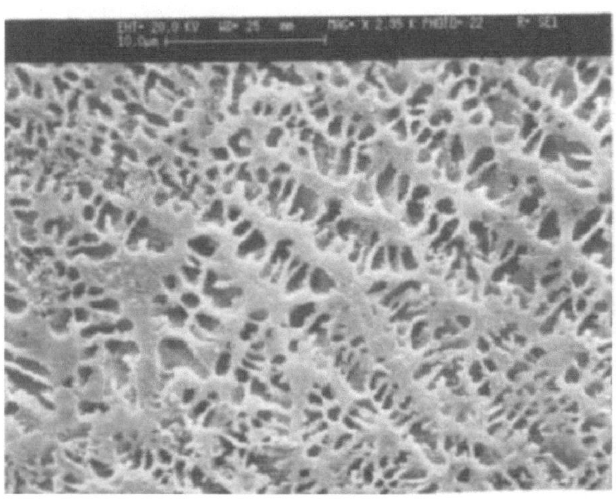

Fig. 19. SEM micrograph showing the columnar dendritic region in the melted zone in Al-Si-W-Ni alloy. (traverse speed, 100 mm/min) (Leech, 1989)

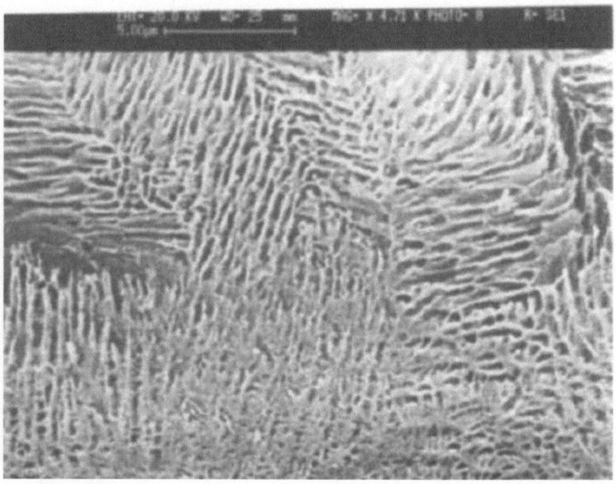

Fig. 20. SEM micrograph of the region of lamellar in Al-Si alloy at a traverse speed of 413 mm/min (Leech, 1989)

An interpretation of the microstructures involves reference to the phase-under cooling diagram shown in figure 21. After laser melting, a rapid extraction of heat from the liquid adjacent to the substrate will produce direct cooling into the α+ eutectic region, the resulting nucleation and growth of α columnar dendrites causing a rejection of silicon into the remaining melt. As the silicon content of the melt increased and with rise in temperature due to latent heat, it is proposed that the composition-cooling line moved to the right into the eutectic-coupled zone. The formation of the lamellar region in the laser melt zone thereby corresponded to the zone of couple growth (Leech, 1989).

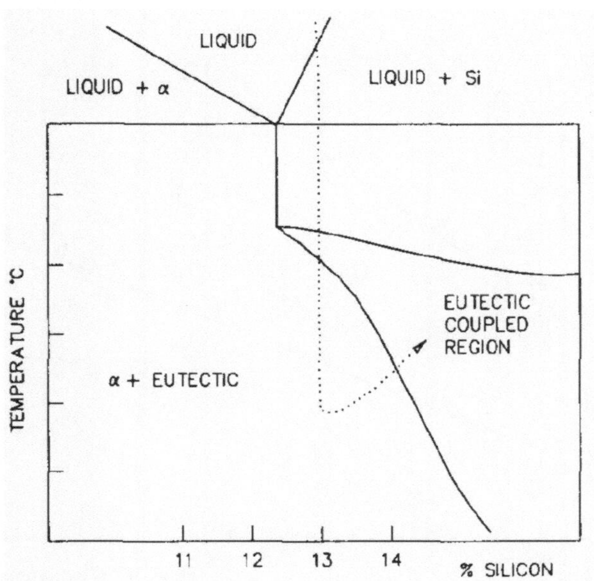

Fig. 21. Schematic phase diagram illustrating the micro structural–undercooling relations during quenching (Leech, 1989)

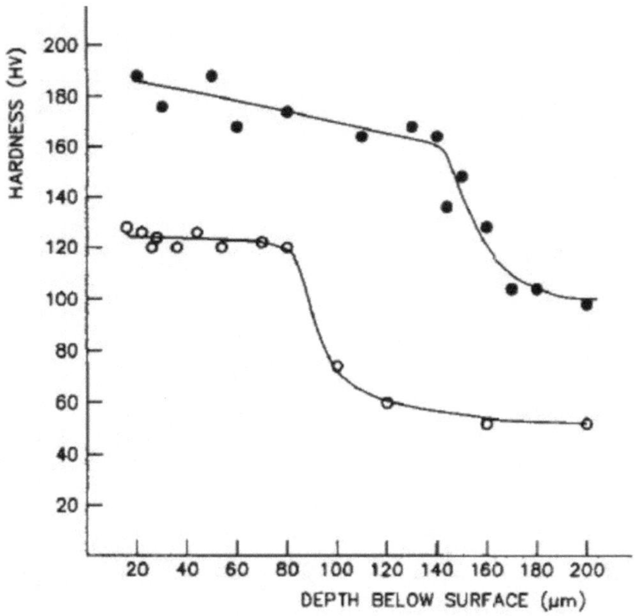

Fig. 22. Hardness profiles taken across the laser-melted regions in the Al-13.6%Si-2.23%Cu-1.94%Ni (•) and Al-13.0%Si (○) alloys at a scan rate of 10 mm s^{-1} (Leech, 1989)

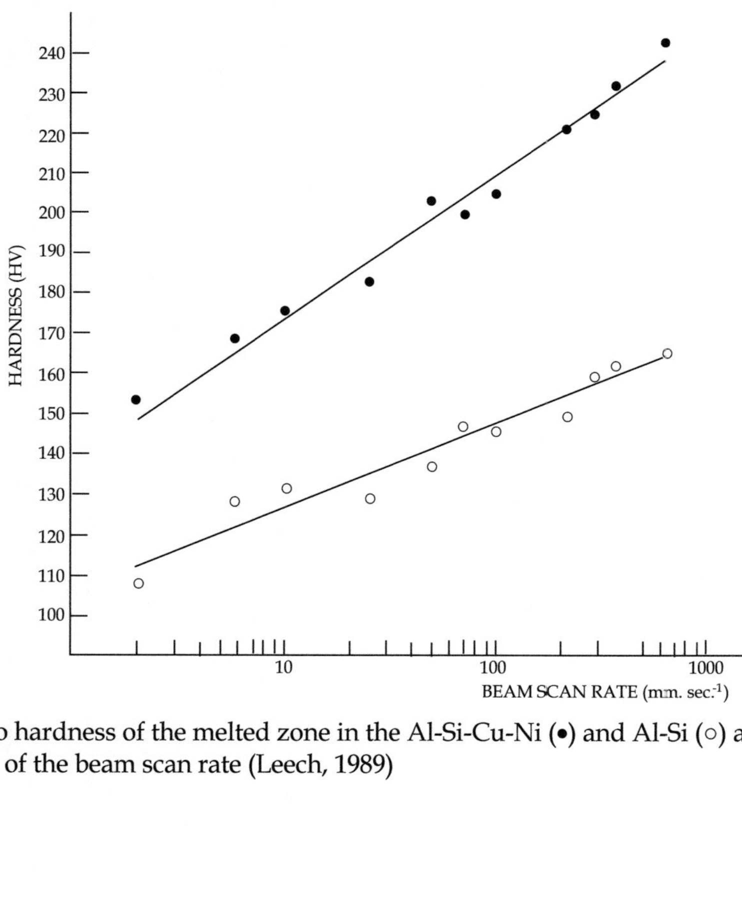

Fig. 23. Micro hardness of the melted zone in the Al-Si-Cu-Ni (•) and Al-Si (○) alloys, plotted as a function of the beam scan rate (Leech, 1989)

The microhardness variation with distance from the melt surface after laser surface melting of Al-Si-Ni alloy and Al-Si alloy is shown in figure 22 (Leech, 1989). Apart from the differences between the alloys in molt zone depth, the curves also illustrate the higher hardness attained throughout the resolidified region of Al-Si-Cu-Ni alloy that in the Al-Si alloy. Leech (Leech, 1989) also reported that micro hardness of laser surface melted layer of Al-Si-Cu-Ni and Al-Si alloys is dependent on laser scan rate (figure 23).

Increasing quenching rates, may promote the formation of finer dispersions of copper and nickel-bearing intermetallic particles (Leech, 1989).

Corrosion resistance is an important matter in aluminum alloys. There are several methods of surface engineering to improving the corrosion behavior of aluminum alloys, that everyone has advantages and disadvantages. Laser surface melting is one of those techniques. There have been a number of studies of the influence of LSM on the corrosion properties of aluminum alloys, and the results achieved have been ambiguous with respect to the benefits of LSM. In some cases, it is severally accepted that laser surface melting can be used for improving the localized corrosion resistance of aluminum alloys as a result of homogenization and refinement of microstructures, and phase transformations. For example, Chong et. al (Chong et al., 2003) studied the corrosion behavior of Al-2014 alloy in T6 and T451 conditions after laser surface melting. After the corrosion tests, they found a large number of pits, randomly distributed on the surface of as-received Al2014 alloy in two conditions (figure 24a). In this instances although Al2014 alloy in both tempers consisted of similar types of intermetallic particles, the copper content in the aluminum matrix for T6 is lower than that for T451. In the NaCl electrolyte, Al_2Cu, and Al-Cu-Mn-Fe-Si particles tend to be cathodic to the matrix (Chong et al., 2003), and pits are likely to initiate and grow in the copper-depleted zone around these particles (Guillaumin & Mankowski, 1999). Mg_2Si particles are anodic to the aluminum matrix, and have a tendency to dissolve and leaving cavities. Figure 24b shows that after LSM, pits formed on the laser-melted surfaces are larger but shallower that in the as- received alloy, with a semi-continuous network, consisting of copper-rich precipitates, remaining within the pits, indicating their cathodic nature. It is proposed that the concentrations of solid solution alloy elements, (particularly copper in Al 2014), are key factors influencing pitting corrosion. Such increase of copper content in the Al 2014 matrix can reduce the potential difference between the Al_2Cu phases and the aluminum matrix, thereby reducing the driving force of pitting corrosion. The reduction in population or the elimination of Mg_2Si particles which are anodic to aluminum matrix may

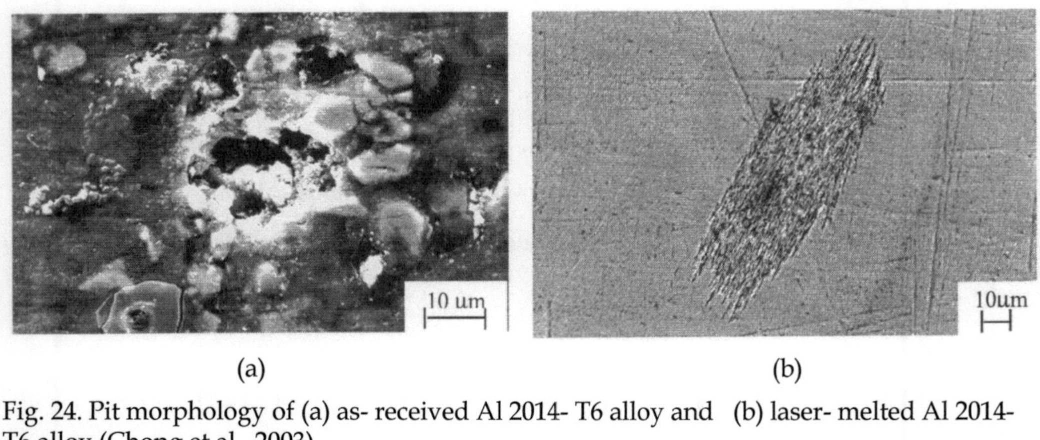

(a) (b)

Fig. 24. Pit morphology of (a) as- received Al 2014- T6 alloy and (b) laser- melted Al 2014-T6 alloy (Chong et al., 2003)

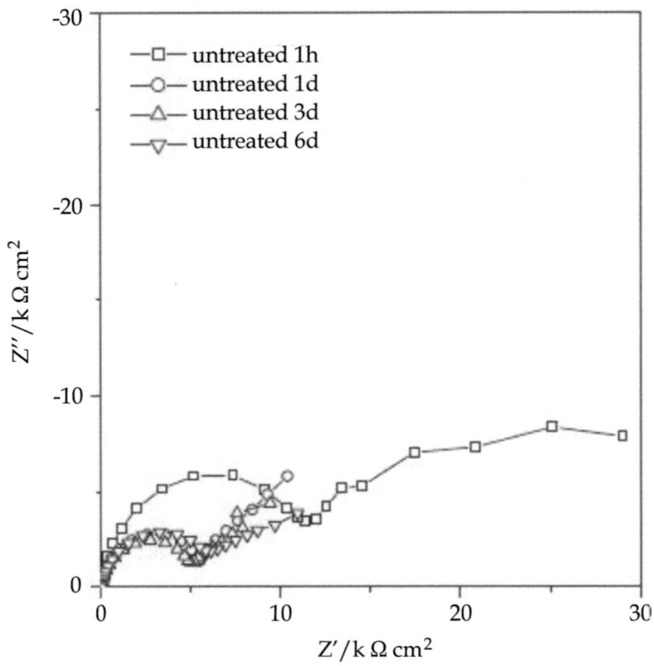

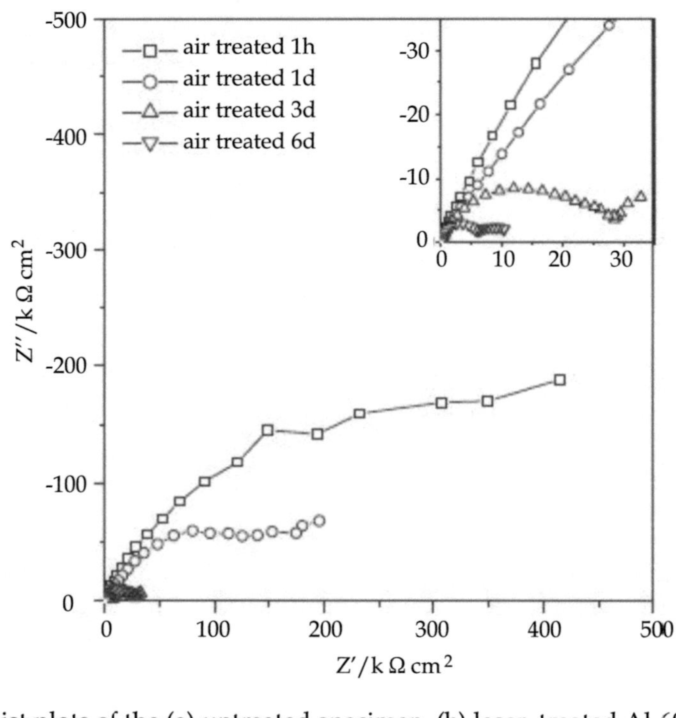

Fig. 25. Nyquist plots of the (a) untreated specimen, (b) laser, treated Al-6013 alloy (Xu et al., 2006)

further improve the behavior by reducing cavities due to the dissolved particles. Regarding influences of preferred orientation, the literature (Guillaumin & Mankowski, 1999) indicates that the pitting potential of aluminum increases in order of $(E_{pit})_{\{001\}} \geq (E_{pit})_{\{011\}} > (E_{pit})_{\{111\}}$, however, the presence of alloyed copper in solid solution reduces the dependence of E_{pit} on surface orientation. Thus, the preferred orientation of α-Al along [200] direction in laser-melted alloy does not appear to play a significant role in the improved pitting behavior.

High- strength aluminum alloys (HSAL) are high susceptible to various forms of corrosion, particularly in the presence of chloride-containing media. Thus, these alloys are very susceptible to pitting corrosion fatigue, and the degradation of HSAL by this phenomenon is a matter of major concern, particularly as many structural parts are inaccessible for inspection and cannot be monitored, thus hiding the defects of corrosion as they approach a critical for fatigue. The improvement in pitting corrosion fatigue behavior of HSAL alloys after LSM is reported by Xu and co-workers (Xu et al., 2006). Figure 25 shows the results of impedance measurements of unmelted and surface melted Al 6013 alloy. These results are displayed in the form of Nyquist plots as a function of immersion time up to a period of 6 days.

The spectra suggest that for untreated and laser-treated Al alloy, corrosion pitting has occurred to various degrees at different times during the immersion test. This is evidenced by the presence of start of the immersion test, with the diameter of the arches decreasing with immersion time (figure 25a).

As for the laser- treated Al 6013 alloy, a compressed capacitive loop with a small diffusion tail at the low-frequency range was seen at the first hour of the test, and a second loop emerged after 1 day of immersion (figure 25b). Figure 26 shows the equivalent circuit of EIS plots to interpret the electrochemical behavior of untreated and laser-treated Al 6013 alloy. The equivalent circuit component values as a function of immersion time are listed in table 1.

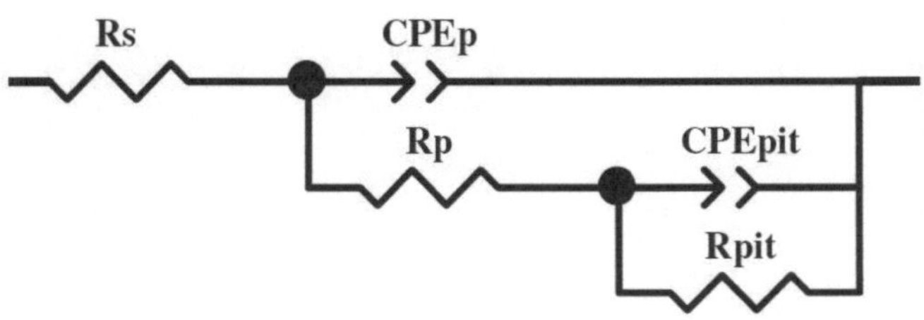

Fig. 26. Equivalent circuits for the untreated and laser-treated Al 6013 alloy (Xu et al., 2006)

Specimen		CPE_p		R_p, Ω cm^2	CPE_{pit}		R_{pit}, Ω cm^2
		Y_0, F cm^{-2} Hz^{1-n}	n		Y_0, F cm^{-2} Hz^{1-n}	n	
Untreated	1 h	5.74e−6	0.93	1.28e4	2.26e−4	0.94	1.85e4
	1 day	1.23e−5	0.91	5.99e3	6.87e−4	0.85	1.53e4
	3 days	1.71e−5	0.89	5.50e3	9.81e−4	0.90	7.00e3
	6 days	2.04e−5	0.88	6.68e3	7.77e−4	0.84	8.47e3
Air-treated	1 h	4.52e−6	0.62	6.54e5			
	1 day	1.69e−6	0.83	1.29e4	4.12e−6	0.65	1.873e5
	3 days	4.80e−6	0.74	2.67e4	4.64e−4	0.53	3.31e4
	6 days	1.84e−5	0.91	6.70e3	8.58e−4	0.94	4.65e3

Table 1. Calculated values of the equivalent circuit components of the impedance plots (Xu et al., 2006)

The magnitude CPE_p (constant phase element), a measure of the capacitance at the surface of laser- melted alloy is much less than that of the as-received alloy, especially up to the immersion time of 3 days. This indicates that less ion adsorption has occurred at the surface of the laser melted alloy. This confirms the good corrosion resistance of the layers containing laser-formed aluminum oxide in reducing the rate of electro chemical reaction at the laser-meted surface. Xu (Xu et al., 2006) reported that the corrosion fatigue life of the laser-surface melted Al 6013 alloy is two times longer than that of the as-received Al alloy (figure 27). Also, the corrosion current for the laser-surface melted Al 6013 alloy is considerably lower than that for the as-received Al 6013 alloy. The improvement in the pitting corrosion of the laser-surface- melted Al alloy. An increase in the corrosion resistance of Al- Si alloys after laser surface melting in both 10% H_2SO_4 and 10% HNO_3 solutions is observed by Wong and co-workers (Wong & Liang, 1997). Also, they reported that, in the 10% HCl and 5% NaCl solutions laser melting has little effect on the corrosion resistance of Al-Si alloys. Because the Cl ions destroy the Al_2O_3 film completely. In the case of 5% NaCl solution, $NaAlO_2$ is formed and the protective oxide film Al_2O_3 is again destroyed, which intensifies the corrosion of the aluminum alloys (Yongqing et al., 1998).
Corrosion resistance of laser surface melted Al 2024 alloy is investigated by Li and co-workers (Li et al., 1996). Free corrosion in naturally aerated chloride electrolyte solution revealed a change in the mechanism of corrosion for the LSM alloy. A small number of large pits, initiated in the α-Al cells and/or dendrites, are found at random over the surface. In contrast, for the as-received alloy where pitting is initiated at Al_2CuMg precipitates, corrosion took the form of intergranular corrosion.

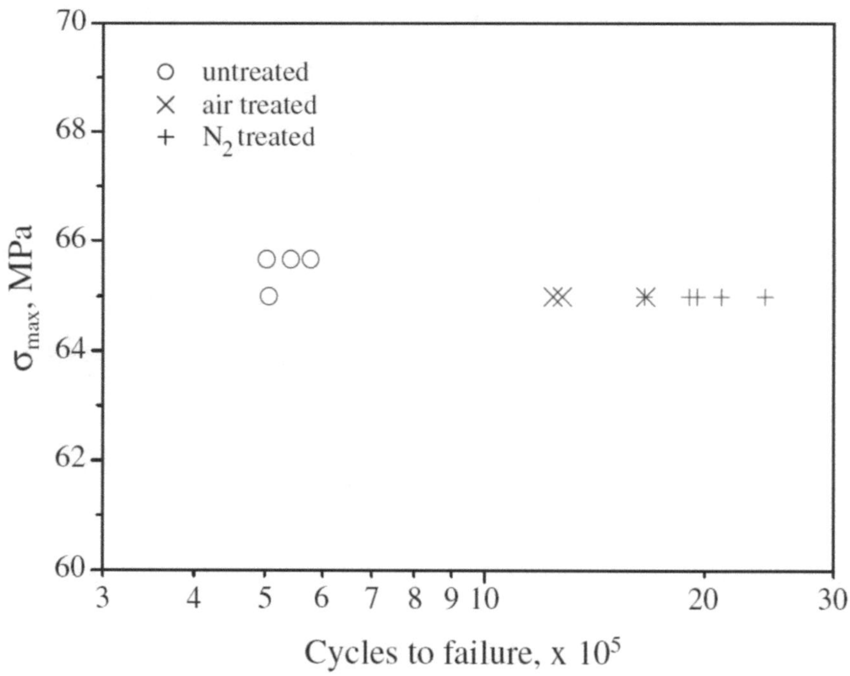

Fig. 27. Test of the fatigue life of the untreated and laser- treated Al 6013 alloy in a 3.5% NaCl solution at a potential of – 675 mv (Xu et al., 2006)

5. Laser surface cladding of aluminium alloys

Aluminium-based metal matrix composites (Al-MMC_s) have high strength, hardness and wear resistance, and find application in various industrial sectors, such as automotive and aerospace industries (Anandkumar et al,. 2007). The major drawbacks of these materials are their high coat and complex production methods compared to conventional alloys, but for many applications, like rapid tooling, the bulk stress levels are compatible with the use of high-strength Al alloy, the required wear resistance being achieved by coating the component with a high wear resistance materials such as a ceramic-reinforced Al-matrix composite (Anandkumar et al,. 2007). Aluminium alloys have been cladded with ceramics such as SiC, B_4C, TiC due to their high hardness and thermal stability and various other metallic materials such as Ti, B, Ni etc. to enhance their surface properties (Anandkumar et al,. 2007). These ceramic reinforcement particles have a low reflectivity; therefore they absorb a considerable amount of laser energy (Anandkumar et al., 2009) and may reach very high temperatures, which will lead to intense reactions between the reinforcement and the liquid metal or to particle dissolution in the melt pool. The tendency of reactivity of reinforcement particles with depends on their temperature, which depends on the interaction time between the particles and the laser beam (Anandkumar et al., 2009).

In this case, the velocity of injected powder is an important factor that affects on the interaction time and particles temperature. The temperature variation of injected powder particles is calculated by several researchers using mathematical modeling. Huang et al. (Huang et al., 2005) calculated the beam attenuation and particle temperature variation due to the interaction of an off-axis powder stream with a laser beam on the basis of Lambert-Beer law and Mie's theory. They found that the temperature of injected powder particles increases with decreasing the angle between the powder jet and the laser beam from 45 to 0°, because the particles trajectory through the laser beam is longer.

Also, a mathematical model for calculation of particles temperature under laser beam irradiation is established by jouvard and co-workers (Jouvard et al., 1997). Figure 28 shows an off-axis blown powder laser cladding process diagram used for jouvard model.

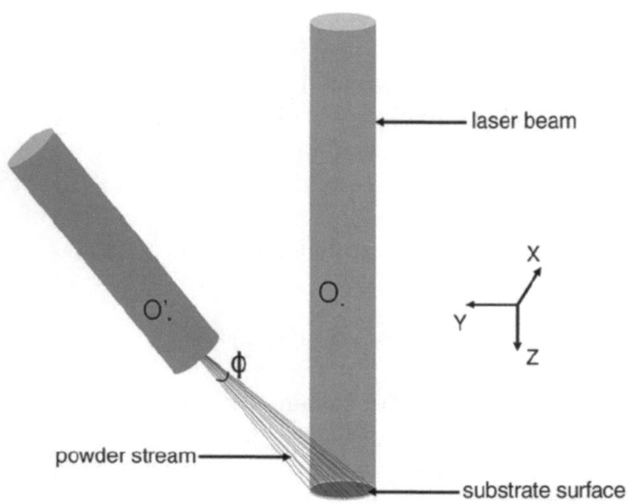

Fig. 28. Diagram of laser beam-powder stream interaction (Anandkumar et al., 2009)

They reported that, the temperature of a powder particle (T) interacting with the laser beam can be calculated by (Jouvard et al., 1997):

$$T = T_0 + \frac{I_{(x,y)}\eta A_p t}{m_p c_p} \tag{7}$$

Where

T_0 initial temperature of the particles (25°C)

$I_{(x,y)}$ laser radiation intensity

η absorptivity of the particle material

A_p cross-sectional area of the particle

m_p mass of the particle

c_p specific heat capacity of the particle material

t laser beam-particle interaction time

In this equation following simpler assumptions is considered:

1. the laser beam is parallel and has a Gaussian energy distribution.
2. the powder particles are spherical and uniformly distributed in the powder stream,
3. energy loss by convection and radiation is negligible (Fu et al., 2002),
4. the effect of gravity and the drag exerted by the surrounding gas on particle movement are negligible and all particles have the same velocity,
5. the shadow effect of the particles on each other is accounted for,
6. the fraction of the laser beam energy absorbed by a particle is given by the absorptivity of the particle material (η) for the laser radiation wavelength,
7. the temperature distribution in each particle is uniform,
8. latent heat effects due to melting are neglected (Anandkumar et al., 2009).

As the interaction time (t) is given by d/v_p, where d is the distance traveled by the particle through the laser beam and v_p its projected velocity component, Eq. (7) can be written as:

$$T = T_0 + \frac{I_{(x,y)}\eta A_p d}{m_p c_p v_p}. \tag{8}$$

The trajectories of the particles are represented by a series of lines diverging from O (Figure 28) and the energy absorbed by particle is calculated as a line trajectory integral through laser beam, because the intensity of the beam depends on x and y and finally also on z. To establish a function describing the laser beam attenuation in the z direction, the interaction region is divided into n layers of thickness Δz and the fraction of radiation intensity (C) absorbed by the particles in each layer is calculated using the following equation:

$$C = I_{(x,y)}\eta A_p \Delta z\left(\frac{N}{V}\right), \tag{9}$$

where (N/V) is the density of the powder stream, which depends on the powder feed rate and the injection velocity (Gingu et al., 1999). The particles in the n-th layer absorb part of the incoming radiation intensity and the remaining intensity is regarded as the input intensity for the n+1 layer and so on. The final temperature of the particles is computed by solving Eq. (10) using Wolfram Research Mathematica®6 software,

$$T = T_0 + \frac{\eta A_p}{v_p m_p c_p} \int_{z_s}^{z_o} \left(I_{(x,y)} - C \right)^{(n-1)} ddz \tag{10}$$

where, the integral limits z_s and z_o are the z coordinates of the pointwhere the particle enters the beam and impinges the substrate, respectively (Anandkumar et al., 2009).

In case of modification using many ceramics, especially carbides, it has been found that they also chemically react with Al and form compounds which decrease the strength of Aluminum alloy. For example, ceramic particles of SiC, tend to react and dissolve in molten Al alloy, and leading to the formation of Al_4C_3 and ternary Al-Si-C carbides during solidification (Viala et al., 1990; Hu et al., 1996) according to the reaction at a temperature range between 940 and 1620 k (Anandkumar et al., 2007):

$$4Al_{(l)} + 3SiC_{(s)} \rightarrow Al_4C_{3\,(s)} + 3Si \tag{11}$$

At higher temperatures (above 1670 k), the reaction product is the ternary carbide Al_4SiC_4, formed by the reaction (Anandkumar et al., 2007):

$$4Al_{(l)} + 4SiC_{(s)} \rightarrow Al_4SiC_{4\,(s)} + 3Si \tag{12}$$

The presense of this phase in microstructure of Al alloy is shown in figure 29a (Anandkumar et al., 2009). The hardness of Al_4C_3 is very lower then Al_4SiC_3 (300 and 1200 Hv, respectively) but, unlike Al_4SiC_4, it is brittle and tends to react with water, forming aluminium hydroxide (Anandkumar et al., 2007).

Accordingly, the presence of Al_4C_3 in the surface microstructure results in poor mechanical properties and low long-term stability (Anandkumar et al., 2007) and its formation must be avoided.

According to the equation 10, Anandkumar et al. (Anandkumar et al., 2009) calculated the SiC particle temperature at two different jet velocities for laser surface cladding of Al-Si alloy. They found that when the particle injection velocity is 5 m/s, the particles are exposed to the laser radiation for a shorter time and they absorb correspondingly less energy. As a result, the temperature of the particles reaching the melt pool is much lower and no significant reactions occur between SiC and molten aluminum, leaving the composition of the melt essentially unchanged. During cooling this liquid solidifies as primary α-Al dendrites and α-Al+Si eutectic (Figure 29b).

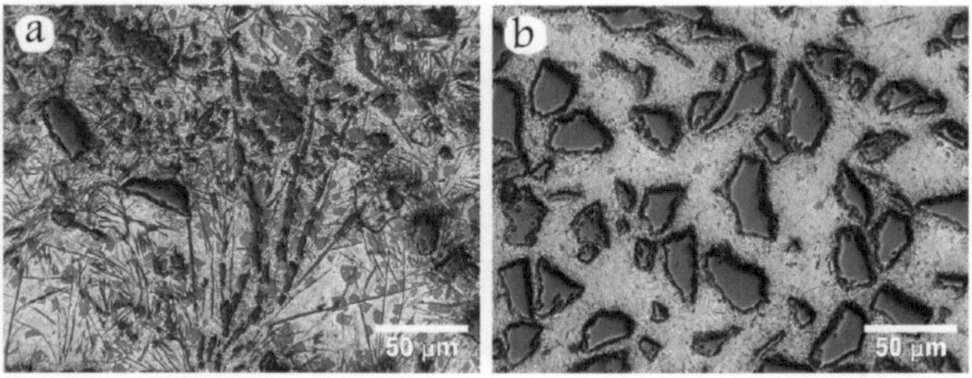

Fig. 29. Microstructure of clad tracks prepared at particle injection velocities of (C) 1 m/s and (d) 5 m/s (Anandkumar et al., 2009)

Due to the lower absorption of laser beam energy by aluminum alloy, the temperature of metallic particles is always much lower than that of ceramic particles. The maximum temperature attained by the particles as a function of their injection velocity is shown in figure 30 (Anandkumar et al., 2009). Also, the temperature distribution of SiC particles injected at 1 m/s along the X axis (y=0) is shown in figure 31. Particles arriving at the X axis traveled the same time through the laser beam: particles reaching the surface near the laser beam axis are subjected to higher radiation intensity and reach higher temperatures, while the temperature decreases towards the periphery of the powder stream as the beam intensity decreases. By contrast, the temperature of the particles increases linearly along the Y axis (Figure 32). Two factors explain this evolution. On one hand, the length of the particle's path through the laser beam varies along the line: it is zero for particles reaching the leading edge of the melt pool and increases with Y up to the trailing edge of the melt pool, where it reaches its maximum value. On the other hand, attenuation of the laser beam by the particles, which decreases from the leading to the trailing edge, further enhances the particle's temperature increase in this direction. The present results show that particle injection velocity is a key parameter in control of the microstructure and properties of metal matrix composite coatings produced from metal–ceramic powder mixtures by laser cladding and laser particle injection (Anandkumar et al., 2009). The particles injection velocity must be kept higher than a certain threshold to avoid excessive heating of the ceramic particles reaching the melt pool and potential reactions between the reinforcement material and the liquid metal.

Other laser parameters such as the power of laser and scanning rate have an important effect on the properties and features of clad layers. Sallamand and Pelletier (Sallamand & Pelletier, 1993), (during laser cladding of aluminium-base alloy with Al-Si and Ni-Al powders), found that at low laser powers or high scanning speeds (or both), some of the injected particle are unmelted and some porosity is sometimes detected as shown in figure 33. Also, with higher power or lower scanning.

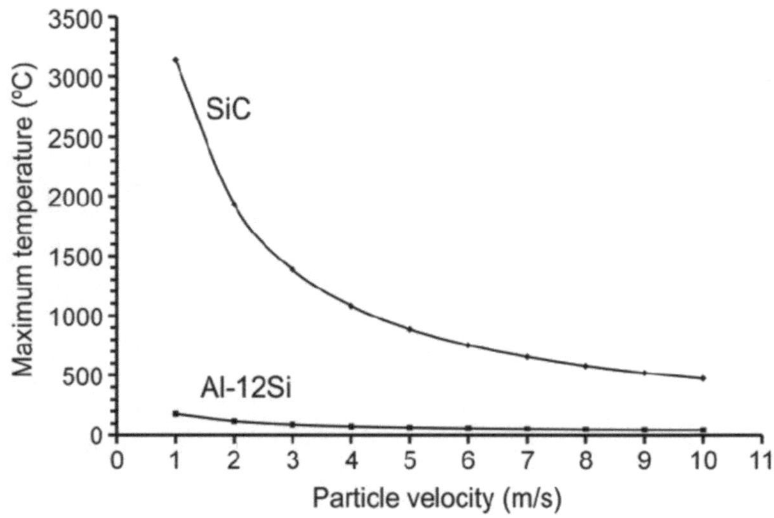

Fig. 30. Maximum temperature attained by the particles as a function of their injection velocity (Anandkumar et al., 2007)

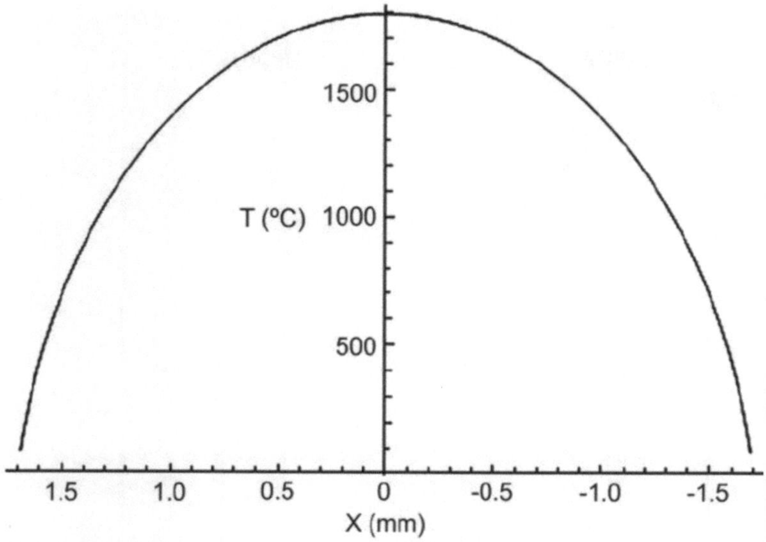

Fig. 31. Temperature distribution of SiC particles injected at 1 m/s along the X axis (y=0) (Anandkumar et al., 2007)

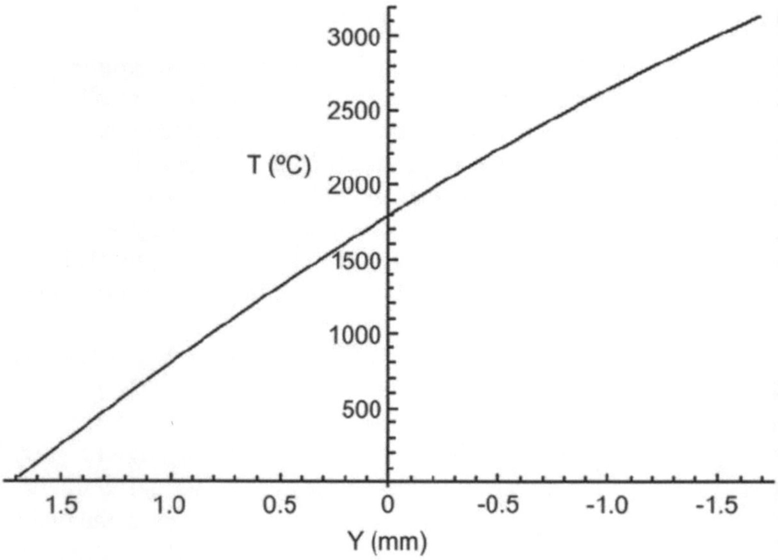

Fig. 32. Temperature distribution of SiC particles injected at 1 m/s along the Y axis (x=0) (Anandkumar et al., 2007)

Speeds, or both, i. e. with a higher interaction time τ and/or a higher absorbed energy, melting of the injected particles occur, as mentioned above. When the power fraction absorbed by the powder is higher than that by the substrate, only limited melting of the substrate occurs and therefore cladding is formed, with a low dilution rate of the incoming powder.

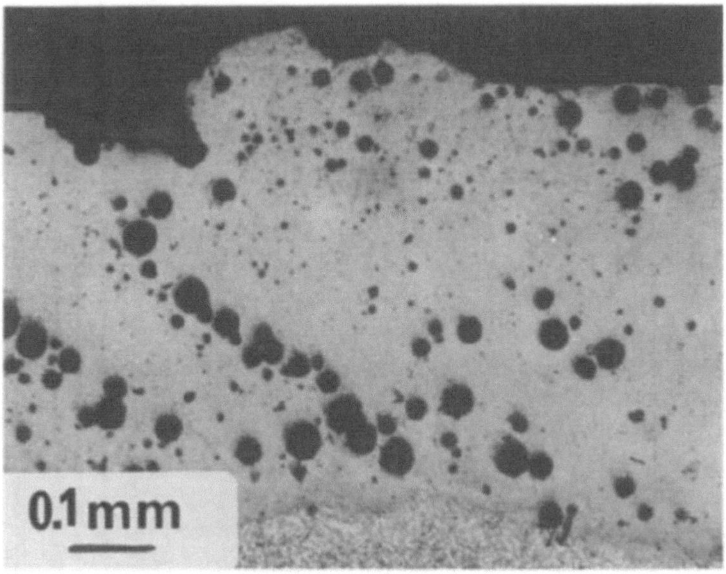

Fig. 33. Micrograph of cladding before optimization of the processing conditions showing that pores are detected. (AI-7at.% Si+injection of (AI, Si,Ni) powders; laser power P= 1900 W; scanning speed v = 1.5 cm s^{-1}; diameter d of the laser beam on the sample 1.25 mm; magnification G = 100) (Sallamand & Pelletier, 1993)

About the Ni powder, when adding the Ni powder into the melted aluminum alloy zone, it is the need for good homogenization of the nickel. The diffusion-controlled process can be enhanced by increasing the temperature, but then vaporization and plasma formation above the sample have to be avoided in order to obtain regular treated zones. It can also be enhanced by increasing the interaction time; however, an increase in the lifetime of the melted pool yields an increase in the melted depth and, consequently, a higher dilution rate of the nickel (Sallamand & Pelletier, 1993). The microstructure of a typical cladding is shown in Figure 34. It appears to be mainly dendritic. The orientations of the dendrites are not very regular; two explanations can be proposed:

1. A cross-section effect of a three-dimensional network occurs, where dendrites are perpendicular to the solidification front which progresses from the bottom to the top of the sample.
2. Convection movements in the melted pool can modify the regularity of the growth direction, since they induce perturbations, both in the thermal gradients and in the chemical composition (Sallamand & Pelletier, 1993). Nevertheless, the main result is the existence of a fine and dendritic microstructre, without cracks, pores or undissolved nickel, aluminium or silicon particles. Therefore the duration of the melted pool was long enough to achieve first complete melting of the injected particles and then good interdiffusion of the different elements. It may be observed in Figure 35 that the geometrical features of the dendrites are progressively modified from the interfacial zone to the surface of the sample; a progressive refinement occurs. This phenomenon is due to the evolution of the solidification rate during the process itself: as shown by many workers this rate starts from zero at the interface and increases to a maximum value at the end of the phenomenon, on the surface of the specimen.

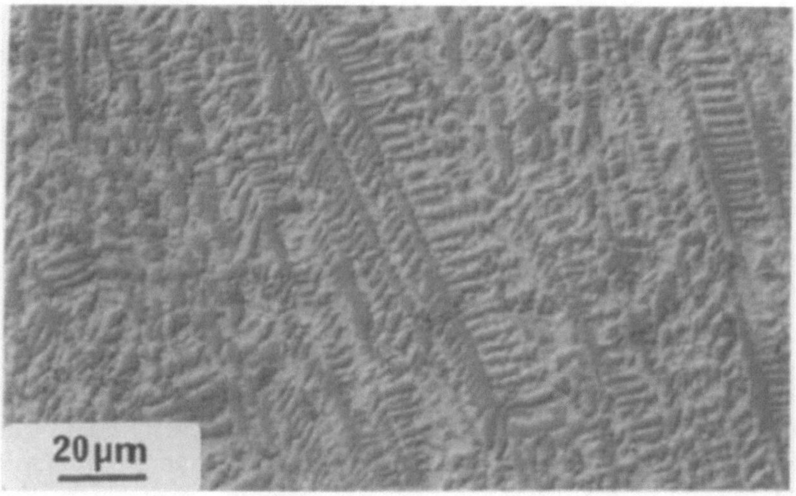

Fig. 34. Microstructure of typical cladding (AI-7at.%Si + injection of (AI, Si, Ni) powders; laser powcr P = 2800 W; scanning speed t, = 1.5 cm s^{-1}; diameter d of the laser beam on the sample, 1.25 mm; magnification G = 80) (Sallamand & Pelletier, 1993)

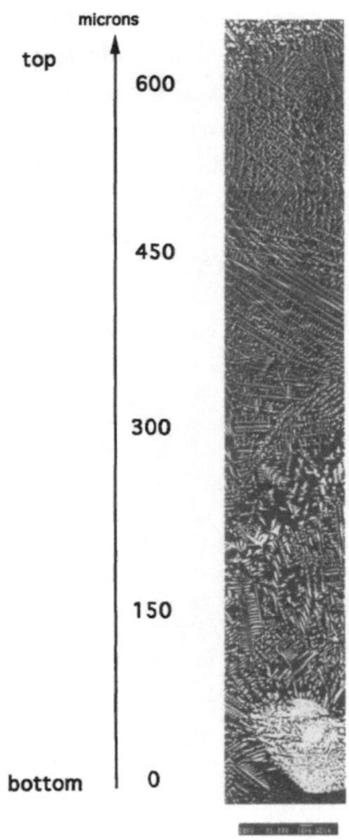

Fig. 35. Evolution of the size of the dendrites from the bottom to the top of the cladding (AI-7at.%Si+injection of (AI,Si,Ni) powders; laser power P= 2800 W; scanning speed v = 2.0 cm s^{-1}; diameter d of the laser beam on the sample, 1.25 mm) (Sallamand & Pelletier, 1993)

6. Laser shock peening of aluminium alloys

Laser shock peening (LSP) is an innovative surface treatment technique, which is successfully applied to improve fatigue performance of metallic components. After the treatment, the fatigue strength and fatigue life of a metallic material can be increased remarkably owing to the presence of compressive residual stresses in the material. The increase in hardness and yield strength of metallic materials is attributed to high density arrays of dislocations and formation of other phases or twins, generated by the shock wave.

The ability of a high energy laser pulse to generate shock waves and plastic deformation in metallic materials was first recognised and explored in 1963 in the USA (Ding & Ye, 2006). A schematic configuration of an LSP process on a workpiece is shown in figure 36 (Dubourg et al., 2005).

When shooting an intense laser beam on to a metal surface for a very short period of time (around 30 ns), the heated zone is vaporised to reach temperatures in excess of 10 000°C and then is transformed to plasma by ionisation. The plasma continues to absorb the laser energy until the end of the deposition time. The pressure generated by the plasma is transmitted to the material through shock waves (Ding & Ye, 2006). Although metals can be highly reflective of light, keeping the constant laser power density and decreasing the wavelength from IR to UV can increase the photon-metal interaction enhancing shock wave generation. However, the peak plasma pressure may decrease because decreasing the wavelength decreases the critical power density threshold for a dielectric breakdown, which in turn limits the peak plasma pressure. The dielectric breakdown is the generation of plasma not on the material surface, which absorbs the incoming laser pulse, limiting the energy to generate a shock wave. In Figure 37, the decrease in the wavelength from IR to green reduces the dielectric breakdown threshold from 10–6GW/cm^2, resulting in maximum peak pressures of approximately 5.5 and 4.5GPa, respectively (Ding & Ye, 2006).

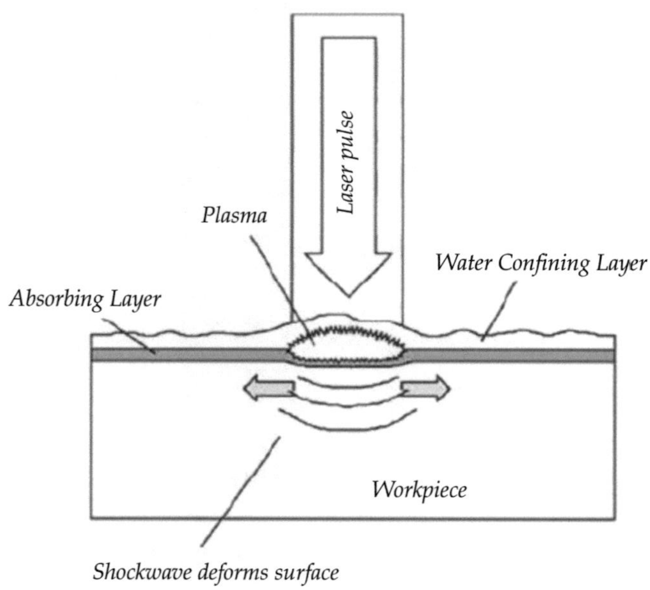

Fig. 36. Schematically principle of laser shock processing (Ding & Ye, 2006)

The transmission of an incident laser pulse throughout a water layer is expected to be controlled significantly by its pulse duration and / or to its rise time. Indeed, the faster energy deposition may generate the better laser-target coupling in plasma confined regime with water (Peyre et al., 2005).
Payer et al. (Peyre et al., 2005) studies the influence of laser intensity, wavelength, and pulse duration on the pressures generated in plasma. Results are presented in figures 38, 39.

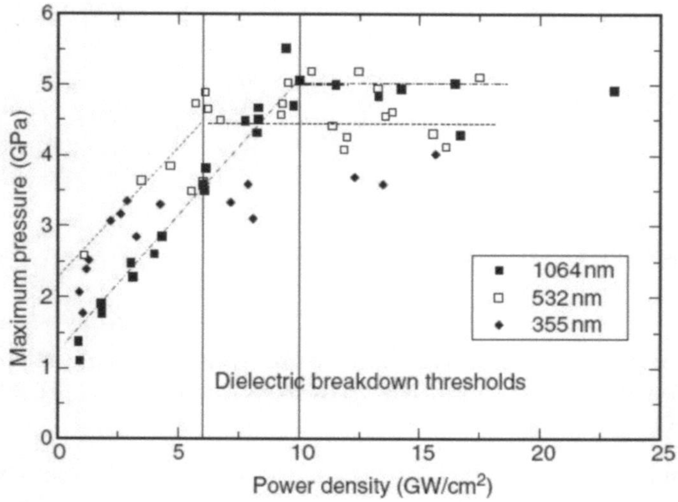

Fig. 37. Peak plasma pressures obtained in WCM as a function of laser power density at 1.064mm, 0.532mm and 0.355mm laser wavelength (Ding & Ye, 2006)

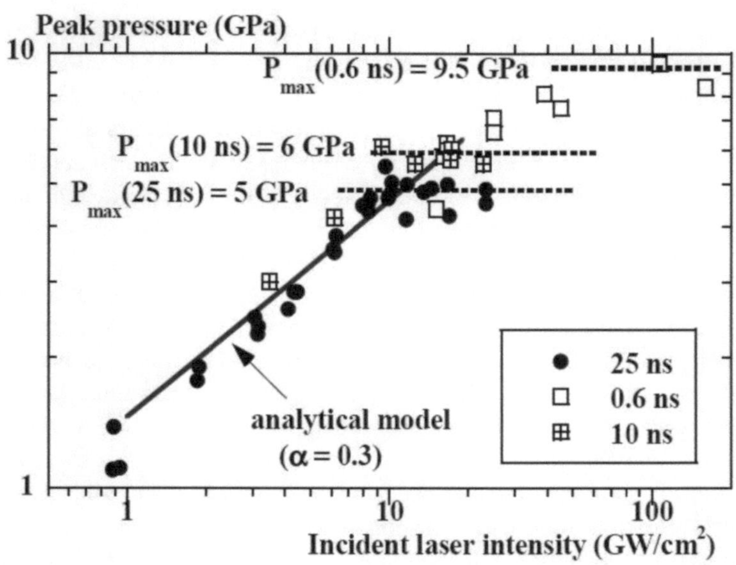

Fig. 38. Influence of laser intensity and pulse durations on the pressures generated in plasma confined with water regime (λ=1.06 μm)-compaison with the analytical model of confinement (25 ns) (Peyre et al., 2005)

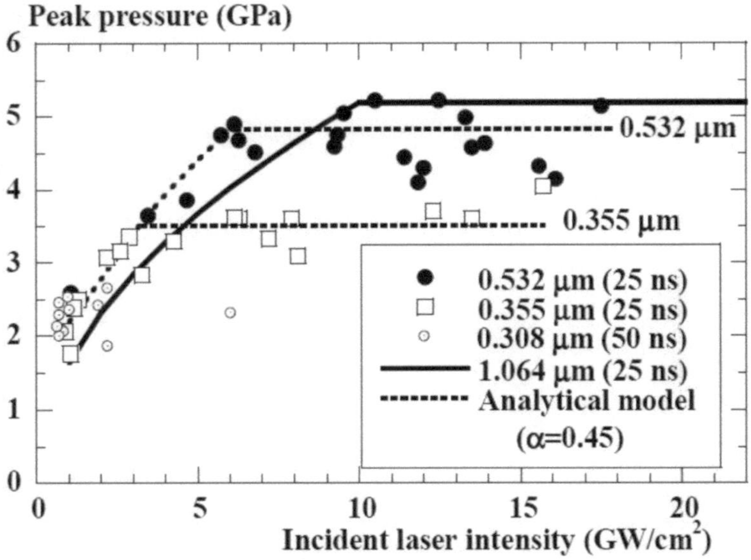

Fig. 39. Influence of laser wavelength on the pressures generated in plasma confined regime with water (all measurements performed at 25 ns pulse durations except 0.308 µm at 50 ns) (Peyre et al., 2005)

It can be seen that the maximum available pressure was saturated to nearly 5-6 GPa above 8-10 GW/cm^2 laser intensity (Fig.1). This saturation was shown to occur because of a parasitic breakdown plasma at the surface of the water which effect was to limit the energy reaching the target and cut temporally the incident laser pulse, thus reducing its effective duration 8. These pressure levels are usually sufficient to harden all the metallic materials but, most of times, impact sizes need to be reduced to reach the convenient power densities (Peyre et al., 2005). Also, the first conclusion to draw from these results is that pressure saturation levels increase with shorter laser pulse durations: from 5-5.5 GPa at 25 ns to 6 GPa at 10 ns and 9.5-10 GPa at 0.6 ns. At shorter durations, the pressure saturation occurs at much higher laser intensity (I_{th} = around 100 GW/cm^2 versus 10 GW/cm^2 at 10-30 ns). This clearly indicates that energy transmissions through the water thickness are improved and that deleterious effects from breakdown plasmas are reduced by the use of shorter durations (Peyre et al., 2005). As can be seen from figure 39 Maximum output pressures P_{max} and intensity thresholds I_{th} tend to be reduced with decreasing wavelengths. At the same pulse duration, maximum pressures decrease from 5.5 GPa at 1.06 µm to 5 GPa at 0.532 µm and 3.5 GPa at 0.355 µm. Intensity thresholds in the UV regime are also reduced to nearly 4 GW/cm^2 versus 10 GW/cm^2 at 1.06 µm. Moreover, the pressure durations (and in turn the transmitted laser pulse durations) decrease much more drastically above the intensity thresholds at lower wavelength. Also, at low intensity (1-4 GW/cm^2) the efficiency of the pressure generation is shown to be improved at 0.532 µm and 0.355 µm. Indeed, according to the analytical model of confinement, the "α" coefficient gives a good fitting with experimental measurement with α = 0.45 versus α = 0.3 in the IR configuration). This could be due to a better target-plasma absorption in the UV range (Peyre et al., 2005). LSP generates compressive residual stresses (CRS) which are known to be the key to enhanced surface Properties (Ding & Ye, 2006).

Residual stresses increase with increasing laser induced pressures until a given pressure level called Psat where a plastic saturation occurs and above which CRS remain nearly constant. Below HEL (Hugoniot Elastic Limit), no plastic deformation occurs and in turn no residual stresses. Maximum RS levels induced by LSP are close to -0.5 σ_Y for one local deformation and -0.7 σ_Y for numerous ones (Figure 40) (Ding & Ye, 2006).

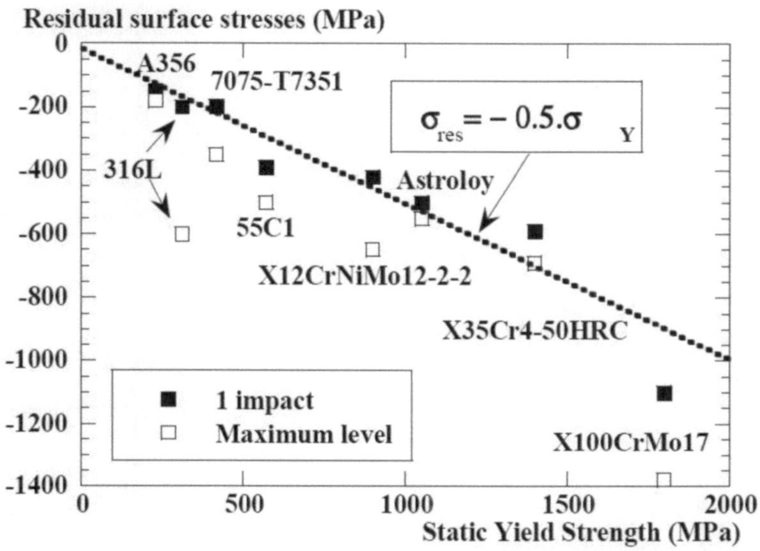

Fig. 40. Influence of the mechanical properties of the targets on the residual stress levels achievable by LSP - Results taken from (Aluminium alloys), (Astroloy Ni superalloy), (X100CrMo17) , (316L and X12CrNiMo12-2-2) (Ding & Ye, 2006)

Many recent studies have evidenced the beneficial influence of LSP on mechanical cyclic properties. On cast and wrought aluminium alloys (Al-7Si and Al12Si, 7075), some 25 to 40 % fatigue limit increases were displayed on notched specimens submitted to R=0.1 bending loadings. These results, superior to shot-peening (+25 % versus +12 % on 7075) were shown to be due to some large improvements in the fatigue crack initiation stage (Peyre et al., 1996).

Lu et al. (Lu et al., 2010) studied the effect of laser shock peening on properties of aluminum alloys. In their report, the residual stress profiles of the treated samples after multiple LSP impacts with the impact time as functions of the distance from the top surface are shown in Figure 41. The substrates are approximately in the zero-stress state, indicating that the effect of initial stress on the shock waves may be ignored (Tan et al., 2004). It can be noted from Figure 41 that the significant compressive residual stresses mainly exist in near-surface regions for all cases and the top surfaces have the maximum values of compressive residual stresses (Lu et al., 2010).

The peak surface compressive residual stress and the depth of compressive residual stress are significantly increased to 116 MPa and to 0.79 mm, respectively, as a result of 3 LSP impacts on the sample surface. After 4 LSP impacts, the peak value of surface compressive residual stress is increased to 123 MPa, and the depth of compressive residual stress reaches about 0.80 mm. It can be seen that the surface compressive residual stress is increased by

25.93% and 13.73% when the impact time increases from 1 to 2 and from 2 to 3, whereas the surface compressive residual stress is increased by 6.89% when the impact time increases from 3 to 4, but the surface compressive residual stress is kept to about 123 MPa after the multiple LSP with 4 and 5 LSP impacts (Lu et al., 2010). It can be seen from Figure 42 that the increasing rate of surface compressive residual stress decreases almost linearly with the impact time, but the increase of surface residual stresses gradually reaches the saturated state when the impact time exceeds 4. The similar results can be seen elsewhere (Masse & Barreau, 1995; Ding & Ye, 2003).

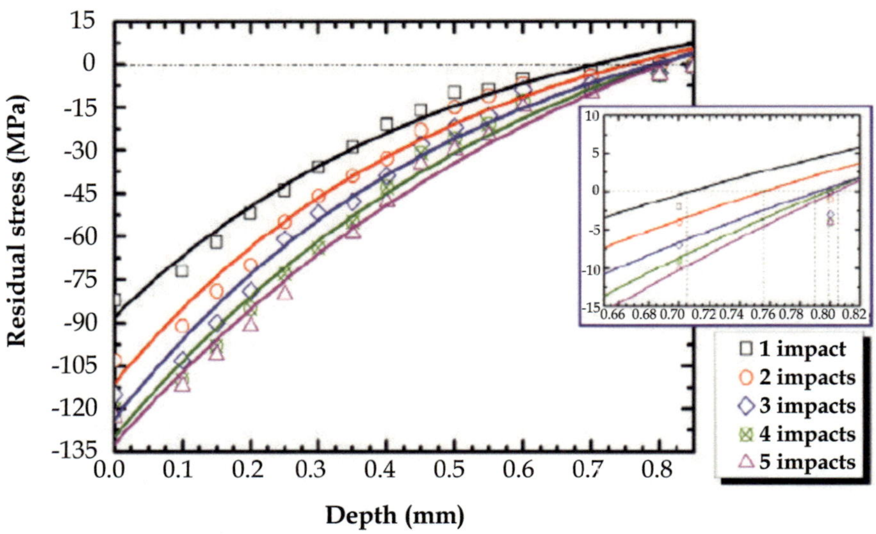

Fig. 41. Residual stress profiles of the hardening layer after multiple LSP impacts with the impact time (Lu et al., 2010)

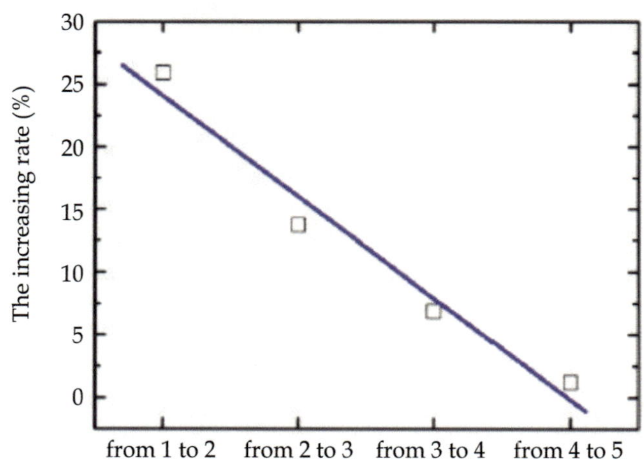

Fig. 42. The comparison between the increasing rate of surface residual stress and the impact time (Lu et al., 2010)

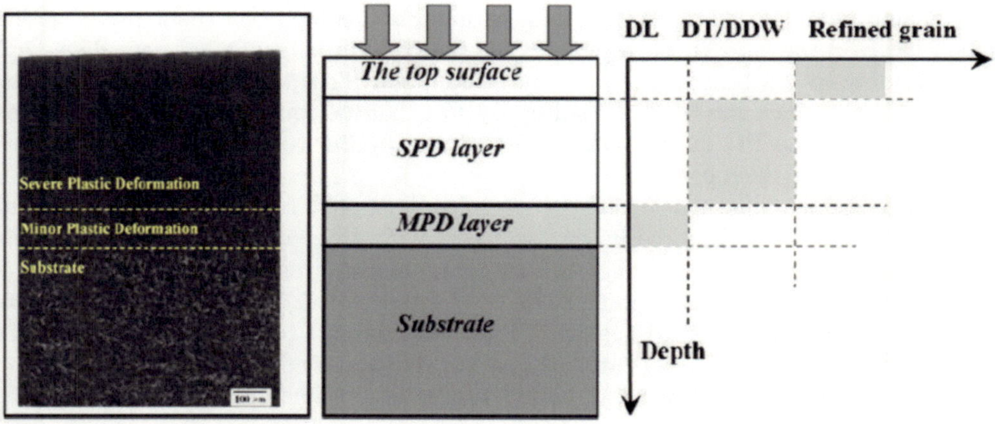

Fig. 43. Schematic illustrations of micro-structure characteristics along depth direction in the hardening layer subjected to 3 LSP impacts (Lu et al., 2010)

It is well known that residual stresses in metal materials are often the result of micro-plastic deformation accompanying the micro-structure changes (Yilbas & Arif, 2007). As a result, it is reasonable to assume that the LSP induced strengthening in metal materials is due to the generation of dislocations. The schematic illustrations of the micro-structure characteristics of the hardening layer subjected to 3 LSP impacts are shown in Figure 43. After 3 LSP impacts, the change of dislocation structure can be also clearly seen at different layers, i.e., it varies from DLs to DTs and DDWs, to subgrains or refined grains as functions of the distance from the top surface. After multiple LSP impacts, the grains in the SPD layer are clearly refined and there are plenty of DLs and DTs with high density in the SPD layer. As a result of the grain refinement, the shocked area is strengthened according to the classical dislocation theory (Chen et al., 2003), where

$$\tau_N = \frac{2\alpha\mu b_N}{D} \tag{13}$$

$$\tau_p = \frac{2\alpha\mu b_p}{D} + \frac{\gamma}{b_p} \tag{14}$$

Here μ is the shear modulus (~35 GPa for Al alloy), γ is the stacking fault energy (104–142 mJ m^{-2} for Al alloy (Lu et al., 2010)), D is the grain size, and b_N and b_P are the magnitudes of the Burgers vectors of the perfect dislocation and the Shockley partial dislocation, respectively. The parameter α reflects the character of the dislocation and contains the scaling factor between the length of the dislocation source and the grain size.
The grain boundaries are taken as dislocation sources, as predicted by computer simulations for subgrains or refined grains. When the grain size becomes smaller than a critical value, D_C, determined by equating Eqs. (13) and (14),

$$D_c = \frac{2\alpha\mu\left(b_N - b_p\right)b_p}{\gamma} \tag{15}$$

The generation of subgrain interfaces and stacking faults offers an alternative interpretation to dislocation pile-up at grain boundaries to explain the continuous grain-size strengthening, as suggested by Eq. (14), and the strain hardening of the metal materials. The reaction between the laser shock wave and the sample will be generated near the sample surface, leading to the generation of the dislocation and the micro-structural deformation near the surface, which can be explained by the fact that the compressive residual stresses are generated in the PD layer, and the magnitude of the compressive residual stress decreases away from the top surface.

The grain refinement mechanism is schematically illustrated in figure 44. Based on the micro-structure features observed in various layers with different strains in the hardening layer, the following elemental states are involved in the grain refinement process: (1) development of DLs in original grains (state (I) in figure 44); (2) the formation of DTs and DDWs due to the pile-up of DLs (state (II) in figure 44); (3) transformation of DTs and DDWs into subgrain boundaries (state (III) in figure 44); and (4) evolution of the continuous dynamic recrystallization (DRX) in subgrain boundaries to refined grain boundaries (states (IV) and (V) in figure 44).

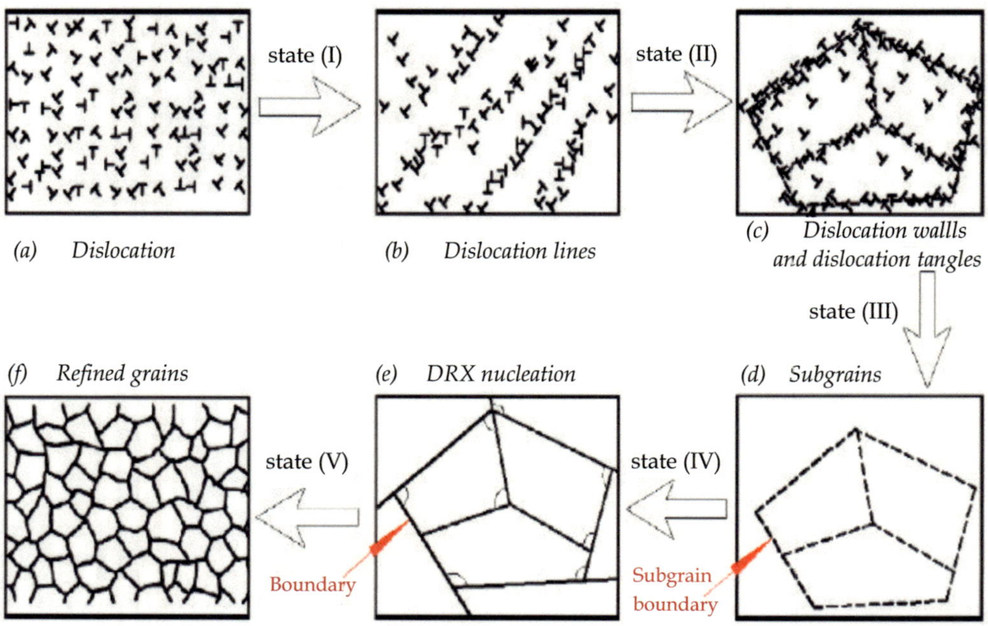

Fig. 44. Schematic illustration showing micro-structural evolution process of LY2 Al alloy induced by multiple LSP impacts (Lu et al., 2010)

The comparison of effect sot peening and laser sock peening on fatigue behavior of Al- alloy was investigated by Gao (Gao, 2011). To determine the effect of surface enhancement on fatigue property and get the optimum parameters, the FLPF analysis under the same stress load or strain load conditions is usually employed. The FLPF is calculated as:

$$FLPF = \frac{N_{\text{mod}\,ifiedspecimen}}{N_{baselinespecimen}} - 1 \tag{16}$$

For the different surface conditions, the fatigue lives of specimens and FLPF are listed in Table 2.

Surface treatment	Minimum fatigue life	Maximum fatigue life	Average fatigue life	FLPF
Machined baseline	2.41×10^4	2.69×10^4	2.55×10^4	0
SP-GB150	5.18×10^4	5.52×10^4	5.49×10^4	1.153
SP-S110	4.74×10^4	5.27×10^4	4.84×10^4	0.898
SP-Z150	8.84×10^4	9.46×10^4	9.23×10^4	2.620
SP-S110+Z150	8.65×10^4	9.21×10^4	8.97×10^4	2.518
SP-S110+GB150	1.15×10^5	1.24×10^5	1.19×10^5	3.667
LP-N=2	9.19×10^4	1.15×10^5	9.92×10^4	2.890
LP-N=4	1.52×10^5	2.14×10^5	1.94×10^5	6.608
LP-N=6	1.46×10^5	1.91×10^5	1.65×10^5	5.471
LP-N=8	9.54×10^4	1.08×10^5	9.98×10^4	2.914

Table 2. Fatigue lives of specimens and FLPF under 300MPa stress (Gao, 2011)

The compressive residual stress distribution along surface layer for laser-peened and shot-peened specimens under different regimes are shown in Figures 45, 46.

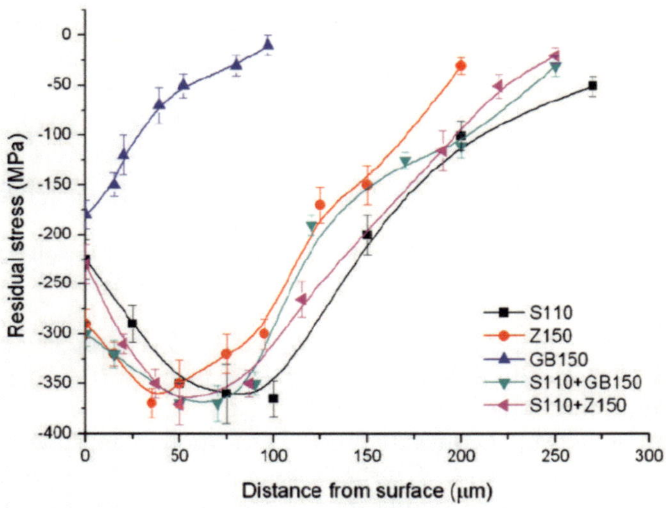

Fig. 45. Compressive residual stress field caused by shot peening (Gao, 2011)

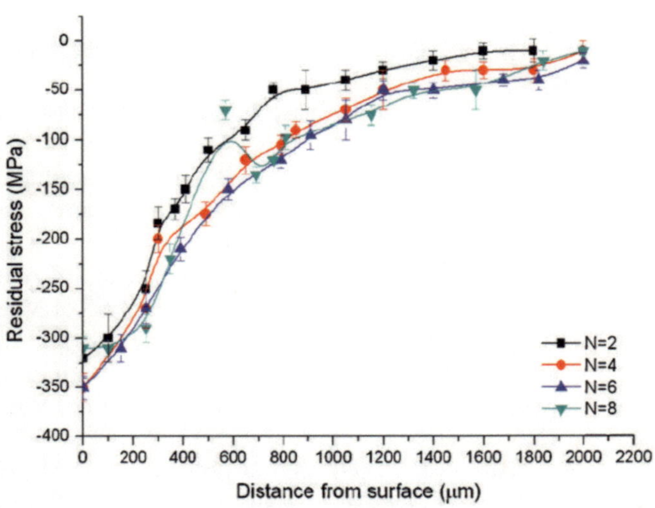

Fig. 46. Compressive residual stress field caused by laser peening (Gao, 2011)

The fatigue strength for 1×107 cycles of 7050–T7451 aluminum alloy was increased by shot peening and laser peening. Fatigue strength of the best-laser-peened specimens is 42% higher than as-machined specimens and the fatigue strength of the best shot-peened specimens is 35% higher than as-machined (Gao, 2011).

7. References

Aleksandrov, V. D. (2002). Modification of the surface of aluminium alloys by laser treatment", *Metal Science and Heat Treatment,* vol. 44, No.3-4, pp.168-171.

Almeida, A., Petrov, P., Nogueira, I., & Vilar, R. (2001). Structure and properties of Al-Nb alloys produced by laser surface alloying. *Materials Science and Engineering A,* Vol. 303, pp.273-280.

Anandkumar, R., Almeida, A., Colaco, R., Vilar, R., Ocelik, V., Th, J., & De Hosson, M. (2007). Microstructure and wear studies of laser clad Al-Si/SiC composite coatings. Surface and Coating Technology, Vol. 201, pp. 9497-9505.

Anandkumar, R., Almeida, A., Vilar, R., Ocelík, V., & De Hosson, J.Th.M. (2009). Influence of powder particle injection velocity on the microstructure of Al–12Si/SiCp coatings produced by laser cladding. *Surface & Coatings Technology,* Vol. 204, pp. 285–290.

Chen, M. W., Ma, E., Hemke, K. J., Sheng, H. W., Wang, Y. M., Cheng, X. M. (2003). Deformation Twinning in Nanocrystalline Aluminum. Science, Vol. 300, pp.1275-1277.

Chong, P.H., Liu, Z., Skeldon, P., & Thompson, E. (2003). Corrosion behavior laser surface melted 2014 aluminium alloy in T6 and T451 tempers. The journal of corrosion science and engineering, Vol. 6, paper 12.

Costa, L. & Vilar, R. (1996). Diffusion-limited layer growth in spherical geometry: A numerical approach. *Journal of Applied Physics,* Vol. 80, No. 8, pp.4350-4353.

Damborenea, J. de. (1998). surface modification of metals by power lasers. *Surface coatings technology,* Vol. 100-101, pp.377-382.

Ding, K., & Ye, L. (2003). Three-dimensional Dynamic Finite Element Analysis of Multiple Laser Shock Peening Processes. Surface Engineering, Vol. 19, p.351-358.

Ding, K. & Ye, L. (2006). *Laser shock peening Performance and process simulation.* Woodhead Publishing Limited, USA.

Draper, C. W., & Poate, J. M. (1985). Laser surface alloying. *International Metals Reviews,* Vol. 30, pp. 85-108.

Dubourg, L., Ursescu, D., Hlawka, F., & Cornet, A. (2005). Laser cladding of MMC coatings on aluminium substrate: influence of composition and microstructure on mechanical properties. *Wear,* Vol. 258, pp. 1745–1754.

Fu, Y.C., Loredo, A., Martin, B., & Vannes, A.B. (2002). A theoretical model for laser and powder particles interaction during laser cladding. Journal of Materials Processing Technology, Vol. 128, pp. 106-112.

Gao, Y. K. (2011). Improvement of fatigue property in 7050–T7451 aluminum alloy by laser peening and shot peening. *Materials Science and Engineering,* Vol. A528, pp. 3823–3828

Gingu, O., Mangra, M., & Orban, R. L. (1999). In-Situ production of Al/SiCp composite by laser deposition technology. *Journal of Materials Processing Technology,* Vol. 89-90, pp.187-190.

Guillaumin, V., & Mankowski, G. (1999). Localized corrosion of 2024 T351 aluminium alloy in chloride media. *Corrosion Science*, Vol. 41, pp.421-438.

Hu, C., Xin, H., & Baker, T.N. (1996). Formation of continuous surface Al-SiCp metal matrix composite by overlapping laser tracks on AA6061 alloy. Materials Science and Technology, Vol. 12, pp. 227-232.

Huang, Y.L., Liang, G.Y., Su, J. Y., & Li, J.G. (2005). Interaction between laser beam and powder stream in the process of laser cladding with powder feeding. Model Simul. Mater. Sci. Vol. 13, pp.47-56.

Jouvard, J.M., Grevey, D.F., Lemoine, F., & Vannes, A.B. (1997). Continuous Wave Nd:YAG Laser Cladding Modeling: A Physical Study of Track Creation During Low Power Processing. Journal of Laser Application, Vol. 9, pp. 43-50.

Leech, P. W. (1989). The Laser Surface Melting of Aluminium-Silicon-Based Alloys. *Thin Solid Films*, Vol. 177, pp. 133 140.

Li, R., Ferreira, M.G.S., Almeida, A., Vilar, R., Watkins, K. G., McMahon, M.A., & Steen, W.M. (1996). Localized corrosion of laser surface melted 2024-T351 aluminium alloy. *Surface and coatings technology*, Vol. 81, pp.290-296.

Lu, J.Z., Luo, K.Y., Zhang, Y.K., Cui, C.Y., Sun, G.F., Zhou, J.Z., Zhang, L., You, J., Chen, K.M., & Zhong, J.W. (2010). Grain refinement of LY2 aluminum alloy induced by ultra-high plastic strain during multiple laser shock processing impacts. *Acta Materialia*, Vol. 58, pp. 3984–3994.

Majumdar, D. J. & Manna, I. (2002). A Theoretical Model for Predicting Microstructure during Laser Surface Alloying. *Lasers in Engineering*, Vol. 12, pp. 171-190.

Masse, J. E., & Barreau, G. (1995). Surface modification by laser induced shock waves. Surface Engineering, Vol. 11, p.131-132.

Munitz, A. (1985). Microstructure of rapidly solidified laser-molten Al- 4.5 wt % Cu surfaces. *Metall Trans B*, Vol. 16, pp.149- 161.

Nagarathnam, K. & Taminger, K.M.B. (2001). Technology Assessment of Laser-Assisted Materials Processing in Space. CP 552, *Space Technology and Applications International Forum*, paper edited by M.S. El-Genk, pp. 153-160.

Peyre, P., Fabbro, R., Berthe, L., Scherpereel, X., & Bartnicki, E. (2005). Laser shock processing of materials and related measurements", CLFA/LALP, 94114 Arcueil, France.

Peyre, P., Fabbro, R., Merrien, P., & Lieurade, H.P. (1996). Laser shock processing of aluminium alloys. Application to high cycle fatigue behavior. *Materials Science & Engineering*, Vol. A210, pp.102-113.

Pinto, M. A., Cheung, N., Ierardi, M.C.F., Garcia, A. (2003). Microstructural and hardness investigation of an aluminum–copper alloy processed by laser surface melting. *Materials Characterization*, Vol. 50, pp. 249- 253.

Rams, J., Padro, A., Urena, A., Arrabal, R., Viejo, F., & Lopez, A.J. (2007). Surface treatment of *aluminium matrix* composites using a high power diod laser. Surface and Coatings Technology, Vol. 202, pp. 1199-1203

Renk, T. J., Buchheit, R. G., Sorensen, N. R., Cowell Senft, D., Thompson, M. O., & Grabowski, K. S. (1998). Improvement of surface properties by modification and alloying with high-power ion beams", *Phys. Plasmas*, Vol. 5, pp. 2144-2150.

Sallamand, P., & Pelletier, J. M. (1993). Laser cladding on aluminium-base alloys: microstructural features", *Materials Science and Engineering*, Vol. A 171, pp. 263-270.

Senthil Selvan, J., Soundararajan, G., & Subramanian, K. (2000). Laser alloying of aluminium with electrodeposited nickel: optimisation of plating thickness and processing parameters. *Surface and Coatings Technology*, Vol. 124, pp. 117–127.

Tan, Y., Wu, G., Yang, J. M., & Pan, T. (2004). Laser shock peening on fatigue crack growth behaviour of aluminium alloy. Fatigue Fracture Engineering and Materials Structure, Vol. 27, pp.649-656.

Tomlinson W.J. & Bransden A.S. (1996). Cavitation erosion of laser surface alloyed coatings on Al-12%Si", International Journal of Multiphase Flow, Vol. 22, No. 1, pp. 152-152.

Tomida, S., Nakata, K. (2003). Fe–Al composite layers on aluminum alloy formed by laser surface alloying with iron powder. *Surface and Coatings Technology,* Vol. 174–175, pp. 559–563.

Viala, J.C., Fortier, & Bouix, P. (1990). Stable and metastable phase equilibria in the chemical interaction between aluminium and silicon carbide. *Journal of Materials Science,* Vol. 25, pp. 1842-1850.

Wong, T.T., Liang, G.Y., Tang, C.Y. (1997). The surface character and substructure of aluminum alloys by laser-melting treatment", *Journal of materials processing Technology,* Vol. 66, pp.172-178.

Wong, T.T., & Liang, G.Y. (1997). Effect of laser melting treatment on the structure and corrosion behavior of aluminium and Al-Si alloys. *Journal of Materials Processing Technology,* Vol. 63, pp.930-934.

Xu, W.L., Yue, T.M., Man, H.C., & Chan, C.P. (2006). Laser surface melting of aluminium alloy 6013 for improving pitting corrosion fatigue resistance. *Surface & Coatings Technology,* Vol. 200, pp. 5077–5086.

Yilbas, B. S., Arif AFM , (2007). Laser shock processing of aluminium: model and experimental study. J Phys D, Appl Phys 40, pp.6740-6747.

Yongqing, F., Batchelor, A.W., Yanwei, G., Khor, K.A., & Huting X. (1998). Laser alloying of aluminum alloy AA 6061 with Ni and Cr. Part 1. Optimization of processing parameters by X-ray imaging. *Surface and Coatings Technology*, Vol. 99, pp. 287-294.

Yue, T. M., Yan, L.J., Chan, C.P., Dong, C.F., Man, H.C., & Pang, G.K.H. (2004). Excimer laser surface treatment of aluminium alloy AA7075 to improve corrosion resistance. *Surface and Coatings* Technology, Vol. 179, No. 2-3, pp.158-164.

Zimmermann, M., Carrard, M., Kurz, W. (1989). Rapid solidification of Al –Cu eutectic alloy by laser remelting. *Acta Metallurgica*, Vol. 37, No. 12, pp.3305–13.

7

Microstructural Changes of Al-Cu Alloys After Prolonged Annealing at Elevated Temperature

Małgorzata Wierzbińska and Jan Sieniawski
Rzeszow University of Technology, Rzeszow, Poland

1. Introduction

The precipitation–strengthened 2xxx series Al–Cu alloys are one of the most important high-strength aluminium alloys. They have been employed extensively in the aircraft and military industries, in which materials are frequently subjected to elevated temperature. The aluminium casting alloys, based on the Al–Cu system are widely used in light–weight constructions and transport applications requiring a combination of high strength and ductility.

Al-Cu alloys are less frequently used than Al-Si-Cu grades due to technological problems in production process (e.g. high propensity to microcracking during casting). However they are the basis for development of multicomponent alloys. Typical alloys for elevated temperature application are Al-Cu-Ni-Mg alloys (containing about 4,5% Cu, 2% Mg and 2%Ni). Their good properties at elevated temperature result from formation of intermetallic phases Al_6Cu_3Ni and Al_2CuMg, both during crystallization and precipitation hardening (El–Magd & Dünnwald, 1996; Martin, 1968; Mrówka-Nowotnik et al., 2007).

Mechanism of precipitation hardening in cast and wrought binary Al-Cu alloys is well known and widely covered in literature. There are some suggestions that decomposition of supersaturated α(Al) solid solution in other precipitation hardened alloys like Al-Cu-Mg, Al-Si-Cu, Al-Mg-Si follows the same route as in the Al-Cu alloys with some specific features of the particular stages of the process (Martin, 1968;). The interest in course and kinetics of the aging process has the practical meaning as the early stages of aging leads to significant improvement of mechanical properties of the alloys. Maximum hardening effect in Al-Cu alloy is a result of in situ transformation of GP zones into transient phase θ''. Increase in aging temperature leads to decrease of the hardness of solid solution α(Al) due to precipitation of equilibrium θ phase on the grain boundaries or on the θ'/matrix phase boundaries. Prolonged aging may lead to microstructure degradation related to coagulation and/or coalescence of the highly dispersed hardening phase precipitates resulting in decrease of hardening effect (Mrówka-Nowotnik et al., 2007; Wierzbińska & Sieniawski, 2010). Therefore development of the chemical composition of the alloy, especially intended for long term operation at elevated temperature, requires taking into account factors resulting in deceleration of the coagulation process and obtaining stable microstructure consisting of solid solution α grains and highly dispersed precipitates of the second phase (Wierzbińska & Sieniawski, 2010).

2. Material and methodology

The investigation was performed on the two casting alumium alloys AlCu4Ni2Mg and AlCu6Ni. AlCu4Ni2Mg is the standard alloy used currently for highly stressed structural elements of engines and AlCu6Ni1 is an experimental alloy that was chosen to investigate the influence of the increased content of Cu on the phase composition, microstructure morphology and mechanical, technological and operational properties. The alloys were cast into metal moulds and subjected to X-ray inspection in order to exclude the presence of porosity or oxide films.

The alloys were subjected to heat treatment T6 followed by annealing at 523 K and 573 K for 150 and 500 hours. After analysis of the results of preliminary tests it was found, that it is advisable to apply additional annealing times at particular temperature, i.e. 100, 300 and 750 hours.

Heat treatment conditions were established on the basis of the phase equilibrium diagrams Al-Si and Al-Cu and available heat treatment data for the alloys with similar chemical composition (both from literature and used in industry practice). The consideration was also given to requirements concerning mechanical properties of the alloys resulting from operation condition of the structural elements made of these alloys. The chemical composition of the investigated alloys and heat treatment parameters are presented in table 1.

Element	Element content, wt.%	
	AlCu4Ni2Mg	AlCu6Ni
Mn	<0.10	0.90
Ni	2.10	1.10
Cu	4.30	6.36
Zr	–	0.01
Fe	0.10	0.20
Si	0.10	0.10
Mg	1.50	0.05
Zn	0.30	–
Al	balance	balance
solution treatment	$793^{\pm5}$K/5h/ water cooling	$818^{\pm5}$K/10h/ water cooling
artificial ageing	$523^{\pm5}$K/5h/ air cooling	$498^{\pm5}$K/8h/ air cooling

Table 1. Composition of AlCu4Ni2Mg and AlCu6Ni alloys and heat treatment parameters

Examination of the alloys microstructure was carried out using light microscope (LM), as well as scanning (SEM) and transmission (TEM) electron microscopes.

3. Results and discussion

Figs. 1 to 4 show the results of microscopic observations of AlCu4Ni2Mg and AlCu6Ni alloys (in T6 condition). In both of investigated alloys large, irregular shaped precipitates of

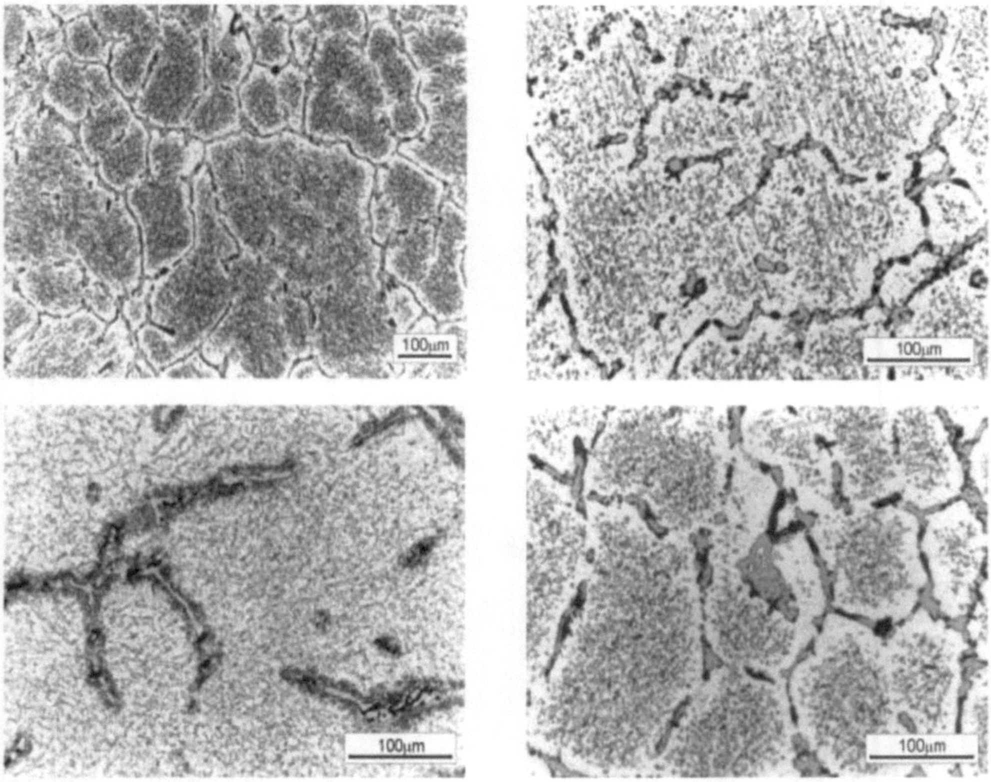

Fig. 1. Microstructure of AlCu4Ni2Mg alloy in T6 condition (LM)

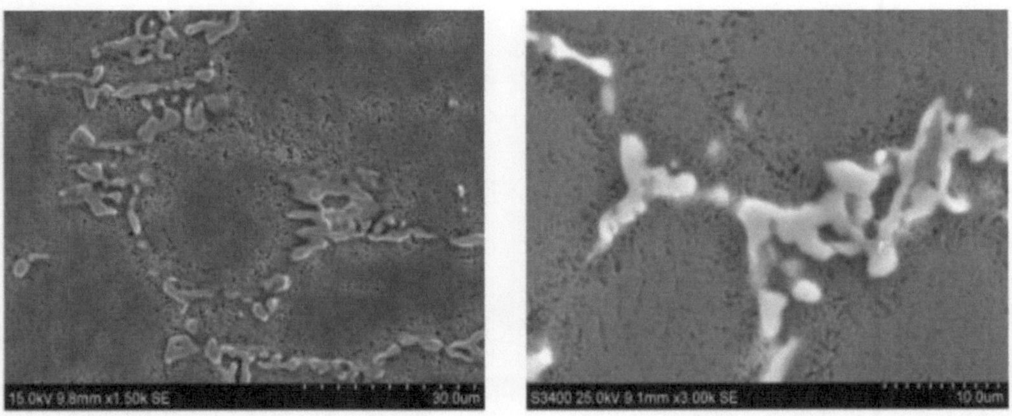

Fig. 2. Microstructure of AlCu4Ni2Mg alloy in T6 condition: precipitations of intermetallic phases in interdendritic areas (SEM)

intermetallic phases, located on the dendrite boundaries of solid solution α-Al and dispersive, spheroidal and strip shaped hardening phase precipitates homogenously distributed throughout the solid solution were observed.

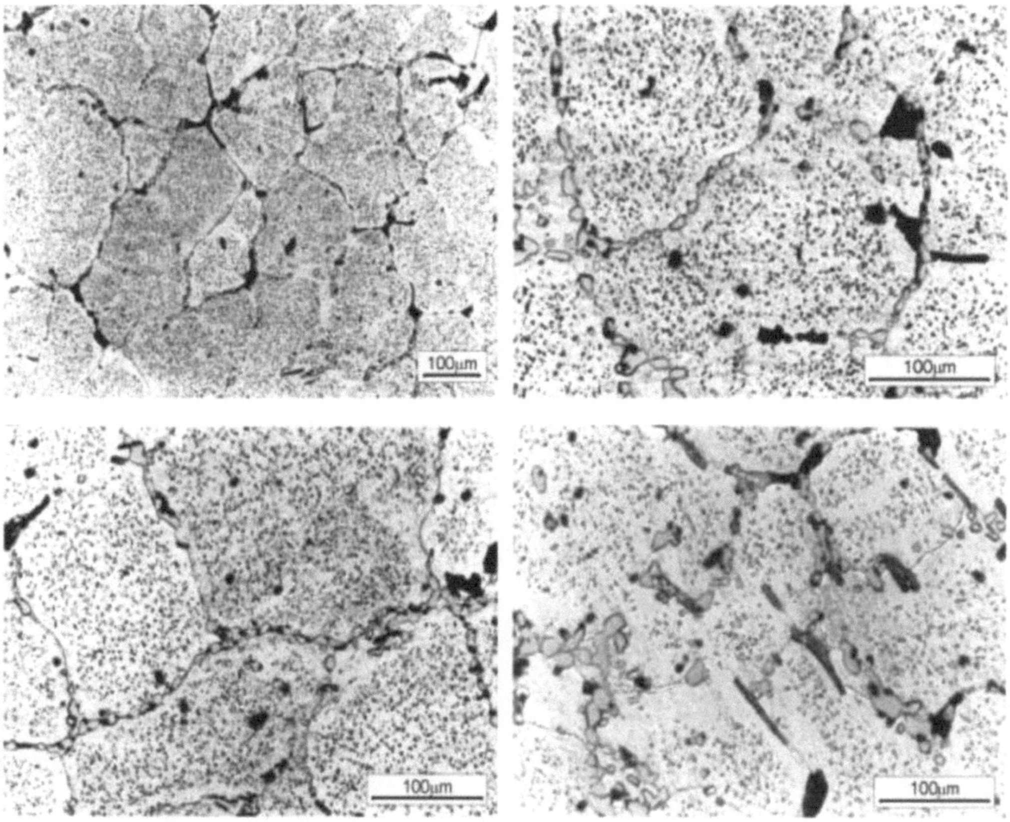

Fig. 3. Microstructure of the AlCu6Ni alloy in T6 condition (LM)

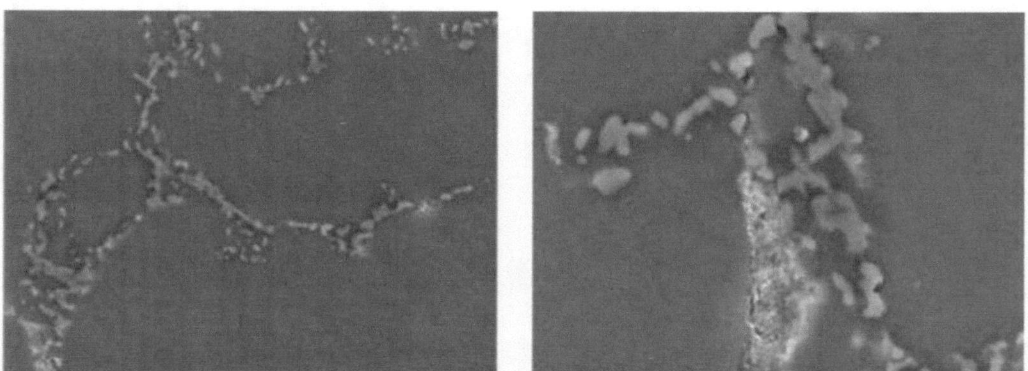

Fig. 4. Microstructure of the AlCu6Ni1 alloy in T6 condition (SEM)

Based upon the EDS results the phases forming large size particles was identified as Al-Cu-Ni, Al-Cu-Ni-Fe and Al-Cu-Mn (fig. 5–6) (Mrówka-Nowotnik et al. 2007, Wierzbińska & Sieniawski 2010).

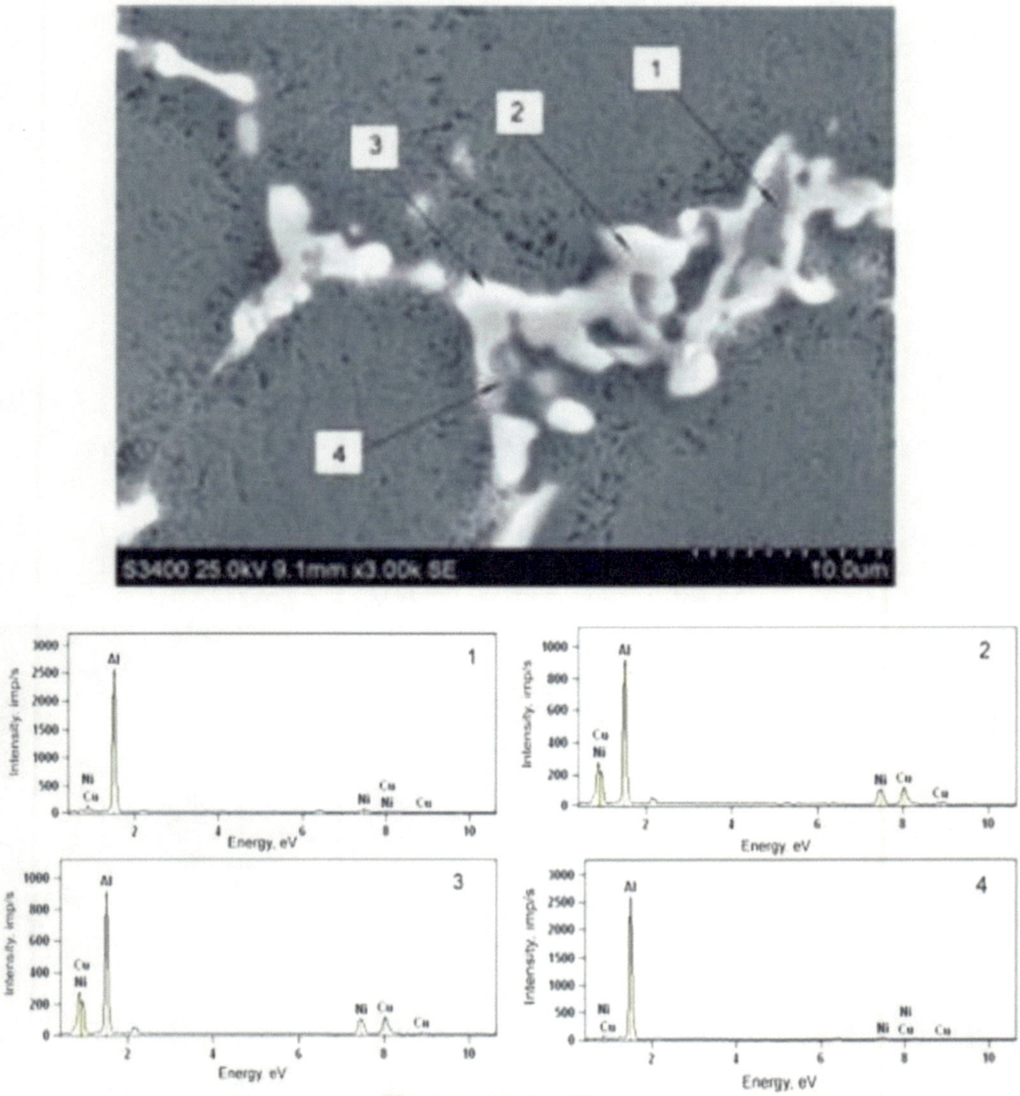

Element content, wt.%

	Al-K	Fe-K	Ni-K	Cu-K
AlCu4Ni2Mg – pt 1	76,32	5,95	13,39	4,35
AlCu4Ni2Mg – pt2	47,26		23,27	29,48
AlCu4Ni2Mg – pt 3	54,05		23,14	22,81
AlCu4Ni2Mg – pt 4	85,32	3,18	9,08	2,43

Fig. 5. AlCu4Ni2Mg alloy – EDS analysis of the areas 1–4

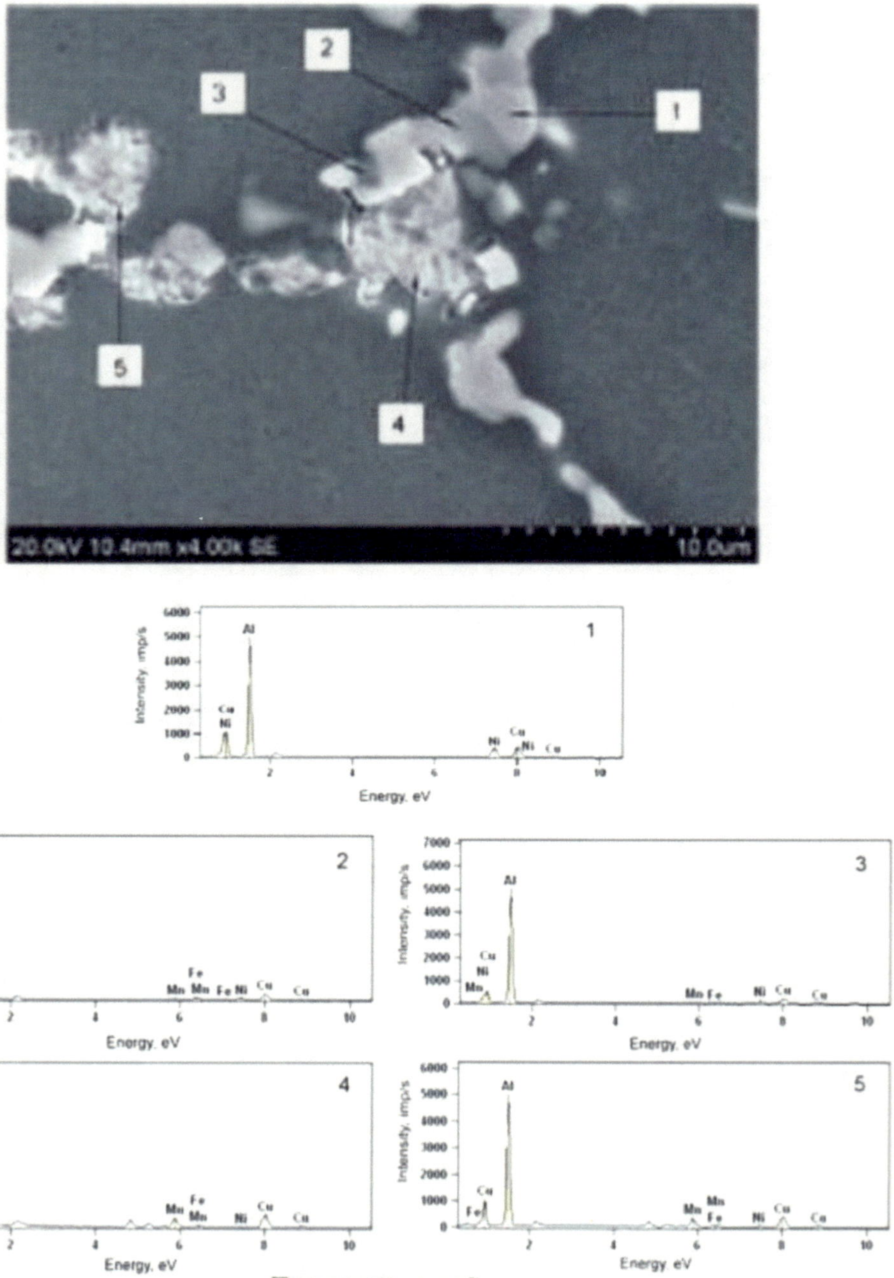

Element content, wt.%

	Al-K	Mn-K	Fe-K	Ni-K	Cu-K
AlCu6Ni1 - pt 1	53,07			19,43	27,51
AlCu6Ni1 - pt 2	60,00	2,38	4,65	6,68	26,29
AlCu6Ni1 - pt 3	65,90	3,94	1,29	8,44	20,43
AlCu6Ni1 - pt 4	51,14	10,42	1,19	2,24	35,02
AlCu6Ni1 - pt 5	55,12	10,06	1,87	2,19	30,76

Fig. 6. AlCu6Ni alloy – EDS analysis of the areas 1–5

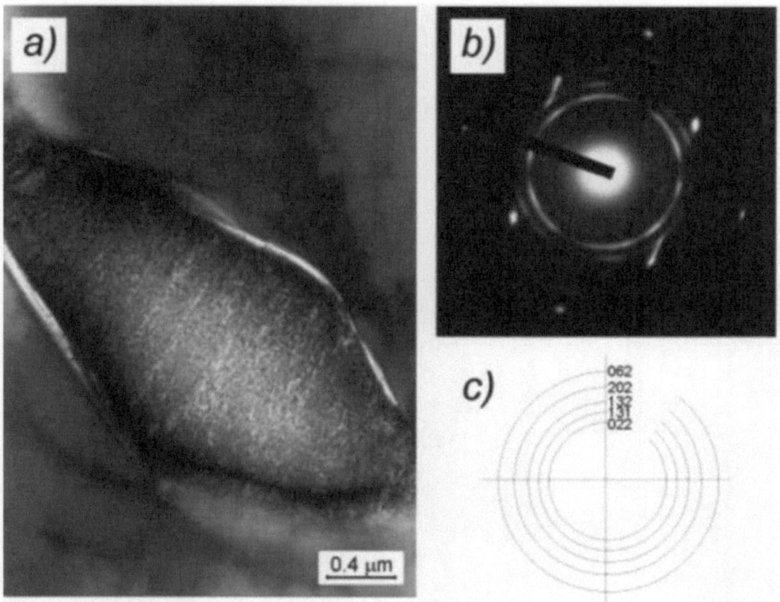

h k l	022	131	132	202	062
d (nm)	28,73	23,50	20,24	17,34	14,12
d (nm): *S-Al₂CuMg*	28,29	13,13	20,18	17,49	14,16

Fig. 7. Microstructure of the AlCu4Ni2Mg alloy: a) precipitates of *S-*Al_2CuMg phase, b) electron diffraction pattern obtained from the precipitate, c) solution of the diffraction pattern

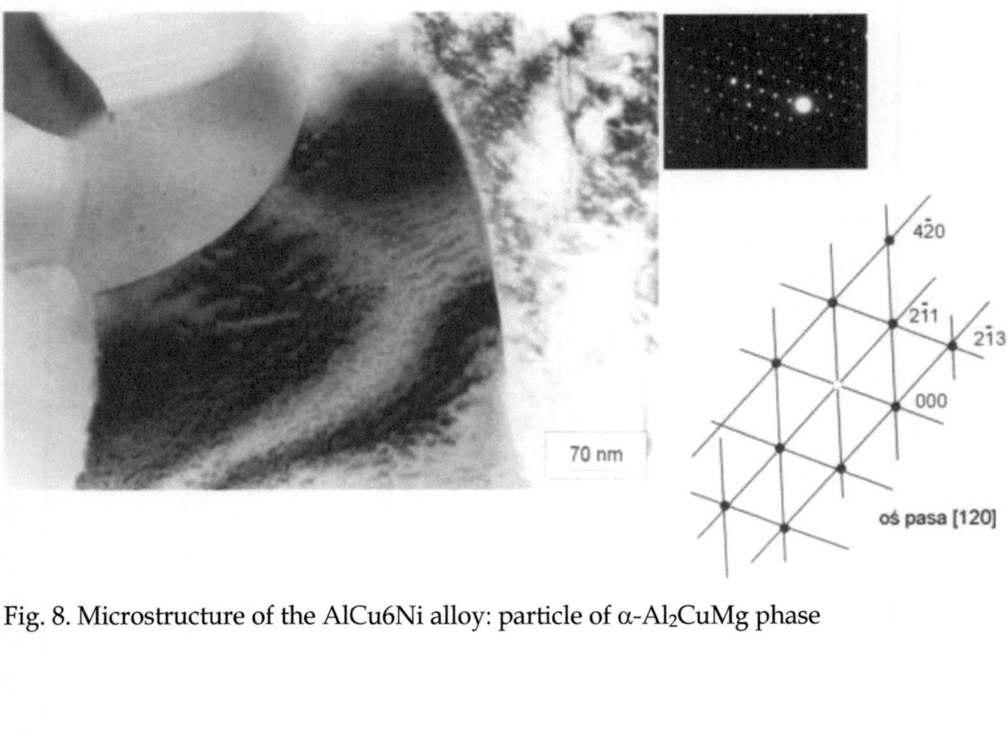

Fig. 8. Microstructure of the AlCu6Ni alloy: particle of α-Al_2CuMg phase

The intermetallic phases S-Al_2CuMg (fig. 7) and α-Al_2CuMg (fig. 8) as well as hardening phase θ'-Al_2Cu (fig. 9-11) were identified in the alloy microstructure by electron diffraction analysis (Pearson, 1967).

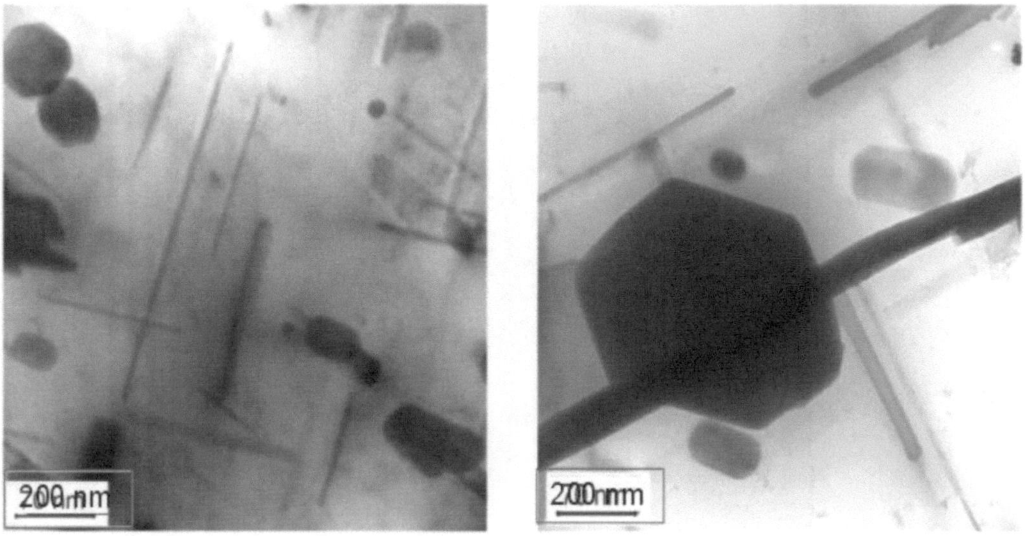

Fig. 9. Microstructure of the AlCu4Ni2Mg alloy in T6 condition (TEM – thin foil). The precipitates of hardening phase θ'-Al_2Cu

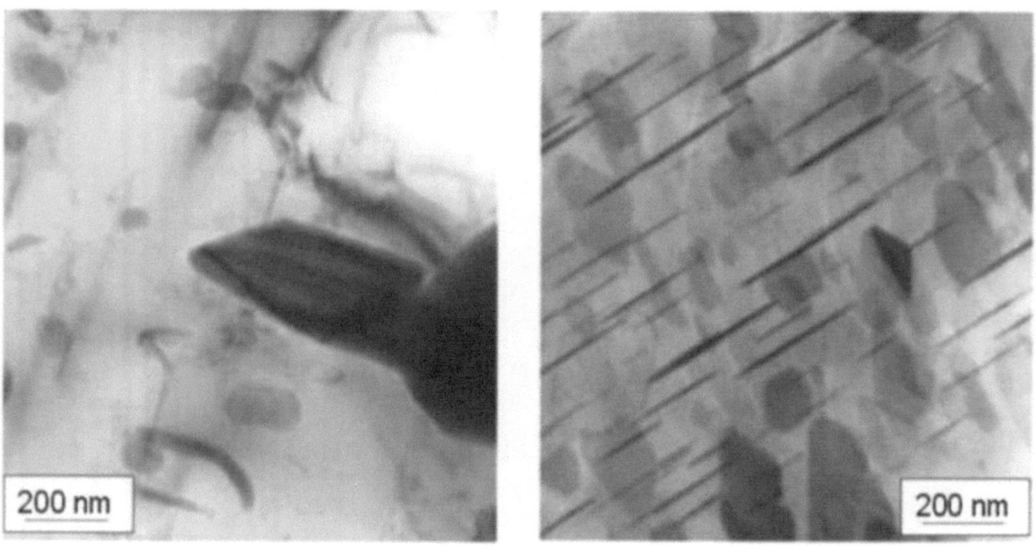

Fig. 10. Microstructure of the AlCu6Ni alloy in T6 condition (TEM – thin foil). The precipitates of hardening phase θ'-Al_2Cu

The shape of Al_2Cu particles was diversified from nearly regular polygons - „crystallites" to strongly elongated - "rod-shaped" (fig. 11–12).

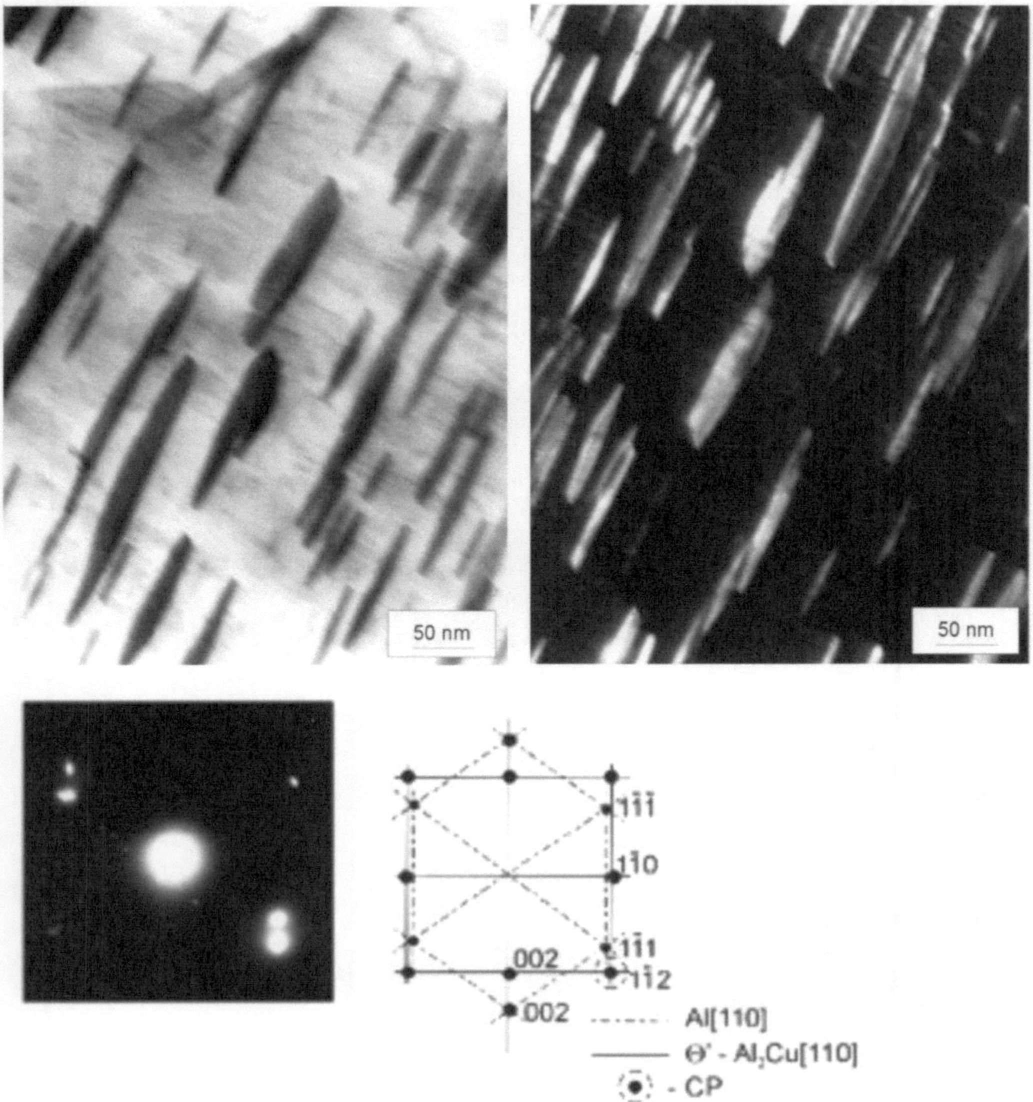

Fig. 11. Microstructure of the AlCu6Ni alloy – precipitates of θ'-Al_2Cu phase in the shape of plates (TEM – thin foil)

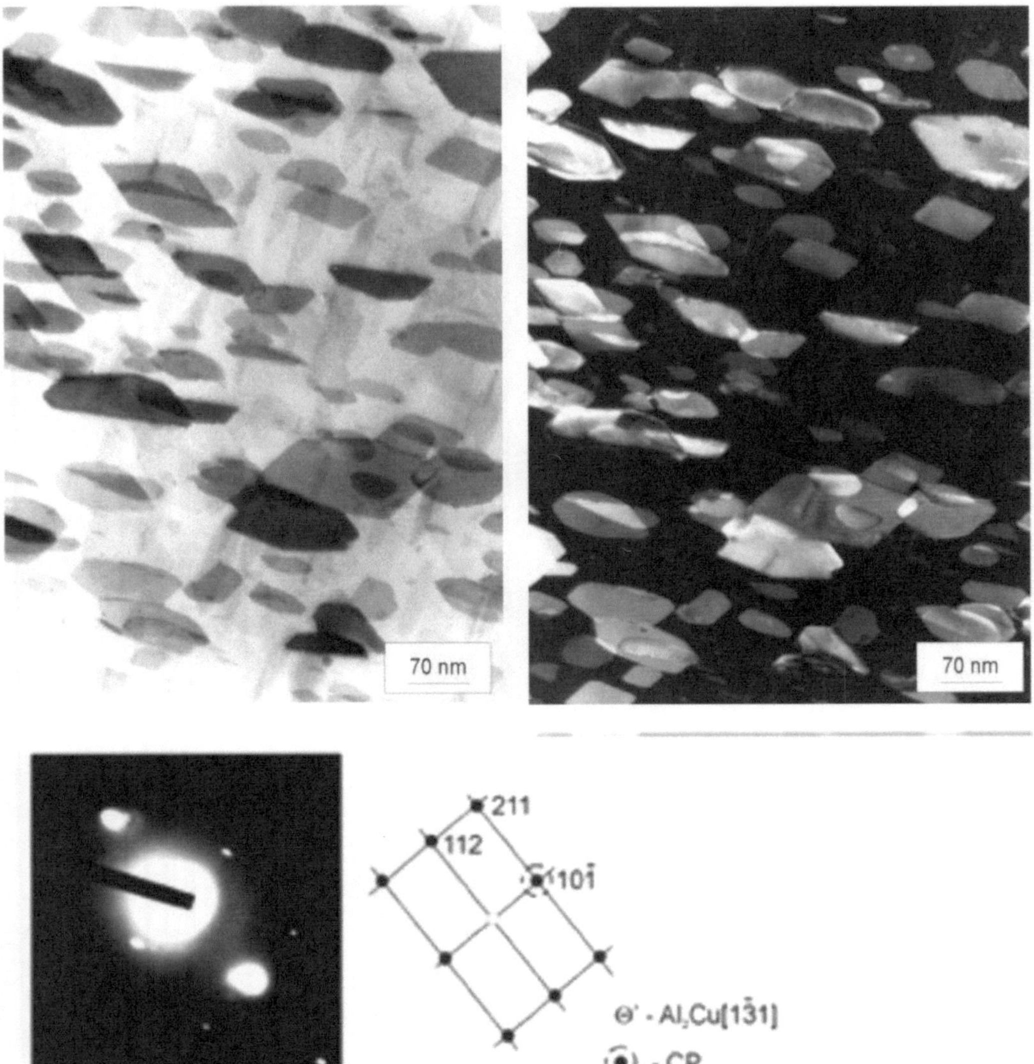

Fig. 12. Microstructure of the AlCu6Ni alloy – precipitates of θ'-Al_2Cu phase in the shape of crystallites (TEM – thin foil)

Examination of the alloys microstructure after prolonged annealing revealed that the precipitates of Al_6Fe and S-Al_2CuMg phases and large precipitates of intermetallic phases at the dendrite boundaries practically did not change (fig. 13–14) even after very long time of annealing (750h). Whereas, significant increase in size of dispersive particles of θ'-Al_2Cu hardening phase was observed (fig. 15-18).

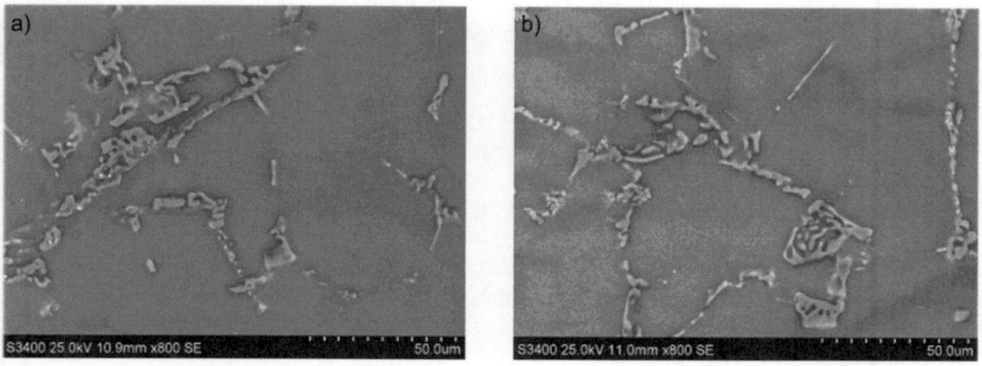

Fig. 13. Microstructure of the AlCu4Ni2Mg alloy after annealing: a) 523K/100h, b) 573K/750h

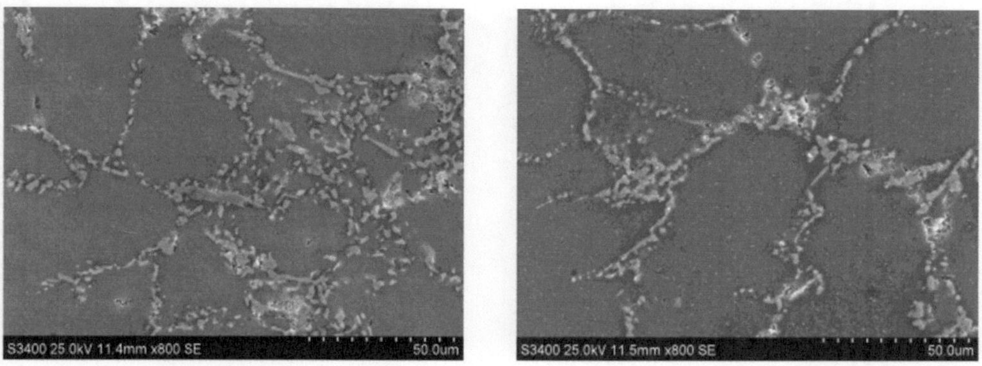

Fig. 14. Microstructure of the AlCu6Ni1 alloy after annealing: a) 523K/100h, b) 573K/750h

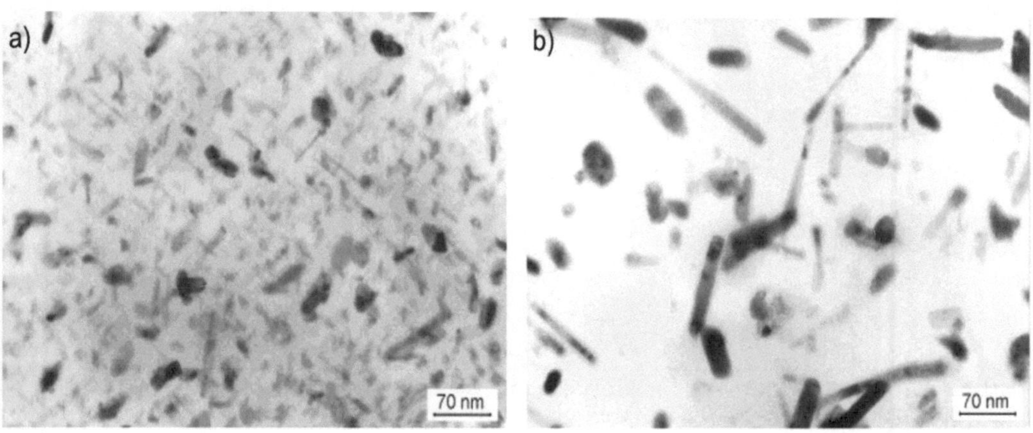

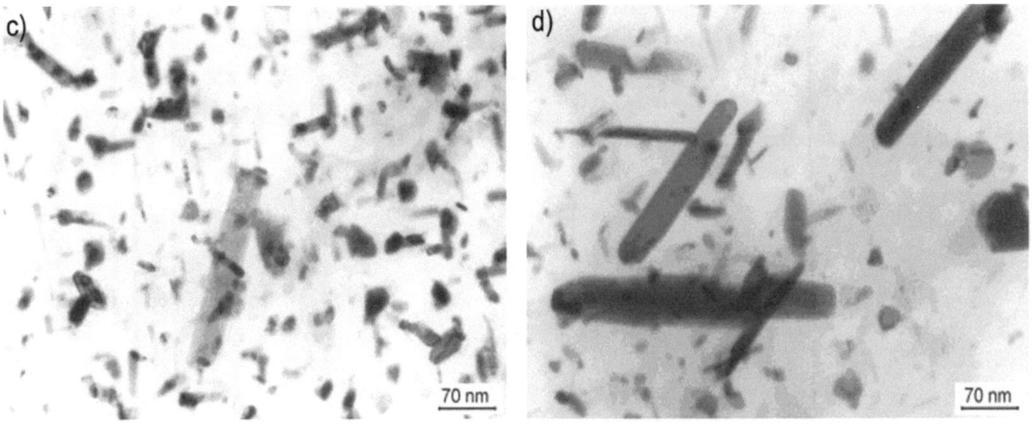

Fig. 15. Microstructure of the AlCu4Ni2Mg alloy – precipitates of the θ'-Al_2Cu phase after annealing at 523 K for: a)100 h, b) 300 h, c) 500 h, d) 750 h

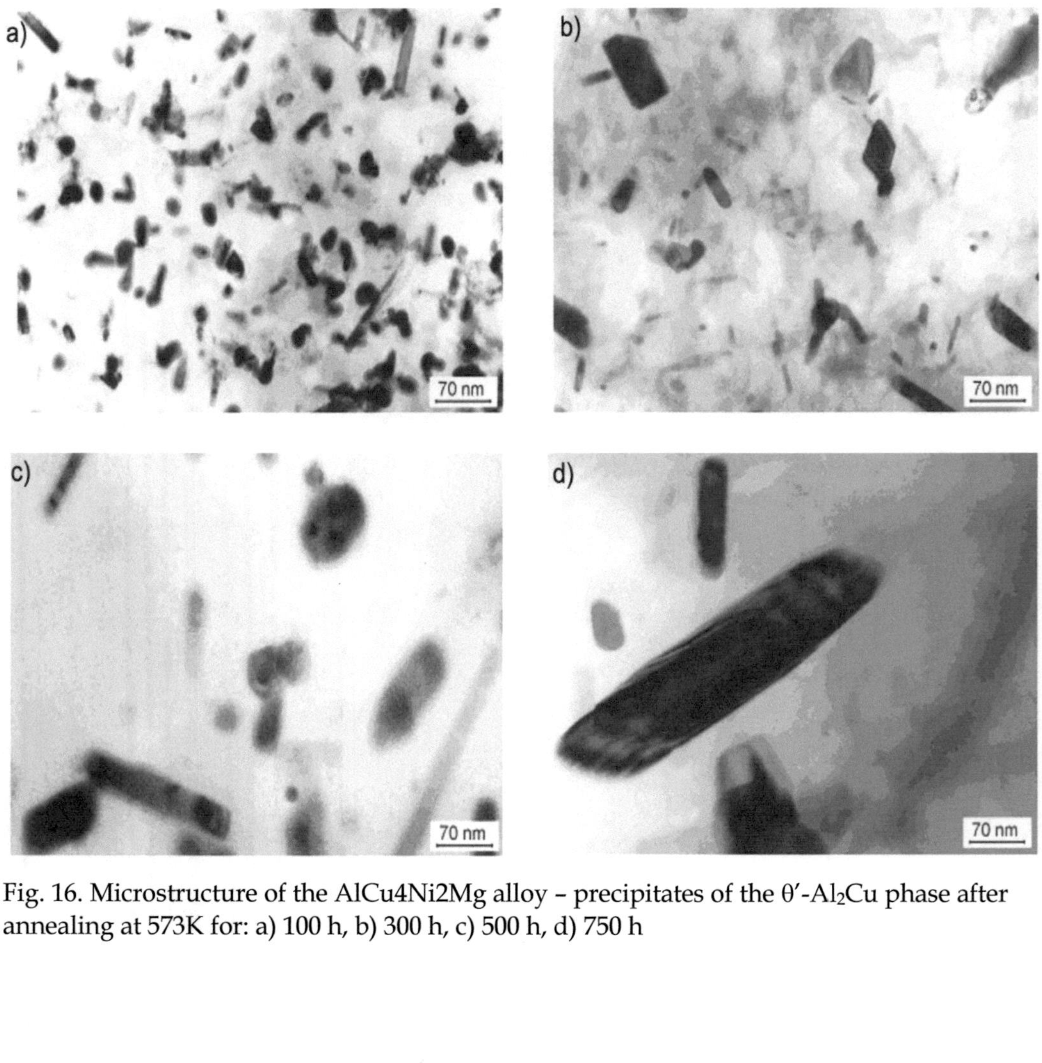

Fig. 16. Microstructure of the AlCu4Ni2Mg alloy – precipitates of the θ'-Al_2Cu phase after annealing at 573K for: a) 100 h, b) 300 h, c) 500 h, d) 750 h

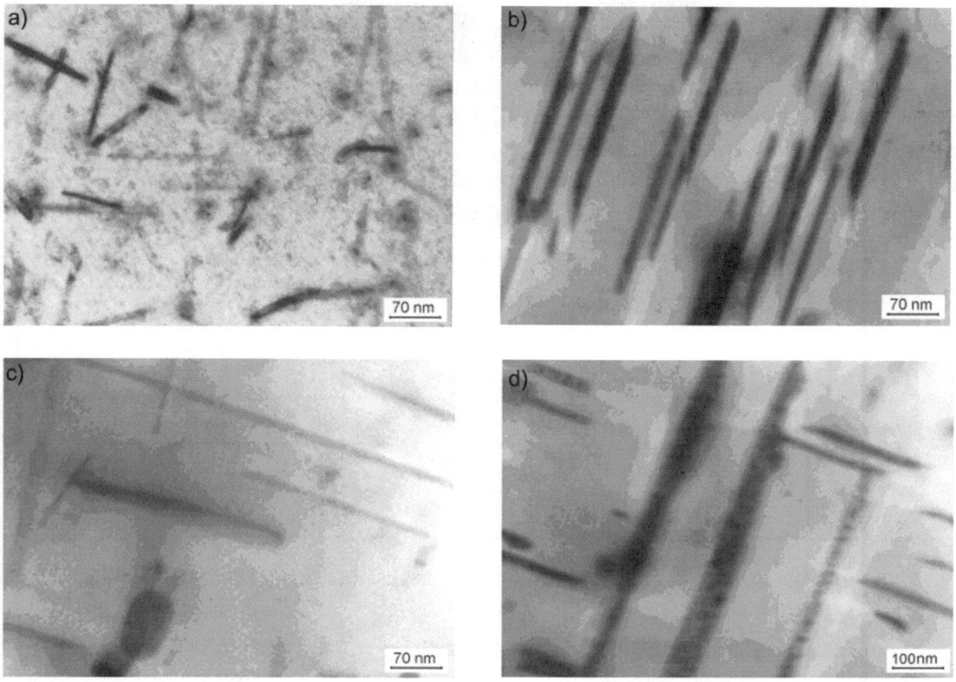

Fig. 17. Microstructure of the AlCu6Ni alloy – precipitates of the θ′-Al_2Cu phase after annealing at 523 K for: a)100 h, b) 300 h, c) 500 h, d) 750 h

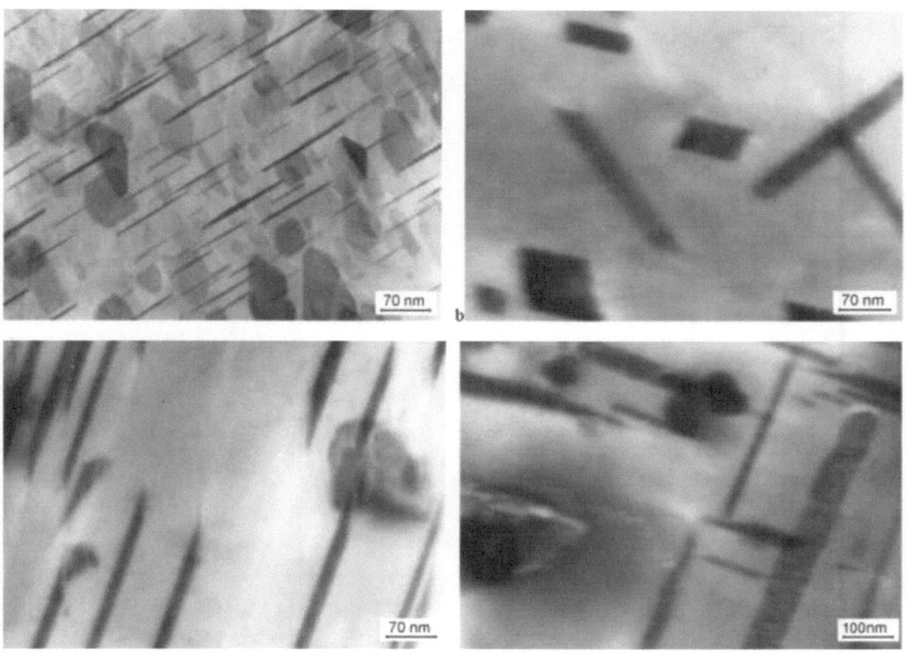

Fig. 18. Microstructure of the AlCu6Ni – precipitates of the θ′-Al_2Cu phase after annealing at 573K for: a) 100 h, b) 300 h, c) 500 h, d) 750 h

Microstructure examination revealed that in both alloys i.e. AlCu4Ni2Mg and AlCu6Ni growth of the hardening phase precipitates occured as a result of long-term thermal loading, which was proportional to the temperature and time of annealing. However higher coarsening propensity was found for Al6CuNi alloy which arose from higher content of the element forming hardening phase (6% Cu). It was confirmed by analysis of the change of shape and size of the θ'-Al_2Cu precipitates in both alloys after annealing at 573K for 150 and 750 h comparing to the standard T6 condition (table 2).

Alloy	Shape parameters of θ'-Al_2Cu precipitates	Heat treatment conditions		
		T6	T6 + annealing at 573 K	
			150 h	750 h
AlCu4Ni2Mg	length, l (nm)	75,12	650,28	887,45
	width, w (nm)	25,20	131,15	158,19
	shape factor l/w	2,98	4,95	5,61
AlCu6Ni	length, l (nm)	55,82	4465,60	6255,05
	width, w (nm)	10,30	115,36	149,11
	shape factor, l/w	5,42	38,71	41,95

Table 2. Evolution of θ'-Al_2Cu precipitates in AlCu4Ni2Mg i AlCu6Ni alloys during annealing at 573K for 150 and 750h

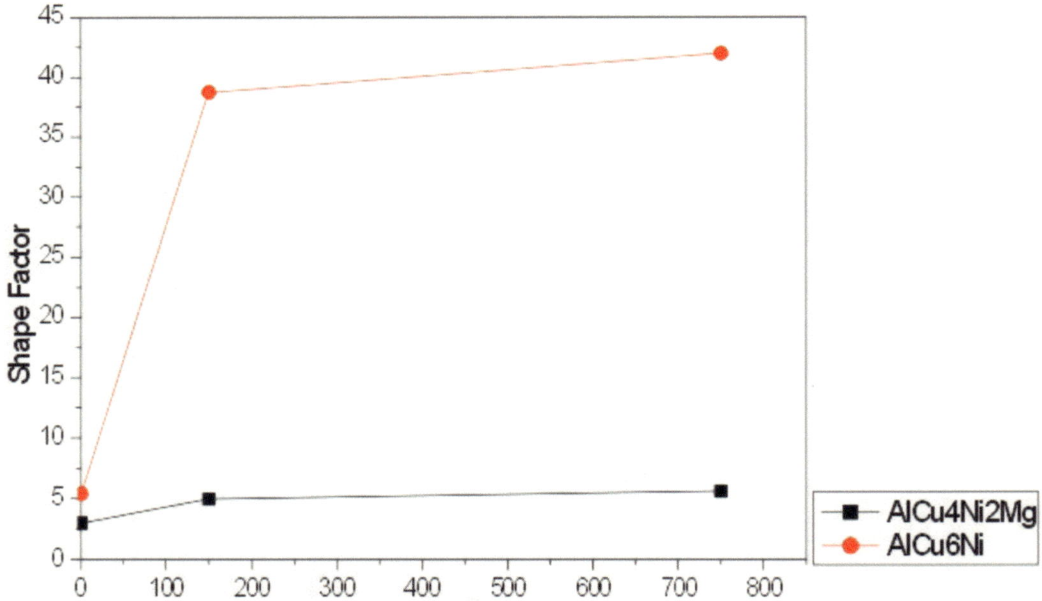

Fig. 19. Change of shape factor of the θ'-Al_2Cu precipitates in AlCu4Ni2Mg and AlCu6Ni as a result of annealing at 573K for 150 and 750h

Results of the measurements showed that annealing of the alloys studied at 573K led to significant growth of hardening θ′-Al_2Cu phase precipitates already after 150h. The biggest change both of size and shape factor of the particles (sevenfold increase) was observed in AlCu6Ni alloy. In the AlCu4Ni2Mg alloy precipitates growth was not so substantial - shape factor was only doubled. Increase in annealing time (750h) resulted in further growth of precipitates. However the process was not so dynamic as in the initial stages of annealing (table 2, fig. 19) - only minor changes of shape factor were observed.
Microstructure examination indicated that growth of the hardening phase precipitates is the main symptom of the microstructure degradation caused by long-term thermal loads. Coarsening and change of the shape of hardening phase particles lead to change of mechanism of their interactions with dislocations and as a consequence of that decrease of strength properties of the alloys (Hirth & Lothe, 1968).
Results of the static tensile test for the alloys studied in T6 condition and after additional annealing at 523 and 573K for 100, 150, 300, 500 and 750h are presented in table 3 and in figures 20 and 21.

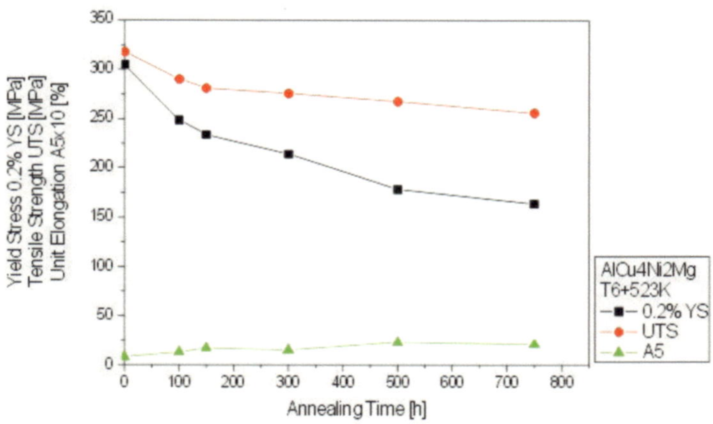

(a)

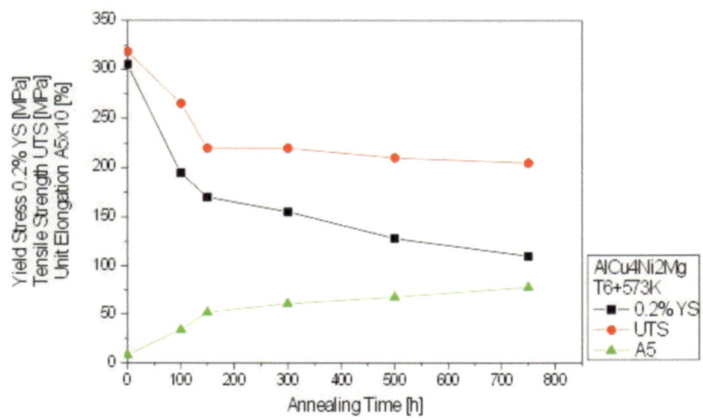

(b)

Fig. 20. Ultimate tensile strength, 0.2% offset yield strength and elongation A_5 for AlCu4Ni2Mg alloy as a function of annealing time at the temperature of a) 523K and b) 573K

Mechanical properties	Heat treatment – temperature and time of annealing										
	T6	T6+523 K					T6+573 K				
		100h	150h	300h	500h	750h	100h	150h	300h	500h	750h
AlCu4Ni2Mg alloy											
0.2%YS, MPa	305	249	234	214	178	164	195	170	155	128	110
UTS, MPa	318	290	281	276	268	256	265	220	220	210	205
A_5, %	0,8	1,3	1,7	1,5	2,3	2,1	3,4	5,2	6,1	6,8	7,8
AlCu6Ni alloy											
0.2%YS, MPa	285	225	215	185	168	142	180	147	140	118	104
UTS, MPa	323	305	290	277	263	243	245	240	216	192	177
A_5, %	0,7	1,3	1,9	2,6	3,1	3,8	2,4	4,2	5,6	6,8	7,0

Table 3. Mechanical properties of the AlCu4Ni2Mg and AlCu6Ni alloys in standard T6 condition and after additional annealing at 523 and 573K

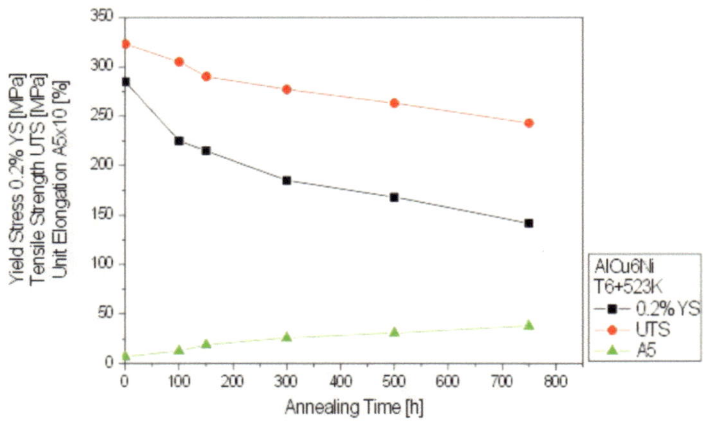

(a)

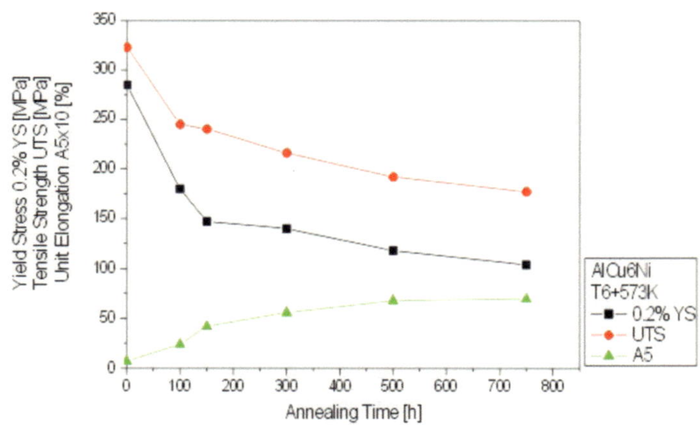

(b)

Fig. 21. Ultimate tensile strength, 0.2% offset yield strength and elongation A_5 for AlCu6Ni alloy as a function of annealing time at the temperature of a) 523K and b) 573 K

Annealing temperature	[(UTS-UTS(T)) / UTS] × 100%					[(YS-YS(T)) / YS] × 100%				
	100h	150h	300h	500h	750h	100h	150h	300h	500h	750h
	AlCu4Ni2Mg									
523 K	9	12	13	16	19	18	23	30	42	46
573 K	17	31	31	34	35	36	44	49	58	64
	AlCu6Ni									
523 K	5	10	14	19	24	21	25	35	41	50
573 K	24	26	33	40	45	37	48	51	59	64

Table 4. Relative decrease of ultimate tensile strength and 0.2% offset yield strength of the AlCu4Ni2Mg and AlCu6Ni1alloys after annealing at 523 and 573K

It was found that both alloys subjected to long-term annealing exhibit significant reduction of mechanical properties. This tendency was characterized by the coefficient calculated according to the formula $[(R - R_{(T)}) \times R^{-1} \times 100\%]$ where: R – UTS or YS in T6 condition, $R_{(T)}$ – UTS or YS after annealing at 523/573K (table 4). The analysis of the dependence of that

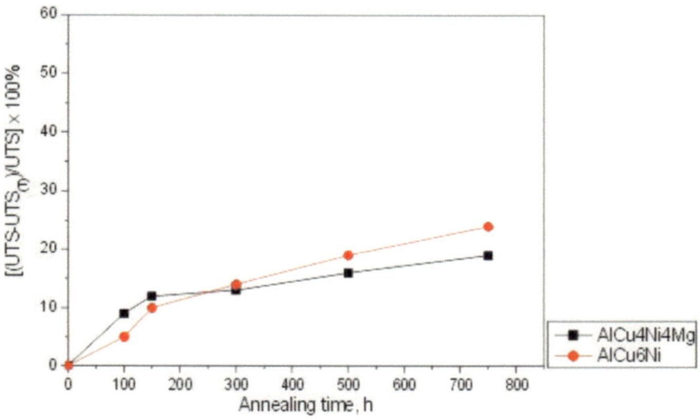

(a)

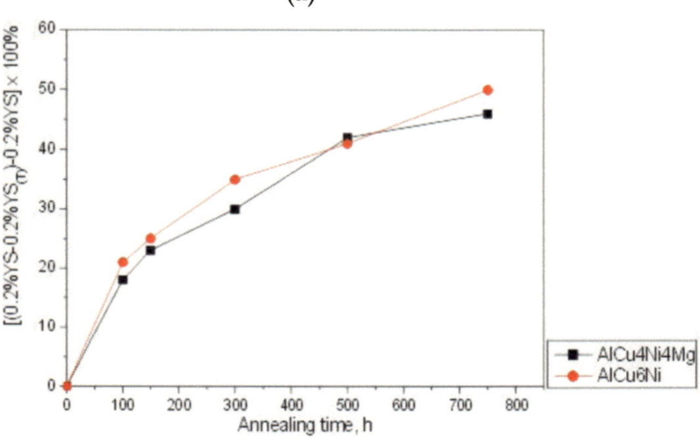

(b)

Fig. 22. Relative change of ultimate tensile strength (a) and 0.2% offset yield strength (b) of the AlCu4Ni2Mg and AlCu6Ni alloys as a function of time of annealing at 523 coefficient value on time of annealing enabled comparison of stability of mechanical properties of the investigated alloys (fig. 22 and 23).

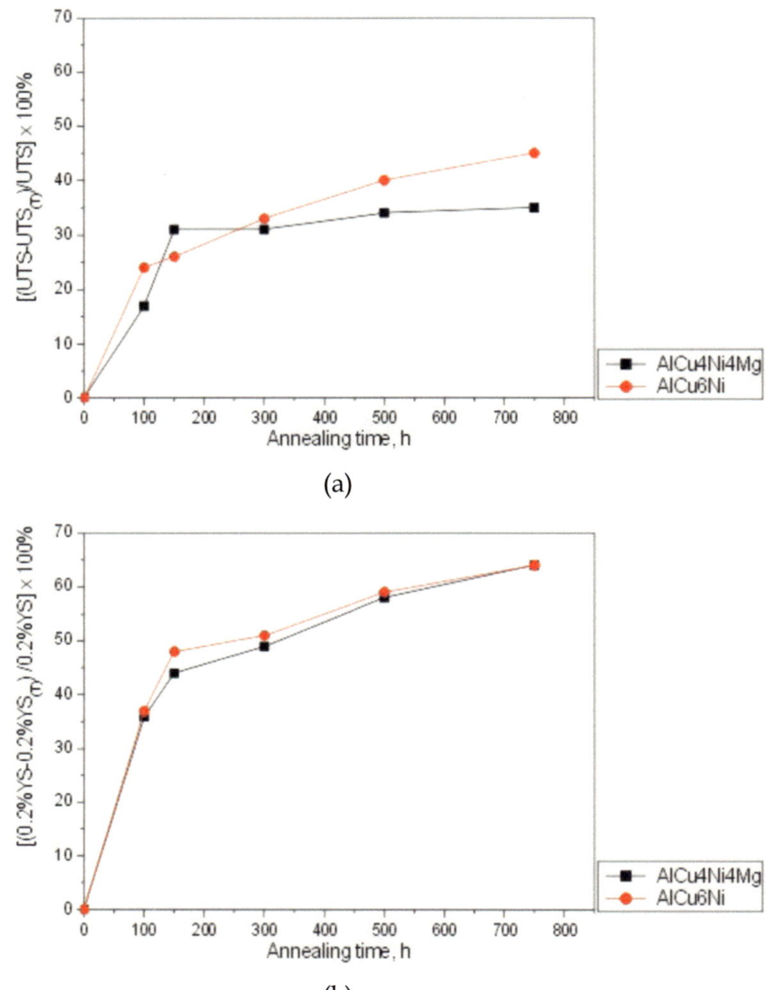

Fig. 23. Relative change of ultimate tensile strength (a) and 0.2% offset yield strength (b) of the AlCu4Ni2Mg and AlCu6Ni alloys as a function of time of annealing at 573K

Repeatability of the mechanical properties of AlCu4Ni2Mg and AlCu6Ni alloys after long-term annealing was determined on the basis of variation of the static tensile test results (table 5). Five specimens were tested for each temperature and time of annealing. Coefficient of variation was calculated using formula:

$$W_z = \frac{s}{\overline{x}} \times 100 \quad (1)$$

where: s – standard deviation, $\overline{x}$ – average value

Values of the ultimate tensile strength and 0.2% offset yield strength of the alloy subjected to long-term thermal loads (573K/750h) characterize its ability to preserve strength properties in operation condition of the castings (table 5).

Annealing temperature, K	Alloy	0.2% offset yield strength		W_z,%
		$YS_{(max)}$, MPa in T6 condition	$YS_{(min)}$, MPa after annealing for 750h	
523	AlCu4Ni2Mg	305	164	18
	AlCu6Ni	285	142	11
573	AlCu4Ni2Mg	305	110	25
	AlCu6Ni	285	104	18

Table 5. Minimum values of the 0.2% offset yield strength of the AlCu4Ni2Mg and AlCu6Ni alloys after annealing at 523/573K for 750h and maximum values obtained for T6 condition

Both alloys exhibit similar repeatability of tensile test results, however AlCu6Ni alloy shows slightly better stability of strength properties (table 5). However AlCu4Ni2Mg alloy is superior to AlCu6Ni alloy in terms of maximum and minimum yield strength after particular heat treatment. It has also higher ultimate tensile strength.

4. Conclusions

In the AlCu4Ni2Mg and AlCu6Ni alloys degradation of the microstructure takes place as a result of long-term thermal loading. It consists largely in coarsening and the change of the shape of hardening phase particles (θ'-Al_2Cu). The changes are proportional to the annealing time and temperature and lead to significant decrease of the mechanical properties of the alloys. The alloys studied are characterized by different content of Cu - primary element forming hardening phase. Increased Cu content in AlCu6Ni alloy caused only slight improvement of the stability of its strength properties. The AlCu4Ni2Mg alloy containing less Cu but with addition of Mg is characterized by better strength properties than AlCu6Ni alloy in T6 condition and preserves relatively high tensile strength and good ductility after long-term thermal loading. Taking into account criterion of mechanical properties and their stability both alloys studied can be successfully applied for highly stressed elements of aircraft structures operating in the temperature range of 523-573K.

5. References

El-Magd, E. & Dünnwald, J. (1996). Influence of constitution on the high-temperature creep behavior of AlCuMg alloy. *Metallkunde*, Vol.506, pp.411-414

Hirth, J.P. & Lothe, J. (1968). Theory of dislocations. McGraw-Hill, New York-London

Martin, J.W. Preciptation Hardening. (1968). Pergamon Press, Oxford

Mrówka-Nowotnik, G., Wierzbińska, M., & Sieniawski J. Analysis of intermetallic particles in AlSi1MgMn aluminium alloy. (2007). *Journal of Archieves in Materials and Manufacturing Engineering*, Vol.1-2, No.20, pp.155-158

Person, W.B. (1997). A Handbook of Lattice Spacing and Structures of Metals and Alloys, Vol.2, Pergamon Press, Oxford-London-Edinburgh-New York-Toronto-Sydney-Paris-Braunschweig

Wierzbińska, M. & Sieniawski, J. (2010). Microstructural changes to AlCu6Ni1 alliy after prolonged annealing at elevated temperature. *Journal of Microscopy*, Vol.237, No.3, pp.516–520

8

PIII for Aluminium Surface Modification

Régulo López-Callejas et al.*
Instituto Nacional de Investigaciones Nucleares, Plasma Physics Laboratory
A.P. 18-1027, 11801, México D. F.
México

1. Introduction

Aluminium is the third more abundant element in the Earth crust. The metal exhibits useful properties such as low density, high strength, good formability and a high resistance to corrosion. Aluminium can gain significant mechanical strength by means of alloying, whereby it is the most used metal after steel. In this sense, aluminium properties depend on its purity and its crystalline structure is face centred cubic [Wang et al, 1999].

Aluminium is, among other characteristics, malleable, easily machined and very ductile. Its high sensitivity to oxidation endows it with a waterproof passivation layer, typically 5-20μm thick according to the prevailing humidity, considerably adherent, which contributes to corrosion tolerance and general durability. The passivation layer consists of the amphoteric aluminium oxide Al_2O_3, often known as *alumina* or *aloxite* in mining and materials science. As corrosion is a major source of failure in Materials Engineering, aluminium is an obvious choice to face aggressive environments, including the atmospheric one.

Aluminium as a pure element has a low mechanical resistance which prevents its application under deformation and fracture conditions. Thus, low density combined with good resistance make aluminium alloys very attractive in design considerations. The properties of these alloys depend on a complex interaction among chemical composition, microstructural failures in solidification, thermal treatments, etc. although an increase in the alloy content tends, in general, to diminish the tolerance to corrosion. That is why quenching processes have been developed to improve the response to corrosion of highly alloyed materials. It is essential to select the precise alloy to match the resistance, ductibility, formability, solubility, corrosion tolerance, etc., required by an application.

Modifying aluminium composition by the adding nitrogen in an ion implantation process provides the treated samples with surface hardness and improved tribological properties by heating them in a nitrogen rich atmosphere. In this way, at low doses, aluminium nitride (AlN) becomes structured in the shape of clusters, the nitride content clearly increasing with the dose. Ion implantation is applied to pieces subjected to major friction and load forces such as rolling tracks, cylinder sleeves, etc., which require some core plasticity enabling

---

* Raúl Valencia-Alvarado[1], Arturo Eduardo Muñoz-Castro[1], Rosendo Peña-Eguiluz[1], Antonio Mercado-Cabrera[1], Samuel R. Barocio[1], Benjamín Gonzalo Rodríguez-Méndez[1] and Anibal de la Piedad-Beneitez[2]
[1]*Instituto Nacional de Investigaciones Nucleares, Plasma Physics Laboratory A.P. 18-1027, 11801, México*
[2]*Instituto Tecnológico de Toluca A.P. 890, Toluca, México*

them to absorb vibrations and impacts, but maintaining their high surface hardness against wearing and deformation [Manova et al, 2001].

AlN was first synthesised in 1877. It is mostly formed by covalent bonds and exhibits a hexagonal crystalline structure which is isomorphic to the wurzite form of zinc sulphide. AlN is stable at very high temperatures in inert atmospheres. Its surface oxidation in air takes place at 700°C, although 5-10 nm layers developed at room temperature have been detected [Selvaduray and Sheet 1993]. This layer protects the material above 1370°C. AlN is stable in hydrogen and carbon dioxide atmospheres even at 980°C. It dissolves slowly in mineral acids, which attack its grain borders, and in strong alkalis that react with the grain itself. AlN is gradually hydrolysed but it is resistant to several fused salts such as chlorides and cryolites.

The experimentation discussed in the present chapter concerns the alloy composition of aluminium 6061 containing Mg and Si as shown in Table 1. With a view to evaluating the content per element and assessing the crystalline phases identified by DRX in each aluminium 6061 sample, micographs (figure 1) and corresponding spectra (figure 2) were obtained by SEM. The diffractogramme of the control (reference) sample, seen in figure 3, presents aluminium peaks at the 2θ values: 38.47°, 44.74°, 65.13°, 78.23° and 82.43° (JCPDS 4-0787 standard) the last peak showing a greater intensity than that reported on Table 2. Aluminium oxide (Al_2O_3) can be detected at the 2θ angles 34.60°, 36.49°, 40.22° (JCPDS 12-0539 standard) and at 42.76° (JCPDS 24-0493 standard).

	%Si	%Fe	%Cu	%Mn	%Mg	%Zn	%Ti	%Cr	%other	%Al
6061	0.4-0.8	0.7	0.15-0.40	0.15	0.8-1.2	0.25	0.15	0.04-0.35	0.15	Balance

Table 1. Al 6061 composition

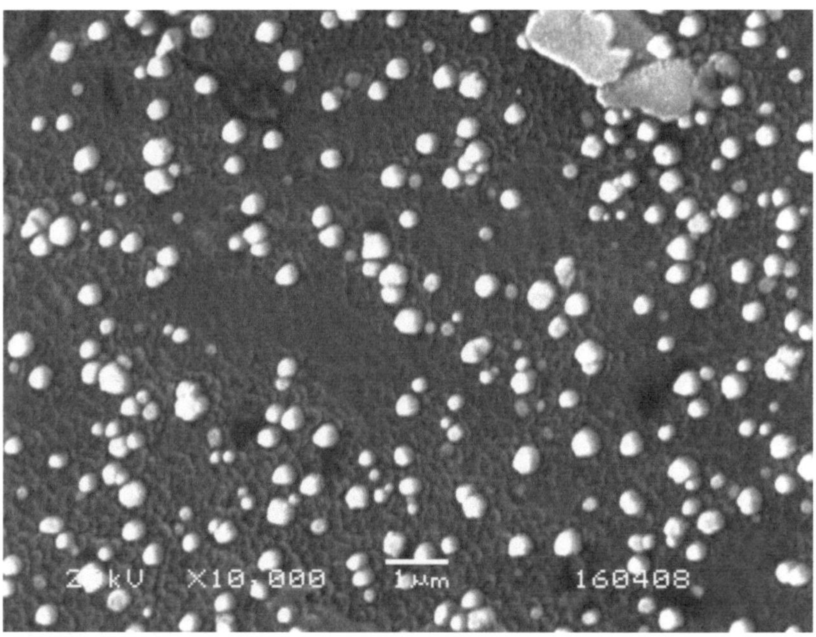

Fig. 1. Reference simple micrograph

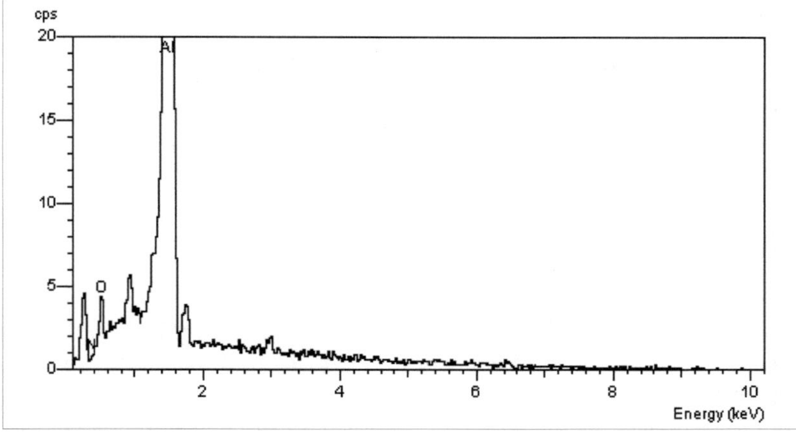

Fig. 2. Reference simple spectrum

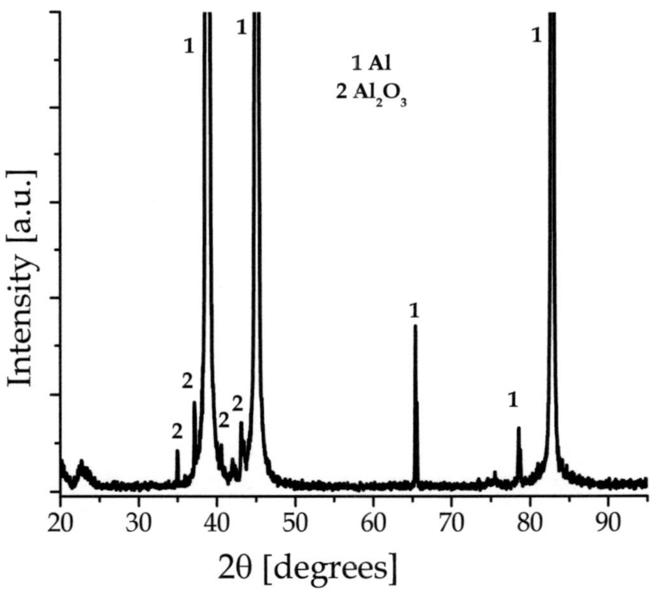

Fig. 3. Reference simple diffractogramme

Angle (2θ)	Relative intensity
38.47°	100
44.74°	47
65.13°	22
78.23°	24
82.43°	7

Table 2. Al peaks (JCPDS 4-0787 standard)

All in all, Aluminium and its alloys are attractive materials to the car, aviation, food and chemical industries as much as the pharmaceutical research. However, these materials lack surface hardness and other tribological qualities which limit their application nitriding is an effect time surface modification used to enhance corrosion tolerance in addition to wear resistance. In this chapter, a study of the formation of aluminium nitride (AIN) by means of the Plasma Implantation Ion Immersed (PIII) is presented.

2. Plasma immersion ion implantation process

Ion implantation based on linear accelerator technology has been long developed to modify material surfaces. The large implantation areas required by industrial applications and the extended processing times demanded by the treatment have made this implantation modality both expensive and complex, limiting its usefulness. By contrast, PIII technology [Conrad et al. 1987] overcomes many of the linear accelerator shortcomings providing a high ion density in a simple, fast, effective and economical way.

The PIII process has been amply described in several papers [Conrad et al. 1987], [Anders, 2000]. A brief description of it can follow. The sheath that normally surrounds an unbiased conductor (sample) submerged in plasma is characterised by an excess of electrons, no matter that the plasma is initially (t=0) in quasi-neutrality. When a voltage pulse, typically a few microseconds long, is applied to the "sample", the sheath is drastically altered and can even vanish momentarily. The comparatively small inertial mass of the electrons allows them to be expelled from the close vicinity of a cathode sheath, or negatively biased "sample", in a very short time. Consequentially, the ion array, or matrix, becomes exposed thanks to the ion's greater inertia. Later on, this charge distribution is enhanced as further electrons are repelled to the point in which the electric field of the biased "sample" is completely shielded. Thus, few centimetres away from the close vicinity of the "sample", the plasma remains unaltered, with the possible exception of the plasma waves created by the bias pulse.

The "sample" bias originates a short distance positive charge gradient and, with it, a potential gradient, namely, the electric field which impulses the ions towards the "sample". Once the ion matrix appears, a steady ion current flow onto the piece, to the extent of the availability of ions in the matrix. As the ions are implanted, the piece emits additional electrons according to its work function and, clearly, to the ion energy. The loss of these electrons extends the sheath by uncovering more ions. The bias pulse width and plasma density are usually adjusted in order, for as many sheath ions as possible, to be implanted into the piece, which is kept, therefore, immerse in the plasma. At the same time, the plasma represents a load to the high voltage pulsed energy supply which bias the work piece. By the end of the few microsecond pulse, the ion matrix is depleted and the system returns to very much the same initial conditions previous to t=0.

Conventional beam-line ion implantation has proven to modify significantly the surface properties of different materials. Nevertheless, PIII offers an alternative to conventional beam-line ion implantation. It has shown the advantages of relative simplicity, high ion fluence, the possibility of implanting complex three-dimensional objects, achieving an area treatment independent of the processing time and providing safe low temperature processing. By contrast, PIII is limited by the lack of charge to mass separation, having an implant energy distribution non homogeneous and the generation of X-rays from the production of secondary electrons.

3. Instrumentation for PIII process

This section contains a detailed description of the instrumentation used to carry out the PIII on the vacuum chamber with biased electrode, a specific high voltage modulator, and the diagnostic systems enabling to estimate the plasma parameters.

3.1 Vacuum chamber

To accomplish the PIII process, the device was designed and constructed as shown schematically in figure 4. The plasma is produced in a stainless steel cylindrical vacuum chamber 0.6 m high, 0.3 m in diameter and 5 mm thick in the wall; thus the vacuum chamber volume is 0.042 m^3. It has been provided with different ports for: a) pressure sensor, b) gas injection, c) electrode bias, d) target support, f) electrical probe diagnostics, g) spectroscopy diagnostics, along with other ports for future needs. The vacuum system consists of a turbo pump with a 500 l/s capacity.

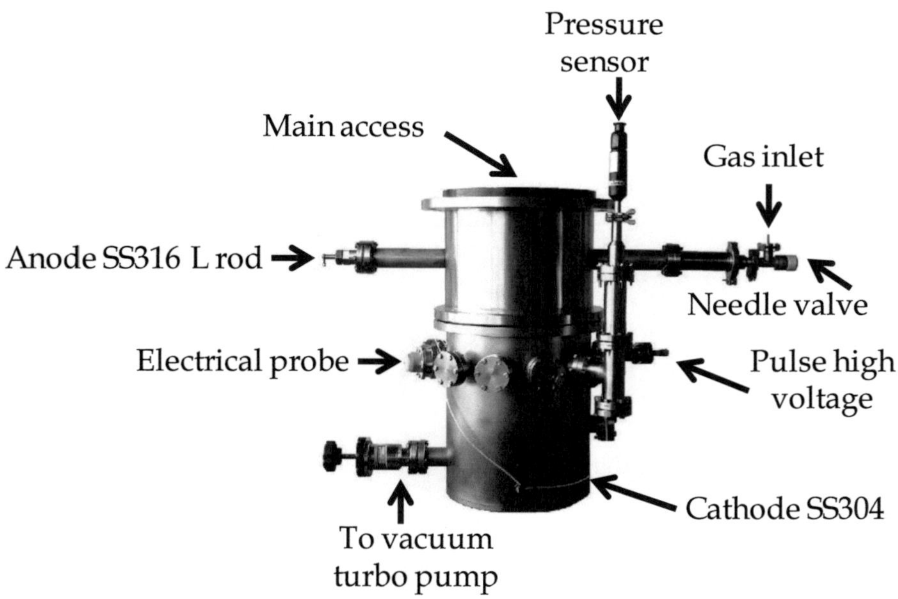

Fig. 4. General view of the vacuum chamber discharge and its main accessories

3.2 Plasma bias

The plasma is generated by a DC glow discharge between a stainless steel solid cylinder, acting as an anode, and the vacuum chamber as a cathode (figure 4). The DC power supply ranged within 0-1200 V/2A. The gas admission control to the vacuum chamber drives a gas dosing valve and the work gases being nitrogen and argon. The whole device is typically operated at a 1×10^{-6} Torr as base pressure and, during the PIII process, the work pressure falls into the 10^{-2} to 10^{-1} Torr interval.

3.3 Pulse generator chamber

The pulse CD supply (see figure 5) consists of a three phase full wave rectifying circuit (D1-D6) coupled to a Variac which enables the user to select the CD output voltage level. Later

on, the voltage is filtered so to supply the high voltage pulse transformer with a CD signal bearing the least possible ripple. The selected commutation device is an IGBT SKM200GB125D by SEMIKRON™ that is driven by an M57959L module. This is a high speed component endowed with a voltage logic level input and insulated by a high speed opto-coupler protected against the event of a short circuit. A commuted converter in a flyback configuration was chosen to build it, due to its relative simplicity and low cost.
The control stage has been implemented by a SG3527A pulse wave modulator (PWM) which imposes the width and repetition rate of the pulses applied to the M57959L module. According to the circuit configuration, the PWM work cycle can vary between 0 and 49% at a 55-2600 Hz rate. Such characteristics can be set up by the P1 and P2 potentiometers.

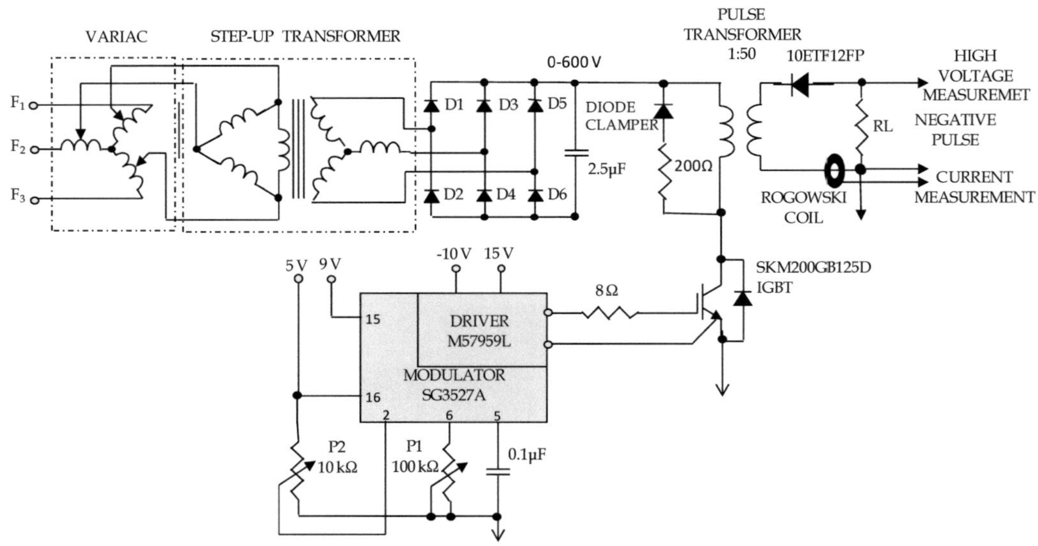

Fig. 5. High voltage pulse generator diagram

3.4 Electric probe diagnostics

In order to measure plasma parameters such as electron temperature (T_e) and plasma density (n) in a simple way, a double Langmuir probe is used. In order to increase the lifetime of the probe, a mechanical system (guard) was designed and constructed. This guard protects the electric probe within the chamber when the diagnostics is not being carried out. The probe is exposed to the plasma discharge for short intervals and only when the diagnostic system is activated. A probe was built out of an alumina rod with two perforations as the insulating element between two tungsten conductors and the metallic capsule that gives support to the probe (figure 6). The tungsten filaments are 0.195 mm in diameter and 4.3 mm long. The probe, inside the guard, is made of a stainless steel pipe, 0.95 cm in diameter and 25 cm long, intended to couple to the engine system that shifts the guard on and off.
This double probe was biased by means of a specifically designed and constructed triangular and sawtooth waveform generator operating in either modality thanks to the SW2 switch (see figure 7). An XR2206 function generator, and associated electronic components, was also used at the low voltage (±15V) stage so that a two scale frequency

output is obtained, due to the SW1 switch, between 0.2 and 20Hz and between 20 and 2000 Hz. The voltage is applied to the double probe in the order of ±150V by two STK4050V high voltage operational amplifiers connected in a differential way. Each amplifier was individually configured as an inverter with a gain factor of 47, given the relationships R17/R16 and R27/R26. Consequentially, the output peak value is ±75 V when 1.6Vp are applied at the amplifier input.

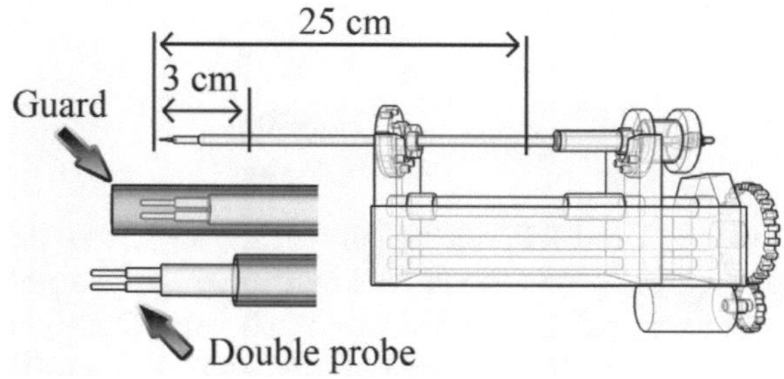

Fig. 6. Longitudinal positioning mechanism of the probe

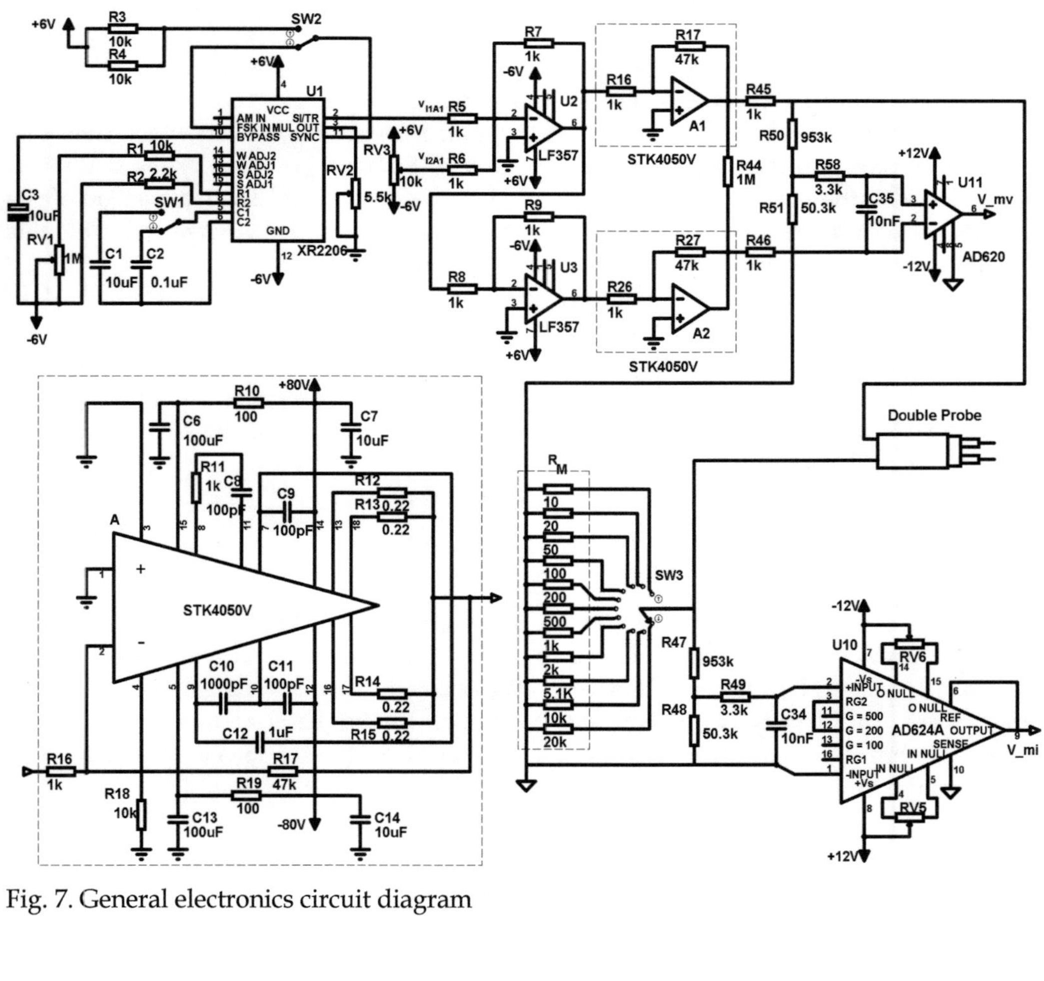

Fig. 7. General electronics circuit diagram

The data acquisition of the voltage and current signals from the signal conditioning stage was carried out with a DAQ PCI-6023E National Instruments board. The measurement of the current circulating through the plasma immerse probe was achieved by the use of an R_M, resistor. Thus, when the current varies between $\pm I_M$, a $\pm V_M$ voltage will be obtained in a directly proportional way. This voltage is located at the amplification stage input with a gain factor G_{AMP}=200. Then, assuming a maximal voltage limit at the board $V_{OA(MAX)}$, a sensor resistance R_M and an attenuation factor F_A, the expression of the maximal current interval $I_{M(MAX)}$ can be given by:

$$I_{M(MAX)} = \frac{V_{OA(MAX)}}{R_M F_A G_{AMP}} \tag{1}$$

The attenuation network (R47 and R48) of the current sensor where the low pass filter (R49 and C34) output signal is connected to the AD624 instrumentation amplifier. In this way, the I_M proportional V_M voltage signal is increased to a 200 gain. Once conditioned, the signal complies with the voltage specifications of the DAQ PCI-6023E acquisition board.

The probe displacement has been achieved by means of a longitudinal positioning mechanism allowing the guard, containing the double probe, to be introduced to the reactor as much as 25 cm while allowing exposing or retracting a 3 cm long tip of the probe with respect to the guard (figure 6).

A LabVIEW™ compatible program was specifically designed to operate the system. The software was applied to process, visualise and storage: the applied voltage, response current and probe positioning. The latter is set at the graphic interface (figure 8.a) which transmits the advance, stop and retreat signals, through an 8 bit terminal provided by the DAQ PCI-6023E board, to the power electronics associated to the mechanism. Then, a fraction of the collected current is selected in order to be plotted against the applied voltage and, from this characteristic curve, determining the main plasma parameters (figure 8b).

The saturation current and the electron temperature provide valuable information in determining the plasma parameters. Two values are calculated from the locus of the V-I plot, both from its positive and negative parts. The V-I double symmetrical cylindrical probe characteristic curve can be approached by the nonlinear function (Equation 1). The Levenberg-Marquardt fit method was implemented so to determine the coefficients of it [Herman and Gallimore, 2008]:

$$I(V) = I_{isat} \tanh\left(\frac{V}{2T_{eV}}\right) + A_1 V + A_2 \tag{2}$$

here, T_{eV} is the electron temperature [eV], I_{isat} is the saturation ion current [A] of each one of the probe ends, A_1 is an account of the expanded ion saturation current sheath depth, whereas A_2 refers to the reflection and displacement currents resulting from stray capacitances. The density calculation was performed on the basis of two types of data analysis: Bohm Approximation and Orbital Motion Limit (OML), given respectively by [Herman and Gallimore, 2008]:

$$n_i = \frac{I_{isat}}{0.61 A_S e}\sqrt{\frac{m_i}{kT_{eV}}} \tag{3}$$

$$n_{OML} = \sqrt{\frac{[-\Delta(I_i^2)/\Delta V_p]m_i}{0.2e^3 A_p^2}} \tag{4}$$

where A_S is the area of sheath [m2], e is the electron charge [C], m_i is the mass of ion [kg], k is Boltzmann´s constant [J/K], I_i is the ion current [A], V_p is the probe bias voltage [V], and A_P is the exposed probe electrode surface area [m2].

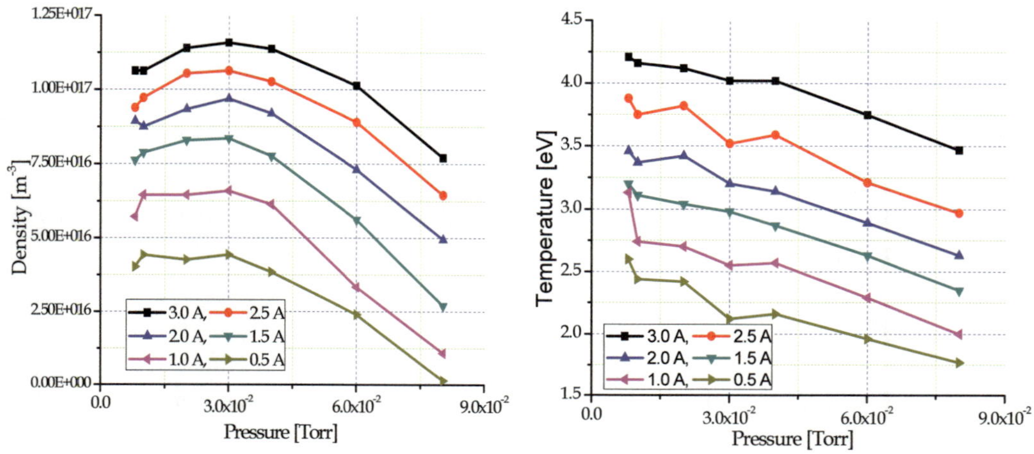

(a) Density (b) Temperature

Fig. 8. Nitrogen plasma parameters in DC

Instrumentation hardware and software have been calibrated on DC plasmas from argon, nitrogen and gas mixtures. Figure 8 displays some results obtained when the probe reached the centre of the reactor filled with nitrogen. The DC current supply went from 500mA to 3000mA, both cases being a function of the work pressure.

In a typical experiment under the previously specified plasma parameters, the following steps are conducted by means of the electric probe. First, the guard and electrical probe array are positioned inside the vacuum vessel, and, as a second step, the probe is moved outside the guard, putting it in contact with the plasma. When the electric probe stops, it is biased by one cycle of the sawtooth signal. With the electric probe system in position, it is possible to measure the electron density and temperature at different locations inside the vacuum chamber. The graphics shown in figure 8 correspond to the centre of the chamber.

3.5 The non collisional ion sheath model

Plasmas are ionised gases and, therefore, electrically conductive to some extent. In this way, plasmas are capable to shield regions of a scale estimated by the Debye length (λ_{De}) provided that the number of charged particles within the Debye sphere is far greater than one and that their motion obeys forces essentially electromagnetic. Other assumptions of the model are [Anders, 2000]:

- The ion flow is not collisional which apply to low gas pressures.
- Electrons are massless so that they respond instantly to the applied potentials given that the implantation time scale is much greater than the plasma cyclotron frequency ω_e.

- A bias $-V_0$ is applied to the piece at t=0, where $V_0 >> T_e$ (with T_e clearly expressed in volts) whereby $\lambda_{De} << s_0$ the latter being the initial sheath depth.
- A quasi-static matrix is formed instantly demanding a current which satisfies Child's law and is provided by the uncovered ions nearby.
- The transit time through the matrix is null, i.e., the implantation current is identical to the amount of ion charge uncovered per second.
- All charged particles are singly ionised.

In this manner, the Child's law current density at a voltage V_0, through a sheath of thickness s, can be expressed as

$$j_c = \frac{4}{9}\frac{\varepsilon_0 (V_0)^{3/2}}{s^2}\sqrt{\frac{2e}{M}} \tag{5}$$

where ε_0 is the vacuum permittivity, e is the electron charge and M the ion mass. By equating j_c with the amount of charge per unit of time and per unit of area that crosses the sheath border, $en_0(ds/dt)$, one can find the expansion speed of this border:

$$\frac{ds}{dt} = \frac{2}{9}\frac{{s_0}^2 u_0}{s^2} \tag{6}$$

where $s_0 = \sqrt{(2\varepsilon_0 V_0)/en_0}$ is the ion matriz thickness and $u_0 = (2eV_0/M)$ is the characteristic ion speed. Equation (6) becomes, after integration,

$$s(t) = s_0\left(1 + \frac{2}{3}\omega_i t\right)^{1/3} \tag{7}$$

where $\omega_i = \sqrt{e^2 n_0/\varepsilon_0 M} = u_0/s_0$ is the ion frequency of the plasma.
Estimating the ion matrix thickness during a PIII process is crucial as its size must not approach the reactor dimensions (see section 3.1) in order to have enough plasma to collect and implant ions from. Figure 9 illustrates the dynamic evolution of the ion sheath in the case of nitrogen (M=28) when the bias potential $-V_0$ ranges from 1kV to 8 kV, provided that the plasma density is 9×10^{16} m^{-3} (cf. figure 8.a).
By integrating Eq. (5), the ion fluence F impinging on the aluminium piece can be calculated. In a planar geometry with a maximum sheath width $s(t)$ and a voltage V_0 during a pulse of length t_p, the fluence is:

$$F = n\left(s(t) + t_p\sqrt{\frac{kT_e}{M}}\right) \tag{8}$$

Typically, in a 50 μs pulse, a plasma density of 9×10^{16} m^{-3}, T_e=3eV and voltage of 5 kV, the fluence density can reach up to 1.7×10^{16} ions per m^2. With a 500 Hz repetition rate, it is possible to implant doses in the order of 10^{21} ions/m^2 in ~1 hour.

4. Results and discussion

The experimentation was carried out on a commercially pure aluminium rod (6061-T6) sectioned into cylindrical pieces, 10 mm in diameter and 5 mm thick. The samples were mirror

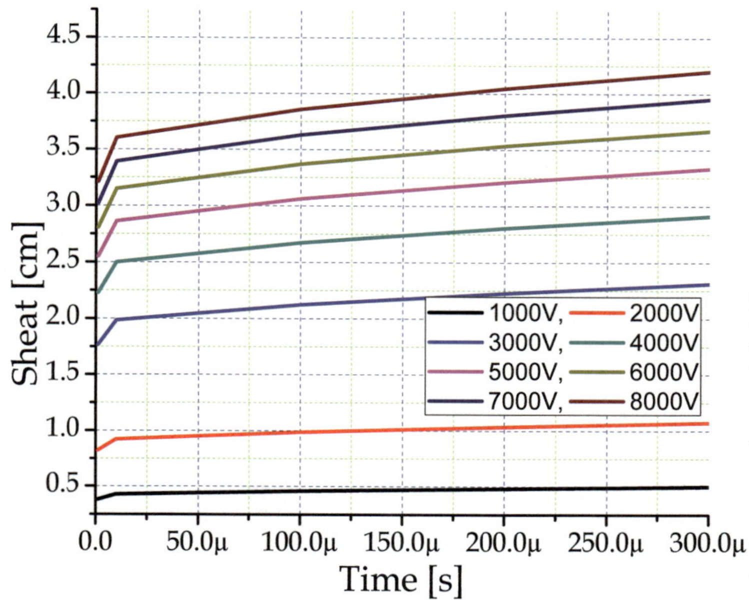

Fig. 9. Ion sheath evolution at a 9×10^{16} m^{-3} plasma density and biasing between -1 and -8 kV

polished and ultrasonically cleansed in acetone. A base pressure ~10^{-6} Torr was achieved with a turbo-molecular vacuum pump, and then the work pressure was established at 3×10^{-2} Torr (see figure 8.a) by admitting nitrogen of a 99.998% purity and its mixtures. Each sample receives a previous 30 min cleansing stay in Argon plasma to be finally implanted for 1-1.5 h periods. The PIII process was conducted with -2 to -5.5 kV bias squared pulses, with ion doses in the order of 1.7×10^{16} ions/cm^2 per pulse, while the electron plasma density is kept at about $9\text{x}10^{16}$ m^{-3}. The gas admission mixture was calibrated by using flow regulators. All the specimens were treated at 400°C. The samples were treated in four separate groups under previously optimised controlled conditions: the first one in 99.998%pure nitrogen (N), the second one in 70%Nitrogen and 30%Argon (70N-30A), the third one in 50%Nitrogen-50%Argon (50N-50Ar) and the last one in 30%Nitrogen-70%Argon (30N-70Ar) mixtures.

4.1 Aluminium treated at 2kV and 150 μs width pulse

The first treatment was applied during 1h periods at a 500 Hz repetition rate. The resulting micrographs are shown in figure 10 and the corresponding EDX spectra in figure 11. O, N and Al are always present in this specimen.

The micrograph in figure 10.d (N30-Ar70) presents a smoothed surface due to the intense bombardment with Ar, which is corroborated by the respective rugosity plot (figure 12). Likewise, the smoothing of the surface in micrograph 10.a follows from the N treatment. In the case of the N50-Ar50 mixture (figure 10.c) the grain size appears particularly inhomogeneous, with an average magnitude of 0.35 μm. Spiked grains of different sizes confirm the highest rugosity (R_a) values occurring in this lot of specimens. As the nitrogen bombardment intensity decreases, the hardness declines, except in the case of N50-Ar50 which displays the highest hardness (figure 13). This result could be explained from the absence of aluminium oxide in the N70-Ar30 and N50-Ar50 treated samples with respect the pure nitrogen and N30-Ar70 cases. All the treated samples improve their hardness with respect to that of the untreated one, with the exception of the N30-Ar70 case which exhibits only the cubic crystalline phase, while the rest do both the cubic and hexagonal phases.

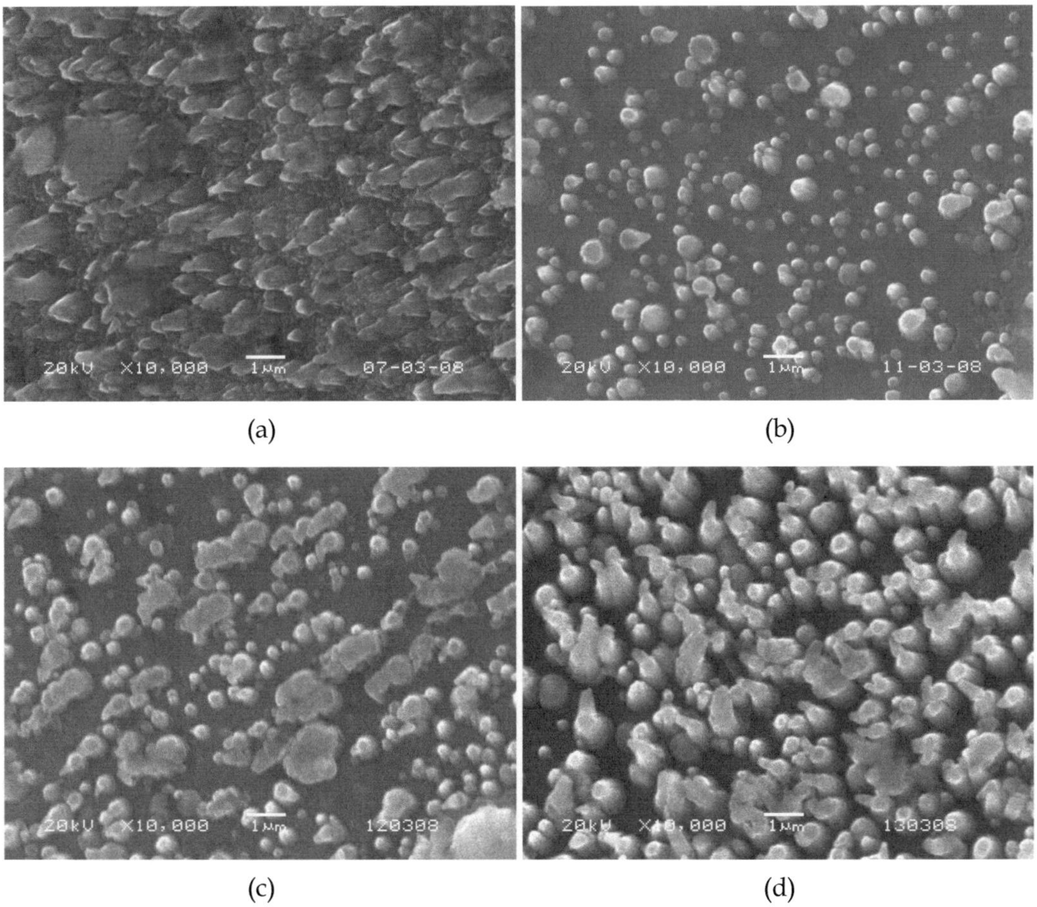

Fig. 10. Treatment micrographs: (a) Ar(N), (b) Ar/(N70-Ar30), (c) Ar/(N50-Ar50), (d) Ar/(N30-Ar70)

Diffractogramme 14.a, corresponding to N, shows the highest presence of AlN, both in cubic and hexagonal phases. Figure 14.c identifies the highest content of AlN in the cubic phase, as the presence of Ar seems to promote this phase and to inhibit the hexagonal one. This fact is due, perhaps, to the catalytic potential of argon, even though there is a competition between sputtering and implantation. In figure 14.b and 14.d, one observes only a small peak of the cubic phase of AlN. Thus, the high treatment temperature (400°C) may have changed the surface microhardness unfavorably, compared with the untreated case. The insignificant content of phases may be due to the relatively high sputtering produced by Ar in contrast with the nitrogen implantation.

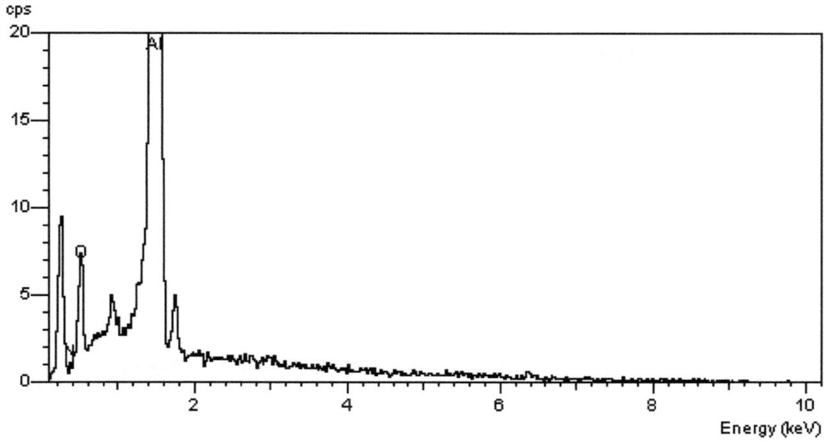
cps
20
15
10
5
0
Al
O
2
4
6
8
10
Energy (keV)

(a)

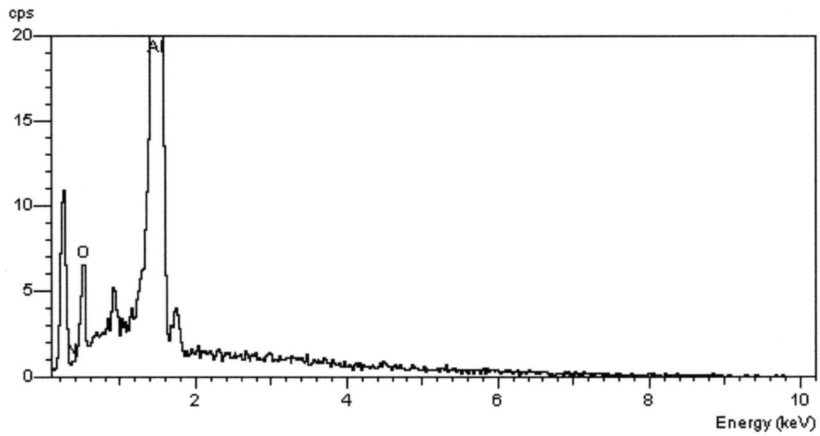
cps
20
15
10
5
0
Al
O
2
4
6
8
10
Energy (keV)

(b)

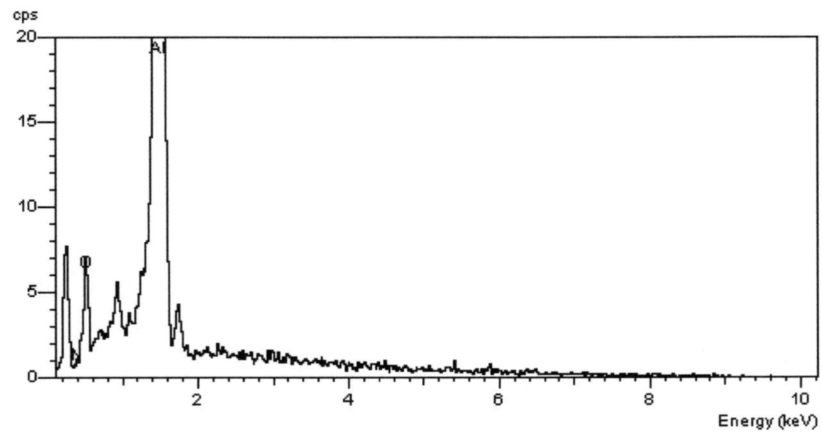
cps
20
15
10
5
0
Al
O
2
4
6
8
10
Energy (keV)

(c)

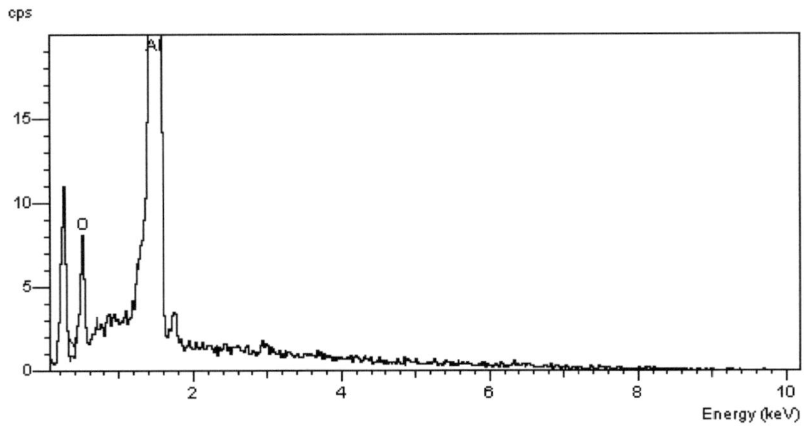

(d)

Fig. 11. Aluminium spectra: (a) Ar(N), (b) Ar/(N70-Ar30), (c) Ar/(N50-Ar50), (d) Ar/(N30-Ar70)

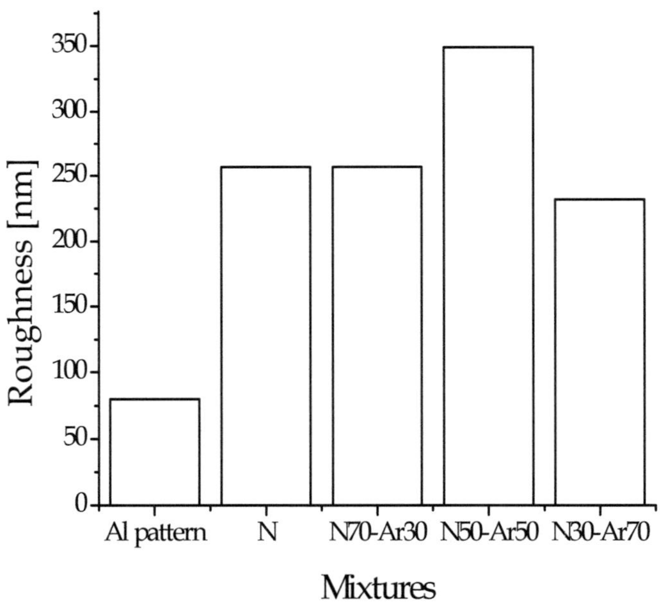

Fig. 12. Rugosity

4.2 Aluminium treated at 3.5kV and 75 µs width pulse

These tests were aimed at attracting the ions with greater energy by increasing the bias voltage while reducing the pulse width. The resulting micrographs are presented in figure 15. The maximal rugosity is identified in the sample treated with the Ar/(N30-Ar70) mixture (figure 15.c) given the respective rugosity tests (see figure 17). The changes in the morphology of the Ar/(N70-Ar30) and Ar/(N50-Ar50) treated samples, given that of the

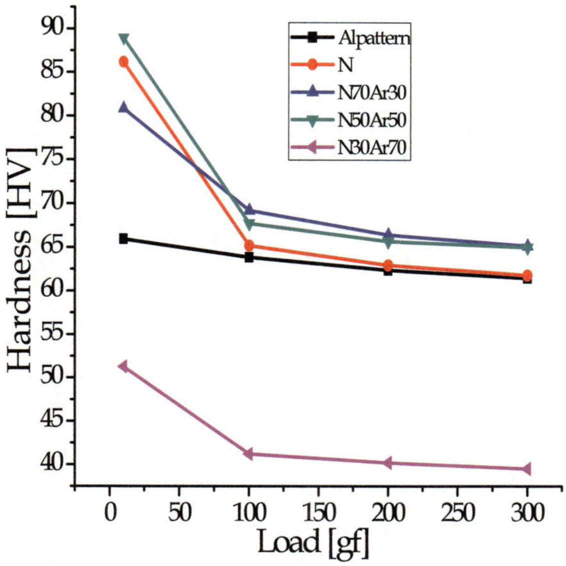

Fig. 13. Hardness

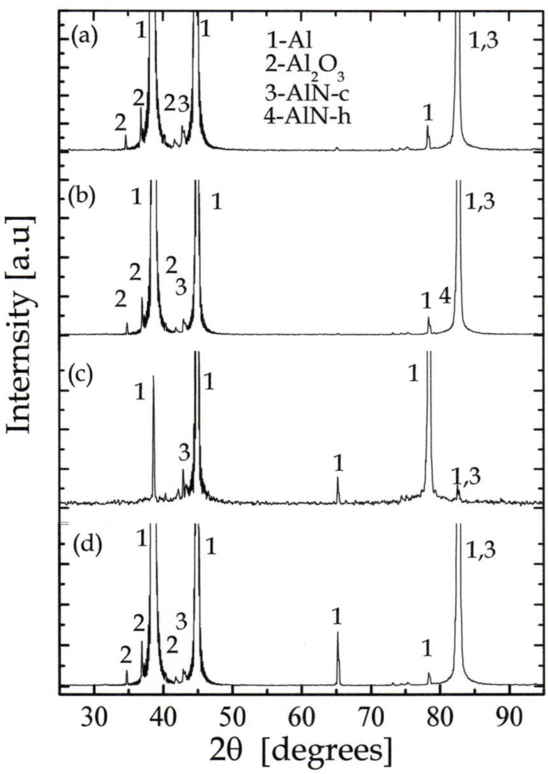

Fig. 14. Difractogrammes from: (a)Ar/(N), (b) Ar/(N70-Ar30), (c) Ar/(N50-Ar50) and (d) Ar/(N30-Ar70)

control one (figure 1) are evident. Figure 16 displays the EDS outcome suggesting the presence of N and an increase in O.

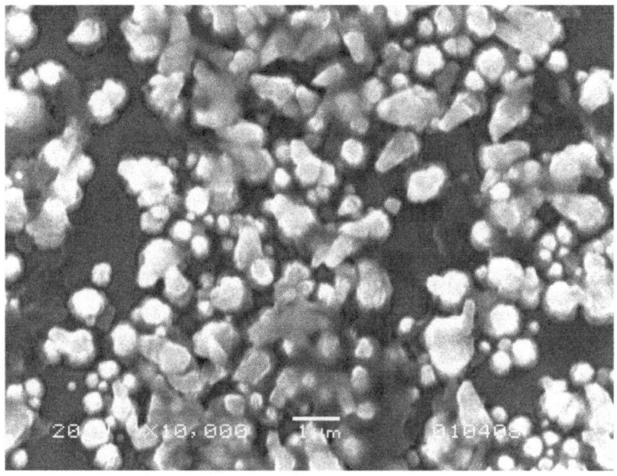

(a)

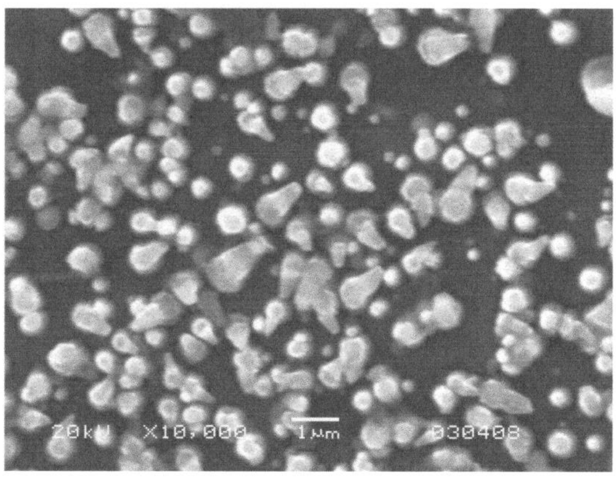

(b)

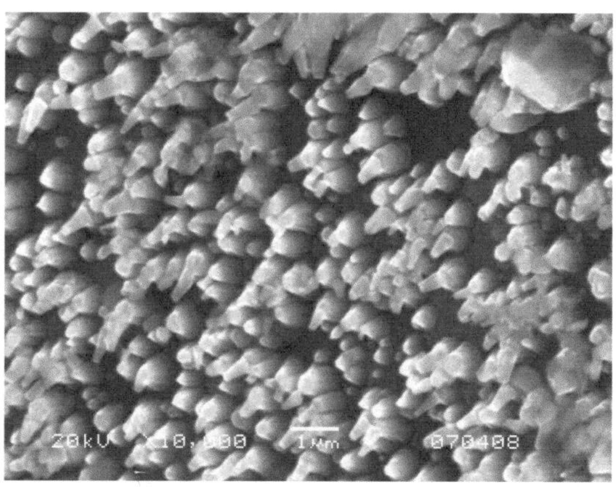

(c)

Fig. 15. Micrographs from: a) Ar/(N70-Ar30), b) Ar/(N50-Ar50), c) Ar/(N30-Ar70)

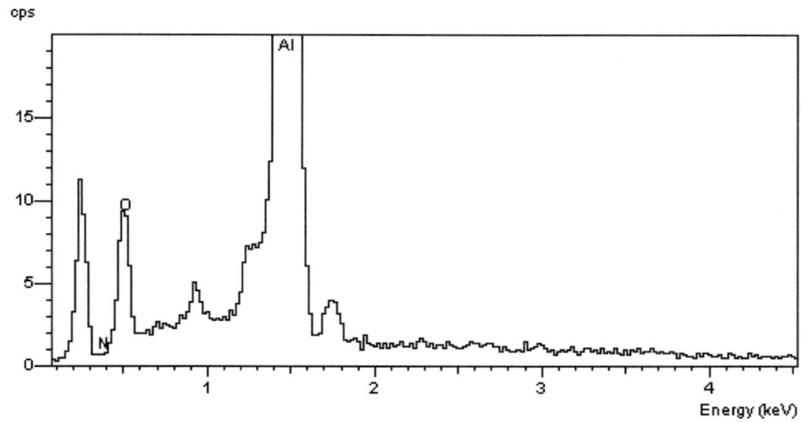

(a)

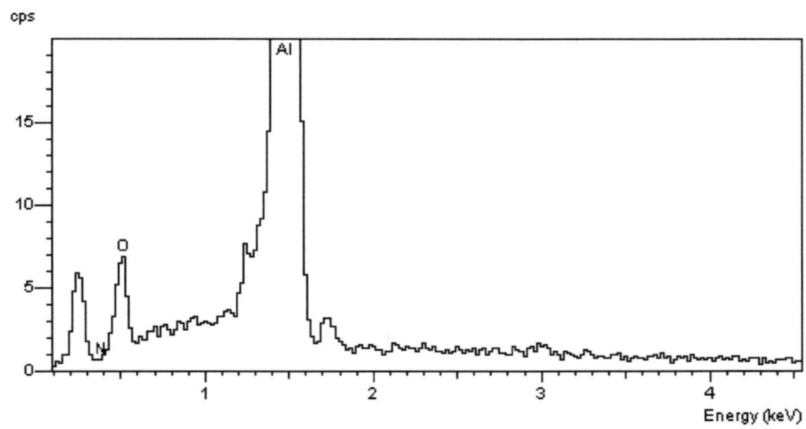

(b)

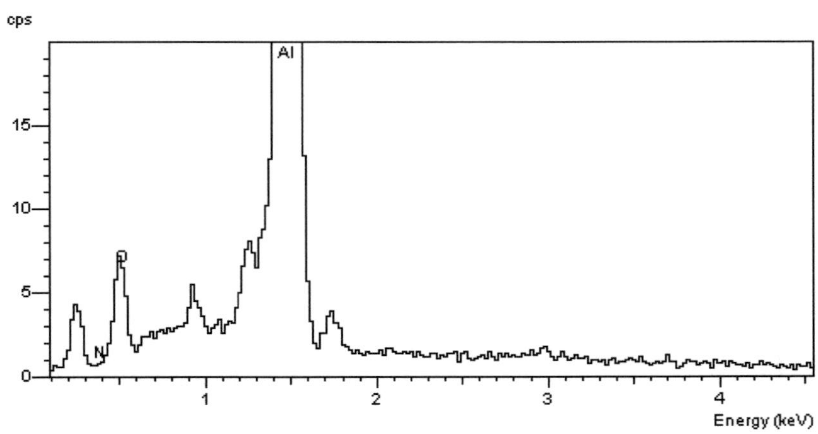

(c)

Fig. 16. Aluminium spectra from: a) Ar/(N70-Ar30), b) Ar/(N50-Ar50), c) Ar/(N30-Ar70)

Figure 17 shows that the greatest rugosity is achieved by the Ar/(N50-Ar50) sample which, at the same time, presents the lowest (290 nm) hardness (figure 18). The Ar/(N70-Ar30) sample reached an average value of 267 nm despite the maximal N concentration in the mixture and the consequent ion impact on the piece. The Ar/(N30-Ar70) sample obtained an R_a value of 195 nm: the lowest in the present experiment.

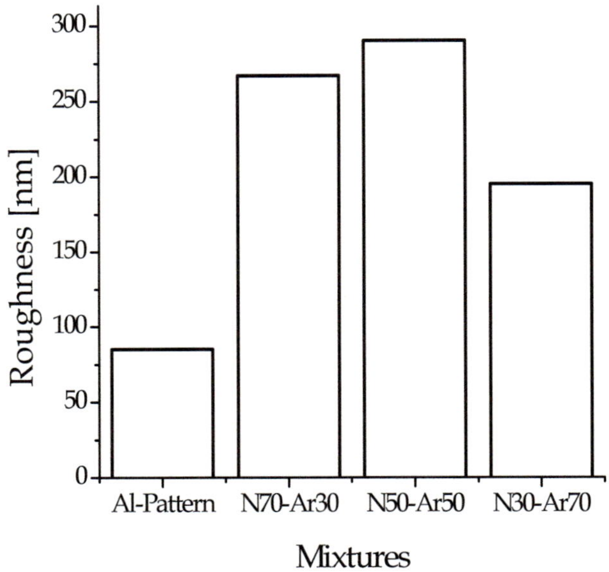

Fig. 17. Rugosity

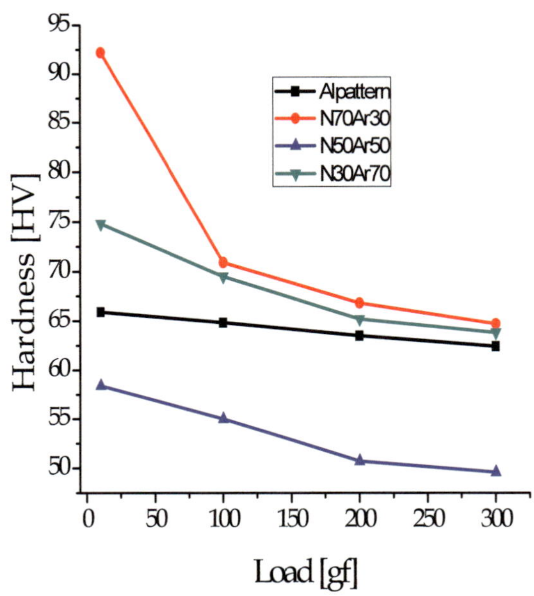

Fig. 18. Hardness

As follows from the hardness plot in figure 18, the maximal enhancement, up to 62HV0.1 at a 100 g load, was achieved with the highest nitrogen concentration mixture Ar/(N70-Ar30). By contrast, the Ar/(N30-Ar70) and Ar/(N50–Ar50) treated samples measured 60HV0.1 and 35HV0.1, respectively.

A comparative view of the diffractogrames shown in figure 19 indicates that the relative intensity of the Al peak at 2θ = 38.47° attained with the Ar/(N50-Ar50) mixture, decreases when the Ar/(N70-Ar30) and Ar/(N30-Ar70) ones are used (figures 19.a and 19.c). Quite the opposite with respect to the peak at 78.23° (figure 19.b) whose intensity is greater than those seen in figure 19.a y 19.c. The latter may be due to a low concentration of compounds like AlN and Al_2O_3 and to Al peaks either intrisically moderate or missing (such as the one expected at 2θ = 65.13°). Likewise, the Ar/(N50-Ar50) diffractogramme seen in figure 19.b, does not show the 2θ = 82.43° Al peak while the main Al peak is particularly reduced wich is ultimately attributable to the implantation process itself. The diffractogramme of figure 19.c points to the fact that a low N concentration favours the cubic phase of AlN at 2θ=41.80° (87-1053 JCPDS standard), 78.41° and 82.62° (46-1200 JCPDS standard).

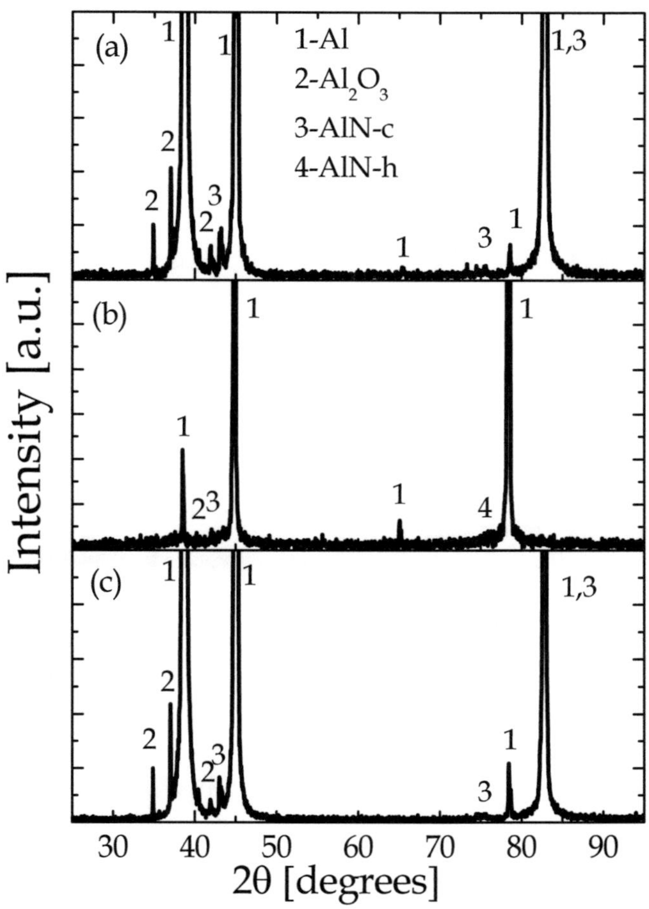

Fig. 19. Diffractogrammes a) Ar/(N70-Ar30), b)Ar/(N50-Ar50) c)Ar/(N30-Ar70)

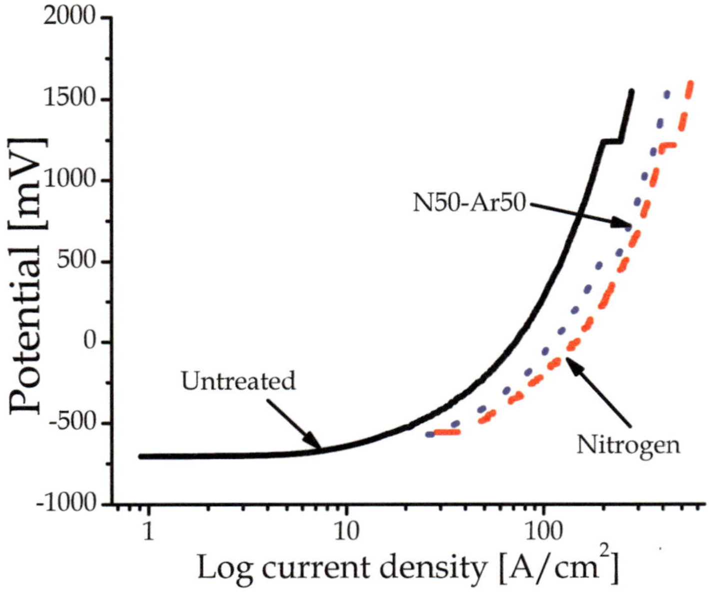

Fig. 20. Potentiodynamic polarization curves

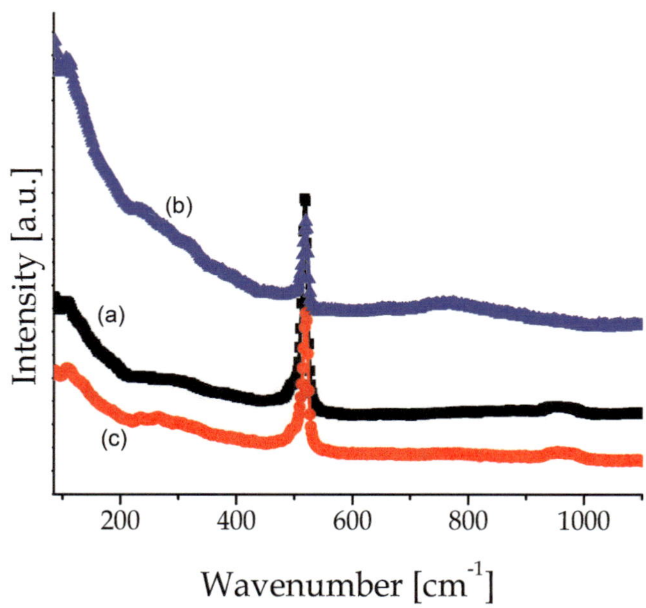

Fig. 21. Raman spectra (a) Nitrogen, (b) 50%N/50%Ar mixture and (c) 70%N/30%Ar

4.3 Corrosion analysis

Electrochemical potentiodynamic polarization tests were carried out within a cell containing 1 l of de-aerated 1.0N solution of H_2SO_4 as electrolyte. The measured results for untreated as well as treated samples are shown in figure 20. The more positive corrosion potential of the treated samples indicates a more electrochemically noble surface and enhanced tolerance to corrosion. The treated sample response is quite similar, irrespective of the temperature and duration of the treatment.

4.4 Raman analysis

The samples treated with a mixture of nitrogen-argon were analyzed by means of Raman Spectroscopy in order to validate the results obtained from XRD (He-Ne laser at a 632.8 nm wavelength). Figure 21 shows the resulting Raman spectra for AlN. The 514 cm^{-1} peak, typical of AlN, is always visible, which coincides with the information provided by XRD diffractogrammes. Likewise, the spectra displayed in figure 4.a., where the absence of aluminium nitride is evident, are confirmed. The results of XRD and Raman spectroscopy show that the general sample improvement depends on the amount of N in the Ar/N mixture concentration.

5. Conclusions

The mechanical improvement of aluminium alloys by conventional nitriding techniques is considerably complex whereas plasma immersion ion implantation provides a simple and effective way to enhance the wear resistance and corrosion tolerance of these alloys. The present study has contributed to the knowledge of the AlN structure in 6061-T6 aluminia following a low energy (2-6 KeV) PIII at a 75-150 μs pulse width. Such a process approaches the conditions of a glow discharge where the anode is a 15 cm long stainless steel rod, 3.5 cm in diameter, placed horizontally at the top of the cylindrical 304 stainless steel vacuum chamber 60 cm high and 30 cm in diameter which, in turn, plays the role of cathode of the discharge. The latter is fed by a DC power supply with a maximum output power of 1200 W, specifically designed and constructed from a current-source converter operating in a resonant mode. Several previously optimized work gas compositions were used: pure nitrogen, 30% argon/70% nitrogen, 50% argon/50% nitrogen, 70% argon/30% nitrogen providing ion doses in the order of 10^{21} ions/cm^2. Samples were implanted at ~400°C for ~1 h periods. X-ray diffractometry, scanning electron microscopy, Vickers microhardness tests, profilometry, corrosion and Raman spectroscopy methods were applied to evaluate the treatment outcome. The highest microhardness values were achieved with the equal part gas mixture and a voltage bias. The greatest roughness was obtained by increasing the implantation pulse width up to 150 μs with the same mixture. The roughness seems to remain invariant when pure nitrogen is used provided that longer time implantation periods are completed. Increasing the surface microhardness of aluminium without jeopardizing its average rugosity depends critically on selecting the correct Ar and N proportions to be used in a very low voltage PIII.

An ostensible improvement on the AlN microhardness results from the presence of the hexagonal crystalline phase and the elimination of the cubic one, the characteristic peaks of AlN in the near surface having been confirmed by Raman spectroscopy. A compromise is to be established between rugosity and microhardness through the several variables in the process.

The optimal conditions characterised in the present study for the PIII treatment of 6061T6 aluminium samples can be summarised as plasmas made out of mixtures of argon with, at least, 50% nitrogen, applied once the sample temperatures reach around 450°C. Thus, the implantation of the aluminium samples results in the surface formation of nitrides. X ray

diffraction of the implanted pieces reveals the presence of AlN in the cp and hcp crystalline phases where the peak intensities increase along with the nitrogen content. The presence of the hexagonal phase has been detected when either pure nitrogen or a 50% mixture have been used, suggesting a correlation between the h phase and the enhanced microhardness. Raman spectroscopy has confirmed the signature peak of AlN and, in addition to XRD, shows that the general surface improvement is enhanced with the N proportion in the Ar/N mixture concentration.
A compromise between high hardness and low roughness in pure nitrogen is observed due to a competition between sputtering and nitriding after, at least, 1 hour of treatment. In particular, maximal microhardness values were found in samples treated with the equal part mixture. The best roughness was achieved with this gas mixture in all cases, although increasing along with the implantation pulse width up to a 300 nm peak at 150 μs. Such a performance can be maintained in a pure nitrogen plasma, provided that longer 1 hour implantation periods are performed.

6. Acknowledgment

The authors are grateful to the technical collaboration received from Israel Alejandro Rojas Olmedo, Hannalí Millán Flores, Everardo Efrén Granda Gutiérrez, María Teresa Torres Martínez, Pedro Angeles Espinoza and Isaías Contreras Villa,.

7. References

Anders André, editor. (2000), "Handbook of plasma inmersion implantation ion and deposition". Ed. John Wiley and Sons, ISBN 0-471-24698-0, USA

Conrad, J. R., (1987) Sheath thickness and potential profiles of ion-matrix sheaths for cylindrical and spherical electrodes, *Journal of Applied Physics*, Vol.: 62, No. 3, pp 777 - 779

Conrad J. R., Radtke J. L., Dodd R. A., Worzala Frank J. and Tran Ngoc C. (1987) Plasma source ion-implantation technique for surface modification of materials, *Journal of Applied Physics*, Vol 62, No. 11, pp. 4591-4596

Herman D. A., Gallimore A. D. (2008) An ion thruster internal discharge chamber electrostatic probe diagnostic technique using a high-speed probe positioning system, *Review Scientific Instruments*, Vol. 79, 013302, 10 pages.

Manova D., Mändl S. and Rauschenbach B. (2001), Oxygen behaviour during PIII-nitriding of aluminium, *Nuclear Instruments and Methods in Physics Research Section B: Beam Interactions with Materials and Atoms*, Vol. 178, No. 1-4, pp 291-296.

Selvaduray G., Sheet L. (1993) Aluminium nitride: review of synthesis methods Source, *Materials Science and Technology*, Vol. 9, No. 6, pp 463-473

Wang J. A., Bokhimi X., Morales A., Novaro O., López T. and Gómez R. (1999). Aluminum Local Environment and Defects in the Crystalline Structure of Sol-Gel Alumina Catalyst, *J. Phys. Chem. B*, Vol. 103, pp. 299-303

9

Optimizing the Heat Treatment Process of Cast Aluminium Alloys

Andrea Manente[1] and Giulio Timelli[2]
[1]*Cestaro Fonderie Spa*
[2]*University of Padova, Department of Management and Engineering*
Italy

1. Introduction

The unfailing increased use of light alloys in the automotive industry is, above all, due to the need of decreasing vehicle's weight. The same need has to be taken into account in order to face up also both energetic and environmental requirements (Valentini, 2002). In terms of application rates, Al and its alloys have an advantage over other light materials. The reduced prices, the recyclability, the development of new improved alloys and casting processes, the increased understanding of design criteria and life prediction for stressed components and an excellent compromise between mechanical performances and lightness are the key factors for the increasing demand of Al alloys. A consolidated example of aluminium alloy employment regards the production of wheels, which, together with an improved aesthetic appearance, guarantees an improvement of driving, like directed consequence of the inertia reduction. These critical safety components are somewhat unique as they must meet, or exceed, a combination of requirements, from high quality surface finish, as wheels are one of the prominent cosmetic features of cars, to impact and fatigue performance. Due to their excellent castability and good compromise between mechanical properties and lightness, AlSiMg alloys are the most important and widely used casting alloys in wheel production (Conserva et al., 2004). Further, the increasing application of these alloys has been driven by the possibility to improve the mechanical properties of cast components through the use of heat treatments. Various heat treatments, e.g. different combinations of temperatures and times, have been standardized by Aluminium Associations and they are used in Al foundry depending on the casting process, the alloy type and the casting requirements (ASM Handbook, 1990). Standard T6 heat treatment is generally applied in wheel production. This heat treatment provides two beneficial effects for cast aluminium alloy wheels: an improved ductility and fracture toughness through spheroidization of the eutectic silicon particles in the microstructure and a higher alloy yield strength through the formation of a large number of fine precipitates which strengthen the soft aluminium matrix (Zhang et al., 2002). The T6 heat treatment comprises three stages (ASM Handbook, 1991): solution heat-treating, quenching and artificial aging.

Solution heat-treating at relatively high temperature is required to activate diffusion mechanisms, first, to dissolve Mg-rich phases formed during solidification and, then, to homogenize the alloying elements, such as Mg and Si, so as to achieve an elevated yield stress subsequent ageing (ASM Handbook, 1991). Further, the solution heat treatment changes the morphology of eutectic Si from polyhedral, or fibrous morphology in the modified alloys, to globular structure. Various efforts have been made to investigate the effects of solution temperature and time on microstructure and mechanical properties of AlSiMg foundry alloys (Zhang et al., 2002; Rometsch et al., 1999; Pedersen & Arnberg, 2001; Shivkumar et al., 1990a; Dwivedi et al., 2006; Taylor et al., 2000; Langsrud & Brusethaug, 1998; Cáceres et al., 1995; Cáceres & Griffiths, 1996; Wang & Cáceres, 1998).

Quenching is usually carried out to room temperature to obtain a supersaturated solid solution of solute atoms and vacancies, in order to achieve an elevated strengthening subsequent ageing (ASM Handbook, 1991; Liščič et al., 1992; Komarova et al., 1973; Totten et al., 1998; Totten & Mackenzie, 2000). The most rapid quench rate gives the best mechanical properties, but it can also cause unacceptable amounts of distortion or cracking in components (Auburtin & Morin, 2003). Thus, parts of complex shape, often with both thin and thick sections, are commonly quenched in a medium that provides a slower cooling. This quenchant can be hot water, an aqueous solution of polyalkylene glycol, or other fluid medium such as forced air or mist. In this way the heat transfer coefficient between the piece and the quenchant is reduced, the heat transfer from the surface is delayed and a more uniform temperature between the surface and the centre is obtained (Liščič et al., 2010; Totten et al., 1998; Totten & Mackenzie, 2000; Bates, 1987; Bates, 1993). Therefore, a balance between fast cooling and distortion minimization is required in quenched components.

Artificial ageing consists of further heating the casting at relatively low temperatures (120-210°C) and it is during this stage that the precipitation of dissolved elements occurs. These precipitates are responsible for the strengthening of the material. In AlSiMg alloys, the decomposition of the supersaturated solution begins with the clustering of Si atoms. This clustering leads to the formation of coherent spherical GP zones, consisting of an enrichment of Mg and Si atoms, that elongate along the cube matrix direction to develop into a needle shape coherent β'' phase. With prolonged ageing, the needle shaped GP zones grow to form rods of an intermediate phase, β', which is semicoherent with the matrix. The final stable β-Mg_2Si phase forms as an incoherent platelets on the α-Al matrix and has ordered face-centered-cubic structure. Several studies have been made to investigate the effect of artificial ageing temperature and time on strengthening mechanism of cast AlSiMg alloys. Ageing in the temperature range 170-210°C gives comparable peak yield strength (Rometsch & Schaffer, 2002; Alexopoulos & Pantelakis, 2004), and, with higher temperatures, the time to peak can be shortened. At ageing temperatures higher than 200°C, the β'' phase is substituted by the β', which contributes less to strengthening (Eskin, 2003).

It is of vital importance to consider both the foundry process and the T6 heat treatment on the whole, in order to achieve the required performances and specific properties (Merlin et al., 2009). While the benefit of T6 heat treatment is accepted, the additional cost and production time associated with such a heat treatment are substantial. Considering the whole production cycle of a standard automotive aluminium alloy wheel made by a low-pressure die-casting process (LPDC), the casting process normally takes less than 6 min, while a typical T6 heat treatment cycle may take more than 10 h. This means that shortening the total time of the T6 heat treatment cycle has a great impact on productivity and manufacturing cost.

In the present work, some process variables, which play a key role in production cycle of wheels have been investigated and improved. An integrated methodology for developing and optimizing the production and the final quality of A356-T6 18-inch wheels, in terms of casting distortion and hardness, is proposed. This study focuses on examining both the effect of cooling rate on wheel distortion and hardness during the post-cast and quenching steps, and the influence of the solutionizing temperature and time, and the powder coating cycles on the microstructure and mechanical properties of the 18-inch wheels.

2. Material and experimental techniques

An approach for optimizing wheel production has been applied on A356-T6 18-inch wheels, which are 5-spoke wheels in the T6 temper, with a diameter of 457 mm and a rim width of 203 mm. Fig. 1 shows a sketch of the analysed wheel, which is generally cast by LPDC. The casting has a weight of about 18 kg.

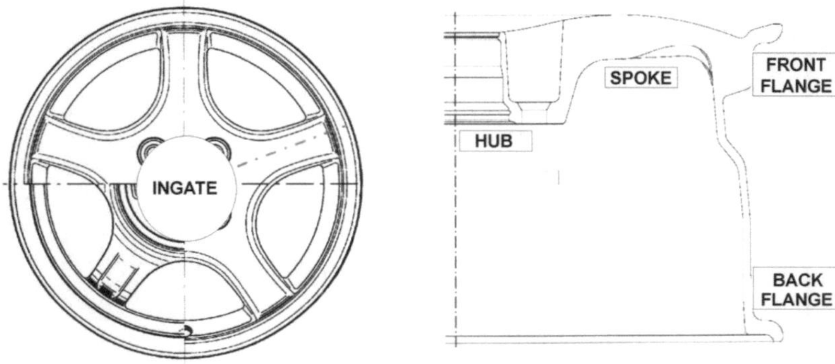

Fig. 1. Sketch of the low-pressure die-cast wheel analysed; the ingate is located in the hub region

2.1 Alloy and casting parameters

The cast wheels were produced with an AlSi7Mg alloy (EN AC-42100, equivalent to the US designation A356), whose composition is indicated in Table 1. The material was melted in a furnace set up at 730 ± 5°C. The melt was degassed with a rotary impeller by using nitrogen and modified with Sr-containing master alloy. AlTi5B1 rod type grain refiner was also added to the molten metal. The hydrogen level was evaluated before casting through a Reduced Pressure Test (RPT).

Alloy	Al	Si	Fe	Cu	Mn	Mg	Zn	Ti	Sr
A356	bal.	7.20	0.135	0.009	0.010	0.265	0.004	0.126	0.0279

Table 1. Chemical composition of A356 alloy used in the present work (wt.%)

The die cavity is geometrically complex and is comprised of four sections: a bottom die, two side die sections, and a top die. These die sections are made by an AISI H13 tool steel. The temperature in the die, measured with thermocouples, was in the range of 450-520 ± 10°C.
The casting process is cyclic and begins with the pressurization of the furnace, which contains a reservoir of molten aluminium. The excess pressure in the holding furnace forces the molten aluminium to fill the die cavity in 60 ± 4 s with a final pressure of 0.4 ± 0.015 bar. An overpressure of 1.2 ± 0.03 bar, reached after 10 ± 2 s from the end of the filling, was then applied for 210 ± 5 s. During solidification, cooling rates are controlled by forcing air (2–3 bar) through internal channels in the top and bottom dies, at various times during casting

cycle. On the side dies, cooling can be ensured by air jets, aimed at various sections of the exterior face. After the complete solidification, the side dies open and the top die is raised vertically. The wheel remains fixed to the top die prior to be ejected onto a transfer tray rolled under the top die. The die is then closed and the cycle begins again. Typical cycle times are 5-6 min. The wheel was then automatically picked up by a robot and cooled. To obtain a set of different cooling rates, water in the temperature range of 30-90°C was adopted. Slow cooling rate in air was also used.

2.2 Heat treatment and powder coating cycle

The wheels were T6 heat treated in an industrial plant, whose lay-out is shown in Fig. 2 (Manente, 2008). The lay-out consists of a one-way line, where the wheels, loaded in suitable steel frame (handling unit), follow and complete the whole heat treatment cycle.

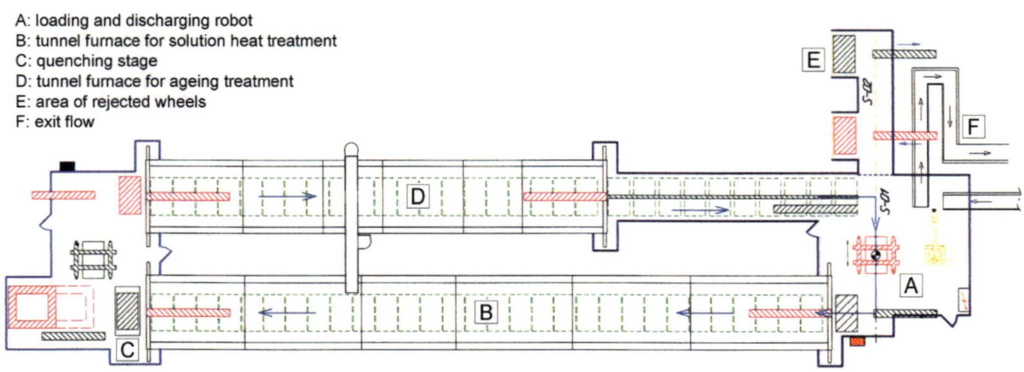

Fig. 2. Lay-out of the T6 heat treatment plant used in the present work (Manente, 2008)

A robot provides for loading 30 wheels in a five plane basket (Fig. 2 - Stage A). The basket is then moved into an air circulating tunnel furnace, where it is driven forward in 30 consecutive steps (Fig. 2 - Stage B). In the first 6 steps, the wheels are heated up to the set up solution temperature, while in the further steps they are maintained at temperature. The wheels were solution treated at 540 ± 5°C for 4, 5, 6, 7 and 8 hours (including heat up time) and immediately quenched (Fig. 2 - Stage C). The quenched delay was measured to be 20 s. To obtain a set of different quench rates, water at different temperature was adopted as quenchant. The water temperature ranged from 50 to 95°C. Slow quenching in air was also used. Table 2 shows the targeted and achieved quench water temperatures.

	Water temperature (°C)							
Targeted	50	60	70	75	80	85	90	95
Achieved	48	58	67	75	81	86	89	94

Table 2. Targeted and achieved temperature of water quenching

The wheels are subsequently transferred to an air circulating tunnel furnace, where they artificial aged (Fig. 2 - Stage D). This stage consists of 20 steps, where in the first 4 steps, the wheels are heated up to the set up ageing temperature, while in the further steps they are maintained at temperature. The wheels were artificially aged at 145 ± 5°C for 4 hours after solutionizing and water quenching (T6). This is a typical underageing treatment used in the manufacture of wheels. The rejected or sound wheels are finally moved to Stages E or F respectively, as indicated in Fig. 2.

After machining and cleaning operations, the wheels are generally powder coated and left inside an air electric furnace at 170 ± 5°C for 1 hour, including the heat-up time. Fig. 3. shows a typical thermal cycle of the wheels during powder coating. In the present work the effect of coating cycles has been studied by varying the number of cycles from 1 to 3.

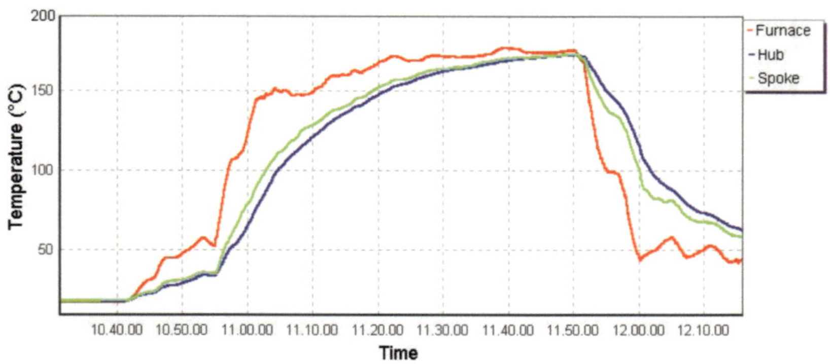

Fig. 3. Thermal cycle used for powder coating wheels; thermocouples are placed directly into the furnace chamber and embedded into the hub and the spoke region of the wheel

2.3 Microstructural characterization

Detailed microstructural characterisation of the as-cast and T6 heat treated wheels was carried out using an optical microscope and a scanning electron microscopy (SEM) equipped with an energy-dispersive spectrometer (EDS). The quantitative analysis of various phases in the microstructure were characterised using an image analyser software. The samples, drawn from the hub, the spoke and the rim region of the wheels, were mechanically prepared to a 3-μm finish with diamond paste and, finally, polished with a commercial fine silica slurry. Average secondary dendrite arm spacing (SDAS) values were obtained using the linear intercept method. A series of at least 10 photographs of each specimen were taken and several measurements were done, in order to obtain reliable mean values. To quantify the microstructural changes during solution heat treatment, the image analysis was focused on the size and shape factor of the eutectic Si particles. Size is defined as the equivalent circle diameter (d); the shape factor (α) is the ratio of the maximum to the minimum Ferets. To obtain a statistical average of the distribution, a series of at least 15 photographs of each specimen were taken; each measurement included more than 700 particles. The secondary phases, such as the Mg-rich particles and the Fe-rich intermetallics, were excluded from the analysis. Further, the polished specimens were chemically etched in a Keller etchant (7.5 mL HNO_3, 5 mL HCl, 2.5 mL HF and 35 mL H_2O).

2.4 Distortion and hardness testing

Brinell hardness measurements were carried out throughout the casting, on well defined locations, by using a load of 250 kgf, according to the standard ASTM E92-82. An average over 15 measurements was taken to evaluate the hardness of each wheel. Target hardness values after complete T6 heat treatment range between 90 and 95 HB.

The amount of distortions of the wheels was carried out after post-cast cooling (ε) and after quenching (ε_t), by using a circular gauge, which allows to calculate the maximum variation of the diameter of the wheel along the rim. Generally, the maximum accepted distortion of a wheel is 1.5 mm, while wheels with higher distortions are normally rejected. This is a typical standard used for wheel manufacturing (Manente, 2008).

3. Results and discussion

The methodology to analyse and optimize the quality of A356-T6 18-inch wheels, in terms of casting distortion and hardness, and to optimise the whole process manufacturing is based on different steps:

- analysis of as-cast wheels;
- analysis of solution heat treated wheels;
- analysis of quenched wheels;
- analysis of powder coated wheels.

3.1 As-cast wheel

3.1.1 Thermography measurements

A series of infrared (IR) thermographs was taken during wheel ejection from the top die and just prior the wheels were water cooled, to obtain 2D temperature maps of the casting surface. Fig. 4 shows an IR image of the wheel surface and the top die. The wheel stays on the transfer tray set under the top die. The top die shows a temperature between 435 and 495°C and the highest temperature is concentrated at the ingate, i.e. the hub region. These thermal values are comparable with the reading of the thermocouples which are located few mm under the die surface. The small differences can be related to the emissivity coefficient set up in the IR camera or some small variations in the experimental process parameters, influencing the thermal evolution and distribution of the die. High temperatures are localised in the die around the thickest regions of the casting, i.e. the hub and the spoke (to some extent), where the die receives a great quantity of solidification heat. Contrary, the surface temperature of the wheel shows a temperature range of 340-420°C, with the lowest temperature in the rim and the highest in the zone of the wheel between the spoke and the rim (Fig. 4). Under these conditions, the wheel was then automatically picked up by a robot and cooled.

Fig. 4. Infrared thermal mapping of the wheel during extraction from die

3.1.2 Microstructural observations of as-cast wheels

The microstructure of the modified A356 alloy consists of a primary phase, α-Al solid solution, and an eutectic mixture of aluminium and silicon. The α-Al precipitates from the liquid as the primary phase in the form of dendrites. The scale of microstructure in different zones of the wheel was characterized by means of SDAS measurements. The coarseness of the microstructure varied inversely with the casting thickness, i.e. the solidification rate. Typical microstructure of the as-cast wheel is shown in Fig. 5, referred to the hub, spoke and rim zones, corresponding to 55, 36 and 22 μm in SDAS respectively. Local solidification times (t_f) were estimated by means of SDAS measurements through the following relationship (Dantzig & Rappaz, 2009):

$$\mathrm{SDAS} = 5.5\left(-\frac{\Gamma_{sl}\, D_l \ln\left(\dfrac{C_{eut}}{C_0}\right)}{m_l(1-k_0)(C_{eut}-C_0)}\, t_f \right)^{1/3} \qquad (1)$$

where Γ_{sl} is the Gibbs-Thomson coefficient, D_l the diffusion coefficient in liquid, m_l the slope of the liquidus curve, k_0 the partition coefficient, C_0 and C_{eut} are the initial alloy concentration and the eutectic composition respectively. The solidification time was estimated to be 184 s in the hub, 52 s in the spoke and 12 s in the rim zone. The solidification sequence is approximately directional, starting at the outermost point of the wheel (rim) and continuing toward the centre of the wheel (hub), where the ingate is located.

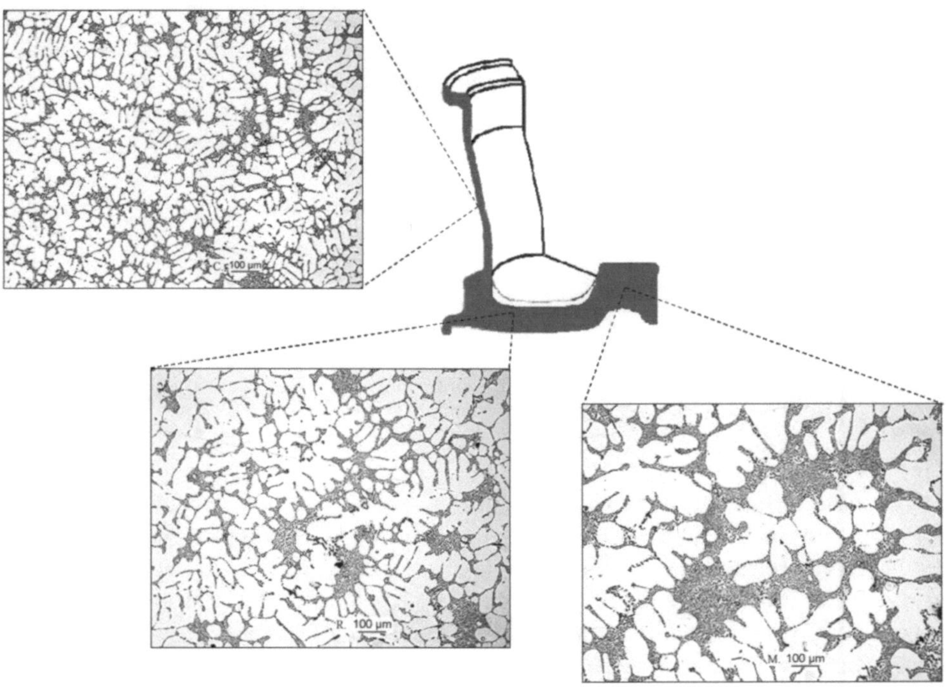

Fig. 5. Microstructure of as-cast wheel with reference to the different positions analysed

Coarse intermetallics compounds, such as Mg-rich particles and Fe-rich intermetallics, both in the form of coarse α-Al(FeMnSi) particles and needle-shaped β-Al_5FeSi, were also observed, especially in the hub region where the solidification rate is lower (Fig. 6).

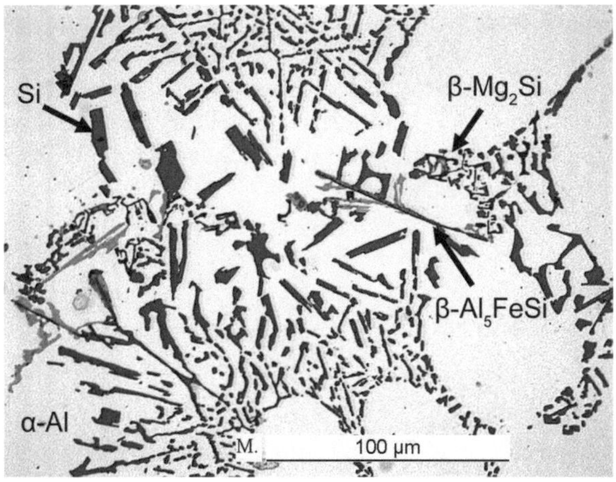

Fig. 6. Optical micrograph showing secondary phase particles in hub region; the eutectic silicon particles are present in the interdendritic channels, β-Al_5FeSi phase appear with typical needle shape and β-Mg_2Si particles as Chinese script

As it has been well established (Apelian et al., 2009; Kashyap et al.,1993), the eutectic Si phase in the microstructure of the Sr-modified alloy exhibits a fibrous morphology under as-cast conditions. The mean equivalent diameter *d* of eutectic Si particles increases approximately from the rim (~0.8 μm) toward the spoke (~1.6 μm) and the hub region (~2 μm). It was established that by reducing the cooling rate, the microstructure is characterised by coarse eutectic Si particles, while by reducing the solidification time the formation of a high number of fine Si particles is predominant. Further, the size distribution of Si particles was investigated in several studies (Tiryakioglu, 2008; Shivkumar et al., 1989; Grosselle et al., 2009) and found to follow the three-parameters lognormal distribution as follows

$$f(d) = \frac{1}{(d-\tau)\sigma\sqrt{2\pi}} \exp\left[\frac{-(\ln(d-\tau)-\mu)^2}{2\sigma^2}\right] \tag{2}$$

where d is the diameter of Si particles, τ the threshold, σ the shape and μ is the scale parameter.

3.1.3 Distortion behaviour in the as-cast temper

The different cooling media produced different amount of distortions in the 18-inch wheels. Generally, the distortion was in the range between 0.6 and 1.1 mm. Fig. 7 compares the wheel distortion induced by air or water cooling. It is evidenced how water temperature higher than 70°C produces similar distortions as air cooling (ε ~0.6 mm).

By increasing the water temperature, the amount of distortions decreases. This relationship has been estimated by linear regression analysis, using the coefficient of the determination R^2 to evaluate the quality of the least-squares fitting (Fig. 7). When R^2 is equal to 1, the fit is

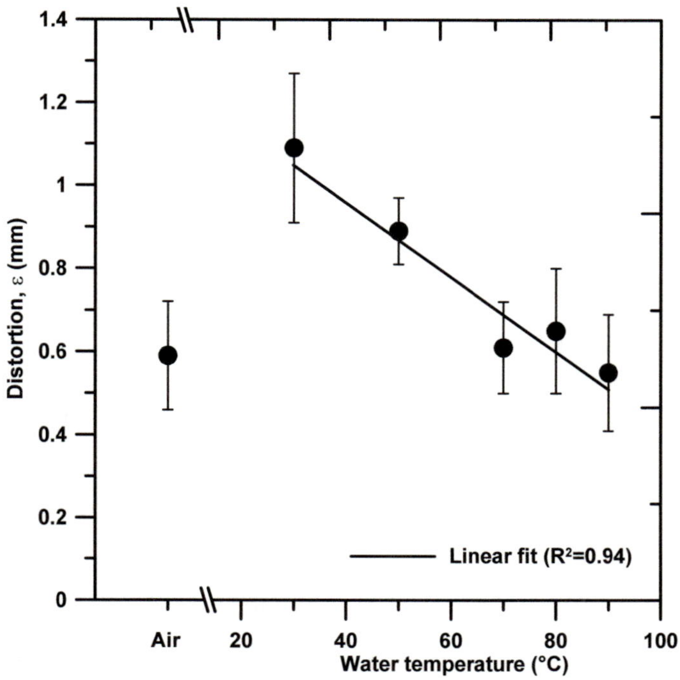

Fig. 7. Wheel distortion as a function of the cooling medium, i.e. air and water at different temperature, after ejection from the die; standard deviations are given as error bars

perfect. In the considered range of water temperature, the distortion ε can be described according to the following regression model ($R^2 = 0.94$):

$$\varepsilon = -0.009 \cdot T + 1.342 \quad (3)$$

where T is the water temperature in °C. Every wheel cooled in air and water shows a certain warp degree, which is more or less evident. The amount of distortions can reach critical level that compromises the functionality of the wheel too, as shown in Fig. 8. Residual

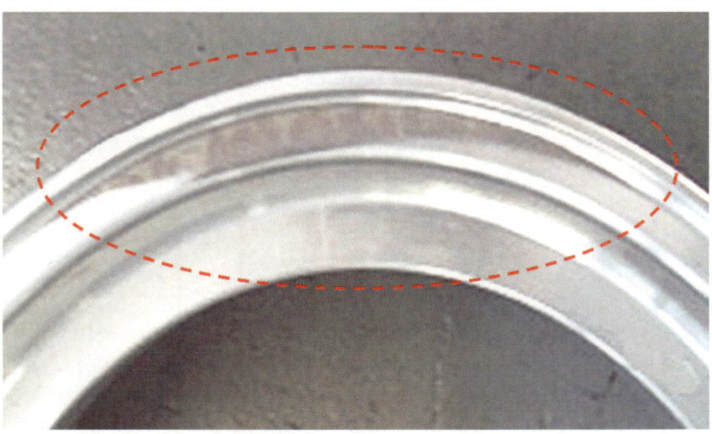

Fig. 8. Wheel distortion after post-cast cooling in water at a temperature of 30°C

stresses originate from differential thermal gradient and contraction during post-cast cooling. The wheel is extracted from the die at high temperature, as previously shown, and rapidly cooled. Therefore, the stress is so high that plastic deformation in the casting, free from the die, occurs. Generally, the casting distortion is more pronounced in casting ejected from the die at high temperature and in components showing drastic thickness changes (ASM Handbook, 1991). Further, higher the temperature difference between the casting and the cooling medium, greater will be the residual stresses and the casting distortion (Bates, 1987).

3.2 Solution heat treatment

3.2.1 Evolution of eutectic silicon particles

The influence of the solution heat treatment time on the microstructure of 18-inch wheels is shown in Fig. 9. The micrographs refer to the hub, which is the thickest zone of the wheel with a coarse microstructure, SDAS ~55 µm. In the range of solution temperature and times used, and due to the high diffusion rate of Mg in the α-Al matrix, the Mg bearing phases are completely dissolved and not more evident even in the coarse microstructure of the hub. These findings are in agreement with the results reported elsewhere (Rometsch et al., 1999; Zhang et al., 2002).After 4 h of solution heat treatment at 540°C, the Si particles become coarser and the interparticle distance increases (Fig. 9b). Rayleigh instability occurs; silicon particles undergo necking and are broken down into fragments. Due to the instability of the interfaces between the two different phases and a reduction in the total interface energy, spheroidization and coarsening processes occur. A prolonged solution treatment leads to extensive coarsening of the particles, with a small effect on the spheroidization level (Fig. 9c-f). The interparticle spacing increases too. Because the coarsening and spheroidization are diffusion-controlled processes (Greenwood, 1956), they are directly proportional to the solution temperature and time. These findings are in agreement with Meyers (Meyers, 1985). Further, previous results (Zhang et al., 2002) showed there exists a decrease in average Si crystal size after short solution heat treatment, before the average size increases. From the literature, the most severe coarsening of eutectic Si particles takes place between 25 and 400 minutes of solution treatment of the unmodified alloy, while the average particle size increased more evenly in the modified alloy (Pedersen, 1999). It has been stated that the typical growth rates for gravity die castings are in the range of 0.02 to 0.07 µm/h (Pedersen, 1999).

The results of the Pedersen's work on the quantitative variation in the Si particle size and shape factor of an AlSi7Mg0.3 alloy with similar microstructural scale as the hub of the wheel (SDAS ~54 µm) as a function of solution time are reported in Fig. 10. The Si growth is estimated in terms of variation of the equivalent radius with respect to $t^{1/3}$, as defined by the ordinary Lifshitz-Slyozow-Wagner model (Liftshitz & Sloyozov, 1961):

$$R^3 - R_0^3 = \frac{8}{9}\frac{D C_0 \gamma V^2}{R_{gas} T} t \tag{4}$$

where T and t are the temperature and time, respectively; R is the radius of the particle; R_0 is the initial radius at t=0; R_{gas} is the gas constant; V is the molar volume; C_0 is the equilibrium concentration of structures in matrix; γ is the surface energy of the particle; and D is the diffusion coefficient. The regression analysis leads to R^2 equal to 0.97, indicating the reliability of the model.

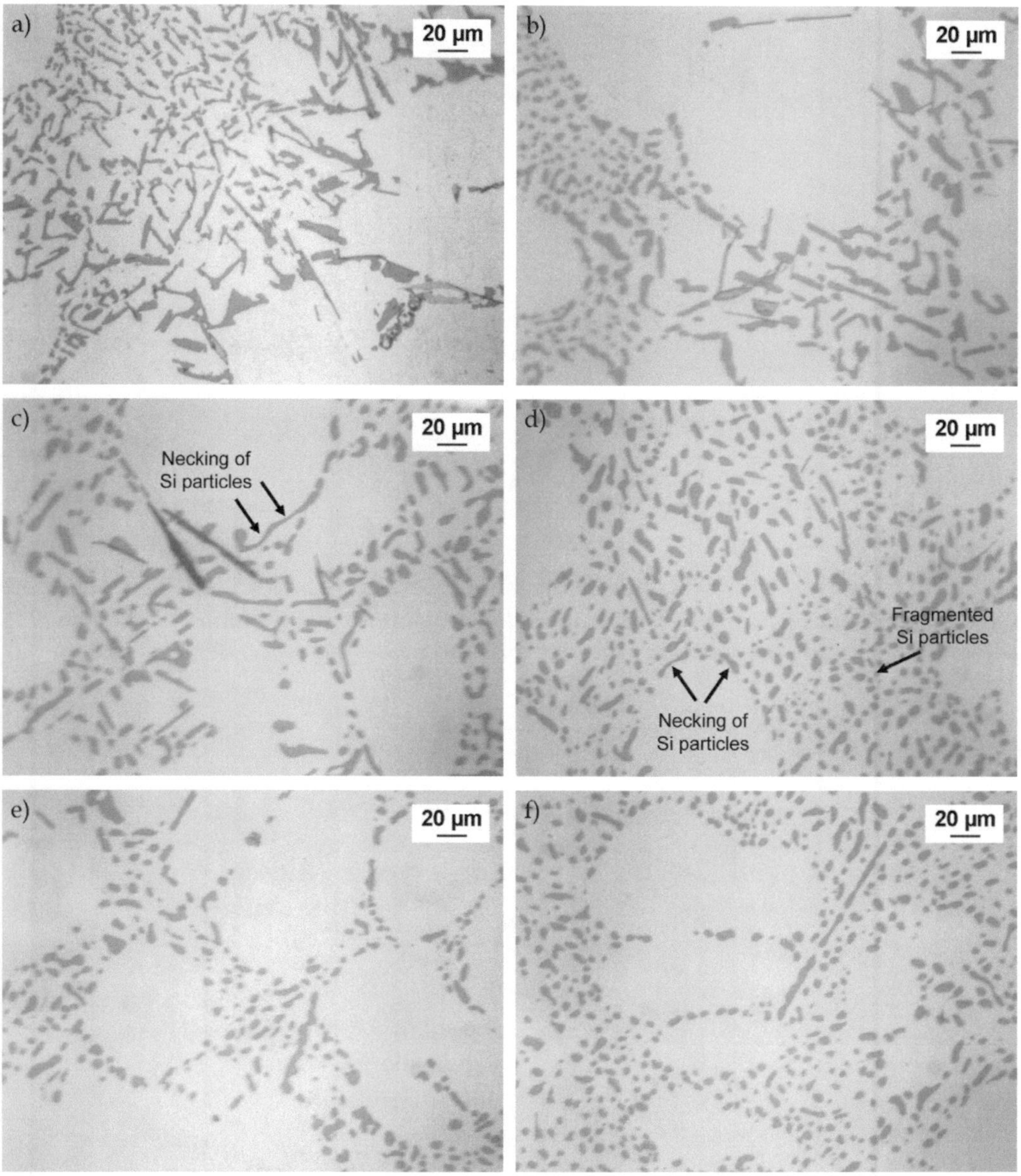

Fig. 9. Eutectic Si particles in the hub region of A356-T6 18-inch wheels; the alloy has been solubilised at 540°C for various time: (a) as-cast, (b) 4, (c) 5, (d) 6, (e) 7 and (f) 8 hours. Silicon particles undergo necking and are broken down into fragments, then, spheroidization and coarsening mechanisms occur

Fig. 10b shows the distribution of the eutectic Si particles as a function of the shape factor for samples heat treated at 540°C for various time. Pedersen observed how the particles undergo great changes in shape factor *a* distribution after short times (30 minutes) of solution heat treatment; the fraction of particles with a smaller *a* parameter is immediately reduced, while the number of particles with a greater *a* parameter is increased. Similar

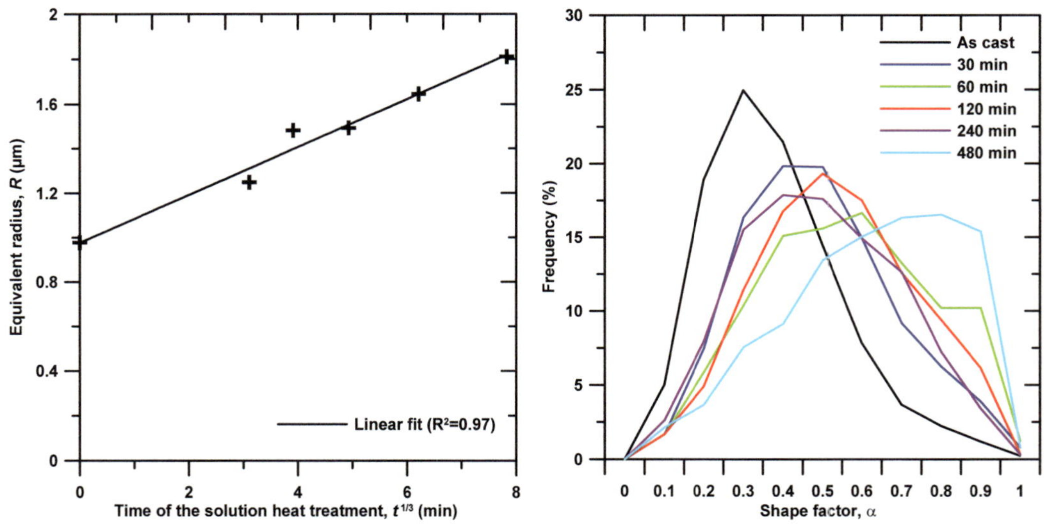

(a) (b)

Fig. 10. (a) Linear regression analysis of eutectic Si equivalent radius with $t^{1/3}$; the point zero in the time axis represents the as-cast condition; (b) frequency distribution of the shape factor *a* after solution treatment at 540°C for different times (Pedersen, 1999)

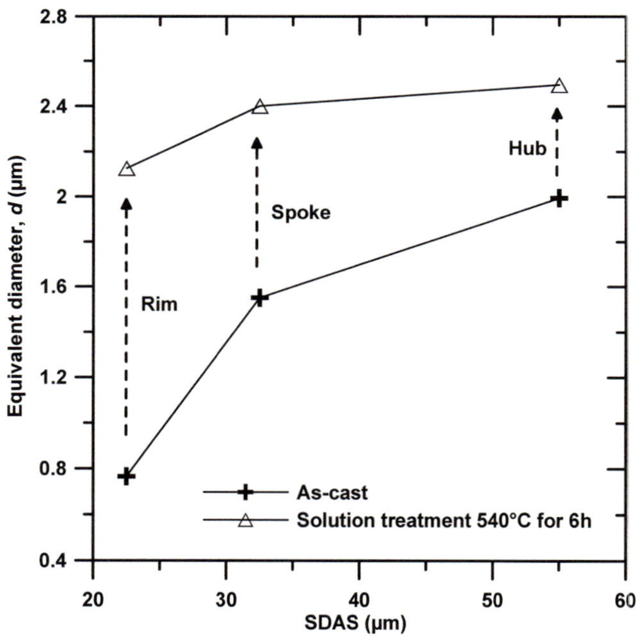

Fig. 11. Average diameter *d* of the eutectic Si particles as a function of SDAS; data refer to the different positions of the as-cast and solution heat treated wheels

changes in particle distribution are not observed by increasing the solution times within 4 hours, even if the distribution curves flatten with solution time and their peaks move to the

right toward higher *a* values. Only after 8 hours solution time, the shape factor distribution moves to higher *a* values. The eutectic Si particles in AlSi7Mg gravity-cast alloys crack progressively with increasing applied plastic deformation, and the crack is favourable for the larger and longer particles, even if the progression of particle cracking is more gradual in a finer microstructure (Cáceres & Griffiths, 1996). In addition, it was observed that the population of cracked particles is distributed according to the $a{\cdot}d$ parameter and is characterized by its average $a{\cdot}d$ value.
Since solidification rate has a dramatic effect on the size and morphology of eutectic Si particles, it is important to be aware of the influence of the solidification rate on the required minimum solution time for realizing the required coarsening and spheroidization. It was reported (Shivkumar et al., 1990c) that 3-6 h at 540°C is the optimal time for a Sr-modified sand-cast A356 alloy; while 30 min at 540°C is needed for a low-pressure die-cast Sr-modified A356 alloy with SDAS of 25 µm (Zhang et al., 2002). Fig. 11 shows the effect of a solution treatment at 540 °C for 6 h on the Si particle size in the different positions of the wheels ,where different microstructural scales were observed. The coarsening mechanism is faster in the rim and spoke region, where SDAS is about 22 and 36 µm respectively. While the coarse microstructure of the hub presents slower coarsening of Si particles, as indicated by the values of equivalent diameter in the as-cast and solution heat treated temper.

3.2.2 Partial melting

The increase of solution temperatures for the heat treatment of the wheels would be desirable since it increases the diffusion rate of Si atoms in the Al matrix, leading to rapid fragmentation and coarsening mechanism of eutectic Si particles, and, therefore, to shorten the total time of the T6 heat treatment cycle. It was demonstrated that for a given short solution treatment time of 9.5 minutes, increasing the temperature from 540 to 550°C the number fraction of Si particles with a diameter of greater than 1 µm increases by more than 10%. Similar changes in the distribution of the shape factor for Si particles are observed by increasing the solution temperature, that is the number fraction of the particles with a shape factor of greater than 0.5 increases by approximately 10% (Zhang et al., 2002). Earlier works (Shivkumar et al., 1990b) showed that extremely high coarsening occurred at temperatures greater than 540°C for A356.2 alloys. However, the major problem associated with higher heat treatment temperatures remains the liquid phase formation, which increases with temperature.
In the present work, the possibility to heat the wheels at higher solution temperature was evaluated. A Fourier thermal analysis was carried out to determine the evolution of the solid fraction during solidification of the A356 alloy used for wheel production. A detailed description of the equipment, the casting procedure, and the process parameters is given elsewhere (Piasentini et al., 2005). The relationship between fraction of solid (f_s) and temperature of solidifying A356 alloy is shown in Fig. 12 for a cooling rate of 1°C/s. With increasing solution temperature above 540°C (final solidification point), the amount of liquid phase (100 f_s) increases slowly at first and then rapidly near the Al-Si eutectic reaction of ~560°C, at which point the fraction of liquid (100-f_s) is about 15%.
At relatively lower solution temperatures, melting starts at grain boundaries and interdendritic regions. In alloys with a dendritic structure, local melting starts generally at interdendritic channels, since these often contain high concentrations of alloying elements/impurities. At higher solution temperatures, local melting may also start at grain

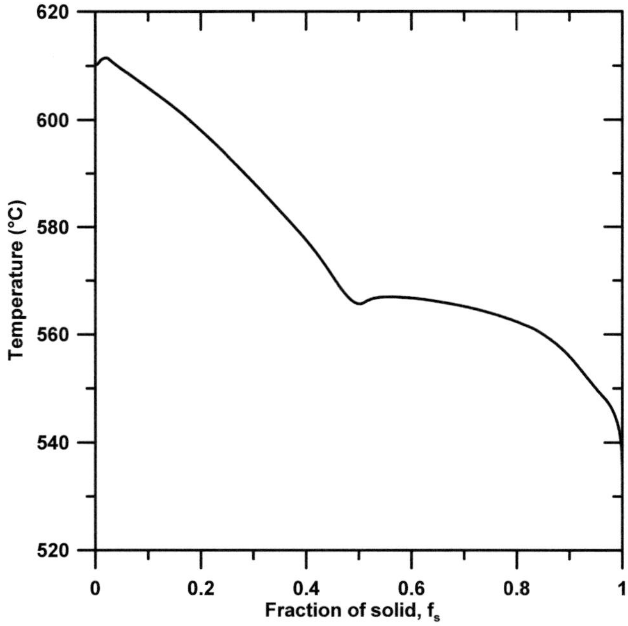

Fig. 12. Solid fraction versus temperature of the A356 alloy used in the present work

boundaries. However, it is difficult to distinguish between interdendritic and grain boundary melting in the microstructure. Interdendritic and grain boundary melting is shown in Fig. 13. The Fe-rich intermetallics melt at solution temperatures above 550°C leading to formation of spherical liquid droplets within the dendrites/grains. At high solution temperatures the width of the grain boundary melted zone increases, and spherical interdendritic liquid droplets enlarge and coalesce to form a large network of interdendritic liquid. On quenching this liquid, reprecipitation of silicon and other intermetallic particles may occur, and the average size increases. Quenching also leads to a large amount of shrinkage porosity adjacent to melted regions, which can coalesce and lead to the complete fracture of the casting, as seen in Fig. 14. The amount of liquid phase formed with high temperature solution treatment depends greatly on the initial solidification rate.

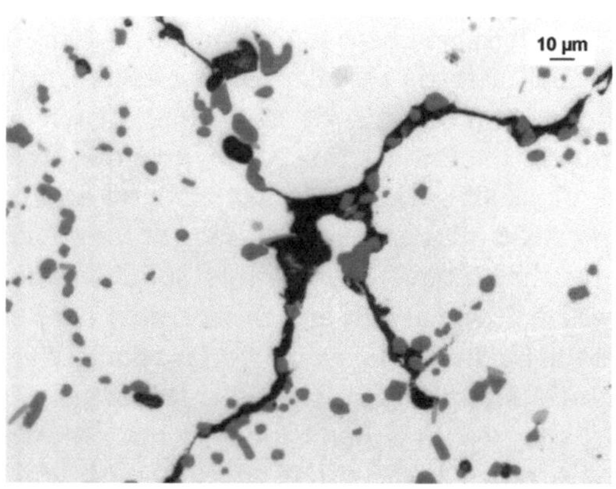

Fig. 13. Interdendritic and grain boundary incipient melting

Therefore, regions of the wheel solidified at high cooling rate, such as the rim, show large amounts of liquid phase formation as compared to those solidified at lower cooling rate, such as the spoke, presumably due to greater segregation of solute elements at interdendritic regions and grain boundaries.

Fig. 14. Fracture path developed by coalescence of shrinkage porosity due to quenching of liquid phase

3.3 Quenching

3.3.1 Microstructural observations

Fig. 15 shows the microstructure of artificial aged wheels, which were water quenched at 45 and 95°C. The size and shape of the eutectic Si particles were not influenced by the quenching condition used in the present work. The different quenching media influenced probably the Mg_2Si and Si precipitates in the α-Al matrix obtained by subsequent artificial ageing. Detailed TEM investigations on A356 alloy, reported elsewhere (Zhang & Zheng, 1996), revealed that, at the peak-aged condition and with a water quench at 25°C, the α-Al matrix consists of a large number of needle-shaped and coherent β″-Mg_2Si precipitates. The size of the precipitates is approximately 3 to 4 nm in diameter and 10 to 20 in length. With a

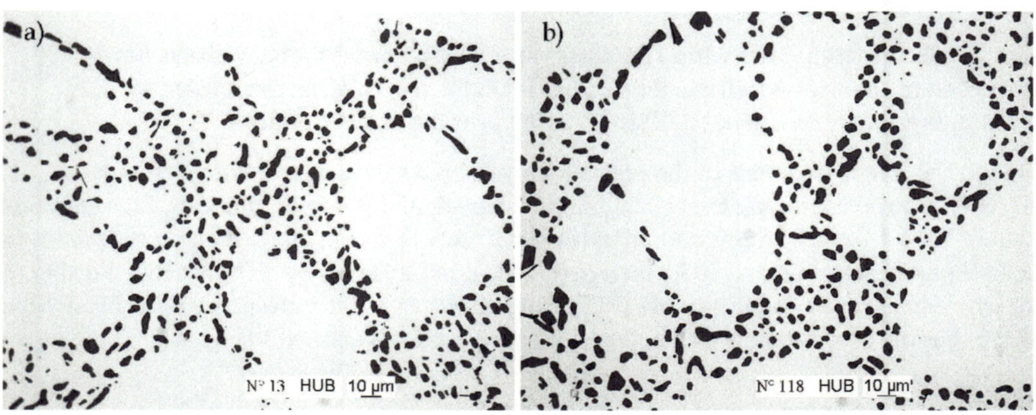

Fig. 15. Microstructure in the hub of the wheels; the micrographs refer to artificial aged A356 alloy solubilised at 540°C for 6 h and water quenched at (a) 45 and (b) 95°C

water quench at 60°C, Zhang and Zheng observed how the density of the precipitates decreases and the size of the precipitates increases slightly; at the same time a significant number of fine Si precipitates resulting from precipitation of excess Si could be observed in the α-Al matrix.

With a slow quenching in air, very different precipitation features are normally evidenced. By air quenching, the material remains at high temperatures for a longer period, which enhances the diffusion of silicon and magnesium. Besides a high density of fine β″-Mg_2Si precipitates, the α-Al matrix also contained a large number of areas with coarse rods β′-Mg_2Si grouped parallel to each other (Zhang & Zheng, 1996). While the first precipitates have an average size approximately 2 to 3 nm in diameter and around 40 nm in length, the latter show an average size ~15 nm in diameter and 300 nm in length.

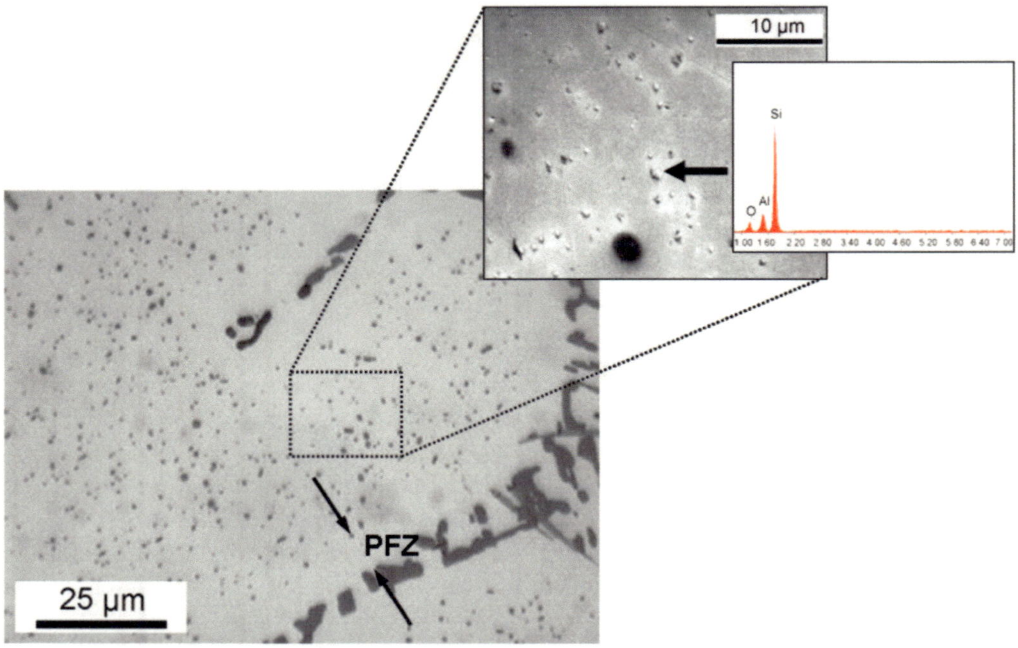

Fig. 16. Silicon precipitates within dendrites in A356-T6 wheels that have been slowly quenched in air; arrows indicate the Si particles in the α-Al matrix, as revealed by EDS spectra. Precipitate-free zone (PFZ) is indicated near the eutectic regions

Due to the low Mg content in the present alloy, a high excess Si concentration is present in the α-Al matrix. Assuming the stoichiometric formation of β′-Mg_2Si, this alloy concentration should form 0.3 wt.% Mg_2Si and an excess of 1 wt.% Si in the alloy, which precipitates as coarse particles within the α-Al matrix (Fig. 16), as revealed by EDS spectra. Further, a clearly visible precipitate-free zone (PFZ) can be seen near the eutectic regions, illustrating that Si has diffused towards existing crystals; such region is marked in Fig. 16.

3.3.2 Distortion behaviour of quenched wheels

The overall distortion ε_t on 18-inch wheels was measured after quenching in water at different temperature. The different quenching rates obtained using water at different temperatures lead to different amount of distortions. Fig. 17 shows the wheel distortion after quenching as a function of the water temperature. By increasing the water temperature, the amount of distortion is reduced; for instance, water at 95°C produces an overall distortion of 1.1 mm, while the wheel distortion is increased up to 1.9 mm with water quenching at 45°C. In the present work, the relationship between the overall casting distortion after quenching and the water temperature has been estimated by linear regression analysis (Fig. 17). The distortion ε_t can be described according to the following regression model:

$$\varepsilon_t = -0.022 \cdot T + 3.101 \tag{5}$$

where T is the water temperature in °C. The regression analysis leads to R^2 equal to 0.96.

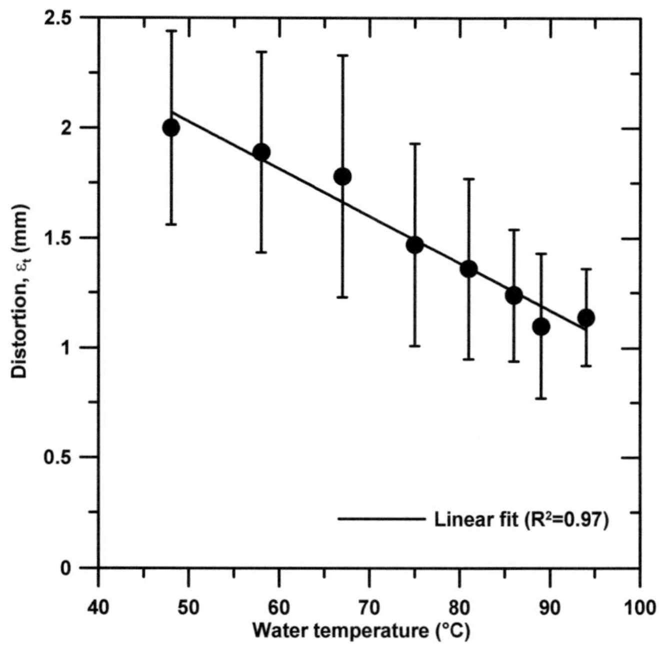

Fig. 17. Wheel distortion after quenching as a function of water quenching temperature; standard deviations are given as error bars

It has to be mentioned, that the distortion measurements after quenching was carried out on a batch of wheels that were previously cooled in water at 30°C at the exit of the LPDC machine. As previously seen, this operation produces an average distortion of about 1.1 mm. Thus, the "real" distortion caused by water quenching ε' was calculated by removing the effect of post-cast cooling (Fig. 18). Again, the wheel distortion progressively reduces by increasing the water temperature, and with a temperature higher than 80°C is approximately zero. This behaviour is explained considering the cooling history and the heat transfer condition of an isothermal mass being quenched from a high initial temperature (solution temperature) in a stagnant bath of liquid. Bath quenching starts with a relatively slow rate of cooling, apparently due to a very rapid development of a thin vapour layer which prevents from the contact of "new" water. The film boiling regime persists from elevated surface temperatures down to a lower temperature limit commonly referred to as the minimum heat flux or Leidenfrost temperature. Below this temperature limit there exists the transition boiling regime, in which the droplets begin to effectively wet the surface resulting in higher heat transfer rates and a faster decrease in the surface temperature. As the surface temperature decreases in the transition boiling regime from the Leidenfrost temperature, the heat transfer rate increases. At the lower temperature boundary of the transition boiling regime, the heat transfer rate reaches a maximum and the temperature of the mass drops rapidly (Liščič et al., 2010; Bernardin et al., 1997).

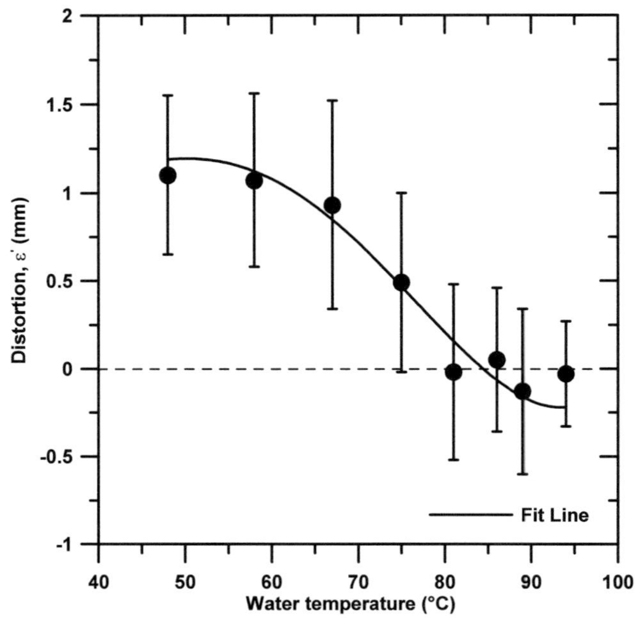

Fig. 18. Effective wheel distortion ε′ caused by quenching as a function of temperature of water quenching; standard deviations are given as error bars

By using warm water, the Leidenfrost temperature shifts to lower values and the film boiling regime is stable in a greater temperature range. In the range of stable film boiling the temperature falls slowly, almost independent of the bath temperature. Therefore, a uniform cooling is obtained throughout the wheel and the amount of distortion is reduced. Contrary, if the temperature falls soon below the Leidenfrost temperature, the film boiling collapses and the temperature drops rapidly. The higher the Leidenfrost temperature is, that is the sooner the film collapses, the shorter is the total quenching time. Therefore, the 18-inch wheels quenched in water bath at a temperature higher than 80°C keep the initial distortion caused by rapid cooling after casting process.

Even if the non-homogeneous cooling of the casting during quenching remains the main cause of the distortions, another important feature to be considered is the non-homogeneous heat exchange of the batch of wheels inside the water tank. In automotive wheel production, generally, several wheels are contemporary quenched by using a steel basket. In this work, batches of 30 wheels, automatically loaded in a five plane steel frame, are quenched. The different heat transfer conditions created in the water bath influence the distortion behaviour of the wheels in the basket. Fig. 19 shows the average distortion of the wheels at the different planes of the basket as a function of the water quenching temperature.

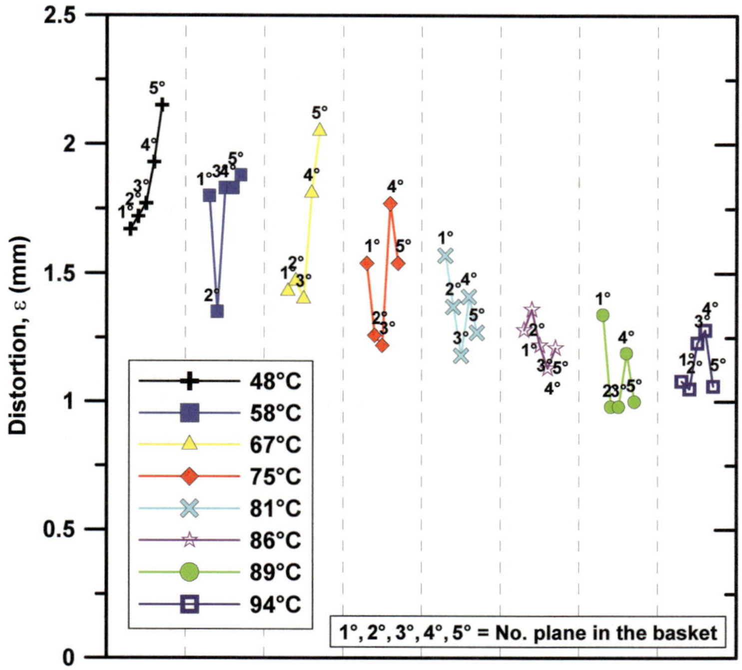

Fig. 19. Average distortion of the wheels in the five planes of the basket as a function of the temperature of water quenching

Generally, the first and the last planes of the frame present the extreme values of distortion. This can be explained considering the quenching operation. The wheels at the first planes of the basket are the first to enter in the water bath and their immersion produces a strong water evaporation with the formation of large vapour pockets, which go up toward the bath surface. The amount of vapour increases progressively at the top of the water bath, as the basket is immersed in water (Fig. 20).

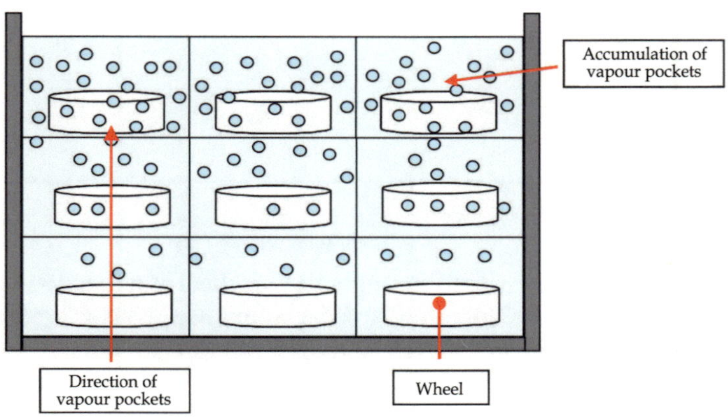

Fig. 20. Draft of the vapour accumulation at the top of the water bath; as the supporting basket is progressively immersed in water, the wheels produce strong water evaporation with the formation of large vapour pockets, which go up toward the bath surface

The vapour pockets may collapse on the casting surface and locally change the heat transfer coefficient between the piece and the quenchant by preventing from the contact of "new" water. Once again, a non-homogeneous quenching rate is established throughout the wheel. The wheels at last planes of the basket undergo different quenching conditions than those at the first planes.
The influence of water temperature on hardness of wheels after ageing at 145°C for 4 h is shown in Fig. 21. The different water temperature, in the range between 40 and 95°C, doesn't influence (to some extent) the hardness properties of the A356 alloy, that is the hardness fluctuates slightly around 92 HB. Generally, the hardness of A356 alloy decreases by lowering quench rates. It has been studied that with a quench rate higher than 110°C/s, obtained with water at temperature lower than 60°C, the peak hardness of A356 alloy is not influenced by the quench rate (Zhang & Zheng, 1996); nevertheless, a little difference (~4 HB) occurs by water quenching in the temperature range between 60 and 100°C (Fracasso, 2010). Furthermore, the time to peak hardness increases for extremely slow quench rates (0.5°C/s), while for faster quench rates, above 20°C/s, no shift is seen in the time to the peak. Therefore, by increasing the temperature of water quenching up to 95°C, the target hardness of the wheels after a complete T6 heat treatment is achieved and the wheel distortion is reduced.

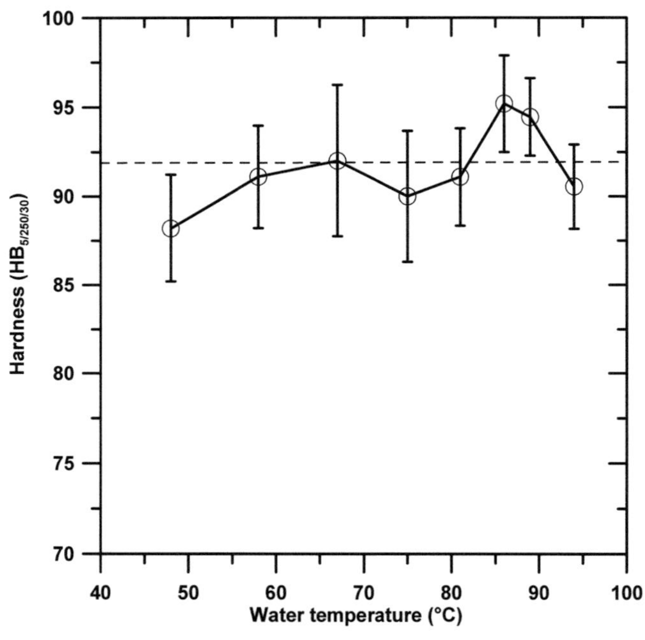

Fig. 21. Brinell hardness measured throughout the wheel as a function of the different temperature of water quenching; standard deviations are given as error bars. Data refer to wheels solution treated at 540°C for 6 h and aged at 145°C for 4 h

3.4 Powder coating

Most aluminium wheels are clear coated for corrosion resistance and aesthetic appearance. Unprotected aluminium wheels quickly corrode and pit when exposed to road salt and excessive moisture. If the corrosion continues unchecked for too long, the cosmetic damage may be too great to reverse. Generally, several coats are applied to aluminium wheels to guarantee a suitable corrosion resistance. After each coat the wheels are left inside an air electric furnace for drying at 170 ± 5°C for 1 hour. From the heating curve in Fig. 3, it is observed that it takes approximately 20 minutes to heat the wheels from room temperature to 145°C. Due to slow heating, the coating treatment effect experienced by the wheels during the heating stage is not negligible. Then, the wheels are maintained for 35 minutes in a range of temperature between 145 and 170°C. The temperature and time used in the present work for powder coating activate the diffusion mechanism of the solute atoms, such as Mg and Si, leading to the precipitation of dissolved elements and the coarsening of existing precipitates, i.e. the bake hardening effect. The influence of powder coating cycles on the hardness of T6 heat treated wheels is shown in Fig. 22. The hardness increases progressively after each coating cycles of about 3%. The average hardness of wheels after machining is around 92 HB, while after 3 coating cycles the hardness increases up to 98 HB.

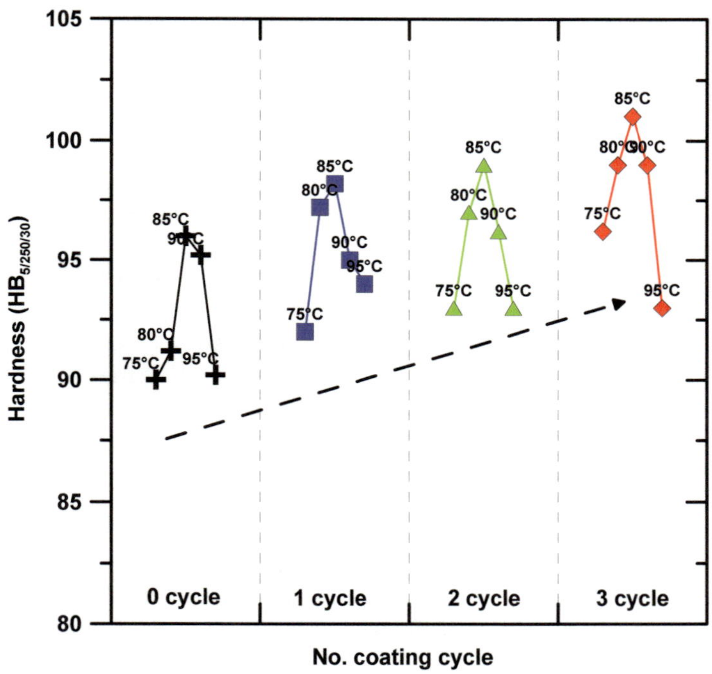

Fig. 22. Effect of coating cycles on hardness of wheels, which were solution treated at 540°C for 6 h, water quenched and aged at 145°C for 4 h; data refer to water quenching at different temperature

4. Conclusions

In the present work, some process variables, which play a key role in production cycle of wheels have been investigated and improved. An integrated methodology for developing and optimizing the production and the final quality of A356-T6 18-inch wheels, in terms of casting distortion and hardness, has been proposed. This study has focused on examining both the effect of cooling rate on wheel distortion and hardness during the post-cast and quenching steps, and the influence of the solutionizing temperature and time, and the powder coating cycles on the microstructure and mechanical properties of the 18-inch wheels.

Based on the results obtained in the present study, it can be drawn that the different cooling rate of the wheels, ejected from the die at high temperature, produces different amount of distortions. By increasing the water temperature, the amount of distortions linearly decreases. Water cooling at a temperature higher than 70°C produces similar distortion as air cooling.

Considering the T6 heat treatment applied to the wheel production, a solution heat treatment of 6 h at 540°C is sufficient to dissolve completely the Mg-rich phases and to achieve a homogeneous solid solution. This solution treatment causes spheroidization and coarsening of the eutectic Si particles, leading to substantial changes in the microstructure throughout the 18-inch wheel. Higher solution temperatures lead to incipient melting at grain boundary and in the interdendritic regions. On quenching this liquid, reprecipitation of silicon and other intermetallic particles occur, and the average size increases. Quenching also leads to a large amount of shrinkage porosity adjacent to melted regions, which can coalesce and lead to the complete fracture of the wheel.

Furthermore, quenching is usually carried out from solution temperature to room temperature to obtain a supersaturated solid solution of solute atoms and vacancies, in order to achieve an elevated strengthening subsequent ageing. Here, the wheel distortion progressively reduces by increasing the temperature of water quenching, and a temperature higher than 80°C is sufficient to avoid distortion, allowing to achieve at the same time the required mechanical properties.

Finally, the powder coating of the wheels influences the final mechanical properties by activating the diffusion mechanism of the solute atoms, such as Mg and Si. This leads to the precipitation of dissolved elements and the coarsening of existing precipitates. The result is an increase of the hardness of about 3% after each coating cycle. This means that the powder coating can be integrated into the whole T6 heat treatment cycle of wheels, with a great impact on productivity and manufacturing cost of wheels.

5. References

Alexopoulos N.D. & Pantelakis S.G. (2004). Quality evaluation of A357 cast aluminium alloy specimens subjected to different artificial aging treatment. *Materials and Design*, Vol.25, No.5, pp. 419-430, ISSN 0261-3069

Apelian D., Shivkumar S. & Sigworth G. (1989). Fundamental aspects of heat treatment of cast Al-Si-Mg alloys. *AFS Transactions*, Vol.97, pp. 727-742

ASM Metals Handbook (1990). *Properties and Selection: Nonferrous Alloys and Special-Purpose Materials*, Vol.2, ASM International, ISBN 978-087-1703-78-1, Materials Park, OH, USA

ASM Metals Handbook (1991). *Heat treating*, Vol.4, ASM International, ISBN 978-087-1703-79-8, Materials Park, OH, USA

Auburtin P. & Morin N. (2003). Thermo-mechanical modeling of the heat treatment for aluminium cylinder heads. *Mécanique & Industries*, Vol. 4, No.3, pp. 319-325, ISSN 1296-2139

Bates C.E. (1987). Selecting quenchants to maximize tensile properties and minimize distortion in aluminium parts. *Journal of Heat Treating*, Vol.5, No.1, pp. 27-40, ISSN 0190-9177

Bates C.E. (1994). Quench Optimization for Aluminum Alloys. *AFS Transactions*, Vol.101, pp. 1045-1054

Bernardin J.D., Stebbins C.J. and Mudawar I. (1997). Mapping of impact and heat transfer regimes of water drops impinging on a polished surface. *International Journal of Heat and Mass Transfer* , Vol.40, No.2, pp.247-267, ISSN 0017-9310

Cáceres C.H. & Griffiths J.R. (1996). Damage by the cracking of silicon particles in an Al-7Si-0.4Mg casting alloy. *Acta materialia*, Vol.44, No.1, pp. 25-33, ISSN 1359-6454

Cáceres C.H., Davidson C.J. & Griffiths J.R. (1995). Deformation and fracture behaviour of an Al-Si-Mg casting alloy. *Materials Science and Engineering A*, Vol.197, No.2, pp. 171-179, ISSN 0921-5093

Conserva M., Bonollo F. & Donzelli G. (2004). *Alluminio, manuale degli impieghi*, Edimet, ISBN 978-888-6259-27-9, Brescia, Italy

Dantzig J.A. & Rappaz M. (2009). *Solidification*, CRC Press, Taylor & Francis Group, ISBN 978-084-9382-38-3, Boca Raton, USA

Dwivedi D.K., Sharma R. & Kumar A. (2006). Influence of silicon content and heat treatment parameters on mechanical properties of cast Al-Si-Mg alloys. *International Journal of Cast Metals Research*, Vol.19, No.5, pp. 275-282, ISSN 1364-0461

Eskin D.G. (2003). Decomposition of supersaturated solid solutions in Al-Cu-Mg-Si alloys. *Journal of Materials Science*, Vol.38, No.2, pp. 279-290, ISSN 0022-2461

Fracasso F. (2010). Influence of quench rate on the hardness obtained after artificial ageing of an Al-Si-Mg alloy. Master Thesis, University of Padova, Padova, Italy

Greenwood G.W. (1956). The growth of dispersed precipitates in solutions. *Acta Metallurgica*, Vol.4, No.3, pp. 243-248

Grosselle F., Timelli G., Bonollo F., Tiziani A. & Della Corte E. (2009). Correlation between microstructure and mechanical properties of Al-Si cast alloys. Metallurgia Italiana, Vol.101, No.6, pp. 25-32, ISSN 0026-0843

Kashyap K.T., Murali S., Raman K.S. & Murthy K.S.S. (1993). Casting and heat-treatment variables of Al-7Si-Mg alloy. *Materials Science and Technology*, Vol.9, No.3, pp. 189-203, ISSN 0267-0836

Komarova M.F., Buynov N.N. & Kaganovich L.I. (1973). Influence of quenching rate and small alloying additions on the kinetics and morphology of precipitations in aluminium-silicon-magnesium alloys. *Physics of Metals and Metallography*, Vol.36 No.3, pp. 72-79, ISSN 0031-918X

Langsrud Y. & Brusethaug S. (1998). Age hardening response of AlSiMg foundry alloys, In: *ICAA6 – Aluminum Alloys, Their Physical and Mechanical Properties*, T. Sato, S. Kumai, T. Kobayashi and Y. Murakami, (Ed.), 733-738, The Japanese Institute of Light Metals, Japan

Liftshitz I.M. & Sloyozov V.V. (1961). The kinetics of precipitation from supersaturated solid solutions. *Journal of Physics and Chemistry of Solids*, Vol.19, No.1-2, pp. 35-50

Liščič B., Tensi H.M., Canale L.C.F. & Totten G.E. (2010). *Theory and Technology of Quenching*, CRC Press, Taylor & Francis Group, ISBN 978-084-9392-79-5, Boca Raton, USA

Manente A. (2008). *La fonderia di alluminio nella pratica quotidiana*, Edimet, ISBN 88-86259-35-1, Brescia, Italy

Merlin M., Timelli G., Bonollo F. & Garagnani G.L. (2009). Impact behaviour of A356 alloy for low-pressure die casting automotive wheels. *Journal of Materials Processing Technology*, Vol.209, No.2, pp. 1060-1073, ISSN 0924-0136

Meyers C.W. (1985). Solution heat treatment effects in A357 alloys. *AFS Transactions*, Vol.112, pp. 741-750

Pedersen L. & Arnberg L. (2001). The effect of solution heat treatment and quenching rates on mechanical properties and microstructures in AlSiMg foundry alloys. *Metallurgical and Materials Transactions A*, Vol.32, No.3, pp. 525-532, ISSN 1073-5623

Pedersen L. (1999). *Solution heat treatment of AlSiMg foundry alloys*. Doctoral Thesis, Norwegian University of Science and Technology, ISBN 82-471-0409-1, Trondheim, Norway

Piasentini F., Bonollo F. & Tiziani A. (2005). Fourier thermal analysis applied to sodium eutectic modification of an AlSi7 alloy. *Metallurgical Science and Technology*, Vol.23, No.2, pp. 11-20

Rometsch P.A. & Schaffer G.B. (2002). An age hardening model for Al-7Si-Mg casting alloys. *Materials Science and Engineering A*, Vol.325, No.1-2, pp. 424-434, ISSN 0921-5093

Rometsch P.A., Arnberg L. & Zhang D.L. (1999). Modelling dissolution of Mg_2Si and homogenisation in Al-Si-Mg casting alloys. *International Journal of Cast Metals Research*, Vol.12, No.1, pp. 1-8, ISSN 1364-0461

Shivkumar S., Keller C. & Apelian D. (1990a). Aging behavior in cast Al-Si-Mg alloys. AFS Transactions, Vol.98, pp. 905-911

Shivkumar S., Ricci Jr. S. & Apelian D. (1990b). "Influence of solution and simplified supersaturation treatment on tensile properties of A356 alloy. *AFS Transactions*, Vol.98, pp. 913-922

Shivkumar S., Ricci Jr. S., Steenhoff B., Apelian D. & Sigworth G. (1989). An experimental study to optimize the heat treatment of A356 alloy. *AFS Transactions*, Vol.97, pp. 791-810

Shivkumar S., Ricci S., Keller C. & D. Apelian (1990c). Effect of solution treatment parameters on tensile properties of cast aluminum alloys. *Journal of Heat Treating*, Vol.8, No.1, pp. 63-70, ISSN 0190-9177

Taylor J.A., StJohn D.H., Barresi J. & Couper M.J. (2000). Influence of Mg content on the microstructure and solid solution chemistry of Al-7%Si-Mg casting alloys during solution treatment. *Materials Science Forum*, Vol.331, pp. 277-282, ISSN 0255-5476

Tiryakioglu M. (2008). Si particle size and aspect ratio distributions in an Al-7%Si-0.6%Mg alloy during solution treatment. *Materials Science and Engineering A*, Vol.473, No.1-2, pp. 1-6, ISSN 0921-5093

Totten G.E. & Mackenzie D.S. (2000). Aluminum quenching technology: a review. *Materials Science Forum*, Vol.331, pp. 589-594, ISSN 0255-5476

Totten G.E., Webster G.M. & Bates C.E. (1998). Cooling curve and quench factor characterization of 2024 and 7075 aluminum bar stock quenched in type 1 polymer quenchants. *Heat Transfer Research*, Vol.29, No.1, pp. 163-175, ISSN 1064-2285

Valentini G. (2002). Application of die-castings in automotive industry: a review, *Proceedings of HTDC 2002 International Conference High Tech Diecasting (Al and Mg alloys)*, pp. 237-250, ISBN 88-85298-43-5, Vicenza, Italy, February 22, 2002

Wang Q.G. & Cáceres C.H. (1998). Fracture mode in Al-Si-Mg casting alloys. *Materials Science and Engineering A*, Vol.241, No.1-2, pp. 72-82, ISSN 0921-5093

Zhang D.L. & Zheng L. (1996). Quench sensitivity of cast Al-7 wt pct Si-0.4 wt pct Mg Alloy, *Metallurgical and Materials Transaction A*, Vol.27, No.12, pp. 3983-3991, ISSN 1073-5623

Zhang D.L., Zheng L.H. & StJohn D.H. (2002). Effect of a short solution treatment time on microstructure and mechanical properties of modified Al-7wt.%Si-0.3wt.%Mg alloy. *Journal of Light Metals*, Vol.2, No.1, pp. 27-36, ISSN 1471-5317

Part 4

Mechanical Behavior of Aluminium Alloys and Composites

10

High Strength Al-Alloys: Microstructure, Corrosion and Principles of Protection

Anthony E. Hughes[1], Nick Birbilis[2], Johannes M.C. Mol[3a], Santiago J. Garcia[3b], Xiaorong Zhou[4] and George E. Thompson[4]
[1]CSIRO Materials Science and Technology, Melbourne
[2]Department of Materials Engineering, Monash University, Clayton
[3]TU Delft, Department of Materials Science and Engineering[a] and Novel Aerospace Materials, Aerospace Engineering[b], Delft
[4]School of Materials, The University of Manchester, Manchester
[1,2]Australia
[3]Netherlands
[4]United Kingdom

1. Introduction

Aluminium alloys have highly heterogeneous microstructures compared to many other metal alloys. This heterogeneity originates from alloy additions and impurities which combine to produce both the desired microstructure as well as undesired, large particles, called constituent particles (and residual impurity particles) which have a range of compositions. In corrosion science these latter particles are commonly referred to as intermetallic (IM) particles. The heterogeneous nature of aluminium alloys is most evident in members of the high strength alloys of the 2xxx, 6xxx, 7xxx and 8xxx and most particularly the 2xxx series alloys where alloy additions are required to obtain the high strength to weight ratio properties of these materials.

For many years now, the study of corrosion in these alloys was, and in many instances continues to be, a phenomenological exercise. So the literature on this subject largely involves studies of a small number of intermetallic (IM) particles under a variety of conditions which are difficult to relate to each other in order to form a more general model of corrosion in highly heterogeneous aluminium alloys. This is particularly true for the 2xxx series of alloys which lacks a system to unambiguously categorise these IM particles compositional variation makes it difficult to relate these particles with well know composition, crystallography and electrochemistry. The difficulty in devising such a system should not be underestimated since the intermetallic particles form at various stages during manufacture, individual particles have compositionally different phase domains and their distribution including the spatial relationship to one another is often dictated by the processing route. Nevertheless, in recent years there have been significant advances in the understanding of both the microstructure of some high strength alloys as well as its influence on corrosion. These advances have their foundations in the wider accessibility to a range of newer electrochemical and physico-chemical characterisation techniques. The use of advanced electrochemical techniques has led to a greater understanding of the properties of the intermetallic particles themselves and their roles in corrosion of alloys. The physico-chemical studies have led to a better characterisation of the composition of intermetallic particles, and, importantly of their spatial distributions. The convergence of the electrochemical and physicochemical approaches, in combination with modelling, is leading to a statistical basis for understanding the influence of the intermetallic particles on corrosion of aluminium alloys.

This chapter therefore aims to summarise our current understanding of the microstructure of high strength aluminium alloys, particularly the more microstructurally complex alloys such as the 2xxx, and 7xxx series alloys. It then looks at how different components of the microstructure contribute to corrosion processes and finishes by examining the principles of protection of aluminium alloys using traditional and newer techniques used to assess the degree of protection.

2. Microstructure

2.1 Aluminium alloys in general

While aluminium alloy microstructure, for some specific alloys, is relatively well known, the microstructures for some high strength aluminium alloys, particularly the older AA2*xxx* alloys, is not well described or understood in the scientific literature, particularly the corrosion literature. This is partially due to manufacturing processing conditions which do not realise the intended microstructure and partially due to quasi or non-equilibrium microstructure existing in real alloys because of the difficulty of obtaining full thermodynamic equilibrium. Typical examples of common high strength alloys used in aircraft manufacture, for example, include AA2024-T3, AA7075-T6 and AA6061-T6. This section, therefore, provides a general overview of the relationship between processing and microstructure.

Processing can significantly alter the bulk microstructure, resulting in microstructural gradients and zones with different characteristics. A good example of these changes can be found in wrought alloy sheet product. First there is a gradient in grain size and constituent particle size across the sheet. Second, shear deformation, resulting from rolling, creates a surface layer called a near surface deformed layer (NSDL) with a very fine microstructure which may have a different degree of precipitation compared to the bulk depending on the heat treatments.

2.2 Aluminium production

Bauxite production has increased 50% in the past decade to an all time high of over 200 million tonnes worldwide; with Australia the largest producer, followed by China and Brazil. Four tonnes of bauxite are used to produce two tonnes of alumina, which can then produce one tonne of aluminium. Recycling of aluminium requires 95% less energy than for primary aluminium production. In order to meet the mechanical and corrosion performance requirements for many alloys as required under performance specifications, much of the recycled metal must be blended or diluted with primary metal to reduce impurity levels. The result is that, in many cases, recycled metal tends to be used primarily for lower grade casting alloys and products (Polmear 2004), however with ~35% of Al being produced from recycled material, the future ramifications for corrosion will need addressing.

2.3 Physical metallurgy of aluminium alloys

The functional properties of aluminium alloys (mechanical, physical, and chemical) depend on alloy composition and alloy microstructure as determined by casting conditions and thermomechanical processing. Only a small number of metals have sufficient solubility to serve as major alloying elements (Das 2006) and alloys derived from these few form the basis of the present classes of commodity Al-alloys. Magnesium, zinc, copper and silicon have significant solubility, whilst additional elements (of <1% solubility) are also used to confer improvements to alloy properties, namely grain refinement, and such elements include manganese, chromium, zirconium, titanium and less commonly (due to cost) scandium (Hatch 1984; Polmear 2004). Alloying of Manganese with Fe-containing intermetallic particles reduces the electrochemical activity of these Fe-containing particles thus improving the corrosion resistance of the alloy (Polmear 1995)

The low strength of pure aluminium (~10 MPa) mandates alloying. The simplest strengthening technique is solution hardening, whereby alloying additions have appreciable solid solubility over a wide range of temperatures and remain in solution after many thermal cycles.

The most significant increase in strength for aluminium alloys is derived from age hardening (often called precipiation hardening) which can result in strengths as high as 800 MPa. The principal of age hardening requires that the solid solubility of alloying elements decreases with temperature. The age hardening process can be summarised by the following stages:

i. solution treatment at a temperature within a single phase region to dissolve the alloying element(s)
ii. quenching of the alloy to obtain what is termed a supersaturated solid solution
iii. decomposition of the supersaturated solid solution at ambient or moderately elevated temperature to form finely dispersed precipitates.

The fundamental aspects of decomposition of a supersaturated solid solution are complex (Raviprasad, Hutchinson et al. 2003; Kovarik, Miller et al. 2006; Winkelman, Raviprasad et al. 2007). Typically however, Guinier-Preston (GP) zones and intermediate phases are formed as precursors to the equilibrium precipitate phase (Hatch 1984) (Figure 1 reveals a

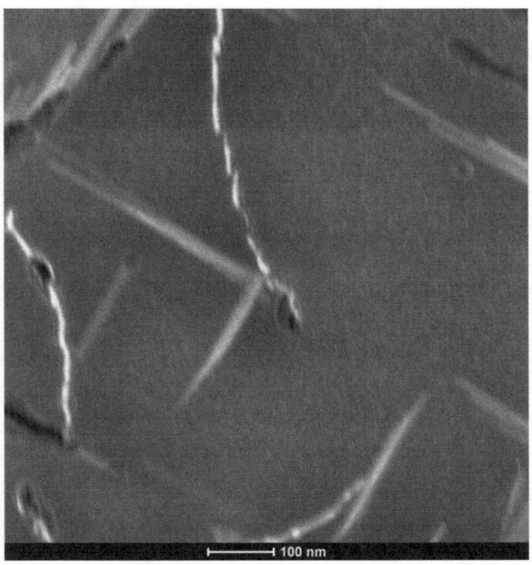

Fig. 1. Dark field scanning transmission electron micrograph of coarse Al_2CuMg precipitate particles in an AA2xxx (Al-Cu-Mg) alloy - imaged down <100> zone axis

typical micrograph showing precipitate particles). GP zones are formed when solute atoms (e.g. Cu, Zn and Mg) accumulate along preferred crystal directions in the Al lattice and form a strengthening phase.

Properties can be enhanced further by careful thermo-mechanical processing that may include heat treatments like duplex aging and retrogression and re-aging. Maximum hardening in commercial alloys is often achieved when the alloy is cold worked by stretching after quenching and before aging, increasing dislocation density and providing more heterogeneous nucleation sites for precipitation. Whilst only moderate increases in strength can be obtained in Al-alloys by exploiting the Hall-Petch[1] relationship (Polmear 1995), refinement is important for a range of properties including fracture and toughness. Grain refinement in aluminium alloys is achieved by additions of small amounts of low solubility elements such as Ti and B to provide grain nuclei, and by recrystallisation control using precipitates called dispersoids (typically 40 x 200 nm) which are formed from aluminium and alloying additions such as Cu, Cr, Zr or Mn to promote insoluble particles which subsequently can restrict or pin grain growth.

The microstructures developed in aluminium alloys are complex and incorporate a combination of equilibrium and non-equilibrium phases. Non-equilibrium phases exist in essentially all high-strength alloys, and as such, their properties are very temperature dependent.

Typical commercial alloys can have a chemical composition incorporating as many as ten alloying additions (with a number of these additions being unavoidable impurities). As such, from a corrosion perspective, one must understand the role of impurity elements on microstructure. Whilst not of major significance to alloy designers, impurity elements such as Fe, Mn and Si can form insoluble compounds called constituent particles. These are comparatively large and irregularly shaped with characteristic dimensions ranging from 1 to ~ 50 µm. These particles are formed during alloy solidification and are not appreciably dissolved during subsequent thermo-mechanical processing. Rolling and extrusion tend to break-up and align constituent particles within the alloy. Often constituents are found in clusters made up of several different intermetallic compound types. Because these particles are rich in alloying elements, their electrochemical behaviour can be significantly different to the surrounding matrix phase. In most alloys pitting is associated with specific constituent particles present in the alloy (Buchheit 1995; Liao, Olive et al. 1998; Wei, Liao et al. 1998; Guillaumin and Mankowski 1999; Park, Paik et al. 1999; Ilevbare, Schneider et al. 2004; Schneider, Ilevbare et al. 2004; Lacroix, Ressier et al. 2008; Lacroix, Ressier et al. 2008; Boag, Taylor et al. 2010. These are discussed below.

2.4 Alloy classification

The International Alloy Designation System (IADS) gives each wrought alloy a four-digit number of which the first digit is assigned on the basis of the major alloying element(s) (Polmear 1995; Winkelman, Raviprasad et al. 2007)). The main alloying element for AA2xxx is Cu and for AA7xxx is Zn, with Mg playing a important role is both classes of alloys.

For cast aluminium alloys, alloy designations principally adopt the notation of the Aluminium Association System. The casting compositions are described by a four-digit system that incorporates three digits followed by a decimal (described in more detail in

[1] The Hall-Petch relationship states that the yield strength is proportional to the inverse square root of the grain size.

(Hatch 1984)). The temper designation system adopted by the Aluminium Association is similar for both wrought and cast aluminium alloys.

2xxx

Copper is one of the most common alloying additions, since it has appreciable solubility and a significant strengthening effect by its promotion of an age hardening response. These alloys were the foundation of the modern aerospace construction industry and, for example AA2024 (Al-4.4Cu-1.5Mg-0.8Mn), can achieve strengths of up to 520MPa depending on temper. The microstructure of this series is considered further below. Cu, however, is one of the nobler alloying elements and therefore supports a high rate of oxygen reduction which drives one half of the galvanic reaction. The cell is completed by the dissolution of any element less noble, particularly Al thereby facilitating the onset and propagation of corrosion.

7xxx

The Al-Zn-Mg alloy system provides a range of commercial compositions, primarily where strength is the key requirement. Al-Zn-Mg-Cu alloys have traditionally offered the greatest potential for age hardening and as early as 1917 a tensile strength of 580MPa was achieved, however, such alloys were not suitable for commercial use until their high susceptibility to stress corrosion cracking could be moderated. Military and commercial aerospace needs led to the introduction of a range of high strength aerospace alloys of which AA7075 (Al-5.6Zn-2.5Mg-1.6Cu-0.4Si-0.5Fe-0.3Mn-0.2Cr-0.2Ti) is perhaps the most well known, and which is now essentially wholly superseded by AA7150 (or the 7x50 family). The high strength 7xxx series alloys derive their strength from the precipitation of η-phase ($MgZn_2$) and its precursor forms. The heat treatment of the 7xxx series alloys is complex, involving a range of heat treatments that have been developed to balance strength and stress corrosion cracking performance, comprising secondary (or more) heat treatments that can include retrogression and re-aging (Sprowls 1978).

2.5 Processing of aluminium alloys

The surface layers of aluminium alloys can be altered during processing and storage environments, which adds complexity to the surface finishing and corrosion performance (Fishkis and Lin 1997). These effects include the formation of near surface deformed layers (NSDL) during mechanical processing, the elongation of crystalline structure during rolling and extrusion, breakup of brittle intermetallic particles, differences in surface roughness and porosity, and the segregation of specific alloying elements to the surface.

Casting from the melt is the first processing step. The three most commonly used processes are sand casting, permanent mould casting and die casting. Sand moulds are gravity fed whereas the metal moulds used in permanent mould casting are either gravity fed or by using air or gas pressure to force metal into the mould. In high pressure die castings, parts up to approximately 5 kg are made by injecting molten aluminium alloy into a metal mould under substantial pressure using a hydraulic ram.

For large production scale, direct chill (DC) casting is a semi-continuous process used for the production of rectangular ingots or slab for rolling to plate, sheet, foil and cylindrical ingots or billet for extruded rods, bars, shapes, hollow sections, tube and wire. DC casting is the first step in the production of Al alloys prior to the thermomechanical treatments, and whilst it may appear to be a topic not requiring discussion in such a chapter, it is important to realise that corrosion performance of an alloy is dictated by each processing step, starting with the solidification of the molten alloy during DC casting.

DC casting starts by pouring molten metal into a water-cooled aluminium or copper mould. Accumulation of alloying elements at the surface can occur through segregation processes where mobile elements diffuse from the bulk and from grain boundaries. In general, the surface enriched elements have a high negative free energy for oxide formation and high diffusion coefficient through aluminium metal. Such elements include lithium, magnesium and silicon (Carney, Tsakiropoulos et al. 1990). The segregation occurs during forming and heat treatments and has been demonstrated to influence the corrosion and wear properties of the alloys (Nisancioglu 2004).

Thermodynamic considerations often fail to correctly predict the phase and solid solution content of an as-cast microstructure because of the non-equilibrium nature of solidification during DC casting. This is important as alloy corrosion properties are controlled by solid solution levels and intermetallic phase crystallography and morphology, which depend on complex kinetic competitions during nucleation and growth. Most importantly, the constituent particles do not appreciably dissolve during subsequent solution heat treatment, and will thus persist into the final product.

2.6 Surface microstructures

Fabrication processes, including rolling, machining and mechanical grinding, produce aluminium products of the required gauge thickness and shape for various applications. Rolling blocks or slabs that are up to many tonnes, requires heating to temperatures up to 500°C and passing through a breakdown mill using heavy reductions per pass to reduce the slab gauge from as large as 5000 mm down to 15 to 35 mm. The slab surface undergoes intense shear deformation during this process and a NSDL develops. The shearing process also influences IM particles below the surface resulting in a larger number of IM particles in the vicinity of the surface than in the body of the material. This is not as a result of precipitation processes but is due to fragmentation of brittle particles during rolling. Hence the particle number density at the surface is higher, but the percentage of surface area is the same indicating particle fracture rather than new particle formation (Hughes, Boag et al. 2006). The characteristics of the surface of sheet AA2024-T3 with respect to the body of the material (obtained by polishing) are compared in Table 1 (Hughes, Boag et al. 2006). The slab from the breakdown mill is then typically hot rolled on a multistand tandem mill down to a gauge of 2.5 to 8 mm. Hot rolling deforms the original cast structure with the grains being elongated in the rolling direction. The elongated microstructure developed during hot

IM Particle Characteristic	Body	Surface
Number Density:	$5.3x10^5cm^{-2}$	$11.7x10^5cm^{-2}$
Average Particle Size:	$6.66\mu m^2$	$1.98\mu m^2$
Median Particle Size:	$1.6\mu m^2$	$1.2\mu m^2$
%Surface Area:	2.89%	2.82%
Total Particles per $1mm^2$:	5300	11690
Minimum Particle Size:	$0.40\mu m^2$	$0.34\mu m^2$
Maximum Particle Size:	$327\mu m^2$	$114\mu m^2$

Table 1. IM particle size (area) for polished (body) and as-rolled (surface) AA 2024

rolling can have a profound effect on corrosion properties like stress corrosion cracking and exfoliation corrosion. For example, the exfoliation corrosion of 7xxx alloys was shown to be due to manganese segregation to interfaces (grain boundaries and the external surface) during DC casting (Evans 1971).

Typically, a NSDL is characterised by ultra-fine, equiaxed grains, with grain boundaries decorated by nano-sized oxide particles (Fishkis and Lin 1997; Leth-Olsen, Nordlien et al. 1998; Plassart 2000; Scamans 2000; Afseth, Nordlien et al. 2001; Zhou, Thompson et al. 2003; Liu, Frolish et al. 2010; Scamans, Frolish et al. 2010; Thompson 2010; Zhou 2011). It is associated with the susceptibility of several aluminium alloys to filiform corrosion (Zhou, Thompson et al. 2003; Liu, Zhou et al. 2007; Liu, Laurino et al. 2010).The depth of the modified surface region ranges from a few nanometers (after polishing) to 8 μm (during rolling). In the latter case the thickness varies with each rolling pass (Fishkis and Lin 1997). Further, a transition region, characterized by microbands that consist of elongated grains aligned parallel to the working surface, may be sandwiched between the surface regions and the bulk alloy. The deformed layers are stable at ambient temperature, associated with the local presence of a large fraction of high angle grain boundaries. The structure is also stabilized through pinning of the grain boundaries by oxide particles and precipitates (Figure 2). NSDLs with grain boundaries decorated by oxide can survive typical annealing and solution heat treatment processes. As a result, metal finishing and surface treatments are

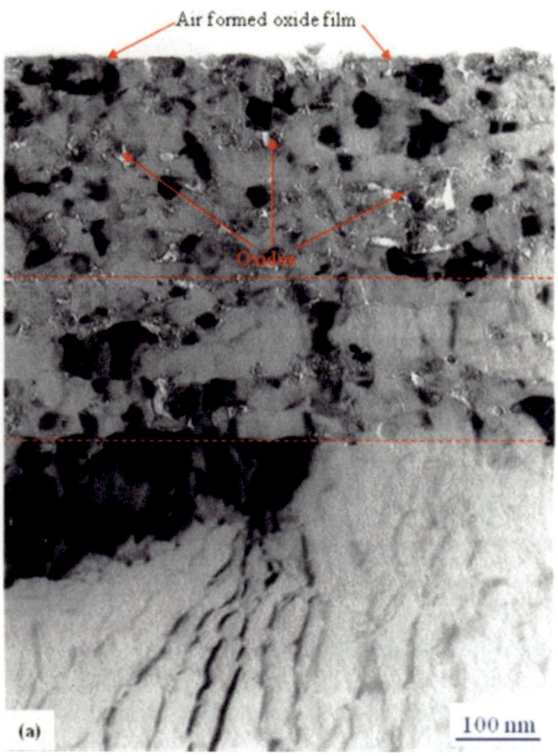

Fig. 2. Transmission electron micrographs of ultramicrotomed sections displaying the surface/near-surface regions of an AA5754 H19 aluminium alloy: (a) as cold rolled, transverse to rolling direction

required to remove these electrochemically active layers (Leth-Olsen, Nordlien et al. 1997; Mol, Hughes et al. 2004; Hughes, Mol et al. 2005). However, the presence of fine grains alone in the deformed layer, with grain boundaries free of oxide particles, is insufficient to hinder grain coarsening during typical annealing treatments (Zhou 2011).

Importantly, the NSDL has significant influence on properties such as the electrochemical and corrosion behaviour as well mechanical properties, material joining and optical properties. The high population of grain boundaries and severe deformation in the deformed layer promote precipitation of intermetallic particles during subsequent heat treatment (Liu, Zhou et al. 2007). For example, a near-surface deformed layer on AA6111 automotive closure sheet alloy can be generated by mechanical grinding during rectification, as shown in Figure 3 (left). Subsequent paint baking, i.e. thermal exposure at 180°C for 30 minutes, promotes the precipitation of Q phase (with various compositions: $Al_5Cu_2Mg_8Si_6$ (Pan, Morral et al. 2010), $Al_4CuMgSi_4$ (Hahn and Rosenfield 1975)) particles, ~20 nm diameter, at preferred grain boundaries within the deformed layer (Figure 3 centre), but with no precipitates being formed in the underlying bulk alloy. The presence of Q phase precipitation in the near-surface deformed layer increases dramatically the susceptibility of the alloy to cosmetic corrosion that propagates intergranularly, with micro-galvanic coupling between the Q phase precipitates and the adjacent aluminium matrix providing the driving force (Figure 3(right)).

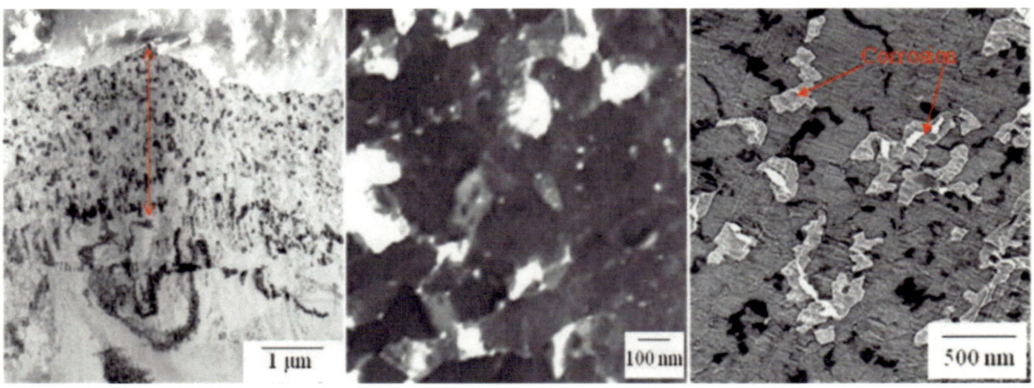

(a) (b) (c)

Fig. 3. Transmission electron micrographs of ultramicrotomed section of AA6111 aluminium alloy after SHT, mechanical grinding and 30 minutes at 180°C: (a) bright field image, revealing a near-surface deformed layer and (b) dark field image at increased magnification, revealing grain boundary precipitates. (c) transmission electron micrograph showing intergranular corrosion

2.7 High strength aluminium alloys AA2xxx and AA7xxx

Microstructural variation in the high strength Al-alloys exists over a range of scales as reported in Table 2. At the atomic and nanoscopic scale the microstructure is related to the mechanical properties of the alloy. This microstructure involves defect structures, hardening precipitates and dispersoid particles. The high strength of the 2xxx and 7xxx series alloys is due to the hardening precipitates with dispersoids playing a secondary role. Dispersoids can pin grain growth limiting grain size thus making a small contribution to increased

strength. Defects are generally undesirable since some types of defects give rise to poorer creep resistance. As will be seen below they appear to enhance corrosion in the form of grain etchout and have an influence on intergranular attack.
IM particles such as constituent and impurity particles exist at larger scales with minimum sizes generally between 0.5 to 1.0 µm. Some types of these particles can achieve local thermodynamic equilibrium during ingot formation. However, processing anomalies may mean that an equilibrium structure is not achieved in practice. Other constituents such as compositions in the Al-Cu-Mg ternary subphase field for 2xxx alloys and Al-Mg-Zn for 7xxx can redissolve during subsequent heat and other treatments.

Microstructural Feature and when formed	Size	Associated Corrosion
Atomic Defects (At any time during processing)	Point Defects <1Å Line defect - tens of nm long Dislocations	Grain etchout associated with higher grain stored energy.
Grain Boundaries (At any time during processing)	tens of nm wide (including the zone of influence such as depleted zones)	Intergranular attack. Some evidence from misorientation angle for preferred corrosion, generally only facilitated by second phase precipitates.
Hardening Precipitates (Ageing after Solution treating)	20 nm x 200 nm	Can facilitate intergranular attack.
Dispersoids (Ageing after Solution treating)	50 nm x 400 nm	Under some conditions undergo preferential attack
Constituent Particles and Impurity Particles (Primary Ingot Production)	Generally 0.5 µm to 50 µm	Localised attack of particle if anodic wrt the matrix and trenching in surrounding matrix if cathodic to the matrix
Clusters of particles (Ingot Working)	50 µm to 500 µm	Associated with pitting attack that propagates into the surface.

Table 2. Microstructural features in high strength aluminium alloys

AA2xxx

The AA2xxx series of alloys are among the most complicated to analyse. While there have been several reports of the compositions of different phases within this group, most have focused on the legacy alloy AA2024-T3, which, unfortunately, is one of the most complex of the 2xxx series of alloys. Perhaps one of the better known of these works was published in 1950 by Phragmen (Phragmen 1950). He examined all the binary, ternary, quaternary quinternary and senary compositions in order to under stand the IM particles in AA2024. Unfortunately the classification of these particles relied heavily on metallographic techniques (etches) and optical microscopy meaning that assignment of IM particles was, in many cases made on appearance and not on composition. These types of studies, however, form the basis for modern assignment of IM particles. From the perspective of obtaining the desired mechanical properties at the nanometer scale, characterisation has focused on the evolution of the alloy microstructure. Corrosion *initiation*, however, is much more closely related to the large constituent particles whose compositions are based on major alloying elements. Corrosion propagation involves all scales of the alloy microstructure.

Copper and magnesium are the two major alloying additions in AA2xxx wrought alloys and because of the Cu, this series is less resistant to corrosion than alloys of other series. Much of the thin sheet made of these alloys is produced as an Al-clad composite, with a relatively pure aluminium alloy as the outer layer, but thicker sheet and other products in many applications have no protective cladding. Electrochemical effects on corrosion can be stronger in these alloys than in alloys of many other types because of two factors: larger variations in electrochemical activity with variations in amount of copper in solid solution and, under some conditions, the presence of non-uniformities in solid solution concentration. The decrease in resistance to corrosion with increasing copper content is exacerbated by the corrosion process itself by the formation of minute copper particles or films deposited on the alloy surface as a result of corrosion.

To begin to understand the microstructure and its influence on corrosion it is important to know compositions of second phase intermetallic particles. However, the characterisation of these IM particles in the corrosion literature is poor with virtually no studies that relate composition with crystal structure. Hence, there is a need for a detailed identification system which links compositional variation within IM particles, with crystallography and electrochemical character. The objective is to move from a purely phenomenological description of corrosion to a level of understanding where the corrosion process can be predicatively modelled. This level of understanding is primarily aimed at developing structural health management algorithms for maintenance management and several approaches to this are already outlined in the literature (Hughes, Hinton et al. 2007; Cavanaugh, Buchheit et al. 2010; Ralston, Birbilis et al. 2010), (Trueman 2005). The determination of appropriate metrics for the compositional and electrochemical characteristics for IM particles is not so straight forward and is addressed below.

IM particles in AA2024-T3 are currently identified and categorised by one or a mixture of the following:

i. composition
ii. electrochemistry
ii. crystallography
iv. shape

Composition and electrochemistry are the most useful categories for corrosion studies and the convergence of these two systems is desirable and has already been achieved for alloys with simpler microstructure. This means that in many alloys a particular composition can be associated with a specific electrochemical behaviour. This is not the case, for example, in AA2024-T3 where there is large compositional variation as described below. The crystallography of IM phases is not so useful in corrosion studies since it is likely that specific crystal structures do not have a one to one relationship with either composition or the electrochemistry in AA2024-T3. What generally happens in the corrosion literature is that once the composition of an IM particle is determined then the particle is assigned a standard stoichiometry which is derived from crystallographic studies. So there is an assumption that a particular composition has a specific crystallographic structure which may not be valid. For example, Wei and coworkers (Gao, Feng et al. 1998; Wei, Liao et al. 1998) used TEM to study IM particles in AA2024-T3 and AA7075-T6 and found that compositions containing Al, Cu, Fe, Mn and Si, had a rhomohedral structure which did not match the hexagonal structure previously reported for particles of this type for example Al_8Fe_2Si or $Al_{10}Mn_3Si$. Classifying IM particles by shape is used both in metallurgy and corrosion. It is commonly used to distinguish between phases that contain Al-Cu-Fe-Mn-Si which tend to be angular and S (Al_2CuMg) and θ (Al_2Cu) phase which tend to be rounded. This distinction is alluded to in standard texts describing the microstructure of AA2024-T3 (Hatch 1984). This is the least reliable method of identification since it has recently been demonstrated that Al_7Cu_2Fe also has a rounded structure and is a similar size to S and θ-phase constituent particles suggesting that assignement by shape could easily lead to mis-identification.

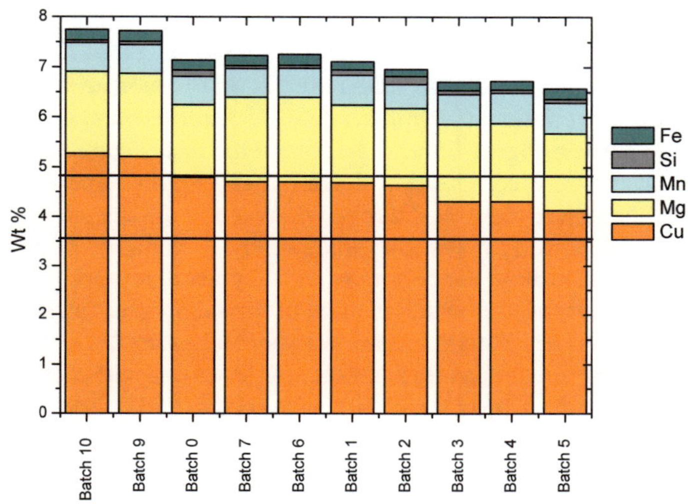

Fig. 4. Analyses in wt% for Cu, Mg, Mn, Si and Fe in ten different batches of AA2024-T3(51) purchased over the period 1995 to late 2000

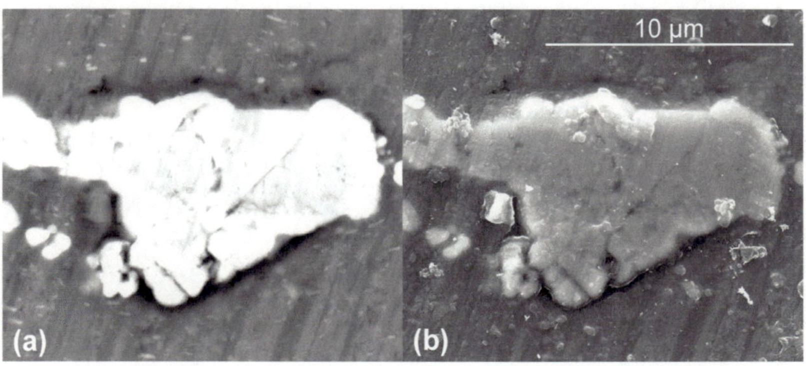

Fig. 5. (a) backscattered and (b) secondary electron images of an IM particle with compositional domains. Corroded in 0.1 M NaCl for 30 minutes at room temperature

To understand the complexity of the AA2024-T3 microstructure we begin with the compositional variation between several batches of sheet product (Figure 4). The major alloying components in the 2xxx series are Cu and Mg and they generally have similar mole fractions. Cu, Mn and Mg have specified compositional bands and the bands for Cu are superimposed on Figure 4. Clearly, most batches fall within specification, although two batches breach the upper specification limit by 0.3 to 0.4 wt%. The IM particles are formed from these elements and the variation in composition from batch to batch manifests itself in compositional variation of the IM particles. The origin of this variation is described above and is related to the source material.

The reason why compositional variation from batch to batch represents a difficulty is that the transition elements can be substitutional in many intermetallics. For example Ayer et al. (Ayer, Koo et al. 1985) found Zn and Ni in Al_7Cu_2Fe, and Gao et al. (Gao, Feng et al. 1998) found considerable compositional variation within a phase nominated as $(Fe,Mn)_xSi(Al,Cu)_y$ with some particles containing mainly Fe and Cu with small amounts of Mn and others had significant Mn and Si; similar results were reported by Boag et al.(Boag, Taylor et al. 2010). Gao et al. (Gao, Feng et al. 1998) even found small amounts of Cr and Zn in Al_2CuMg. These variations can make it difficult to classify the particles on the basis of composition alone. In addition to composition variation, many individual IM particles often contain compositional domains within the particle. Figure 5 shows backscattered and secondary electron images of an IM particle with different composition domains. The bright regions in the backscatter images are Cu and Fe rich as described above whereas the darker parts of the IM particle contain more Mn and Si. It is not know whether these domains have different crystal structures. The dark band around the bottom of the IM particle represents the beginning of a form of corrosion called trenching which is often observed around cathodic IM particles. It is clear that this corrosion has initiated in the matrix adjacent to the Cu and Fe rich part of the IM particle indicating greater electrochemical activity of this part of the particle.

From the literature, standard texts such as Hatch (Hatch 1984) lists the IM particles in AA2024-T3 and AA7075-T6 as presented in Table 3. These compositions are derived from metallurgical studies. At the top of the Table are the 'as-cast" compositions and in the bottom the wrought compositions. According to these standard texts, heat treatment dissolves much of the Al_2Cu and Al_2CuMg whereas all Fe-containing IM particles convert to Al_7Cu_2Fe. Thus the wrought composition contains the IM particles listed in Table 3. Compositional analyses in several other studies have some overlap with the composition of phases (based on stoichiometry) listed in Table 3. Compositional analyses from the three most comprehensive studies in the literature are presented Tables 4 to 6.

Treatment	**Phases**	
	AA2024-T3	AA7075-T6
Cast Ingot formation	$(Mn,Fe)_3SiAl_{12}$ Mg_2Si Al_2Cu (θ-phase) Al_2CuMg (s-phase) $Al_3(Fe,Mn)$ $Al_6(Fe,Mn)$	$(Fe,Cr)_3SiAl_{12}$ Mg_2Si Zn_2Mg ($(Zn,Cu,Al)_2Mg$)
Wrought	$A_{12}CuMg$ Unreacted $(Mn,Fe)_3SiAl_{12}$ Al_7Cu_2Fe $Al_{20}Mn_3Cu_2$ (Dispersoid)	$(Fe,Cr)_3SiAl_{12}$ Al_7Cu_2Fe $Al_{18}Mg_3Cr_2$ (Dispersoid)

Table 3. Typical breakdown of constituent particles in AA2024-T3 and AA7075-T6

Particle Type	Measured Stoichiometry	Other elementse
Al_2CuMg	$Al_{51.62}Cu_{24}Mg_{23.52}$	Fe(0.08), Mn(0.05), Cr(0.09) Zn(0.75)
Al_2CuMg (EPMA)	$Al_{52.8}Cu_{24.5}Mg_{22.7}$	-
Al_2Cu	$Al_{62.98}Cu_{33.83}Mg_{2.82}$	Fe(0.11), Mn(0.04), Cr(0.17) Ni(0.06)
$(Al,Cu)_ySi\ (Fe,Mn)_x$	$Al_{67.6}\ Cu_{3.36}Fe_{14.89}\ Mn_{6.87}Si_{6.27}$	Mg (0.43)

Table 4. Composition of IM particles in AA2024-T3 determined by Gao et al. (Gao, Feng et al. 1998)

Wei and co-workers(Gao, Feng et al. 1998) identified S-phase, θ-phase and the remainder were categorised as $(Fe,Mn)_xSi(Al,Cu)_y$ indicating no Al_7Cu_2Fe. Electron diffraction of these phases found an unidentified rhombohedral structure and particles with the general composition of $(Fe,Mn)_xSi(Al,Cu)_y$ were reported as variants of Al_8Fe_2Si or $Al_{10}Mn_3Si$. Buchheit et al.(Buchheit, Grant et al. 1997) performed an electron microprobe analysis of 652 particles and identified S-phase and a range of phases containing Al, Cu, Fe and Mn as listed in Table 5. In the most recent study, Boag et al. (Boag, Hughes et al. 2009), examined around 82,000 compositional domains in 18,000 IM particles and identified the compositions in Table 6. An example of these compositional domains within IM particles is shown in Figure 6. $(Al,Cu)_{21}(Mn,Fe)_4Si$ are widespread in the alloy and were assigned to the $(Fe,Mn)_xSi(Al,Cu)_y$ phase identified by Wei and co-workers. The S-phase and θ-phase predominantly occur as domains within individual, but composite particles. Al_7Cu_3Fe, in most instances, are a third group of particle and have domains, generally within their centre of Al_{10}(Cu, Mg). A periphery phase was observed around S/θ phase composite particles, which appeared to coincide with a precipitate free zone (Boag 2009; Boag, Hughes et al. 2009). There was also a periphery phase surrounding many particles.

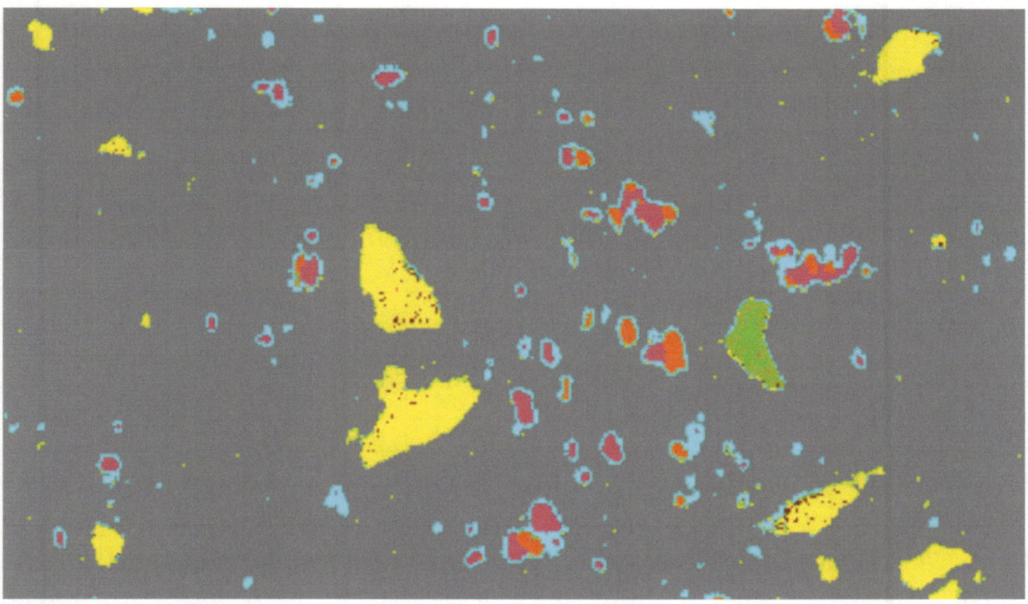

Fig. 6. Microprobe image IM particles in AA2024-T3 showing compositional domains. ■ = $(Al,Cu)_{21}(Mn,Fe)_4Si$, ■ = Al_2CuMg, ■ = Al_2Cu, ■ = Al_7Cu_3Fe, ■ = Al_{10}(Cu, Mg), ■ = Al_3(Cu,Fe,Mn)

What is clear from these studies is that there is no definitive composition for particles that contain Al, Cu, Mn, Fe, and Si (with small additions of other elements) and it is clear from observations such as those in Figure 5 that these regions of compositional variation have different electrochemical activity. In addition to the compositional variation there is evidence of considerable microstructural variations within the compositional field defined for AA2024-T3. For example θ-phase has been reported recently in studies of AA2024-T3, whereas no θ-phase was detected by Buchheit et al. (Buchheit, Grant et al. 1997) and Hughes and co-workers have examined different batches of AA2024-T3 sheet product and detected some batches with only S-phase and some with S-phase/ θ-phase composite particles(Boag, Taylor et al. 2010).

AA7xxx

In AA7xxx wrought alloy the major alloying element is zinc along with magnesium or magnesium plus copper in combinations that develop various levels of strength. Those containing copper have the highest strengths and have been used as structural materials, primarily in aircraft applications.

Particle Type	Number percent	Area percent
Al_2CuMg	61.3	2.69
$Al_3(Cu,Fe,Mn)$	12.3	0.85
Al_7Cu_3Fe	5.2	0.17
$(Al,Cu)_6Mn$	4.3	0.11
Indeterminate	16.9	0.37

Table 5. Composition of IM particles in AA2024-T3 determined by Buchheit et al. (Buchheit, Grant et al. 1997)

Phase Label	Measured Stoichiometry	Area (% of total)	Particle Density (number/cm²)	Mean Particle Diameter (μm)
Matrix	$Al_{96}Cu_2Mg_5$	Residual	-	
$(Al,Cu)_21(Mn,Fe)_4Si$	$Al_{77}Cu_5Mn_5Fe_{10}Si_4$	0.742	22052	5.19
Al_2CuMg	$Al_{61}Cu_{20}Mg_{15}$	0.381	22412	4.52
Al_7Cu_3Fe	$Al_{70}Cu_{18}MnFe_6$	0.089	22076	1.84
$(Al,Cu)_{93}(Fe,Mn)_5(Mg,Si)_2$	$Al_{90}Cu_3MgMn_2Fe_3$ Si	0.252	140296	1.46
$Al_{10}(Cu, Mg)$	$Al_{90}Cu_7Mg_2$	0.983	81856	5.38
$Al_3(Cu,Fe,Mn)$	$Al_{73}Cu_{11}Mn_4Fe_{10}Si$	0.062	17728	1.97
Periphery	$Al81Cu_{12}Mg_4MnFe$	0.018	3868	2.26
Al_2Cu	$Al_{70}Cu_{27}$	0.298	17568	4.60
Total		2.83%	320,000	N/A

Table 6. Composition of IM particles in AA2024-T3 determined by Boag et al. (Boag, Hughes et al. 2009)

The AA7xxx wrought alloys are anodic to AA1xxx wrought aluminium and to other aluminium alloys. Resistance to general corrosion of the copper-free wrought AA7xxx alloys is good, approaching that of the wrought AA3xxx, AA5xxx and AA6xxx alloys. The copper-containing alloys of the AA7xxx series, such as 7049, 7050, 7075, and 7178 have lower resistance to general corrosion than those of the same series that do not contain copper

(Meng and Frankel 2004). All AA7xxx alloys are more resistant to general corrosion than AA2xxx alloys, but less resistant than wrought alloys of other groups. The AA7xxx series alloys are among the aluminium alloys most susceptible to SCC and Cu is beneficial from the standpoint of resistance to SCC.

While the total weight of alloying components in AA7075-T6 is higher than AA2024-T3 by around 1 to 2% the microstructure tends to be simpler, in terms of the number and identification of IM particle types. Hardening precipitates are generally of the family η-phase (Zn_2Mg) and dispersoids are of the composition $Al_{20}Cu_2Mn_3$ and $Al_{18}Mg_3Cr_2$. Like AA2024-T3 reports of constituent particle compositions vary. Gao et al.(Gao, Feng et al. 1998) report two phases: $Al_{23}Fe_4Cu$ and SiO_2. However, constituent particles compositions reported by others authors suggest Al_7Cu_2Fe, Al_2Zn, Al_3Zr and Mg_2Si (Birbilis and Buchheit 2005; Wloka and Virtanen 2008).

2.8 Clustering

Clustering of IM particles is an emerging area of importance in understanding pit initiation and stabilisation (Chen, Gao et al. 1996; Park, Paik et al. 1996; Park, Paik et al. 1999), (Ilevbare, Schneider et al. 2004), (Liao, Olive et al. 1998; Schneider, Ilevbare et al. 2004; Harlow, Wang et al. 2006), (Cawley and Harlow 1996; Hughes, Boag et al. 2006; Mao, Gokhale et al. 2006; Hughes, Wilson et al. 2009; Hughes, MacRae et al. 2010). Clustering may be important at several different length scales and perhaps even times scales (for corrosion processes). Clustering at length scales similar to the IM particle size can be attributed to IM particle fracture during mechanical processing and, in some instances to non-equilibrium microstructures. An explanation of IM particle clustering reported for larger scale of a few hundred microns is not clear. A study by Mao et al., (Mao, Gokhale et al. 2006) revealed both short range (size similar to the particle dimensions) and long range (few hundred times the particle size) clustering in AA7075 alloy plate material. Clearly the clustered structures are elongated in the rolling direction and they have a range of different sizes. Hughes and co-workers (Hughes, Boag et al. 2006; Hughes, Wilson et al. 2009; Hughes, Muster et al. 2010) reported significant clustering in AA2024-T3 alloy sheet between phase domains within IM particles, as well as between IM particles themselves. In their study they identified strong clustering behaviour between S-phase and θ-phase, S-phase and the Al_7Cu_2Fe phase and to a lesser extent between S-phase and IM particles with an average stoichoimetry of $(Al,Cu)_{21}(Fe,Mn)_4Si$. In that particular study the microstructure consisted of individual particles which had compositional domains of S and θ, which represented a the highest degree of clustering.

Clustering behaviour has also been reported for a number of other aluminium alloys including AA6061-T6, AA7075-T6 and AA5005 (Cawley and Harlow 1996; Hughes, Boag et al. 2006). Coupling between IM particles types of different electrochemical activity has been observed at stable pit sites and attributed to their initiation (Liao, Olive et al. 1998; Boag, Taylor et al. 2010). On the other hand Wei and co-workers (Chen, Gao et al. 1996; Liao, Olive et al. 1998) and Ilevbare et al. (Ilevbare, Schneider et al. 2004) have concluded that clustering in AA2024-T3 and AA7075-T6 alloys leads to large stable pits, primarily through excessive lateral trenching. Cawley and Harlow (Cawley and Harlow 1996) found that IM particles in AA2024-T3 alloys tended to be clustered whereas the pits tended to be randomly distributed because the spatial relationships between IM particles is lost during excessive corrosion.

In the studies above, clustering was assessed on a statistical basis to determine the average properties of clusters, i.e. lateral size of the cluster, number of particles, types of particles. This raises an interesting question of how these results should be interpreted for modelling applications. The data reported to date tends to describe average clustering behaviour but severe corrosion events might more appropriately be assigned to the extreme properties of the clusters i.e., the densest collection of particles or the most active collection of particles. In this context Boag et al. (Boag, Taylor et al. 2010) observed that the clusters with the highest density of IM particles were those associated with active corrosion on AA2024-T3. These studies suggest that once average IM properties have been assessed for any particular sample then it might also be necessary to determine extreme values.
To conclude this section, it is evident that the compositions of IM particles in AA2xxx and AA7xxx alloys is extremely varied and not well described by any particular classification system for any particular alloy.

3. Corrosion and microstructure

3.1 Corrosion fundamentals

Corrosion in aluminium alloys is generally of a local nature, because of the separation of anodic and cathodic reactions and solution resistance limiting the galvanic cell size. The basic anodic reaction is metal dissolution ($Al \rightarrow Al^{3+} + 3e^-$) and the cathodic reactions are oxygen reduction ($O_2 + 2H_2O + 4e^- \rightarrow 4OH^-$) and hydrogen reduction ($2H^+ + 2e \rightarrow H_2$) in acidified solution such as in a pit environment as a result of aluminium ion hydrolysis.
It is the interaction between local cathodes and anodes and the alloy matrix that leads to nearly all forms of corrosion in aluminium alloys. These include trenching, intermetallic particle etchout, pitting corrosion, intergranular attack and exfoliation corrosion. Surface and subsurface grain etchout is dictated more by grain energy which is derived from grain defect density as described above. Grain etchout, has a significant role in exfoliation corrosion since the volume of hydrated aluminium oxide generated during dissolution is larger than the original volume of the grain.
Relatively pure aluminium presents excellent corrosion resistance due to the formation of a barrier oxide film that is bonded strongly to its surface (passive layer) and, that if damaged, re-forms immediately in most environments (re-passivation). This protective oxide layer is especially stable in aerated solutions of most non-halide salts leading to an excellent pitting resistance. Nevertheless, in open air solutions containing halide ions, with Cl- being the most common, aluminium is very susceptible to pitting corrosion. This process occurs, because in the presence of oxygen, the metal is readily polarized to its pitting potential, and, because of the presence of chlorides, forms a very soluble chlorinated aluminium (hydr)oxide that does not allow the formation of a stable oxide on the aluminium surface.
On the other hand, industrial alloys surfaces are almost as heterogeneous materials. The surface of a wrought or cast alloy is likely to contain a either mixed Al-Mg oxide (for alloys with Mg (Harvey, Hughes et al. 2008)) or aluminium oxide, almost regardless of the alloy type. This is primarily because of the heat of segregation of Mg is high and it has a favourable free energy for the formation of the oxide. Aluminium, readily oxidises both in IM particles as well as from the matrix. If the surface was mechanically undisturbed then this oxide would be relatively protective, However, most real surfaces have some sort of mechanical finishing which results in the formation of the NSDL and shingling. Shingling

occurs where the alloy matrix is spread across the surface including IM particles since the IM particles are harder than the surrounding matrix and less susceptible to deformation (Zhou 2011). Even on polished surfaces, the matrix and the IM particles rapidly form different oxide structures (Juffs, Hughes et al. 2001; Juffs, Hughes et al. 2002). This is almost certainly due to different chemical environments due to different electrochemical reactions over the IM particles compared to the matrix. Furthermore, the morphology and the oxide are not continuous from the IM particles to the matrix and this represents a significant defect site.

The solution potential of an aluminium alloy is primarily determined by the composition of the aluminium rich solid solution, which constitutes the predominant volume fraction and area fraction of the alloy microstructure. While the solution potential is not affected significantly by second phase particles of microscopic size, these particles frequently have solution potentials differing from that of the solid solution matrix resulting in local (micro-) galvanic cells, leading to a variety of local types of corrosion, such as pitting, exfoliation etc. Since most of the commercial aluminium alloys contain additions of more than one type of alloying element, the effects of multiple elements on solution potential are approximately additive. The amounts retained in solid solution, particularly for more highly alloyed compositions, depend on production and thermal processing so that the heat treatment and other processing variables influence the final electrode potential of the product.

Solution potential measurements are useful for the investigation of heat treating, quenching, and aging practices, and they are applied principally to alloys containing copper, magnesium, or zinc. By measuring the potentials of grain boundaries and grain bodies separately, the difference in potential responsible for local types of corrosion such as intergranular corrosion, exfoliation, and stress corrosion cracking (SCC) can be quantified (Guillaumin and Mankowski 1999; Zhang and Frankel 2003). Solution-potential measurement of alloys containing copper also show the progress of artificial aging as increased amounts of precipitates are formed and the matrix is depleted of copper. Potential measurements are valuable with zinc-containing (AA7xxx) alloys for evaluating the effectiveness of the solution heat treatment, for following the aging process, and for differentiating among the various artificially aged tempers. These factors can affect corrosion behaviour significantly.

3.2 Effects of microstructure on corrosion

From a corrosion perspective, the dominant features of alloy microstructure are the grain structure and the distribution of second phase IM particles including constituent and impurity particles, dispersoids and precipitates. At the largest scale, corrosion is observed around clusters of constituent and impurity particles which results in severe pitting attack (Chen, Gao et al. 1996; Liao, Olive et al. 1998; Boag, Taylor et al. 2010; Glenn, Muster et al. 2011; Hughes, Boag et al. 2011). Attack around isolated intermetallic particles is now relatively well understood and more on this will be said below. Dispersoids and precipitates have electrochemical characteristics that differ from the behaviour of the surrounding alloy matrix, which is the cause of localized forms of corrosion attack that is often termed micro-galvanic corrosion; however it is also now appreciated that such a term does not cover the full complexity of corrosion on Al-alloys. For example Figure 7 shows the co-existance of fine precipitates in the matrix with a coarse constituent particle embedded within the low grain.

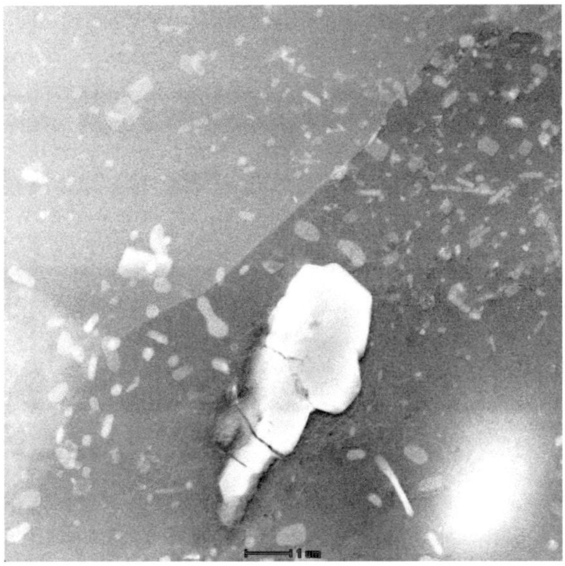

Fig. 7. Dark field scanning transmission electron micrograph of constituent particle co-existing with S-phase (Al_2CuMg) precipitate particles in AA2024-T3 sheet

Over the years there have been a number of studies that have assessed the effect of intermetallic particles on the corrosion susceptibility of specific aluminium alloys (Scully, Knight et al. 1993; Birbilis and Buchheit 2005),(Zamin 1981; Mazurkiewicz and Piotrowski 1983; Scully, Knight et al. 1993; Seri 1994). In the 1990s, Buchheit collected the corrosion potential values for intermetallic phases common to aluminium alloys mainly in chloride containing solutions (Buchheit 1995). More recently various groups have focussed on the electrochemical properties of Fe containing intermetallics (Pryor and Fister 1984; Afseth, Nordlien et al. 2002), and Cu containing intermetallics (Searles, Gouma et al. 2001; Birbilis, Cavanaugh et al. 2006; Birbilis and Buchheit 2008) which has been expanded into a comprehensive treatise covering a variety of common intermetallics present in aluminium alloys (Frankel 1998; Birbilis and Buchheit 2005; Birbilis and Buchheit 2008). A summary of the results of these studies is shown below in Table 7.

3.3 Second phases

3.3.1 Attack around isolated IM particles

IM particles may be either anodic or cathodic relative to the matrix under any particular set of solution conditions. As a result, two main types of pit morphologies are typically observed around isolated IM particles. Circumferential pits generally appear as trenches around a more or less intact particle and the corrosion attack is mainly in the matrix phase. In the case of clusters, the development of the trench around the cluster is a secondary corrosion reaction following extensive grain boundary attack. Trenching around isolated particles does not lead to more severe corrosion (Chen, Gao et al. 1996; Ilevbare, Schneider et al. 2004; Schneider, Ilevbare et al. 2004; Boag, Hughes et al. 2011). This can be understood in terms of a quantity called the pit stability product first reported by Galvele. This quantity is defined as *i.r* where *r* is the depth of the pit and *i* is the current density. This quantity must be greater than 10^{-2} Acm^{-1} for a nascent pit to be able to grow rapidly enough to establish a long enough diffusion path for an oxygen gradient to be established. This results in oxygen reduction near the mouth of the pit at these early stages and an acidic salt solution at the active pit face. Studies have shown that this product is too low for trenching events to develop into stable pits around isolated cathodic particles even for S-phase dealloying (Schneider, Ilevbare et al. 2004).

The second type of pit morphology is due to the selective dissolution of the constituent particle. Pits of this type are often deep and may have remnants of the particle in them. Figure 8 shows a model of dealloying of an S-phase IM particle which leads to a Cu-enriched remnants as well as non-faradaic liberation of the Cu. Under neutral pH conditions magnesium and aluminium are preferentially dissolved from the Al_2CuMg phase, leaving a Cu-enriched and high surface area remnant, which then exhibits solution potentials noble to the matrix (Buchheit, Grant et al. 1997). Ultimately, the form that copper takes on the surface is thought to be important in determining the corrosion-performance of alloys such as AA2024. The redistribution of copper has been demonstrated to enhance the kinetics of oxygen reduction processes and negatively affect corrosion. In dealloying from a bulk phase the physical structure of a surface has been predicted by percolation theory to be dependent upon the dissolution rate and concentration of the noble elements in the phase (Sieradzki 1993; Newman and Sieradzki 1994). Rapid dissolution rates lead to more porous network structures, where there is a possibility that unoxidized fragments enriched in the more noble metal will be released into solution, whereas slow dissolution allows surface diffusion and relaxation processes to maintain a stable surface structure. Also, if the noble metal content is sufficient, dealloying will not lead to an isolation of the percolation network. Theory suggests the copper concentration of 25 at% contained in the Al_2CuMg phase allows it to dealloy and form both porous copper-rich networks and also to release clusters of both oxidized and unoxidized copper into an electrolyte. It is also noted that hydrodynamic forces may assist in the release of fragments (Buchheit, Martinez et al. 2000; Vukmirovic, Dimitrov et al. 2002; Muster 2009).

Phase	Corrosion Potential (mV_{SCE})				
	0.1M NaCl pH 2.5	0.01M NaCl pH 6	0.1M NaCl pH 6	0.6M NaCl pH 6	0.1M NaCl pH 12.5
Al_3Fe	-510	-493	-539	-566	-230
Al_2Cu	-546	-592	-665	-695	-743
Al_6Mn	-	-839	-779	-779	-
Al_3Ti	-	-620	-603	-603	-
$Al_{32}Zn_{49}$	-	-1009	-1004	-1004	-
Mg_2Al_3	-	-1124	-1013	-1013	-
$MgZn_2$	-1007	-1001	-1029	-1029	-1012
Mg_2Si	-1408	-1355	-1538	-1538	-1553
Al_7Cu_2Fe	-535	-549	-551	-551	-594
Al_2CuMg	-750	-956	-883	-883	-670
$Al_{20}Cu_2Mn_3$	-	-550	-565	-565	-
$Al_{12}Mn_3Si$	-	-890	-810	-810	-
Al-2%Cu	-	-813	-672	-744	-
Al-4%Cu	-	-750	-602	-642	-

Table 7. Summary of corrosion potentials for intermetallic particles common to Al alloys

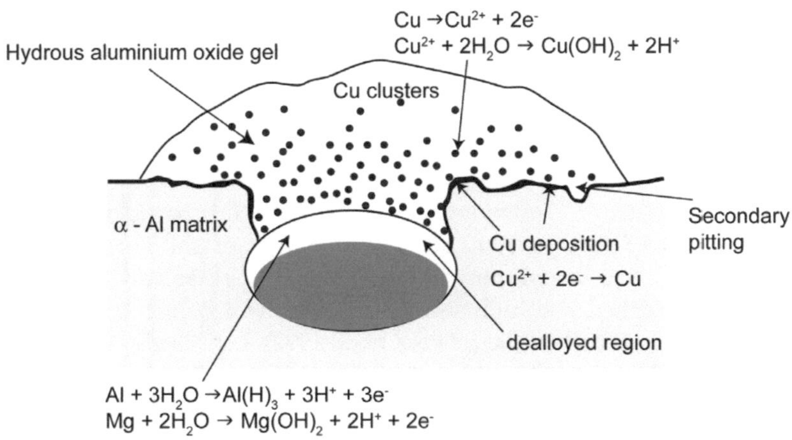

Fig. 8. Schematic of dealloying of Al_2CuMg phase contained within an aluminium alloy matrix. Anodic polarisation of the particle results in preferential loss of Al and Mg. A Cu-rich network, which coarsens with age, and is susceptible to break-up during hydrodynamic flow, releases small Cu-rich particles (diameter approximately 10-100 nm). The electrically isolated particles are dissolved, making Cu ions available for replating as elemental Cu onto cathodic sites. The replated cathodic sites serve as efficient local cathodes that stimulate secondary pitting (after Buchheit et al., 2000)

Localised corrosion activity is, however, a complex phenomenon that is still under active research. Localised corrosion leads to local pH gradients as recently studied in detail by Ilevbare and Schneider (Ilevbare, Schneider et al. 2004; Schneider, Ilevbare et al. 2004). Enhanced oxygen reduction at cathodic sites generate hydroxyl ions promoting local pH increases, which can then modify the subsequent rate and morphology of corrosion propagation. Recent work by Boag et al. revealed the timescales on which various forms of attack occurred on isolated IM particles in AA2024-T3. In this study the IM particles in AA2024-T3 were divided into S-phase, AlCuFeMn phases and $(Al,Cu)_x(Fe,Mn)_ySi$ since these compositions appeared to have separated timescales for attack (Figure 9). These compositions follow the electrochemical activity and provide one method for categorising the IM particles which brings together composition and electrochemistry. Whether this is sufficient for modelling purposes remains to be answered since it does not distinguish, for example, between compositional domains within the one particle that have different corrosion activities as seen in Figure 5. With respect to Figure 9, attack began at the S-phase particles which underwent dealloying indicating anodic behaviour, but then switched to trenching indicating active cathodic behaviour. This was followed by trenching around AlCuFeMn particles after 30 minutes and finally by trenching around $(Al,Cu)_xSi(Fe,Mn)_y$ at 120 minutes.

The precise morphology of particle-induced pitting is important for emerging damage accumulation models. For these models to be predictive, it is necessary to develop a comprehensive, self-consistent accounting of this type of pitting. In cases where the electrochemical characteristics of constituent particles have been rigorously characterized, they have been found to have much more complicated behaviour than categorised by simple characterizations like "noble" or "active". In recent years, Buchheit and co-workers have attempted an accounting of such phenomena (Figure 10). In such work, electrochemical

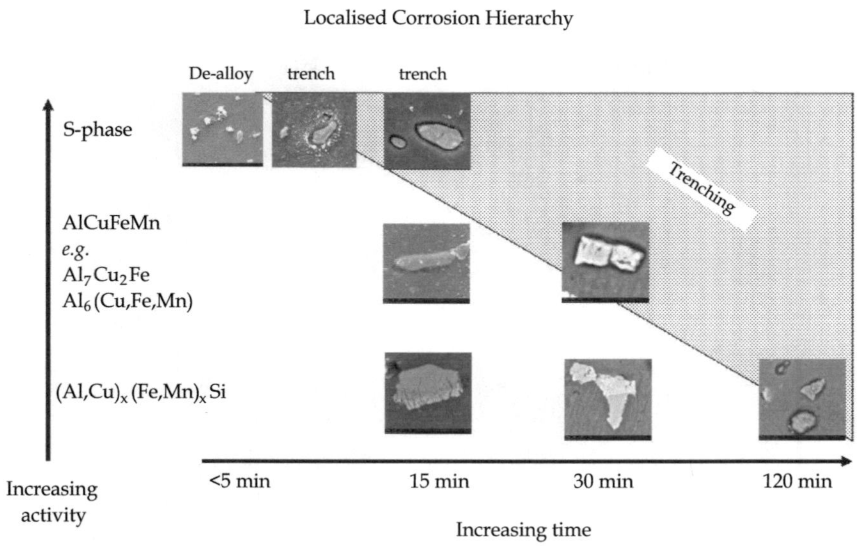

Fig. 9. Hierarchy of localised corrosion attack in 0.1 M NaCl for AA2024-T3 from Boag et al. (Boag, Hughes et al. 2011)

information relating to microstructural features (from the work used to generate Table 7) was overlaid spatially above real micrographs of AA7075. The data represents the given anodic or cathodic current sustained (at a potential corresponding to the open circuit potential) for each microstructural feature. It is seen that, depending on environment, the currents realised are rather different; and this reconciles very well with 'ground truth' observations from SEM of exposed specimens.

An assumption behind the electrochemical work is that stable pits are on the extreme end (in an extreme value statistics sense) of the metastable pitting events as can be seen in Table 8 for total charge passed, pit stability product and the pit size. There have been other studies that suggest that the current transients are fundamentally different involving activation - repassivation events within active pits (Sasaki, Levy et al. 2002). Some studies have also tried to connect phenomenological measurements to the current transients such as through measuring the volume of the trench around an IM particle to the total charge passed. While the total charge passed is similar for both cases, it can be seen (Figure 9) that S-phase dealloying is complete with 5 minutes which means that there are likely to be 75 trenching events occurring simultaneously/cm^2 within the first 5 minutes. However, the frequency of metastable current transients on AA2024-T3 was close to 0.02 s which is nearly four orders of magnitude smaller. Transient currents associated with η-phase have been observed in AA7075 (Wloka and Virtanen 2008).

3.3.2 Corrosion and particles clusters

As described earlier, the reports of clustering of intermetallic particles in Al-alloys is an emerging area of research in aluminium corrosion. While severe pitting has for many years been attributed to clustering (Chen, Gao et al. 1996; Gao, Feng et al. 1998; Liao, Olive et al. 1998; Schneider, Ilevbare et al. 2004), it is only recently that detailed studies of the connections between clusters and severe pitting have emerged. Part of this work is based on electrochemical studies of current transients measured on corroding surfaces to try to

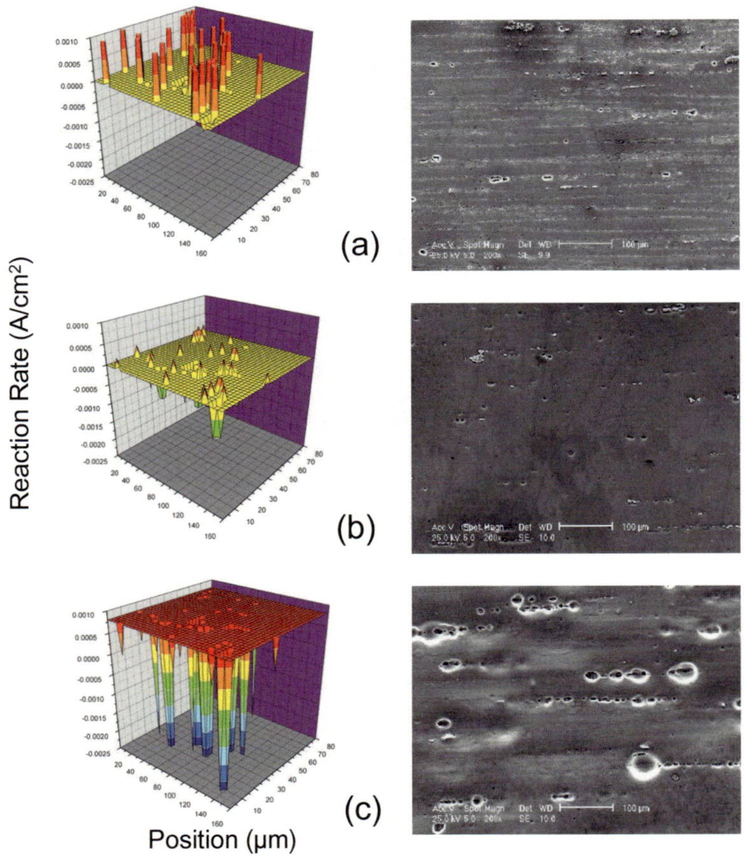

Fig. 10. Left: 3-D representations of the spatial variation in electrochemical reaction rate associated with the 7075 alloy microstructure at a potential of -0.8Vsce in aerated 0.1M NaCl solution at 23°C and pH 2.5 (a), pH 6 (b), and pH 12 (c). Right: SE images of AA7075 after a 24 h exposure under free corrosion conditions in aerated 0.1M NaCl solution at 23°C and pH 2.5 (a), pH 6 (b), and pH 12 (c)

establish the boundary between the *i.r.* product for metastable and stable pits. Such current transients manifest in three ways (i) spontaneous nucleation/passivation events which have very short timescales, (ii) metastable pitting events which have lifetimes up to a several seconds on AA2024-T3 before decaying back to background currents and (iii) stable pitting events, which increase to a transient peak but then decay to a constant current indicating ongoing electrochemical activity. The literature in this area largely comes from a phenomenological or electrochemical perspective, using different terminologies, Table 8, summaries the activities from these different perspectives.

In terms of the relationship of clustering to stable or severe pitting, it can be seen clearly from Figure 10 that a higher density of IM particles, means a greater level of electrochemical activity. Excessive trenching is the most common explanation for the development of stable pitting from clusters of IM particles. However, in terms of developing models for corrosion initiation, it is important to understand all the pathways to the establishment of stable pits and the mechanisms for those pathways. Recently, Hughes and co-workers reported

	Metastable	Stable
Phenomenological		
Attack descripition	Pitting (Chen, Gao et al. 1996; Liao, Olive et al. 1998) Trenching/S-phase etch [(Kolics, Besing et al. 2001) (Liao, Olive et al. 1998; Schneider, Ilevbare et al. 2004) (Lacroix, Ressier et al. 2008), (Suter 2001; Blanc, Gastaud et al. 2003)	Severe pitting (Chen, Gao et al. 1996; Liao, Olive et al. 1998) Co-operative corrosion (Hughes, Boag et al. 2011)
Attack Type	Individual IM particles	Clusters
Attribution	S-phase dealloying and etchout, Trenching	Coupling of different IM particle types Clustered trenching (Liao, Olive et al. 1998) Corrosion Rings (Liao, Olive et al. 1998)
Propagation	Virtually none	GBA, IGA Subsurface S-phase etchout Grain etchout
H_2 evolution	None?	H_2 Evolution from within corrosion rings: It can be continuous or irregular
Electrochemical		
Total charge (μC)	< 13 (Trueman 2005) to 18 (Pride, Scully et al. 1994)	> 13 (Trueman 2005) to 18 (Pride, Scully et al. 1994)
i.r	$<10^{-2}$A/cm	$>10^{-2}$A/cm

Table 8. Corrosion Attack on A2024-T3

corrosion initiation within rings of corrosion product on AA2024-T3. Rings of corrosion product were observed a early as 1952 by Pearson et al. in 2S aluminium alloy and developed as early as 15 minutes after immersion. These rings were also observed in the work of Wei's group (Chen, Gao et al. 1996). Hughes et al. reported that the early stages of corrosion within these rings is characterised by H_2 evolution and extensive grain boundary attack on the surface which happens prior to, or at the same time as trenching around IM particles (Figure 11). Boag et al. showed that only the most clustered sites on the surface of AA2024-T3 were associated with chloride signals drawing a strong link between clustering and stable pitting. The subsurface attack at these sites was almost exclusively intergranular, penetrating as much as 60 μm in 120 minutes exposure to 0.1M NaCl. The substantial intergranular attack extending deep into the AA2024-T3 and preceded the development of open pits, suggesting that intergranular attack proceeds prior to any substantial grain etchout, even at the surface.

These results suggest that trenching is either a secondary corrosion process or that there are different mechanisms of corrosion initiation which contribute to stable pitting. To further complicate the picture, Zhou, Thompson and co-workers observed that in some alloys there were large S/θ composite particles and that stable pits were established by direct penetration through the composite particle into the subsurface region of the alloy. This type of attack may be responsible for the widely accepted view that severe pitting can start at S-phase particles. Clearly there may be several avenues to the establishment of stable pits and this is still an area of active research.

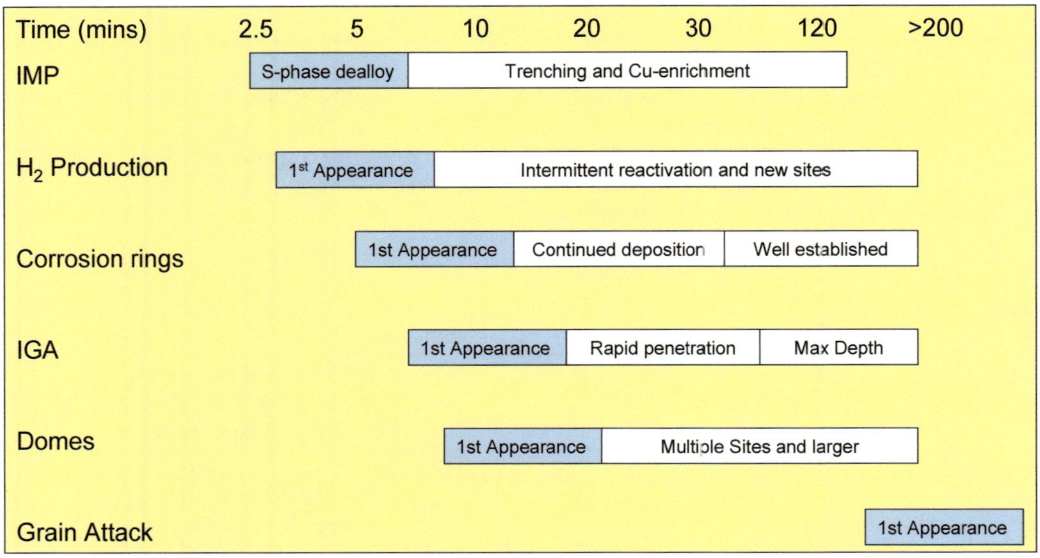

Fig. 11. Stages to stable Pitting after Hughes et al. (Hughes, Boag et al. 2011)

3.4 Intergranular attack

Intergranular corrosion is a phenomenon of which the precise mechanisms have been under debate for almost half a century (Hunter 1963). Whilst in a simple view intergranular corrosion can be considered as a special form of microstructurally influenced corrosion, intergranular corrosion can be summarised as a process whereby the grain boundary 'region' of the alloy is preferentially attacked, this is most often because it is anodic to the bulk or adjacent alloy microstructure. Intergranular corrosion can initiate at second phase IM particles in the surface, from pits and from grain boundaries at the surface. Intergranular corrosion penetrates more rapidly than pitting corrosion, and whilst both may have a deleterious effect on corrosion fatigue, the sharper tips produced by intergranular attack are drastic stress concentrators which may reduce the number of cycles to failure.

Corrosion activity may develop because of some heterogeneity in the grain boundary structure. In aluminium-copper alloys, precipitation of Al_2Cu particles at the grain boundaries leaves the adjacent solid solution anodic and more prone to corrosion. With aluminium-magnesium alloys the opposite situation occurs, since the precipitated phase Mg_2Al_3 is less noble than the solid solution. Serious intergranular attack in these two alloys may however be avoided, provided that correct manufacturing and heat treatment conditions are observed.

The distribution of second phase material has a significant influence on the corrosion behaviour of high strength aluminium alloys. Thus, if second phases are located preferentially at grain boundaries, they may promote intergranular corrosion (IGC) due to their compositional, and, hence, electrochemical differences with respect to the adjacent alloy matrix. Further, as a result of precipitation at grain boundaries, the formation of a narrow band on either side of the grain boundary, the precipitation free zone (PFZ), also influences the corrosion behaviour of aluminium alloys since the PFZ is depleted of particular alloying elements. Thus, intergranular corrosion of high strength aluminium alloys is often attributed to compositionally different features at the grain boundary due mainly to anodic dissolution of (i) the precipitation free zone where noble alloying elements such as copper are depleted, (ii) anodic second phase precipitates at the grain boundary, or (iii) grain boundaries with segregated alloying elements, such as magnesium, or impurity elements. In the case of the Al-Cu-Mg system, for example, numerous previous studies have suggested that in AA2024 aluminium alloy, when copper-rich precipitates ($CuAl_2$) formed at grain boundaries, copper-depleted regions develop adjacent to the boundaries, which are anodic with respect to copper-rich grain boundaries and the grain matrix. A lot of IGC in AA2024 aluminium alloy can be explained as micro-galvanic corrosion of copper-depleted regions driven by cathodic areas of copper-rich grain boundaries and grain matrix. Further, when S phase ($CuMgAl_2$) particles are preferentially precipitated at grain boundaries (typically <100 nm), their anodic nature with respect to the adjacent grain matrix, results in their preferential dissolution.

Not all IGA is so well understood. In AA2024-T3, for example, IGC has been observed in the absence of second phase precipitates, and penetrates up to 60 μm within 120 min of immersion in 0.1 M NaCl at ambient temperature (Glenn, Muster et al. 2011; Hughes, Boag et al. 2011). While some of the attacked grain boundaries were decorated with $CuAl_2$ or $CuMgAl_2$ precipitates, many grain boundaries that were subject to intergranular attack were not associated with such precipitates, as shown in Figure 12. The TEM image of the corrosion front reveals two parts of a grain boundary, (i) the attacked grain boundary region behind the corrosion front and (ii) the intact grain boundary ahead of the corrosion front. Clearly, precipitation at the grain boundary is absent. Interestingly, although corrosion propagated in a confined region along the grain boundary and the corrosion front followed the grain boundary, corrosion development across the grain boundary is uneven with the interior of grain B being preferentially attacked. Corrosion has developed more than 70 nm into grain B, with comparably little attack on grain A.

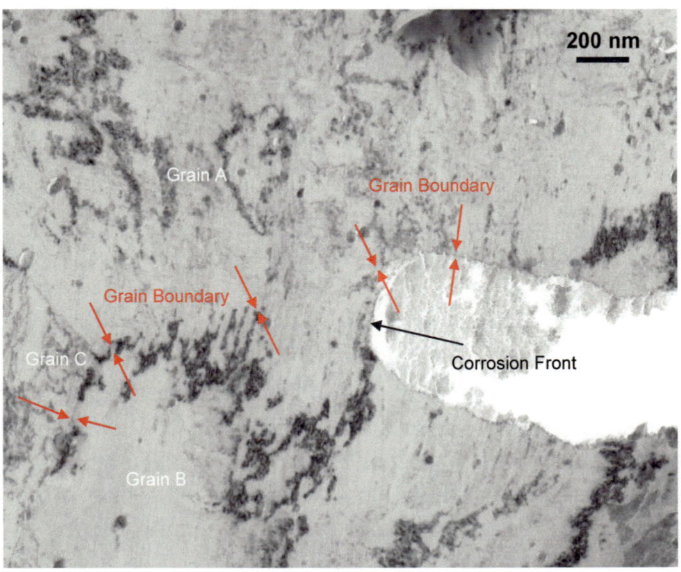

Fig. 12. Transmission electron micrograph of the corrosion front, revealing two parts of a grain boundary: the attacked grain boundary region behind the corrosion front and the intact grain boundary ahead the corrosion front

Further investigation of the relationship between grain boundary misorientation and its susceptibility to corrosion revealed that the distribution of grain boundary misorientation for the attacked grain boundaries was similar to that in the as-fabricated alloy (Glenn, Muster et al. 2011; Luo to be published). Thus, there must be other factor(s) that influence the corrosion susceptibility of grain boundaries and its zone of influence.

It was revealed using electron backscatter diffraction (EBSD) that grain stored energy plays a significant role in grain boundary attack. The stored energy of a pixel was determined from the misorientation of a pixel from its neighbours within grains. For each pair of pixels, with misorientation above a selected threshold value, a mean boundary energy was calculated using the Read-Shockley equation:

$$\gamma_s = \gamma_0 \, \theta \, (A - \ln \theta)$$

where θ is the misorientation angle, γ_0 and A are constants. The stored energy of an individual grain/subgrain is determined by averaging the sum of the mean boundary energies over the area of individual grains/subgrains. Then, the spatial distribution of the stored energy is represented as a map, which reflects the average population density of dislocations in individual grains/subgrains.

Figures 13(a) and (b) display a scanning electron micrograph of an AA2024-T3 aluminium alloy surface after immersion in 0.5 M NaCl solution for 4 h at ambient temperature and the stored energy map of the same area (Luo to be published). Brighter grains/subgrains have a higher level of stored energy. The corrosion product has been removed from the alloy surface which has been further cleaned using an argon plasma. The attacked local regions along grain boundaries are clearly evident. Comparing the SEM image with the stored energy map of Figure 13(b), it is evident that the regions of attack are located along the grain boundaries that surround grains of relatively high stored energy. Corrosion is not confined within the region immediately adjacent to the grain boundaries, but has developed 1-2 μm into the grains of relatively high stored energy, suggesting that grains with relatively high level of defects are more susceptible to corrosion. The cold work applied to the alloy to achieve the T3 temper resulted in relatively high population density of dislocations in the alloy. The population density of dislocations may vary from grain to grain since grains with different orientations to the rolling direction are subjected to higher levels of strain. Thus, more defects may be introduced in certain grains which have experienced more deformation than other grains. Consequently, some grains are more susceptible to corrosion than other.

Figure 14 displays a backscattered electron micrograph of the cross section of AA2099-T8 aluminium alloy after polarization to 0.824 V (SCE) in a 0.5 M NaCl solution. Grain boundary attack is evident to a depth up to ~55 μm. Individual subgrains were selectively attacked, as indicated by arrows. Some attacked subgrains are close to the surface region, with others being relatively deep into the bulk alloy. It is believed that the subgrains close to the surface region were exposed to the solution earlier than those far away from the surface region. Thus, the attack to certain subgrains within the inner regions suggested that the selective attack is determined by the intrinsic microstructure. EBSD indicates that the attacked sites tend to be the grains of relatively high stored energy compared with the intact grains that are surrounding the attacked grains, suggesting that the grains of high stored energy have relatively high corrosion susceptibility (Luo to be published).

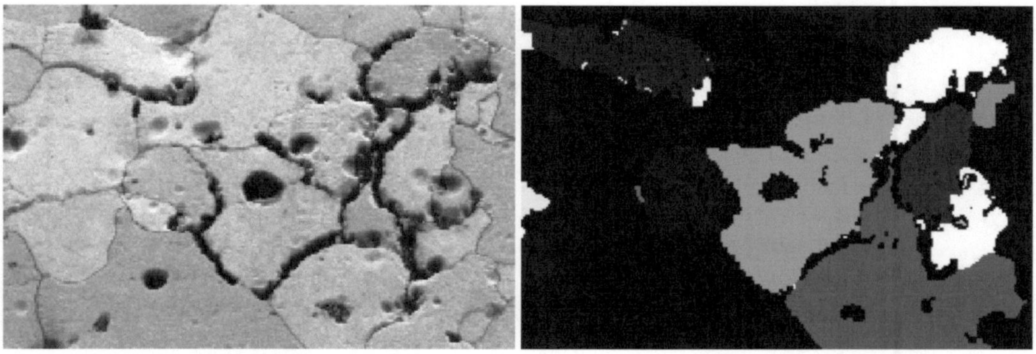

Fig. 13. a) Scanning electron micrograph of the alloy surface after 4 h of immersion in 0.5 M NaCl solution at ambient temperature, revealing attacked local regions along grain boundaries; and b) grain stored energy map of the same area as a), obtained with threshold value for misorientation set at 1.3

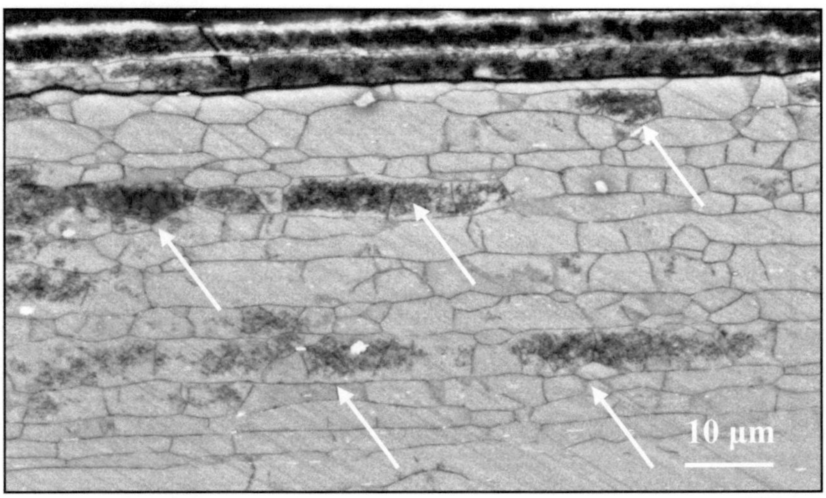

Fig. 14. Scanning electron micrograph of the cross section of AA 2099-T8 aluminium alloy after polarization to 0.824 V (SCE) in 0.5 M NaCl solution

In AA7xxx aluminium alloys, when anodic precipitates, such as η-$MgZn_2$ phase, are formed at the grain boundaries, then these are relatively active with respect to the grain matrix. IGC occurs, with micro-galvanic coupling between the η-$MgZn_2$ phase precipitates and the aluminium matrix adjacent to the particles providing the driving force (Wadeson, Zhou et al. 2006). Again however susceptibility to intergranular corrosion is strongly dependent on the heat treatment condition and its effect on grain boundary solute segregation and the morphology and composition of the grain boundary precipitate and the surrounding alloy matrix. (Knight 2003). The most resistant heat treatments are based on the use of over-aging to the T7 treatment or more complex heat treatments which involve retrogression and re-aging to minimise the trade off between alloy strength and intergranular corrosion resistance (Polmear 1995; Davies 1999). The images in Figure 15 help rationalise the origins of intergranular corrosion, in the Al-Zn-Mg system.

Exfoliation corrosion (Davies 1999; Zhao and Frankel 2007) of aluminium alloys is also frequently due to intergranular corrosion. It generally occurs where the alloy microstructure has been heavily deformed (i.e. by rolling) and the grain structure has been flattened and extended in the direction of working. Intergranular corrosion attack from transverse edges and pits then run along grain boundaries parallel to the alloy surface. Exfoliation is characterised by leafing off of layers of relatively uncorroded metal caused by the swelling of corrosion product in the layers of intergranuar corrosion. Exfoliation is observed on aircraft components, for example, around riveted or bolted components.

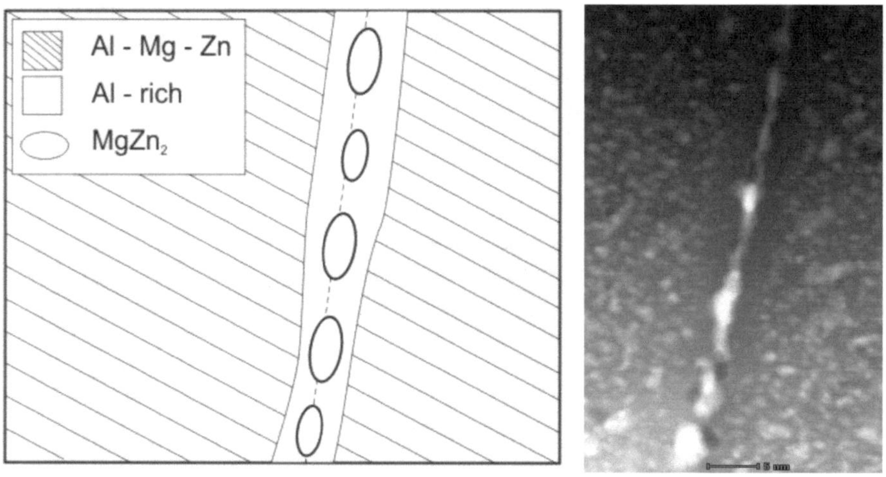

Fig. 15. (a) Schematic of hypothetical grain boundary in an Al-Zn-Mg alloy. This schematic indicates the different chemistry that exists in the grain interior, solute depleted zone (precipitate free zone) and grain boundary precipitates - giving rise to electrochemical heterogeneity localised at the grain boundary region. (b) Conventional bright field TEM image of high angle grain boundary in AA7075-T651, revealing grain boundary precipitates ($MgZn_2$) and a distinguishable precipitate free zone

3.5 Accumulation

The accumulation of noble metals such as iron and copper at the surface of aluminium alloys is problematic even in the absence of specific intermetallic phases such as Al_2CuMg. The accumulation of Cu in almost all Al alloys (even those with low Cu content) has been an issue in metal finishing for many years and was recently extensively reviewed by Muster et al. (Muster 2009). Copper accumulation at corrosion sites has also been investigated extensively. Vukmirovic et al. (Vukmirovic, Dimitrov et al. 2002) showed that copper accumulation on the surface of AA2024 also arises from the copper in the aluminium solid solution. In a range of corrosive environments, aluminium has been shown to preferentially oxidise, resulting in the build-up of copper within a layer approximately 2 - 5 nm thick at the alloy surface (Jung, Dumler et al. 1985; Habazaki, Shimizu et al. 1997). The behaviour of alloying elements in this sense is somewhat dependent upon the Gibbs free energy of oxide formation, which controls the enrichment of elements at the alloy surface and in the surface oxides during corrosion processes. Copper and other more noble elements (i.e. gold) have high Gibbs free energies of oxide formation per equivalent ($\Delta G^{o}/n$) relative to that of alumina and therefore show extensive enrichment at the metal/oxide interface. In contrast, elements such as magnesium and lithium for example, have a lower $\Delta G^o/n$ value than aluminium and, therefore, are more likely to appear in the oxide or electrolyte solution following corrosion processes (Muster 2009).

In environments where aluminium alloys continually experience anodic dissolution it has been suggested that alloys with a wide range of copper concentrations (0.06 to 26 at%) can display copper enrichment at the metal-oxide interface (Blanc, Lavelle et al. 1997; Habazaki, Shimizu et al. 1997; Garcia-Vergara, Colin et al. 2004). Once a certain level of enrichment occurs, copper atoms (and most likely other noble alloying elements) are thought to arrange themselves into clusters through diffusion processes and eventually protrude from the alloy surface due to undermining of the surrounding aluminium matrix (Sieradzki 1993; Habazaki, Shimizu et al. 1997). These copper clusters may be released as elemental copper into the oxide layer by being undermined or copper ions may be oxidized directly from the protruding clusters. It has also been demonstrated that the level of copper enrichment is also influenced by grain orientation. In terms of general corrosion performance, the enrichment of copper at the alloy surface is also likely to increase the number of flaws that exist in the aluminium oxide.

4. Corrosion and protection

This section covers general approaches to protection of aluminium alloys in view of recent advances in the understanding of alloy microstructure. It includes an overview of pretreatment processes such as anodising, conversion coating and organic coatings (barrier and inhibitor combinations). It will examine recent advances in inhibitor design such as building in multifunctionality and touch on self healing coating systems. Approaches using multifunctionality can target anodic and cathodic reactions more effectively than using individual monofunctional inhibitors.

Standard metal finishing processes, which have been used for many years, are likely to continue to be used into the future unless they contain chemicals that are targeted for replacement such as chromium. The function of these coatings is primarily to provide better adhesion properties for paint coatings and a secondary role is to provide corrosion protection. The general approach for applying these coatings relies on metal finishing treatments (treatment prior to painting involving immersion in acidic and alkaline baths)) with the objective of reducing the heterogeneous nature of the metal surface such as removing the NSDL and second phase particles (Muster 2009). This is achieved in multistep treatment processes for metal protection (Twite and Bierwagen 1998; Buchheit 2003; Muster 2009) as for instance:

1. selective deoxidation (IM particle removal and surface etching);
2. deposition or growth of a manufactured oxide via electrochemical (anodising) or chemical (conversion coating) means;
3. use of an organic coating for specific applications, normally including a primer and a top-coat.

On aluminium, most anodised coating processes produce an outer oxide with a cellular structure on top of a thin barrier layer that provides some protection against corrosion. Inhibitors can be incorporated into the outer porous layer of the anodized layer during formation or as a seal after formation to offer some extra protection upon damage. Chromic acid anodizing is one of a number of processes that are available for electrochemical growth of surface protective oxides. More environmentally-friendly alternatives to chromic acid anodizing such as sulfuric, sulfuric-boric, sulphuric-tataric and phosphoric based processes have been available for a long time. There have been a number of recent advances in reducing the energy consumption of anodizing processes as well as improving coating properties. These advances are based on an improved understanding of the alloy microstructure described above and involve selective removal of second phase particles as part of the anodising process.

An alternative approach to anodizing is to precipitate a coating on a surface through chemical means called conversion coatings. For high strength Al-alloys such as 2xxx and 7xxx series chromate conversion coating (CCC) is still the preferred process. Replacements for chromate-based conversion coatings include a range of treatments based on self-assembled monolayers, sol-gel chemistries, Ti/Z oxyfluorides, rare earth, cobalt, vanadates, molybdates and permanganate processes (Twite and Bierwagen 1998; Buchheit 2003; Kendig and Buchheit 2003). These processes are widely developed for chemically pretreated surfaces that have nearly all the IM particles removed and are not specifically designed to address electrochemical and compositional variations found for a heterogeneous surface such as when the IM phases are present. Work like that in Figure 10 depicting the reaction rate variation across the surface, however, opens an avenue to start designing inhibitors where the initial reaction rate distribution across a surface can be significantly reduced to limit the overall activity of the surface. In this context reaction of inhibitive phases with manufactured IM compounds as well as IM particles within the alloy have been studied for a number of systems (Juffs, Hughes et al. 2001; Juffs 2002; Juffs, Hughes et al. 2002; Birbilis, Buchheit et al. 2005; Scholes, Hughes et al. 2009).

Once the anodised or conversion coating is applied, the surface is ready to receive the organic coating. There are many different types of organic coatings, however because of the focus on 2xxx and 7xxx alloy used in the aerospace industry this section will only deal with that application area. The organic coating system usually consists of a primer and a topcoat. The primer is the main protective layer including corrosion inhibitors that can be released when corrosive species or water reach the metal. From the perspective of providing protection for the underlying aluminium alloy, the inhibitor needs to be available during a corrosion event at a concentration higher than the minimum concentration at which the inhibitor stops corrosion (critical concentration). While this sounds obvious, the critical inhibitor concentration needs to be maintained over many years for structures such as airframes, where maintenance may not be possible in parts of the aircraft because of poor access. The chromate systems itself provide continuous protection and repair to the surface for as long as the dose of chromate remains above the critical concentration. This mechanism of inhibitor release and metal protection is recognized as a self-healing mechanism, since the release of the active species recovers the protective layer on top of the metal.

The search for green inhibitors as replacements for chromate has been driven by legislative imperatives for a number of years. Needless to say, replacement inhibitors do not have the same intrinsic inhibitive power at low solubility as chromate. Thus solubility, inhibitive power and transport within the primer system (which consists of a number of inorganic phases as well as the epoxy) ultimately means that finding a replacement for chromate is difficult. This means that alternatives must be present at higher concentration leading to the use of more soluble compounds and consequently encapsulation as a method of regulating the response to external or internal triggers emerges as a prospective way to achieve this objective.

Many current inhibitors are water soluble salts and thus ionic. Consequently, they exist as either anions or cations in solution and perform the single function of anodic or cathodic inhibition. So the simplest improvement to inhibitor design is to increase the functionality by finding compounds which play both a cationic and anionic inhibitive role. A large range of cations including Zn, Ca, and rare earths (Bohm, McMurray et al. 2001; Du, Damron et al. 2001; Kendig and Buchheit 2003; Taylor and Chambers 2008; Muster, Hughes et al. 2009) have been combined with either organic (Osborne, Blohowiak et al. 2001; Sinko 2001) (Voevodin, Balbyshev et al. 2003; Khramov, Voevodin et al. 2004; Blin, Koutsoukos et al. 2007; Taylor and Chambers 2008; Muster, Hughes et al. 2009) or inorganic (oxyanions, carbonates, phosphates, phosphites, nitrates, nitrites, silicate (Bohm, McMurray et al. 2001; Sinko 2001; Blin, Koutsoukos et al. 2007; Taylor and Chambers 2008)) compounds.

Anions with dual functionality, such as some of the transition metal oxyanions which are both oxidants and anions, have been investigated extensively. The oxidizing oyxanions or some organophosphates have some degree of bio-inhibition required for some applications. Substitution of different organophosphates into rare earth-based inhibitors provide versatility in designing inhibitors for specific applications (Birbilis, Buchheit et al. 2005; Hinton, Dubrule et al. 2006; Ho, Brack et al. 2006; Blin, Koutsoukos et al. 2007; Markley, Forsyth et al. 2007; Markley, Hughes et al. 2007; Forsyth, Markley et al. 2008; Deacon, Forsyth et al. 2009; Scholes, Hughes et al. 2009). Thus Ce(di-butyl phosphate)$_3$ is a good inhibitor and relatively green whereas Ce(di-phenyl phosphate)$_3$ is also a good inhibitor, but the diphenyl phosphate also has strong bio-inhibition characteristics (García 2011). However, good bio-inhibition usually means that there are increased environmental and health risks. Obviously the number of cathodic and anodic inhibitors means that there is an enormous number of possible combinations, particularly if ternary and quaternary combinations are considered. Hence high-throughput techniques are being used to assess new inhibitor.

As pointed out above, the kinetics of inhibitor release are of the utmost importance since the inhibitor should be available at levels above the critical inhibitor concentration. Optimization of the release kinetics by novel delivery systems become integral to incorporation of new inhibitors.

There are a number of different approaches to release mechanisms for release of healing agents or corrosion inhibitors which can be incorporated into organic coatings. Both mechanical damage and water are triggers for inhibitor release. In the former case mechanical damage breaks capsules containing water soluble inhibitors. In the latter case water dissolves inhibitor directly incorporated in the primer. Droplet formation within defects such as scratches means that the inhibitor is only released when required i.e., when the defect is moist (Furman, Scholes et al. 2006). There is some evidence to suggest that initial high release of inhibitors may be facilitated through atmospheric exposure of the intact paint where penetration of water into the film "prepares" the inhibitor, probably via surface hydrolysis reactions, within the paint, for diffusion and release into the defect (Du, Damron et al. 2001; Furman, Scholes et al. 2006; Scholes, Furman et al. 2006; Souto, González-García et al. 2010). The presence of water in the film allows soluble inhibitor species to be released into the paint system and diffuse to the metal/coating interface to provide in-situ corrosion prevention or repair called pre-emptive healing (Mardel, Garcia et al. 2011) (Zin, Howard et al. 1998; Osborne, Blohowiak et al. 2001). Thus it has been

demonstrated that water can trigger cerium dibuthylphosphate ($Ce(dbp)_3$) release into an epoxy matrix resulting in improved adhesion and resistance to filiform corrosion attack through interfacial modification (Mardel, Garcia et al.).
In terms of delivery systems, hard capsules, which have been used in polymer healing (Dry 1996; White, Sottos et al. 2001; Mookhoek, Mayo et al. 2010) need to be smaller for paint systems particularly in the aerospace industry where coatings are typically 20 µm or less (Yin, Rong et al. 2007; Fischer 2010; Mookhoek, Mayo et al. 2010) (Hughes, 2010). In polymer applications, capsules up to a few hundred microns can be accommodated (Yin, Rong et al. 2007; Wu, Meure et al. 2008; Tedim, Poznyak et al. 2010). The concept of encapsulation has already been successfully applied to protective organic coatings under different concepts: i) liquids filling completely the void created by the damage by adopting a bi-component systems where one component is encapsulated and the other distributed in the matrix (Cho, White et al. 2009), or single based components with water reactive oils like linseed and tung oils (Suryanarayana, Rao et al. 2008; Samadzadeh, Boura et al. 2010)and ii) liquids (i.e. silyl esters) forming a hydrophobic and highly adhesive layer covering the metallic surface by reaction with the underlying metal and the humidity in air (Garcia, Fischer et al. 2011). One adaption for capsules is to increase the volume of self healing material by manufacturing rods instead of spheres. Rods with the same cross-sections as spheres can deliver larger volumes of material (Bon, Mookhoek et al. 2007; Mookhoek, Fischer et al. 2009). For inhibitors, their role is to prevent a surface reaction (corrosion) and therefore, the volume of material required is much smaller than that required to actually fill the defect. Consequently, there has been considerable effort looking at "nano-containers" (Voevodin, Balbyshev et al. 2003; Raps, Hack et al. 2009; Tedim, Poznyak et al. 2010).
Water is the most obvious trigger since it can permeate most polymers. pH variations are more specific and respond to the pH excursions that occur in corrosion reactions and by an understanding reactions that occur at different sites in the alloy microstructure. The presence of chloride ions (and other anions) within the coating can be used as specific triggers for the release of corrosion inhibitors and uptake of corrodents using anion exchange materials, such as layered double hydroxides (e.g. hydrotalcites) (Tedim, Poznyak et al. 2010) (Bohm, McMurray et al. 2001; Buchheit, Guan et al. 2003; Williams and McMurray 2003; Zheludkevich, Salvado et al. 2005; Mahajanarn and Buchheit 2008). In this context hydrotalcites have been loaded with vanadate, chromate, nitrate and carbonate which exchange for chloride ions and prevent interfacial damage (Bohm, McMurray et al. 2001; Williams and McMurray 2003; Mahajanarn and Buchheit 2008). The incorporation of Mg particles into paint acts as sacrificial anodes to protect Al alloys and steels (Battocchi, Simoes et al. 2006).

5. Conclusions

The broad range of microstrucutral characteristics associated with the high strength Al-alloys have been examined in detail. Perhaps the most attention has been paid to the AA2024-T4 legacy alloy where the microstructure is complicated by the broad compositional variation of second phases, particularly the constituent particles. The role of these features on corrosion has been described and areas where the role of the microstructure is still not clearly understood have been identified and discussed. In the light of a continually emerging understanding of alloy microstructure some general principles of inhibitor design and incorporation into paints systems have been explored.

6. References

Afseth, A., J. H. Nordlien, et al. (2001). "Effect of heat treatment on filiform corrosion of aluminium alloy AA3005." Corrosion Science 43(11): 2093-2109.

Afseth, A., J. H. Nordlien, et al. (2002). "Filiform corrosion of AA3005 aluminium analogue model alloys." Corrosion Science 44(11): 2543-2559.

Allen, C. M., K. A. Q. O'Reilly, et al. (1998). "Intermetallic phase selection in 1XXX Al alloys." Progress in Materials Science 43(2): 89-170.

Ayer, R., J. Y. Koo, et al. (1985). "Microanalytical Study of the Heterogeneous Phases in Commercial Al-Zn-Mg-Cu Alloys." Metallurgical Transactions a-Physical Metallurgy and Materials Science 16(11): 1925-1936.

Birbilis, N. and R. G. Buchheit (2005). "Electrochemical characteristics of intermetallic phases in aluminum alloys - An experimental survey and discussion." Journal of the Electrochemical Society 152(4): B140-B151.

Birbilis, N. and R. G. Buchheit (2008). "Investigation and discussion of characteristics for intermetallic phases common to aluminum alloys as a function of solution pH." Journal of the Electrochemical Society 155(3): C117-C126.

Birbilis, N., R. G. Buchheit, et al. (2005). "Inhibition of AA2024-T3 on a phase-by-phase basis using an environmentally benign inhibitor, cerium dibutyl phosphate." Electrochemical and Solid State Letters 8(11): C180-C183.

Birbilis, N., M. K. Cavanaugh, et al. (2006). "Electrochemical behavior and localized corrosion associated with Al7Cu2Fe particles in aluminum alloy 7075-T651." Corrosion Science 48(12): 4202-4215.

Blanc, C., S. Gastaud, et al. (2003). "Mechanistic studies of the corrosion of 2024 aluminum alloy in nitrate solutions." Journal of the Electrochemical Society 150(8): B396-B404.

Blanc, C., B. Lavelle, et al. (1997). "The role of precipitates enriched with copper on the susceptibility to pitting corrosion of the 2024 aluminium alloy." Corrosion Science 39(3): 495-510.

Blin, F., P. Koutsoukos, et al. (2007). "The corrosion inhibition mechanism of new rare earth cinnamate compounds - Electrochemical studies." Electrochimica Acta 52(21): 6212-6220.

Boag, A. (2009). The Relationship Between Microstructure and Stable Pitting Initiation in Aerospace Aluminium Alloy 2024-T3. Melbourne, RMIT. PhD.

Boag, A., A. E. Hughes, et al. (2011). "Corrosion of AA2024-T3 Part I. Localised corrosion of isolated IM particles." Corrosion Science 53(1): 17-26.

Boag, A., A. E. Hughes, et al. (2009). "How complex is the microstructure of AA2024-T3?" Corrosion Science 51(8): 1565-1568.

Boag, A., R. J. Taylor, et al. (2010). "Stable pit formation on AA2024-T3 in a NaCl environment." Corrosion Science 52(1): 90-103.

Bockris, J. O. M., Reddy, A.K.N., Gamboa-Aldeco, M.E. (2000). Modern Electrochemistry. New York, Kluwer.

Bohm, S., H. N. McMurray, et al. (2001). "Novel environment friendly corrosion inhibitor pigments based on naturally occurring clay minerals." Materials and Corrosion-Werkstoffe Und Korrosion 52(12): 896-903.

Bon, S. A. F., S. D. Mookhoek, et al. (2007). "Route to stable non-spherical emulsion droplets." European Polymer Journal 43(11): 4839-4842.

Buchheit, R. G. (1995). "A Compilation of Corrosion Potentials Reported for Intermetallic Phases in Aluminum-Alloys." Journal of the Electrochemical Society 142(11): 3994-3996.

Buchheit, R. G., R. P. Grant, et al. (1997). "Local dissolution phenomena associated with S phase (Al2CuMg) particles in aluminum alloy 2024-T3." Journal of the Electrochemical Society 144(8): 2621-2628.

Buchheit, R. G., H. Guan, et al. (2003). "Active corrosion protection and corrosion sensing in chromate-free organic coatings." Progress in Organic Coatings 47(3-4): 174-182.

Buchheit, R. G., Hughes, A.E. (2003). Chromate and Chromate-Free Coatings. Corrosion: Fundamentals, Testing and Protection. C. Moosbrugger. Mterials Park, Oh, USA, ASM International. 13A: 720 -735.

Buchheit, R. G., M. A. Martinez, et al. (2000). "Evidence for Cu ion formation by dissolution and dealloying the Al2CuMg intermetallic compound in rotating ring-disk collection experiments." Journal of the Electrochemical Society 147(1): 119-124.

Carney, T. J., P. Tsakiropoulos, et al. (1990). "Oxidation and Surface Segregation in Rapidly Solidified Al-Alloy Powders." International Journal of Rapid Solidification 5(2-3): 189-217.

Cavanaugh, M. K., R. G. Buchheit, et al. (2010). "Modeling the environmental dependence of pit growth using neural network approaches." Corrosion Science 52(9): 3070-3077.

Cawley, N. R. and D. G. Harlow (1996). Journal of Material Science 31: 5127-5134.

Chen, G. S., M. Gao, et al. (1996). "Microconstituent-induced pitting corrosion in aluminum alloy 2024-T3." Corrosion 52(1): 8-15.

Cho, S. H., S. R. White, et al. (2009). "Self-Healing Polymer Coatings." Advanced Materials 21(6): 645-+.

Das, S. K. (2006). "Designing Aluminium Alloys for a Recycle-Friendly World." Light Metal Age June.

Davies, J. R. (1999). Corrosion of Aluminium and Aluminium Alloys. Columbus, OH, ASM International.

Deacon, G. B., M. Forsyth, et al. (2009). "Synthesis and Characterisation of Rare Earth Complexes Supported by para-Substituted Cinnamate Ligands." Zeitschrift Fur Anorganische Und Allgemeine Chemie 635(6-7): 833-839.

Dry, C. (1996). "Procedures developed for self-repair of polymer matrix composite materials." Composite Structures 35(3): 263-269.

Du, Y. J., M. Damron, et al. (2001). "Inorganic/organic hybrid coatings for aircraft aluminum alloy substrates." Progress in Organic Coatings 41(4): 226-232.

Engelberg, D. L. (2010). Shreir's Corrosion. R. J.A. Amsterdam, Elsevier.

Evans, U. R. (1971). "Inhibition, Passivity and Resistance. A Review of Acceptable Mechanisms." Electrochimica Acta 16(11): 1825-1840.

Fischer, H. R. (2010). natural Science 2: 873-901.

Fishkis, M. and J. C. Lin (1997). "Formation and evolution of a subsurface layer in a metalworking process." Wear 206(1-2): 156-170.

Forsyth, M., T. Markley, et al. (2008). "Inhibition of corrosion on AA2024-T3 by new environmentally friendly rare earth organophosphate compounds." Corrosion 64(3): 191-197.

Frankel, G. S. (1998). "Pitting corrosion of metals - A review of the critical factors." Journal of the Electrochemical Society 145(6): 2186-2198.

Furman, S. A., F. H. Scholes, et al. (2006). "Corrosion in artificial defects. II. Chromate reactions." Corrosion Science 48(7): 1827-1847.

Gao, M., C. R. Feng, et al. (1998). "An analytical electron microscopy study of constituent particles in commercial 7075-T6 and 2024-T3 alloys." Metallurgical and Materials Transactions a-Physical Metallurgy and Materials Science 29(4): 1145-1151.

Garcia-Vergara, S., F. Colin, et al. (2004). "Effect of copper enrichment on the electrochemical potential of binary Al-Cu alloys." Journal of the Electrochemical Society 151(1): B16-B21.

Garcia, S. J., H. R. Fischer, et al. (2011). "Self-healing anticorrosive organic coating based on an encapsulated water reactive silyl ester: Synthesis and proof of concept." Progress in Organic Coatings 70(2-3): 142-149.

García, S. J., Mol, J.M.C., Muster, T.H., Hughes, A.E., Mardel, J., Miller, T., Markely, T., Terryn, H., de Wit, J.H.W. (2011). Advances in the Selection and use of Rare-Earth-Based Inhibitors for Self Healing Organic Coatings, Accepted for publication in Self-Healing Properties of New Surface Treatments. Green Inhibitors. L. Fedrizzi, EFC-Maney Publishing. 58.

Glenn, A. M., T. H. Muster, et al. (2011). "Corrosion of AA2024-T3 Part III: Propagation." Corrosion Science 53(1): 40-50.

Guillaumin, V. and G. Mankowski (1999). "Localized corrosion of 2024 T351 aluminium alloy in chloride media." Corrosion Science 41(3): 421-438.

Habazaki, H., K. Shimizu, et al. (1997). "Nanoscale enrichments of substrate elements in the growth of thin oxide films." Corrosion Science 39(4): 731-737.

Hahn, G. T. and A. R. Rosenfield (1975). "Metallurgical Factors Affecting Fracture Toughness of Aluminum-Alloys." Metallurgical Transactions A 6(4): 653-668.

Harlow, D. G., M. Z. Wang, et al. (2006). Metall. mat. Trans A 37A: 3367-3373.

Harvey, T. G., A. E. Hughes, et al. (2008). "Non-chromate deoxidation of AA2024-T3: Sodium bromate-nitric acid (20-60 degrees C)." Applied Surface Science 254(11): 3562-3575.

Hatch , J. E. (1984). Aluminium: Porperties and Physical Metallurgy, ASM International.

Hinton, B. R. W., N. Dubrule, et al. (2006). Raman, EDS and SEM studies of the interaction of corrosion inhibitor $Ce(dbp)_3$ with AA2024-T3. 4th International Symposium on Aluminium Surface Science and Technology. Beaune, France.

Ho, D., N. Brack, et al. (2006). "Cerium dibutylphosphate as a corrosion inhibitor for AA2024-T3 aluminum alloys." Journal of the Electrochemical Society 153(9): B392-B401.

Hughes, A. E., A. Boag, et al. (2011). "Corrosion of AA2024-T3 Part II Co-operative corrosion." Corrosion Science 53(1): 27-39.

Hughes, A. E., A. Boag, et al. (2006). Statistical Approach to Determine Spatial and Elemental Correlations of Corrosion Sites on Al-Alloys. Aluminium Surface Science and Technology Conference Beaune, France, ATB Metallurgie, 45 (1-4): 551-556.

Hughes, A. E., Cole, I.S., Muster, T.M. and Varley, R.J. (2010). "Combining Green and Self Healing for a new Generation of Coatings for Metal Protection." Nature Asia Materials 2(4): 143-151.

Hughes, A. E., B. Hinton, et al. (2007). "Airlife - Towards a fleet management tool for corrosion damage." Corrosion Reviews 25(3-4): 275-293.

Hughes, A. E., C. MacRae, et al. (2010). Surface Interface Analysis 42: 334-338.

Hughes, A. E., J. M. C. Mol, et al. (2005). "A morphological study of filiform corrosive attack on cerated AA2024-T351 aluminium alloy." Corrosion Science 47(1): 107-124.

Hughes, A. E., T. H. Muster, et al. (2010). Corrosion Science Accepted.

Hughes, A. E., N. Wilson, et al. (2009). Corrosion Science 51: 1565-1568.

Hunter, M. S. F., G.R., Robinson, D.L. (1963). 2nd International Conference on Metallic Corrosion: 66.

Ilevbare, G. O., O. Schneider, et al. (2004). Journal of Electrochemical Society 151: B453-B464.

Ilevbare, G. O., O. Schneider, et al. (2004). "In situ confocal laser scanning microscopy of AA 2024-T3 corrosion metrology - I. Localized corrosion of particles." Journal of the Electrochemical Society 151(8): B453-B464.

Juffs, L. (2002). Investigation of Corrosion Coating Deposition on Microscopic and Macroscopic Intermetallic Phases of Aluminium Alloys. Melbourne, RMIT. Master of Science.

Juffs, L., A. E. Hughes, et al. (2002). "The use of macroscopic modelling of intermetallic phases in aluminium alloys in the study of ferricyanide accelerated chromate conversion coatings." Corrosion Science 44(8): 1755-1781.

Juffs, L., A. E. Hughes, et al. (2001). "The use of macroscopic modelling of intermetallic phases in aluminium alloys in the study of ferricyanide accelerated chromate conversion coatings." Micron 32(8): 777-787.

Jung, D. Y., I. Dumler, et al. (1985). "Electronmicroscopic Examination of Corroded Aluminum-Copper Alloy Foils." Journal of the Electrochemical Society 132(10): 2308-2312.

Kendig, M. W. and R. G. Buchheit (2003). "Corrosion inhibition of aluminum and aluminum alloys by soluble chromates, chromate coatings, and chromate-free coatings." Corrosion 59(5): 379-400.

Khramov, A. N., N. N. Voevodin, et al. (2004). "Hybrid organo-ceramic corrosion protection coatings with encapsulated organic corrosion inhibitors." Thin Solid Films 447: 549-557.

Knight, S. P. (2003). "A review of HEat Treatments." Australasian Corrosion Association.

Kolics, A., A. S. Besing, et al. (2001). "Interaction of chromate ions with surface intermetallics on aluminum alloy 2024-T3 in NaCl solutions." Journal of the Electrochemical Society 148(8): B322-B331.

Kovarik, L., M. K. Miller, et al. (2006). "Origin of the modified orientation relationship for S(S")-phase in Al-Mg-Cu alloys." Acta Materialia 54(7): 1731-1740.

Lacroix, L., L. Ressier, et al. (2008). "Combination of AFM, SKPFM, and SIMS to study the corrosion behavior of S-phase particles in AA2024-T351." Journal of the Electrochemical Society 155(4): C131-C137.

Lacroix, L., L. Ressier, et al. (2008). "Statistical study of the corrosion behavior of Al2CuMg intermetallics in AA2024-T351 by SKPFM." Journal of the Electrochemical Society 155(1): C8-C15.

Leth-Olsen, H., J. H. Nordlien, et al. (1997). "Formation of Nanocrystalline Surface Layers by Annealing and Their Role in Filiform Corrosion of Aluminum Sheet." Journal of the Electrochemical Society 144(7): L196-L197.

Leth-Olsen, H., J. H. Nordlien, et al. (1998). "Filiform corrosion of aluminium sheet. III. Microstructure of reactive surfaces." Corrosion Science 40(12): 2051-2063.

Liao, C. M., J. M. Olive, et al. (1998). "In-situ monitoring of pitting corrosion in aluminum alloy 2024." Corrosion 54(6): 451-458.
Liu, Y., M. F. Frolish, et al. (2010). "Evolution of near-surface deformed layers during hot rolling of AA3104 aluminium alloy." Surface And Interface Analysis 42(4): 180-184.
Liu, Y., A. Laurino, et al. (2010). "Corrosion behaviour of mechanically polished AA7075-T6 aluminium alloy." Surface And Interface Analysis 42(4): 185-188.
Liu, Y., X. Zhou, et al. (2007). "Precipitation in an AA6111 aluminium alloy and cosmetic corrosion." Acta Materialia 55(1): 353-360.
Luo, C., Hughes, A.E., Zhou X., Thompson G.E. (to be published). Corrosion Science.
Mahajanarn, S. P. V. and R. G. Buchheit (2008). "Characterization of inhibitor release from Zn-Al- V10O28 (6-) hydrotalcite pigments and corrosion protection from hydrotalcite-pigmented epoxy coatings." Corrosion 64(3): 230-240.
Mao, Y., A. M. Gokhale, et al. (2006). Computational Materials Science 37: 543-556.
Mardel, J., S. J. Garcia, et al. (2011). "The characterisation and performance of Ce(dbp)3-inhibited epoxy coatings." Progress in Organic Coatings 70(2-3): 91-101.
Markley, T. A., M. Forsyth, et al. (2007). "Corrosion protection of AA2024-T3 using rare earth diphenyl phosphates." Electrochimica Acta 52(12): 4024-4031.
Markley, T. A., A. E. Hughes, et al. (2007). "Influence of praseodymium - Synergistic corrosion inhibition in mixed rare-earth diphenyl phosphate systems." Electrochemical and Solid State Letters 10(12): C72-C75.
Mazurkiewicz, B. and A. Piotrowski (1983). "The Electrochemical-Behavior of the Al2cu Intermetallic Compound." Corrosion Science 23(7): 697-&.
Meng, Q. J. and G. S. Frankel (2004). The effect of Cu content on the localized corrosion resistance of AA7xxx-T6 alloys. Corrosion and Protection of Light Metal Alloys. R. G. Buchheit, R. G. Kelly, N. A. Missert and B. A. Shaw. 2003: 62-81.
Mol, J. M. C., A. E. Hughes, et al. (2004). "A morphological study of filiform corrosive attack on chromated and alkaline-cleaned AA2024-T351 aluminium alloy." Corrosion Science 46(5): 1201-1224.
Mookhoek, S. D., H. R. Fischer, et al. (2009). "A numerical study into the effects of elongated capsules on the healing efficiency of liquid-based systems." Computational Materials Science 47(2): 506-511.
Mookhoek, S. D., S. C. Mayo, et al. (2010). "Applying SEM-Based X-ray Microtomography to Observe Self-Healing in Solvent Encapsulated Thermoplastic Materials." Advanced Engineering Materials 12(3): 228-234.
Muster, T. H., Hughes, A.E., Thompson. G.E. (2009). Cu Distributions in Aluminium Alloys. New York, Nova Science Publishers.
Newman, R. C. and K. Sieradzki (1994). "Metallic Corrosion." Science 263(5154): 1708-1709.
Nisancioglu, K. S., O., Yu, Y., Nordlien, Y.K. (2004). 55th Annula Meeting of the International Society of Electrochemistry, Thessaloniki.
Osborne, J. H., K. Y. Blohowiak, et al. (2001). "Testing and evaluation of nonchromated coating systems for aerospace applications." Progress in Organic Coatings 41(4): 217-225.
Pan, X., J. E. Morral, et al. (2010). "Predicting the Q-Phase in Al-Cu-Mg-Si Alloys." Journal of Phase Equilibria and Diffusion 31(2): 144-148.
Park, J. O., C.-H. Paik, et al., Eds. (1996). Critical Factors in Localised Corrosion II., The Electrochemical Society, Pennington, NJ.

Park, J. O., C.-H. Paik, et al. (1999). Journal of the Electrochemical Society 146: 517-523.

Phragmen, G. (1950). "On the Phases Occurring in Alloys of Aluminium with Copper, Magnesium, Manganese, Iron and Silicon." The Journal of the Institute of Metals 77: 489-553.

Plassart, G., Aucouturier, M. (2000). 2nd International Conference on Aluminium Surface Science and Technology. Manchester: 29 - 35.

Poelman, M., M. G. Olivier, et al. (2005). "Electrochemical study of different ageing tests for the evaluation of a cataphoretic epoxy primer on aluminium." Progress in Organic Coatings 54(1): 55-62.

Polmear, I. J. (1995). Light Allopys: Metallurgy of the Light Metals. London, Arnold.

Polmear, I. J. (2004). "A Century of Age Hardening." Materials Forum 28: 1 - 14.

Pride, S. T., J. R. Scully, et al. (1994). "Metastable Pitting of Aluminum and Criteria for the Transition to Stable Pit Growth." Journal of the Electrochemical Society 141(11): 3028-3040.

Pryor, M. J. and J. C. Fister (1984). "The Mechanism of Dealloying of Copper Solid-Solutions and Intermetallic Phases." Journal of the Electrochemical Society 131(6): 1230-1235.

Ralston, K. D., N. Birbilis, et al. (2010). "The effect of precipitate size on the yield strength-pitting corrosion correlation in Al-Cu-Mg alloys." Acta Materialia 58(18): 5941-5948.

Raps, D., T. Hack, et al. (2009). "Electrochemical study of inhibitor-containing organic-inorganic hybrid coatings on AA2024." Corrosion Science 51(5): 1012-1021.

Raviprasad, K., C. R. Hutchinson, et al. (2003). "Precipitation processes in an Al-2.5Cu-1.5Mg (wt. %) alloy microalloyed with Ag and Si." Acta Materialia 51(17): 5037-5050.

Samadzadeh, M., S. H. Boura, et al. (2010). "A review on self-healing coatings based on micro/nanocapsules." Progress in Organic Coatings 68(3): 159-164.

Sasaki, K., P. W. Levy, et al. (2002). "Electrochemical noise during pitting corrosion of aluminum in chloride environments." Electrochemical and Solid State Letters 5(8): B25-B27.

Scamans, G. M., Afseth, A., Thompson, G.E., Zhou X (2000). 2 nd International Conference on Aluminium Surface Science and Technology. Manchester: 9 - 15.

Scamans, G. M., M. F. Frolish, et al. (2010). "The ubiquitous Beilby layer on aluminium surfaces." Surface And Interface Analysis 42(4): 175-179.

Schneider, O., G. O. Ilevbare, et al. (2004). Journal of Electrochemical Society 151(B$65-B472).

Schneider, O., G. O. Ilevbare, et al. (2004). "In situ confocal laser scanning microscopy of AA 2024-T3 corrosion metrology - II. Trench formation around particles." Journal of the Electrochemical Society 151(8): B465-B472.

Scholes, F. H., S. A. Furman, et al. (2006). "Chromate leaching from inhibited primers - Part I. Characterisation of leaching." Progress in Organic Coatings 56(1): 23-32.

Scholes, F. H., A. E. Hughes, et al. (2009). "Interaction of Ce(dbp)(3) with surface of aluminium alloy 2024-T3 using macroscopic models of intermetallic phases." Corrosion Engineering Science and Technology 44(6): 416-424.

Scully, J. R., T. O. Knight, et al. (1993). "Electrochemical Characteristics of the Al2cu, Al3ta And Al3zr Intermetallic Phases and Their Relevancy to the Localized Corrosion of Al-Alloys." Corrosion Science 35(1-4): 185-195.

Searles, J. L., P. I. Gouma, et al. (2001). "Stress corrosion cracking of sensitized AA5083 (Al-4.5Mg-1.0Mn)." Metallurgical and Materials Transactions a-Physical Metallurgy and Materials Science 32(11): 2859-2867.

Seri, O. (1994). "The Effect of Nacl Concentration on the Corrosion Behavior of Aluminum-Containing Iron." Corrosion Science 36(10): 1789-&.

Sieradzki, K. (1993). "Curvature Effects in Alloy Dissolution." Journal of the Electrochemical Society 140(10): 2868-2872.

Sinko, J. (2001). "Challenges of chromate inhibitor pigments replacement in organic coatings." Progress in Organic Coatings 42(3-4): 267-282.

Souto, R. M., Y. González-García, et al. (2010). "Examination of organic coatings on metallic substrates by scanning electrochemical microscopy in feedback mode: Revealing the early stages of coating breakdown in corrosive environments." Corrosion Science 52(3): 748-753.

Sprowls, D. O. (1978). "High Strength Aluminium Alloys with Improved Resistance to Corrosion and Stress Corrosion Cracking." Aluminium 54(6): 214 - 217.

Suryanarayana, C., K. C. Rao, et al. (2008). "Preparation and characterization of microcapsules containing linseed oil and its use in self-healing coatings." Progress in Organic Coatings 63(1): 72-78.

Suter, T., Alkire, R.C. (2001). "Microelectrochemical Studies of Pit Initiation at Single Inclusions in Al 2024-T3." jJournal of the Electrochemical Society 148(1): B36-B42.

Taylor, S. R. and B. D. Chambers (2008). "Identification and characterization of nonchromate corrosion inhibitor synergies using high-throughput methods." Corrosion 64(3): 255-270.

Tedim, J., S. K. Poznyak, et al. (2010). "Enhancement of Active Corrosion Protection via Combination of Inhibitor-Loaded Nanocontainers." Acs Applied Materials & Interfaces 2(5): 1528-1535.

Thompson, G. E. a. Z. X. (2010). Mater. World 18: 26-27.

Trueman, A. R. (2005). "Determining the probability of stable pit initiation on aluminium alloys using potentiostatic electrochemical measurements." Corrosion Science 47(9): 2240-2256.

Twite, R. L. and G. P. Bierwagen (1998). "Review of Alternatives to Chromate for Corrosion Protection of Aluminum Aerospace Alloys." Progress in Organic Coatings 33(2): 91-100.

Voevodin, N. N., V. N. Balbyshev, et al. (2003). "Nanostructured coatings approach for corrosion protection." Progress in Organic Coatings 47(3-4): 416-423.

Vukmirovic, M. B., N. Dimitrov, et al. (2002). "Dealloying and corrosion of Al alloy 2024-T3." Journal of the Electrochemical Society 149(9): B428-B439.

Wadeson, D. A., X. Zhou, et al. (2006). "Corrosion behaviour of friction stir welded AA7108 T79 aluminium alloy." Corrosion Science 48(4): 887-897.

Wei, R. P., C. M. Liao, et al. (1998). "A transmission electron microscopy study of constituent-particle-induced corrosion in 7075-T6 and 2024-T3 aluminum alloys." Metallurgical and Materials Transactions a-Physical Metallurgy and Materials Science 29(4): 1153-1160.

White, P. A., A. E. Hughes, et al. (2009). "High-throughput channel arrays for inhibitor testing: Proof of concept for AA2024-T3." Corrosion Science 51(10): 2279-2290.

White, S. R., N. R. Sottos, et al. (2001). "Autonomic healing of polymer composites." Nature 409(6822): 794-797.

Williams, G. and H. N. McMurray (2003). "Anion-exchange inhibition of filiform corrosion on organic coated AA2024-T3 aluminum alloy by hydrotalcite-like pigments." Electrochemical and Solid State Letters 6(3): B9-B11.

Winkelman, G. B., K. Raviprasad, et al. (2007). "Orientation relationships and lattice matching for the S phase in Al-Cu-Mg alloys." Acta Materialia 55(9): 3213-3228.

Wloka, J. and S. Virtanen (2008). "Detection of nanoscale eta-MgZn2 phase dissolution from an Al-Zn-Mg-Cu alloy by electrochemical microtransients." Surface and Interface Analysis 40(8): 1219-1225.

Wu, D. Y., S. Meure, et al. (2008). "Self-healing polymeric materials: A review of recent developments." Progress in Polymer Science 33(5): 479-522.

Yin, T., M. Z. Rong, et al. (2007). "Self-healing epoxy composites - Preparation and effect of the healant consisting of microencapsulated epoxy and latent curing agent." Composites Science and Technology 67(2): 201-212.

Zamin, M. (1981). "The Role of Mn in the Corrosion Behavior of Al-Mn Alloys." Corrosion 37(11): 627-632.

Zhang, W. L. and G. S. Frankel (2003). "Transitions between pitting and intergranular corrosion in AA2024." Electrochimica Acta 48(9): 1193-1210.

Zhao, X. Y. and G. S. Frankel (2007). "Quantitative study of exfoliation corrosion: Exfoliation of slices in humidity technique." Corrosion Science 49(2): 920-938.

Zheludkevich, M. L., I. M. Salvado, et al. (2005). "Sol-gel coatings for corrosion protection of metals." Journal of Materials Chemistry 15(48): 5099-5111.

Zhou, X., G. E. Thompson, et al. (2003). "The influence of surface treatment on filiform corrosion resistance of painted aluminium alloy sheet." Corrosion Science 45(8): 1767-1777.

Zhou, X. L., Y., Thompson, G.E., Scamans, G.M., Skeldon, P., Hunter, J.A. (2011). "Near-Surface Deformed Layers on Rolled Aluminium Alloys." Metallurgical and Materials Transactions A: Physical Metallurgy and Materials Science.

Zin, I. M., R. L. Howard, et al. (1998). "The mode of action of chromate inhibitor in epoxy primer on galvanized steel." Progress in Organic Coatings 33(3-4): 203-210.

11

Mechanical Behavior and Plastic Instabilities of Compressed Al Metals and Alloys Investigated with Intensive Strain and Acoustic Emission Methods

Andrzej Pawełek
Aleksander Krupkowski Institute of Metallurgy and Materials Science
Polish Academy of Sciences, Kraków,
Poland

1. Introduction

The aim of the present chapter has been to substantiate and explain the relation among the behavior of acoustic emission (AE) parameters, the course of external load, evolution of microstructure and the dislocation mechanisms of slip and the localization of deformation connected with twinning, formation of slip and shear bands. The problem is of fundamental meaning, when qualitative and quantitative relations between the rate of AE events, amplitude and energy of AE signals and other AE descriptors in relation to micro-processes occurring in a material are to be discussed.

It is commonly believed that twinning is the most efficient source of acoustic emission (Bidlingmaier et al., 1999; Boiko, 1973; El-Danaf et al., 1999; Heiple & Carpenter, 1987; Tanaka & Horiuchi, 1975) due to fast release of great amount of elastic energy. It is connected with the fact that the velocity of twinning dislocations is higher than this of slip dislocations (Boiko, 1973), which results in the increase of contribution of accelerating effects in the recorded AE impulses.

One of the first AE investigations concerned the tensile test of titanium and its alloys (Tanaka & Horiuchi, 1975), in which it was established, that the AE activity in Ti was bound with twinning from the beginning, while in Ti alloys the AE impulses from twinning appeared after a high degree of deformation. During compression of the γ-TiAl alloy, AE sources were identified as generally coming from slip, twinning and the propagation of microcracks. It was reported, however, that the detailed mechanisms by which moving dislocations create elastic waves are still not fully understood (Bidlingmaier et al., 1999).

Moreover, the problem of twinning in Al has still remained controversial. It is believed quite commonly, that at least in simple uniaxial strain state, like in a tensile test, twins do not appear in Al. One of the aim of this chapter is to demonstrate, that there are numerous proofs, that in a complex strain state, which occurs in the channel-die compression of single Al crystals at temperature of liquid nitrogen the twinning processes do occur.

The reasons for undertaking such a research are numerous. At first, there is lack in literature of the experimental data on the AE behavior during channel-die compression of single crystals. Secondly, in the contemporary materials science, one of the basic problems of plastic deformation of metals are questions of strain localization due to the formation and development of slip lines and slip bands as well as shear banding and twinning processes.

In the last decade the methods of intensive strain have become more and more widely used to obtain microstructure refinement and finally ultra fine-grained (UFG), nanocrystalline materials which have the excellent mechanical properties, such as great strength and plasticity or even superplasticity occurring in the conditions of relatively not too high temperatures (Vinogradov, 1998). They allow obtaining massive samples of metals ready for a further treatment. This refers in particular to the packet rolling with bonding, so called ARB (Accumulative Roll-Bonding) method (Saito et al., 1999; Pawełek et al., 2007). There are also known products obtained on industrial scale by the method of channel compression ECAP (Equal Channel Angular Pressing), (Kuśnierz, 2001). The method of torsion under high pressure HPT (High Pressure Torsion), (Valiev et al., 2000) has been the least known since obtaining the high pressure itself is a difficult problem.

The subject concerns the Al alloys of AA6060, AA2014 and AA5182 type as well as AA5754 and AA5251 ones. The examinations of Al alloys of AA6060 and AA2014 types were carried out applying the HPT method as well as ECAP technique with circular cross section of the channel. The anisotropy of Portevin-Le Châtelier (PL) and AE effects was described and the relation between the PL and AE effects in UFG (nanocrystalline) Al alloy after intensive strain processing was reported here for the first time.

On the other hand the results of the examinations of Mg-Li and Mg-Li-Al alloys, presented here for comparison, were carried out applying HPT method (Kúdela et al., 2011) as well as ECAP technique (Kuśnierz et al., 2008) with squared cross section of the channel.

The aim of this chapter has been also an attempt to present the correlations between the mechanisms of plastic deformation and the AE phenomenon and the discussion of the connection of AE with the possible phenomenon of superplasticity in UFG (nanocrystalline) aluminium alloys.

2. Materials and methods

2.1 Production of single and bi-crystals

The Al crystals of several different crystallographic orientations were obtained using a standard Bridgman method, while the Al bi-crystals of crystallite orientations {100}<011>/{110}<001> (Goss/shear) were produced applying a modified Bridgman technique of horizontal crystallization.

The samples of Al single- and bi-crystals, of dimensions 10x10x10mm of various orientations were cut out and subjected to tests of channel-die compression at room and liquid nitrogen temperature (77K) using an INSTRON testing machine equipped with channel-die (Fig. 1), which ensured plastic flow merely in the normal direction (ND) and in the elongation direction (ED), parallel to the channel axis, since the deformation in the transverse direction (TD) was held back by the channel walls.

The samples were deformed in a multi-stage way in order to obtain appropriate values of intermediate and final degrees of deformation. After each deformation stage they were trimmed to satisfy the slenderness ratio. In each case the traverse speed of the testing machine was 0.05 mm/min.

An apparatus recording the AE descriptors was coupled with the testing machine. Both systems are unique, fully computerized set for a simultaneous measurement of external compressive force and AE descriptors. A broad-band piezoelectric sensor recorded AE signals in the range from 100 kHz up to 1 MHz. The contact of the sensor with the sample was maintained with the aid of steel rail which served as a washer in the channel-die, as well as an acoustic wave-guide. In order to eliminate unwanted effects of friction against the wall of the channel, each sample was covered with a Teflon foil. The more detailed description of the AE method is presented in the next section 2.2.

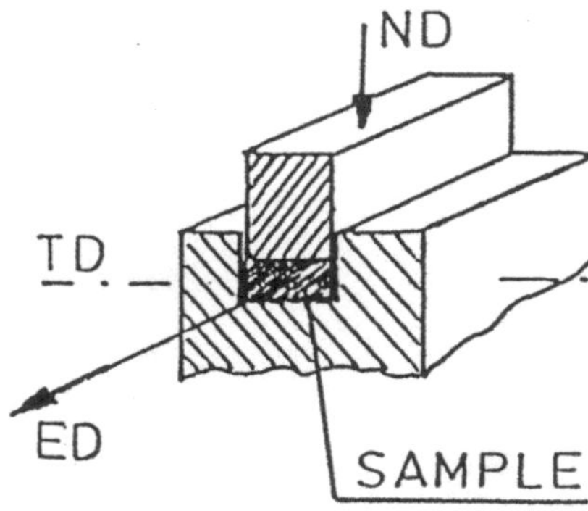

Fig. 1. Scheme of the device for the channel-die compression test

2.2 Acoustic emission method

The AE phenomenon takes place during a rapid release of elastic energy, accumulated in the material as a result of acting external or internal conversions, which can be emitted in the form of elastic waves which the frequency is contained between several kilohertz and a few megahertz. In metals and alloys, in general, they are generated as the effect of plastic deformation and particular dislocation strain mechanisms. The AE method enables sensitive monitoring effects in real time, even in considerable volume of investigated elements.

Considering simplifying assumptions, that the function of amplitude of strain field changes in the AE source has a form of elementary shock, the point of observation is in a distant area, and the elastic wave propagates in a homogeneous medium, the elementary equation of the signal propagation distance given in literature (Resnikoff & Wells, 1998) takes the form:

$$G(x,\ t'-t;x)=\frac{1}{4\pi\rho v_p^2}\gamma_i\gamma_j\frac{1}{r}\delta(t'-t-r/v_p)-\frac{1}{4\pi\rho v_s^2}(\gamma_i\gamma_j-\delta_{ij})\frac{1}{r}\delta(t'-t-r/v_s) \qquad (1)$$

where: G_{ij} $(\boldsymbol{r}',t'-t;\boldsymbol{r})$ is Green function for the displacement in directions x_i', y_i', z_i' in point $\boldsymbol{r}'$, in time t', in the case, when a local disturbance of strain field in point $\boldsymbol{r}$ in time t becomes the source of the displacement, ρ – medium density, v_p – velocity of dilatation wave, v_s – velocity of shear wave, γ_i, γ_j - for i=1, 2, 3 and j=1, 2, 3 are directional cosines source-receiver and receiver-source, r – distance between AE source and sensor, δ_{ij} – Kronecker delta, $\delta(x)$ – delta function equal $+\infty$, for x=0 and equal 0 for the remaining values x.

Apart from the AE signal the apparatus registers also a noise of acoustic background and that generated during the processing of the recorded signal. In the course of its processing from the analogue form into digital one, so called *quantization noise* occurs, resulting from the process of rounding the instantaneous value of the signal to the levels, which are the components of a binary form of the record. The decrease of quantization noise was attained through the use of modern analogue-numerical processors of high linearity of processing and resolution of 12 bytes in an optimal range of input voltages of about ±5V.

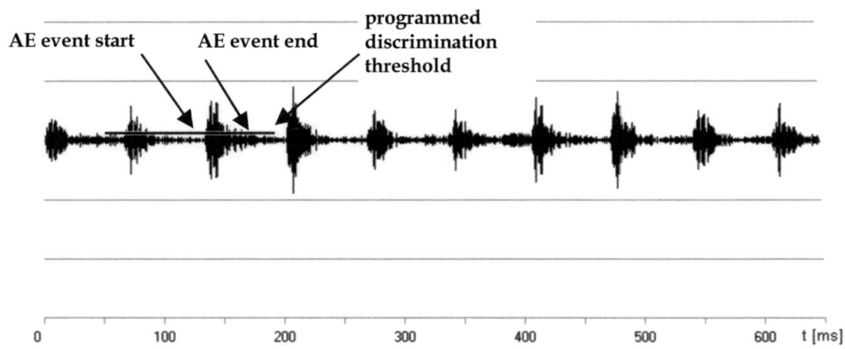

Fig. 2. The principles of AE event detection

The line determining the maximum level of noise voltage of surrounding background is shown in Fig. 2. The level is taken as a discrimination voltage. The occurrence of AE is defined as a moment of increase of instantaneous signal value above the discrimination voltage. The duration of the AE event is determined to the moment of the decrease of instantaneous signal value below the discrimination voltage. The method takes into account the detection of events with the possibility of software increase/decrease of the discrimination level voltage (Paupolis, 1980). The applied algorithm of AE event detection enables a program implementation of numerical records of signals containing even a few hundred of megabites. By means of such an algorithm, it is possible to detect the events lasting at least three times the sampling period of the applied analogue-numerical processor. For example, for the frequency of sampling used at present in long-lasting examinations, which is 88.2kHz, the minimum duration time of an event is 34 microseconds. A 9812 ADLINK fast card of analogue-numerical processor was used for the analysis of AE signals. Such a device enables the increase of instrument sensitivity and detection of AE events differing by an order of magnitude. The indexes of start and end of AE event recorded in the program table can serve to the determination of its duration. The E energy of AE event can be derived from the approximate formula:

$$E = 0.5\,(v_{max})^2\,\Delta t, \tag{2}$$

where v_{max} denotes maximum value of AE signal during the event, Δt - time of AE event.
To characterize the material subjected to the examinations, values of arithmetic means of all measured values E, v_{max} and Δt are needed. The AE instrument is also equipped with an analogue system, which allows obtaining an effective value of the signal.
The transformation of the set of instantaneous values of the measured $v(t)$ signal into effective value V_{RMS} for time T is realized according to formula:

$$V_{RMS} = \sqrt{\frac{1}{T}\int_0^T v^2(t)dt}. \tag{3}$$

The AE signals generated by different sources in the examined object can be analyzed inspecting the changes of its *spectral characteristic*. A continuous AE signal $v(t)$ in a selected finite range of time can be demonstrated as a function of its spectral characteristic $A(\omega)$, where ω is a frequency pulsation f, defined as $\omega=2\pi f$. Assuming absolute integrality, function $v(t)$ is linearly transformed into the function of spectral density $A(\omega)$ according to the Fourier transformation:

$$v(t) = \frac{1}{\pi}\int_{0}^{\infty} A(\omega)\exp(j\omega)\,d\omega. \tag{4}$$

In consequence, the procedure of spectral density function determination $A(\omega)$ for consecutive segments of discrete set of AE signal samples was elaborated together with a corresponding graphic presentation of results in the form of acoustic maps i.e. *acoustograms* or *spectrograms*. A numerical method of Windowed Fourier Transformation (WFT) is applied here. Next, the discrete form of spectral density function is determined using several thousands of signal samples adjacent to the central sample of the recorded AE event. The algorithm, which transforms the set of signal samples into a set of spectral density coefficients $c_n : v(m) \Rightarrow c_n(\omega)$ is similar to the approximate formula (Scott, 1991):

$$c_n \approx \frac{1}{N}\sum_{m=0}^{N-1} v(m)\cdot \mathrm{mod}(\, e^{jn2\pi m/N}), \tag{5}$$

in which j denotes $\sqrt{-1}$ and mod is the module of complex expression.
The acoustogram of AE event set presented earlier in dependence of signal amplitude on time in Fig. 2 has now been shown in Fig. 3. The spectral characteristic of signal is illustrated every 0.5ms. The discrete equivalent of the $A(\omega)$, spectral density function is presented in the form of color code.
The AE analyzer is equipped with an additional measurement channel enabling simultaneous recording of sample load by the computer as well as the registration of AE parameters in the form of AE event rate together with their duration, amplitude and effective value connected with conventional value of the event energy and in consequence, the distribution of events number versus their energy.

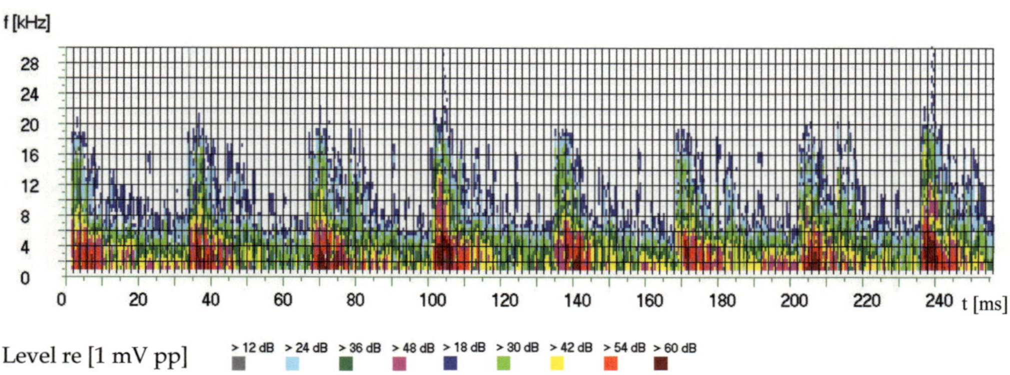

Fig. 3. Acoustogram of AE event set, previously presented in Fig. 2 in the form of dependence of signal amplitude on time

The effective value of noises at the inlet of the preamplifier is about 20÷30mV, depending on the selected frequency band. During the signal processing the value undergoes about four-fold decrease due to the application of a band-pass filtration. An active upper-pass filter of 8th order of cut off frequency 5kHz is joined to the preamplifier. Another filter of 4th order of 20kHz frequency can be additionally switched on. Thanks to that, the signals of a vibroacoustic background, which do not originate from the processes occurring because of sample loading, are eliminated from the further processing. The signal is next passed to a lower-pass filter of cut-off frequency 1MHz. The total amplification of AE signal at the outlet of instrument is 70dB and the threshold voltage is 0.5V. The effective voltage value of the AE signal recorded is derived through the second exit. The analysis of energy and duration of individual AE events is possible with an appropriate program, which determines time of the AE event occurrence, its maximum amplitude, sum of recorded amplitudes and duration time of the event up to significant decrease of its amplitude.

2.3 Methods of microstructure observations

After each stage of a compressive test, microstructure observations were carried out using a standard technique of optical microscopy at NEOPHOT instrument. The further observations were performed using transmission (TEM) and scanning (SEM) electron microscopes. The techniques of *convergent beam electron diffraction* (CBED) and *electron back scattered diffraction* (EBSD) as well as SEM with *field emission gun* (FEG) were applied in the examinations of bi-crystals.

2.4 Methods of intensive deformation

Fig. 4a shows the scheme of the ECAP method. The parameters of the installation have the following values: b=10mm, a=30mm, angle α=31.3° or α=90°. Equivalent strain (for square cross-section) is equal to $\varepsilon_n = 0.5922n$, where n – number of passes. For angle Φ= 90° and α=0 it amounts to $\varepsilon_n = 0.9069n$.

Fig. 4b shows the scheme of the HPT method of torsion under high pressure. The sample is in the form of a roll with R radius and the height l. Dilatation strain γ after N rotations is equal to $\gamma = (2\pi RN)/l$, and the equivalent strain $\varepsilon_N = \gamma/1.73$.

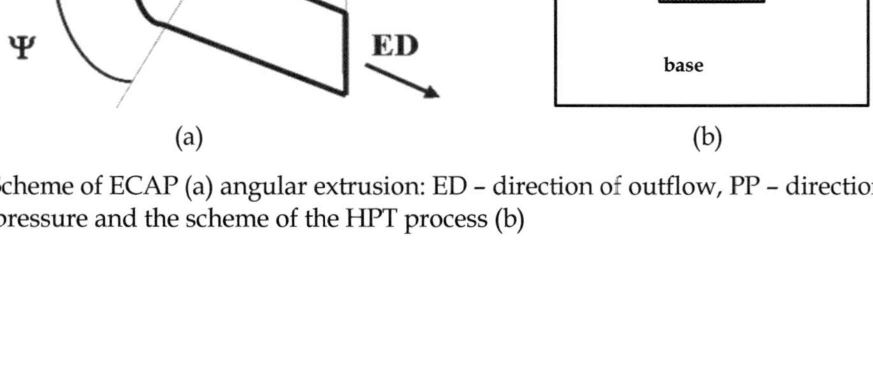

Fig. 4. Scheme of ECAP (a) angular extrusion: ED – direction of outflow, PP – direction of punch pressure and the scheme of the HPT process (b)

Fig. 5 presents the scheme of the ARB technique. Purified and degreased surfaces of two sheet plates are folded and fastened, next heated and rolled to reduction $z = 50\%$. The sheet obtained after rolling is cut into halves and subjected to the same procedure as before. The procedure may be repeated several times. For example, a sheet plate with thickness g_o, subjected to rolling in succession n time to the reduction of z=50%, i.e. after n passes, has thickness $g_n=g_o/2^n$, and the total reduction is equal to $z_n=1- g_n/g_o=1-1/2^n$. The tensile tests were carried out with ten-fold plane specimens using the standard INSTRON machine. The rate of the traverse of the testing machine was 2mm/min. Each specimen was of gauge length l_0=90mm (overall length l_c=105mm), b_0=20mm wide and a_0=3.50mm thick.
The samples for ECAP tests (for circular cross-section) had the shape of rolls with diameter 20mm and height 30÷40mm while that for the ECAP processing in the channel of square cross section was in the shape of rectangular prisms of dimensions 10x10x40mm. The samples to be used in HPT tests had the shape of discs of diameter 10mm and thickness 3÷5mm. The samples intended for compression had the shape of cubes with the edge not greater than 10mm in the case of ECAP or the shape of square plates with side 10mm and thickness 1mm, cut in the case of ARB from the samples prepared earlier for the tensile tests. The discs after HPT, intended for the compression, had the thickness of the order of 1÷2mm.

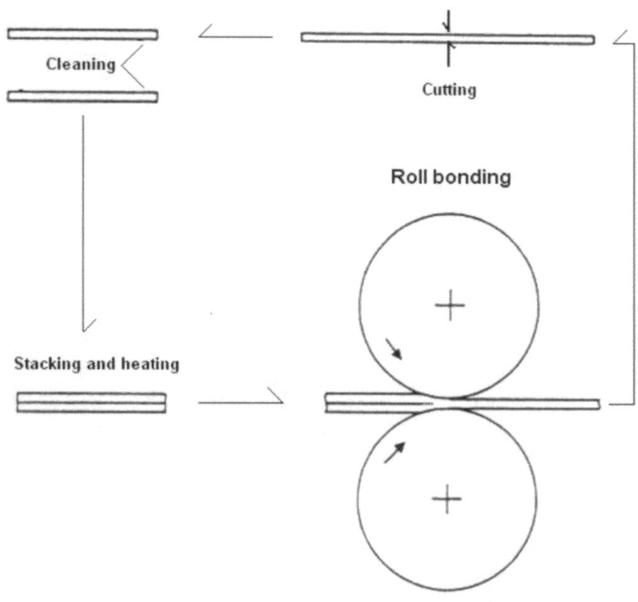

Fig. 5. Scheme of the ARB process

Simultaneously with the registration of the external force F and the sample elongation, the basic AE parameters in the form of AE event rates, or as RMS – the effective value of voltage of the registered AE signal were measured.
The aim of the research has been the documentation and interpretation of correlation of the AE descriptors during compression tests of Al alloys before and after pre-deformation by intensive strain methods. The evolution of micro- and/or nanostructure in dependence on dislocation mechanisms of deformation as well as slip processes occurring along grain boundaries, responsible for possible superplastic flow, is also considered.

3. AE in Al mono- and bi-crystals compressed at liquid nitrogen temperature

3.1 Twinning and shear band formation in Al single crystals

The examinations of AE in Al mono- and bi-crystals were aimed at the identification of unstable plastic flow mechanisms connected with the deformation twinning and the formation of shear bands. The compression test were carried out mainly at temperature of liquid nitrogen (77K). Moreover, the still controversial problem of deformation twinning in Al crystals is also considered here.

The course of AE and the external force in the Al crystal of Goss orientation {110}<001> is shown in Fig. 6a whereas Fig. 7 comprises, for the purpose of comparison, the results of AE and the course of external force for the crystal of {112}<111> orientation.

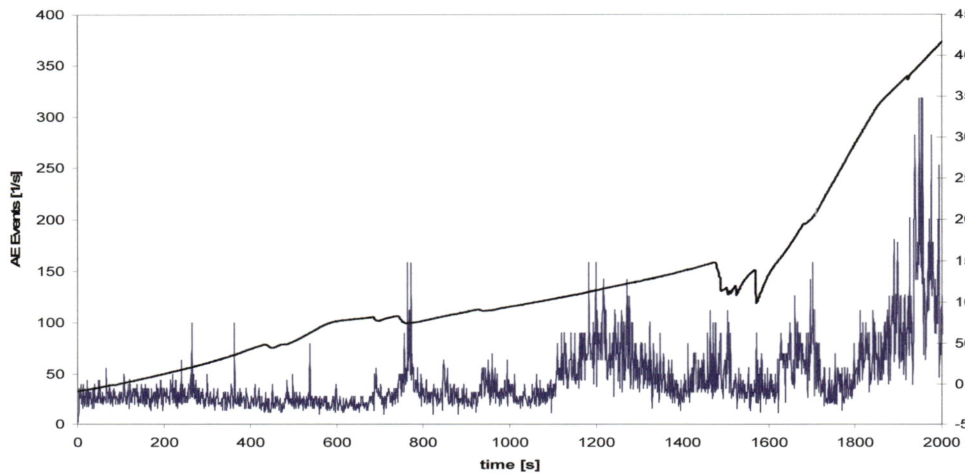

(a)

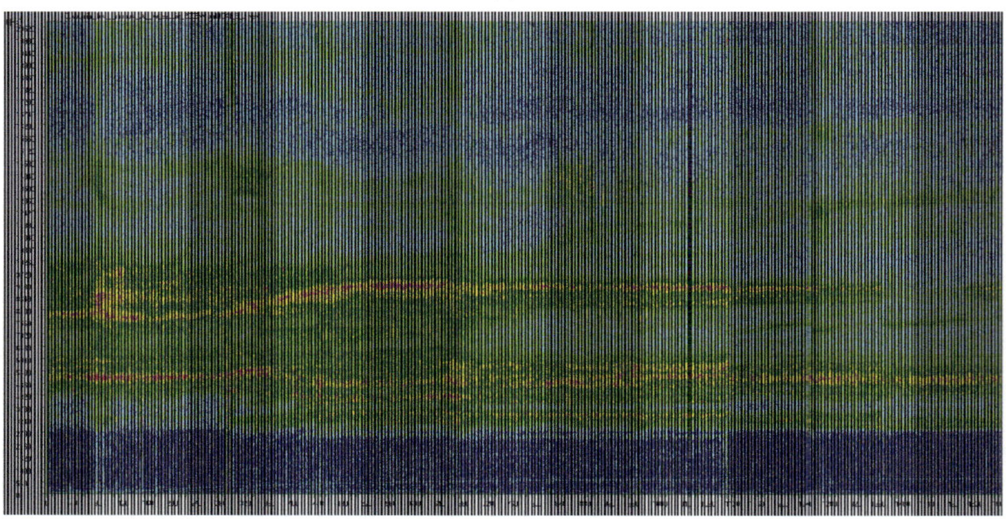

(b)

Fig. 6. Course of AE and compressive force in Al single crystal of Goss orientation (a) together with corresponding acoustogram (b)

When analyzing Fig. 6, it can be stated, that evident correlation exists between the course of force and behavior of AE. All the local decreases of the force curve are accompanied by more or less distinct areas of elevated AE activity. It seems that these strong plastic instabilities on the compression curve correspond to the occurrence of shear bands. Vertical areas in the acoustogram containing a broad range of frequency spectrum visible in Fig. 6b.

On the other hand, low temperature courses of AE impulses together with external compressive force in dependence on time are presented in Figs. 8 and 9a and 10a for the Al single crystals of two selected orientations: {112}<111> (Fig. 8) and {531}<231> (Fig. 9a for reduction z≅27.1% and Fig. 10a for reduction z≅51.4%). Attention should be drawn to some characteristic features of the recorded courses.

Moreover, the experimental {111} pole figures (EXP) presented in Figs.9b and 10b, as well as the calculated orientation distribution functions (ODF) referred to in Figs.9c and 10c, illustrate explicitly the existence of twin orientation after compression to z≅51.4% (Fig. 10b and 10c), and suggest strongly the possibility of deformation twinning also in the Al single crystals channel-die compressed at the liquid nitrogen temperature. In {111} pole figure (Fig. 10c), the component of twin orientation $(\bar{4}\,4\,1)\,[\bar{1}\,\bar{3}\,8]$, appearing after reduction z≅51.4% (initial matrix orientation $(\bar{1}\,3\,5)[\bar{1}\,3\,\bar{2}]$, Figs. 9b and 9c) is now given by orientation $(\bar{2}\,2\,5)[\bar{3}\,7\,\bar{4}]$ (Fig. 10c) – corresponds to the twinning on the active co-planar slip system.

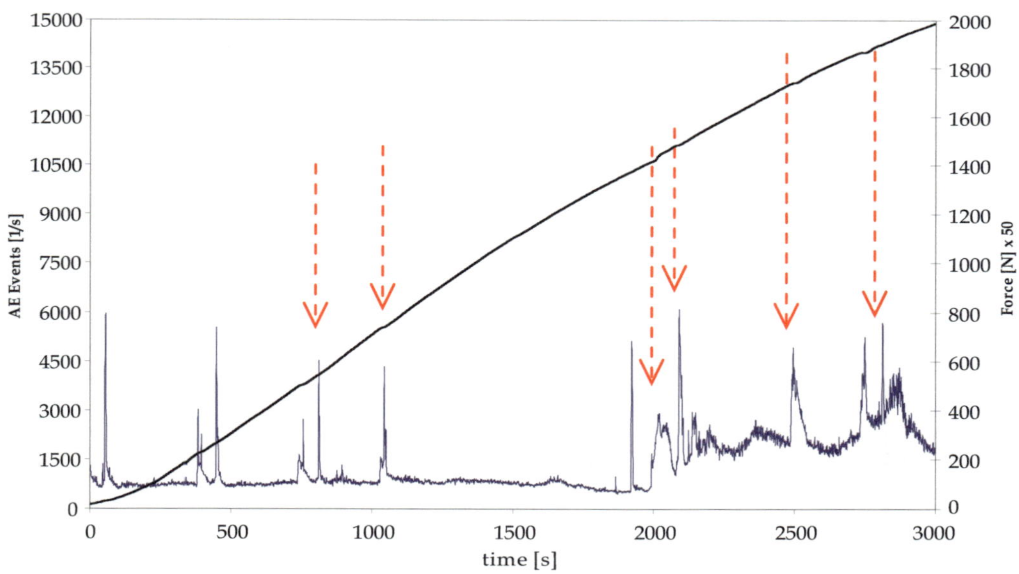

Fig. 7. Acoustic emission and compressive force in Al single crystal of orientation type C≡{112}<111>. The arrows indicate the correlations between AE and drops of force

Based on the dislocation dynamics and the AE model (Jasieński et al., 2010; Pawełek, 1988a; Pawełek et al., 2001; Ranachowski et al., 2006) a number of AE impulses, which were generated due to the appearance of an individual twin lamella can be estimated. It was assumed, that the twins formed as a result of the pole mechanism action. It was also accepted, that an individual AE impulse occurred, when a partial twinning dislocation, which moved in the area of a single atomic plane, approached the surface. This suggestion is in agreement with the results and concepts reported by Boiko et al. (1973, 1974, 1975) for the relationship between AE and elastic twins generated in calcite crystals. It was also assumed for simplification that distance *a* between the atomic planes was equal to the value of Burgers vector of dislocation i.e. $a \cong b \cong 1.0 \times 10^{-4} \mu m$.

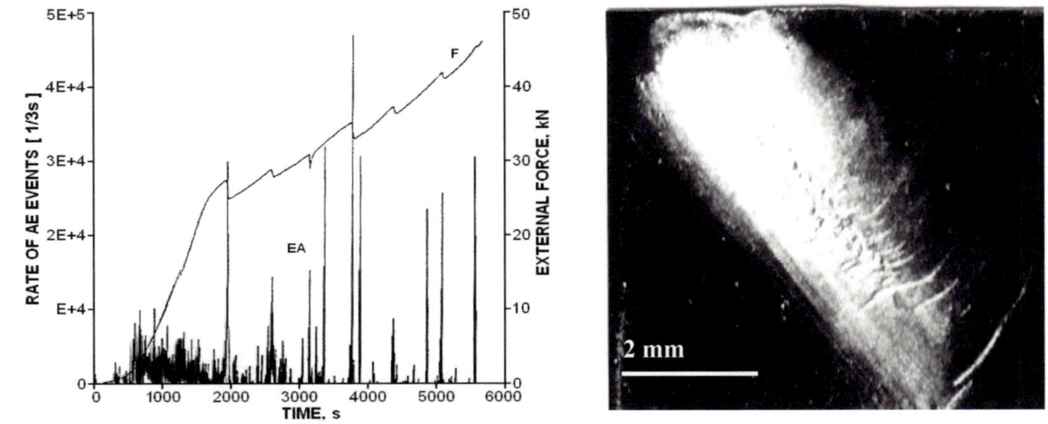

Fig. 8. Courses of AE and external force of Al single crystal of orientation {112}<111> channel-die compressed at T=77 K up to reduction z=61.6%; microstructure inserted nearby illustrates shear band

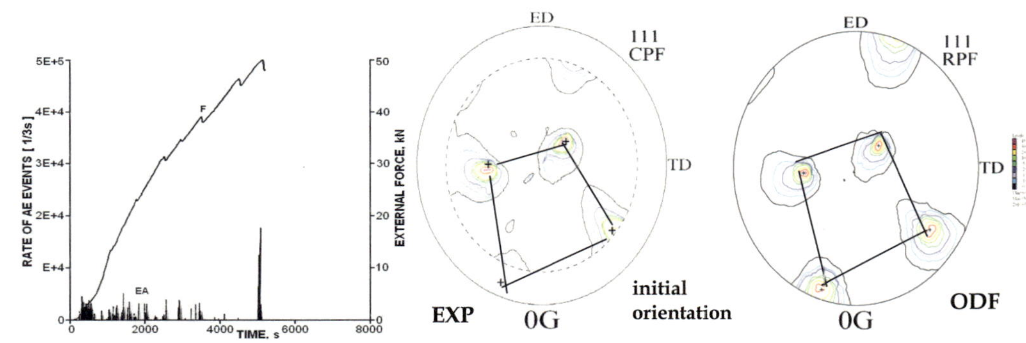

(a) (b) (c)

Fig. 9. Courses of AE and external force (a) of Al single crystal of initial orientation {531}<231> channel-die compressed at T=77K up to reduction z=27.1% and corresponding {111} pole figures: experimental EXP (b) and recalculated ODF (c)

The thickness of twin lamella, estimated visually from microstructure images, was in the range from 0.1μm up to 1.0×10^{3}μm, which, at magnifications of order 10x÷100x, used the most frequently, gave the thickness of a real twin lamella, at first approximation, in the range from 1 up to 100μm. Hence the number of atomic planes engaged and the number of elementary impulses completing the AE peak from an individual twin was of order $10^{4} \div 10^{6}$, which was in satisfactory agreement, as far as the order of value was concerned, with the value observed.

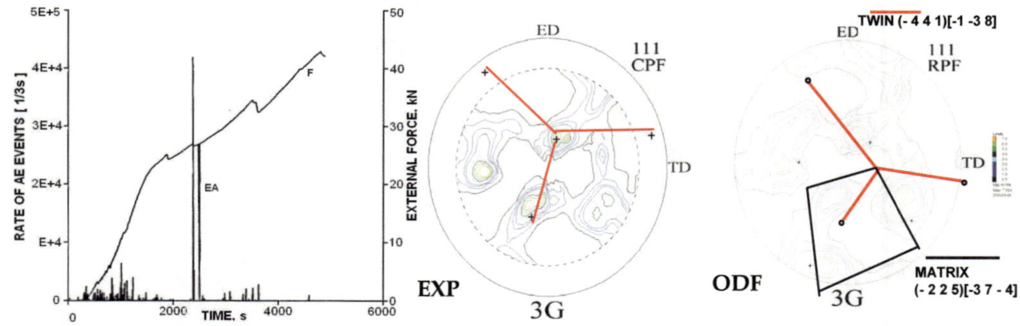

Fig. 10. Courses of AE and external force (a) of Al single crystal of initial orientation {531}<231> channel die-compressed at T=77K up to reduction z=51.4% and corresponding {111} pole figure experimental EXP (b) and recalculated ODF (c)

3.2 AE during channel-die compression of Al bi-crystals

The examinations of AE tests in Al bi-crystals subjected to low-temperature channel-die compression were also performed. Fig. 11 show the behavior of AE and force together with corresponding acoustogram for the Al bi-crystals of {110}<100>/{110}<011 hard/Goss orientation. After the first stage of deformation (z≅30%), the sample was trimmed and compressed again until it was reduced by z≅50%. Courses of AE and external force for a bi-crystal of {110}<001>/{100}<011> Goss/shear orientation are illustrated in Fig. 12 for the first stage of deformation (z≅20%). For such an orientation also structural examinations were performed using TEM and CBED and SEM-FEG/EBSD techniques.

3.3 Development of Al bi-crystal dislocation structure

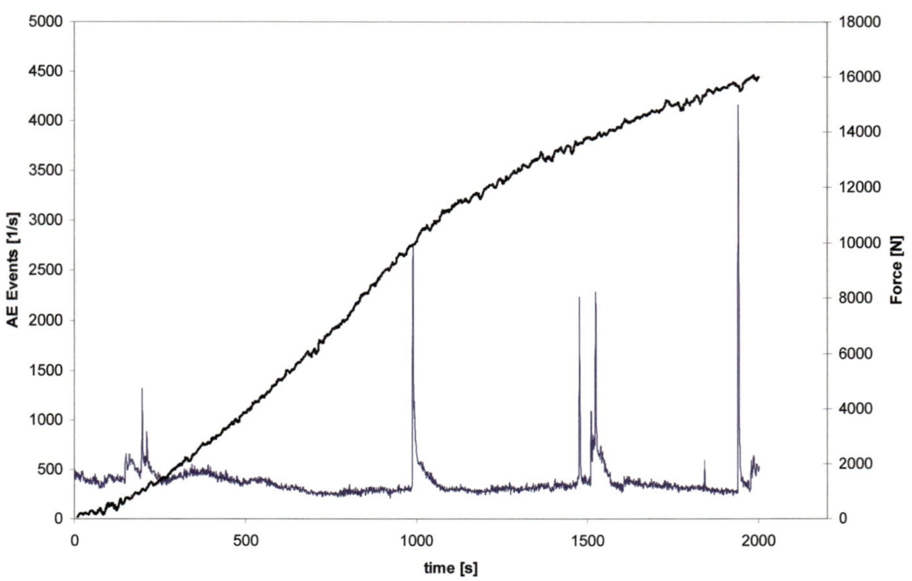

(a)

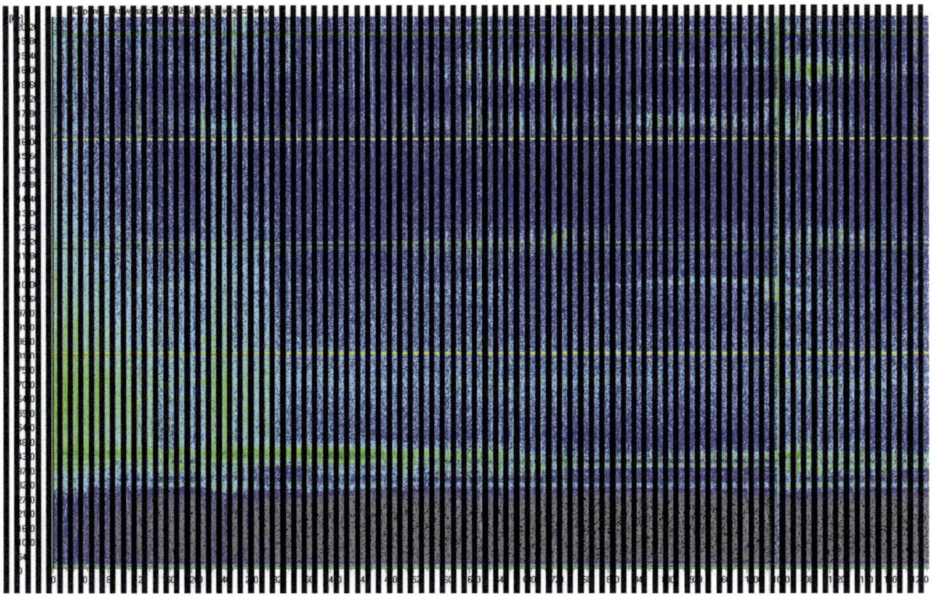

(b)

Fig. 11. AE and force of channel-die compressed Al bi-crystal of {110}<100>/{110}<011> hard/Goss orientation reduced by z≅50%: (a) and corresponding acoustogram (b)

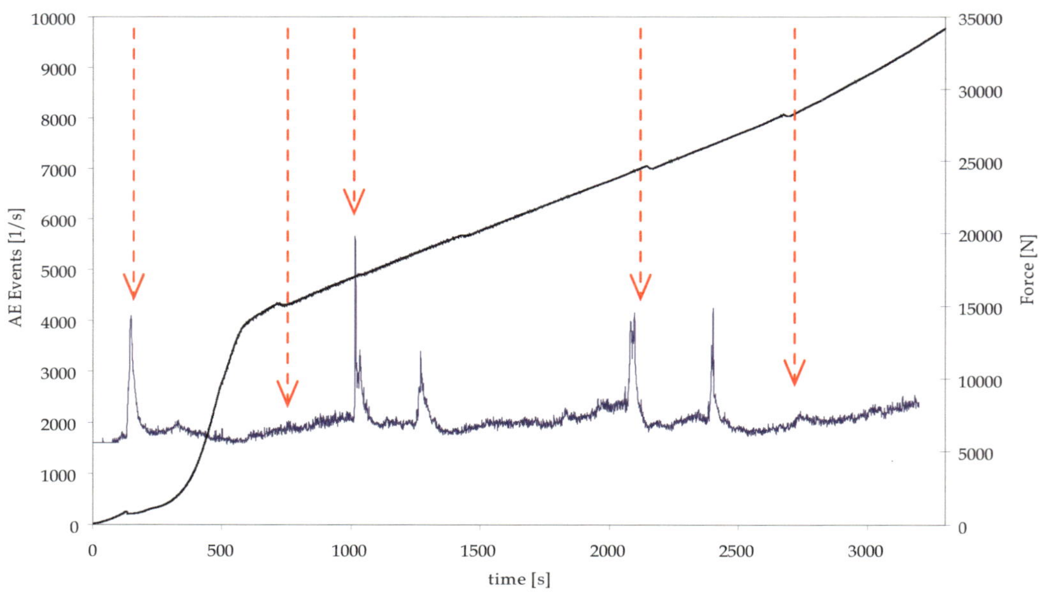

Fig. 12. Behavior of AE and compressive force of bi-crystal of {110}<001>/{100}<011> Goss/shear orientation. The arrows indicate the correlations between AE and drops of force

The Goss {110}<001> orientation remained stable during compression in a broad range of deformations, while the shear {100}<011> orientation was strongly unstable and underwent decomposition from the initial stages of deformation by rotation around the transverse direction (TD).

3.3.1 TEM microstructure observations

The examinations in the micro and mezzo- scale were focused on the structure and texture analyses carried out using transmission (TEM) and scanning (SEM) electron microscopy. In both cases, the observations were correlated with measurements of local orientations changes with the use of *convergent beam electron diffraction* (CBED) and *electron back scattered diffraction* (EBSD) techniques. The TEM orientation measurements were performed using semi-automatic system, while the SEM instrument equipped with field emission gun (FEG) disposed of fully automated system Channe 5™ of HKL Technology firm.

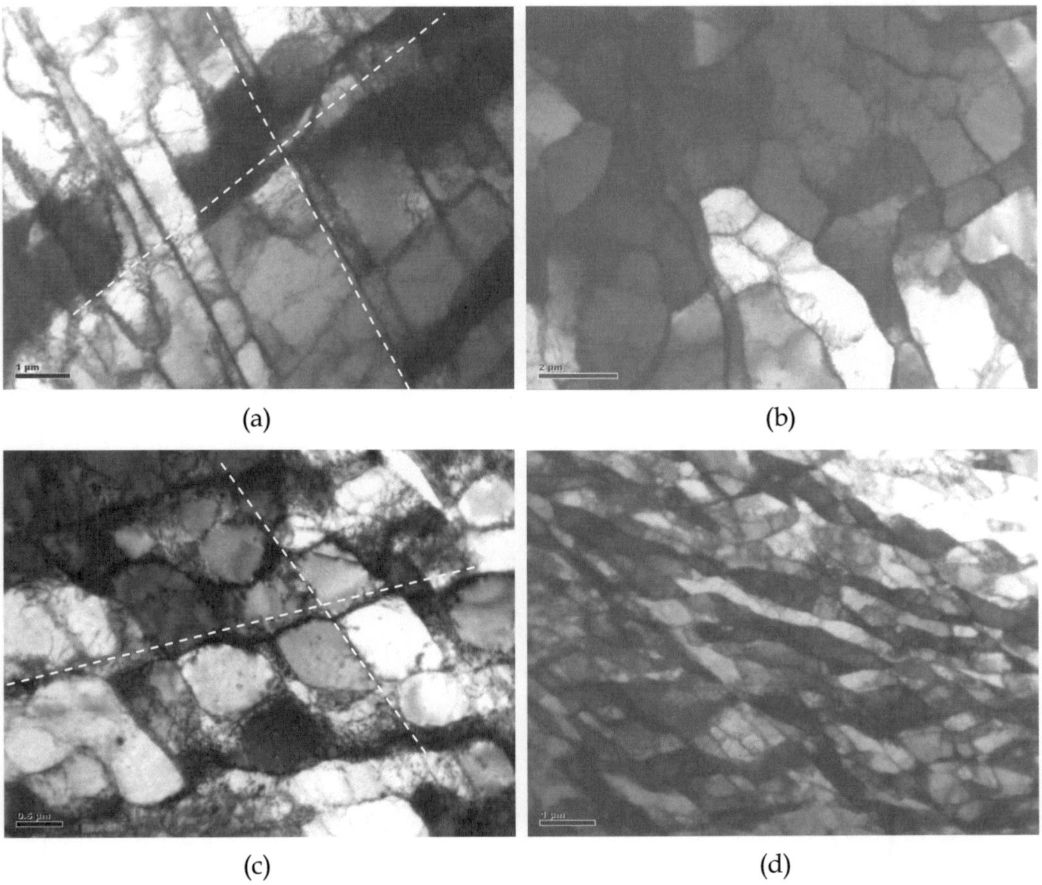

Fig. 13. Development of dislocation structure observed in TEM in crystallites of Al bi-crystal; (a) and (c) Goss orientation, (b) and (d) shear orientation; total deformation of bi-crystal z≅20%(c) and z≅70%(d). In the Goss orientation traces of planes, on which active slip systems operated are marked

The observations of dislocation structure in TEM in each crystallite of the bi-crystal deformed by z≅20% and z≅70% were correlated with the local orientation measurements in a high-resolution scanning microscope using SEM-FEG/EBSD techniques. The development of dislocation structure observed in TEM is presented in Fig. 13 for each crystallite and for two deformation degrees. The advance of structure refinement together with the increase of deformation was observed in the crystallites with the Goss orientation {100}<011>. In crystallites of shear orientation {100}<011>, two pairs of coplanar slip systems were active what led to the decomposition of the crystal resulting in the change of initial crystallite orientation to two symmetrically situated positions of {112}<111> orientations. The areas, in which different slip systems dominated were separated by intermediate bands.

3.3.2 SEM examinations of local texture changes

The texture-structure examinations were performed using the SEM-FEG/EBSD system at a mezzo-scale, which allows reproducing the "electron" image of structure as regards the crystallographic orientation changes.

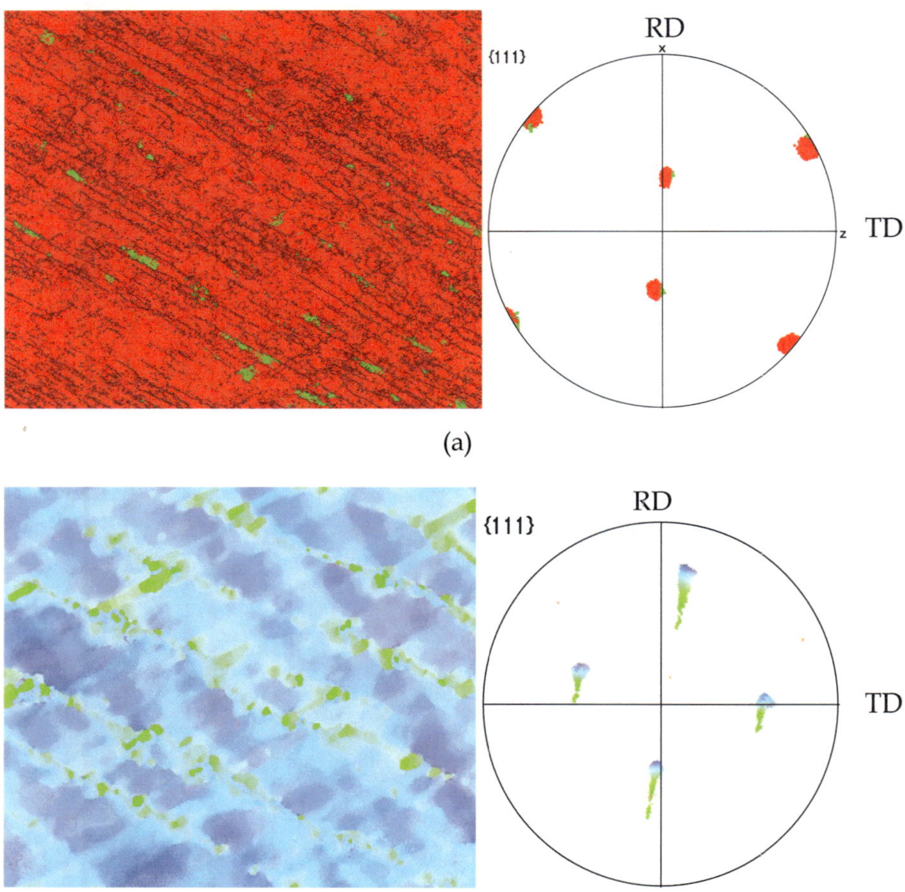

(a)

(b)

Fig. 14. Orientation maps (shown as a „function" of IPF colors) for the part of bi-crystal of Goss (a) and shear (b) orientation and corresponding {111} pole figure. Measurement step 100nm. Reduction z≅20%

The obtained orientation maps of Goss and shear type presented in Fig. 14 illustrate well the observed tendencies to broadening of the initial orientation of the crystallites in the bi-crystal. In the Goss orientation (Fig. 14a), after an applied deformation degree, initial orientation remains stable; only weakly visible tendency to broadening of the {111} plane poles mainly by a rotation around ND is observed. In the case of crystallite of orientation shear (Fig. 14b) a strong tendency to rotation of crystalline lattice through the rotation around TD, towards two complementary positions of the {112}<111> orientation. The presented orientation map shows the structure-texture changes in the area, in which one pair of coplanar systems dominates.

3.4 Comparison with other fcc single crystals

In order to document better the correlation between the AE behavior and the localization of deformation connected with twinning processes and the formation of shear bands, experiments at temperature of liquid nitrogen (77 K) were carried out on single crystals of Ag and Cu. The selected results for the Ag single crystals are presented in Fig. 15, and for copper in Fig. 16. The results refer to the same orientation {112}<111> and two subsequent

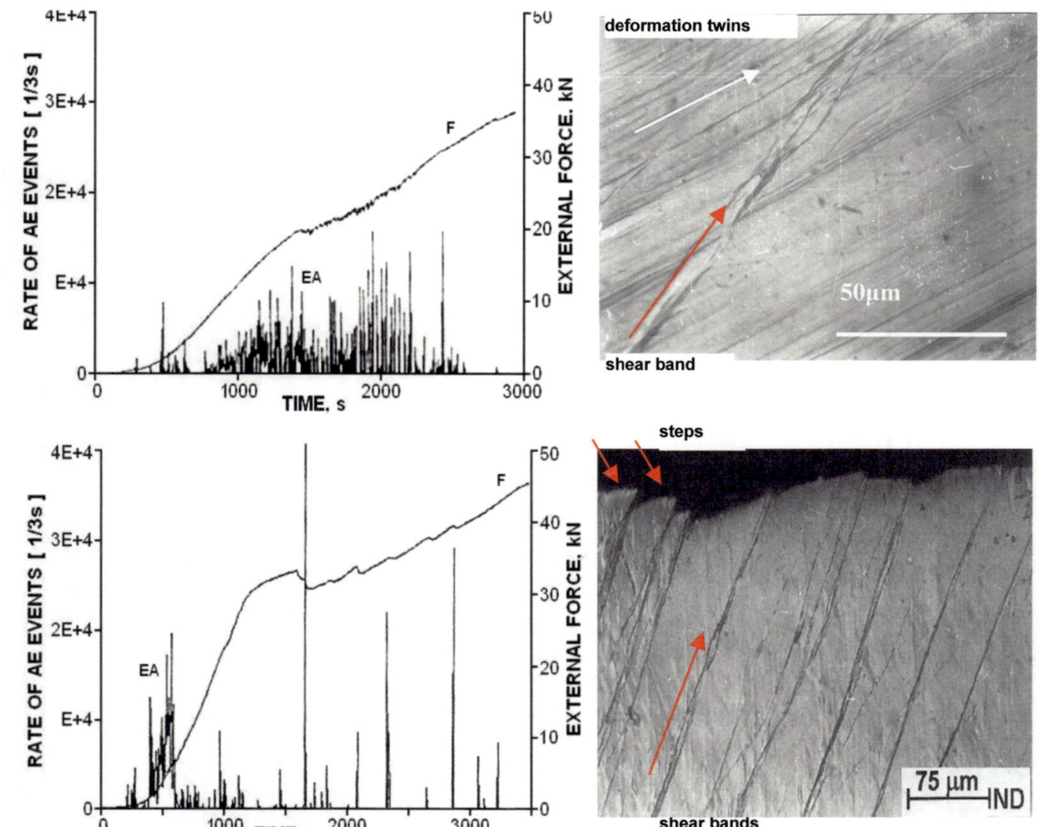

Fig. 15. Courses of AE and external force and corresponding microstructures of Ag single crystals of orientation {112}<111> channel-die compressed at T=77K: (a) – reduction z≅33% and (b) – reduction z≅63.4%

stages of compression, comprising conventionally intermediate reductions in the range from about 30% to 65%. It is visible, that the behavior of AE and its correlations with external loads are qualitatively very similar. It confirms that the deformation mechanism changes from an ordinary slip through strong twinning (Fig. 15a and 16a and appropriate optical microstructures) to the mechanism of shear band formation (Fig. 15b, 16b and corresponding optical images).

The considerable drop of AE event rate is a characteristic feature of twinning → shear bands transition, while corresponding high AE peaks are distinctly correlated with abrupt drops of the external load, which is the most evidently caused by the appearance and development of individual shear bands, which belong to the same primary family. For example, the last high AE peak visible in Fig. 16a at about 2200s may originate from an already forming shear band.

Comparing the courses of force and AE and the microstructure (Fig. 8) for the Al crystal of {112}<111> orientation with respective plots and images for the Ag and Cu single crystals of the same orientation (Fig. 15 and 16, respectively) and analyzing the courses of force and AE for the {531}<231> orientation in the Al single crystals (Figs.9a and 10a) a similarity to a large extent can be noticed, which lets us state, that also in the Al single crystals compressed at low temperatures, the transition of the type twinning → shear bands after initial slip deformation is quite probable.

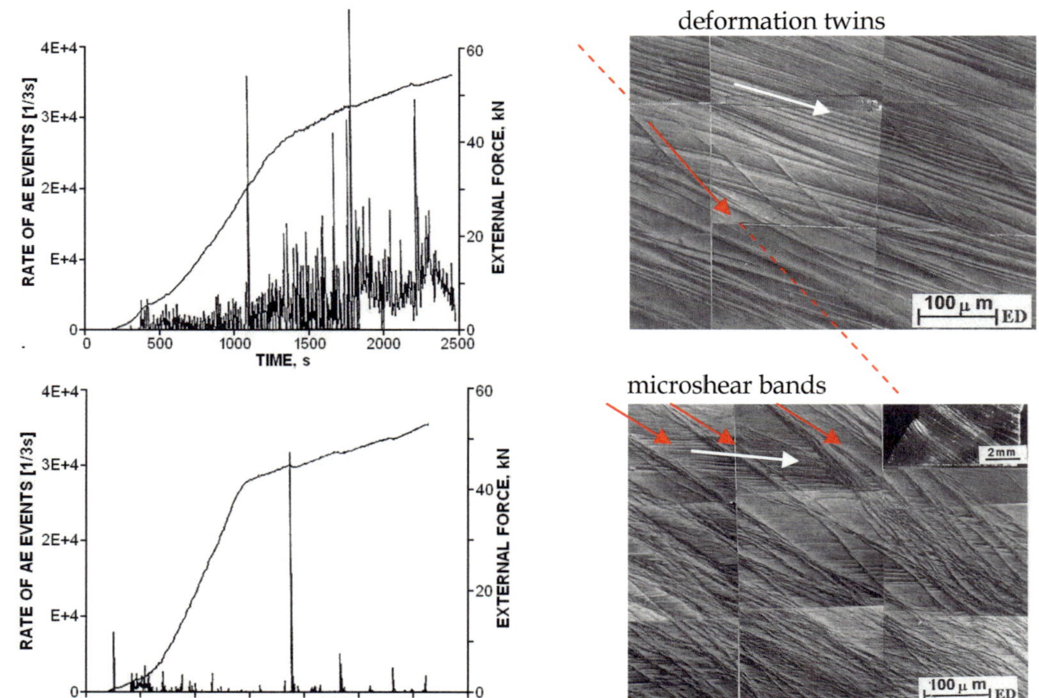

Fig. 16. Courses of AE and external force and corresponding microstructures of Cu single crystals of orientation {112}<111> channel-die compressed at T=77K: (a) - reduction z≅41% and (b) - reduction z≅53.4%

3.5 Acoustic emission vs. twinning in Al crystals

The presented results helped to establish a scheme of the microstructure evolution and mechanisms of deformation during channel-die compression of single crystals of fcc metals. The substantial element of the model is, that in the range of **intermediate reductions** (from about 30% to 65%) in the initial stages of compression, a change of the deformation mechanism from intensive twinning resulting in high AE of big activity of AE sources into the generation and localization of primary family of shear bands takes place. In the range of **small reductions** (up to about 35%) the dislocation mechanisms of ordinary slip dominate and processes of twinning can be initiated, while in the range of **high reductions** (above about 65%) the formation of another family of shear bands begins in the secondary slip systems, not coplanar with respect to the primary systems (Pawełek et al., 1997).

Based on the above considerations it can be stated, that the presented results directly indicated the correlation of the following four elements: high peak of AE event rate, abrupt decrease of external force, the formation of twin lamella or the nucleation of shear band as well as the appearance of a step at the surface of deformed crystal.

However, the twins were not observed neither in the Al crystals nor Al bi-crystals using the accessible methods of optical and electron microscopy. On the other hand the presented pole figures in Figs. 9 and 10 surely do not exclude the possibility of twinning. Moreover, they become a kind of proof that the process of deformation twinning in fact has occurred. It should be stressed that in this kind of discussion an argument is often raised, that the existence of twin orientations itself is not a proof that the process of deformation twinning has taken place. Similarly, the microstructures obtained using the TEM technique (Paul et al., 2001) may certify the fact that the deformation twinning occur also in single crystals of Al, although - it should be impartially said - they are not too convincing.

There is also another kind of confirmation of such a statement: it is the audible effect. In many cases, during the compression tests knocks typical for twinning were heard in the frequency range audible for the human ear. Hence, it is probable, that the difficulties in the documenting the twins in microstructure images are due to very high stacking fault energy of Al. Very fast processes of recovery or even recrystallization taking place in the sample being moved from the liquid nitrogen to the ambient temperature "blurr" the possible twins formed during deformation. In general the problem has not been definitely solved so far, although the results obtained may contribute, to some extent, to its full solution.

The observed correlations between AE and the mechanisms of deformation can be explained in terms of highly synchronized and collective behavior of groups of many dislocations, particularly in reference to the processes of dislocation annihilation at the free surface of the sample. Moreover, the description of dislocation annihilation based on the soliton properties of dislocation (Pawełek, 1988a; Pawełek et al., 2001) is closer to the reality than the description resulting from the application of the theory of continuous media (& Burkhanov, 1972; Natsik & Chishko, 1972, 1975).

4. AE in polycrystalline Al alloys deformed before and after intensive strain operations

4.1 AE in AA6060 and AA2014 alloys compressed before and after using the ECAP method

The measurement of AE were carried out for Al alloys of AA6060 and AA2014 type subjected to compression in a channel-die after the ECAP processing in a channel of circular cross-section. The AE behavior and the courses of compressive force of Al alloy of AA6060 type subjected to compression tests after 2- and 4-fold processing in the ECAP circular channel are presented in Figs. 17a and 17b, respectively, in which a significant decrease of AE level measured with the RMS parameter was observed. The observation confirms the

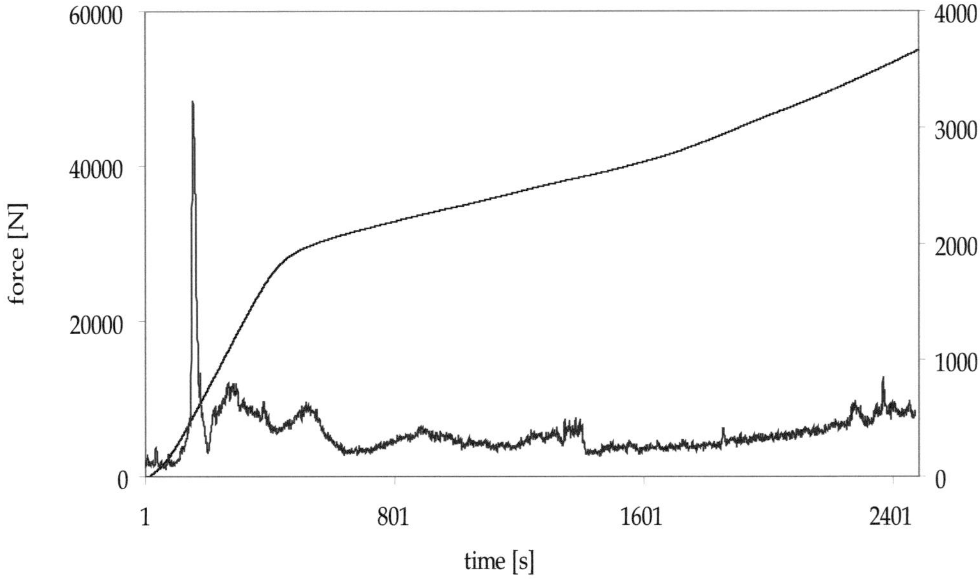

(a)

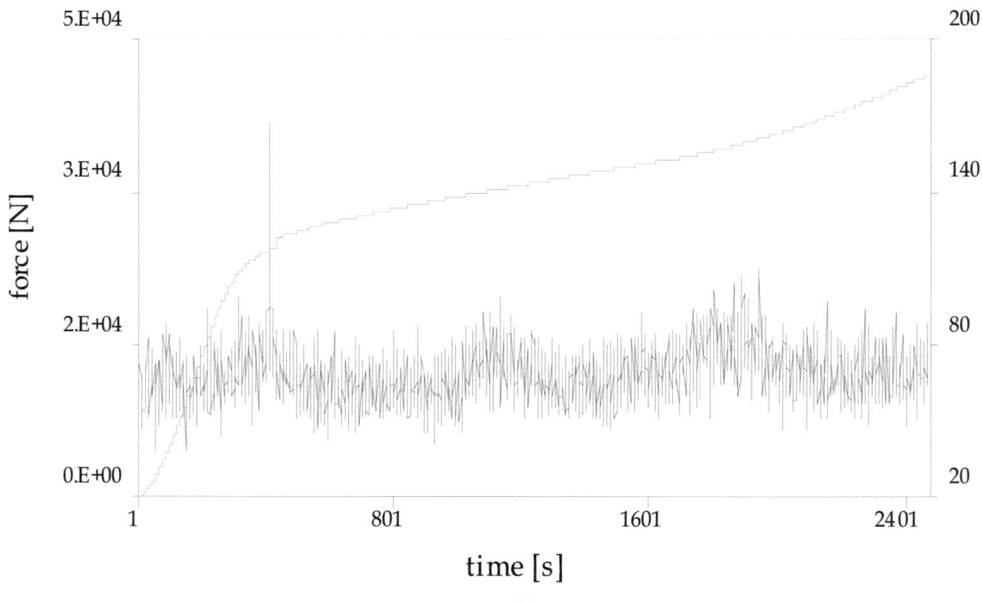

(b)

Fig. 17. AE and compressive force in Al alloy of AA6060 type subjected to tests of compression after two-fold (a) and four-fold (b) processing in the ECAP angular channel of circular cross section tendency of AE to decrease reported earlier (Kúdela et al., 2011; Kuśnierz et al., 2008) in the samples of Mg-Li and Mg-Li-Al alloys compressed after processing with intensive deformation, and for comparison, presented here in the section 4.3 of this chapter. The next Figs. 18a and 18b show a TEM microstructure of the AA6060 alloy on horizontal and cross sections observed after 4-fold ECAP operation in the circular channel.

The results of AE examinations of the Al AA2014 alloy during compression tests are shown in Fig. 19 in which the courses of AE and external force in the sample after 2-fold processing in the angular circular channel ECAP (Fig. 19a) as well as the corresponding TEM microstructure (Fig. 19b) allow the conclusion, that the average level of AE is lower than in the AA6060 alloy, compressed also after 2-fold ECAP operation.

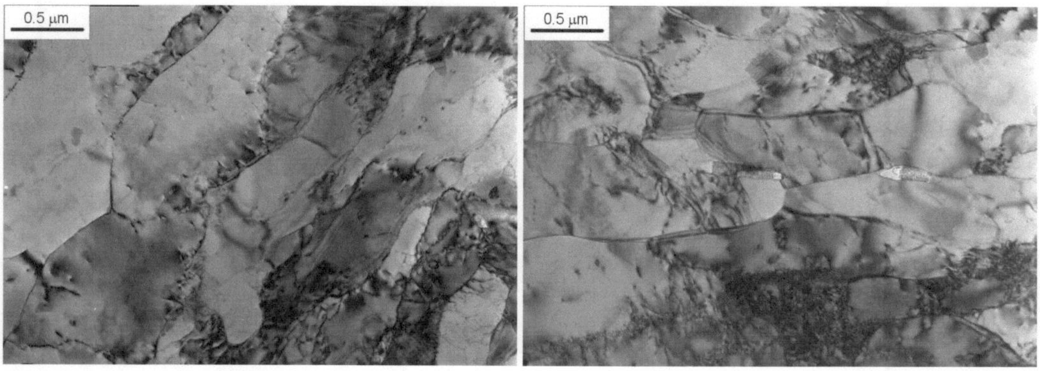

(a) (b)

Fig. 18. TEM microstructure of Al alloy of AA6060 type after 4-fold processing in the ECAP angular channel of circular cross section: (a) - horizontal section, (b) - cross section

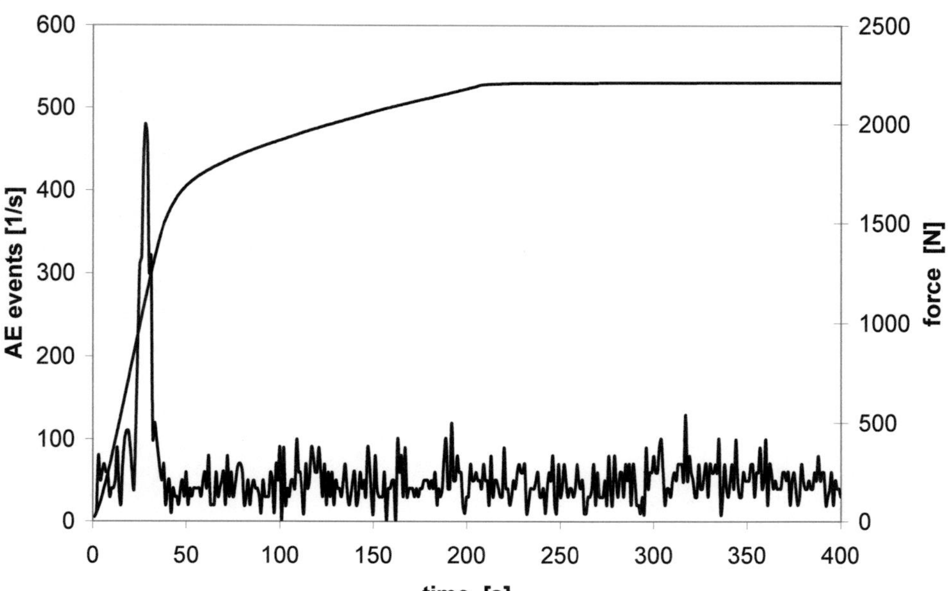

(a)

(b)

Fig. 19. Courses of AE and compressive force (a) in AA2014 Al alloy compressed after 2-fold ECAP processing of circular cross section and its corresponding TEM microstructure (b)

4.2 AE in Al AA6060 and AA2014 alloys compressed before and after HPT treatment

The AE and compressive force courses were examined on the Al AA6060 alloys subjected to compression tests before the HPT operation, after one and two HPT operations (Figs. 20a and 20b, respectively). It is visible, that the AE distinctly decreases after two-fold HPT processing compared with the AE level after one HTP operation. However, from Fig. 21, an evident decrease of the AE rate in the sample already after one HPT rotation (Fig. 21b) compared with the initial state of AA2014 alloy sample (Fig. 21a) can be observed.

4.3 Comparison to Mg-Li and Mg-Li-Al alloys

Fig. 22 presents one of more important results obtained so far on the Mg8Li alloy compressed before and after the application of ECAP technique in squared cross section channel (Kuśnierz et al., 2008). These results are even more pronounced in the case of Mg10Li and Mg10Li5Al alloys compressed after the application of HPT technique (Kúdela et al., 2011). The result of the applied HPT process, is presented in Fig. 23 for the Mg10Li alloy after a three-fold, whereas the relation of AE activity versus compression force for the Mg10Li5Al alloy is shown in Fig. 24 also after a three-fold HPT treatment.

The decrease of AE activity of compressed samples subjected to the both ECAP and HPT processes is closely connected with the size of grains. The initial microstructure of Mg8Li alloys (Fig. 22a) consists of size grains of order of several hundreds of micrometers (10^2μm), whereas the microstructure after a large plastic deformation by ECAP is visible in Fig. 22b. After the ECAP and/or HPT processes, applying only a few rotations, the grain size decreased three orders of magnitude to hundreds of nanometers (10^{-1}μm, Fig. 22b, 23b and 24b). These microstructures of Mg-Li and Mg-Li-Al alloys, presented in are the most pronounced examples of the refining effect of intensive strain processes arriving at UFG and/or nanocrystalline structure.

On the other hand the decrease of the AE level in Mg10Li and Mg10Li5Al (Figs. 23a and 24a) is of about two order of magnitude, as in the case before HPT operation, not presented here, since it is about 10^4/s, similarly as in the case of Mg8Li alloys (Fig. 22a).

The similar statement we can refer to the microstructures of Al alloys of AA6060 and AA2014 type, presented in Figs. 18 and 19, respectively. The phenomenon of the decrease of

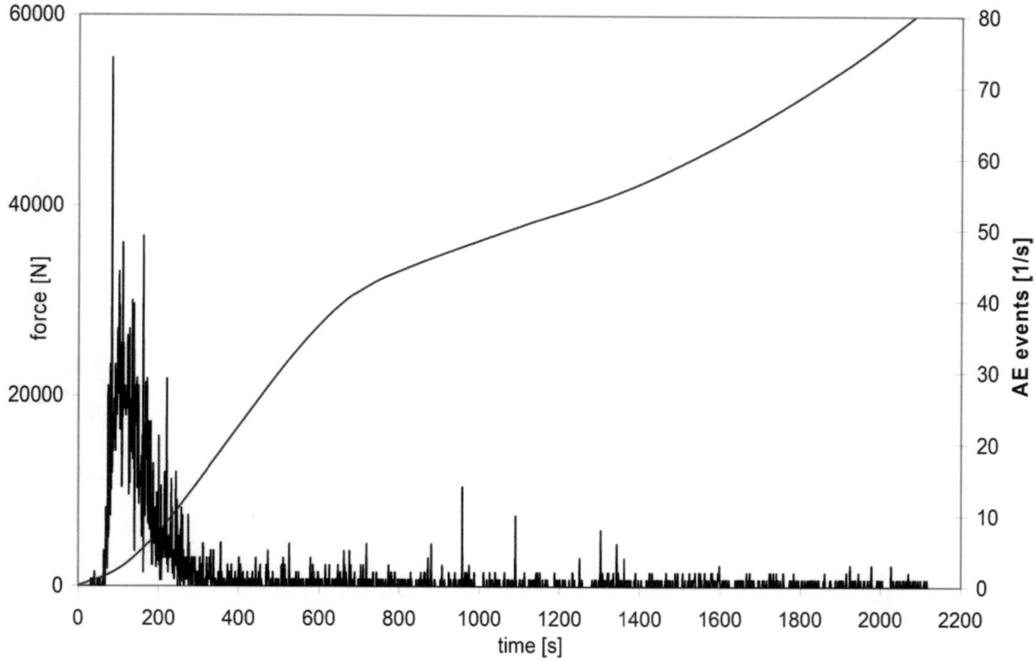

(a)

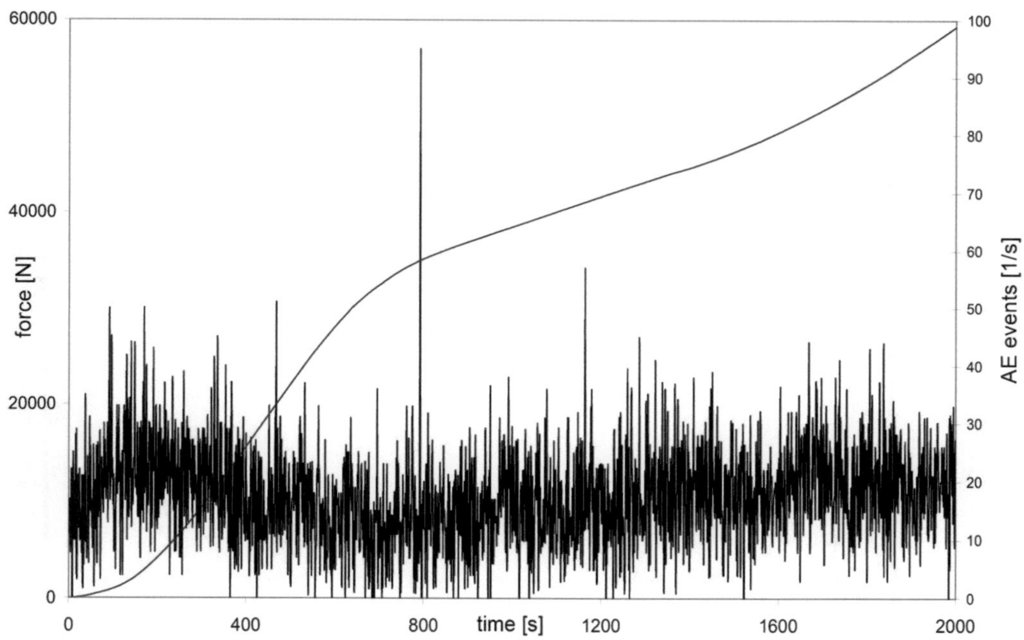

(b)

Fig. 20. AE and compressive force in dependence on time in Al alloy of AA6060 type subjected to tests of compression after one (a) and two (b) HPT processes

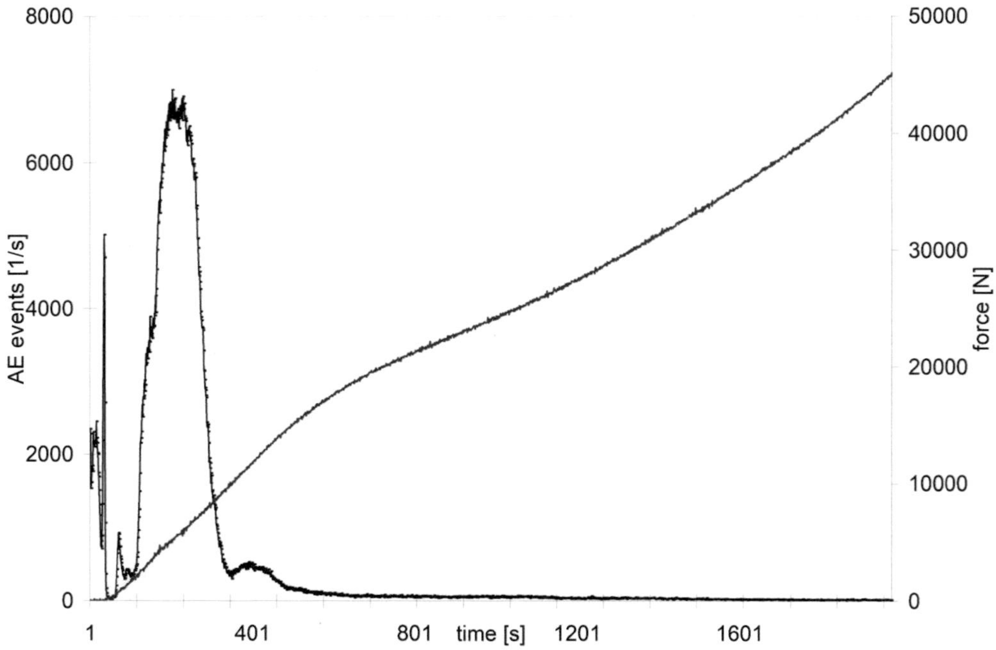

(a)

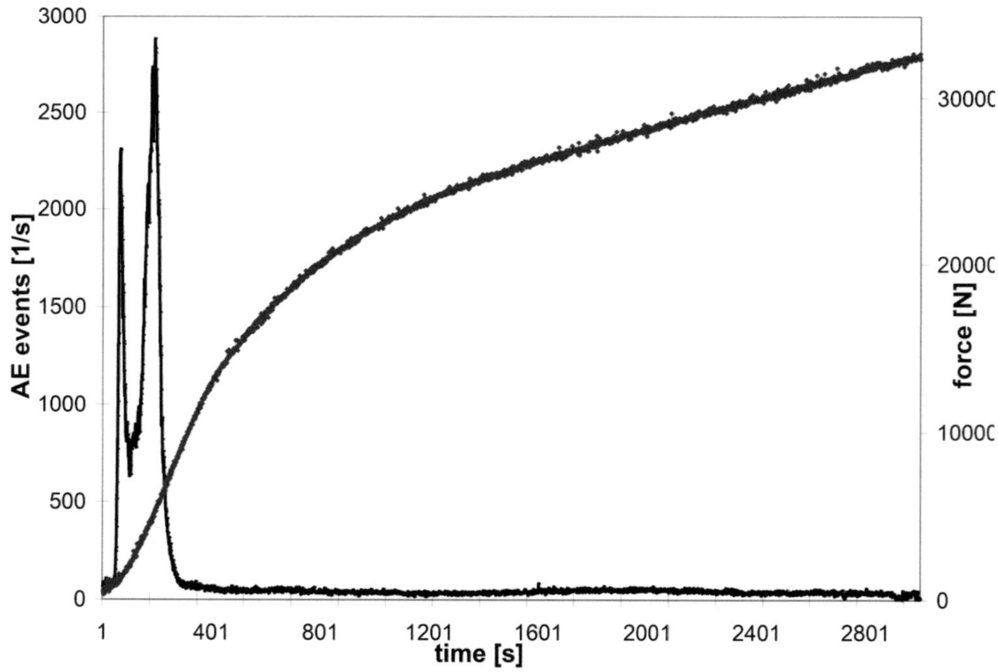

(b)

Fig. 21. Behavior of AE and compressive force in dependence on time in Al alloy of AA2014 type subjected to tests of compression before (a) and after one HPT process (b) intensity and activity of AE in the materials subjected to intensive strain processing, may be explained here based on the consideration of two vital processes. The first one is connected with the strengthening mechanism resulting from the intensive deformation, because a significant growth of dislocation density compared with the initial state takes place after the processing. In this way a collective motion of dislocations generated during the compression is strongly limited due to intensive interaction of mobile dislocations, e.g. with the forest dislocations or precipitate particles and solute atoms. Another process is bound with the tendency to the growth of plasticity (or even superplasticity) in intensively deformed materials. The contribution in the AE decrease after the intensive processing occurs, when on the expense of typical dislocation slips along the favored planes of the crystalline lattice within individual grains, the start of the grain boundary slips begins, which is probably less acoustically effective compared with the effective mechanism of collective and synchronized annihilation of many dislocations.

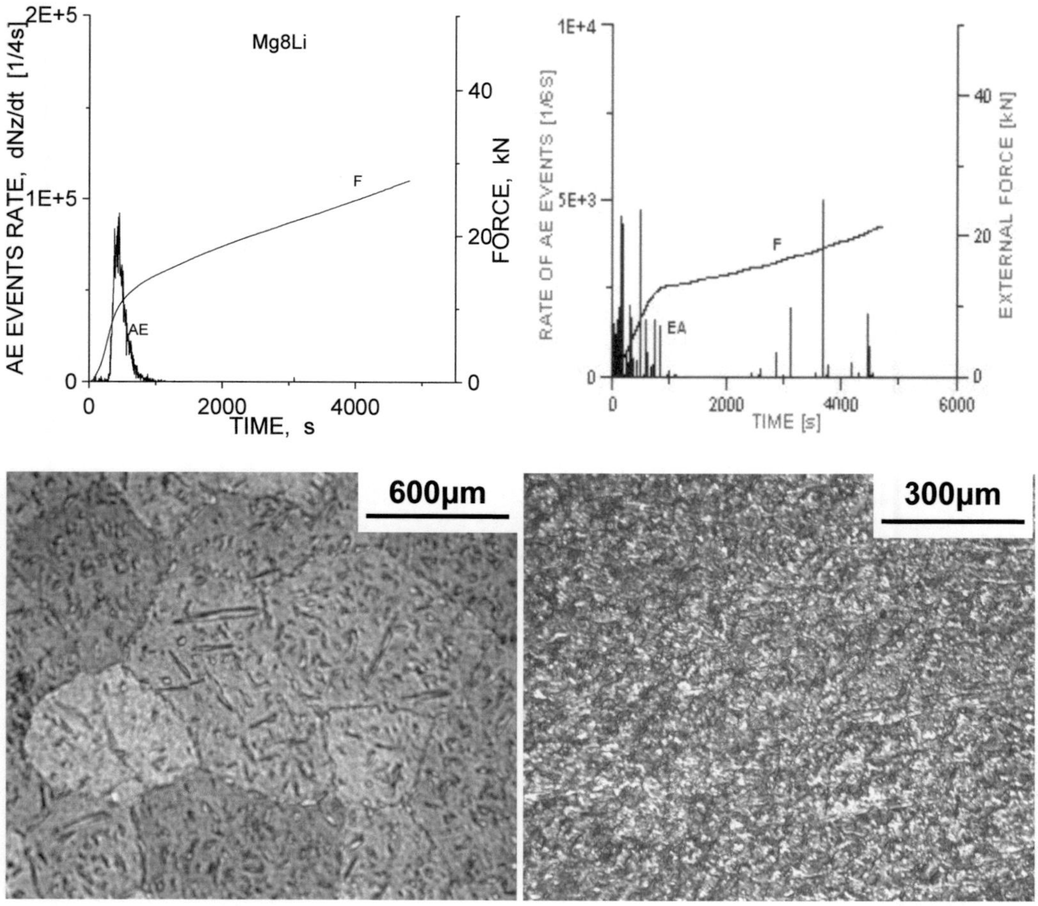

(a) (b)

Fig. 22. AE and external force in two-phase Mg8Li alloys before (a) and after (b) four-fold ECAP processing. At the bottom the corresponding optical microstructures

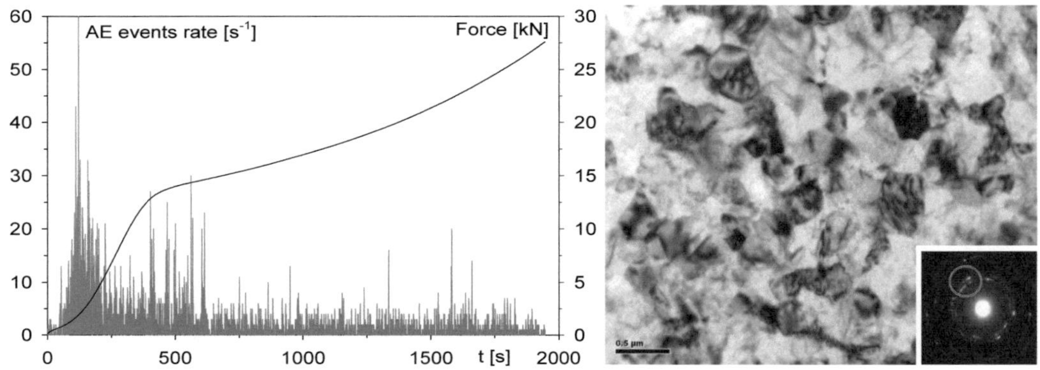

(a) (b)

Fig. 23. Courses of AE events rate and force versus time during compression of Mg10Li alloy after application of three-fold HPT rotations (a) and corresponding TEM microstructure (b)

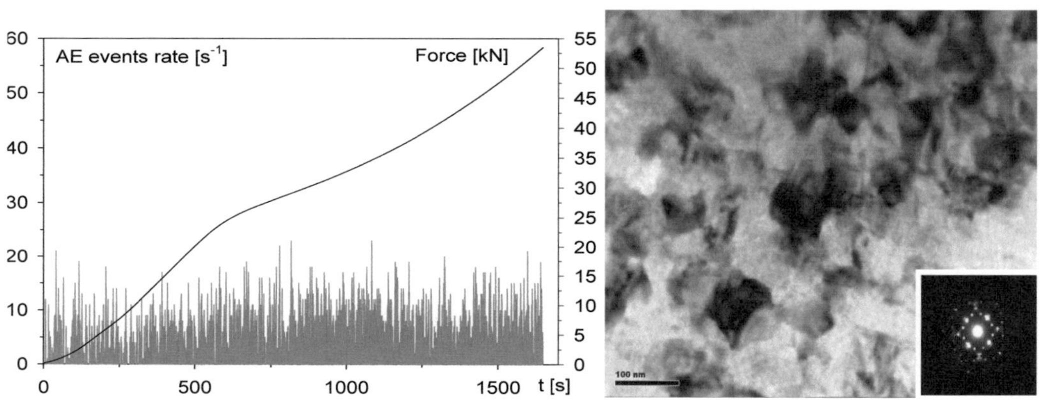

(a) (b)

Fig. 24. Courses of AE events rate and force versus time during compression of Mg10Li5Al alloy after three-fold HPT operation (a) and corresponding TEM microstructure (b)

4.4 AE and the Portevin–Le Châtelier effects in Al alloys

The AE effect, which accompany the Portevin–Le Châtelier one (PL effect - known also as discontinuous or serrated yielding or jerky flow), are quite well documented (van den Beukel, 1980; Caceres & Bertorello, 1983; Cottrell, 1953; Korbel et al., 1976; Pascual, 1974; Pawełek, 1989). Pascual (Pascual, 1974), as one of the first showed, that strong correlations occurred between the AE behavior and plastic flow instabilities, resulting from the in-homogenous deformation are typical for the PL phenomenon.

It was established that the local peaks of yielding corresponded to the increases of AE and that they resulted from the dislocation breakaway from the atmospheres of foreign atoms (Cottrell atmospheres) as well as the multiplication of dislocations at the front of propagating deformation band, similar to the well known Lüders' band. The results presented below will be shortly discussed further on the basis of a simple dislocation-dynamic (DD) model of PL effect (Pawełek, 1989), described slightly in section 4.4.2.

4.4.1 Anisotropy of AE and PL effects in Al alloys

The phenomenon of PL effect anisotropy was observed for the first time in works (Mizera & Kurzydłowski, 2001; Pawełek et al., 1998). The present research was carried out in order to confirm the anisotropy of the both AE and PL phenomena as well as to study the possibility of the occurrence of PL and/or AE effects also in materials processed with intensive deformation techniques (Pawełek et al., 2007, 2009).

Al alloys of AA5754 type. The examinations of PL and AE effects were performed in fact for 5 orientations of samples cut out at angles β=0°, 22.5°, 45°, 67.5° and 90° with respect to the rolling direction. Fig. 25, shows the AE rate and courses of external force during the tensile tests only for three samples of Al alloys of AA5754 type.

orientation (cut out angle *β*)	0°	22.5°	45°	67.5°	90°
total number of events Σ_C for AA5754 alloy	3400	3500	8020	2520	4500

Table 1. The total sum of AE events in Al AA5754 alloy in dependence on cut out angle β

Moreover, when analyzing the plots in Fig. 25a-c, it can be found, that anisotropy of AE in AA5754 alloy is connected with the maximum quantities Σ_c (about 8000), which occur for cut out angles β=45° whereas the minimum of Σ_c (about 2500) is for β=67.5°. It is illustrated in Table 1, where maximum Σ_c (red color) and minimum ones (blue) are given.

Al alloys of AA5182 type. Cold rolled sheets of Al AA5182 alloy were the subject of plastic deformation anisotropy analysis connected with the PL effect. The samples were cut out of the rolled sheet along the rolling direction (RD), transverse direction (TD) and at angle 45° between them. The investigated sheets were subjected to uniaxial tension at ambient temperature using a static QTEST testing machine at constant strain rate $5.3x10^{-4}s^{-1}$ to the moment of their failure. In Fig. 26, the corresponding collection of intensity of AE signal counts recorded during the tensile test are showed in the form of histogram.

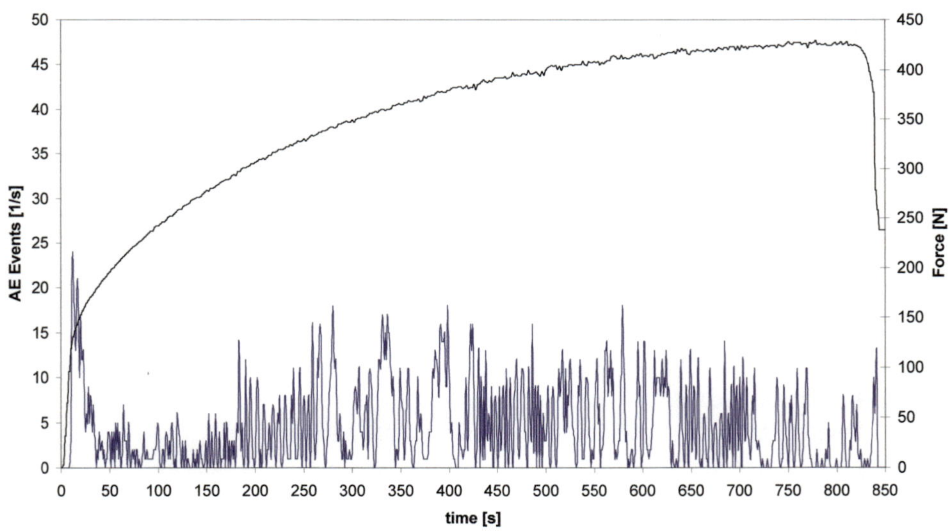

(a)

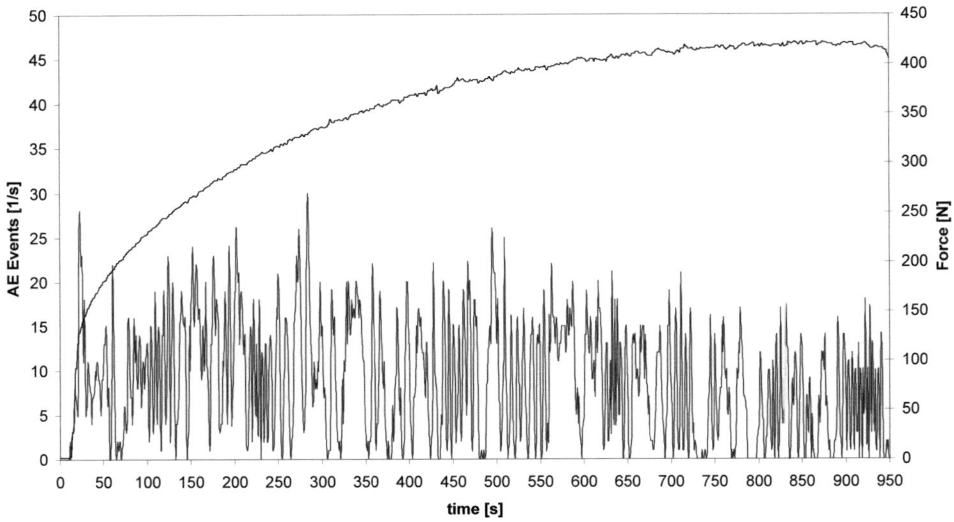

(b)

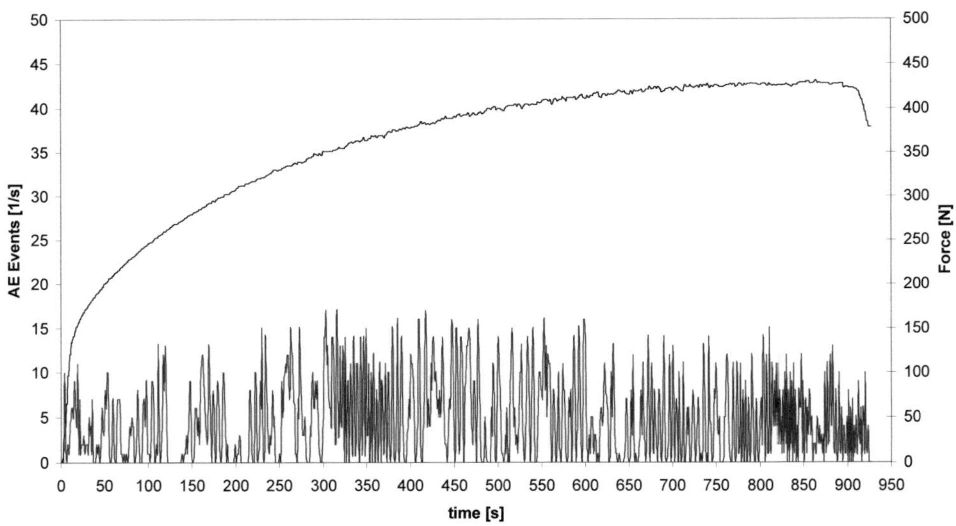

(c)

Fig. 25. Anisotropy of AE and PL effects in tensile test of Al AA5754 alloy. AE and external force for individual cut out angles: (a) – β=0°, (b) – β=45° and (c) – β=90°

The anisotropy of PL and AE effects in these alloys resulted from the fact, that the highest number of AE signal counts were recorded during the tension of samples, which were cut out in the TD direction perpendicular to the rolling direction RD.

The correlations of amplitude of AE signals with the tensile curves of samples in the rolling direction, transverse direction and inclined 45° to them are shown in Fig. 27. The analysis of the results showed that during plastic deformation of the Al AA5182 alloy the AE intensity bound with the motion of dislocation occurs at a defined level of load dependant on microstructure of materials. As it was suggested in previous works (Pawełek et al., 1998) the different distribution of grain orientations, i.e. the differentiation of sample textures were found to be reasons for the anisotropy of AE and PL. Generally, it means that the maximum AE, for example in the Al AA5754 sample for β=45°, is the result of the fact that the number of privileged slip systems of {111} type is greater than in the sample for other values of β, and, in consequence the number of active dislocation sources generating the AE events is greater. Moreover, the reasons for the AE generation during the effect of PL are related with the collective behavior of dislocation groups generated by the sources formerly blocked by the Cottrell atmospheres. Based on Cottrell idea, an own model of PL effect was proposed in (Pawełek, 1989). This model is presented schematically in Fig. 30 and discussed in short in the next section.

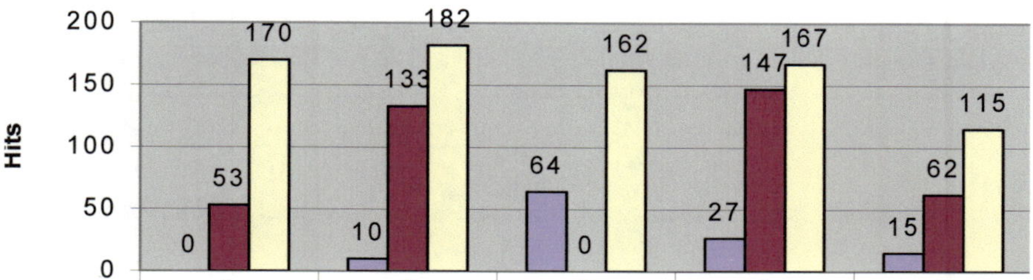

	1	2	3	4	5
RD	0	10	64	27	15
45	53	133	0	147	62
TD	170	182	162	167	115

Fig. 26. Intensities of AE signals during tensile test in the form of histogram

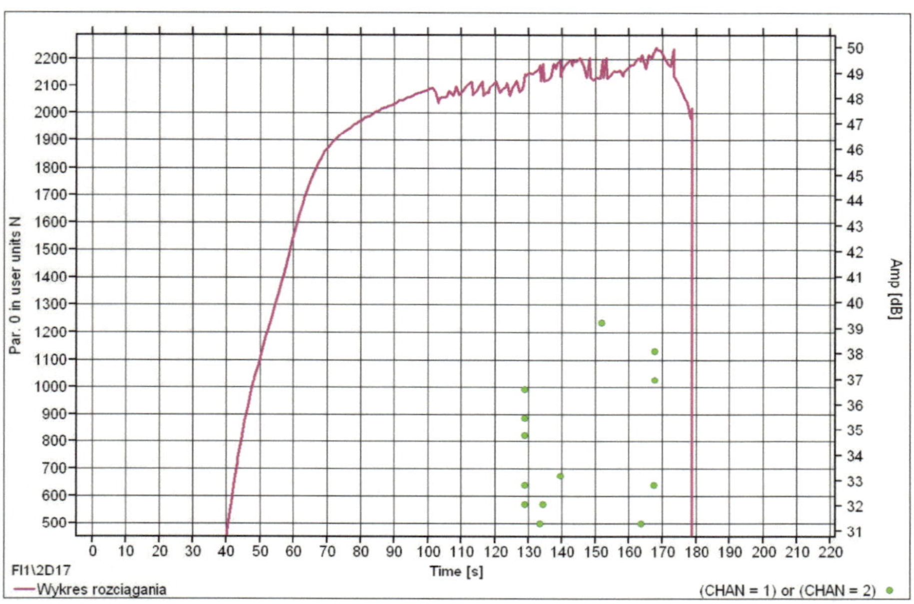

(a)

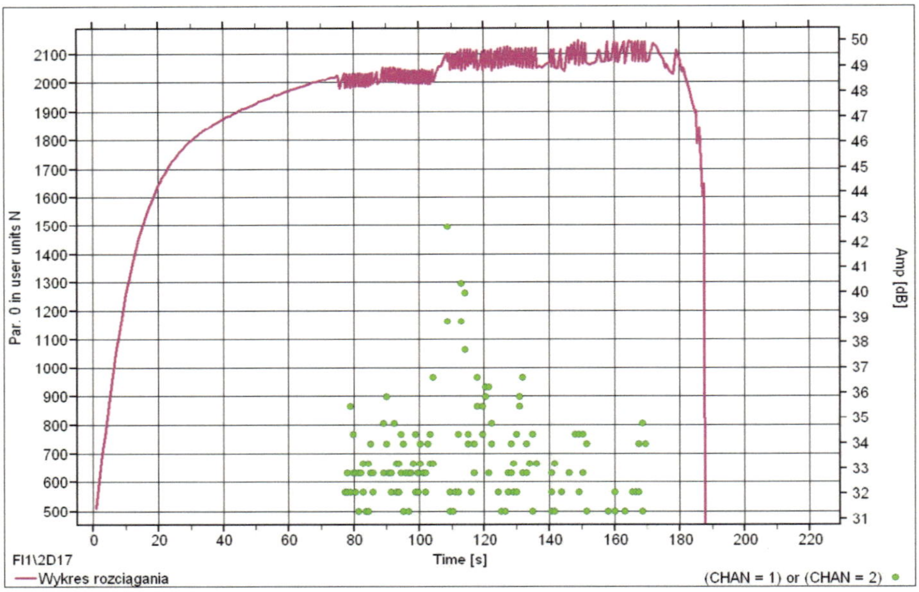

(b)

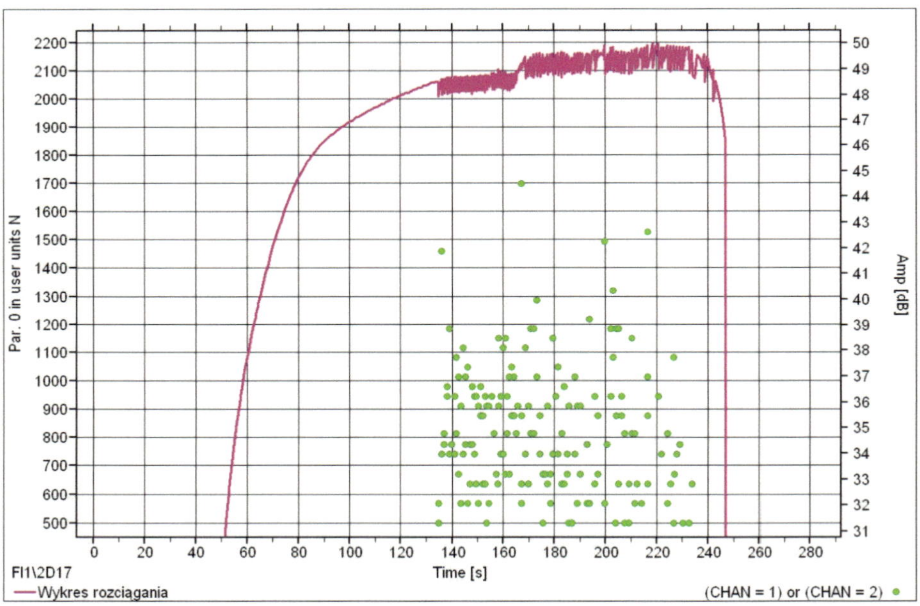

(c)

Fig. 27. Correlations of amplitude of AE signals with the tension curve of samples in the rolling direction (a); transverse direction (b) and inclined 45° to them (c)

4.4.2 AE and PL effects in Al AA5251 alloys processed with the ARB method

The results of the first investigations of the relations between the mechanical properties, PL effect and the AE signals generated in a tensile test of Al alloys of AA5251 type before and after ARB process are presented in Fig. 28. It is shown that in the case of not pre-deformed alloy (Fig. 28a) more essential correlations between the AE and the PL effects appear than in the case of alloy, pre-deformed with the ARB method (Fig. 28b). The behavior of force and AE during tension of AA5251 alloy obtained after n=6 passes of ARB. It can be seen that the correlations between the PL and AE effects continue to occur: local drops of force, characteristic for the PL effect correspond to the peaks of the rate of AE events. However, both the activity and the intensity of AE as well the values of the local drops of force are no longer so distinct as in the case of not pre-deformed samples. Thus, it can be said that both the PL and the AE effects in samples of more refined grain size (UFG, nanocrystalline) show the tendency to disappear.

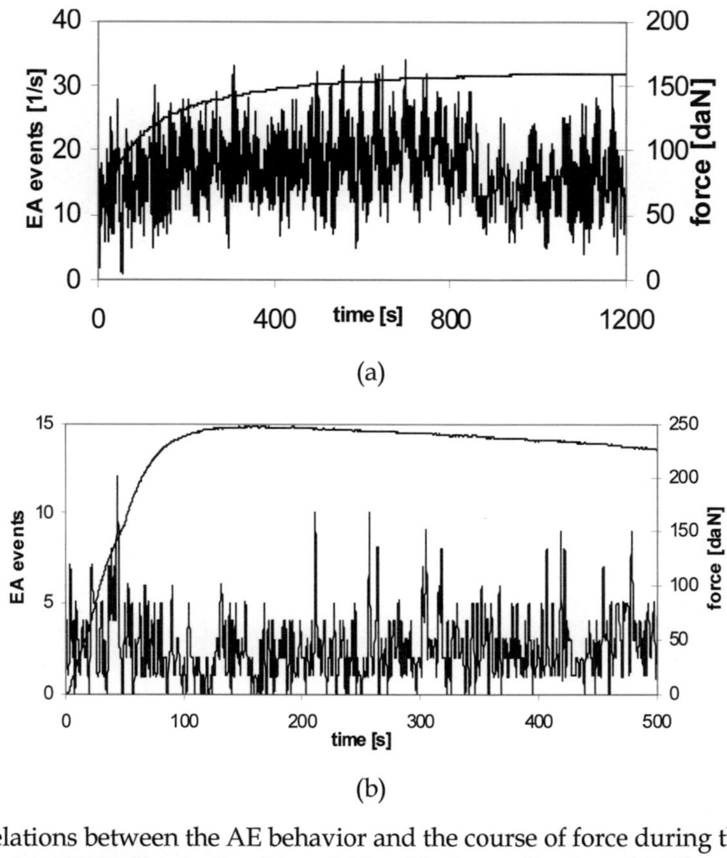

Fig. 28. Correlations between the AE behavior and the course of force during the PL effect in a tensile test of AA5251 alloy before (a) and after (b) the application of ARB operation

The application of modern software enabled the spectral analysis of AE signals in the preparation of acoustic maps (acoustograms). Fig. 29 shows, by the way of example, such an acoustogram for a sample after n=6 passes by the ARB method. It should be especially noticed that it is clearly shown here that the correlations between the PL and AE effects occur in the frequency range of AE signals above 17kHz (except the line at about 360s), which seems to be a very characteristic, never noticed earlier, feature of the PL effect. In all cases where AE and PL effects were examined in not pre-deformed state, this frequency range was considerably lower - most often below 8 kHz.

Most of the models of PL effect (e.g. van den Beukel, 1980; Král & Lukáč, 1997; Onodera et al., 1997) are of phenomenological character and none of them explain clearly the physical mechanisms of the formation and propagation of the related deformation bands and which would be coherent with the models of the sources of AE. The presented results are briefly discussed below in the context of the dislocation models of the PL effect reported in literature (e.g. Pawełek, 1989; Pascual, 1974) and the theoretical concepts concerning the source of AE generation during plastic deformation of metals (e.g. Kosevich, 1979; Natsik & Burkhanov, 1972; Natsik & Chishko, 1972, 1975; Pawełek, 1988a; Pawełek et al., 2001).

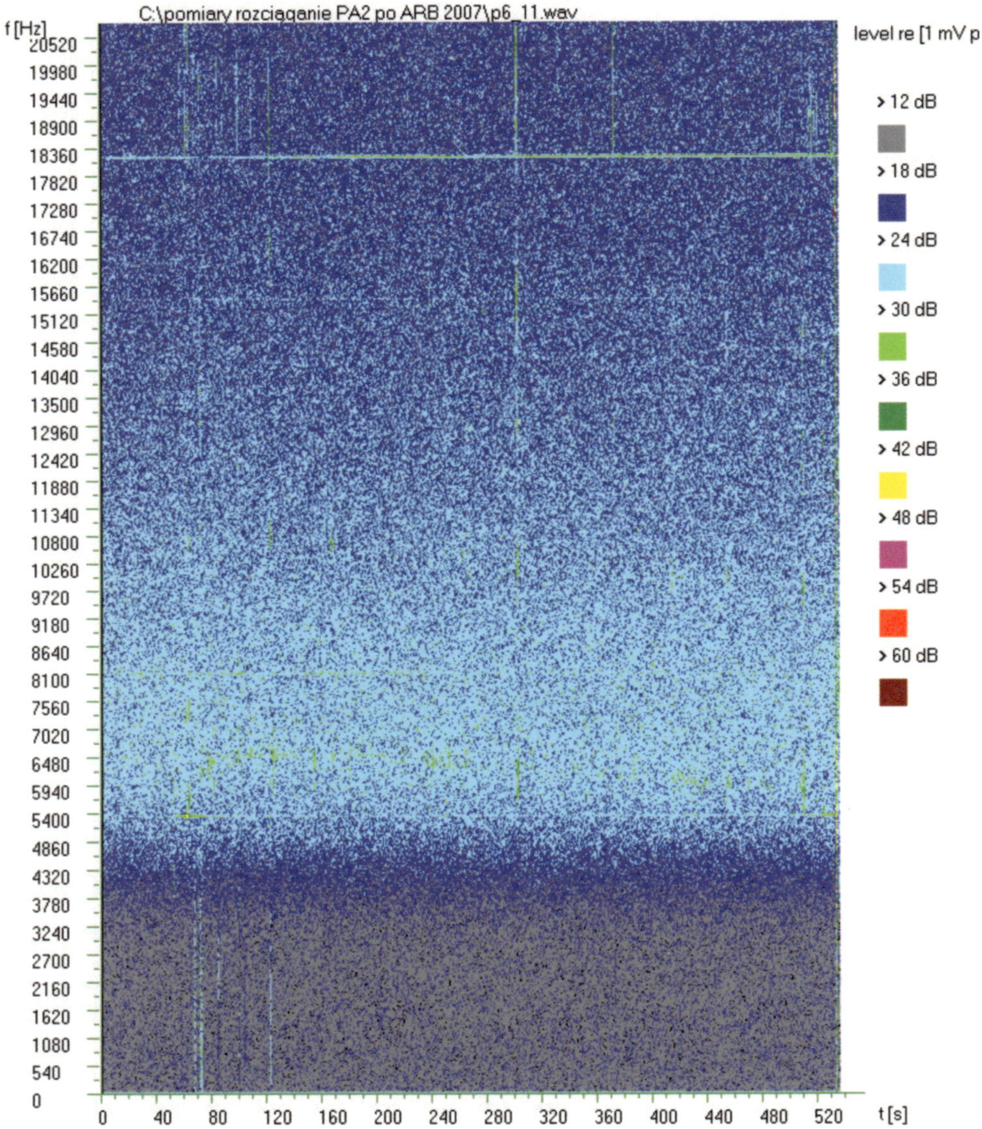

Fig. 29. Acoustic map of AE signals generated during PL effect in tensile test of AA5251 alloy after six repetitions of ARB operation

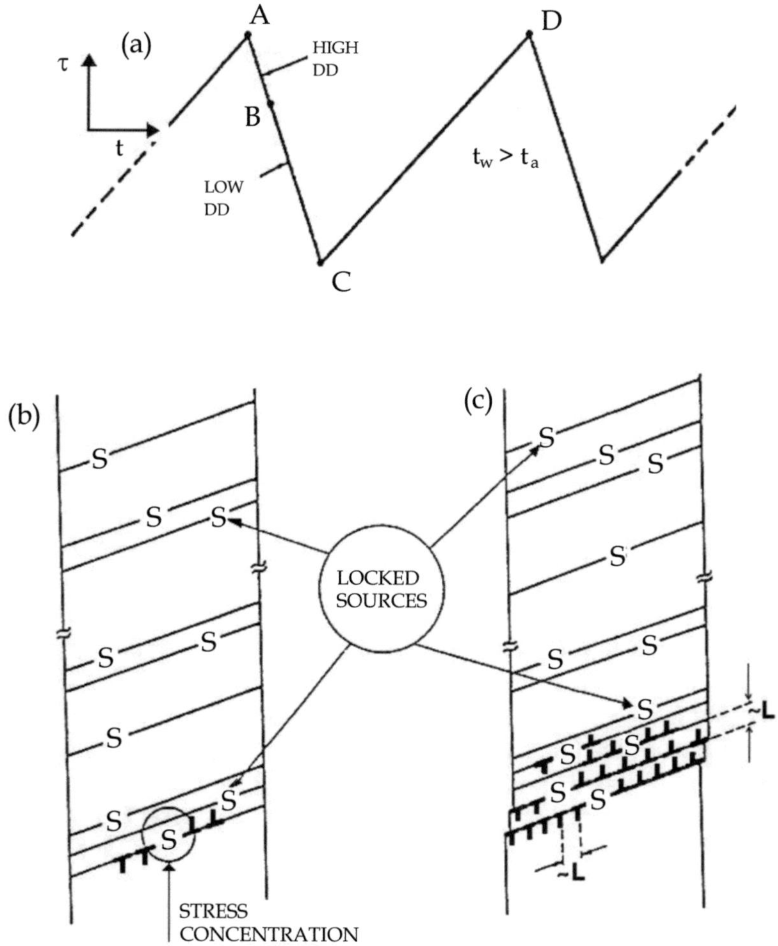

Fig. 30. Simple dislocation-dynamic (DD) model of the PL effect: (a) jump-like drop of force, (b) localization and nucleation of a band and (c) propagation of a slip band

In accordance with a simple dislocation-dynamic (DD) model of the PL effect (Pawełek, 1989), each local drop of the external force on the work-hardening curve (Fig. 30a) is connected with unlocking of the dislocation sources in a certain localized area of the sample. The consequence is the formation of a slip band (Fig. 30b) which continues to propagate (Fig. 30c) until the waiting time t_w reaches again the value of the aging time t_a. The strain rate in the slip band $\dot{\varepsilon}_d$ is greater than the rate of the homogeneous strain $\dot{\varepsilon}$ due the high DD. Accordingly, on force-time curve a local drop must occur (Fig. 30a), since, according to the known equation of Penning: $K^{-1}d\sigma/dt + \dot{\varepsilon}_d = \dot{\varepsilon}$ there occurs relation $d\sigma/dt<0$ for $\dot{\varepsilon}_d > \dot{\varepsilon}$; K is the coefficient of rigidity of the system of machine–sample.

The discussion was carried out in respect with collective properties of motion of many dislocation groups as well as the internal and surface synchronized annihilation of dislocations. The dislocation model of AE event generation was the starting point, which was based on soliton properties of dislocations (Pawełek, 1985, 1987, 1988a,b; Pawełek & Jaworski, 1988; Pawełek et al., 2001).

It was suggested, that the AE increases accompanying the local flow peaks were bound to highly dynamic sources of overstress, which acted in the state of strong overstress (Pawełek et al., 1985) due to an abrupt breakaway from the Cottrell atmospheres. It is thus very probable, that in this case, apart from the annihilation of dislocations, the contribution of the enhancement of dislocation dynamics, which, due to the effect of overstress of the dislocation sources (e.g. Frank-Read (FR) type), may be significantly higher that in the case of usual action of FR sources. Also the results of one of the latest works on the PL and AE effects (Chmelík et al., 1993) do confirm that the dominating factors binding both AE and PL phenomena result from the multiplication of dislocations during the action of FR sources and the breakaway of dislocations from the Cottrell atmospheres.
Simultaneously with the above process there takes place the generation of AE events both due to the acceleration as well as annihilation of dislocations. Dislocations generated from the FR sources may attain very great accelerations resulting from the interactions of the dislocation-dislocation type. However, there are more premises (Boiko et al., 1973, 1974, 1975) maintain that the contribution to AE signals due to annihilation is considerably higher than that resulting from acceleration. These authors carried out the calculations showing that the expression for AE included three terms related to the dislocation annihilation, the rate of dislocation generation and to the dislocation acceleration. However, at the same time the two last terms are always considerably less important than the dislocation annihilation term. Moreover, the contribution from the annihilation of the dislocation segments when the dislocation loops are bearing off from the FR source is intensified and dominated by the processes of the surface annihilation of dislocations, such as it takes place e.g. in the case of the formation of dislocation steps on the sample surface due to the formation of slip lines and slip bands or the shear microbands. This observation is in accordance with the results obtained in another work (Merson et al., 1997), in which the strong influence of the surface on the AE generated due to plastic deformation of metals was clearly demonstrated. Moreover, the anisotropy occurrence of PL and AE effects was confirmed on the example of AA5754 and AA5182 type of Al alloys. The tendency to the decrease of intensity of the AE and PL effects in UFG (nanocrystalline) Al alloy of AA5251 type was observed (Pawełek, 2009) for the first time.

5. Conclusions

The more important results obtained in this chapter may be formulated in the form of the following, more detailed conclusions:

- The AE method shows, that low temperature transition of the twinning → shear bands observed in Cu and Ag single crystals is also observed in Al single crystals.
- The formation of an individual twin lamella, a micro shear band and step on the surface is related with the generation of AE peak due to surface annihilation of $10^4 \div 10^6$ dislocations.
- The intensity and activity of AE in alloys subjected to compression tests after processing with the methods of intensive deformation (HPT, ECAP and ARB) distinctly fell in respect to the unprocessed alloys.
- The decrease of AE in intensively processed alloys is connected with a significant increase of refinement of microstructure and the tendency to the plasticity growth.

- The correlations between the AE and the mechanisms of deformation may be considered in terms of collective synchronized acceleration and annihilation of many dislocations.
- A hypothesis, that the decrease of AE in alloys compressed after intensive strain is due to strengthening processes and beginning of slip along grain boundaries was put forward.
- The anisotropy of AE and PL effects is bound with the maximum value of total sum of AE events and maximum abrupt drops of external force.
- The PL and AE effects in alloy after ARB treatment reveal the tendency to disappear.
- The relation of AE and PL effects is in good accordance with a simplified dislocation-dynamic model of the PL phenomenon.
- Correlations between the PL and AE effects occur in the frequency range above 17kHz, whereas in metals, not generating the PL effect, they occur in a lower range - below 8kHz.

6. Acknowledgement

The studies were financially supported by the research projects of the Polish Ministry of Science and Higher Education No N N507 598038 and No N507 056 31/128 as well as by the research project of the Polish Committee for Scientific Research No 3 T08A 032 28.
I would like to thank also my friends and coworkers: dr Andrzej Piątkowski, prof. Zbigniew Ranachowski, dr Stanislav Kúdela, prof. Henryk Paul and prof. Zdzisław Jasieński, for their contribution to the presented paper and valuable discussion.

7. References

Beukel, van den, A. (1975). Theory of the Effect of Dynamic Strain Aging on Mechanical Properties. *Physica Status Solidi A*, Vol.30, pp.197-206, ISSN 0031-8965.

Bidlingmaier, T.; Wanner, A.; Dehm, G. & Clemens H. (1999). Acoustic Emission during Room Temperature Deformation of a γ-TiAl Based Alloy. *Zeitschrift für Metallkunde*, Vol.90, No.8, pp.581-587, ISSN 0044-3093.

Boiko, V.S. (1973). Dislocation Description of Twin Dynamic Behavior. *Physica Status Solidi B*, Vol.55, pp.477- 482, ISSN 0370-1972.

Boiko, V.S.; Garber, R.I.; Krivenko, L.F. & Krivulya, S.S. (1973). *Fiz. tverd. Tela*, Vol.15, p.321.

Boiko, V.S.; Garber, R.I. & Krivenko, L.F. (1974). *Fiz. tverd. Tela*, Vol.16, p.1233.

Boiko, V.S.; Garber, R.I.; Kivshik, V.F. & Krivenko L.F. (1975). *Fiz. tverd. Tela*, Vol.17, p.1541.

Caceres, C.H. & Bertorello H.R. (1983). Acoustic emission during non-homogeneous flow in Al Mg alloys. *Scripta Metallurgica*, Vol.17, No.9, pp.1115-1120.

Cottrell, A.H. (1958). Dislocations and Plastic Flow in Crystals, Oxford University Press.

Chmelík, F.; Trojanowá, Z.; Převorovský, Z. & Lukáč P. (1993). The Portevin-Le Châtelier effect in Al-2.92%Mg-0.38%Mn alloy and linear location of acoustic emission. *Materials Science & Engineering A*, Vol.164, Nos.1-2, (May 1993), pp.260-265, ISSN 0921-5093.

El-Danaf, E.; Kalidindi, S.R. & Doherty, R.D. (1999). Influence of Grain Size and Stacking-Fault Energy on Deformation Twinning in Fcc Metals. *Metallurgical and Materials Transactions A*, Vol.30, (May 1999), pp.1223-1233, ISSN 1073-5623.

Heiple, C.R. & Carpenter, S.H. (1987). *Journal of Acoustic Emission,* Vol.3, p.177.

Jasieński, Z.; Pawełek, A. & Piątkowski, A. (2010). Low temperature deformation twinning in channel-die compressed aluminium single crystals evidenced by acoustic emission. *Materials Science-Poland,* Vol.31, No.3, pp.528-530, ISSN 0208-6247.

Kuśnierz, J. (2001). Microstructure and texture evolving under Equal Channel Angular (ECA) processing. *Archives of Metallurgy,* Vol.46, No.4, pp.375-384, ISSN 0860-7052.

Kúdela, S.; Pawełek, A.; Ranachowski, Z.; Piątkowski, A.; Kúdela, S., Jr. & Ranachowski, P. (2011). Effect of Al alloying on the Hall-Petch strengthening and Acoustic Emission in compressed Mg-Li-Al Alloys after HPT processing. *Kovové Materiály – Metallic Materials,* Vol.49, No.4, pp.271-277, ISSN 0023-432X.

Kuśnierz, J.; Pawełek, A.; Ranachowski, Z.; Piątkowski, A.; Jasieński, Z.; Kudela, S. & Kudela, S., Jr. (2008). Mechanical and Acoustic Emission Behavior Induced by Channel-Die Compression of Mg-Li Nanocrystalline Alloys Obtained by ECAP Technique. *Reviews on Advanced Materials Science,* Vol.18, pp.583-589,

Korbel, A.; Zasadziński J. & Sieklucka Z. (1976). A new approach to the Portevin-LeChatelier effect. *Acta Meallurgica,* Vol.24, No.10, (October 1976), pp.919-923, ISSN 0001-6160.

Král, R. & Lukáč P. (1997). Modelling of strain hardening and its relation to the onset of Portevin-Le Chatelier effect in Al-Mg alloys. *Materials Science & Engineering A,* Vol. 234-236, pp.786-789, ISSN 0921-5093.

Kosevich, A.M. (1979). In: *Dislocations in Solids,* F.R.N. Nabarro, (Ed.), Vol.1, p.33, North-Holland Publ. Co., Amsterdam, The Netherlands.

Mizera, K. & Kurzydłowski, K.J. (2001). On the anisotropy of the Portevin-Le Chatelier plastic instabilities in Al-Li-Cu-Zr alloy. *Scripta Materialia,* Vol.45, No.7, (October 2001), pp.801-806, ISSN 1359-6462.

Merson, D.; Nadtochiy, M.; Patlan, V.; Vinogradov, A. & Kitagawa, K. (1997). On the role of free surface in acoustic emission. *Materials Science & Engineering A,* Vol.234-236, pp.587-590, ISSN 0921-5093.

Natsik, W.D. & Burkhanov, A.N. (1972). *Fiz. tverd. Tela,* Vol.14, p.1289.

Natsik, W.D. & Chishko, K.A. (1972). *Fiz. tverd. Tela,*Vol.15, p.3126; (1975). ibid, Vol.17, p.341.

Onodera, R.; Morikawa, T. & Higushida, K. (1997). Computer simulation of Portevin-Le Chatelier effect based on strain softening model. *Materials Science & Engineering A,* Vol.234-236, pp.533-536, ISSN 0921-5093.

Paupolis, A. (1980). *Circuits and Systems. A Modern Approach,* Holt, Rinehart & Winston, (Ed.), New York, USA.

Pawełek, A.; Piątkowski, A. & Jasieński, Z. (1997). Nonlinear and dislocation dynamic aspects of acoustic emission and microstructure evolution during channel-die compression of metals. *Molecular and Quantum Acoustics,* Vol.18, 321-358, ISSN 0208-5151.

Pawełek, A.; Piątkowski, A.; Jasieński, Z. & Pilecki S. (2001). Acoustic Emission and Strain Localization in FCC Single Crystals Compressed in Channel-Die at Low Temperature. *Zeitschrift für Metallkunde,* Vol.92, No.4, pp.376-381, ISSN 0044-3093.

Paul, H.; Darrieulat, M. & Piątkowski, A. (2001). Local Orientation Changes and Shear Bending in {112}<111>-Oriented Aluminium Single Crystals. *Zeitschrift für Metallkunde,* Vol.92, No.11, pp.1213-1221, ISSN 0044-3093.

Pascual, R. (1974). Acoustic emission and dislocation multiplication during serrated flow of an aluminium alloy. *Scripta metallurgica.,* Vol.8, No.12, pp.1461-1466.

Pawełek, A. (1989). On the Dislocation-Dynamic Theory of the Portevin-Le Chatelier Effect. *Zeitschrift für Metallkunde,* Vol.80, No.9, pp.614-618, ISSN 0044-3093.

Pawełek, A.; Piątkowski, A.; Jasieński, Z.; Litwora, A. & Paul, H. (1998). Acoustic emission and Portevin-Le Chatelier effect during tensile deformation of polycrystalline α-brass. *Molecular and Quantum Acoustics,* Vol.19, pp.201-215, ISSN 0208-5151.

Pawełek, A.; Kuśnierz, J.; Jasieński, Z.; Ranachowski, Z. & Bogucka, J. (2009). Acoustic emission and the Portevin - Le Châtelier effect in tensile tested Al alloys before and after processing by accumulative roll bonding (ARB) technique, *Proceeding of 10th French-Polish Colloquium,* Paris, France, May 20-21, 2008; *Archives of Metallurgy and Materials,* Vol.54, No.1, pp.83-88, ISSN 1733-3490.

Pawełek, A.; Bogucka, J.; Ranachowski, Z.; Kudela, S. & Kudela, S., Jr. (2007). Acoustic emission in compressed Mg-Li and Al alloys processed by ECAP, HPT and ARB methods. *Archives of Acoustics,* Vol.32, No.4 (Supplement), pp.87-93, ISSN 0137-5075.

Pawełek, A. (1988a). Possibility of a soliton description of acoustic emission during plastic deformation of crystals. *Journal of Applied Physics,* Vol.63, No.11, (June 1988), pp.5320-5325, ISSN 0021-8979.

Pawełek, A. (1988b). An attempt at a soliton approach to plastic flow of crystals. *Physics Letters A,* Vol.128, No.1,2, (March 1988), pp.61-65, ISSN 0375-9601.

Pawełek, A. (1987). Density of kiks on a dislocation segment in thermodynamic equilibrium and the interaction between solitons. *Journal of Applied Physics,* Vol.62, No.6, (September 1987), pp.2549-2550, ISSN 0021-8979.

Pawełek, A. & Jaworski, M. (1988). A moving dislocation kink as the soliton on a background of quasi periodic process in unbounded sine-Gordon system. *Journal of Applied Physics,* Vol.64, No.1, (July 1988), pp.119-122, ISSN 0021-8979.

Pawełek, A. (1985). Soliton-soliton and soliton-antisoliton interaction in the Frenkel-Kontorova model of dislocation. *Acta Physica Polonica A,* Vol.68, No.6 (December 1985), pp.815-831, ISSN 0587-4264.

Pawełek, A.; Stryjewski, W.; Bochniak, W. & Dybiec, H. (1985). Mobile Dislocation Density Variation during Strain Rate Change Evidenced by Acoustic Emission. *Physica Status Solidi A,* Vol.90, pp.531-536, ISSN 0031-8965.

Ranachowski, Z.; Piątkowski, A.; Pawełek, A. & Jasieński, Z. (2006). Spectral analysis of acoustic emission signals generated by twinning and shear band formation in silver single crystals subjected to channel-die compression tests. *Archives of Acoustics,* Vol.31, No.4 (Supplement), pp.91-97, ISSN 0137-5075.

Resnikoff, H. & Wells, R. (1998). *Wevelet Analysis,* Springer, New York, USA.

Saito, Y.; Utsunomiy, H.; Tsuji, A. N. & Sakai, T. (1999). Novel ultra-high straining process for bulk materials - development of the accumulative roll-bonding (ARB) process. *Acta Materialia,* Vol.47, No.2, (January 1999), pp.579-583, ISSN 1359-6454.

Scott, I. G. (1991).*Basic Acoustic Emission,* Gordon and Breach, New York, USA.

Tanaka, H. & Horiuchi, R. (1975). Acoustic Emission due to deformation twinning in titanium and Ti - 6Al - 4V alloy. *Scripta Met.,*Vol.9, No.7, pp.777-780.

Vinogradov, A. (1998). Acoustic emission in ultra-fine grained copper. *Scripta Mateialia,* Vol.39, No.6, (August 1998), pp.797-805, 1359-6462.

Valiev, R.Z.; Ismagaliev, R.K & Alexandrov, I.V. (2000). Bulk nanostructured materials from severe plastic deformation. *Progress in Materials Science,* Vol.45, No.2, pp.103-189, ISSN 0079-6425.

12

Aluminum Alloys for Al/SiC Composites

Martin I. Pech-Canul
Centro de Investigación y de Estudios Avanzados del IPN Unidad Saltillo
Ramos Arizpe Coahuila,
México

1. Introduction

Aluminum has played and continuous to play a key role in the development of metal matrix composites (MMCs) reinforced with a variety of ceramic materials including Al_2O_3, TiC, B_4C, and SiC. From the wide range of MMCs systems studied thus far and on account of the attractive properties of SiC, Al/SiC composites have drawn the attention of a plethora of research scientists and technologists. Like with any other composite material, the materials behavior lies much in the matrix characteristics as in the reinforcement properties. Several aspects are to be considered with regard to the metallic matrix, namely, composition, response to heat treatments, mechanical and corrosion behavior. And since aluminum offers flexibility in terms of these aspects, accordingly, a number of aluminum alloys have been used in studies intended for research and technological applications. The choice, however, for one or another alloy depends also on other factors as the composite processing route, which in turn can be dictated by the volume fraction of the reinforcement in the composite. For instance, the stir casting route is more suitable for low volume fractions (< 20%), whilst the infiltration routes are more appropriate for high volume fraction of the reinforcement (> 40%). Another important factor for selection of the aluminum alloy is the composites application and specific requirements in service. For instance, one composite may behave better under certain loads or in corrosive environments.

The aim of this chapter is to provide the readers with an insight into the factors that affect the properties of Al/SiC composites and the most important response parameters, associated to mechanical, heat-treatment, and corrosion behavior. The chapter is organized based on a hierarchical concept, starting with the role of alloy composition, followed by the resulting mechanical properties and its dependence in heat treatments, closing with the corrosion behavior. At the same time, this is derived from the central paradigm of materials science and engineering, based on the correlation: *processing* → *microstructure* → *properties* → *performance.* A review of the main findings in studies related to mechanical properties, heat treatments and corrosion behavior is presented. In view of that, the chapter is provided with references from earlier-to-the most recent studies on the behavior of Al/SiC composites, on the basis of the importance of aluminum alloy characteristics.

2. Factors and response variables related to Al alloys for Al/SiC composites

Aluminum alloy composition is one of the factors that most notoriously influence the properties of Al/SiC composites. Other factors are processing time and temperature, atmosphere type, and reinforcement characteristics. The alloy chemistry impinges on the mechanical properties and corrosion behavior, and when applicable, on the composite response to solution treatment and age hardening. It is interesting to note, however, that the wide variety of Al/SiC composites reported thus far have been prepared both with commercial and experimental alloys.

2.1 Use of commercial and experimental aluminum alloys

Several reasons dictate the need for using one or another alloy, including final application, processing route, and with no doubt, availability and cost. As for final application, and according to the materials science and engineering standpoint, the clear establishment of stress level, temperature and environment (corrosion behavior) related aspects in service is paramount. Associated to all these requirements is the chemical composition of the aluminum alloy used for the manufacture of the composite. Table 1 shows the designations of a variety of aluminum alloys used in systematic studies on Al/SiC composites. They will be evoked in the following sections concerning mechanical property, heat treatments and corrosion response of Al-alloy/SiC composites.

Alloy designation	Related field of study	Reference(s)
2024	Aging behavior Deformation behavior	Ahmad et. el., 2000, Mousavi, 2010; Zhang, 2001
2124	Creep behavior	Li & Mohamed, 1997
2219	Fatigue and projectile penetration performance	Greasley et. al., 1995
A3xx.x A3xx.x	Corrosion in humid environment Pitting corrosion	Pardo, et. al., 2003 Pardo, et. al., 2005
5456 (UNS A95456)	Pit morphology	Ahmad et. al., 2000; Trzaskoma, 1990
6013-T4 and T6	Aging behavior	Guo and Yuan, 2009
6061 UNS(A6061) 6061-T6 6061-T6	Pitting corrosion Pit morphology Galvanic corrosion Residual fatigue life prediction	Ahmad et. al., 2000; Aylor et. al., 1985: Trzaskoma, 1990 Hihara & Latanison, 1992 Shan & Nayeb-Hashemi, 1999
Al-Li, Al-Cu	Fatigue and fracture	King & Bhattacharjee, 1995
7075	Fatigue and projectile penetration performance	Greasley, et. al.,1995
Al-9.42Si-0.36 Mg	Hardness	Sahin & Acilar, 2003
AS7G06	Hardening kinetics	Cottu el al., 1992
Al-4Mg	Corrosion behavior	Candan & Bilgic, 2004
Al-13.5 Si-9 Mg	Activity of SiC in Al MMCs (electrochemical studies)	Díaz-Ballote et. al., 2004
Al–15.52 Mg–13.62 Si	Microhardness and superficial hardness	Montoya-Dávila et. al., 2007
Al–13 Mg–1.8Si Al-Si-Mg (Si/Mg var.)	Pitting behavior The role of Mg_2Si in corrosion behavior	Montoya-Dávila et. al., 2009 Escalera-Lozano, et. al., 2010

Table 1. Commercial and experimental aluminum alloys used in Al/SiC composites

As regards to fabrication route, the alloy composition to be utilized – commercial or experimental – depends strongly on the manner used to incorporate the SiC reinforcement into the alloy matrix, and on other factors such as the SiC volume fraction and shape (fibers, whiskers or particles). In some cases, processing is made with the metal in the liquid state and in others, in solid state, as in the powder metallurgy route. Due to the inherent advantages of using the aluminum alloy in molten state, a copious number of researchers have used liquid aluminum in their work. Hence, this section revolves around the implications of using liquid aluminum for the processing and characterization of Al/SiC composites.

With the metal in liquid state, the SiC reinforcements can be incorporated by way of stirring or mixing, followed by casting into metallic molds to produce ingots, which are then remelted to produce parts by other routes, like squeeze casting. Alternatively, the alloy may be incorporated into a porous preform formed by the SiC reinforcements, via the infiltration route, which in turn can be conducted either under the application of forces or vacuum or by capillary action. The latter is the so-called non-assisted, spontaneous or pressureless infiltration. In both cases – mixing and infiltration – wettability of the SiC reinforcements by the aluminum alloy is a crucial prerequisite, in tandem with an optimum fluidity of the alloy.

One of the main problems faced when processing Al/SiC composites with the metal in molten state is that liquid aluminum tends to attack SiC according to the following reaction (Iseki et. al., 1984):

$$4Al_{(l)} + 3SiC_{(s)} \leftrightarrow Al_4C_{3(s)} + 3Si_{(in\ l\ Al)} \quad (1)$$

The Al_4C_3 compound has deleterious effects within the composite because, firstly, as a brittle phase degrades the mechanical properties, and secondly, it reacts with liquid water or with moisture in the ambient, debilitating even more the composite, according to the following reactions (Kosolapova, 1971; Park & Lucas, 1997):

$$Al_4C_{3(s)} + 18\ H_2O_{(l)} \rightarrow 4Al(OH)_{3(s)} + 3CO_{2(g)} + 12H_{2(g)} \quad (2)$$
$$\Delta G_{25\,^{o}C} = -1746\ kJ/mol$$

$$Al_4C_{3(s)} + 12\ H_2O_{(g)} \rightarrow 4Al(OH)_{3(s)} + 3CH_{4(g)} \quad (3)$$
$$\Delta G_{25\,^{o}C} = -1847\ kJ/mol$$

Several approaches have been proposed in the literature to prevent or retard aluminum carbide formation and overcome its harmful effect; these include: i) modifications to the aluminum alloys compositions, ii) coatings on the SiC reinforcements, iii) varying processing temperature and contact time, iv) artificial or intentional oxidation of the SiC reinforcements, and v) incorporation of silica (SiO_2) powders in the SiC preforms (Rodriguez-Reyes et.al., 2006).

One of the first and most successful approaches to avoid the attack of SiC by liquid aluminum and its consequences, and at the same time improve wetting, was the modification of alloy composition with silicon and magnesium. It is well established that silicon in the aluminum alloy plays a key role, as it lowers the melting point of the alloy and decreases the contact angle between the solid and the liquid, thus enhancing wettability (Pech-Canul et. al., 2000 (a)).

On the other hand, from equation (1) and in the light of Le Chatelier's principle it is simple to see the beneficial effect of silicon, since it will tend to reverse the direction of the reaction. However, an uncontrolled excess of silicon in the alloy may have an adverse effect because above a critical level –the eutectic point – it tends to increase the viscosity of the alloy. In the case of the Al-Si system it corresponds to 12.6 wt. % Si at the temperature of 577 °C.

It is clear, that long processing times and at high temperatures - like those used in SiC oxidation - turn out to be costly to the manufacture of Al/SiC composites. In this regard, within the author's research group, several approaches that prove to be successful have been tested. Initially, wettability studies - using four experimental aluminum alloys - were performed in order to establish the optimum parameters to lower contact angle and surface tension, and in parallel, avert Al_4C_3 formation (Pech-Canul et. al., 2000 (b)). Under optimized conditions, Al/SiC_p composites were fabricated by pressureless infiltration and characterized physically and in mechanical tests, specifically, modulus of rupture and modulus of elasticity. It was shown that Al/SiC composites made with silicon rich aluminum alloys and siliconized SiC show properties that are significantly different from those of similar composites produced with unsiliconized SiC or with aluminum alloys that do not contain silicon. Under optimized infiltration conditions, metal matrix composites with less than 3% porosity, over 200 GPa modulus of elasticity, and about 300 MPa modulus of rupture were routinely produced (Pech-Canul & Makhlouf, 2000).

Another element with a prime importance as an alloying element for aluminum is magnesium. Various investigations have been devoted to study the effect of Mg on the microstructure characteristics and mechanical properties of SiC-reinforced aluminum alloys. A comprehensive review on the role of magnesium in the processing of Al-based composites reinforced with various ceramic materials has been reported by B. C. Pai and co-workers (B. C. Pai et al., 1995; Aguilar-Martínez et. al., 2004). One prominent outcome from wettability tests, studying the effect of magnesium in aluminum alloys for pressureless infiltration is that it lowers the surface tension of the liquid aluminum, thus, enhancing wettability (Pech-Canul et. al., 2000 (b)).

In another approach used by the author's research group, incorporation of SiO_2 powders into the SiC preforms also proved to be beneficial (Rodriguez-Reyes et.al., 2006) to prevent formation of the unwanted phase Al_4C_3. The use of 6 V. % SiO_2 either in the form of quartz or cristobalite powders of ≈ 5 µm average particle size, completely hindered formation of aluminum carbide. Later on, the same group tested a simple method by coating the SiC particles with 0.1 volume fraction of colloidal silica (0.02–0.06 µm particle size) (Montoya-Dávila et. al., 2007). With this method, aluminum carbide was prevented again. In each of these methodologies, several experimental aluminum alloys with unconventional levels of silicon and magnesium were used. A last but not least important factor is the atmosphere used during the infiltration operation. According to the optimization of processing parameters, a change in the atmosphere from argon-to-nitrogen during pressureless infiltration significantly improves the wetting of SiC by the liquid aluminum alloy and consequently, substantially enhances infiltration (Pech-Canul et. al., 2000(b)).

It should be no surprise that with modifications in the levels of Si and Mg in the aluminum alloy, alterations in the matrix microstructure should be observed. These changes are manifested as the appearance or disappearance of phases with specific composition and morphology. In this context, one main secondary phase responsible for the mechanical and corrosion of response Al/SiC composites is magnesium silicide (Mg_2Si). This intermetallic phase is responsible for hardness increase in many commercial and experimental alloys, but at the same time, it may sabotage the integrity of the composite because it may form galvanic couples with the aluminum matrix or with other phases (Wei et. al., 1998; Escalera-Lozano et. al., 2010). The role played by Mg_2Si becomes as important when it is formed in the alloy matrices only, as when it is present in aluminum-based metal matrix composites because although it may be dissolved during composite processing by the liquid state route, it is formed again during the solidification of the alloy as the composite matrix.

In summary, to some extent all factors involved in the processing of Al/SiC composites by the liquid state route influence the contact angle (θ) and liquid-vapor surface tension (γ_{lv}). In the case of composites processed by the infiltration route, the magnitude of θ and γ_{lv} will determine whether the liquid will incorporate with or without the use of external forces. Figure 1 summarizes the factors that affect the magnitude of θ and γ_{lv}.

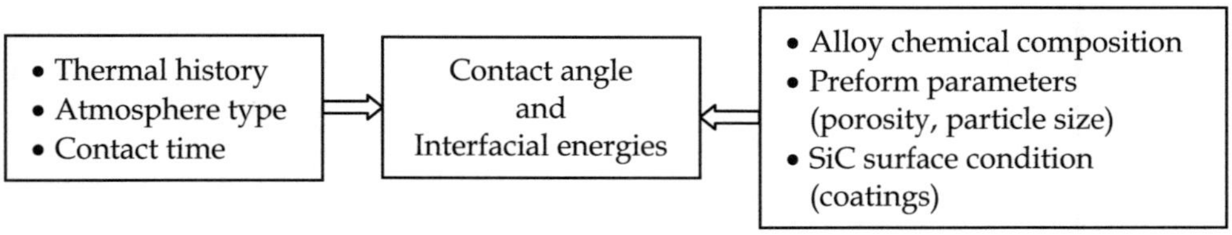

Fig. 1. Factors affecting the contact angle and interfacial energies in the Al/SiC system

From all the above factors, aluminum alloy composition plays a decisive role. However, it is possible to distinguish, chronologically speaking, between its effect *during* processing and *after* the fabrication stage. The resulting microstructure - embracing the matrix alloy and the matrix/reinforcement interface condition - is the first manifestation of the effect of alloy composition. Phase composition, amount and shape are dependent on the aluminum alloy. Then, after processing, they influence the mechanical properties, and when applicable, the heat treatments, and in the end, the corrosion behavior. As shown in Table 1, a broad variety of commercial and experimental aluminum alloys have been used in studies related to the processing and characterization of Al/SiC composites. Their particular effect on the composites' mechanical properties, heat treatment and corrosion behavior will be reviewed in the following sections (2.2, 2.3 and 2.4, respectively). An assessment of the SiC/Al-matrix interface microstructure condition is considered to be a prerequisite before evaluating the mechanical behavior of the composites. A sound interface is more likely to lead to reliable mechanical properties.

2.2 Mechanical behavior

Development of aluminum matrix composites reinforced with silicon carbide has been stimulated by the promise of improving the properties of aluminum alloys, and needless to say, mechanical properties were perhaps the first involved in systematic investigations. Several mechanical property related aspects, including deformation behavior (Zhang et. al., 2001), fatigue and fatigue-life predictions (Kumai et. al., 1991, King & Bhattacharjee 1995, Greasley et. al., 1995, Shan & Nayeb-Hashemi, 1999), fracture mechanisms predictions (Kumai et. al., 1991, King & Bhattacharjee 1995), hardness (Sahin & Acilar, 2003) and creep (Li & Mohamed, 1997) have been involved in earlier and recent studies, using both, theoretical and experimental approaches. Likewise, unconventional approaches have been proposed to strengthen Al/SiC composites (Tham et. al., 2001).

One of the first concerns was the effect of the difference in the coefficient of thermal expansion (CTE) of the matrix and that of the reinforcement on the mechanical behavior of the composite. This is because when a metal matrix composite is cooled down to room temperature from the fabrication or annealing temperature, residual stresses are induced in the composites due to the mismatch of the thermal expansion coefficients between the matrix and the reinforcements (either in the form of fibers or particles). And since alloy composition affects the CTE of the matrix, it is clear that the behavior of one Al-alloy/SiC system cannot be directly inferred from the behavior of any other given aluminum alloy.

Arsenault et. al. performed an investigation on the magnitude of the thermal residual stresses by determining the difference of the yield stress ($\Delta\sigma_y$) between tension and compression resulting from the thermal residual stresses (Arsenault et. al, 1987). A theoretical model based on the Eshelby's method was constructed to predict the thermal stresses and $\Delta\sigma_y$. A good agreement was observed between the theoretical prediction and the experimental results. In previous work, Arsenault and Shi (Arsenault and Shi, 1986) proposed a simple model based on prismatic punching to explain the relative dislocation density due to the differential thermal contraction. According to the model, the presence of SiC particles of platelet morphology in an aluminum metal matrix composite results in the generation of dislocations at the Al/SiC interface when the composite is cooled from the annealing temperature. The intensity of dislocation generation at the Al/SiC interface was found to be related to the size and shape of the SiC particles.

Kumai et. al. presented a review on the fatigue and crack growth behavior of SiC particulate aluminum alloy composite. It was concluded that the improved fatigue life reported in stress-controlled tests results from the higher stiffness of the composite; therefore, it is generally inferior to monolithic alloys at a constant strain level. The role of the particulate reinforcement was examined for fatigue and fatigue crack initiation, short crack-growth and long-crack growth. Crack initiation is observed to occur at the matrix-SiC interface or at the cracked SiC particles in powder metallurgy processed composites depending on particle size and morphology. The da/dN *vs.* ΔK relationship in the composites is characterized by crack growth rates existing within a narrow range of ΔK and this is because of the lower fracture toughness and relatively high threshold values in composite compared with those in monolithic alloys. An enhanced Paris region slope attributed to the monotonic fracture contribution was reported and the extent of this contribution is found to depend on particle size (Kumai et. al., 1991).

King & Bhattacharjee studied the interfacial effects on fatigue and fracture in discontinuously reinforced metal matrix composites (DRMMCs), using Al-Li/SiC_p and Al-Cu/SiC_p systems. It was reported that the presence of weak interfaces leads to static modes of crack growth becoming important in fatigue crack growth resistance. However, whether this is deleterious to damage tolerance depends on the nature of the composite. In particles reinforced MMCs weak interfaces lead to void nucleation and growth, contributing to high m values and an early onset of stage 3 crack growth. In contrast, in whisker reinforced composites, crack deflection, associated with the presence of weak interfaces, can be beneficial in deflecting cracks and hence reducing growth rates. In order to exploit the benefits of crack deflection, it is essential to be able to predict both whether or not it will occur, and if so, the resulting path of the deflected crack. Mixed mode fracture testing approaches appear to offer some solutions in this area. However, when referring to the interface it is important to consider not just a two dimensional interface, but the whole interface zone (King & Bhattacharjee, 1995).

Greasley et. al. used either 2219 or 7075 Al alloys to study fatigue and projectile penetration performance of SiC particle reinforced aluminum matrix composite. It was found that the plastic deformation is the major energy absorption mechanism with a significant input from the melting phenomenon. However, these mechanisms can only be effective if spalling is prevented and the role of the reinforcement in this area is still important. Clearly fracture processes themselves do not absorb much energy and so reinforcement additions should not be made to promote fracture at the expense of enhancing spalling (Greasley et. al., 1995).

A study of the residual fatigue life prediction of 6061-T6 aluminum matrix composites reinforced with 15 vol. % SiC particulates (SiC_p) by using the acoustic emission (AE) technique and the stress delay concept was carried out by Shan et. al. In their work, acoustic emission activity of a 6061/SiC_p composite was monitored during tensile tests in the as received condition and after the specimens were subjected to cycle fatigue loading for a number of cycles. It was found that the AE activity rises quickly once the material is well in the plastic regime. The activity was related to the particle/matrix debonding and linkage of voids. In the high cycle fatigue, the residual tensile strength of the composites was found not to be affected by the prior cyclic loading, since the crack initiation period dominated the life of the composite (Shan et. al, 1999).

Li and & Mohamed studied the creep behavior of Al 2124/SiC composites prepared by powder metallurgy. It was concluded that SiC particles are not directly responsible for the threshold stress behavior in the composite (Li & Mohamed, 1997).

Various attempts have been made to strengthen Al/SiC composites, including unconventional approaches. For instance, Tham et. al. proposed a strengthening approach in which the phase Al_4C_3 is made to form intentionally at the Al/SiC interface, varying the contact time between the SiC particles and the molten aluminum alloy during processing. A layer of aluminum carbide was found to increase the composite yield strength, ultimate tensile strength, work hardening rate, and work-to-fracture, and change the fracture pattern from one involving interfacial decohesion to one where particle breakage was dominant. These changes were attributed to a stronger interface bond, which is thought to result from the tendency of the Al_4C_3 reaction layer to form semicoherent interfaces and orientation relationships with the aluminum matrix and SiC particles and for it to be mechanically keyed-in to both phases. The stronger interface bond also enhanced the levels of plastic constraint which, when coupled with the greater work-hardening, promoted local matrix failure, thereby reducing the composite ductility (Tham et. al., 2001).

The deformation behavior below 0.2% offset yield stress in 2024 Al/SiC_p composites and its unreinforced matrix were investigated experimentally under three heat treatment conditions by Zhang et.al. In the case of annealing, incorporation of SiC particles into aluminum matrix can enhance the plastic flow stress (PFS) in the macroplastic stage, but almost has no effect on PFS in the microplastic stage, suggesting that the strengthening led by the SiC particulate is more effective in a larger stress (or strain) region. Quenching treatment would increase PFS in the microplastic both in micro- and microplastic stages for the unreinforced 2024 Al alloy while slightly lower PFS in the microplastic stage for the composite. Quenching followed by artificial aging shows the highest PFS both in the micro- and macroplastic stages for both materials, implying that like the conventional yield strength PFS in microplastic stage of the composite is also strongly controlled by the precipitates formed in the matrix during aging treatment. The results were attributed to the microstructure features such as the residual thermal stresses, dislocation density and matrix hardness (Zhang et. al., 2001).

The mechanical properties are also affected by processing type, like the extrusion and aging, with the net effect related to the production of dislocations. For instance, the effect of extrusion and sintering temperature on the mechanical properties of SiC/Al-Cu composites was discussed by C. Sun and co-workers (Sun et. al., 2011). The extrusion and increase of the sintering temperature can break up the oxide coating on the matrix powder surfaces, decrease the number of pores, accelerate the elements' diffusion and increase the density and particle interfacial bonding strength, thus significantly improving the mechanical properties of the composite. During aging, new precipitates nucleate and grow from the supersaturated matrix. The increase in the strength and decrease in the elongation at the under-aged stage are due to the increase in the volume fraction of the precipitates. A high number of volume-fractioned precipitates will effectively inhibit the movement of dislocations, generate more geometrically necessary dislocations and reach the critical dislocation density for fracture earlier during deformation, and thus increase the strength and decrease the elongation.

It is well established that the volume fraction of reinforcement affects the mechanical and thermal properties of the metal matrix composites. Al MMCs with high volume fractions (40 > V_f < 70 %) are typically produced when high hardness and stiffness are sought after. This type of composites is usually produced by the infiltration technique. Composites with Young's modulus as high as 225 GPa and 405 MPa flexure strength were produced by Cui (Y. Cui, 2003) using the pressureless infiltration method. Sahin & Acilar studied the effect of SiC_p volume fraction on the physical properties and hardness of Al/SiC_p composites. Composites were produced by the vacuum infiltration technique using an alloy Al-9.42 %Si-0.36 % Mg (wt. %) and up to 55 vol. % SiC. Results showed that hardness and density of the composite increased with increasing load and increasing particle content (Sahin & Acilar, 2003). A different approach was used within the research group of the author in an endeavor to increase the Al/SiC composite's hardness (Montoya-Dávila et. al., 2007). The alloy Al–15.52 Mg–13.62 Si (wt. %) was employed to prepare composites in argon followed by nitrogen at 1100 °C for 60 min. The use of preforms with monomodal, bimodal and trimodal distribution of SiC particulates resulted in an increase of hardness with increase in the reinforcement particle size distribution. Superficial hardness behavior was explained by the combined effect of work-hardening in the alloy matrix and particle-to-particle impingement. As for hardness increase, a demonstrated and successful approach consists of the use of heat treatments, reviewed in the next section.

2.3 Heat treatments (solution and age-hardening)

In order to comply with the requirements for some specific applications, metals are not used as pure materials in the as-cast conditions, and aluminum is not the exception. The precipitation-strengthening (hardening) process is used to increase the strength of many aluminum and other metal alloys. The object of precipitation strengthening is to create in a heat treated alloy a dense and fine dispersion of precipitates in a matrix of deformable metal. The precipitate particles act as obstacles to dislocation movement and thereby strengthen the heat-treated alloy. The precipitation-strengthening process involves three basic steps: 1. Solution heat treatment, 2. Quenching, and 3. Aging (natural or artificial).

One prominent characteristic of age hardening is that the precipitate is *coherent* with the matrix, implying that certain matrix planes match specific planes of the precipitate quite closely in atom spacing and are continuous throughout the precipitate. This small difference in spacing produces a strain field around the precipitate particle, which causes blocking of dislocations for some distance from the actual precipitate. Usually the strain field can exist only while the precipitate is small. Precipitation hardening is commonly employed with high strength aluminum alloys. Although a large number of these alloys have different proportions and combination of alloying elements, the mechanism of hardening has perhaps been studied most extensively for the aluminum-copper alloys (Callister, 1997).

If many aluminum alloys are prone to age hardening and if those alloys are used for the manufacture of Al/SiC composites, then it is reasonable to think that precipitation-strengthening (hardening) processes might also occur in the composites, but with a slightly (or considerably) different mechanism due to the presence of the SiC reinforcement. One foreseeable difference is the solution treatment and/or aging response between the unreinforced alloy and the composite, including aging kinetics and peak hardness. In this particular discussion, it is postulated that the mechanism in the composites might be

influenced by one or more of the following factors: alloy composition, morphology, volume fraction, particle size and properties (physiochemical and thermal) of the reinforcements, and finally, the fabrication route and heat treatment given to the composites. Most of the work done is concerned with heat treatment parameters and alloy chemistry.
It is generally accepted that aging treatment can significantly increase the properties of some aluminum alloys and their respective composites, especially those alloys of the series 2xxx and 6xxx. However, some discrepancies are observed in the literature, because while some authors propose that the addition of reinforcing particles accelerates the aging kinetics, others suggest that it decreases or marginally affects it (Mosuavi & Seyed, 2010). As a result, the accelerated aging phenomenon aroused the interest of various researchers, stimulating both, theoretical and experimental studies to explain the operative mechanism.
For instance, Dutta & Bourell expounded that accelerated aging in metal matrix composites (MMCs) can be attributed to an increase in dislocation density in the vicinity of the reinforcements or to the matrix residual stress field near reinforcements (Dutta & Bourell, 1989). Both mechanisms aid the diffusion of solute atoms, thereby leading to more rapid precipitation. They studied experimentally the precipitation behavior of aluminum 6061 alloy reinforced with 10 vol. % SiC whiskers of variable aspect ratio and compared results with the precipitation behavior of a control aluminum alloy 6061 in the unstrained and plastically strained conditions. It was found that the strained control alloy, with approximately the same expanded plastic work as the composite, showed a similar β' precipitation rate and activation energy as the composite. On the contrary, the unstrained alloy had a much higher activation energy for precipitation. A theoretical model was developed to predict the rate of precipitation in the residual stress field of the matrix. This rate was compared with the rate of precipitation on a regular edge dislocation array. It was found that for realistic values of fiber radii and dislocation densities (about 0.25-1 μm and 10^{13}-10^{14} m^{-2}, respectively), both mechanisms give comparable precipitation rates. However, solute atoms flowing towards the matrix-fiber interface under the influence of residual stress field on encountering matrix dislocations are trapped, thereby lowering the activation energy to that of precipitation on dislocations. It was concluded that for MMCs with large fibers and high dislocation densities, dislocation generation is the principal contribution to accelerated aging while, in MMCs with small fibers and low dislocation densities, the residual stress mechanism predominates. For intermediate fiber radii and dislocation densities, both mechanisms could be important although, in real MMCs, dislocations seem to play the dominant role (Dutta & Bourell, 1989). Similar results were reported later on by Borrego and co-workers using the 6061 aluminum alloy, but with 15 vol. % SiC whiskers (Borrego et. al., 1996). They found that accelerated aging in composite materials with respect to the unreinforced alloy is more accentuated with increasing temperature, and attributed this behavior to the increasing dislocation density with extrusion temperature. The dislocation density of composites was calculated from the τ_{peak} values, observing that as the extrusion temperature goes up the dislocation density increases. This increment also accounts for the increasing hardness of the composite. It was suggested that the principal hardening mechanism of the composites is due to the residual dislocations (Taylor). Other contributions due to grain/subgrain structure (Petch/Hall), particle strengthening (Orowan) and/or initial work hardening would represent a factor of 0.3 (or smaller) that due to the residual dislocations (Borrego et. al., 1996). In a study conducted using the 2024 Al alloy in Al/SiC_p composites, it was found that the presence of SiC particles led to increasing the peak hardness of the alloy. The aging behavior of the 2024 Al alloy and its composite reinforced with 20 vol. % SiC particles was studied after solution treatment at

495 °C for 1-3 h. The suitable solution treatment time was about 2 h for both, the unreinforced alloy and the composite that leads to the fastest aging kinetics and the maximum hardness. The composite reached its peak hardness in shorter time compared with the unreinforced alloy on solution treatment for 2 and 3 h, but reached it in longer time on solution heat treatment for 1 h (Mosuavi & Seyed, 2010). Using the same alloy type, 2124, Thomas and King studied the formation of phases during precipitation by differential scanning calorimetry (DSC) (Thomas & King, 1994).

Although most of the studies on precipitation hardening in aluminum alloys are related to the intermetallic phase Mg_2Si, not in all studies the precipitates responsible for age hardening correspond to Mg_2Si, as reported by Cottu et. al., who identified rod-shape precipitates (atomic ratio Mg/Si= 1) with chemical composition not consistent with Mg_2Si stoichiometry (Cottu et. al., 1992). They showed that age-hardening kinetics of Al-Cu-Mg alloy 10 wt. % SiC fiber composite was enhanced by the presence of the reinforcement during a T6 heat treatment, attributed to the plastic deformation induced during heat treatment due to the difference between coefficients of thermal expansion (CTE) of matrix and reinforcement. In addition to the works by Dutta & Bourell, and Guo & Yuan, Cottu et. el., (Cottu, et. al., 1992) concluded that the hardening kinetics is enhanced by the SiC reinforcement due to the fact that precipitation preferentially develops on dislocation lines. In a study with composites with the aluminum alloy AS7G06 reinforced with chopped SiC fibers (about 10-15 μm diameter and 3-6 mm long), they found that the high-temperature deformation strongly increases the precipitation rate as the material is reinforced.

In summary, several factors are expected to affect the aging behavior of Al/SiC composites, in addition to aluminum alloy composition. They include fabrication route and processing parameters comprising temperature, time, atmosphere, as well as SiC reinforcement size and shape. In the case of fibers, it has been reported that the length and radii play a significant role in the age hardening mechanism (Dutta & Bourell, 1989). Incorporation of a different phase also may play a significant role, like that observed in SiC/Gr/6013Al composites. In the work of Guo and Yuan (Guo & Yuan, 2009) it was reported that the aging behavior of the composites is similar to that of the 6013Al showing two peak hardness during artificial aging at 191 °C. The composite reaches its peak hardness in shorter time and exhibits a smaller increase of hardness. Moreover, during natural aging, the composite reaches stable hardness in longer time than does the matrix alloy, with lower increase of hardness. There was evidence that low fractions of graphite powders affect the aging behavior and mechanical properties of aluminum matrix composites with SiC particulate reinforcements.

Investigations comparing the effect of different aluminum alloys are scarce. It was recently stated that there is still a lack of information regarding the aging behavior of Al alloys and their corresponding composites (Mosuavi & Seyed, 2010). The author's research group is currently focused on studying the effect of silicon in the alloy (Al- 10.33 Si- 17.75 Mg *vs.* Al-21.85 Si- 16.21 Mg, wt. %) on the solution treatment and artificial age hardening of Al/SiC composites. SiC was protected with SiO_2 coatings. Preliminary results suggest that with the high Si alloy, hardness increases, but the peak hardness is not as sensitive - as in the alloy with low Si level - to aging time. With the latter, peak hardness was attained in aging for 3 h at 170 °C, after a solution treatment at 350 °C for 3 h.

2.4 Corrosion behavior

Numerous investigations have been devoted to the study of the corrosion behavior of Al/SiC composites (Ahmad et. al., 2000; Aylor & Moran, 1985; Candan & Bilgic, 2004;

Hihara & Latanison 1992; Pardo et. al., 2003; Pardo et. al., 2005; Kiourtsidis et. al. 1999; Shin et. al., 1997; Trzaskoma, 1990; Wei et. al., 1998). Several factors are known to affect the corrosion behavior of SiC-reinforced composites of aluminum alloys, namely, alloy composition, SiC physicochemical characteristics and volume fraction, and processing route. Factors related to processing are: porosity, the presence of intermetallic phases within the matrix, and the formation of reaction products at the interface between the metallic matrix and the reinforcing phases. The size and shape of the SiC reinforcement is also of great importance, as fine particles and short/thin fibers are more prone to the dissolution.

The intermetallic phases can already be present in the raw aluminum alloy or can be formed in situ during processing of the composite. For that reason, the corrosion behavior of the composites cannot be inferred directly from the response of their respective aluminum alloys. Both, in the aluminum alloy (before composite's processing) and in the composites, the intermetallic phases can form galvanic couples with the matrix or between them, and make the composites susceptible of corrosion. If the composites are manufactured with different aluminum alloys, then each type of Alloy/SiC system would require its own corrosion study. Magnesium silicide (Mg_2Si) is one of the intermetallics that play a dominant role in the corrosion behavior of the composites (Escalera-Lozano et. al., 2010). On the other hand, several corrosion parameters (pitting corrosion, corrosion potential, passivation potential, etc.) can be evaluated in the different studies using various experimental techniques (potentiodynamic polarization, double cyclic polarization, etc.).

Pitting corrosion is believed to be one of the main mechanisms for damage of high-strength aluminum alloys, and usually it initiates by the breakdown of the passive film of the metal surface. Accordingly, the effect of reinforcement on the pitting behavior of aluminum-base metal matrix composites was studied earlier (Aylor & Moran, 1985). Aylor & Moran studied the effect of processing type on the corrosion behavior, utilizing different 6061 Al processing forms (wrought vs. powder metallurgy) and SiC/Al composite heat-treating (as-fabricated vs T-6 temper). It was observed that processing type does not affect the anodic polarization characteristics of SiC/Al and Al and that the presence of SiC in the 6061 Al matrix does not alter the corrosion potential in aerated ocean water. Moreover, SiC does not increase the pitting susceptibility of the Al/SiC composite. The morphology and extent of pitting differs between the SiC/Al and Al materials. SiC/Al composites exhibit pitting concentrated predominantly at the SiC/Al interfaces, with the pitting being greater in number, smaller in size and more shallow in penetration depth, relative to the unreinforced aluminum alloys. The electrochemical behavior of Al/SiC composites was essentially identical to that of the powder metallurgy processed and wrought aluminum alloys; however, the pitting attack on the composites was distributed more uniformly across the surface, and the pits penetrated to significantly less depth. Pit morphology of aluminum alloy and Al/SiC composites was studied by Trzaskoma. Pit morphology of Al 5456 (UNS A95456), Al 6061 (UNS A6061), SiC_w/Al 5456, and SiC_w/Al 6061 was studied in order to compare pitting process of SiC_w/Al metal matrix composites and that corresponding unreinforced alloys. Pits on the composites are significantly more numerous, shallow, and widespread than on the monolithic materials. Studies of pit structure suggest that there are two stages in pit development. The first involves the initial dissolution of metal atoms and opening of the pit, and the second involves pit enlargement or growth. For both materials, pits initiate at secondary particles within the metal matrix. In the case of Al 5456 and SiC_w/Al 5456, it is shown that these particles are intermetallic phases composed of alloying elements Mg, Cr, Mn, and Al, as well as Fe, which is an impurity of the metal. Under equivalent conditions of preparation and processing, a greater number of intermetallic phases form in the composite than the alloy and hence the composite has more pit initiation sites (Trzaskoma, 1990).

In a study of the influence of reinforcement proportion and matrix composition on the corrosion resistance of cast aluminum matrix composites (A3xx.x/SiC_p) in a humid environment, Pardo and co-workers found that the corrosion process was influenced more by the concentration of the alloy elements than by the proportion of the SiC particles (Pardo et. al., 2003). Later on, they confirmed this observation, but investigating the influence of reinforcement grade and matrix composition on pitting corrosion behavior of the same type of composites, using the potentiodynamic polarization technique and four aluminum alloys. The corrosion damage was caused by pitting attack and by nucleation and growth of $Al_2O_3.3H_2O$ on the material surface (Pardo et. al., 2005).

In a study using Al-4 Mg alloy matrix, Candan and Bilgic investigated the corrosion behavior of Al-60 vol.%/SiC in NaCl solution. Experimental results revealed that precipitation of Mg_2Si as a result of the reaction between the Al-Mg alloy and the SiC particles has a beneficial effect on corrosion resistance of Al-4Mg alloy matrix due to the interruption of the continuity of the matrix channels within the pressure infiltrated composite (Candan & Bilgic, 2004). Hihara and Latanison, studied the galvanic corrosion of aluminum matrix composites using ultrapure Al and 6061-T6 electrodes to study the galvanic corrosion current density (CD) of Al-matrix composites. Results indicated that the galvanic corrosion rate of Al is approximately 30 times less when coupled to SiC. Oxygen reduction was the primary cathodic reaction in the aerated solutions. In deaerated 0.5 M Na_2SO_4 and 3.15 wt.% NaCl, galvanic corrosion was negligible (Hihara & Latanison 1992).

Using the double cycle polarization technique Kiourtsidis et. al., conducted a study on the pitting behavior of AA2024/SiC_p composite and found that pitting potential is unaffected by the presence and volume fraction of SiC particles in the composites (Kiourtsidis et. al. 1999).

In a different study, using four alloys (6013, 6061, 5456 and 2024 with heat treatments (T4 and T6)), it has been shown that pit initiation is dependent on the alloy type and heat treatment. Further, microscopic observations show that pit initiation sites are correlated with secondary phase particles, suggesting that secondary phases, rather than SiC particles, contribute to the pitting behavior of the composites (Ahmad et. el., 2000).

Related studies by the author's research group - using the Scanning electrochemical microscopy (SECM) technique in composites processed with the alloy Al-13.5% Si-9% Mg - allowed determining that SiC particles are electrochemically active (Díaz-Ballote, et. al., 2004). The data suggested that the electronic conductivity at these sites is higher than that of the Al_2O_3 film covering the alloy matrix surface. *In situ* SECM images of samples and current *vs.* tip-substrate distance curves were used to investigate the reduction of dissolved oxygen on the silicon carbide particles. Results with samples of SiCp/Al composites immersed in distilled water alone or in either 0.1 M NaCl or boric acid/borax buffer containing ferrocenemethanol as mediator demonstrate that the SiC particles are conductive and act as local cathodes for the reduction of oxygen. More recently, within the same group, the electrochemical behavior of the passive film of an alloy of the same type (Al–17Si–14Mg) was studied in anodic polarization tests and compared to that for pure Al. Results showed that for the alloy, the passive current density increased but the pitting susceptibility decreased. The first effect was ascribed to a significant electrochemical activity of the Mg_2Si intermetallics and the second to improved stability of the oxide film. X-ray photoelectron spectroscopy (XPS) analysis of potentiostatically formed passive film on the alloy showed that it consisted of aluminum oxyhydroxide with incorporation of silicon in its elemental

and two oxidized states (+3 and +4). Mott–Schottky analysis showed that trivalent silicon ion acted as an n-type dopant in the film (Coral-Escobar et. al., 2010).
Regarding the role of secondary phases, it has been reported that particles that contain Al, Cu and Mg tend to be anodic relative to the alloy matrix, while those that contain Al, Cu, Fe and Mn tend to be cathodic relative to the matrix (Wei et. al., 1998). It is generally accepted that Mg_2Si tends to be anodic with respect to the matrix and can act as initiation sites for corrosion. It is believed that this phase dissolves leaving behind a cavity, which can act as a nucleation site for pitting. Accordingly, a study of the role of Mg_2Si in the electrochemical behavior of Al-Si-Mg aluminum alloys was undertaken by the author´s group, using four experimental aluminum alloys with variations in the Si/Mg molar ratio (A1-0.12, A2-0.49, A3-0.89, A4-1.05)(Escalera-Lozano et. al., 2010). The corrosion potential in open circuit (Eoc) and polarization resistance (Rp) were measured. Results show that the augment in Si/Mg molar ratio increases the presence of Mg_2Si intermetallic phase. During immersion tests in neutral aerated chloride solutions the anodic activity of the Mg_2Si intermetallic decreased rapidly, as indicated by a fast ennoblement of open circuit potential. After the immersion period (7 days), higher Rp values for alloys A3 and A4 (21 and 26 $K\Omega cm^2$, respectively) as compared to those for alloys A1 and A2 (5 and 10 $K\Omega cm^2$, respectively), suggest a greater corrosion resistance in Cl- containing environments for alloys A3 and A4. Another major form of degradation of Al/SiC composites is via chemical corrosion, attributed to the aluminum carbide (Al_4C_3) phase, formed by the dissolution of the SiC reinforcement during processing with aluminum in liquid state. A study of the effect of processing methods on the formation of Al_4C_3 in Al 2024/SiC_p composites has been conducted by Shin et. al. (Shin et. al., 1997). In another recent investigation, the author´s group focused on the role of Al_4C_3 in the corrosion characteristics of Al-Si-Mg/SiCp composites with varying Si/Mg molar ratio in neutral chloride solutions (Escalera-Lozano et. al, 2009). Immersion tests in aerated 0.1 M NaCl showed that for composites with Si/Mg molar ratios of 0.12 and 0.49, chemical degradation by hydrolysis of Al_4C_3 was followed by intense anodic dissolution at the matrix reinforcement interface, while composites corresponding to Si/Mg molar ratios of 0.89 and 1.05 did not exhibit intense localized attack.

3. Conclusion

With all the valuable research work conducted thus far in the field, there can be no denying that aluminum plays a pivotal role in the development of composite materials reinforced with SiC. And perhaps, amongst all the Al/reinforcement systems, the Al-SiC one has become a benchmark. It turns out that alloy composition influences the processing route to be employed as well as the mechanical, heat-treatment and corrosion behavior of the composites. The use of aluminum in liquid state has serious implications because attack of SiC and the subsequent phenomena do compromise the integrity of the composite. In this regard, since the aluminum-matrix/reinforcement interface plays a critical role in transferring the load from the matrix to the reinforcing phase, the soundness of the interface is always a crucial aspect to take care of. What's more, wettability studies aimed at optimizing processing conditions are always wise. From heat treatment investigations, there is a consensus in that SiC in aluminum leads to an accelerated age hardening, compared to the unreinforced alloy. Being pitting one of the major concerns in the corrosion behavior of Al/SiC composites, the confluence of results suggest that in the composites, pitting is greater in number, smaller in size and shallower in penetration depth, relative to the unreinforced aluminum alloys. Since pits initiate at secondary particles within the metal matrix, and as a greater number of intermetallic phases form in the composites, these have more pit initiation sites. It is suggested from this discussion that corrosion tests should precede mechanical evaluation involving hardness, creep and fatigue and fracture studies. A thoughtful consideration of the abovementioned factors and response variables involved increases the likelihood for Al/SiC composites to achieve their full potential in a safe and cost-effective way.

4. Acknowledgement

The author gratefully acknowledges "Consejo Nacional de Ciencia y Tecnología" (Conacyt in México) for financial support under project with reference No. 47353-Y.

5. References

Aguilar-Martínez, J. A., Pech-Canul, M. I., Rodríguez-Reyes, M., De La Peña, J. L. (2004), "Effect of Mg and SiC type on the processing of two-layer Al/SiC_p composites by pressureless infiltration", *J. Mater. Sci.*, vol. 39, No. 3, pp. 1025-1028, ISSN 0022-2461

Ahmad, Z., Paulette, P. T., Aleem, B. J. A. (2000). "Mechanism of localized corrosion of aluminum-silicon carbide composites in a chloride containing environment", *J. Mater. Sci.*, Vol. 35, No. 10, pp. 2573-2579, ISSN 0022-2461

Arsenault, R. J., Shi, N. (1986)."Dislocation generation due to differences between the coefficients of thermal expansion". *Mater. Sci. and Eng.*, Vol. 81, pp. 175-187, ISSN 0921-5093.

Arsenault, R. J., Taya, M. (1987)."Thermal residual stress in metal matrix composite". *Acta Metall.*, Vol. 35, No. 3 pp. 651-659, ISSN 0001-6160

Aylor, D. M., Moran, P. J. (1985). "The effect of reinforcement on the pitting behavior of aluminum-base metal matrix composites". *J. Electrochem. Soc.: Electrochemical Sci. and Tech.*, Vol. 132, No. 6, pp. 1277-1281, ISSN 0013-4651

Borrego, A., Ibáñez, J. López, V., Lieblich, M., González-Doncel, G. (1996)."Inlfuence of extrusion temperature on the aging behavior of 6061Al-15 vol %SiC_w composites". *Scripta Materialia*, Vol. 34, No. 3, (1 February 1996), ISSN 1359-6462

Callister Jr. W. D. (1997). Materials Science and Engineering An Introduction (Fourth Edition), John Wiley & Sons, Inc. ISBN 0-471-13459-7, New York.

Candan, S., Bilgic, E. (2004). "Corrosion behavior of Al-60 vil. % SiC_p composites in NaCl solution". *Mater. Lett.*, Vol. 58, No. 22-23, (September 2004), pp. 2787-2790, ISSN 0167-577X

Coral-Escobar, E. E., Pech-Canul, M. A., Pech-Canul, M. I. (2010)."Electrochemical behavior of passive films on Al-17Si-14Mg (wt.%) alloy in near-neutral solutions", *J. Solid State Electrochem.*, Vol. 14, No. 5, pp. 803-810, ISSN: 1385-8947

Cottu, J. P., Coudurec, J.-J., Viguier, B., Bernard, L. (1992). "Influence of SiC reinforcement on precipitation and age hardening of a metal matrix composite". *J. Mater. Sci.*, Vol. 27, pp. 3068-3074, ISSN 0022-2461

Cui, Y., (2003). "High volume fraction SiC_p/Al composites prepared by pressureless melt infiltration: processing, properties and applications", Key Engineering Materials, Vol. 249, pp. 45-48, ISSN 1013-9826

Díaz-Ballote, L., Velva, L., Pech-Canul, M.A., Pech-Canul, M.I., Wipf, D. O. (2004)."The Activity of Silicon Carbide Particles in Al-Based Metal Matrix Composites Revealed by Scanning Electrochemical Microscopy". *Journal of the Electrochemical Society*, Vol. 151, No. 6, pp. B299-B303, ISSN: 0013-4651

Dutta, I., Bourell, D. L. (1989)."A theoretical and experimental study of aluminum alloy 6061-SiC metal matrix composite to identify the operative mechanism for accelerated aging". *Mater. Sci. and Eng. A*, Vol. 112, (June 1989), ISSN 0921-5093

Escalera-Lozano, R., Pech-Canul, M. A., Pech-Canul, M. I., Quintana, P. (2009). "Corrosion characteristics of Al-Si-Mg/SiC_p composites with varying Si/Mg molar ratio in neutral chloride solutions". *J. Materials and Corrosion*, Vol. 60, No. 9, (September 2009), pp. 683-689, ISSN 0947-5117

Escalera-Lozano, R., Pech-Canul, M. I., Pech-Canul, M. A., Montoya-Dávila, M., Uribe-Salas, A. (2010)."The role of Mg_2Si in the corrosion behavior of Al-Si-Mg alloys for pressureless infiltration". The Open Corrosion Journal, Vol. 3, pp. 73-79, ISSN 1876-5033

Greasley, A., Hermann, R., Tomlinson, M. (1995). "Fatigue and projectile penetration performance of ceramic particle reinforced aluminum metal matrix composites", *Materials Science Forum*, Vols. 189-190, pp. 321-328, ISNN 0255-5476

Guo, J., Yuan, X. (2009)."The aging behavior of SiC/Gr/6013Al composite in T4 and T6 treatments". *Mater. Sci. and Eng.* A, Vol. 499, pp. 212-214, ISSN 0921-5093.

Hihara, L. H., Latanison, R. M. (1992). "Galvanic corrosion of aluminum matrix composites", *Corrosion*, Vol. 48, No. 7 (July 1992), pp. 546-552, ISSN 0010-9312

Iseki, T., Kameda, T., Maruyama, T.(1984). "Interfacial reaction between SiC and aluminum during joining", *J. Mater Sci.*, Vol. 19, No. 5, pp. 1692-1698, ISSN 0022-2461

King, J. E. and Bhattacharjee, D. (1995). "Interfacial effects on fatigue and fracture in discontinuously reinforced metal matrix composites". *Materials Science Forum*, Vols. 189-190, pp. 43-56, ISNN 0255-5476

Kiourtsidis, G. E., Skolianos, S. M., Pavlidou, E. G. (1999). "A study on pitting behavior of AA2024/SiC_p composites using a double cyclic polarization technique". *Corrosion Science*, Vol. 41, No. 6, (June 1999), pp. 1185-1203, ISSN 0010-938X

Kosolapova, T. Y. (1971). *Carbides, Properties, Production and Applications*, Plenum Press, ISBN 306-30496-1, New York

Kumai, S., King, J. E., Knott, J. F. (1991). "Fatigue in SiC-particulate-reinforced aluminum alloy composites". *Mater. Sci. and Eng. A*, Vol. 146, pp. 317-326. ISSN 0921-5093

Li, Y., Mohamed, F. A. (1997). "An investigation of creep behavior in an SiC-2124 Al composite". *Acta Mater.*, Vol. 45, No. 11, pp. 4775-4785, ISSN 1359-6454

Montoya-Dávila, M., Pech-Canul, M. A., Pech-Canul, M. I. (2007). "Effect of bi- and tri-modal size distribution on the superficial hardness of SiC_p in Al/SiC_p composites prepared by pressureless infiltration". *Powder Technology* Vol. 176, No. 2-3, pp. 66-71, ISSN 0032-5910

Montoya-Dávila, M., Pech-Canul, M. A., Pech-Canul, M. I. (2009). "Effect of SiC_p multimodal distribution on pitting behavior of Al/SiC_p composites prepared by reactive infiltration". *J. Powder Technology*, Vol. 195, No. 3, pp. 196-202, ISSN 0032-5910

Mousavi Abarghouie, S. M. R., Seyed Reihani, S. M. (2010). "Aging behavior of a 2024 Al alloy-SiC_p composite", *Materials and design*, Vol. 31, No. 6, pp. 2368-2374. ISSN 0261-3069

Pai, B. C., Ramani, G., Pillai, R. M., Satyanarana, K. G. (1995). "Role of magnesium in cast aluminum alloy matrix composites". J. Mater. Sci., Vol. 30 No. 8, pp. 1903-1911, ISSN 0022-2461

Pardo, A., Merino, M. C., Merino, S., López, M. D., Viejo F., Carboneras. M. (2003). "Influence of reinforcement grade and matrix composition on corrosion resistance of cast aluminum matrix composites (A3xx.x/SiC_p) in a humid environment". *Materials and Corrosion* 54, No. 5 (May 2003), pp. 311-317, ISSN 1521-4176

Pardo, A., Merino, M. C., Merino, S., Viejo, F., Carboneras, M., Arrabal, R.. (2005). "Influence of reinforcement grade and matrix composition on pitting corrosion behavior of cast aluminum matrix composites (A3xx.x/SiC_p)". *Corrosion Science*, 47, No. 7, (July 2005), pp. 1750-1764, ISSN 0010-938X

Park, J. K., Lucas, J. P. (1997). "Microstructure effect on SiC on SiCp/6061 Al MMC: dissolution of interfacial Al_4C_3". *Scripta Materialia*, Vol. 37, No. 4, (August 1997), pp. 511-516, ISSN: 1359-6462

Pech-Canul, M. I., Katz, R. N., Makhlouf, M. M. (2000)(b). "Optimum Parameters for Wetting Silicon Carbide by Aluminum Alloys". *J. of Metallurgical and Materials Transactions A*, Vol, 31 A, No. 2, pp. 565-573, ISSN 1073-5623

Pech-Canul, M. I., Katz, R. N., Makhlouf, M. M., Pickard, S. (2000)(a). "The Role of Silicon in Wetting and Pressureless Infiltration of SiC_p Preforms by Aluminum Alloys". *J. of Materials Science*, Vol. 35, No. 9, pp. 2167-2173, ISSN 0022-2461

Pech-Canul, M. I., Makhlouf, M.M. (2000). "Processing of A-SiC_p metal matrix composites by pressureless infiltration of SiC_p preforms". *J. Mater. Synthesis and Processing*, Vol. 8, No. 1, pp. 35-53, ISSN 1064-7562

Rodríguez-Reyes, M., Pech-Canul, M. I., Rendón-Angeles, J. C., López-Cuevas, J. (2006). "Limiting the Development of Al_4C_3 to Prevent Degradation of Al/SiC_P Composites Processed by Pressureless Infiltration". *Composites Science and Technology*, Vol. 66, No. 7-8, pp. 1056-1062, ISSN 0266-3538

Sahin, Y., Acilar, M. (2003). "Production and properties of SiC_p-reinforced aluminum alloy composites". *Composites Part A*, Vol. 34, No. 8, pp. 709-718, ISSN 1359-835X

Shan, D., Nayeb-Hashemi, H. (1999). "Fatigue-life prediction of SiC particulate reinforced aluminum alloy 6061 matrix composite using AE stress delay concept", *J. Mater. Sci.*, Vol. 34, No. 13, pp. 3263-327, ISSN 0022-2461

Shin, D.-S., Lee, J.-C., Yoon, E-P. Lee, H.-I. (1997)."Effect of the processing parameters on the formation of Al_4C_3 in SiC_p/Al 2024 composites", *Mater. Res. Bull.*, Vo. 32, No. 9 (September 1997), pp. 1155-1163, ISSN 0025-5408

Sun, C., Shen, R., Song, M., (2011). "Effects of sintering and extrusion on the microstructures and mechanical properties of a SiC/Al-Cu composite", Journal of Materials Engineering and Performance, DOI:10.1007/s11665-011-9940-1, ISSN 1059-9495

Tham, L. M., Gupta, M., Cheng, L. (2001). "Effect of limited matrix-reinforcement interfacial reactions on enhancing the mechanical properties of aluminum-silicon carbide composites". *Acta Mater.*, Vol. 49, No. 16, pp. 3243-3253, ISSN 1359-6454

Thomas, M. P., King, J. E. (1994)."Comparison of the ageing behaviour of PM 2124 Al alloy and Al-SiC_p metal-matrix composite". *J. Mater. Sci.*, Vol. 29, No. 20, ISSN 0022-2461

Trzaskoma, P. P. (1990). "Pit morphology of aluminum alloy and SiC/Aluminum alloy metal matrix composites". *Corrosion*, Vol. 46, No. 5, pp. 402-409, ISSN 0010-9312

Wei, R. P., Liao, Ch.-M., Gao, M. (1998)." A Transmission electron microscopy study of constituent-particle-induced corrosion in 7075-T6 and 2024-T3 aluminum alloys". *Metall. and Mater. Trans. A*. Vol. 29 (April 1998) pp. 1153-1160, ISSN 1073-5623

Zhang, F., Sun, P., Li, X., Zhang, G. (2001). "A comparative study on microplastic deformation behavior in a SiC_p/2024Al composite and its unreinforced matrix alloy". *J. Mater. Lett.*, Vol. 49, No. 2, pp. 69-74, ISSN 0167-577X

Part 5

Corrosion and Mechanical Damage of Aluminium Alloys

13

Effects of Dry Sliding Wear of Wrought Al-Alloys on Mechanical Mixed Layers (MML)

Mariyam Jameelah Ghazali
Department of Mechanical & Materials Engineering
Universiti Kebangsaan Malaysia
Malaysia

1. Introduction

Aluminium alloys are very attractive compared to other materials like steels, particularly for their mechanical properties. Despite of having a relatively low density (2.7 g/cm^3 as compared to ± 7.9 g/cm^3 of steel), they also possess high ductility (even at room temperature), high electrical and thermal conductivity and resistance to corrosion and high thermal conductivity. However, aluminium by itself exhibits poor tribological properties and their usage, for example in automotive applications, has been limited by their inferior strength, rigidity and wear resistance, compared with ferrous alloys. With respect to friction and wear behaviour, it has been well understood that the tribological behaviour of aluminium alloys is strongly influenced by the mechanical, physical and chemical properties of the near-surface materials. Whether lubricated or dry sliding, there is evidence that substantial work-hardening occurs at the worn surface. Surface strains can be well in excess of those found in conventional mechanical working. Intimate contact between ductile materials in particular, normally involved transferred materials, which may result in the formation of a mechanically mixed layer (MML). The MML was generally found to be comprised of materials from both contact surfaces, and may also include oxygen, and was known to have very different properties to the Al-alloy. Although the formation of an MML was known to modify wear behaviour, the exact manner was not fully understood. Moreover, very little was known about the effect that matrix alloy composition had on MML formation although it was claimed that the MML could improved wear resistance.

2. Backgrounds

2.1 Sliding wear theory

Wherever surfaces move against each other, wear will occur; damage to one or both surfaces generally involves progressive loss of material (ASM International & 1992 Hutchings, 1992). The rate of removal is generally slow. Although the loss of material is relatively small, it can be enough to cause complete failure of large and complex machinery. Hence, it is essential to develop a thorough understanding of the wear process, especially its mechanism and behaviour, in order to optimise performance. In the current work, only dry or unlubricated sliding wear will be further discussed, even though it is often associated with an environment containing appreciable humidity. When two surfaces slide or roll against each other under an applied load, two forces will exist:

1. The load acts normal to the surface areas that in contact will exert a compressive stress on the materials, which has a similarity with cold working and is usually concentrated in the rolling case.
2. A force exerted by the machine in the direction of motion, overcomes the following resistance:
   - The friction force, F, that is proportional to the normal load between contacting surfaces.
   - The static coefficient of friction, that is higher at the start of the motion than the dynamic friction.
   - Adhesion; the tendency of the two mating metals to adhere to each other. It may result in the surfaces being locally bonded together, forming a junction.
   - In extreme cases, resistance to motion is caused by abrasive material.

2.2 Wear of aluminium alloys

Due to their low density and excellent corrosion resistance, aluminium has become a substitute for steels especially in structures that require high performance and weight reduction. As with most other metals, aluminium reacts with oxygen in air. A submicron thick oxide layer is formed to provide effective corrosion protection. Aluminium is also non-magnetic and non-toxic, and can be formed by all known metal working processes. The density of aluminium is 2.7g/cm^3 or approximately one third the density of steel and aluminium alloys have tensile strengths of between 70 and 700N/mm^2. At low temperatures the strength of aluminium and its alloys increases without embrittlement in contrast to most steels (Pollack, 1977). Table 1 shows a comparison of the physical characteristics of some of the most important construction materials.

During the 1980's, about 85% of aluminium was used in the wrought form, that is rolled to sheet, strip or plate, drawn to wire or extruded as rods or tubes (Higgins, 1987 and Polmear, 1989). Some of the alloys may undergo subsequent heat-treatment in order to achieve the desired mechanical properties. The most common methods to increase the strength of aluminium alloys are:

- To disperse any second-phase constituents or elements in solid solution and cold work the alloy; there are known as *non-heat-treatable alloys.*
- To dissolve the alloying elements in solid solution and re-precipitate them. These are also as heat-treatable or precipitation-hardening alloys (originally known as 'age-hardening' alloys).

	Al	Fe	Cu	Zn
Density (g/cm^3)	2.7	7.9	8.9	7.1
Melting Point (YP) (°C)	660	1540	1083	419
Electrical conductivity (%)	63	16	100	30
Specific Heat/Thermal Volume (J/kg, K)	900	450	390	390
Thermal Conductivity (W/m, K)	220	75	390	110
Linear exp. coefficient (10^{-6}/K)	24	12	16	26
Electrical resistance (10^{-9} ohms/m)	27.5	105	17	58
Young's Modulus, (GPa)	70	220	120	93

Table 1. Physical characteristics of some of the most important construction materials

Two major and most common types of wear identified by Eyre (1979) that are relevant to industrial applications of aluminium alloys are abrasive and sliding wear especially for Al-Si alloys. In the case of Al-Si, generally, the hard silicon particles addition will contribute to higher hardness hence increase the wear resistance. Moreover, the particles are surrounded by softer and relatively tough matrix, which then improves the overall toughness of the material. This will lead to wear resistance by favouring more plastic behaviour (ASM Handbook, 1994).

As for aluminum alloys that reinforced with ceramic particles, they have shown significant improvements in mechanical and tribological properties including sliding and abrasive wear resistance (Rittner, 2000). The hard ceramic particles provide protection from further detrimental surface damage. An increase of ceramic hard particles content in alloys may enhance its wear resistance behaviour (Geng et al., 2009). The ageing behaviour of discontinuous reinforced metal matrix composites has been a subject of great interest, which is beneficial to optimise the ageing treatment and providing the experimental and theoretical information for designing the composites properties (Sheu and Lin, 1997). Aluminum nitride (AlN) as a reinforcement material has received much interest in electronic industry because of the need for smaller and more reliable integrated circuit.

For applications, aluminium based alloys have been widely used, for instance Al-Sn alloys as bearing metals in automobile designs. The most important properties of being a bearing metal are that it should be hard and wear-resistant, and have a low coefficient of friction. At the same time, it must be tough, shock-resistant, and sufficiently ductile to allow for *running-in*[1] processes made necessary by slight misalignments.

3. Mechanically Mixed Layers (MML)

3.1 Formation of the MML

In the case of ductile materials like aluminium alloys, most wear mechanism observed are consistent with Archard adhesive wear characterised by plastic ploughing and transfer of material from the counterface. With respect to friction and wear behaviour, numerous authors (Perrin and Rainforth, 1995, Leonard et al., 1997, Jiang and Tan, 1996, How and Baker, 1997 and Rigney, 1998) have concluded that the tribological behaviour is influenced by the mechanical, physical and chemical properties of these near-surface materials. In all cases, a mechanically mixed layer (MML) was present in most dry worn wrought aluminium alloys due to the repetitive sliding. However, significant differences between the MML of each alloy were observed. Their thickness which varied with loads suggested that the subsurface zones of the materials to the sliding and impact wear consisted of 3 zones (Rice et al., 1981) as indicated in Fig. 1.

a. Zones 1 – represents the undisturbed base material or original specimen material in the undeformed state, which experiences elastic deformation and thermal cycling when loaded in tribocontact. Its structure and properties are identical to those prior to the wear test.
b. Zones 2 – consists of the part of the original specimen that has obtained new properties due to repetitive tribocontact. Basically, sufferred deformed intermediate region of the base material. Here, plastic deformation occurs especially in ductile materials, grains are distorted and cracks or voids may nucleate.

---

[1] The process by which machine parts improve in conformity, surface topography and frictional compatibility during the initial stage of use.

c. Zones 3 - is known as tribolayer which forms in-situ, and usually contains chemical species from the counterface and test environment as well as the bulk material.

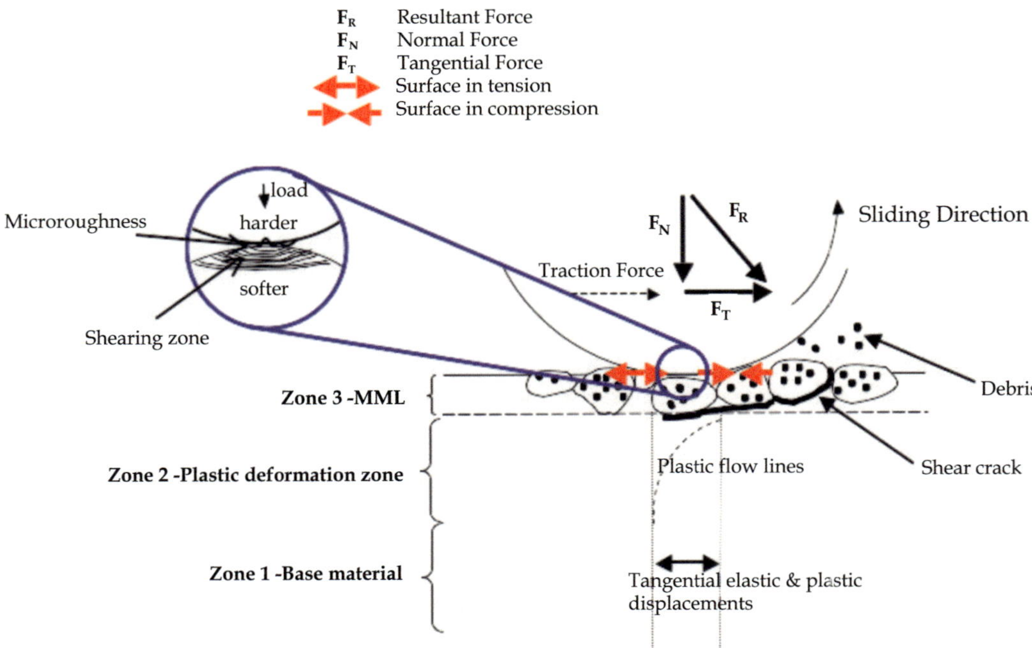

Fig. 1. An illustration of deformation during dry sliding (Ghazali, 2005)

The mixed layer (zone 3), which is commonly known as the mechanically mixed layer (MML), was formed through the incidental mixing between the two materials that statistically occurs at the contact spots under normal pressures. Crack and void formation were generally associated at the zone 2/3 interface and may dictate the dimensions of the wear debris formed (Suh, 1973). The extent and compositional features of these sub-surface zones were found to depend on the conditions of sliding wear, material and environment. Rice et al. (1981-1982) also indicated that these sub-surface zones developed quickly under dry sliding wear conditions. The present work has confirmed that a MML was formed in the sliding wear of the Al-alloys against both counterfaces. Its particles are recognised to have the same physical structure and chemical composition as those of the base pair (Biswas, 2000). The distinctive morphology of the mixed layer has led to a suggestion that its formation was due to a compression of the transfer material and the entrapped debris, which was followed by mechanical mixing during the sliding process. As highlighted by Heilmann et al. (1983), the MML which develops at an early stage (even before loose debris was obtained), is common in both dry and lubricated sliding wear process. In dry sliding condition, a high compressive pressure and large shear strains in the asperities were produced. Heavy plastic deformation and shear strains in the worn surface give rise to dislocation cells and elongated subgrains, as seen in i.e; Fig. 2, which is consistent with Heilmann et al. (1983), Rigney et al. (1981), Chen (1986), Chen and Rigney (1986) and Kuo and Rigney (1992).

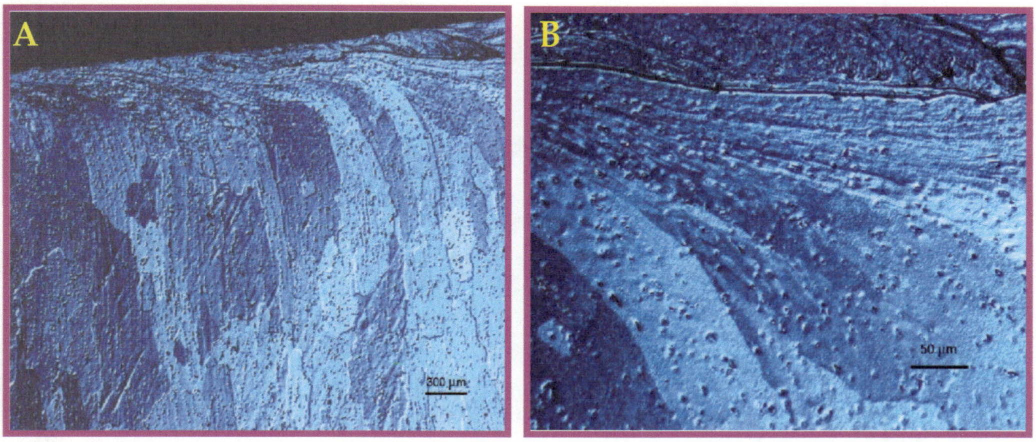

Fig. 2. A is an example of A3004 alloys after 10.8km slid against M2 at 140N. B is the magnified (Normaski) view of a selected area (Ghazali, 2005)

Some MML was very thin and the matrix of the Al alloy almost approached to the top worn surface, is very likely associated with the plastic flow during deformation, where the mixed layer can be replenished with fresh base material by a large plastic flow in the subsurface (Li and Tandon, 1999). Biswas (2000) studied the thickness of the MML appears to be controlled by the depth of crater and abrasion grooves made on the surface. In fact, he further concluded that one of the pre-requisites for the formation of this MML correlates with severe deformation of the top layer of the softer material pair, which in this case is the aluminium alloys. It appears that, when one surface is softer than the other, metal may be transferred from the soft to the hard surface. The material could be transferred back and forth several times during sliding and eventually produce wear debris particles (Heilmann et al., 1983).

3.2 Composition of the MML

In agreement with Rice et al. (1981), Heilmann et al. (1983) and Rigney et al. (1984), the mixed layer is composed of a mixture of two mating materials and from this layer, the loose debris were derived. Based on the results of EDS (Figs. 3), the layer had similar structure and composition to the loose debris. Here, presumably some of the wear debris from the counterface, together with the debris from the pin may have been compacted to form the MML. In other words, it can be concluded that such debris are not derived directly from the base material, with an exception for the case of abrasion, in which microcutting and microploughing are prominent.

In the present study, the MML were found to contain Al, Fe (for Al/M2 case) and O (in the form of oxide), which proves the source of element in the MML obviously originated from the counterface. The oxides were found to be coexisted with other phases in the MML and the wear debris, which is an expected phenomenon since the wear system was exposed to air. They could provide microstructural stability as a second phase in the ultrafine grained structure in debris, as proposed by Rigney et al. (1984). The oxides which have been known to form some protective and some destructive (Fischer, 1997 and Ravikiran et al., 1995) were then fractured and comminuted in further sliding process. The crushed oxides can be dispersed into the mixed surface layer and act as a pinning source of the grain boundaries in the ultrafine mixture in the MML and in the wear debris (Li and Tandon, 1999).

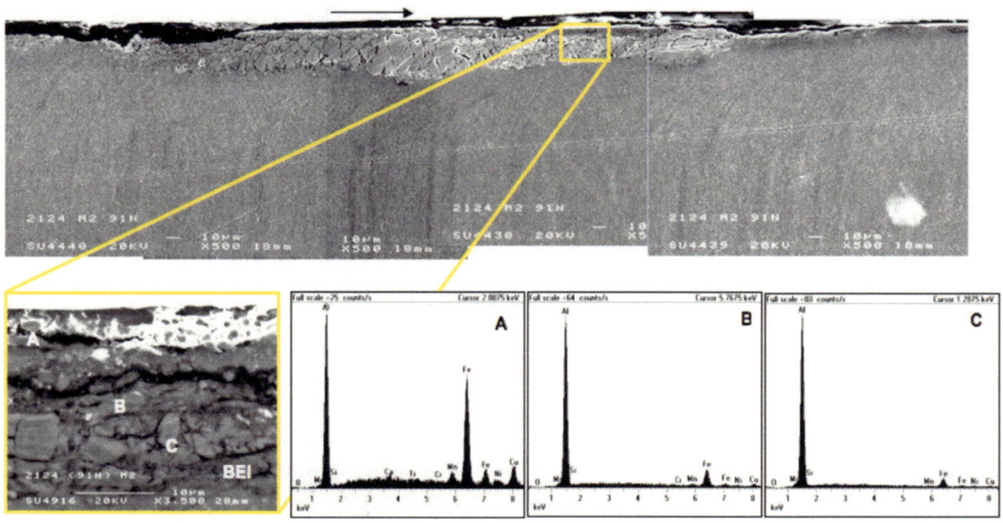

Fig. 3. Subsurface damage of longitudinal cross sections of A2124 alloys against M2 at 140N after sliding 10.8 km. Black arrows indicate the direction of sliding. A corresponding EDS analysis are shown in A, B and C areas (Ghazali, 2005)

3.3 The correlation between the MML and the wear debris

As for wear debris, its formation appeared to occur by two principal mechanisms, namely, the physical displacement of material from the worn surface by the ploughing action of the hard tool steel or alumina asperities, and secondly, delamination of large sheets (up to 1mm in extent) at particularly at high load like 140N. The thickness of the delamination sheets was found broadly consistent with the thickness of the MML, although it could not be defined with certainty whether the delamination occurred within the MML or at the MML/substrate interface. However, the longitudinal cross-sections suggested that both mechanisms were probable. Moreover, the cracks in the MML can give rise to delamination wear as a result of subsurface shear in a manner proposed by Suh (1977) where plate-like wear debris is produced. As one of the main principal wear mechanism in the present study was the delamination of the MML (part or whole), it would be reasonable to expect a correlation between MML thickness and specific wear rate.

3.4 The correlation between MML thickness and specific wear rate

A detailed comparison between several commercial wrought aluminium alloys, namely; A2124, A3004, A5056 and A6092 was carried out for this purpose. For this Al/M2 system, the specific wear rate was relatively insensitive to MML thickness for the A3004 and A5056, although the specific wear rate decreased in a linear manner with increasing MML thickness (refer to Fig. 4a).

In contrast, for the A2124 and particularly the A6092, the specific wear rate was a strong function of the MML thickness. Although a reasonable linear fit was possible for the A2124,

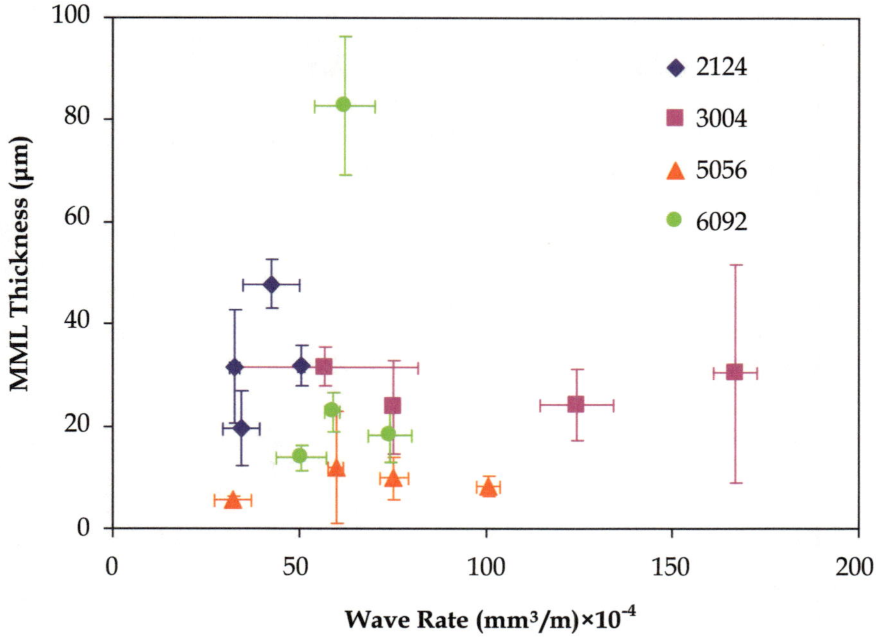

(a)

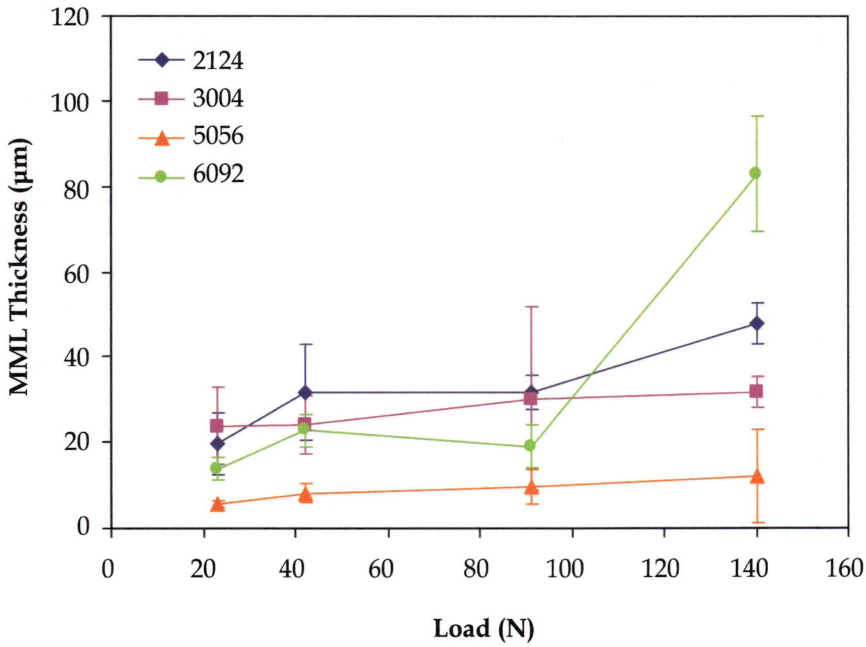

(b)

Fig. 4. The relationship between MML thickness and (a) wear rate and (b) load after 10.8km against M2 counterface (Ghazali, 2005) the A6092 data was better represented by an exponential fit. Since this data did not fit the same trend as the other alloys, the experiment was repeated and measurements re-taken, but with essentially the same result. Thus, the difference in behaviour of this alloy appears to be reproducible. Interestingly, the two alloys where the specific wear rate was relatively insensitive to MML thickness also exhibited MML with the least Fe content (Table 2) and the most homogeneous structure. Conversely, the A6092 exhibited the highest Fe content, the most heterogeneous structure and the greatest influence on the specific wear rate. However, the thickness of the MML cannot explain the dramatic drop in specific wear rate with load observed for the A3004 alloy (Fig. 4b). The thickness of the MML is only one of several potential ways in which the MML can affect wear rate. Clearly for Al/M2 system, the mechanical properties of the MML (in particular hardness and fracture stress) and its adhesion to the substrate are contributing factors.

Element	2124		3004		5056		6092	
	42N	140N	23N	140N	23N	140N	42N	140N
Mg	1.8 ± 0.3	2.0 ± 0.4	-	-	6.8 ± 0.5	6.8 ± 1.1	1.0 ± 0.1	1.7 ± 0.3
Al	85.3 ± 1.9	73.4 ± 5.9	94.9 ± 9.2	95.3 ± 5.6	80.5 ± 7.8	86.8 ± 7.7	61.8 ± 7.6	79.9 ± 5.8
Si	1.0 ± 0.5	-	0.7 ± 0.2	0.8 ± 0.3	1.0 ± 0.4	0.4 ± 0.1	1.6 ± 0.4	1.1 ± 0.3
Mn	1.1 ± 0.1	1.4 ± 0.5	2.5 ± 1.8	1.1 ± 0.2	-	-	0.3 ± 0.1	-
Fe	6.6 ± 1.5	19.1 ± 4.8	1.9 ± 1.5	2.7 ± 1.8	11.7 ± 5.4	6.0 ± 5.5	34.5 ± 7.8	15.9 ± 7.4
Cu	4.3 ± 0.6	4.1 ± 1.1	-	-	-	-	0.8 ± 0.2	1.3 ± 0.6

Table 2. Average quantitative EDS analysis on MML of Al-alloys against M2 (Ghazali, 2005)

In Al/M2 system, the A6092 exhibited the thickest MML and the highest Fe content. Since this was not replicated by the A5056, rather the reverse, it is clear that it is not the Mg content of the A6092 that promotes the formation of a thick MML. Thus, these results imply that stronger adhesion and transfer from the counterface is promoted by the Si in the alloy, while a high Mg content in the Al-alloy reduces adhesion. Similarly, the presence of Cu in the A2124 also appears to have promoted stronger adhesion than an equivalent amount of Mg, although the Cu was not as potent as the Si. The Mn in the A3004 also promoted a relatively thick MML, but one that was more homogeneous than for the A6092. The solubility of these elements in α-Al is in the order Si, Mn, Cu, Mg, which roughly approximates to the thickness of the MML formed. Thus, the observations are in-line with the Archard theory of adhesive wear, as might be expected. However, the level of alloy additions are small (e.g. Si) and it is surprising that the effect was as strong as observed. Thus, the wear performance is largely determined by the properties of the MML.

3.5 Effect of other variables

The atmosphere under an unlubricated wear process can strongly influence sliding wear rates with oxygen content and humidity being probably one of the important factors. In the case of Al-alloys, it is readily combined with oxygen to form a stable oxide layer. Oxidation, may have opposing effects on the wear process; one, it degraded the surface by removing metal atoms and second, it plays protective role in reducing metallic contact and decrease the wear rate (Degnan, 1995). However, whether or not the environment reaction has a beneficial and detrimental effect on wear rate, it depends strongly on the mechanical interaction of the reaction product with the substrate, particularly under surface plasticity condition, (Rainforth et al., 2002), which is in line with the present work.

Moisture in the environment also can have major effect on wear of metals. Endo and Goto (1978) reported the high humidity had a dentrimental effect on the fretting of aluminium alloys but negligible on carbon steels. Moreover, humidity can control the friction at room temperatures, particularly ceramics as higher coefficient of friction may occurr at temperatures above 800°C.

Beside humidity, all wear processes are influenced by temperature. The temperature reached at the surface of the contact is strongly influenced by the width of the contact (Johnson, 1985) and flash temperature is responsible for many wear and friction effects (Gecim and Winer, 1986). Wear occurs in conjunction with the dissipation of frictional energy in the contact and this is always accompanied by a rise in temperature. The frictional energy is generated by the combination of load and sliding speed and its distribution and dissipation is influenced by other contact conditions such as size and relative velocity. In regards to temperature effects on sample size and mass, contact spots have a tendency to remain in one place much longer on the smaller pin (alloy) than the larger side (counterface), causing stronger local heating in the former. Moreover, in this work, the rotating counterface will mostly experience extra cooling convention than the stationary alloy, which constantly hot due to repeated passage during the test. Local heating at contact spots also has other effect. Most obviously, the local hardness is reduced and thereby the load-bearing area is increased (Kuhlmann-Wilsdorf, 1987).

At high loads like the one used in the present study, 140N, friction heating can induce an increase in temperature, resulting a thermal softening beneath the worn surface, and may affect the wear mechanism (Zhang and Alpas, 1997 and Wang and Rack 1991). Maupin et al., (1992, 1993) studied that large grains were replaced by fine nanocrystalline grains which were relatively free of dislocations underneath the worn surface. Such microstructures could develop only if the temperature of the surface due to friction is very high. In addition, the deformed layer beneath the worn surface could result in higher plastic flow and work hardening resulting in increased wear resistance. At such high temperature, oxidation of the surface is also a possibility, as observed in earlier results.

4. Conclusion

In general, the dry sliding of Al/M2 systems showed the following responses as a result of repeated stress and frictional heat cycle:

- Elements present in the Al-alloy with high solubility in steel promoted a thick mechanically mixed layer, with higher Fe content. The effect was marked even for small contents in the Al-alloy.
- The solubility of these elements in α-Fe is in the order of Si, Mn, Cu, Mg, which roughly approximates the thickness of the MML formed.

- MML with high Fe content tended to be comprised of fragmented particulate, while a low Fe content tended to be associated with a more homogenous MML.
- A linear relationship between specific wear rate and the thickness of the MML was observed for 2124, 5056 and 3004, but not for 6092. The specific wear rate was relatively insensitive to MML thickness for the 3004 and 5056. In contrast, for the 2124 and particularly for the 6092, the specific wear rate was a strong function of the MML thickness.

5. Acknowledgment

The financial support of this research through JPA/SLAB (UKM) program, under the guidance of Professor William Mark Rainforth from University of Sheffield is gratefully acknowledged.

6. References

ASM Handbook Volume 18 (1992), Friction, Lubrication, and Wear Technology (ASM International).

ASM Specialty Handbook (1993): Aluminium and Aluminium Alloys, edited by Davis, J.R., ASM International, ISBN 978-0-87170-496- 2, Materials Park, OH.

Biswas, S.K. (2000). Some mechanisms of tribofilm formation in metal/metal and ceramic/metal sliding interactions. *Wear*, Vol. 245, No. 1, pp. 178–189.

Chen, L.H. (1986). Deformation, Transfer and Debris for Mation During Sliding Wear of Metals. *Scripta Metall.*, Vol.24, pp.827-83.

Chen L.H., and Rigney, D. A. (1986). Transfer during unlubricated sliding wear of selected metal system. *Wear*, Vol. 105, pp. 47-61

Degnan, C.C. (1995). The processing and wear behaviour of a W(TiC) reinforced steel matrix composite, PhD Thesis, University of. Nottingham, Nottingham,

Endo K. and Goto H. (1978). Effects of environment on fretting fatigue. *Wear*, Vol. 48, No. 2, pp. 347-367.

Eyre, T.S. (1979), *Treatise on Materials Science and Tech.*, edited by Scott, D., Academic Press, New York.

Fischer, T.S. (1997). New Directions in Tribology, edited by Hutchings, MEP Ltd, London.

Gecim, B., and Winer, W.O. (1986). Effect of surface film on the surface temperature of a rotating cylinder. ASME *J. Tribol.*, Vol. 108, pp. 92-97.

Geng, L. Zhang, B.P. Li, A.B. Dong, C.C. (2009). *Materials Letters* 63 pp. 557.

Ghazali, M.J. (2005), Dry Sliding Wear Behaviour of Several Wrought Aluminium Alloys, PhD. Thesis. University of Sheffield.

Heilman, P., Don, J., Suh, T.C., and Rigney, D.A. (1983). Sliding wear and transfer. *Wear*, Vol. 91, pp. 171-190.

Higgins, R.A. (1987). Materials for the Engineering Technician, 2nd Ed. Edward Arnold, London.

Hutchings, I.M. (1992). *Tribology-Friction and Wear of Engineering Materials*, 1st edn., Edward Arnold, ISBN 0-340-56184-x, London, UK.

How, H.C., and Baker, T.N. (1997). Dry sliding wear behaviour of Saffil-reinforced AA6061 composites. *Wear,* Vol. 210, pp. 263-272.

Jiang, J.Q., and Tan, R.S. (1996). Dry sliding wear of an alumina short fibre reinforced Al-Si alloy against steel. *Wear,* Vol. 195, No. 1-2, pp. 106-111.

Johnson, K.L. (1985). *Contact Mechanics,* Cambridge University Press, Cambridge.

Kuhlmann-Wilsdorf, D. (1987). Demystifying flash temperatures I. Analytical expressions based on a simple model. *Mat. Sci. Eng.,* Vol. 93, pp. 107-118.

Kuo, S.M., and Rigney, D.A. (1992). Sliding Behavior of Aluminum. *Mat. Sci. Tech.,* Vol. 157, pp. 131-143.

Leonard, A.J., Perrin, C., and Rainforth, W.M. (1997). Microstructural changes induced by dry sliding wear of a A357/SiC metal matrix composite. *Mater. Sci. Tech.,* Vol. 13, No. 1, pp. 41-48.

Li, X.Y., and Tandon, K.N. (1999). Mechanical mixing induced by sliding wear of an Al Si alloy against M2 steel. *Wear,* Vol. 225-229, pp. 640-648

Maupin H.E., Wilson R.D., Hawk J.A. An abrasive wear study of ordered Fe3Al. *Wear,* Vol. 159, pp. 241-247.

Maupin, H. E. Wilson, R. D. and Hawk, J. A. (1993). Wear deformation of ordered Fe-Al intermetallic alloys. *Wear,* Vol. 162-164, pp. 432-440.

Perrin, C. and Rainforth, W.M. (1995). The effect of alumina fibre reinforcement on the wear of an Al-4.3%Cu alloy. Wear, Vol. 181-183, No. 12, pp. 312-324.

Pollack, H.W. (1977). *Mat. Sci. and Met.,* 2nd edn., Virginia, Prentice Hall.

Polmear, I.J. (1989). *Metallurgy of the Light Metals,* 2nd edn., Edward Arnold. New York.

Rainforth, W.M., Leonard, A.J., Perrin, C., Bedolla-Jacuinde, A.,. Wang, Y., Jones, H., Luo, Q. (2002). High-resolution observations of friction-induced oxide and its Interaction with the Worn Surface. *Tribol. Int.,* Vol. 35, pp. 731–748.

Ravikiran, A., Nagarajan, V.S., and Biswas, S.K., Pramila Bai, B.N. (1995). Effect of. Speed and Pressure on Dry Sliding Interactions of Alumina against Steel. *J. Am. Ceram. Soc.,* Vol. 78, No. 2, pp. 356-364.

Rice S.L., Nowothy, H., and Wayne, S.F. (1981-1982). Characteristics of metallic subsurface zones in sliding and impact wear. *Wear,* Vol. 74, pp. 131 -142

Rigney, D.A., ed. 1981, in Fundamentals of Friction and Wear of Materials: Am. Soc. for Metals, Metals Park, Ohio.

Rigney, D.A., Chen, L.H., Naylor, M.G., and Rosenfield, A. (1984). Wear processes in sliding systems. *Wear,* Vol. 100, pp. 195-219.

Rigney, D.A. (1998). Large Strains Associated with Sliding Contact of. Metals. *Mater. Res. Innovat.,* Vol. 1, pp 231-234.

Rittner, M. (2000). *Metal matrix composites in 21st. century: markets and opportunities,* BCC, Inc., Norwalk, C.T.

Sheu, C.Y. Lin, S.J. (1997). *Journal of materials science,* 32 pp. 1741.

Suh, N.P. (1973). The delamination theory of wear. *Wear,* Vol. 25, pp. 111-124.

Suh N.P., Jahanmir, S., Flemming, J.R., Pamies-Teixeira, J.J., Saka, N. (1977). Overview of the Delamination Theory of Wear. *Wear,* pp. 44, 1-16.

Wang A., Rack H.J. (1991). Abrasive wear of silicon carbide particulate—and whisker-reinforced 7091 aluminum matrix composites. *Wear,* Vol. 146, pp. 337-348.

Zhang, J., Alpas, T. A. (1997). Transition Between Mild and Severe Wear in Aluminium Alloys, *Acta Metall.*, Vol. 45, pp. 513–528.

14

Comparison of Energy-Based and Damage-Related Fatigue Life Models for Aluminium Components Under TMF Loading

Eichlseder Wilfried, Winter Gerhard,
Minichmayr Robert and Riedler Martin
Montanuniversität Leoben
Austria

1. Introduction

Thermo-mechanical fatigue is generally due to a cyclic thermal load in conjunction with restrained thermal expansion. Because of the considerable amplitude of strain this load leads to local cyclic plastic deformation and thus to material fatigue. Usually several concurrent and complex damage mechanisms are involved in thermo-mechanical fatigue due to the temperatures and stresses attained. In addition to typical fatigue damage caused by plastic deformation, elevated temperature leads to an increase in corrosive effects (e.g. oxidation) and creep damage. The areas of application are manifold: besides thermo-mechanical fatigue of combustion engine components (e.g. cylinder heads, pistons, exhaust elbows) such effects are common with tanks used in the chemical industry, pipelines, braking systems, turbine blades, as well as all machine tool components and components subjected to elevated operating temperatures. All these applications show cyclic thermal load which, for example, is caused by start-up and shutdown procedures, as well as a mechanical load caused by either restrained thermal expansion (e.g. cylinder heads) or considerable centrifugal forces (e.g. turbine blades).

While in 1992 the maximum specific power for a diesel passenger car was 35 kW/l, the typical ignition pressure was about 130 bar, resulting in a maximum piston temperature of 330 °C. Owing to demands targeting reduction of costs, emissions and fuel consumption, an increase in efficiency by means of "Downsizing" is called for. This is realised by reducing the cubic capacity as well as the number of cylinders and along with additional charging, resulting in an increase in firing pressures and temperature the combustion chamber. The specific power thus obtained is in the region of 70 kW/l, together with ignition pressures of 200 bar and a maximum piston temperature of more than 400 °C, (Reichstein, 2005).

Modern cylinder head materials are typically produced out of aluminium-cast alloys, of which aluminium-silicon-magnesium (AlSiMg) and aluminium-silicon-copper (AlSiCu) alloys are most common. Aluminium and silicon form a eutectic at 577°C and 12 weight percent. The Al-solid solution, silicon and additional secondary phases have a eutectic solidification. The cooling rate influences the dendrite arm spacing (DAS) and the morphology of the eutectic silicon. A high cooling rate leads to a low DAS and finer secondary microstructure. The hypoeutectic alloys are used for cylinder heads and hypereutectic alloys are found in pistons and crankcases.

2. Similarities and differences between LCF and TMF

By their very nature, cyclic thermal loads appear with relatively low numbers of cycles in the low cycle fatigue (LCF) region so that the application of strain-based concepts (e.g. strain life diagrams etc.) is self-evident. If the loading is large enough to produce plastic strain, the number of cycles to failure is relatively low, in the order of less than 10,000 cycles. This total strain predominantly consists of plastic strain, which dominates the fatigue life. Widely used methods to determine the material behaviour are total strain based fatigue tests, whereby the resulting cyclic stress-strain hystereses are investigated. The resulting cyclic stress-strain curves as well as strain S-N curves are the basis for further lifetime evaluation where, depending on the material behaviour, softening or/and hardening effects can be found.

Depending on the application, further influences like temperature, mean strain, strain rate, atmosphere or aging-conditions must be considered. The components are primarily obtained by casting and defects such as pores, shrink holes or oxide inclusions ensued during this process have a negative influence on the lifetime. While these influences are extensively studied for isothermal conditions (Fagschlunger et al., 2006, Oberwinkler et al., 2010, Powazka et al, 2010), scientific understanding of the same for TMF is very limited. While LCF tests are always conducted under isothermal conditions, TMF tests are additionally loaded by thermal cycles, normally defined by a minimum and maximum temperature, dwell time and heating/cooling rates.

As TMF experiments are both very cost-intensive and time-consuming, it is often attempted in practice to estimate the fatigue life of components under thermo-mechanical load by means of more common isothermal LCF experiments. However, this approach may lead to non-conservative fatigue life estimates if the cyclic stress-strain behaviour or the effective damaging mechanisms under TMF loading differ considerably from the material behaviour under isothermal conditions. Furthermore LCF and TMF test results might not correlate due to differing methods used for recording and interpreting the deformation behaviour. In order to avoid misinterpretations it is crucial to pay close attention to the locally and temporally fluctuating temperature field, in particular when recording the TMF deformation behaviour. Thus a fundamental examination of the stress-strain behaviour and the predominant damage mechanisms under TMF conditions is crucial in order to enable accurate fatigue life predictions under thermo-mechanical fatigue loading. This approach can also clarify to which extent the employment of isothermal data is justified (Riedler et al., 2004; Riedler, 2005).

Differences may result from the fact that under TMF loading, as opposed to isothermal LCF loading, during every cycle a broad temperature range is experienced, in which the material properties can change and the material response may differ. The key to a comparison of LCF and TMF data thus lies in the evolution of the microstructure, whose integral behaviour is reflected in the shape of the stress-strain hysteresis loops.

3. Damage mechanisms in thermo-mechanical fatigue

The phenomena during thermo-mechanical fatigue are influenced by a variety of processes within different temperature ranges during a thermal cycle, where especially at elevated temperatures the mentioned damage mechanisms can occur either individually or in mutual interaction. Thus the predominant damage processes are thermally activated gliding of

dislocations at low temperatures, cyclic ageing at medium temperatures, and diffusion creep at high temperatures. However, both under IP (temperature and stress cycle are in phase) and OP (temperature and stress cycle are out of phase) TMF loading the microstructure evolution and the oxidation processes are more often than not dominated by the temperature range close to the maximum temperature. The maximum temperature occurs in the tensile stress region in case of an IP-TMF load, and in the compressive stress region in case of an OP-TMF load, see figure 1 (according to Löhe et al., 2004) and figure 2.
On the other hand concerning OP-TMF, crack initiation and growth are linked directly with the processes in the temperature range closest to the minimum temperature where tensile stresses prevail, and are only linked indirectly with processes that occur at the maximum temperature. Nevertheless this indirect influence can be even more distinctive than it would be in isothermal experiments. For example a layer of scale might build up as a result of oxidation which takes effect predominantly at high temperatures. This layer of scale is very brittle at low temperatures and thus causes early crack initiation and accelerated crack propagation under an OP-TMF load.

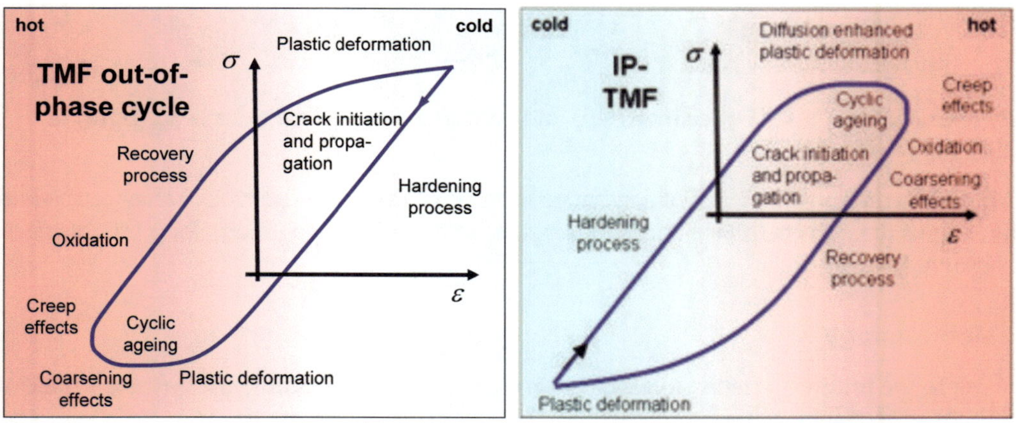

Fig. 1. Active damaging mechanisms during an OP- and IP-TMF cycle (Löhe et al., 2004)

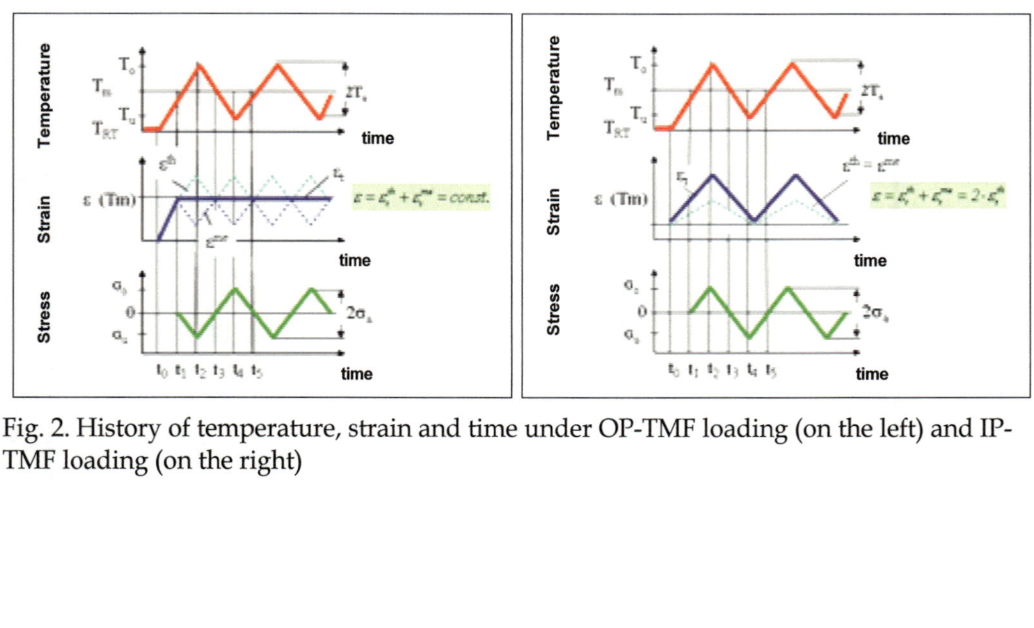

Fig. 2. History of temperature, strain and time under OP-TMF loading (on the left) and IP-TMF loading (on the right)

Crack initiation in various Al-Si-alloys occurs preferentially at the interface between Al-matrix and Si-phase. A meso-scale modelling of the microstructure under thermo-mechanical loading conditions shows stress concentration in the Al-Si-interface (Thalmair, 2009). The repeated recurring thermo-mechanical cycles cause micro-stresses and hence the preferred crack initiation.

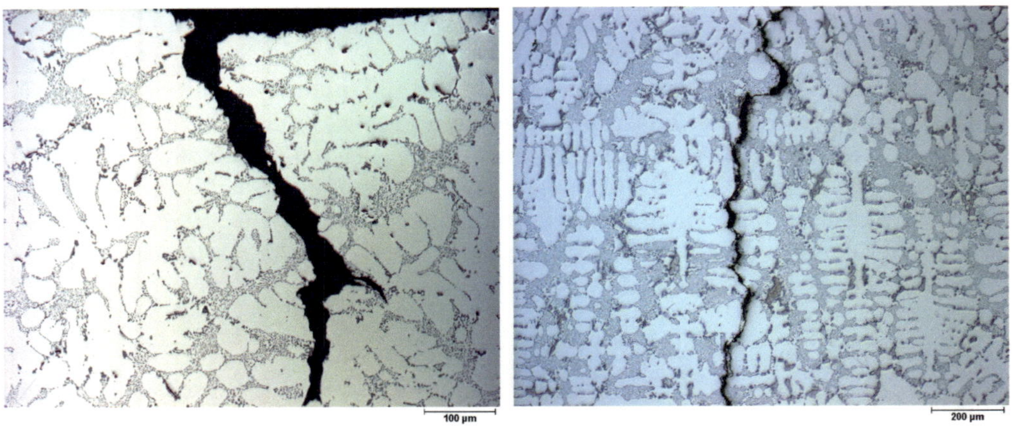

Fig. 3. Crack tip of a TMF-specimen (left) and a thermo-shocked cylinder head (right) of AlSi8Cu3 (Thalmair, 2009)

TMF-test as thermo-shock test of a cylinder head shows very similar crack behaviour, where cracking of the eutectic phase and an inter-dendritic crack-propagation along the interfaces is observed, see figure 3.

4. Methodology

Owing to the multiple effects in components exposed to TMF loading, the developed testing methods are rather varied. Meanwhile, besides LCF experiments and *bi-thermal tests* (Halford et al., 2004) also TMF experiments with combined temperature and strain control are also common. Because of the missing standardisation and the different requirements the experiments differ significantly with respect to heating and cooling temperature application, measurement and control as well as cycle form, strain measurement and consideration of thermal strains. Data from the literature typically allow for little comparability as the information concerning the testing procedures is usually insufficient.

The mechanical loading force application is mostly achieved by means of servo-hydraulic or electro-mechanic testing machines. For this purpose, test equipment manufacturers have started to offer powerful all-in-one systems. However, for special experiments one is still dependent on adaptions or self-made constructions (Riedler & Eichlseder,2004; Minichmayr et al., 2005). Furthermore, it is common to examine specimens with geometries close to the actual component, which only require temperature control. In this case the geometrical constraint is provided for either by the specimen shape or by the mechanical boundary conditions (Simon & Santacreu, 2002; Prillhofer et al., 2005).

It is often tried to simulate the actual behaviour of the component by means of laboratory tests such as for example multiaxial TMF experiments (Otaga & Yamamoto, 2001) or superimposed HCF loading (Minichmayr et al., 2005). By measurement of real components

it is possible to identify actual TMF cycle shapes, which are translated to the test specimen as *industrial cycles* with certain phase shifts between thermal and mechanical strains (Engler-Pinto et al., 1995).
The description of the creep, TMF and LCF testing rigs used for the following experiments as well as a detailed material characterisation can be found in previous papers (Riedler, 2005; Riedler & Eichlseder, 2004; Minichmayr et al., 2005; Minichmayr, 2005).

5. Investigated influences

Firstly, it is important to clarify the governing damage mechanisms that occur in out-of-phase TMF cycles in cylinder heads. Therefore the tests on specimens were specifically designed to take the real circumstances in components as best possible into account - with the aim of using the derived models for lifetime estimations of TMF loaded components. Investigated influences are amongst others (see test matrix in Table 1) mean and local strains, cyclic and constant temperatures, dwell times, pre-aging and aging during service life, HCF-interaction, strain and temperature rates as well as the ratio of mechanical and thermal strain. Further single and multiple step creep tests have been carried out to take into account the stress relaxation phenomena. Additional tests in an argon atmosphere have finally enabled the isolation of the predominating damage mechanism in cylinder heads. All analyses are done in the manner of hysteresis loops, stress-cycle and plastic strain-cycle plots, lifetime diagrams and cyclic deformation behaviour diagrams.

Quasistatic - Creep	LCF	TMF	HCF	Metallographic
Constant temperature	Constant temperature	Maximum temperature	Constant temperature	Fractured surface
Pre-aging	Dwell time	Dwell time	Notch effect	Microstructure
Strain rate	Pre-aging	Pre-aging	Pre-aging	Chem. analysis
Single/multiple step	Mean strain	Mean strain	Dendrite arm spacing	Dendrite arm spacing
Stress relaxation	Strain rate	Temperature rate	Porosity	Porosity
	In lying hole	Rigid clamped - controlled	Mean stress	Precipitations
	Strain amplitude	Strain constraining	Frequency	Grain size
	Argon atmosphere	HCF interaction	Type of loading	Striations
	Incr. step test	Phase shift	Stress amplitude	

Table 1. Test matrix

5.1 Influence of an in lying drilled hole

The aim of this study is to investigate the effects of an in lying drilled hole that is used for an improved quality of the temperature control device, presented in (Riedler & Eichlseder, 2004). The behaviour of the hollow drilled sample is calculated with the finite element method, tested with special LCF test series as well as analyzed by means of fractured surfaces on one wrought and one cast alloy. Whereas the influence on AlCuBiPb is visible, even though, marginally in respect on the lifetime behaviour analyzed with the Manson-Coffin-Basquin (Manson, 1954; Lemaitre & Chaboche, 1985; Basquin, 1910). Approach and the cyclic deformation behaviour analyzed with the Ramberg-Osgood (Ramber & Osgood, 1943) approach, at the Aluminium cast alloy AlSi7MgCu0.5 no difference can be ascertained between the test series of the hollow and solid samples (Riedler & Eichlseder, 2004).

Moreover the analysis by means of fractured surfaces of AlSi7MgCu0.5 of the solid and hollow sample of three LCF strain levels shows assimilable fractured surfaces for each strain level. When decreasing the strain level to lower values, a crack propagation area can be seen beginning at the outside of the specimens. The finite element method shows differences that are of the size of less than one per cent from the maximum axial stress (Minichmayr, 2005).

5.2 Influence of pre-aging

When heat treated aluminium alloys are exposed to elevated or fluctuating higher temperatures in their service life, they show a temperature- and time-dependent aging behaviour which can much decrease the mechanical properties. To investigate these effects on low cycle and thermo-mechanical fatigue, LCF test series at room and higher temperatures, as well as LCF and TMF test series for pre-aged conditions were conducted. Moreover TMF test series with different dwell times at the maximum temperatures were conducted to additionally investigate creep effects.

The first investigation is the separated effect of pre-aging (at an elevated constant temperature) on the deformation and lifetime behaviour by the means of quasi static tests, alternating LCF tests (strain ratio=-1) and temperature-controlled OP-TMF tests (temperature ratio=-1). Figure 4 (left) shows the hysteresis loops for two different total strain levels for non pre-aged and pre-aged specimens at 250°C for 500 hours. At the same LCF strain-level the pre-aged specimens show stress values that are about the half compared to non pre-aged specimens. When investigating the influence of pre-aging on the deformation behaviour by means of tensile tests and LCF tests, at the non pre-aged specimens a high stress hardening tendency can be seen as compared to the tensile test. Pre-aging at 250°C for 500 hours leads to a striking by smaller lifetime in the lower strained LCF region. The deformation behaviour of pre-aged specimens in the manner of stress-cycle or plastic strain-cycle plots shows a nearly straight line without any distinctive hardening or softening, but a markedly higher plastic strain part.

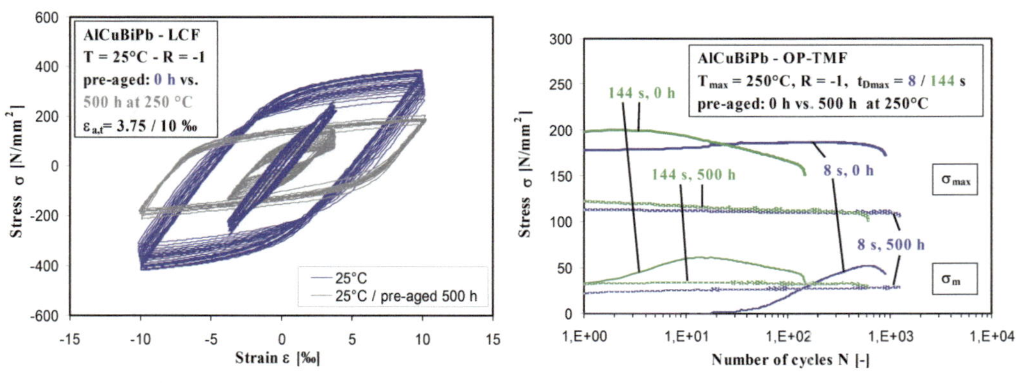

Fig. 4. Influence of pre-aging on the LCF hysteresis loops (on the left) and on the OP-TMF stress-cycle behaviour (on the right)

At the same TMF temperature-level pre-aged specimens at 250°C for 500 hours show the analogous deformation behaviour tendency as obtained at the LCF results, namely a decrease of about 50 per cent compared to non pre-aged specimens, see figure 4 right. The influence of the dwell time decreases with increasing time and temperature of pre-aging.

The differences in lifetime for the dwell time of 8 s and 144 s decreases from a factor more than 6 to a factor of 2, when the specimens are pre-aged for 500 h at 250°C before being tested. After extensive pre-aging the influence of dwell time completely disappears in face of the cyclic deformation and the lifetime behaviour (Riedler & Eichlseder, 2004; Riedler et al., 2005)

5.3 Influence of temperature

A constant elevated temperature influences firstly the quasi static material behaviour and secondly has a time-dependent effect because of hardening vs. softening effects during service life. In this section the time-dependent influence of a constant elevated temperature on LCF is investigated by means of non pre-aged specimens.

A constant elevated temperature of 200°C leads to a higher damage of the material with differences in the lifetime of about one decade compared to the room temperature results. At 250°C the effect is even more drastically, as figure 5 (left) shows. Although at the high strained area the lifetime is a little higher than for 200°C, after the point of intersection at about 100 cycles there is a tremendous drop in the lifetime. Figure 5 (right) shows the summarized presentation of the influences of pre-aging, constant elevated temperature and applied mean strain on the LCF deformation behaviour by means of the plastic strain amplitude part. At a constant temperature of 200°C the stress softening phase starts after a few cycles, what can be seen in an increase of the plastic strain part.

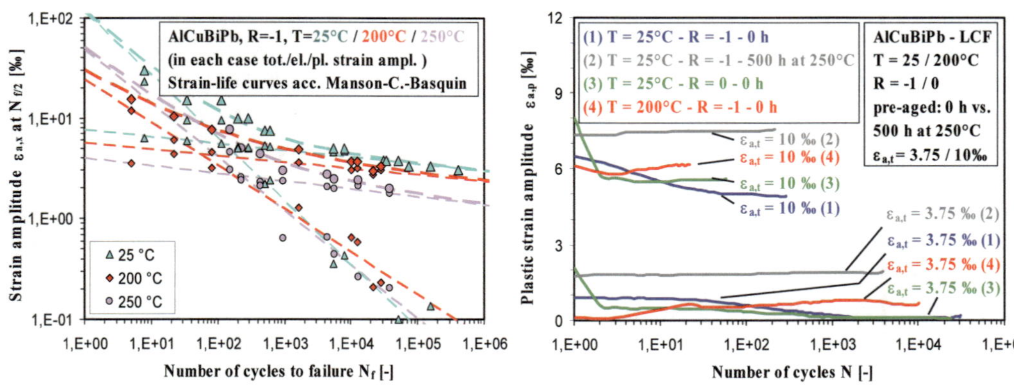

Fig. 5. Influence of elevated temperature on the LCF lifetime (on the left) and influence of pre-aging, elevated temperature and mean strain on the LCF deformation behaviour (on the right)

5.4 Influence of mean strain

Impressed mean strains in the manner of pulsating LCF tests (strain ratio=0) show a visible decrease in lifetime for higher strain levels and only a slight lifetime decreasing effect at lower strain levels. At latest from the half of the number of cycles to failure $N_{f/2}$ on, the cyclic stress deformation curves follow a common progression. Some slight lifetime-decreasing effects ascertainable at pulsating executed test series mostly result from the first few cycles, where the higher tensile stresses and plastic strains (see figure 5 (right)) cause higher damage rates. The comparison of alternating and pulsating executed TMF test series shows the same tendencies as at the LCF results. At latest from $N_{f/2}$ on, the cyclic stress deformation curves follow a common progression and only some slight lifetime-decreasing effects are ascertainable at pulsating executed TMF test series.

5.5 Influence of dwell-time

The typically start-stop-operation of a motor vehicle as well as the alternating fired and non-fired operation causes dwell times. To study this effect out-of-phase TMF tests with four different dwell times at the particular maximum temperatures were conducted (8, 24, 144 and 864 s).

Whereas with the alloy AlCuBiPb, a higher dwell time always causes a lifetime decreasing effect. This is not the case with the alloy AlSi7MgCu0.5, as figure 6 (left) shows, where the strain values are scaled between the minimum and maximum in this range. The TMF strain-life curve for the dwell time of t_{D3}=144 s is quite steep for higher strain values (and therefore temperatures). A point of intersection of the curves for the lower dwell times (8 s and 24 s) is visible at about 1000 cycles. This phenomenon is explained with pronounced softening effects in the first few cycles that occur due to the high aging tendency at the high dwell time and temperature level. The capacious over-aging at this level can mainly be seen in the highly plastic parts, what results in an upward movement of the total strain-life curve. For that reason the high over-aging at the dwell time of 144 s shows that a high mechanical strain amplitude can be endured for a longer time compared to the smaller dwell times. If the maximum temperatures are low, this effect turns around at the dimensioning level for cylinder heads (about 5000 cycles) and a lifetime decreasing effect is visible with a higher dwell time.

Two extreme LCF tests were conducted at the mean value for the TMF maximum temperatures. A LCF test series at a constant higher temperature of 250°C and one tested at room temperature, but pre-aged at 250°C for 500 hours. Figure 6 (right) shows that these two LCF strain-life curves span the TMF range for all dwell times of the materials investigated, if the mechanical strain (and not the thermal strain) is considered. The comparison of the LCF hysteresis loops at 250°C with comparable mechanically strained TMF hysteresis loops for all dwell times also shows a good accordance. Moreover the cyclic deformation behaviour according to Ramberg-Osgood shows the best accordance of the LCF-250°C curve with the TMF curves. This investigation shows that the macroscopic behaviour is comparable if the aging status is similar (Riedler et al., 2004).

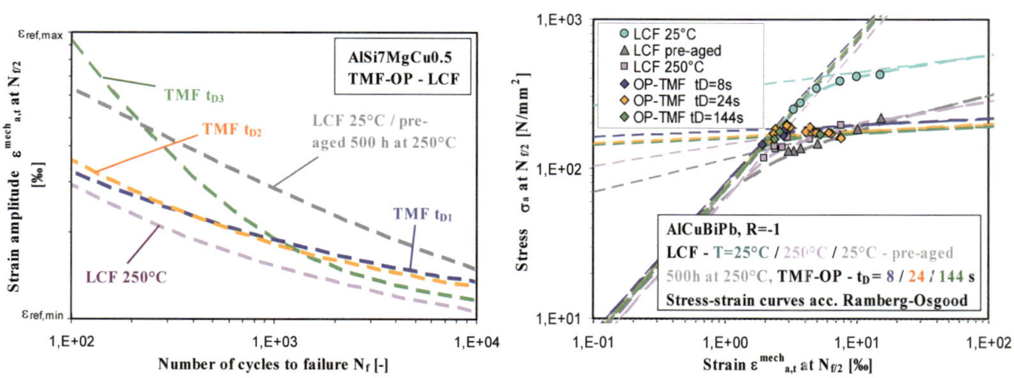

Fig. 6. Influence of TMF dwell-time and LCF-pre-aging and constant elevated temperature on the lifetime (on the left) and on the cyclic deformation behaviour (on the right)

5.6 Influence of phase shifts / strain compensation

In order to investigate the influence of stiffness and phase shifts, different TMF conditions were tested. Besides the ideal OP-TMF situation ($\varepsilon_{t,mech}$=-ε_{th}, K_{TM}=1.0), two overcompensated conditions of the thermal strain (K_{TM}=1.5 and 2.0), a 75% compliance (K_{TM}=0.75, which is near to the real circumstances in cylinder heads) and an ideal in-phase-TMF situation (K_{TM}=-1.0) were tested.

When the local strains are taken into account at rigid clamped specimens, all OP-TMF results can be drawn together in a common strain vs. cycles to failure diagram. Because of creep damage at higher temperatures the IP-TMF lifetime is shorter than the OP-TMF lifetime, as figure 10 shows.

5.7 Influence of strain rates / argon atmosphere

The systematic investigations show that the influence of strain rate on the deformation behaviour is negligible within practical range of temperature rates, but the time and temperature dependent aging behaviour is very important. Unlike the deformation behaviour, the strain rate shows an important influence on the lifetime behaviour due to the additional creep damage involved. These differences and the differences at IP/OP-TMF in the number of cycles to failure allow the separation and investigation of different damage mechanisms. Additional tests in argon atmosphere were executed for this aim, which show a lifetime increasing effect, because of oxidation damage being minimized.

5.8 Influence of HCF interaction

Superimposed HCF-loading has an important influence on the fatigue life of TMF-loaded components. Experiments with superimposed HCF strain amplitudes from 0.01% to 0.1% show a significant decrease in fatigue life depending on the amplitude of the HCF-loads, wherein a small influence of HCF-frequency is found. Metallographic investigations show crack propagation due to HCF-loading and TMF-loading, wherein a combination of HCF-amplitude and the shift in mean-stress causes crack propagation. Good correlation between striations (~crack propagation) and strain amplitude for different loadings is found.

HCF-loadings in a combustion engine occur especially during the heating of the component with maximum ignition pressure. Therefore additional experiments were conducted with superimposed HCF-loadings only during heating period and dwell time. In this case most of the HCF-cycles appear within the compression region of the hysteresis loop. Therefore the effect is obviously reduced. With regard to the typical ignition pressure in a diesel engine, the influence is small compared to the tests without superimposition; see figure 7 (Minichmayr et al., 2005).

5.9 Influence of creep

Because creep effects have to be considered in thermo-mechanical loaded components to take into account stress relaxation phenomena and creep damage, single and multiple step creep tests were carried out. Due to aging effects the decreasing strain rate of the primary creep stage of the single step tests directly merges into the stage of tertiary creep with material softening and therefore increasing creep strain rates. Multiple step tests show, that neither strain nor time are suitable to describe the creep behaviour for that case. Therefore different tests with varying pre-exposure times at test temperature were conducted. The pre-aging time was chosen according to the total time in the multiple step tests. It can be seen, that the minimum creep strain rate of the single step test at 150 MPa is more than 300 times lower compared to the minimum creep strain rate in multiple step test and single step test with pre-aged specimen at the same stress level. Furthermore the test data of the multiple step test and test with pre-aged specimens show a very similar behaviour. Therefore the time at test temperature determines the minimum strain rate independent of strain (Minichmayr et al., 2005).

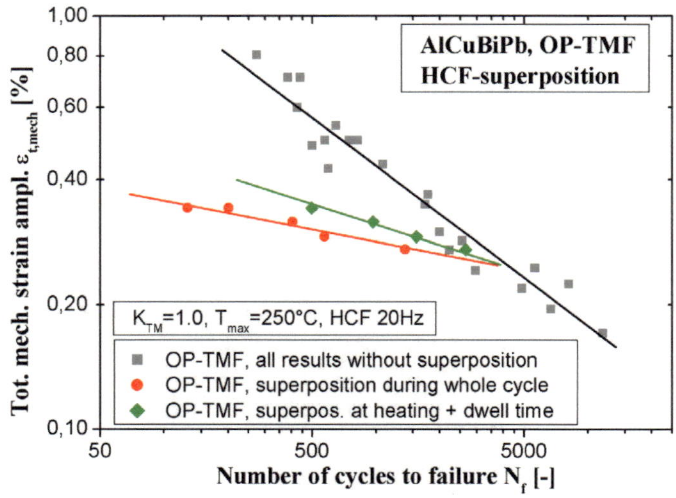

Fig. 7. Influence of HCF interaction on the OP-TMF lifetime

6. Simulation of the cyclic deformation behaviour

6.1 Basics and classification

The description of the elastoplastic deformation behaviour of the construction material forms the basis for assessing the fatigue life of complex components. By means of the finite element method and proper material models it is possible to calculate the local loads (e.g. stress, strain, etc.) under the assumption of adequate boundary conditions.

The most basic material model describes an isotropic plastic hardening independent from the direction of loading. The expansion of the yield surface, which is determined by the *drag stress K* (defined size of the yield surface delimiting the elastic region), can be defined, e.g., in tabular form as a function of the plastic strain. Under cyclic loading each cycle in an isotropic material model leads to further hardening until the maximum strength is obtained and the model shows only elastic, ideal-plastic behaviour. Therefore the isotropic hardening model is adequate only for unidirectional loading. Many materials display the so-called Bauschinger effect under reversed loading (Bauschinger, 1881). The said effect means that plastic deformation occurs already at a significantly lower stress when the load is reversed. The cause of this effect is the formation of dislocation structures, which facilitate plastification in the opposite direction. The Bauschinger effect and the cyclic deformation behaviour, respectively, can only be described by consideration of kinematic hardening, thus using the *back stress* α (which defines the shift of the yield surface in the three-dimensional stress space). In addition high temperatures cause a dependency of stress on the loading rate, which is due to time-dependent processes such as creep. The partitioning

of time-independent plastic deformation and time-dependent creep effects for the deformation behaviour at elevated temperatures is already known from Manson (Manson et al., 1971). At the same time this forms the basis of the *strain range partitioning* concept.
A literature review yields a great number of material models which are able to describe the material behaviour for certain kinds of loading. According to (Christ, 1991) they may be classified according to the underlying approach as follows: empirical models, continuum-mechanical models, physically based models and multi-component models.

6.2 Using the ABAQUS Combinend Hardening Model

The cyclic deformation behaviour is described by the ABAQUS® *Combined Hardening Model* dependent on temperature and ageing. The corresponding variables are the temperature-corrected time and the current temperature. The necessary parameters can be correlated with the $R_{p0,2}$ yield limit from the Shercliff-Ashby model (Shercliff & Ashby, 1990). For the calculation of different states of ageing a user subroutine has been developed. On the one hand it allows for an accumulation of the temperature-corrected time on the basis of the temperature-time curve for separate calculation steps, and on the other hand it is possible to calculate directly the state of ageing for a certain ageing time according to the local maximum temperature.
The calculated hystereses and stress-time curves conform very well to the measured experimental data (fig. 8). Under TMF load both the asymmetry of the stress-strain hystereses and the stress relaxation in the dwell time region are expressed correctly.

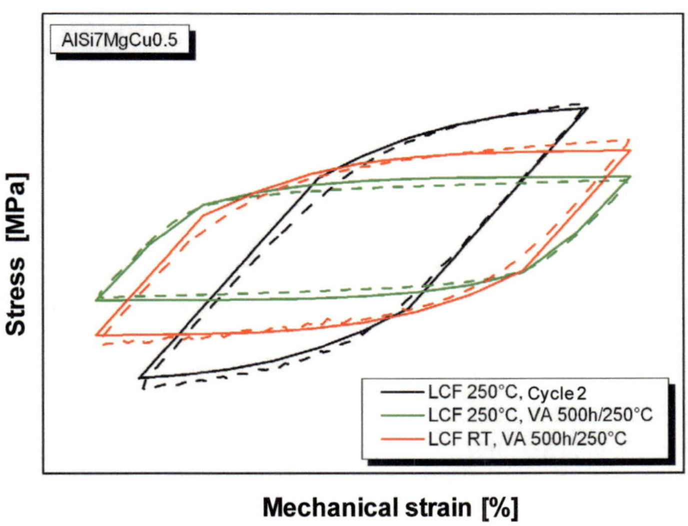

Fig. 8. Comparison of calculated and measured hystereses

7. Simulation of fatigue life behaviour

7.1 Basics and classification

The models available for describing the complex phenomena of thermo-mechanical fatigue range from engineering approaches to physically based models, thereby characterising the combined loading in differing complexity.

Most models for TMF fatigue life calculation are based on linear damage accumulation according to Palmgren-Miner (Palmgren, 1924; Miner, 1945).
Nonlinear cumulative approaches mainly come from Chaboche and Lemaitre (Chaboche & Lesne, 1988; Lemaitre & Chaboche, 1985).
The methods of calculation are mostly based upon local loading parameters such as stress, strain and temperature, which are calculated in complex components by means of elaborate material models. Thus the classification of models describing the fatigue life behaviour under TMF and LCF loading shows the same subdivision as the models describing the cyclic deformation behaviour and shall be briefly explained in the following.

Empirical models

Empirical models represent the major group of this classification. With these models, usually the fatigue life and the parameters of the load cycles are linked. These models are simple, however the partly lacking physical interpretability is a disadvantage.
Furthermore they are effective only for a quite narrow load spectrum as there is no distinction between the individual effects. The empirical models can be set up for various levels of complexity and are divided into approaches based on strain life curves and damage parameters, methods describing creep damage, energy-based approaches, and approaches for partial damage accumulation.

Damage mechanics models

Damage mechanics models usually describe the development of damage by means of methods from continuum mechanics. The origin of these models is found with Kachanov (Kachanov, 1986) and Rabotnov (Rabotnov, 1969) who were concerned with creep damage. Damage mechanics approaches see damage being caused by creep and plastification. As the damage rates are linked to the current damage value, damage accumulation is nonlinear and requires the damage variable to be integrated cycle by cycle.

Physically based models

In general the physically based models play a minor role for practical application, which is due to their complexity and the difficulties in determining their input parameters experimentally. They attempt to characterise the damage development on the basis of atom, vacancy and dislocation movement. At the current state of knowledge and application, physically based models for fatigue life computations are primarily relevant for depicting the physical background of empirical and damage mechanics methods, respectively.

Fracture mechanics models

Fracture mechanics models are linked to the local plastic strains at the crack tip, which can be described, for example, by a ΔJ integral or a modified ΔJ integral, respectively. A link to the physically based models exists with models for micro-crack propagation.

7.2 Application of energetic approaches

An advantage of energy criteria is the significant reduction of parameters in comparison to total strain life curve approaches according to Manson-Coffin-Basquin (Manson, 1954; Coffin, 1954; Basquin, 1910), as energy criteria are able to describe several influences due to the interaction of stress and strain variables. Energy criteria are representative for the cyclic behaviour of materials. They are sound damage indicators which are linked to macroscopic crack initiation and allow for a generalisation to multiaxial loading. The input parameters have to be known, i.e., the local load parameters in consideration of the cyclic deformation behaviour have to be determined in advance. Especially for aluminium alloys the combination of stress and strain variables yields an adequate parameter as the cyclic deformation behaviour is dominantly affected by the ageing effect. If the temperature dependent ageing reduces the stress, the plastic strain increases in a similar way.

Whilst a single-parameter plastic energy approach is an adequate criterion for the highly ductile aluminium alloy AlCuBiPb, for ductile gravity die casting alloys (AlSi7MgCu0.5 and AlSi8Cu3) it is a total strain based energy criterion. For brittle materials such as AlSi6Cu4 lost foam, a fracture mechanics based energy criterion with a cyclic J integral provides the smallest standard deviation (Riedler et al., 2005).

The so-called *unified energy approach* is derived based on this knowledge of material-dependent TMF fatigue life criteria.

$$\Delta W_u = c_u \cdot \Delta W_{u,e} + \Delta W_{u,p} = c_u(\sigma_o \cdot \varepsilon_{a,e}) + (\sigma_a \cdot \varepsilon_{a,p}) \quad (1)$$

$$N_B = A_u \cdot \Delta W_u^{-B_u} \quad (2)$$

The specific hysteresis energy for a representative cycle consists of an elastic and a plastic portion. Whereas the elastic portion is formed by the maximum stress and the elastic strain amplitude, the plastic portion is formed by the amplitude values of stress and plastic strain. The material parameter c_u takes values near 1. The fatigue life is determined by a power law approach according to (2). The quality of computing OP-TMF fatigue life for the examined cylinder head alloy made of aluminium and cast iron by means of the *unified energy approach* can be seen in figure 9. It shows 95% of the data points of the six examined alloys influenced by maximum temperature, average and local strain, pre-ageing as well as ageing in operation lying in a fatigue life scatter band of 2.5, and two-thirds of the data points lying in a scatter band of 1.6.

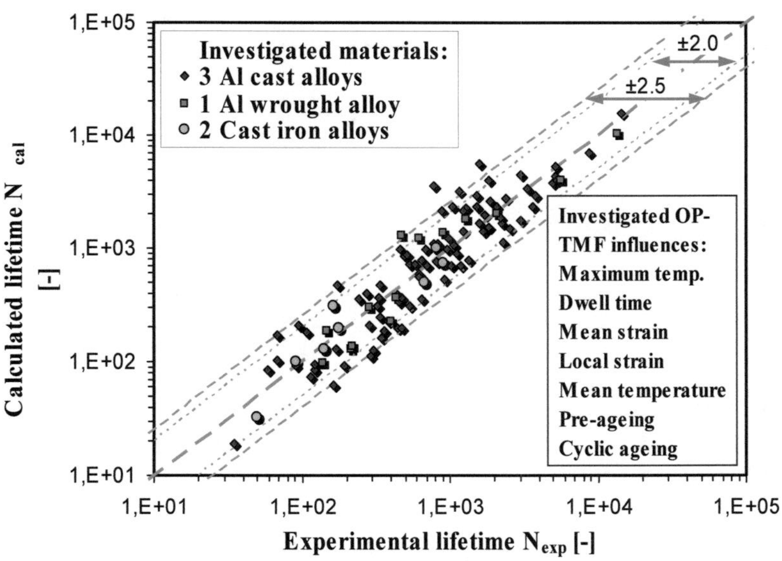

Fig. 9. Quality of the TMF fatigue life simulation by means of the *unified energy approach*

7.3 Application of the damage rate model according to Sehitoglu

The model of Neu-Sehitoglu (Neu & Sehitoglu, 1989) is based on the assumption that overall damage is caused by fatigue, oxidation, and creep:

$$D^{total} = D^{fat} + D^{ox} + D^{creep} \tag{3}$$

or expressed as an equation using the cycles to fracture:

$$\frac{1}{N_B^{total}} = \frac{1}{N_B^{fat}} + \frac{1}{N_B^{ox}} + \frac{1}{N_B^{creep}} \tag{4}$$

The pure fatigue damage portion is described by means of the Manson-Coffin-Basquin approach with the mechanical strain range $\Delta\varepsilon_{mech}$:

$$\frac{\Delta\varepsilon^{mech}}{2} = \frac{\sigma_f^{'}}{E}\left(N_B^{fat}\right)^b + \varepsilon_f^{'}\left(N_B^{fat}\right)^c \tag{5}$$

The parameters E, σ_f', b, ε_f' and c are determined from isothermal fatigue experiments at room temperature. Thereby it is assumed that all experiments at elevated temperature show a similar or shorter fatigue life than at room temperature and furthermore that in these cases the fatigue life reduction is due to the oxidation and creep damage portions.

The oxidation damage portion describes the repeated formation and destruction of an oxide layer at the crack tip as a function of mechanical strain rate, mechanical strain amplitude, temperature and phasing between mechanical strain and temperature:

$$\frac{1}{N_B^{ox}} = \left[\frac{h_{cr}\delta_0}{B\Phi^{ox}K_p^{eff}}\right]^{-\frac{1}{\beta}} \frac{2(\Delta\varepsilon^{mech})^{(2/\beta+1)}}{\dot{\varepsilon}^{1-(\alpha/\beta)}} \tag{6}$$

The temperature dependency of the oxidation is described by means of an Arrhenius approach. The effective oxidation constant is obtained by integration over a complete cycle:

$$K_p^{eff} = \frac{1}{t_C}\int_0^{t_C} D_0 \exp\left(-\frac{Q}{RT(t)}\right)dt \tag{7}$$

The phase factor takes into account that the oxidation damage portion of an OP-TMF load is higher than that of an IP-TMF load:

$$\Phi^{ox} = \frac{1}{t_C}\int_0^{t_C} \phi^{ox} dt \ \text{mit}\ \phi^{ox} = \exp\left[-\frac{1}{2}\left(\frac{(\dot{\varepsilon}^{th} / \dot{\varepsilon}^{mech} + 1}{\xi^{ox}}\right)^2\right] \tag{8}$$

The creep damage portion describes the damage due to pore and intergranular crack formation. The creep damage portion is defined as a function of temperature, equivalent stress, hydrostatic stress and *drag stress*:

$$D^{creep} = \Phi^{creep}\int_0^{t_C} A\exp\left(-\frac{\Delta H}{RT(t)}\right)\cdot\left(\frac{\alpha_1\bar{\sigma} + \alpha_2\sigma_H}{K}\right)^m dt \tag{9}$$

It is made use of the internal variables of the material models according to Slavik-Sehitoglu (Slavik & Sehitoglu, 1987). Creep damage is highest under IP-TMF loading, if the maximum temperature coincides with tensile stress. Isothermal LCF experiments show low creep damage, and for OP-TMF loading it is almost zero. Thus another phase factor is introduced:

$$\Phi^{creep} = \frac{1}{t_C}\int_0^{t_C} \phi^{creep} dt \text{ mit } \phi^{creep} = \exp\left[-\frac{1}{2}\left(\frac{(\dot{\varepsilon}^{th} / \dot{\varepsilon}^{mech} - 1)}{\xi^{creep}}\right)^2\right] \quad (10)$$

In (Neu & Sehitoglu, 1989) the parameters are determined by means of elaborate experiments. In doing so, also experiments for measuring the oxide layer growth at different temperatures are conducted. Furthermore the growth of the oxide layer under repeated break-up is also measured.
The parameter adjustment in (Minichmayr, 2005)] is carried out solely by manual parameter variation and automatic parameter optimisation. In addition to the OP-TMF experiments (classical cylinder head applications, basis of energetic approaches) also LCF experiments with different strain rates, LCF experiments in argon atmosphere as well as in-phase TMF experiments were necessary. By means of non-linear parameter optimisation the corresponding model parameters were determined.
The fatigue life computation by means of the Sehitoglu damage model is slightly more accurate in comparison to the energy criteria; 90% of the data points for the cast alloy AlSi7MgCu0.5 lie within a scatter band of 1.85. The biggest advantage results from several damage mechanisms being active at the same time. Likewise it is possible, for example, to predict the in-phase TMF experiments correctly, which are characterised by a dominant effect of creep damage, see fig. 10 and 11.

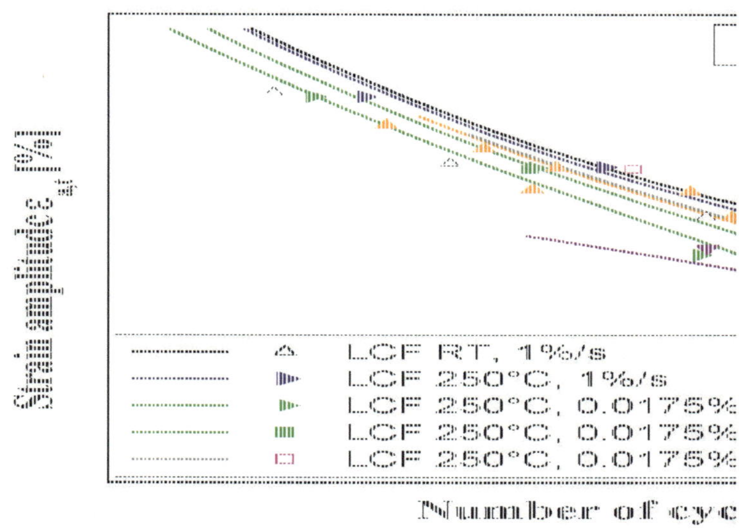

Fig. 10. Fatigue life computing under LCF, OP-TMF, and IP-TMF loading according to the Sehitoglu damage model

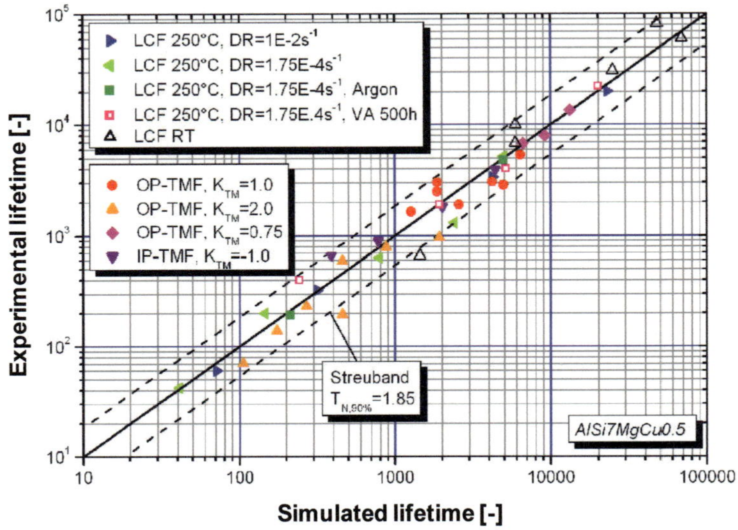

Fig. 11. Quality of the TMF fatigue life calculation using the Sehitoglu damage model

8. Comparative consideration and application to components

Both energy criteria and the model according to Neu-Sehitoglu were used for fatigue life modelling. In contrast to empirical approaches this model distinguishes between different damage mechanisms. The adequate method is chosen according to the type of application; in doing so, it is important to indicate the model limits. Energy criteria seem to represent the best compromise between accuracy and complexity in their application. Major differences occur if the damage mechanisms involved are changing.

Practical application to cylinder heads shows that the fatigue life, calculated on the basis of the dissipated plastic energy, largely depends on the chosen state of ageing. The solution to this problem is an ageing-dependent cumulative damage model.

The commercial fatigue lifetime prediction software FEMFAT (FEMFAT Manual, 2005) features, since version 6.5, a module for calculating the damage under thermo-mechanical loading according to the damage rate model by Neu-Sehitoglu.

With this model it is possible to calculate the local damage portions caused by fatigue, oxidation and creep. The calculation is based on shear strains, which are determined by a critical plane method and are therefore also applicable for multiaxial loading. The predominant part of the overall damage is caused by pure fatigue. The portion of oxidation damage amounts to some 10% in the regions of maximum loading. The fatigue life computation by means of FEMFAT-Sehitoglu provides realistic results concerning the critical areas and fatigue lives.

9. Conclusion

TMF energy criteria is a suitable tool for TMF lifetime assessment of aluminium, provided the limitations of the application are known. They are representative for the cyclic material behaviour and good damage indicators, since they are associated with the macroscopic crack initiation. The damage rate model of Sehitoglu is powerful to describe more influences, albeit with the major disadvantage being the need of an extensive data basis for every specific material. Depending on the application, one specific lifetime calculation method should be preferred.

10. References

Reichstein, S., Hofmann, L. & Kenningley, S. (2005). Entwicklung von Kolbenwerkstoffen für moderne Hochleistungsdieselmotoren, *Giesserei-Praxis*, pp. 380-384, No. 10, Schiele & Schön, Berlin

Fagschlunger, C., Pötter, K., & Eichlseder, W. (2006). Abschätzung der Schwingfestigkeit von porenfreien Randschichten in Al-Gussbauteilen. *MP Materialprüfung*, pp. 142-151, Vol. 48, No. 4, Hanser, München

Oberwinkler, C., Leitner, H., Eichlseder W., Schönfeld, F. & Schmidt, S. (2010). Schädigungstolerante Auslegung von Aluminium-Druckgusskomponenten, *MP Materials Testing*, pp. 513-519, Vol. 52, No. 7-8, Hanser, München

Powazka, D., Leitner, H., Brune, M., Eichlseder, W. & Oppermann, H. (2010). Fertigungsbedingte Einflüsse auf die Schwingfestigkeit von Al-Gussbauteilen, *Giesserei*, pp. 34-42, Vol. 97, No. 7, Gießerei-Verlag, Düsseldorf

Riedler, M.; Eichlseder, W. & Minichmayr, R. (2004). Relationship between LCF and TMF: Similiarities and Varities, *12th International Conference on Experimental Mechanics, ICEM12*, Paper No. 102, Bari

Riedler, M. (2005). TMF von Aluminiumlegierungen - Methodikfindung zur Simulation von thermomechanisch beanspruchten Motorbauteilen aus Aluminiumlegierungen, *Fortschritt-Berichte VDI*, Reihe 5, ISBN 3-18-371805-7

Löhe, D., Beck, T. & Lang, K.-H. (2004). Important aspects of cyclic deformation, damage and lifetime behaviour in thermomechanical fatigue of engineering alloys, pp. 161-175, *Fifth International Conference on Low Cycle Fatigue*, Eds.: Portella, P.D., Sehitoglu, H., Hatanaka, K., DVM, 2004, Berlin

Thalmair, S. (2009). Thermomechanische Ermüdung von Aluminium-Silizium-Gusslegierungen unter ottomotorischen Beanspruchungen, *Dissertation* Univ. Karlsruhe

Halford, G.R., McGaw, M.A.; Bill, R.C. & Fanti, P.D. (1988). Bithermal Fatigue: A Link between Isothermal and Thermomechanical Fatigue, Low Cycle Fatigue pp. 625-637, *ASTM STP 942*, Eds.: Solomon et al., American Society for Testing and Materials, Philadelphia

Riedler, M. & Eichlseder, W. (2004) Temperature control method in elevated and fluctuating temperature fatigue tests, *Materials Engineering*, pp.1-7, Vol. 11, 2004 No. 3, ISSN 1335-0803

Minichmayr, R., Riedler, M. & Eichlseder, W. (2005). Thermomechanische Ermüdung von Aluminiumlegierungen - Versuchstechnik und Methoden der Lebensdaueranalyse, pp. 591-600, *MP Materialprüfung*, Vol. 47, No. 10, Hanser, München

Simon, C. & Santacreu, P.O. (2000). Life Time Prediction of Exhaust Manifolds, pp. 257-267, *Proc. CAMP2002 - High-Temperature Fatigue*, Eds.: Biallas, G., Maier, H.J., Hahn, O., Herrmann, K., Vollertsen, F., Paderborn

Prillhofer, B., Riedler, M. &Eichlseder, W. (2005) Übertragbarkeit von Versuchsergebnissen an Rundproben auf thermomechanisch beanspruchte Bauteile, *1. Leobener Betriebsfestigkeitstage*, Planneralm,

Ogata, T. & Yamamoto, M. (2001). Life Evaluation of IN738LC under Biaxial Thermo-Mechanical Fatigue, pp. 839-847, *Sixth International Conference on Biaxial/Multiaxial Fatigue & Fracture*, Lisboa, Portugal

Engler-Pinto, C.C. Jr., Meyer-Olbersleben, F. & Rézai-Aria, F. (1995). Thermo-Mechanical Fatigue Behaviour of SRR99, Fatigue under Thermal and Mechanical Loading: Mechanisms, pp. 151-157, *Mechanics and Modelling, Kluwer Academic Publishers*, Petten, Eds.: Bressers, J., Rémy, L., Steen, M., Vallés, J.L.

Minichmayr, R. (2005). Modellierung und Simulation des thermomechanischen Ermüdungsverhaltens von Aluminiumbauteilen, Dissertation, Montanuniversität Leoben

Riedler, M. & Eichlseder, W. (2004) Effects of dwell times on thermo-mechanical fatigue, *Zeitschrift Materialprüfung*, Jahrg. 46 11-12, Carl Hanser Verlag, München, S. 577-581.

Bauschinger, J. (1886). Mitt. mech.-techn. Lab., pp. 289, TH München 13, 1886, Zivil-Ing. 27, 1881

Manson, S.S., Halford, G.R. & Hirschberg, M.H. (1971). Creep-Fatigue Analysis by Strain-Range Partitioning, Design for Elevated Temperature Environment, pp. 12-24, Ed.: Zamrik, S.Y., *ASME*, New York

Christ, H.-J. (1991). *Wechselverformung von Metallen*, Springer-Verlag, Berlin

Shercliff, H. R. & Ashby, M. F. (1990). Process Model for Age Hardening of Aluminium Alloys - I. The Model; *Acta metall. Mater*. Vol. 38, No. 10, 1789-1802, 1990.

Palmgren, A. (1924) Die Lebensdauer von Kugellagern, pp. 339-341 *VDI-Zeitung 58*

Miner, M.A. (1945) Cumulative damage in fatigue, Trans. *ASME Journal of Applied Mechanics*, 12-3, pp. A159-A164.

Chaboche, J.L. & Lesne, P.M. (1988). A Non-Linear Continuous Fatigue Damage Model, pp. 1-17 , *Fat. Fract. Engng. Mater. Struct.*, 11

Lemaitre, J. & Chaboche, J.L. (1985). Mecanique des Materieaux Solides, Dunod,

Kachanov, L.M. On Creep Rupture Time, Izv. Akad. Nauk. SSR, Otd Tekh. Nauk, No. 8, pp. 26-31.

Kachanov, L.M. (1986). Introduction to Continuum Damage Mechanics, Martinus Nijhoff Publ., Dodrecht, Holland

Rabotnov, Y.N. (1969). Creep Problems in Structural Members, North Holland Publishing, Amsterdam

Manson, S.S. (1954). Behaviour of materials under conditions of thermal stress, *NACA Report* No. 1170

Coffin, L.F. (1954). A study of the effects of cyclic thermal stresses on a ductile metal, *Trans. ASME 76*, pp. 931-950.

Basquin, O.H. (1910). The exponential Law of Endurance Tests, *Proceedings of the ASTM 10*, pp. 625-630.

Ramberg, W. & Osgood, W.R. (1943). In: *NACA Technical Note* 902.

Riedler, M., Minichmayr, R. & Eichlseder, W. (2005). Methods for the thermo-mechanical fatigue simulation based on energy criterions, pp. 496-503, *6th International Conference of Assessment of reliability of materials and structures: problems and solutions, RELMAS 2005*, St. Petersburg

Neu, R.W.,6 Sehitoglu, H. (1989). Thermo-mechanical Fatigue, Oxidation and Creep, Part II: Life Prediction, *Metal Transactions*, pp. 1769-1783, Vol. 20A

Slavik, D. & Sehitoglu, H. (1987). A Constitutive Model for High Temperature Loading, Part I and II, in Thermal Stress, *Material Deformation and Thermomechanical Fatigue*, Eds.: Sehitoglu, H., Zamrik, S.Y., ASME PVP 123, New York, 1987, pp. 65-82.

FEMFAT: Finite Elemente Methode und Betriebsfestigkeit, *Manual zur Software FEMFAT*, ECS Steyr, 2005.

15

Deformation Characteristics of Aluminium Composites for Structural Applications

Theodore E. Matikas[1] and Syed T. Hasan[2]
[1]*Department of Materials Science and Engineering, University of Ioannina,*
[2]*Faculty of Arts, Computing, Engineering and Sciences, Sheffield Hallam University,*
[1]*Greece*
[2]*United Kingdom*

1. Introduction

Silicon carbide (SiC) particulate-reinforced aluminium matrix composites (AMC) are attractive engineering materials for a variety of structural applications, due to their superior strength, stiffness, low cycle fatigue and corrosion fatigue behaviour, creep and wear resistance, compared to the aluminium monolithic alloys. An important feature of the microstructure in the Al/SiC composite system is the increased amount of thermal residual stresses, compared to unreinforced alloys, which are developed due to mismatch in thermal expansion coefficients of matrix and reinforcement phases. The introduction of the reinforcement plays a key role in both the mechanical and thermal ageing behaviour of the composite material. Micro-compositional changes which occur during the thermo-mechanical forming process of these materials can cause substantial changes in mechanical properties, such as ductility, fracture toughness and stress corrosion resistance.

The satisfactory performance of aluminium matrix composites depends critically on their integrity, the heart of which is the quality of the matrix/particle reinforcement interface. The nature of the interface depends in turn on the processing of the AMC component. At the micro-level, the development of local concentration gradients around the reinforcement can be very different to the nominal conditions. The latter is due to the aluminium alloy matrix attempt to deform during processing. This plays a crucial role in the micro-structural events of segregation and precipitation at the matrix-reinforcement interface.

The strength of particulate-reinforced composites also depends on the size of the particles, interparticle spacing, and the volume fraction of the reinforcement [1]. The microstructure and mechanical properties of these materials can be altered by thermo-mechanical treatment as well as by varying the reinforcement volume fraction. The strengthening of monolithic metallic material is carried out by alloying and supersaturating, to an extent, that on suitable heat treatment the excess alloying additions precipitates out (ageing). To study the deformation behaviour of precipitate hardened alloy or particulate reinforced metal matrix composites the interaction of dislocation with the reinforcing particles is much more dependent on the particle size, spacing and density than on the composition [2]. Furthermore, when a particle is introduced in a matrix, an additional barrier to the movement of dislocation is created and the dislocation must behave either by cutting through the particles or by taking a path around the obstacles [3].

At present, the relationship between the strength properties of metal matrix composites and the details of the thermo-mechanical forming processes is not well understood. The kinetics of precipitation in the solid state has been the subject of much attention. Early work on growth kinetics has been developed for the grain boundary case [4] and for intragranular precipitation [5]. These approaches have been integrated to produce a unified description of the inter- and intra-granular nucleation and growth mechanisms [6, 7]. More recently, successful attempts have been made to combine models of precipitate growth at interfaces with concurrently occurring segregation in aluminium alloys [8]. Studies of the relation between interfacial cohesive strength and structure have only recently become possible. This is due to of remarkable advances in physical examination techniques allowing direct viewing of interface structure and improved theoretical treatments of grain boundary structure.

The ability of the strengthening precipitates to support the matrix relies on the properties of the major alloying additions involved in the formation of these precipitates. The development of precipitates in Al-based alloys can be well characterised through heat treatment processing. Heat treatment affects the matrix properties and consequently the strain hardening of the composite. Furthermore, the distribution and concentration of these precipitates greatly affect the properties of the material where homogenous distribution of small precipitates provides the optimum results.

The role of the reinforcement is crucial in the microdeformation behaviour. The addition of SiC to aluminium alloy increases the strength and results in high internal stresses, in addition to the ones caused by the strengthening precipitates. Furthermore, the SiC reinforced particles are not affected by the heat treatment process. A great deal of attention has been recently devoted to understanding the strengthening mechanisms in metal matrix composites, which are distinguished by a large particulate volume fraction and relatively large diameter. Another important matter in understanding and modelling the strength of particulate MMCs is to consider the effect of particle shape, size and clustering [9-11], as well as the effects of clustering of reinforcement on the macroscopic behaviour and the effects of segregation to the SiC/Al interfaces [12]. Important role also play the effects of casting condition and subsequent swaging on the microstructure, clustering, and properties of Al/SiC composites [13].

Aluminium honeycomb sandwich panel constructions have been successfully applied as strength members of satellites and aircraft structures and also in passenger coaches of high-speed trains such as the TGV in France and the Shinkansen in Japan [14]. However, the cost of producing the all welded honeycomb structure has been a key factor for not using this technology on mass production rate. Recent developments in manufacturing methods have given rise to a range of commercially viable metallic foams, one being Alulight. In comparison to aluminium honeycomb core construction, metallic foams show isotropic properties and exhibit non linear mechanical deformation behaviour. The metallic foams have the potential to be used at elevated temperatures up to 200°C [15]. They also have superior impact energy absorption and improved strength and weight savings. However, the successful implementation of both aluminium honeycomb and metallic foam sandwich panels for aerospace and transportation applications is dependent upon an understanding of their mechanical properties including their resistance to fatigue crack growth and the resistance of aluminium alloys to environmentally induced cracking or stress corrosion cracking.

This chapter discusses first the relationship between the interfacial strength with the thermo-mechanical deformation process and the resulting macroscopic mechanical behaviour of particle-reinforced aluminium matrix composites. Micro-compositional changes which occur during the thermo-mechanical processing of these materials can cause substantial changes in mechanical properties such as ductility, fracture toughness, or stress corrosion resistance. A mico-mechanistic model will be presented for predicting the interfacial fracture strength in AMCs in the presence of magnesium segregation. Finally, the use of powerful nondestructive evaluation tools, such as infrared thermography, will be discussed to evaluate the state of stresses at the crack tip and to monitor fatigue crack growth in particle-reinforced aluminium alloy matrix composites.

In the second part of the chapter the structural integrity of Aluminium Honeycomb (HC) sandwich panels is compared with the new core material concept of aluminium foams. Aluminium Honeycomb sandwich panels are used to reduce weight whilst improving the compressive strength of the structure with the aerospace industry being one of the prime users of HC sandwich panels for structural applications. The cost of producing all welded HC structures has been the key factor for not using this technology on a mass production basis. An alternative to the aluminium honeycomb (HC) sandwich panels is the metallic foam sandwich panel, which has been gaining interest in the same field. These foams are anisotropic, exhibit non-linear mechanical behaviour, and they have the potential for use at temperatures up to 200°C. They have superior impact energy absorption, and improved strength and weight savings. The lower weight as compared to conventional solid wrought aluminium alloys will mean a reduction in fuel consumption thus providing economical savings.

This chapter attempts to investigate whether aluminium honeycomb sandwich panels, with their homogenous hexagonal core can be successfully replaced by metallic foam sandwich panels, which have an inhomogeneous core. A successful replacement would improve the confidence of manufacturers in the exploitation of this new material in replacing traditional materials. Current levels of understanding of cyclic stressing in metallic foam sandwich panels is limited and models of long term understanding of this aspect of failure are very important for both aerospace and automotive sectors. Burman et al [16] suggests that fundamental fatigue models and concepts proven to work for metals can be applied to metallic foam sandwich panels. A study by Shipsha et al [17] investigated experimentally both metallic foam and other cellular foams, using compact tension specimens. Shipsha's et al research is extremely interesting and implies that a sandwich panel should be considered whole and not two separate entities. Banhart and Brinkers has shown that it is very difficult to detect the features leading to fatigue failure in metallic foams due to the metallic foam being already full of micro cracks [18]. However, Olurin [19] investigation suggest that the fatigue crack growth mechanism of Alulight and Alporas foam is of sequential failure of cell faces ahead of crack tip. The main conclusion is that for a given ΔK, the fatigue crack propagation rate, da/dN decreases with increasing density and for a given stress intensity, the fatigue crack propagation rate increases when the mean stress is increased.

Current levels of understanding of cyclic stressing in aluminium foams is limited and models of long term understanding of this aspect of failure are important for both aerospace and automotive sectors. This is particularly important for low-density foam and honeycomb materials which despite thin ligament thickness, have good properties in compression. A method of analysis is proposed to predict life expectancy of aluminium honeycomb and metallic foam sandwich panels.

2. SiC-particulate reinforced aluminium matrix composites

2.1 Materials

Aluminium - silicon - magnesium alloys (A359) are important materials in many industrial applications, including aerospace and automotive applications. The alloys from the Al-Si-Mg system are the most widely used in the foundry industry thanks to their good castability and high strength to weight ratio. Materials based on A359 matrix reinforced with varying amounts of silicon carbide particles are discussed in this chapter.

Four types of material are used: 1) Ingot as received A359/40%SiC, with an average particle size of 19±1 micron, 2) Ingot as received A359/25%SiC, with an average particle size of 17±1 micron, 3) Hot rolled as received A359/31%SiC with an average particle size of 17±1 micron and 4) Cast alloy as received A359/30%SiC with particles of F400grit, with an average particle sizes of 17±1 micron. Table 1, contains the details of the chemical composition of the matrix alloy as well as the amount of silicon carbide particles in the metal matrix composites.

TYPES	Si	Mg	Mn	Cu	Fe	Zn	SiC
INGOT A359	9.5	0.5	0.1	0.2	0.2	0.1	40
INGOT A359	9.5	0.5	0.1	0.2	0.2	0.1	25
CAST A359	9.5	0.5	0.1	0.2	0.2	0.1	30
ROLLE D A359	9.5	0.5	0.1	0.2	0.2	0.1	31

Table 1. Types and composition of the material

The microstructure of such materials consists of a major phase, aluminium or silicon and the eutectic mixture of these two elements. In this system, each element plays a role in the material's overall behaviour. In particular, Si improves the fluidity of Al and also Si particles are hard and improve the wear resistance of Al. By adding Mg, Al - Si alloy become age hardenable through the precipitation of Mg_2Si particulates.

2.2 Heat treatment

Properties in particulate-reinforced aluminium matrix composites are primarily dictated by the uniformity of the second-phase dispersion in the matrix. The distribution is controlled by solidification and can be later modified during secondary processing. In particular, due to the addition of magnesium in the A359 alloy, the mechanical properties of this material can be greatly improved by heat treatment process. There are many different heat treatment sequences and each one can modify the microstructural behaviour as desired [20]. Precipitation heat treatments generally are low temperature, long-term processes. Temperatures range from 110°C to 195°C for 5 to 48 hours. The selection of the time temperature cycles for precipitation heat treatment should receive careful consideration. Larger precipitate particulates result from longer times and higher temperatures. On the other hand, the desired number of larger particles formed in the material in relation to their interparticle spacing is a crucial factor for optimising the strengthening behaviour of the composite. The objective is to select the heat treatment cycle that produces the most

favourable precipitate size and distribution pattern. However, the cycle used for optimising one property, e.g. tensile strength, is usually different from the one required to optimise a different property, e.g. yield strength, corrosion resistance.
Heat treatment of composites though has an additional aspect to consider, the particles introduced in the matrix. These particles may alter the alloy's surface characteristics and increase the surface energies [21].
The heat treatments were performed in Carbolite RHF 1200 furnaces with thermocouples attached, ensuring constant temperature inside the furnace. There were two different heat treatments used in the experiments, T6 and modified-T6 (HT-1) [21, 22].
The T6 heat treatment consists of the following steps: solution heat treatment, quench and age hardening (Fig. 1).

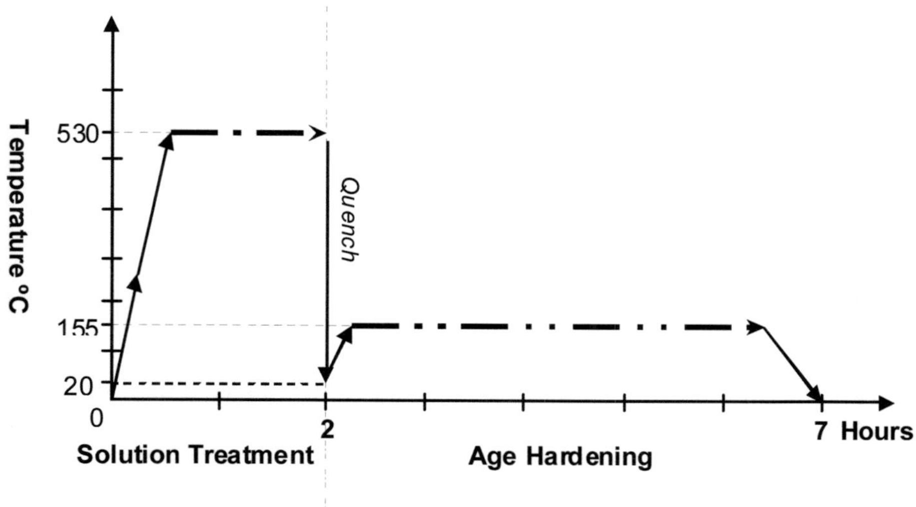

Fig. 1. T6 Heat treatment diagram showing the stages of the solution treatment for 2 hours and artificial ageing for 5h

In the solution heat treatment, the alloys have been heated to a temperature just below the initial melting point of the alloy for 2 hours at 530±5°C where all the solute atoms are allowed to dissolve to form a single phase solid solution. Magnesium is highly reactive with Silicon at this temperature and precipitation of Mg_2Si is expected to be formed. The alloys were then quenched to room temperature. In age hardening, the alloys were heated to an intermediate temperature of 155°C for 5 hours where nucleation and growth of the β' phase. The desired β phase Mg_2Si precipitated at that temperature and then cooled at room temperature conditions. The precipitate phase nucleates within the grains at grain boundaries and in areas close to the matrix-reinforcement interface, as uniformly dispersed particles. The holding time plays a key role in promoting precipitation and growth which results in higher mechanical deformation response of the composite.
The second heat treatment process was the modified-T6 (HT-1) heat treatment, where in the solution treatment the alloys have been heated to a temperature lower than the T6 heat treatment, at 450±5°C for 1 hour, and then quenched in water. Subsequently the alloys were heated to an intermediate temperature of 170±5°C for 24 hours in the age hardened stage and then cooled in air (Fig. 2).

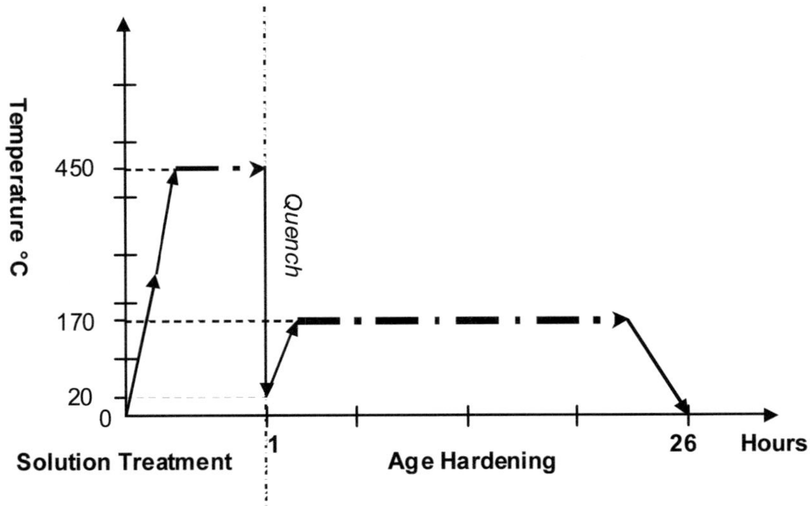

Fig. 2. Modified T6 (HT-1) showing stages of solution treatment for 1 hour and artificial ageing for 24h

In both heat treatments undesired formation of phases, like the Al_4C_3, is a possibility and selection of the solution treatment as well as the age hardening processes should be carefully considered. Temperature and time control, therefore, is extremely important during heat treatment. If the melt temperature of SiC/Al composite materials rises above a critical value, Al_4C_3 is formed increasing the viscosity of the molten material, which can result in severe loss of corrosion resistance and degradation of mechanical properties in the cast composite; excessive formation of Al_4C_3 makes the melt unsuitable for casting. In the A359/SiC composite high silicon percentage added in excess aids to the formation of some oxides (SiO_2) around the SiC reinforcement, something that retards the formation of Al_4C_3, since such oxides prevent the dissolution of SiC particles [22].

2.3 Metallographic examination

In order to analyse the microstructure, a series of sample preparation exercises were carried out, consisted of the cutting, mounting, grinding and polishing of the samples. The microstructures were investigated by SEM, EDAX, and XRD to determine the Al/SiC area percentage, size and count of particulates.

The microstructures of the examined MMCs in the as received condition have four distinct micro phases as clearly marked on the image micrograph, which are as follows: the aluminium matrix, the SiC particles, the eutectic region of aluminium and silicon and the Mg phase (Fig. 3). The distribution of SiC particles was found to be more or less uniform, however, instances of particle free zones and particle clustered zones were observed.

Matrix-reinforcement interfaces were identified by using high magnification Nano-SEM. In the as received hot rolled images the Al Matrix/SiC reinforcement interface is clearly identified (Fig. 4). These interfaces attain properties coming from both individual phases of constituents and facilitate the strengthening behaviour of the composite material.

In the modified T6 (HT-1) condition the microsturucture of the cast 30% SiC has the same phases as in the as received state, plus one rod-shape phase (Fig.5) along the matrix and at the matrix-reinforcement interface has been identified to be Mg_2Si precipitates in an early stage which are not fully grown. This evidence shows that β' phase has been formed with magnesium and silicon reacting together but β phases forming platelets of precipitates have not been formed in this HT-1 heat treatment, and this is probably due to the solution treatment temperature that did not allow enough reactivity time between the main alloying elements.

In the rolled 20% SiC the microstructure of HT-1 heat treatment shows an increase of the Silicon phase as shown in the image (Fig. 6). Silicon has been expanded during solidification and subsequent ageing. This formed round areas around the whole area of the composite. Comparing with the cast 30% SiC sample, in the rolled material the silicon phase is

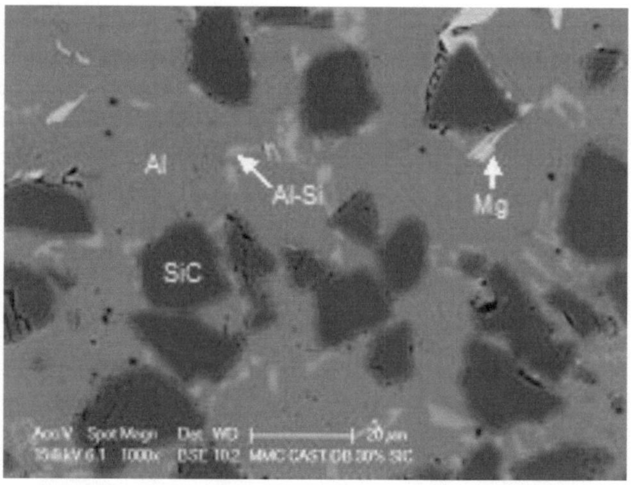

Fig. 3. Microstructure of cast 30% SiC in the as received condition showing four distinct phases: Aluminium matrix, SiC particles, eutectic region of aluminium and silicon and Mg phase

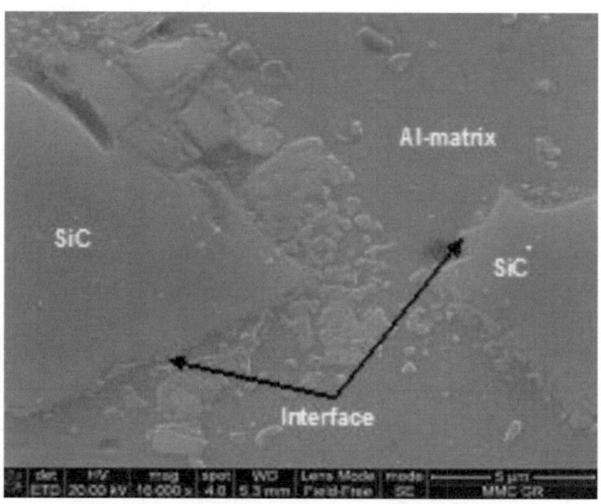

Fig. 4. Microstructure of rolled 31% SiC in the as received condition showing matrix-reinforcement interfaces

increased by 6%. This increase under the same heat treatment conditions is explained by the difference in the percentage of reinforcement in the material. Therefore, it becomes evident that the introduction of SiC reinforcement promotes zone kinetics and phase formation reactions during heat treatment process. The reinforcement, depending on its percentage in the matrix material, accelerates or restrains events such as precipitation and segregation. This is further supported by the fact that precipitation has not been observed in the HT-1 heat treated 20% SiC rolled material, where lower percentage of SiC reinforcement slowed-down the precipitation kinetics and β' phases could not be created in a similar manner as the 30% SiC cast sample.

In the T6 condition the microstructural results showed that in the rolled 31% SiC sample precipitates of Mg_2Si have been formed in the matrix in a platelet shape with a size of around 1-3 μm, as well as in areas close to the interface (Fig. 7). The higher solution temperature and lower age hardening holding time that exist in the T6 heat treatment process, promoted the forming of this type of precipitates which more densely populated

Fig. 5. Microstructure of cast 30% SiC in the HT-1 condition showing rod shape β' phases of Mg_2Si around the matrix and the interface of the reinforcement

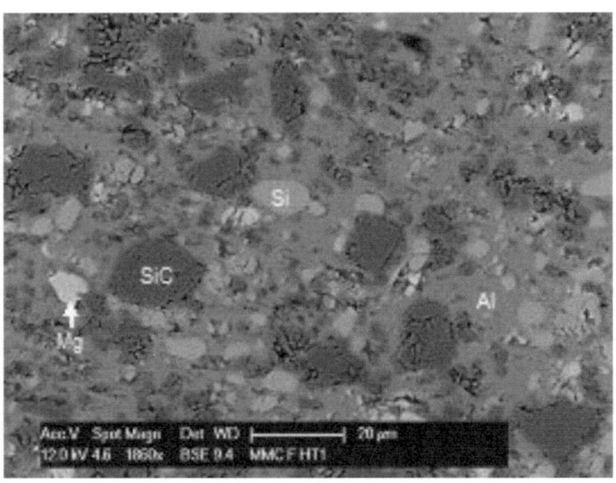

Fig. 6. Hot rolled HT-1 sample showing phases of Aluminium, SiC, Silicon, Mg

the interface region compared to the matrix. In the case of presence of a crack in the matrix, the precipitates act as strengthening aids promoting crack deflection at the interface resulting in an increase of the composite's fracture toughness [20, 23]. Furthermore, the precipitates formed in the matrix act as support to strengthening mechanisms of the reinforcement-matrix interface.

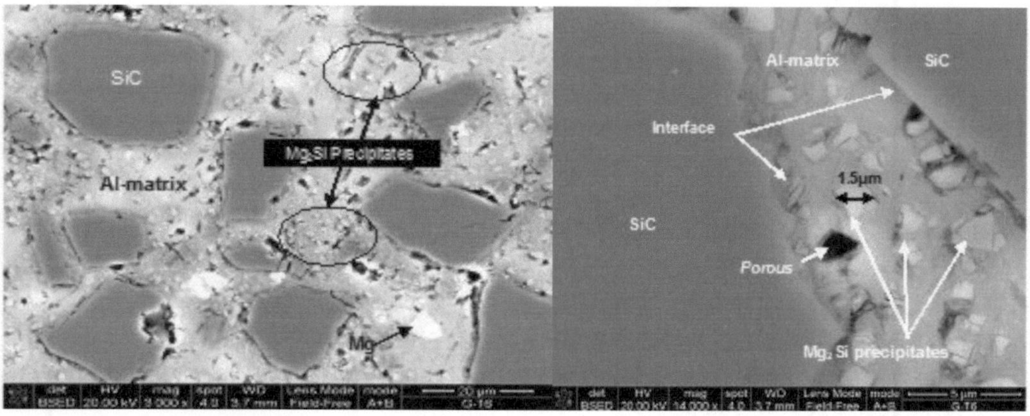

(a) (b)

Fig. 7. (a) Hot rolled 31% SiC –T6 showing precipitate formed around the reinforcement. (b) Hot rolled 31% SiC - T6 showing Mg_2Si precipitates formed between the SiC reinforcement interface in a platelet shape of around 1-3 µm. A porous close to the interface has been identified in a similar size

The X-ray diffraction was carried out on the MMCs in the as received, as well as, in the heat treatment conditions, in samples with 20%, 30% and 31% of SiC particulates. Even though some peaks were superimposed, the results clearly showed the phases present in the microstructures. In particular, in the as received condition and in the heat treatment conditions the results showed existence of aluminium matrix material, eutectic silicon, SiC, Mg_2Si, SiO_2 phases as the distinct ones, and also $MgAl_2O_4$ and Al_2O_3 phases. $MgAl_2O_4$ and Al_2O_3 oxides give good cohesion between matrix and reinforcement when forming a continuous film at the interface. The presence of $MgAl_2O_4$ (spinel) shows that low percentage of magnesium reacted with SiO_2 at the surface of SiC and formed this layer in the interphacial region between the matrix and the reinforcement.

$$2SiO_2 + 2Al + Mg \rightarrow MgAl_2O_4 + 2Si \tag{1}$$

The layers of $MgAl_2O_4$ protect the SiC particles from the liquid aluminium during production or remelting of the composites. This layer provides more than twice bonding strength compared to Al_4C_3. Furthermore, the layer of Al_2O_3 oxide is formed as a coating when SiO_2 is reacting with liquid aluminium.

$$3SiO_2 + 4Al \rightarrow 2Al_2O_3 + 3Si \tag{2}$$

The same phases have been identified in the HT-1 modified condition. In the T6 condition XRD results showed one more phase present which is the spinel-type mixed oxide $MgFeAl0_4$ showing that Fe trace reacted with Mg and in the presence of aluminium and oxygen formed this oxide.

2.4 Micro-hardness testing

The three samples have been compared in relation to their microhardness performance based on the reinforcement percentage, the heat treatment conditions and the different manufacturing forming processes. Microhardness of the three composites has been measured in order to get the resistance of the material to indentation, under localized loading conditions. The microhardness test method, according to ASTM E-384, specifies a range of loads using a diamond indenter to make an indentation, which is measured and converted to a hardness value [21, 22].

Measuring the different phases in the micro-level it is quite challenging, as the SiC reinforcement of ≈17μm in size was not easy to measure, due to small indentation mark left when a small load on the carbide is applied. When introducing higher values of load, the indentation was not localized in the carbide but covered some of the matrix area too. The load was finally set to 50 grams in order to obtain valid measurements coming from different areas of the samples: SiC, aluminium matrix, and the overall composite i.e. areas superimposing matrix and reinforcement.

There are many factors influencing the microhardness of a composite material, including the reinforcement percentage, interparticle spacing and also particle size. Moreover, manufacturing forming processes influence material's microhardness behaviour in relation to the reinforcement percentages in the composites.

The cast sample in the as received condition has the highest MMC microhardness, where the rolled 20% SiC with lower percentage of reinforcement has the lowest values. By altering the microstructure with modified T6 (HT-1) heat treatment all values of the three samples show an increase between 20-45% from the initial state (Fig. 8). This shows the effect of the heat treatment in the micro-deformation of the matrix-reinforcement interface due to the presence of precipitates and other phases and oxide layers.

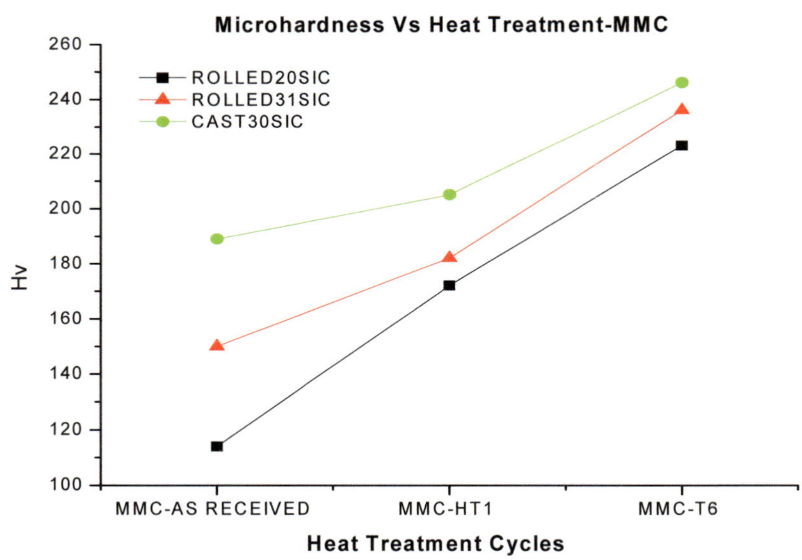

Fig. 8. Microhardness values Vs. Heat treatment cycles for the MMC areas

In the T6 condition it was observed the larger increase in microhardness values from the as received state, ranging from 20% to 90% depending on the reinforcement percentage and manufacturing process. In particular, in the rolled 20% SiC material the increase in microhardness values is in the order of 90%.

Furthermore, variability in microhardness values was observed when comparing cast and rolled materials with different percentage of SiC. However, this variability varied when samples processed at different heat treatment conditions were compared. Highest variability showed samples in the as received condition, whereas lowest variability showed samples in the T6 condition, with samples in the HT-1 condition in between. This can be explained by the fact that precipitates act as strengthening mechanisms and affect the micromechanical behaviour of the composite material.

In the absence of precipitates (in the as received condition), the volume percentage of SiC and the manufacturing processing play a significant role in micromechanical behaviour of the composite. As precipitates are formed due to heat treatment process they assume the main role in the micromechanical behaviour of the material. In the HT-1 heat treatment condition there is presence of β' precipitates which affect the micromechanical behaviour in a lesser degree than in the case of T6 heat treatment condition where fully grown β precipitates are formed. It becomes clear that after a critical stage, which if related to the formation of β precipitates in the composite the dominant strengthening mechanism is precipitation hardening.

While Figure 8 shows results in areas that include the interface region (where precipitates are concentrated) Figure 9, shows results on microhardness values in the aluminium matrix (where precipitates are dispersed). In Figure 9 there is similar variability for all three materials processing states, as received, HT-1, and T6. This implies that in the matrix material the percentage of the reinforcements, the manufacturing process, as well as the precipitation hardening, are strengthening mechanisms of equal importance.

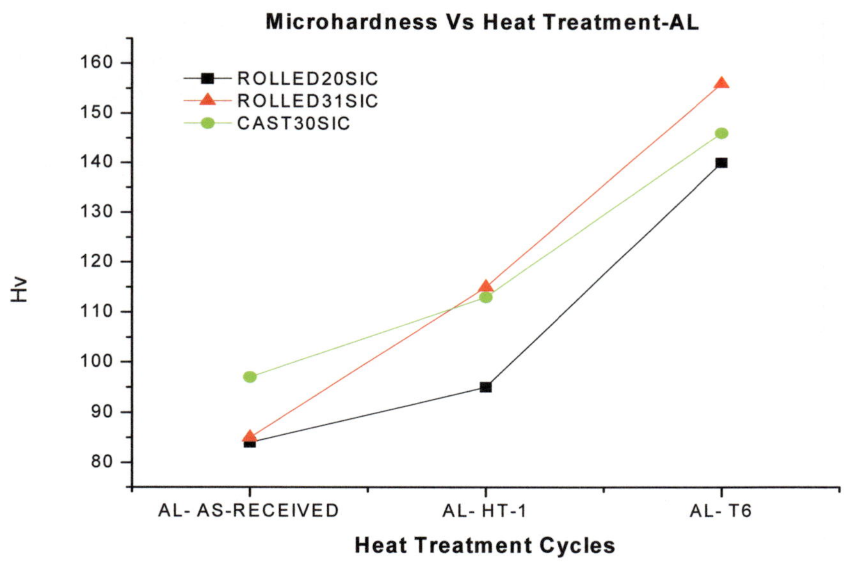

Fig. 9. Microhardness Vs. Heat treatment cycles for Aluminium areas

Figure 10 shows microhardness measurements obtained from areas around the matrix-reinforcement interface in a composite heat treated in the T6 condition. The microhardness values are higher in the close proximity of the interface. It is observed that cast material has higher values than the rolled material. In the case of rolled material, the microhardness raises as the percentage of reinforcement increases.

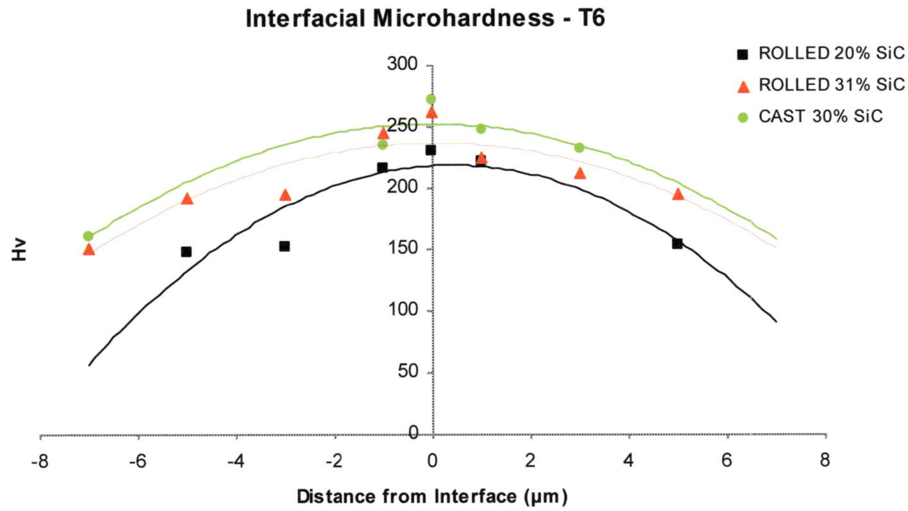

Fig. 10. Interfacial microhardness showing measurements obtained from areas close to the matrix- reinforcement interface in the T6 condition

2.5 Tensile testing

Aluminium - SiC particulate composite samples were tested in tension for two different volume fractions, 20% and 31%, in reinforcement [23]. The dog-bone coupons were tested according to ASTM E8-04 in the as received and, following two different heat treatments, modified T6 (HT-1) and T6 heat treatment conditions.. The mechanical properties of the composites are presented in Table 2.

Material	Condition	$\sigma_{0.2}$(MPa)	σ_{uts}(MPa)	ε(%)	E	$HV_{0.5}$
Rolled Al A359-SiC-20p	T1	146	157	1.5	100	114
	HT-1	147	190	4	102	172
	T6	326	360	2.1	112	223
Rolled Al A359-SiC-31p	T1	158	168	1	108	150
	HT-1	155	187	2	110	182
	T6	321	336	1.3	116	236

Table 2. The mechanical properties of Al/SiC Composites

The engineering stress/strain curves of the composite are shown in Figure 11. As can be clearly seen in Figure 11, the HT-1 heat treatment improved both the strength and strain to failure than the untreated composites for both volume fractions. Furthermore, the failure strain for this temper is considerably higher than for the T6 heat treatment; this may be attributed either to the nucleation of the β` precipitate phases which although not yet visible, may lead to the increase of the plastic deformation through crack deflection mechanisms and/or to annealing which acts competitively to the precipitation leading to the toughening of the composite. However, the T6 heat treatment exhibits the highest strength followed by the HT-1 and the as received state. Finally, as was expected, the "as received" composites behaviour in tension deteriorates with increasing filler concentration.

The experiments showed that for the same range of conditions tested, the yield and the ultimate tensile strengths of the SiC/Al composites were mainly controlled by the percentage of reinforcement as well as by the intrinsic yield/tensile strengths of the matrix alloys. The addition of the SiC reinforcement created stress concentrations in the composite, and thus the aluminium alloy could not achieve its potential strength and ductility due to the induced embrittlement. Composites in the as-received condition failed in a brittle manner with increasing percentage of reinforcement. As a result, with increasing reinforcement content, the failure strain of the composites was reduced as shown in Figure 11. From the above postulations it is obvious that the phase that dominates the mechanical behaviour of the composite is the precipitation phase created by age hardening while the reinforcement phase plays a secondary role.

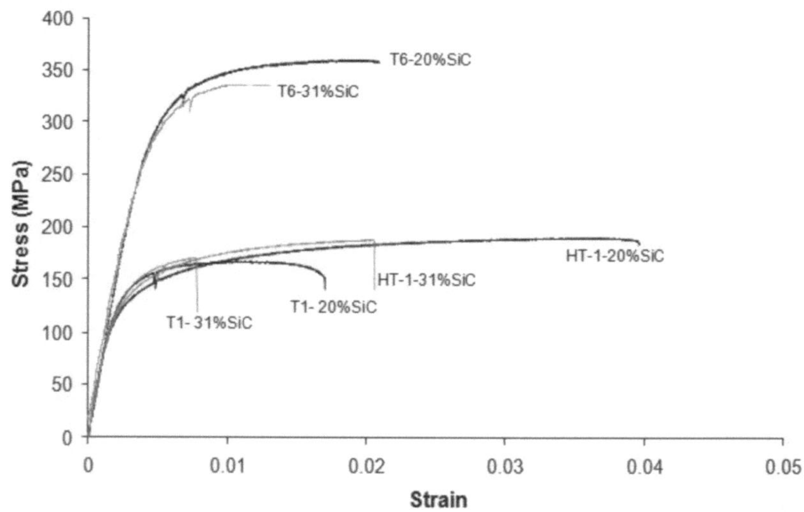

Fig. 11. Stress / Strain curves of Al/SiC Composites

The heat treatment affected the modulus of elasticity of the composites by altering the transition into plastic flow (see Table 2 and Fig. 12). Composites in the T6 condition strained elastically and then passed into a normal decreasing-slope plastic flow. Composites tested in the HT-1 condition exhibit a greater amount of strain than the as received and those heat treated in the T6 condition. The failure strain increasing from about 1.5% strain to about 4% but the greater influence was a sharper slope of the stress-strain curve at the inception of plastic flow.

This increase in elastic proportional strain limit and the steepening of the stress-strain curve were reflected by the higher yield and ultimate tensile strengths observed in the heat-treated composites. The increase in flow stress of composites with each heat-treatable matrix probably indicated the additive effects of dislocation interaction with both the alloy precipitates and the SiC reinforcement. The combination increased the strain in the matrix by increasing the number of dislocations and requiring higher flow stresses for deformation, resulting in the higher strengths observed. Ductility of SiC/Al composites, as measured by strain to failure, is again a complex interaction of parameters. However, the prime factors affecting these properties are the reinforcement content, heat treatment and precipitation hardening.

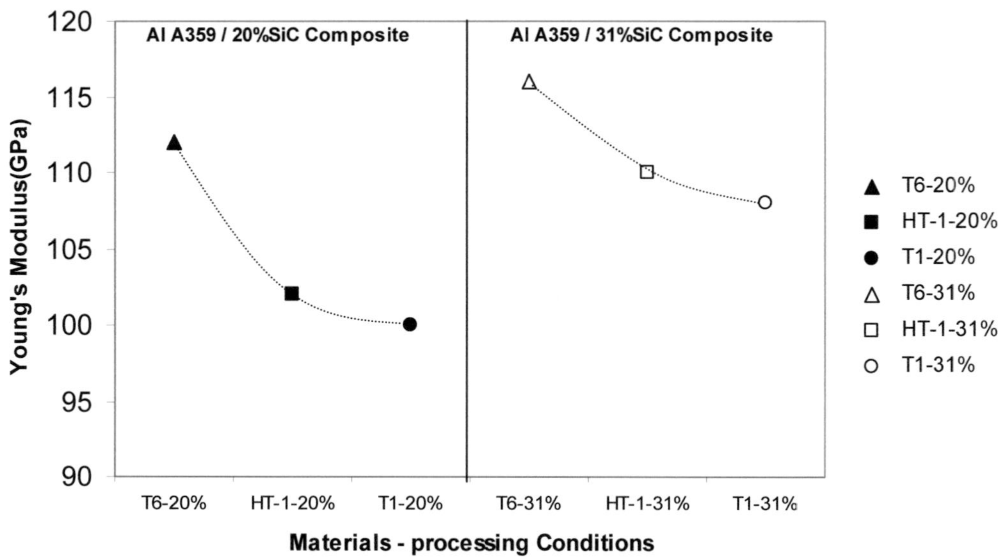

Fig. 12. Young's Modulus vs. Processing Conditions curves showing T6 treated composites having the highest modulus

2.6 Fracture toughness KIC testing

The plane strain fracture toughness test involves the loading to failure of fatigue pre-cracked, notched specimens in tension or in three-point bending. The calculation of a valid toughness value can only be determined after the test is completed, via examination of the load-displacement curve and measurement of the fatigue-crack length. The provisional fracture toughness value, K_Q, is first calculated from the following equation:

$$K_Q = \left(\frac{P_Q}{BW^{1/2}} \right) \cdot f\left(\frac{a}{W} \right) \tag{3}$$

where P_Q is the load corresponding to a defined increment of crack length, B is the specimen's thickness, W is the width of the specimen, and $f(\alpha/W)$ is a geometry dependent factor that relates the compliance of the specimen to the ratio of the crack length and width, expressed as follows:

$$f\left(\frac{a}{W} \right) = \frac{(2 + a / W)(0.86 + 4.64a / W - 13.32a^2 / W^2 + 14.72a^3 / W^3 - 5.6a^4 / W^4)}{(1 - a / W)^{3/2}} \tag{4}$$

Only when specific validity criteria are satisfied, the provisional fracture toughness, K_Q, can be quoted as the valid plane strain fracture toughness, K_{IC}. The standard used for conducting this experiment, ASTM E399, imposes strict validity criteria to ensure that the plane strain conditions are satisfied during the test. These criteria include checks on the form and shape of the load versus displacement curve, requirements on specimen's size and crack geometry, and the 0.2% proof strength values at the test temperature. Essentially, these conditions are designed to ensure that the plastic zone size associated with the pre-crack is small enough so that plane strain conditions prevail, and that the linear elastic fracture mechanics approach is applicable.

Fracture toughness tests were conducted using a servo-hydraulic universal testing machine with data acquisition controller. The system was operated on load control for the fatigue pre-cracking stage, and on position control for the crack opening displacement (COD) testing. The fatigue test for pre-cracking was conducted at a frequency of 1 Hz, at a load ratio R = 0.25 and load range of 3.7 - 4.5 KN according to the materials' ultimate tensile strength. The COD was monitored by a clip gauge attached to the specimen with a testing rate set at 1 mm/min. Moreover, a thermal camera was set for thermographic monitoring of the crack opening displacement. Compact tension (CT) specimens were prepared for fracture toughness tests according to ASTM E399. The thickness B of the specimens was 9.2 mm for the MMC, and 5 mm for the unreinforced aluminium alloys.

Provisional K_Q values were calculated according to ASTM E399 standard for all specimens according to Equations (1) and (2), where Pq = Pmax. Load versus displacement curves for Al/SiCp composites and unreinforced aluminium alloys are shown in Fig. 13. Fracture toughness data for Al/SiCp and unreinforced aluminium alloys are summarised in Table 2.

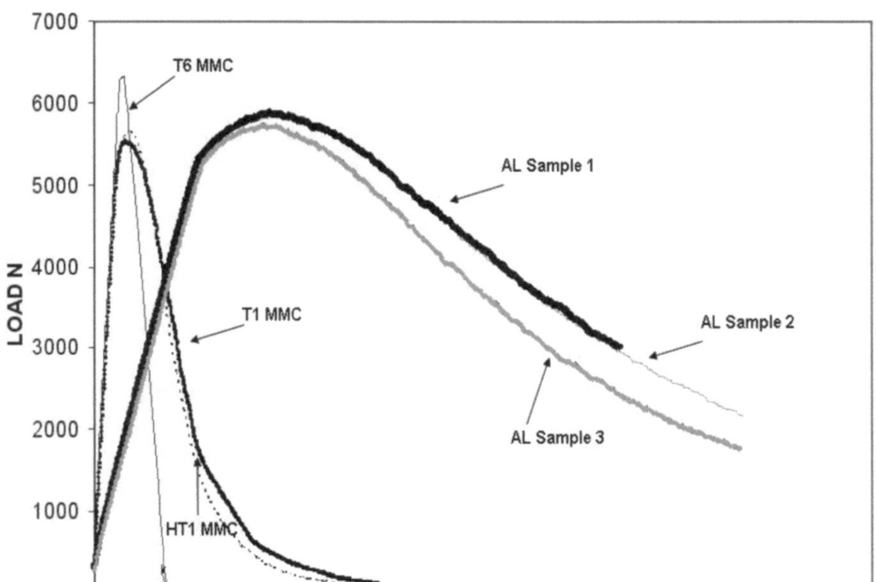

Fig. 13. Load – Displacement curves for Al/SiCp composites subjected to T1, T6 and HT-1 heat treatment conditions and three unreinforced aluminium alloy samples

Material	Heat Treatment	E (GPa)	$Rp_{0.2}$ (MPa)	B (mm)	a/W	a_{eff} (mm)	K_Q (MPa $\sqrt{m}$)	Valid	Reason
2000 series Al	AR	71	75	5.10	0.552	27.62	55,36	No	2**
2000 series Al	AR	71	78	5.13	0.555	26.76	56,00	No	2**
2000 series Al	AR	71	72	5.00	0.558	28.43	58,48	No	2**
A359/SIC/31p	T1	106	117	9.20	0.456	20.79	19,28	Yes	-
A359/SIC/31p	T6	108	120	9.21	0.462	20.12	22,05	Yes	-
A359/SIC/31p	HT1	116	157	9.20	0.467	21.33	20.75	Yes	-
A357/SIC/20p [16]	-	-	215	-	-	-	18.60	-	-
A359/SIC/10p [16]	-	-	300	-	-	-	17.40	-	-

**Validity criteria:
1 Excessive crack curvature
2. Thickness criteria not satisfied
3. Excessive plasticity
4. a/W out of range
5. Non-symmetrical crack front
6. In plane crack propagation

Table 2. Fracture toughness data for Al/SiCp and Al alloys and test validity

As is shown in Table 2, Al/SiC_p composites exhibited lower provisional K_Q values than the reference unreinforced aluminium alloys. In addition, heat treatment processing, and especially T6 treated specimen, had the highest K_Q values compared to the other two heat treatment conditions. According to the load-displacement curves in Figure 3, composites clearly showed more brittle behaviour than the unreinforced aluminium alloys. T6 heat treated composites have the highest strength, but the lowest ductility compared to the other materials. Although these results provide some insight regarding the fracture behaviour of the materials examined, specific validity criteria have to be satisfied in order to obtain K_{IC} values.

2.7 Examination by infrared thermography

Nondestructive evaluation techniques are powerful tools for monitoring damage in composite materials [24]. Infrared thermography was used to monitor the plane crack propagation behaviour of particulate-reinforced AMCs [25, 26]. The deformation of solid materials is almost always accompanied by heat release. When the material becomes deformed or is damaged and fractured, a part of the energy necessary to initiate and propagate the damage is transformed in an irreversible way into heat [26]. The heat wave, generated by the thermo-mechanical coupling and the intrinsic dissipated energy during mechanical loading of the sample, is detected by the thermal camera. By using an adapted detector, thermography records the two dimensional "temperature" field as it results from the infrared radiation emitted by the object. The principal advantage of infrared thermography is its noncontact, non-destructive character.

A rectangular area on the specimen, located just in front of the initial pre-cracking region, was selected, as shown in Fig. 14a.

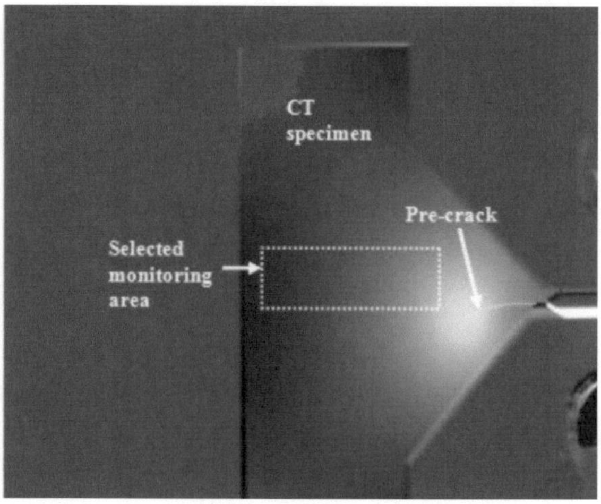

Fig. 14.a CT specimen showing the selected area for thermographic monitoring

The development of fracture was monitored in that area using infrared thermography. The mean temperature in this area versus time during crack growth was calculated using the recorded thermal imprint. As the specimen was stretched in tension, stresses were accumulating in the specimen, and the temperature increased as a function of time. When the accumulated energy became sufficient to propagate the crack, crack growth was observed, resulting in the stress relief. This corresponded to a peak in the temperature-time curve followed by a sudden decrease in temperature. As shown in Fig. 14b, 14c and 14d this behaviour was recurrent until the failure of the specimen. In these figures the thermographic monitoring of Aluminium 2xxx alloy, Al/SiCp T6 composite, and Al/SiCp HT1 composite samples is presented respectively. The different stages of crack growth for each material up to the final fracture of the specimen can be clearly observed. Just prior to fracture, the plasticity zone was clearly delineated on the specimen's surface as a heated region, which may be readily attributed to local plastic deformation. Furthermore, as seen in all figures, the crack was propagated in-plane throughout the experiment.

A comparison of the thermography graphs in Figs. 14b, 14c, and 14d leads to the conclusion that the aluminium alloy exhibited different crack propagation behaviour than the Al/SiCp composites. For the aluminium alloy, the temperature versus time curve in Fig. 5b showed extended plasticity behaviour before final fracture occurred. This behaviour was evidenced by the constant increase in temperature between the temperature picks at the 60th and 140th second (figure 14b). This behaviour may be attributed to the small specimen thickness. However, for the T6 heat treated composite material in Fig. 14c, fracture was more elastic as the multiple temperature peaks indicated a confinement of the plasticity zone. Also, plasticity was formed in a more balanced way regarding the overall fracture process. It was also observed that T6 heat treated composites exhibited fewer picks compared to the HT1 heat treated specimens (Fig. 14d). This was attributed to the presence of a stronger interface in the T6 material as the accumulation of precipitates near the interface, resulted in the improvement of the fracture toughness of the material.

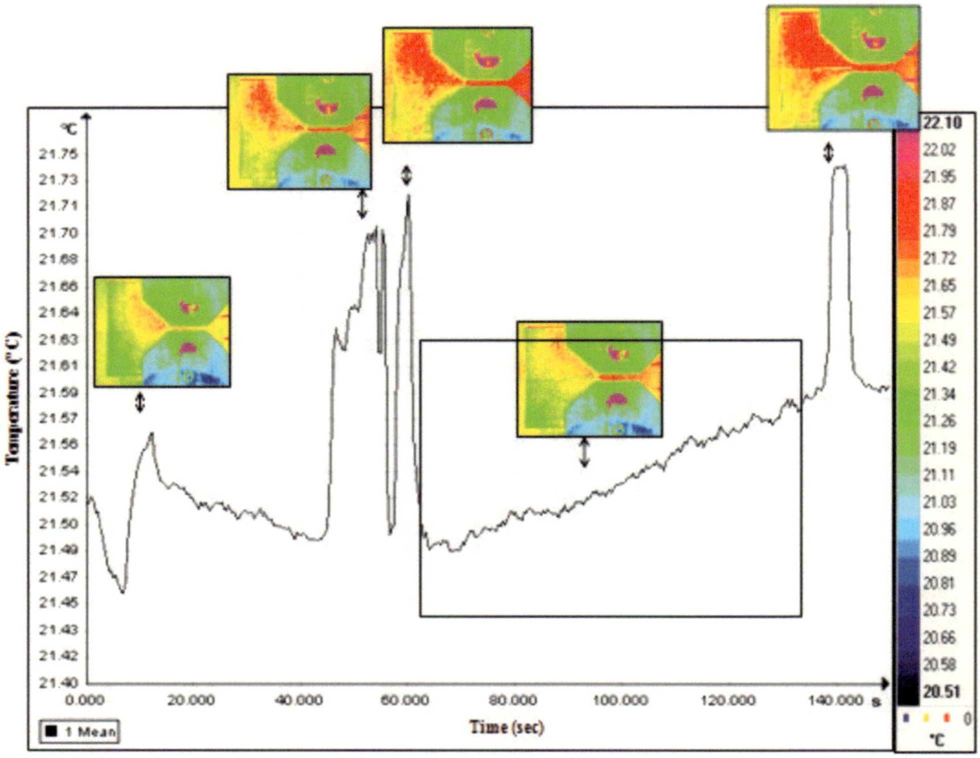

(b)

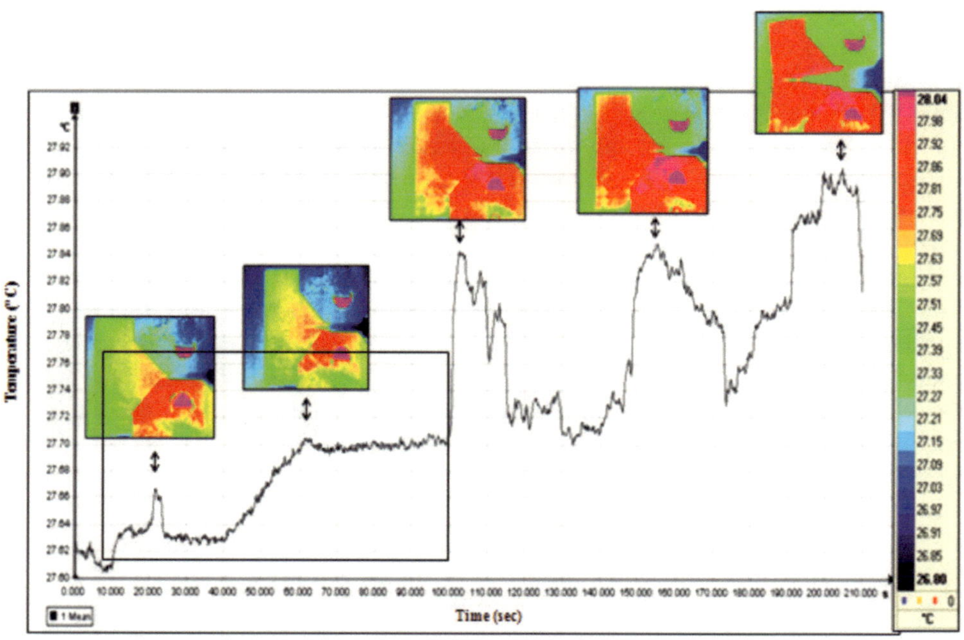

(c)

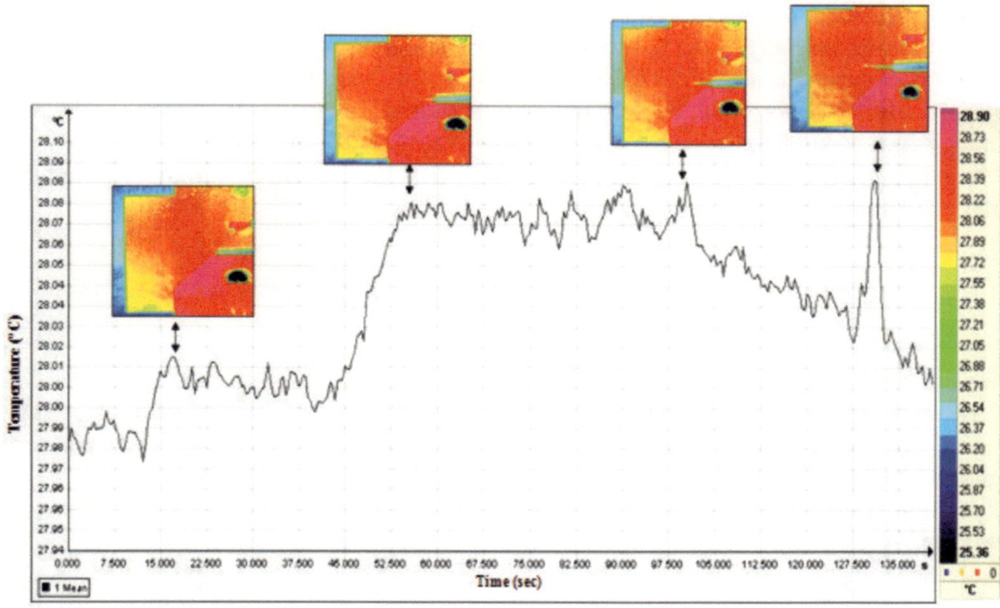

(d)

Fig. 14. Thermographic monitoring of various CT specimens showing the different stages of crack growth up to the specimen's final fracture: (b) Aluminium 2xxx, (c) Al/SiCp T6 composite, (d) Al/SiCp HT1 composite

2.8 A model for predicting interfacial strengthening behaviour of particulate reinforced AMCs

A model proposed by McMahon and Vitek [27] predicts the fracture resistance of a ductile material that fails by an intergranular mechanism. Based on this model, an effective work parameter can be developed to predict fracture strength of an interface at a segregated state using Griffith crack-type arguments. The Griffith's equation, which was derived for elastic body, is applied here because it is assumed that the yielding zone size ahead of the crack is small enough and the fracture is governed by the elastic stress field. The model further assumes that small changes in interfacial energy caused by segregation of impurities at the interface will result in a much larger change in the work of fracture. This is due to the fact that the work of fracture must be provided by a dislocation pile-up mechanism around the advancing crack-tip on the interface. This implies that additional work must be provided to deform the material at the crack-tip in addition to the work needed to overcome the interface energy and to replace it with two surfaces. The definition of interfacial fracture strength, σ_{int}, is then given by:

$$\sigma_{\text{int}} = \sqrt{\frac{100\varepsilon_P E_{\text{int}}}{\pi d}} \tag{5}$$

where,

E is Young's modulus, d is the particle thickness, since it is assumed that cracks of the order of the particle size are present when considering crack propagation through the interface

and the particulate, ε_p is the energy required to create two fracture surfaces = $2\varepsilon_s - \varepsilon_{gb}$ (= ε_o), with ε_s, the surface energy, and ε_{gb}, the grain boundary energy.
The 100 ε_p component allows for dislocation interaction and movement ahead of the crack-tip in ductile materials. This refers to the work required for a total separation of the lattice planes, which is equal to the area under the force-extension curve.
From equation (6) εp can be estimated if Kint (Interface fracture toughness) and Eint (Interface Young's Modulus) are known.

$$\frac{K_{\text{int}}^2}{100E_{\text{int}}} = \varepsilon_p \left(1 - \frac{ZRT\ln(1-c+Bc)}{\varepsilon_p}\right)^n \tag{6}$$

Where,
Z, describes the density of interface sites which are disordered enough to act as segregation sites (= D ρ_S), with D the thickness of the interface region, and ρ_S the density of the interface region (D=300 nm) (ρ= 2.6889 g/cm³ for Aluminium and 3.22 g/cm³ for SiC),
R, is the gas constant (= 8.314472(15) J•K-1•mol-1),
T, is the absolute temperature (= 803.15 K for T6, = 723.15 K for HT1),
c, is the segregate concentration needed to cause embrittlement (= 0.1 for pure aluminium),
B, describes the modification of the boundary energy by impurities using the Zuchovitsky equations,
n, is the work hardening exponent (= 10 for FCC aluminium).
In hard particle reinforced metal matrix composites the stress transfer from the matrix to the particles is mainly controlled by the misfit of the elastic constants between the two phases [28]. To measure the stress transfer to the particle, in an homogeneous material subjected to tensile loading, the stress carrying capability of the particle is defined as the ratio of the normal stress σ_N to the particle in the loading direction to the macroscopic tensile stress, σ_T, i.e. the ratio L = σ_N / σ_T. By using Eshelby's theory, the stress carrying capability of a spherical inhomogeneity can be written as [29]:

$$L = \frac{9x(2+3x)}{(1+2x)(8+7x)} \tag{7}$$

where, x = E_i / E_m, and E_i and E_m are Young's moduli for inhomogeneity and matrix, respectively.
Furthermore, the shear lag model, originally developed by Cox [30] modified by Llorca [31], can be used to estimate the stress carrying capability of a particulate, assuming that the volume fraction of reinforcement is small and the average stress in the matrix is approximately equal to the applied stress:

$$L = 1 + \frac{a}{\sqrt{3}} \tag{8}$$

where $a = \bar{h}/2\bar{r}$ is the aspect ratio of the reinforcement, with $\bar{h}$ and $\bar{r}$ the average length and the average diameter of the particle.
A model has been proposed to estimate the effects of particle volume fraction on fracture toughness in SiC particle-reinforced aluminium alloy matrix composites. This model assumes that SiC particles are uniformly distributed in the matrix and that the pattern of

particle distribution is similar to FCC structure in metals. The fracture toughness of the composite can then be written as [32]:

$$K_{IC} = \frac{K_p}{L_p} V'_m + \frac{2K_{\text{int}}}{L_p + L_m} (V_m - V'_m) + \frac{K_m}{L_m} 2V_m + K_m (1 - 3V_m) \tag{9}$$

where K_{IC}, K_p = 3 MPa m-1/2, K_m = 35 MPa m-1/2, and K_{int} is the fracture toughness of the composite, SiC particulates, A359 aluminium alloy matrix, and interface, respectively. L_p and L_m are the stress carrying capabilities of a particulate and the matrix, respectively. On average, for SiC particles and aluminium alloy matrix, L_p ~ L_m ~ 2. The value of L_m = 1 is applicable for clean surfaces. However, due to processing conditions and the physical interaction at the matrix/reinforcement interface the interface contains partially contaminated surfaces, therefore L_m = 2 since it cannot be considered as a "clean surface". V_m and (V_m-V'_m) are the area fractions for particle cracking and interface failure, respectively. These area fractions though are not accurately known. However Wang and Zhang [33] found that the ratio of particle cracking over interface failure V_m/ (V_m-V'_m) was about 0.13 (= 1.4%/10.7 %) in a SiC particle-reinforced aluminium alloy composite.
Young's modulus of matrix has been obtained for A359 aluminium matrix. The particles E_p, matrix E_m, and interface E_i shown in equation

$$E_C = E_p v_f^{2/3} + E_m \left(1 - V_f'^{2/3}\right) + E_i \left(V_f'^{2/3} - V_f^{2/3}\right) \tag{10}$$

Due to the fact that the difference $\left(V'_f - V_f\right)$ is very small, a good approximation is to consider that the Young's modulus of the interface region is close to that of the matrix; $E_i \cong E_m$ [32].
The parameter B describes the modification of the boundary energy by impurities using the Zuchovitsky equations [34, 35], given by:

$$B = e^{\left(\frac{\varepsilon_1 - \varepsilon_2}{RT}\right)} \cong e^{\left(\frac{0.75\varepsilon_F}{RT}\right)} \tag{11}$$

where ε_2-ε_1 is the difference between the formation energy in the impurity in the bulk and the interface region. It is assumed that the values of the surface energy and the impurity formation energy in the bulk are close, and therefore the numerator in the exponential term depends on the impurity formation energy in the interface region, which is assumed to be 0.75 ε_f, where ε_f is the formation energy of the impurity in the bulk.
Using Faulkner's approach [36], to the derivation of impurity formation energy,

$$\varepsilon_f = \varepsilon_s + \varepsilon_e \tag{12}$$

where, ε_s is the surface energy required forming the impurity atom and ε_e is the elastic energy involved with inserting an impurity atom into a matrix lattice site. This is given by:

$$\varepsilon_f = \frac{0.5\varepsilon_S}{1.94} + \frac{8\pi G}{3e} a_m (a_i - a_m)^2 eV \tag{13}$$

where,
ε_S is the surface energy (1.02 J m-2)

e is the electronic charge (1.60217646 *10^19 Coulomb)
a_i is the impurity atomic radius (0.118 nm for Si)
a_m is the matrix atomic radius (0.143 nm for aluminium)
G is the shear modulus (26 GPa for aluminium)
By performing the calculations the impurity formation energy, ε_f, for A359 aluminium alloy (Al-Si-Mg) can be determined and then substituted in equation (11) to calculate B.
The micro-mechanics model described above is based on thermodynamics principles and is used to determine the fracture strength of the interface at a segregated state in aluminium matrix composites. This model uses energy considerations to express the fracture toughness of the interface in terms of interfacial critical strain energy release rate and elastic modulus. The interfacial fracture toughness is further expressed as a function of the macroscopic fracture toughness and mechanical properties of the composite, using a toughening mechanism model based on stress transfer mechanism. Mechanical testing is also performed to obtain macroscopic data, such as the fracture strength, elastic modulus and fracture toughness of the composite, which are used as input to the model. Based on the experimental data and the analysis, the interfacial strength is determined for SiC particle-reinforced aluminium matrix composites subjected to different heat treatment processing conditions and the results are shown in table 2. It is observed that K_{int} values are close to the K_{1c} values of the composites. Furthermore, σ_{int} values found to be dependent on the heat treatment processing with T6 heat treatment composite obtain the highest interfacial fracture strength.

2.9 Fatigue testing and crack growth behaviour

Tension-tension fatigue tests were conducted using a hydraulic testing machine. The system was operated under load control, applying a harmonic tensile stress with constant amplitude. By specifying the maximum and the minimum stress levels, the other stress parameters could be easily determined. These were the stress range, σ_r, stress amplitude, σ_a, mean stress, σ_m, and fatigue stress ratio, R ($=\sigma_{min}/\sigma_{max}$). Throughout this study, all fatigue tests were carried out at a frequency of 5 Hz and at a stress ratio R = 0.1. Different stress levels between the ultimate tensile strength (UTS) and the fatigue limit were selected, resulting in so-called Wöhler or S-N curves. Tests exceeding 10^6 cycles without specimen failure were terminated. Specimens that failed in or close to the grips were discarded. The geometry of the samples was the same as those used for the tensile characterisation, i.e. rectangular strips of 12.5mm width, and 1.55mm thickness.
The normalised "S-N" curves of the fatigue response of the Al/SiC composites is shown in Fig. 15. The stress was normalised over the UTS of each material and plotted against the number of cycles to failure. As can be observed, whereas in the untreated T1 condition the composite retains at least 85% of its strength as fatigue strength, the corresponding value for the T6 heat treatment is falling to the 70% of UTS. The HT1 heat treatment is exhibiting an intermediate behaviour, with its fatigue strength falling to 75% of the corresponding UTS value. It can be concluded that aggressive heat treatment reduces the damage tolerance of the composites.
A direct comparison of the fatigue performance of the composite with the corresponding quasi static performance in tension reveals that the T6 heat treatment improved the strength of the composite. This can be attributed to a dominant mechanism related to the changes in the microstructure of the composite. This mechanism relates to the precipitations appearing

in the microstructure of the composite at the vicinity of the interphase area, which results to the composite hardening. The creation of the interphase together with the improved stress transfer may be regarded as the main contributing parameters to the improved mechanical properties of the particulate reinforced composite. The improved static strength is followed by a less spectacular performance in fatigue, with the fatigue limit of the material falling to the 70% of the UTS.

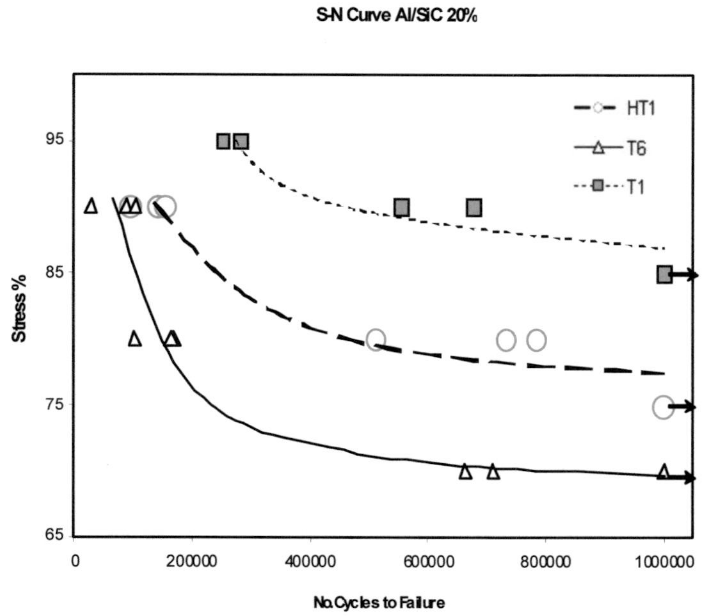

Fig. 15. S-N Curve of Al/SiC 20% Composite

2.9.1 Crack growth rate vs. range of stress intensity

To study the crack growth rate (da/dN) vs. stress intensity range (ΔK) data for aluminium SiC_p composites and aluminium 2xxx series specimens, the materials were subjected to cyclic loading. Fatigue crack growth tests were conducted according to the ASTM E647 standard using a servo hydraulic testing machine. The tests were conducted under load control. Compact tension (CT) specimens were prepared for the fatigue crack growth experiments. The fatigue tests for the monolithic aluminium specimens were performed at a standard frequency of 5 Hz. However, a lower frequency of 1 Hz was selected for the composite specimens in order to minimize the effect of sudden failure due to the brittle nature of these materials. The experiments were performed at a load ratio R = 0.25 and maximum load ranges of 3.7 - 4.5 KN, keeping the maximum stress at about 70% of the material's ultimate tensile strength.

The technique used for determining the crack growth rate versus stress intensity range during the cyclic loading tests was based on non-contact monitoring of the crack propagation by lock-in thermography. This new technique deals with mapping the crack growth nondestructively. Lock-in thermography is based on remote full field monitoring of thermal waves generated inside the specimen by cyclic loading that caused an oscillating temperature field in the stationary regime. Lock-in refers to the necessity for monitoring the exact time dependence between the output signal (thermal wave) and the reference input signal (fatigue cycle). This is done with a lock–in amplifier so that both phase and magnitude images become available.

The detection system included an infrared camera. The camera was connected with the lock-in amplifier, which was then connected to the main servo hydraulic controller. Therefore, synchronization of the frequency through the lock-in amplifier and the mechanical testing machine could be achieved and lock-in images and data capture during the fatigue testing were enabled.

The camera was firstly set at a distance close to the specimen, in order to have the best possible image capture. Then, the fatigue pre-cracking started while synchronizing, at the same frequency, fatigue cycles and infrared camera through the lock-in amplifier.

In order to determine the crack growth rate and calculate the stress intensity factor using thermographic mapping of the material undergoing fatigue a simple procedure was used:

a. The distribution of temperature and stresses at the surface of the specimen was monitored during the test. Therefore, thermal images were obtained as a function of time and saved in the form of a movie.
b. The stresses were evaluated in a post-processing mode, along a series of equally spaced reference lines of the same length, set in front of the crack-starting notch. The idea was that the stress monitored at the location of a line versus time (or fatigue cycles) would exhibit an increase while the crack approaches the line, then attain a maximum when the crack tip was on the line. Due to the fact that the crack growth path could not be predicted and was not expected to follow a straight line in front of the notch, the stresses were monitored along a series of lines of a certain length, instead of a series of equally spaced points in front of the notch. The exact path of the crack could be easily determined by looking at the stress maxima along each of these reference lines.

Four lines of the same length, equally spaced at a distance of 1 mm, were set on the thermal images of the CT specimen at a distance in front of the specimen's notch.

In Figure 16, the crack growth rate for the heat treated composite specimens and the reference aluminium alloy samples are plotted on a logarithmic scale as a function of the stress intensity range. The results showed that the heat treatment processing influences crack growth behaviour of the composite materials. Specimens subjected to T6 heat treatment condition exhibited the highest crack growth rate vs. stress intensity range slope compared to the other composite systems. Moreover, the crack growth rate vs. stress intensity range line of specimens subjected to T6 heat treatment was shifted towards higher ΔK values compared to that from specimens subjected the other two heat treatment conditions. This implies that in order to attain the same crack growth rate, higher stress intensity factor is required for specimens subjected to T6 condition compared to those subjected to T1 and HT-1 conditions. The need for higher stresses for a crack to propagate reveals the material's microstructural strength, where micro-mechanisms such as precipitation hardening promote high stress concentrations at the crack tip, resulting in the toughening of the crack path. The above postulations agree with previews results, where the T6 heat treated composites showed superior strength but the lowest ductility compared to T1 or HT-1 heat treated specimens. Results, shown in Figure 16, indicate that at intermediate values of crack growth rate (10^{-2} to 10^{-5} mm/cycle) the Al/SiCp composites have fracture properties comparable to those of the unreinforced matrix alloys. It is obvious that in these composites crack propagation rate seems more balanced and takes more time than the aluminium alloy obtaining crack growth rate values from around 10^{-1} to 10^{-4} mm/cycle.

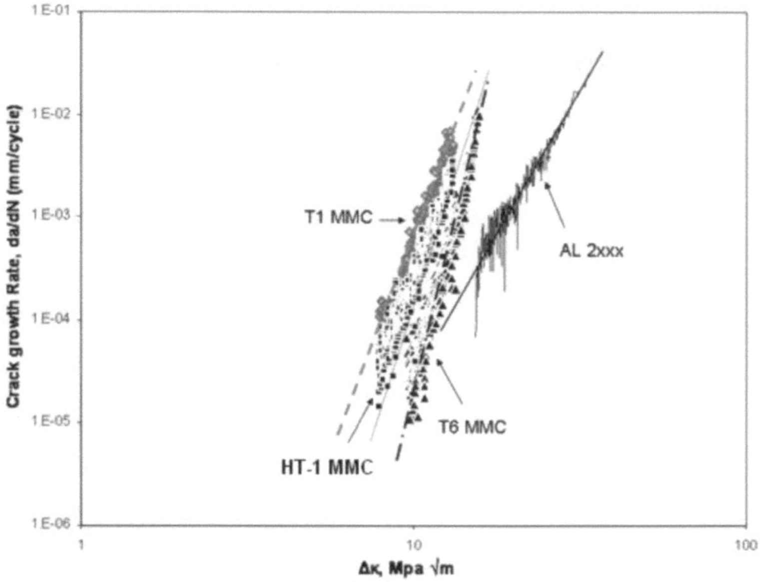

Fig. 16. da/dN vs. ΔK plots of Al/SiCp composite and monolithic aluminum 2xxx specimens

2.9.2 Estimation of da/dN vs. ΔK using thermography and compliance methods

Using the procedure described above based on thermographic mapping, the local stress versus time was measured for the T6 heat treated Al/SiCp along each of the four reference lines placed in front of the CT specimen's notch. The maximum value of stress versus the number of fatigue cycles was then plotted for those four lines (Figure 17). As expected, Figure 17 shows that the local stress, monitored at the location of each line, increases as the crack is approaching that line, then attains a maximum when the crack tip is crossing the line. Finally, after the crack has crossed the line, the local stress measured at the location of the line decreases. This is also expected, since the stress values shown in Figure 17 are stress maxima from all the locations along the particular line. At the exact position on a line where the crack has just crossed, the local stress is null as expected.

From the stress maxima versus fatigue cycles curves, for each reference line, shown in Figure 17, the crack lengths versus the number of fatigue cycles were determined for A359/SiC$_P$ composites in all three different thermal treatment conditions: T1, T6, and HT-1 (Fig. 18). As it is shown in Figure 18, the crack growth rate was found to be quite linear for all heat treatments. Also, there is a small change in the linear slope for the HT-1 heat treated sample, showing increased ductility, which indicates that more time (i.e. cycles) is needed for the crack to grow in this case. For the T6 heat treatment, the results depict a brittle behaviour, as the crack starts to grow earlier than in the other two cases, supporting evidence of brittle fracture.

The stress intensity range was further calculated by the data shown in Figure 17. ΔK values have been estimated from the stress maxima versus fatigue cycles curves for each reference line, shown in Figure 17. Each of the four lines provides a stress intensity range and a da/dN value. The data obtained using lock-in thermography, shown in Figures 18 and 19, were correlated with crack growth rate values obtained by the conventional compliance method and calculations based on the Paris law. Furthermore, the da/dN vs. ΔK curves steaming from the compliance method were plotted in the same graph, for comparison purposes, with those obtained using lock-in thermography (Fig. 20). It can be seen in Figure 20 that the two different methods are in agreement, demonstrating that lock-in thermography is a credible nondestructive method for noncontact evaluation of the fracture behaviour of materials.

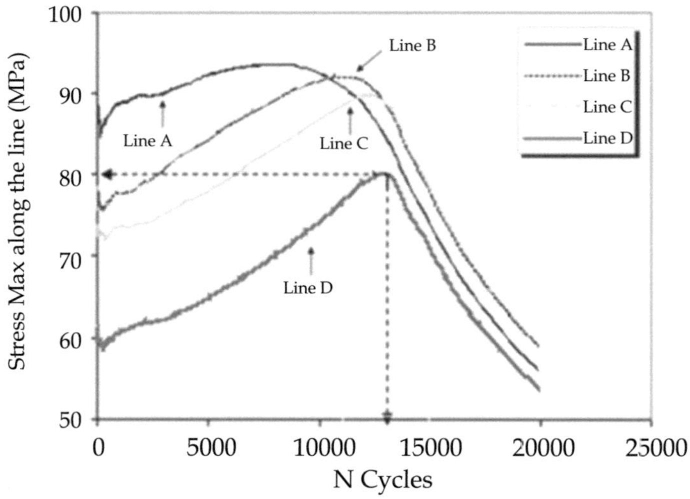

Fig. 17. T6 al/SiCp stress maxima along the four reference lines vs. number of fatigue cycles

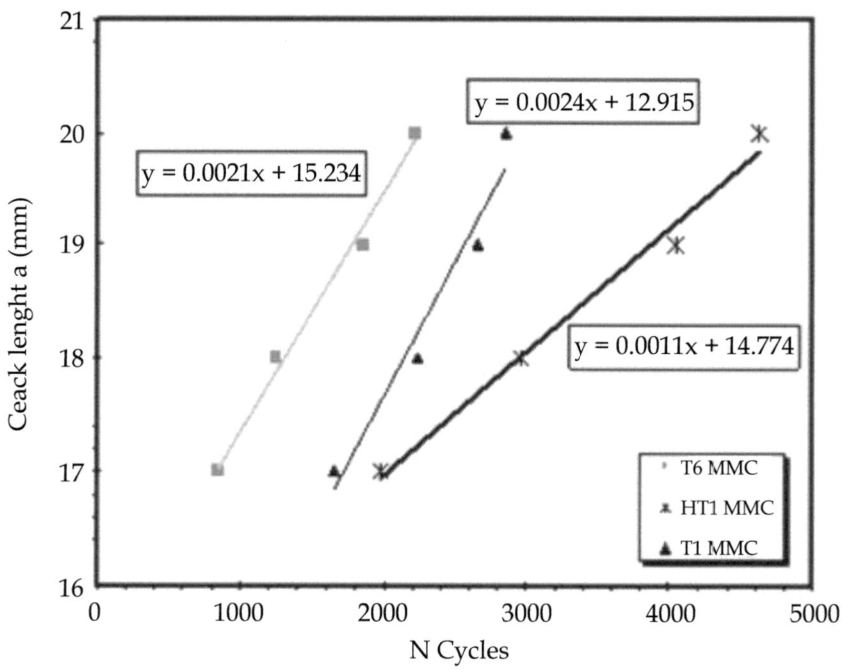

Fig. 18. Crack length vs. cycles obtained from lock-in thermography data for A359/SiCp composites subjected to three different heat treatment conditions

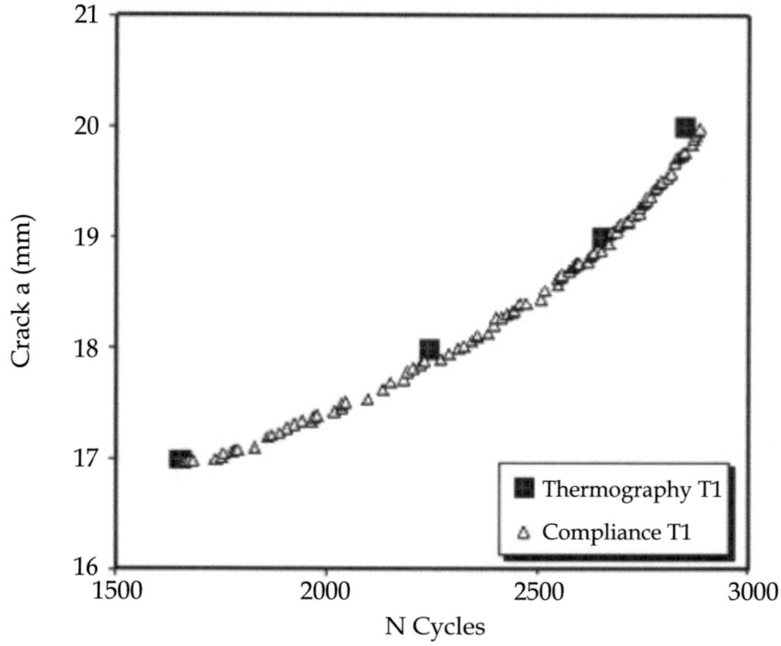

Fig. 19. Crack growth rate determined by compliance vs. thermography for A359/SiCp composite

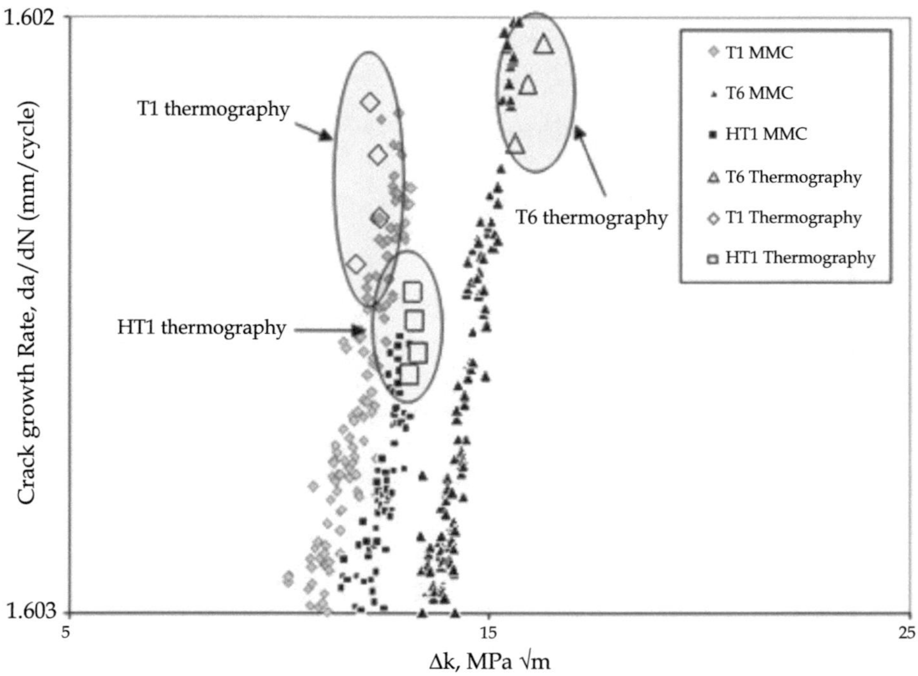

Fig. 20. da/dN vs. ΔK for Al/SiCp specimens - Thermography vs. Compliance method

3. Aluminium honeycomb sandwich panels and metallic foam

3.1 Material and experimental procedure

The compact tension specimens were manufactured from rectangular plates conforming to BS 7448-1: 1991 as shown in Fig. 21.

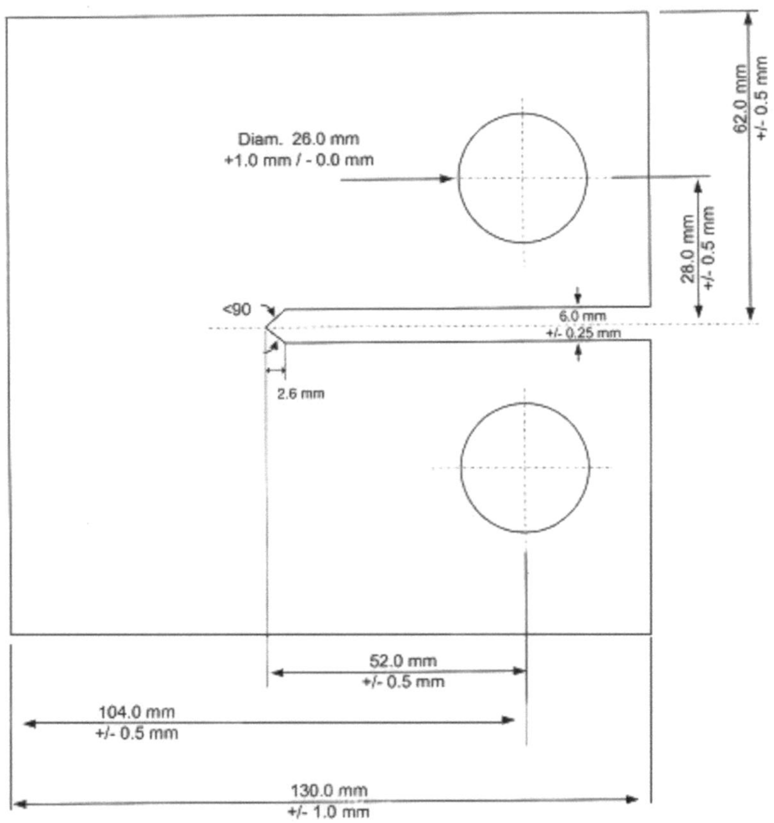

Fig. 21. Compact tension specimen

The fatigue tests were performed on a standard Mayes machine attached to a Pd system. The Pd system was switched on with the current circulating for 30 minutes before any readings were taken; the fatigue cycle was then begun. Data from the Pd system was exported into an excel spreadsheet and the number of cycles versus crack propagation was then plotted. From these results, values of stress intensity (ΔK) versus crack growth rate were then calculated. In order to calculate the stress intensity factor, a standard equation was used.

$$\Delta K = \frac{\Delta P Y}{B\sqrt{W}} \qquad (14)$$

Where, Y = geometry factor, ΔP = change in cyclic load (P_{max} - P_{min}), B = sample thickness and W = sample width. Fatigue tests were conducted in fully tension - tension and at a constant frequency.

3.2 Results and discussion

The results of fatigue testing of aluminium honeycomb sandwich panel both in air and in 3.5% sodium chloride solution, are plotted in Figure 22. A total of twenty cyclic deformation tests were conducted in fully tension-tension at a constant frequency of 2HZ, which is equivalent to two cycles per second. The results of fatigue testing of metallic foam sandwich panel both in air and in 3.5% sodium chloride solution, are plotted in Figure 23.

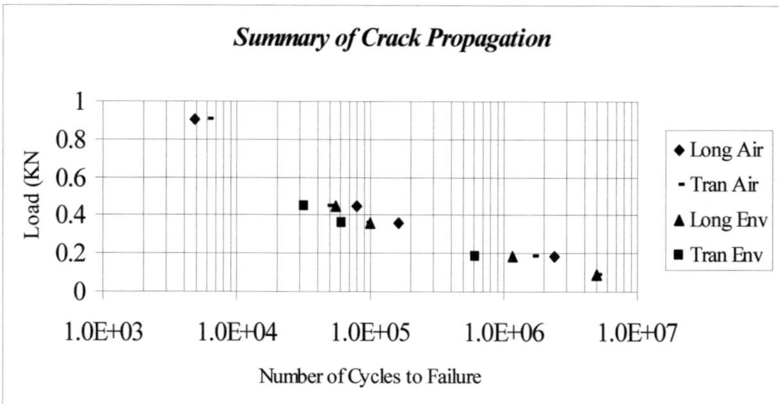

Fig. 22. Aluminium honeycomb fatigue data

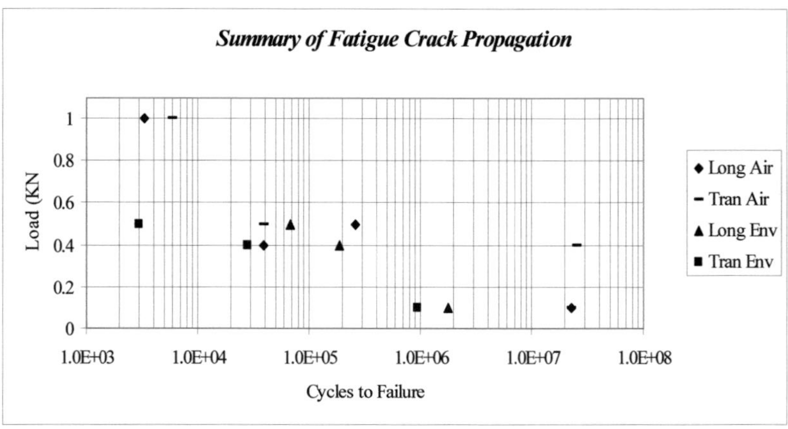

Fig. 23. Metallic foam fatigue data

3.2.1 Aluminium honeycomb

Cyclic deformation data reveals that honeycomb sandwich panel samples do produce consistent results with acceptable repeatability of results even though the honeycomb core is not a conventional structure due to its complex geometry, but because of its homogeneity, it does compare well to the consistent results we would expect from a less complex conventional aluminium solid sample. The results, also, revealed that samples taken from a longitudinal direction constantly have a longer life expectancy, of approximately 40%, then those samples taken from a transverse direction regardless of environmental exposure.

Plotting of fatigue data versus stress intensity for aluminium honeycomb sandwich panels shows that samples tested in a corrosive environment are inferior in performance when compared to samples tested in air. Evidence from crack propagation testing establishes that crack propagation takes place, firstly, within the side plate, leading to some fracture but mainly tearing of the honeycomb structure, only a small amount of crack propagation is evident in the honeycomb structure. The weakest part of the sandwich panel structure appears to be the interface between the aluminium side plate and honeycomb core, with the adhesive used being epoxy resin. Crack propagation testing shows that crack growth is not equal on both sides of the sandwich panel structure; this effect must be due to the complex geometry of the hexagonal core and is a potential difficulty when considering the commercial applications of the aluminium honeycomb sandwich panels.
This research produces a valid method of calculating the Paris exponent, m, with the aluminium honeycomb sandwich panel having a Paris exponent, m, of 1.9. This value is similar when compared to typical values for aluminium alloys of between 2.6 to 3.9.

3.2.2 Metallic foam

Examination of the metallic foam sandwich panel revealed that a consistent form of failure could not be established, with size and position of voids within the metallic foam core having a detrimental effect on failure. Cyclic deformation data revealed that samples tested in air produced inconsistent results showing that the voids within the metallic foam play an important part in crack propagation. However, when samples are tested in an environment, samples taken from the longitudinal direction are superior. This leads to the conclusion that in an environment precipitates within the outer skin have a significant effect on crack propagation. The crack deformation data suggests that due to the complexity of the metallic foam structure and the scatter of results the life of samples exposed to a corrosive environment cannot be correlated with data produced in air. Analysis shows that this is simply untrue and if the two sets of data are plotted a lower and upper trend can be produced, independent of environment and rolling direction, and it is possible to establish a trend of crack growth data within the two bands. This research produced a valid method of calculating the Paris exponent, m. The metallic foam sandwich panel had a Paris exponent, m, of 7.41.

3.3 A proposed method of analysis to predict the fatigue life of sandwich panels

The method of analysis is formed by acquiring the experimental data for aluminium honeycomb and metallic foam sandwich panels. This experimental data is then compared to data produced by calculating the number of cycles to failure. The aim is to calculate the fatigue lives observed experimentally for both aluminium honeycomb and metallic foam sandwich panels. The calculated data will then be used to produce an equation that will predict experimental fatigue life for the complex metallic foam sandwich panels. Conventionally, crack growth rate can be related to the stress intensity factor range using equations (15) and (16).

$$\Delta K = \mathrm{Y}(\Delta P)\sqrt{\pi a} \tag{15}$$

$$\frac{da}{dN} \equiv C(\Delta K)^{m} \tag{16}$$

By rearranging equation (15) and (16) and separating variables and integrating for m ≠ 2 gives,

$$N_f = \frac{2}{(m-2)C(Y\Delta P\sqrt{\pi})^m}\left(\frac{1}{ai^{(m-2)/2}} - \frac{1}{a_f^{(m-2/2}}\right) \tag{17}$$

Before equation (17) could be solved to calculate residual life from a crack size (a_i) one must know the final or critical crack size (a_f). For the critical crack condition when $a = (a_f)$ equation (15) can be rewritten as;

$$a_f = \frac{1}{\pi}\left(\frac{K_c}{YP_{max}}\right)^2 \tag{18}$$

Where; K_c= Fracture Toughness

The fracture toughness of aluminium honeycomb sandwich panels is 0.91 MPa $\sqrt{m}$ and 0.85 MPa $\sqrt{m}$ for Alulight foam. The calculated crack propagation life versus experimental crack propagation life is shown in Figure 24. In Fig. 24 data can be seen for aluminium honeycomb sandwich panels. The graph clearly shows an excellent correlation between calculated and experimental results.

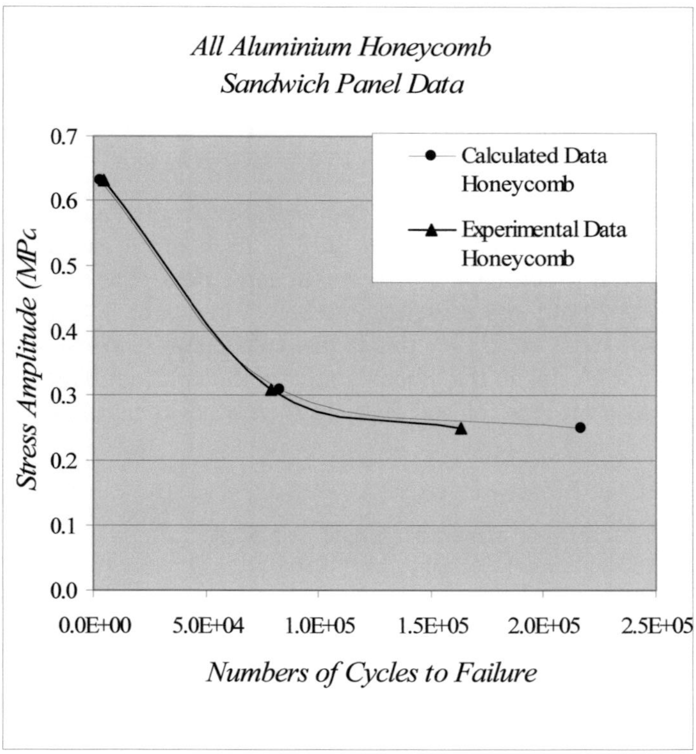

Fig. 24. Calculated Crack Propagation Life versus experimental Crack Propagation Life for Aluminium Honeycomb Sandwich Panel

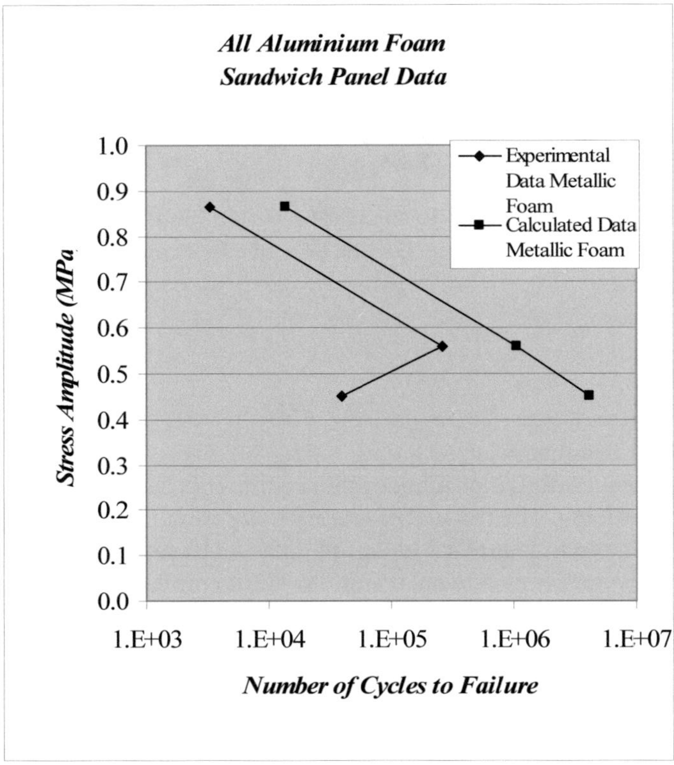

Fig. 25. Calculated Crack Propagation Life versus experimental Crack Propagation Life for Metallic Foam Sandwich Panels

However, the calculated crack propagation life versus experimental crack propagation life for metallic foam sandwich panel is shown in Fig. 25 where it can be clearly seen that calculated data does not correlate with experimental data. The graph illustrates that calculated data always produces a higher number of cycles to failure for metallic foam sandwich panels. The main reason for this is that calculation of stress within the metallic foam structure is complex due to the inhomogenity of the voids within the foam. Equations produced using data from Figs. 24 and 25 were then used to calculate experimental data equation for foam:

Foam Calculated:

$$\sigma = -0.072 \ln N_f + 1.5518 \tag{19}$$

Honeycomb calculated:

$$\sigma = -0.0912 \ln N_f + 1.3589 \tag{20}$$

Therefore, to plot experimental foam data:

$$\sigma = \frac{(-0.1632 \ln N_f - 2.9107)}{2} \tag{21}$$

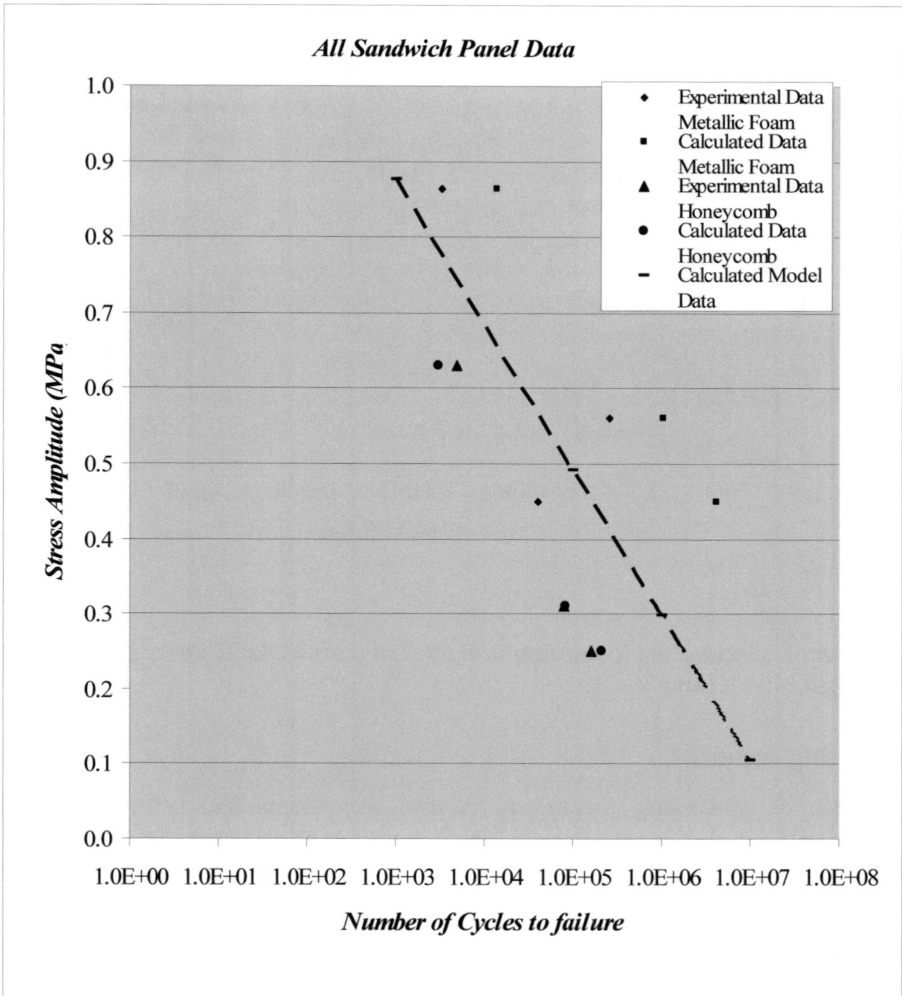

Fig. 26. Calculated Fatigue Life versus Experimental Fatigue Life for Metallic Foam and Aluminium Honeycomb Sandwich Panels

The results of both aluminium honeycomb and metallic foam data for both calculated and experimental cycles to failure is shown in Figure 26. An equation was calculated for each set of data, equations 19 for metallic foam and 20 for aluminium honeycomb respectively. The equations 19 and 20 can then be used to develope an equation 21 to predict the fatigue life of a close cell metallic foam sandwich panel. Using equation 21, calculated life for aluminium honeycomb and metallic foam sandwich panels are compared with original experimental data. What is clear from the Fig. 28 is that all of the experimental data for the two types of specimens correlate with the predicted values. The data correlation proves that a successful model is produced to calculate fatigue life for metallic foam sandwich panels. This model is of extreme importance because it shows that from a structural point of view, metallic foam sandwich panels can successfully replace aluminium honeycomb sandwich panels.

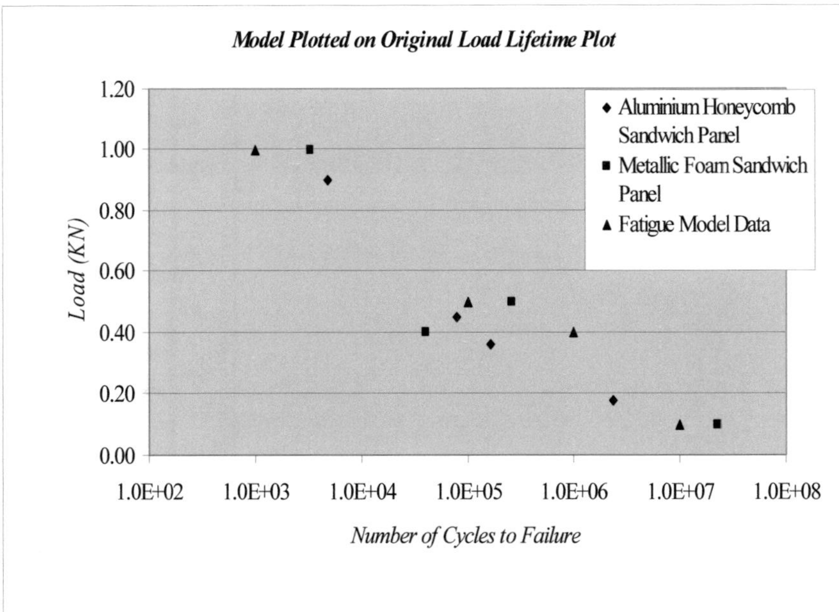

Fig. 27. Showing Fatigue Data for Aluminium Honeycomb, Metallic Foam Sandwich Panel and Calculated Model Data

4. Concluding remarks

The influence of processing conditions in the micromechanical behaviour of Al/SiC composites has been discussed. Two different manufacturing processes (cast and rolled), three reinforcement percentages (20%, 30%, 31%) and three processing states (as received, HT-1, T6 heat treated) have been compared.

The importance of processing conditions in the micro-structural events of segregation and precipitation has been depicted at the micro/nano level using microhardness measurements and nano-scale phase identification of the matrix-reinforcement interface, and the developments of strengthening mechanisms in the composite have been identified. The HT-1 heat treatment condition clearly showed an increase in the microhardness, due to β precipitates as well as other phases and oxides formed in the composite. T6 heat treatment showed the highest microhardness values due to formation of β precipitates, which contribute to strengthening of the interface.

Microhardness and tensile testing results show that the composite micro-mechanical behaviour is influenced by certain factors. In the absence of precipitates (as received state) or in the case of dispersed precipitates (aluminium matrix) the dominant parameters influencing the micromechanical behaviour of the composite are the reinforcement percentage, the interparticle distance and the mean size of particulates. However, when precipitates are concentrated in the areas close to the interface (T6 condition) these precipitates contribute to the strengthening of the composite material.

The thermographic examination of the materials show that heat treated composite samples exhibit regular crack propagation behaviour. Stress concentration, due to the presence of particle reinforcements, produced controlled crack growth and higher stresses, which were related to regular energy release by the material during fracture, indicative of brittle fracture behaviour. On the other hand, the large plastic deformation of the aluminium alloy can be associated with the absence of stress-peaks in conjunction with the monotonic temperature rise for a large part of the temperature / time curve prior to the specimen failure.

A model has been applied to predict the interfacial fracture strength of aluminium in the presence of silicon segregation. This model considers the interfacial energy caused by segregation of impurities at the interface and uses Griffith crack-type arguments to forecast the energy change in terms of the coincidence site stress describing the interface and the formation energies of impurities at the interface. Based on Griffith's approach, the fracture toughness of the interface was expressed in terms of interfacial critical strain energy release rate and elastic modulus. The interface fracture toughness was determined as a function of the macroscopic fracture toughness and mechanical properties of the composite using two different approaches, a toughening mechanism model based on crack deflection and interface cracking and a stress transfer model. The model shows success in making prediction possible of trends in relation to segregation and interfacial fracture strength behaviour in SiC particle-reinforced aluminium matrix composites. The model developed here can be used to predict possible trends in relation to segregation and the interfacial fracture strength behaviour in metal matrix composites. The results obtained conclude that the role of precipitation and segregation on the mechanical properties of Al/SiCp composites is crucial, affecting overall mechanical behaviour.

The tension-tension fatigue properties of Al/SiC composites as a function of heat treatment have been discussed as well as the associated damage development mechanisms. The composites exhibited endurance limits ranging from 70% to 85% of their UTS. The T6 composites performed significantly better in absolute values but their fatigue limit fell to the 70% of their ultimate tensile strength. This behaviour is linked to the microstructure and the good matrix-particulate interfacial properties. In the case of the HT1 condition, the weak interfacial strength led to particle/matrix debonding. In the T1 condition the fatigue behaviour is similar to the HT1 condition although the quasi static tensile tests revealed a less ductile nature.

The crack growth behaviour of particulate-reinforced metal matrix composites was also investigated. Aluminium A359 reinforced with 31% of SiC particles subjected to two different thermal treatments, as well as wrought aluminium 2xxx series specimens, have been examined using thermographic mapping. Heat treated composites, and especially those samples subjected to T6 aged condition, exhibited different behaviour of crack propagation rate and stress intensity factor range than the as-received composite specimens. Furthermore, the composite specimens exhibited different fatigue crack growth rate characteristics than the base aluminium alloy samples. It becomes evident that the path of fatigue crack growth depends on the heat treatment conditions, where crack propagation relies on strengthening mechanisms, such as precipitation hardening. The microstructure of the interphase region was also found to play a significant role in the crack growth behaviour of particulate-reinforced composites. In this sense, T6 heat treated Al/SiCp composite samples exhibits better interphase bonding behaviour than the other composite systems.

The fatigue crack growth curves reveal an approximately linear, or Paris law region, fitting the function $da/dN = C \Delta K$. Crack growth rate vs. stress intensity range curves have been obtained using lock-in thermography. These results are in agreement with crack growth rate measurements using the conventional compliance method and calculations based on the

Paris law. It becomes, therefore, evident that lock-in thermography has a great potential for evaluating nondestructively the fracture behaviour of metallic composite materials.
Finally, cyclic deformation data reveals that metallic foam sandwich panel samples do not produce consistent results with acceptable repeatability of results but by using calculated crack propagation life data and experimental data for both aluminium honeycomb and metallic foam sandwich panels a method of analysis has been proposed to predict fatigue life of metallic foam sandwich panels.

5. References

[1] K.G. Kreider, Composite Materials, in: Metallic Matrix Composites, Volume 4, Academic Press, New York and London (1974).
[2] D.J. Lloyd, Particle Reinforced Aluminium and Magnesium Matrix Composites, Int. Mater. Rev. 39, 1-23 (1994).
[3] N.F. Mott and F.R.N. Nabarro, An attempt to estimate the degree of precipitation hardening, with a simple model, Proc Phys Soc 52, 85 (1940).
[4] H.B. Aaron and H.I. Aaronson, Growth of grain boundary precipitates in Al-4% Cu by interfacial diffusion, Acta Meter. 16, 789 (1968).
[5] H.B. Aaron, D. Fainstein and G.R. Kotler, Diffusion-Limited Phase Transformations: A Comparison and Critical Evaluation of the Mathematical Approximations, J. Appl. Phys. 41(11), 4404-4410 (1970).
[6] H.R. Shercliff and M.F. Ashby, A process model for age hardening of aluminium alloys - I. The model, Acta Mater. 38, 1789-1802 (1990).
[7] R.A. Carolan and R.G. Faulkner, Grain boundary precipitation of M23C6 in an austenitic steel, Acta Mater. 36, 257-266 (1988).
[8] S.T. Hasan, J.H. Beynon and R.G. Faulkner, Role of segregation and precipitates on interfacial strengthening mechanisms in SiC reinforced aluminium alloy when subjected to thermomechanical processing, J. Mater. Process. Technol. 153-154, 757-763 (2004).
[9] M. Manoharan and J.J. Lewandowski, In-situ Deformation Studies of an Aluminum Metal-Matrix Composite in a Scanning Electron Microscope, Scr. Metall. 23, 1801-1804 (1989).
[10] M. Manoharan and J.J. Lewandowski, Effect of Reinforcement Size and Matrix Microstructure on the Fracture Properties of an Aluminum Metal-Matrix Composite, Mater. Sci. Eng. A 150, 179-186 (1992).
[11] S.T. Hasan, Effect of heat treatment on interfacial strengthening mechanisms of second phase particulate reinforced aluminium alloy, 14th International Metallurgical and Materials Conference (Metal 2005), Hradec nad Moravici, Czech Republic, (2005).
[12] J.J. Lewandowski, C. Liu and W.H. Hunt Jr., Effects of Microstructure and Particle Clustering on Fracture of an Aluminum Metal Matrix Composite, Mater. Sci. Eng. A 107, 241-255 (1989).
[13] G. Rozak, J.J. Lewandowski, J.F. Wallace and A. Altmisoglu, Effects of Casting Conditions and Deformation Processing on A356 Aluminum and A356-20% SiC Composites, J. Compos. Mater. 26(14), 2076-2106 (1992).
[14] Bitzer T.1997 Honeycomb Technology: materials, design, manufacturing, applications and testing. Chapman & Hall.
[15] Alulight International GmbH

[16] Burman, M and Zenkert D, 1997, Fatigue in Foam Core Sandwich Beams International Journal of Fatigue, Vol 19, Issue 7, pp 551 -561
[17] Shipsha, A, Burman, M and Zenkert, D. 1999, On Mode I Fatigue Crack Growth in Foam Core Materials for Sandwich Panels. Journal of Sandwich Structures and Materials Vol 2, Issuse 2, pp 103-116
[18] Banhart, J and Brinkers, W, 1999, Fatigue Behaviour of Aluminium Foams, Journal of Materials Science Letters, 18:pp 617-619
[19] Olurin, OB, McCullough, Fleck NA and Ashby MF, 2001, Fatigue crack Propagation in Aluminium Alloy Foams International Journal of Fatigue, Vol 23, Issue 5, pp 375-382
[20] Myriounis D. P., S. T. Hasan, N. M. Barkoula, A. Paipetis, T. E. Matikas, "Effects of heat treatment on microstructure and the fracture toughness of SiCp/Al alloy metal matrix composites", Journal of Advanced Materials, vol. 41(3), pp. 18-27 (2009).
[21] Myriounis D. P., S. T. Hasan and T. E. Matikas, "Influence of Processing Conditions on the Micro-Mechanical Properties of Particulate-Reinforced Aluminium Matrix Composites", Advanced Composites Letters, vol. 17(3), pp. 75-85 (2008).
[22] Myriounis D. P., S. T. Hasan and T. E. Matikas, "Microdeformation behaviour of Al-SiC Metal Matrix Composites", Composite Interfaces, vol. 15(5), pp. 485-514 (2008).
[23] Myriounis D. P., S. T. Hasan and T. E. Matikas, "Heat Treatment and Interface Effects on the Mechanical Behaviour of SiC-Particle Reinforced Aluminium Matrix Composites", Journal of ASTM International - JAI, vol. 5(7), published on-line, DOI: 10.1520/JAI101624 (2008).
[24] D. G. Aggelis, E. Z. Kordatos, T. E. Matikas, "Acoustic Emission for Fatigue Damage Characterization in Metal Plates", Mechanics Research Communications", Mechanics Research Communications, vol. 38, pp. 106–110 (2011).
[25] E. Z. Kordatos, D. P. Myriounis, S. T. Hasan, T. E. Matikas, "Monitoring the fracture behavior of SiCp/Al alloy composites using infrared lock-in thermography", Proceedings of SPIE - The International Society for Optical Engineering, Vol. 7294, Article number 72940X, 2009.
[26] Myriounis D. P., E.Z. Kordatos, S. T. Hasan, T. E. Matikas, "Crack-tip stress field and fatigue crack growth monitoring using infrared lock-in thermography in SiCp/Al alloy composites", Strain, published on-line, DOI: 10.1111/j.1475-1305.2009.00665.x (2010).
[27] C. J. McMahon Jr., V. Vitek, Effects of segregated impurities on intergranular Fracture Energy, Acta Metall, 27(4), 507-513 (1979).
[28] L. Shoyxin, S. Lizhi, L. Huan, L. Jiabao, W. Zhongguang, Stress carrying capability and interface fracture toughness in SiC/6061 Al model materials, Journal of materials science letters 16, 863–869 (1997).
[29] X. Q. Xu, D. F. Watt, Basic role of a hard particle in a metal matrix subjected to tensile loading, Acta Metal. Mater., 42(11), 3717-3729 (1994).
[30] H. L. Cox, The elasticity and strength of paper and other fibrous materials, Br. J. Appl. Phys.,3, 72-79 (1952).
[31] J. Llorca, An analysis of the influence of reinforcement fracture on the strength of discontinuously- reinforced metal matrix composites, Acta Metall. Mater., 43(1), 181-192 (1995).
[32] Myriounis D. P., S. T. Hasan, T. E. Matikas, "Predicting interfacial strengthening behaviour of particulate reinforced MMC - A micro-mechanistic approach", Composite Interfaces, vol. 17(4), pp. 347–355 (2010).

[33] Z. Wang, R. J. Zhang, Microscopic characteristics of fatigue crack propagation in aluminium alloy based particulate reinforced metal matrix composites, Acta metal. mater., 42(4), 1433-1445 (1994).
[34] R. G. Faulkner, Impurity diffusion constants and vacancy binding energies in solids, Mater. Sci. Technol., 1(6), 442–447 (1985).
[35] Hasan ST, Beynon JH, Faulkner RG., Role of segregation and precipitates on interfacial strengthening mechanisms in SiC reinforced aluminium alloy when subjected to thermomechanical processing, Journal of Materials Processing Technology 153-154:758-764 (2004).
[36] R. G. Faulkner, L. S. Shvindlerman, Grain Boundary Thermodynamics, Structures and Mechanical Properties, Materials Science Forum, 207-209, 157-160 (1996).

16

Corrosion Behavior of Aluminium Metal Matrix Composite

Zaki Ahmad[1], Amir Farzaneh[2] and B. J. Abdul Aleem[1]
[1]Mechanical Engineering Department,
King Fahd University of Petroleum & Minerals, Dhahran,
[2]Department of Metals, International Center for Science,
High Technology and Environmental Sciences, Kerman,
[1]Saudi Arabi
[2]Iran

1. Introduction

Metal matrix composite (MMC) is a material which consists of metal alloys reinforced with continuous, discontinuous fibers, whiskers or particulates, the end properties of which are intermediate between the alloy and reinforcement (Schwartz, 1997). These materials have remained the focus of attention of aerospace, automobile and mineral processing industry because of the several advantages they offer which include high strength to weight ratio, elevated temperature toughness, low density, high stiffness and high strength compared to its monolithic counterpart (the original alloy). The particle reinforced metal matrix composites (PRMMC) satisfy many requirements for performance driven applications in aerospace, automobile and electrical industry. The particle reinforced composites can be tailored and engineered with specific required properties for specific application. The commonly used reinforcing materials are silicon carbide, aluminium oxide and graphite in the form of particles and whiskers. Nominal compositions of some well known alloys which are reinforced with whiskers, fibers or particulate is shown table 1. Figure 1 shows that microhardness increases with an increase in filler content of the composites.

	Si	Fe	Cu	Mn	Mg	Cr	Zn	Ti	Al
Al 6061	0.62	0.23	0.22	0.03	0.84	0.22	0.10	0.1	Bal
Al 7075	0.40	0.50	0.60	0.30	2.5	0.15	5.5	0.2	Bal
Al 6013	0.6	0.50	1.1	0.2	0.8	0.1	0.25	01	Bal

Table 1. Nominal composition of some well known alloys reinforced with whiskers and particles

MMC can be continuous or discontinuous. Discontinuous MMC can be isotropic and can be worked with standard metal working techniques such as extrusion, forging or rolling.

Continuous reinforcement uses monofilament fibers, wires or fibers such can carbon fibers. The reinforcement materials commonly used are graphite SiO_2, SiC, TiC, Al_2O_3 and glasses.

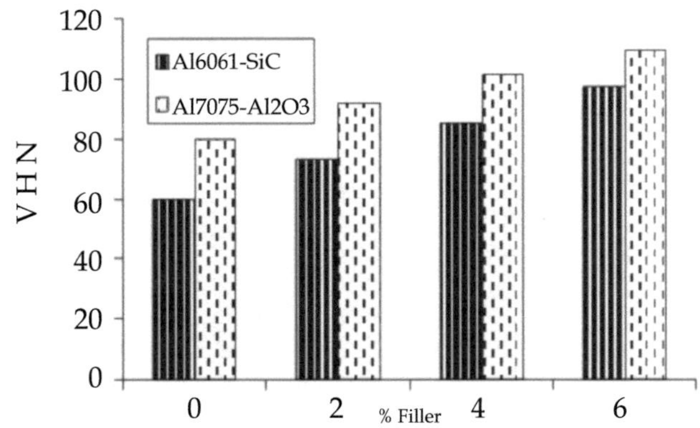

Fig. 1. Microhardness of Al6061-SiC and Al7075-Al2O3 composites (Vaeeresh et al., 2010)

2. Mechanical and physical properties

Metal matrix composites have been shown to exhibit significant improvements in certain physical and mechanical properties over their monolithic metallic counterpart, However, the mechanical properties are strongly dependent on micro structural parameters, in particular, size, shapes volume fraction and orientation of the particles and the composition of matrix.

Parameters	Al 6061	Al 7075	SiC	Al2O3
Flastic Modulus	70 - 80	70 - 80	410	300
Density	2.7	2.81	3.1	3.69
Poisson's Ratio	0.33	0.33	0.14	0.21
Hardness (HB - 500)	30	60	28 W	1175
Tensile Strength(MPA)	115	220	3900	2100

Table 2. Properties of Al 6061 and Al 7075 with and without reinforcement

It is a general observation that the Vickers microhardness observed is greater than the matrix alloy. This is a exemplified by composites, 6061/SiC(p), 6013 SiC(p) and 7075/Al2O3(p) Figure 1 shows the effect of Vol.% of particulates (SiC) on the modulus of elasticity of Al 6061 / SiC, and Al 7075/ Al2O3 composites (Vaeeresh et al., 2010).
The development of metal matrix composites has been a major breakthrough in the last twenty years. The quantum leap in recent years has established their potential for weight critical application in engineering components and structures in aerospace.
It is shown that the tensile strength is increased with increasing volume fraction of SiC particulates. This applies to all metal matrix composites including discontinuously reinforce composite reinforced with SiC particulates or whiskers Figure 2 and 3. The effect of strength may be attributed to the generation of dislocations on cooling of the metal matrix

composites. Such dislocations have been observes by TEM. A high dislocation density was observed on Al 6013/SiC (p) interface.
In a TEM experiment, the generation of dislocations started only at 500 K (Vogelsang et al. 1986). It has also been suggested that dislocation were generated in Al - 6061/ 20 SiC MMC below 573 K.

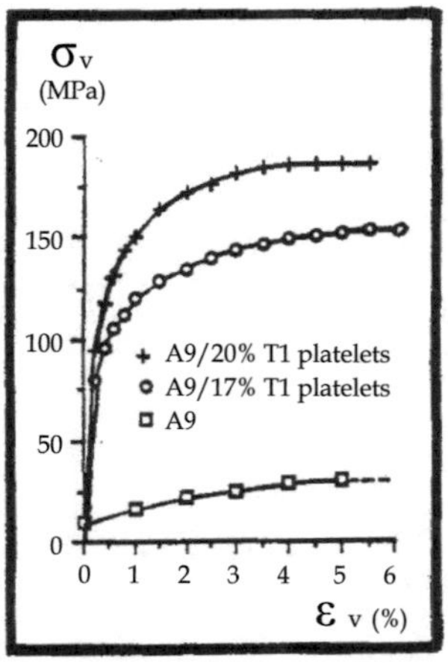

Fig. 2. Effect of the size of the platelets (Massardier et al., 1993)

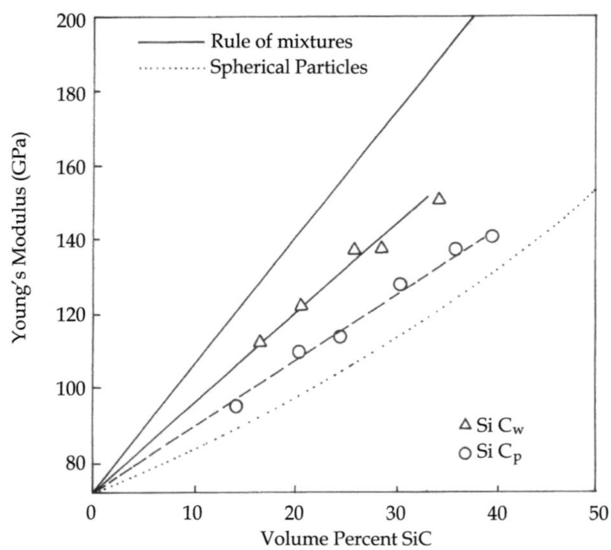

Fig. 3. Young's modulus vs volume percent of SiCw, SiCp and reinforcement (Zaki, 2001)

The elongation (%) of the MMC decreased with increased particulate contents as shown by Al 6061 / 20 SiC (p) - The mechanism of fracture toughness is not fully understood. The presences of large clusters of particles promote crack propagation whereas their uniform distribution retards crack propagation. The fracture toughness values of selected alloys are given in Table 3.

Alloy Designation	Toughness Value
Al2009/ SiC/15(%) W (T 8)	51 mpa √m
Al6061 - 40% SiC(p)	122 mpa √m
Al6013-29SIC(p)	19.5 KSi√2

Table 3. Fracture toughness of selected MMCS

The strains to failure (%) for different Al_2O_3 reinforcement are shown in Table 4 Strain to failure decreases with increase of volume fraction of reinforcement.

Vf %	Percent strain to failure
	(Ef)mm/mn*100
Al 6061	29.26
Al 6061 / 10 vol.% Al2O3	4.72
Al 6061 / 20 vol.% Al2O3	2.29
Al 6061 / 30 vol% Al2O3	1.42

Table 4. Strain to failure of Alloy Al 6061 with increasing volume fractions (Dehlan and Syed, 2006)

Al MMC are finding increasing applications as rotor material in automotive brake systems (Shorowords et al., 2004). Effect of Studies on the effect of sliding velocity on wear friction and tribochemistry of MMC reinforced with 13% SiC or B4C have shown that sliding velocity leads to lower wear rates and lowers friction coefficient for both MMCs.. Studies on interaction between MMC and phenolic brake pads showed that the transfer layer consisting of phenolic pad material acted as a protective layer and reduced wear rates and coefficient of friction. Honda has used aluminum metal matrix cylinder liners in some of their engines including B21Al and H23A, F20 C and F22C.

The effect of cutting speed on tool wear has been investigated. The cutting tool wear increased with increased reinforcement ratios. At constant speed and feed rate, the lowest wear rate has been found in 5 Wt % SiC (p) and the highest wear with 15 Wt % SiC_p increased cutting speed increased the tool wear rate.

From the above description, it may be concluded that the development of MMC has been a big breakthrough in search for stiff high strength materials for aerospace and automotive industry particularly. Whereas the mechanical properties of MMC have remained the focus of attention, the work on corrosion behavior of MMC did not proceed hand in hand with the mechanical and tribological properties. The work on corrosion was undertaken the last decade and a considerable progress has been made in the understanding of corrosion behaviour of metal matrix composites in recent years.

3. Corrosion behavior of Aluminum metal matrix composites

The corrosion behaviour of alloys in sea water 3.5 Wt % NaCl represents an adequate measure of its corrosion resistance. Important results of corrosion studies undertaken in the last decade would be discussed under the following categories.

a. Immersion and long term exposure tests in sea water or 3.5 wt % NaCl.
b. Localized corrosion studies
c. Flow induced corrosion and Erosion corrosion
d. Corrosion inhibition
e. Corrosion mechanism

3.1 Immersion & long term exposure studies

The above studies were conducted in accordance with ASTM designation G 31 - 72 (ASTM, 2004). The results of studies on Al6092 - T6, Al/*B*4C/20P, Al 6092 - T6 /2oSiC(p), and 6092 - T6 20vol%Al2O3 and monolithic 6061-T6 Al, immersed for 90 days in air exposed 0.5 Na_2SO_4 solution, 3.5 wt% NaCl, ASTM sea water and real sea water were recently described (Hongho et al. 2009). In alloy 6092 - T6 Al/B 4C/20P MMC specimen in ASTM Sea water bubbles were observed. The current over most of the area was found to be anodic. The solution at the anode site was found to be acidic (PH 6.4). Corrosion products were formed as observe after monitoring for three days and the area became more alkaline (PH 8.4). A similar phenomena occurred with alloy 6092 reinforced with 20 Vol. % SiC (p) and gradually the alkalinity increased because of its change of area from and anodic to cathodic. The corrosion rates of MMCS in sea water and ASTM sea water were lower than those in 0.5 M Na_2SO_4 and 3.5 wt % NaCl. The rates of monolithic 6061 - T6 Al in both real and ASTM sea water were significantly lower than those in 3.5 wt % NaCl. The surface morphology after the test showed similar general features, one major feature of the surface morphology was the presence of intermetallic precipitates on the surface. The EDS studies suggested these precipitates to contain Al, Mg, O, and C. Mg and HCO_3 irons as the main species corrosion products.

The formation of precipitates is a greater concern in MMC, as localised corrosion is controlled by the formation of such precipitates. The role of precipitates would be discussed in the relevant section of the paper. In general the corrosion rate of Al MMC decreased with time due to the formation of precipitates.

3.2 Localled corrosion of ALMMC's

If is generally accepted that MMC are in general more prone to corrosion than their monolithic counterparts (Berkely et al., 1998; Turnbull and Corros, 1992; Trzskoma, 1991). Conflicting views have been presented on the causes of the localised corrosion. The results of the studies showed that galvanic corrosion between the matrix and the reinforcement occurs. However, this is related to the machining conditions. Three different machining process; Wielding Electrical Discharge Machine (WEDM), Cemented Carbide Turning and Single Point Diamond Turning were employed for investigation. The test results for different process are shown in Table 5 (Yue et al., 2002).

	E_{Corr} (mV)	E_{pitl} (mV)	$E_{Pil} - E_{corr}$ (mV)	I_{Corr} (Am^{-l})
WED	-761.4	- 633 v	128.4	3.80 TE - 4
Carbide Turning	- 673.6	- 655	186	3.194 E - 2
Diamond Turning	- 928.3	- 655	288.3	1.052 E - 3

Table 5. Electrochemical parameters for different machining conditions (Yue et al., 2002)

The electrical discharge machining showed the highest value of pitting potential. The resolidified layer did not show any extensive pitting. The results show that surface conditions have a major effect on pitting potential and the resistance to pitting may be shown by E_{pit} - $E_{Corrosion}$. The difference above is not sufficient to predict corrosion susceptibility. It may be observed that silicon carbide is an insulator and there is hardly any possibility of cathodic reaction occurring on the surface of particles. The theory that Al/SiC is sensitive to corrosion because of micro galvanic coupling applies to some intermetallic compounds, cathodic to the matrix such as $CuAl_2$ which is formed. So far there is no general agreement on the role of SiC particulates on the mechanism of localized corrosion. The electrochemical behaviour of Al2024/AlSiC has been also investigated by scanning micro reference electrode imaging system (Feng et al., 1981; Isacs & Vyas, 1981). The results of investigations on Al2024/Al SiC (A) are given in Table 6.

Volume Fraction	$E_{pitting}$			$E_{protection}$			$E_{corrosion}$	
	0.01 m NaCl	6.1 m NaCl	0.5 m NaCl	0.01 m NaCl	0.1 m NaCl	0.5 m NaCl	0.5 NaCl	0.1 NaCl
0	- 430	- 497	- 565	- 653	- 620	- 612	- 612	- 574
5	- 460	- 528	- 597	- 750	- 700	- 670	- 670	- 610
10	- 485	- 555	- 625	- 740	- 765115(T)	- 720	- 725	- 688
15	- 538	- 632	- 662	- 700	- 720	- 720	- 750	- 671
20	- 550	- 650	- 692	- 670	- 670	- 775	- 775	- 671

Table 6. Summary of electrochemical data (Feng et al., 1981; Isacs & Vyas, 1981)

It was observe that pitting potential E_p decreased as the volume fraction of SiC particulate reinforcement increased. The relation between the volume fraction and $E_{Protection}$ It was clearly observed that the pitting attack occurred at SiC/Al interface which contained intermetallic Cu and Al precipitates. The presence of Mg, Cu, and Fe compounds in Al6013/20% Vol. of SiC has been confirmed also in another work in recent years (Zaki et al., 2000). The interfacial regions may act as active centers for localized corrosion on immersion in sodium chloride solution. The EDS spectrum of Al_2Cu is shown in Figure 4. The pits on Al 2024/SiC interface are shown in Figure 5. In Al 2024/SiC MMC, Mg may segregate in addition to the precipitates of Al_2Cu Mg and Al_2Cu. The segregated magnesium may form active galvanic couple with Al matrix (Jamaludin et al., 2008). There is also the possibility of the intermetallic precipitates to act as local anodes or cathodes because of the difference between the open circuit potentials of these intermetallic with Al matrix. As seen above the role of the precipitates and inclusion is not clearly understood. However, the evidence of localized corrosion of Al MMC suggests, that the Al/SiC interface in active and responsible for localized corrosion. This is also confirmed by studies on (Al 2009/SiC W) (W = whisker). In the rolled material extensive pitting occurred, and on removing the corrosion products it was observed that the pits contained particles $CuAl_2$ (Rohatgi, 2003). On heat treatment the amount of $CuAl_2$ particles was significantly reduced (Rohatgi, 2003) and the rate of corrosion also diminished which suggested that the heat treatment diminished Mg, Fe and $CuAl_2$ precipitates Figure 6 shows the effect of heat treatment on the corrosion behaviour of T6 and as rolled Al 2009/Sic (w) composite. The corroded surface of as rolled specimens is shown in Figure 6.

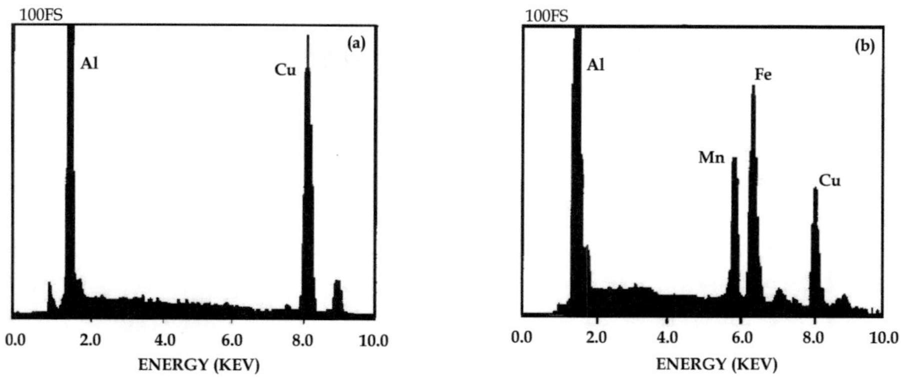

Fig. 4. EDS spectrums of (a) Al2Cu and (b) (CuFeMn) Al6 inclusions (Feng et al., 1998)

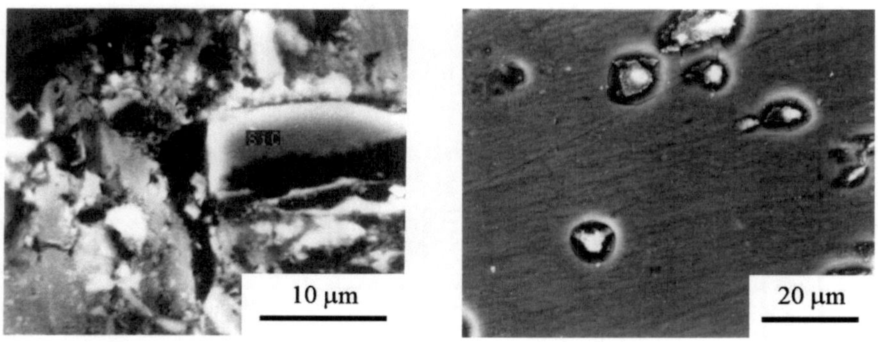

Fig. 5. Scanning electron micrographs of pits on interfaces of (a) SiCp-2024 Al matrix, and (b) inclusions-2024 Al matrix (Feng et al., 1998)

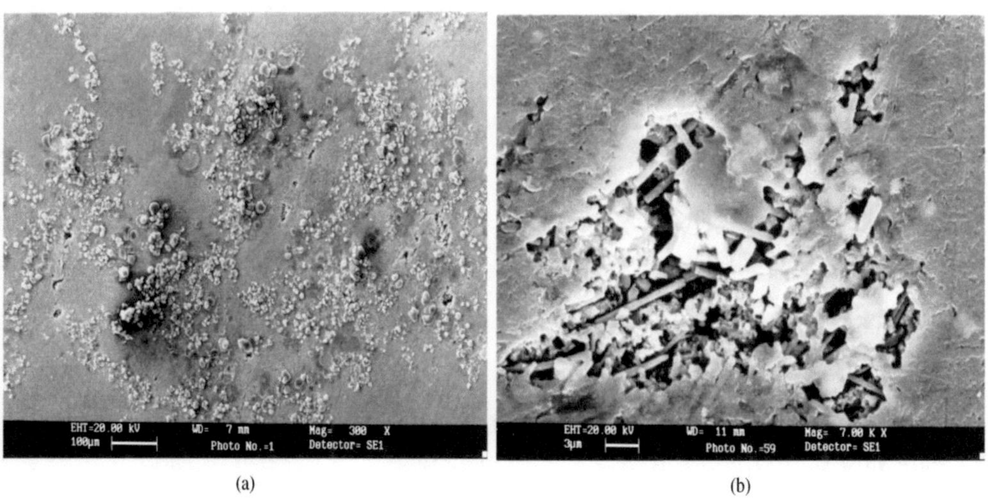

Fig. 6. Corroded surface of the as-rolled specimen after the polarization test (a), (b) showing pit morphology (Yue et al., 2000)

3.3 Flow induced corrosion and Erosion corrosion

The resistance of metallic equipment and structures to the impact of flow induced corrosion is extremely important as it affects their operational life and integrity of equipment. Whereas the effect of velocity on the erosion/corrosion of steel copper, and aluminium alloys are widely reported in literature the information on the metal matrix composite is scanty (Rohatgi, 2004; Griffen &Turnbull, 1994; Lin et al., 1992; Mansfield & Jeanjagnet, 1984; Chen & Mansfeild, 1997; Hihara, 2010; Colman et al., 2011). Studies on Al 6013-20 SiC were conducted in a customized recirculation loop as shown in Figure 7. It consisted of entry valves, a manometer, a centrifugal water pumps, a flow meter and several specimen holders to accommodate flat specimens. Each specimen holders contained four specimens which were housed in an outside container. The velocity was varied by varying the chamber of the specimen holders. Three tempers of Al6013-20 SiC (p) were investigated in the loop. In which a solution of 3.5wt%% NaCl was flowing at velocities ranges from 1-4 ms^{-1}.

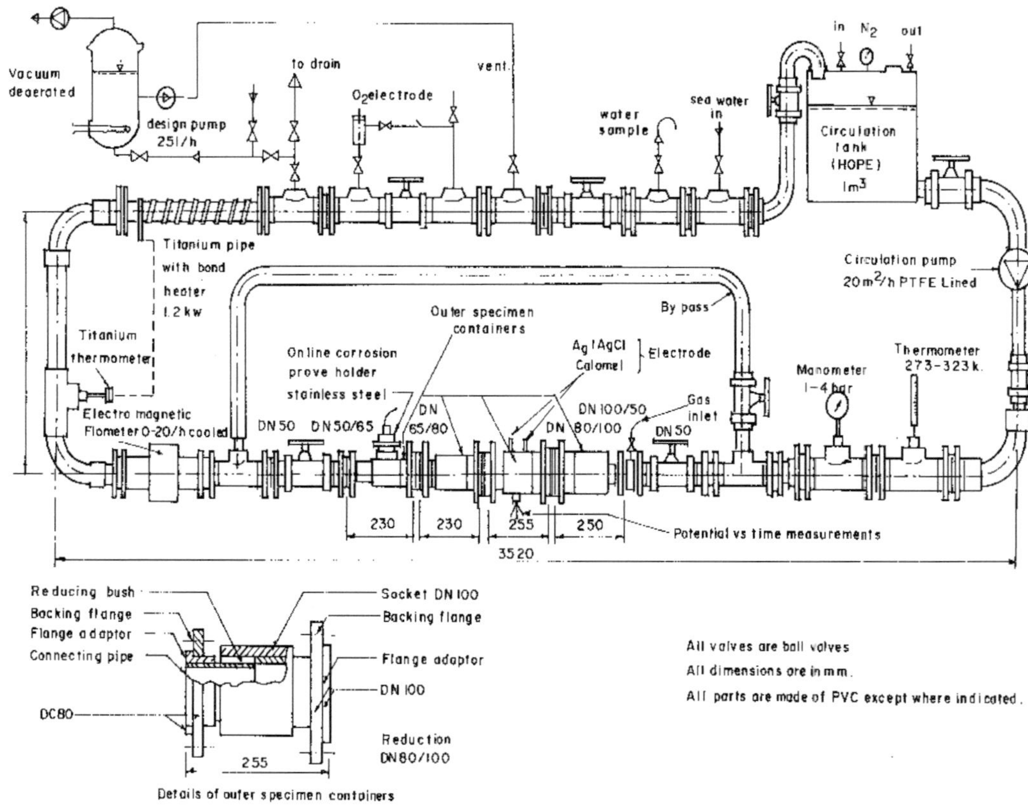

Fig. 7. Schematic diagram of PVC recirculating loop (Zaki, 2001)

After exposure of 100 hours it was shown that temper (0) annealed, and temper F, as fabricated, showed a lower resistance to corrosion in 3.5 wt% NaCl with and without polystyrene suspended particles. Upon increasing the temperature form 30 to 50 and 90 C, the erosion corrosion rate increased as shown in Table 7 and 8 (Zaki, 2007).

	Corrosion Rate(In 3 weight% NaCl + Vol% Polystyrene(mpy)		
Velocity	Temper(0)	Temper (F)	Temper T4
1.0	11.8	9.9	9.6
2.7	12.6	10.8	10.1
3.8	12.9	11.3	11.4

Table 7. Variation of Erosion-Corrosion Rate with Velocity in 3.5wt%% NaCl + 2%Vol Polystyrene (Zaki, 2007)

		Erosion Corrosion Rate(mpy)		
Temperature (°C)	Velocity	Temper(O)	Temper(F)	Temper(T4)
50	1.0	12.1	10.3	9.9
	1.9	3.6	11.2	10.1
	2.7	14.2	12.1	11.4
	3.8	14.9	13.6	10.3
75	1.0	11.9	11.9	117
	1.9	15.5	15.5	161
	2.7	172	17.2	148
	3.8	19.6	19.6	159
90	1.0	13.3	13.3	12.3
	1.9	13.1	15.1	13.6
	2.7	17.8	17.8	163
	3.8	19.7	19.7	17.6

Table 8. The effect of temperature on the erosion - corrosion behavior of Al 6013 - 2051 C (p) in 3.5 wt % NaCl + 2%Vol Polystyrene (Zaki, 2007)

The erosion-corrosion rate increased, linearly with velocity in the presence of SiC particles. It was also found that Temper (T4) of the alloy showed the best resistance to corrosion. The rate of erosion corrosion varied also with temperature. The best resistance offered by T4 may be attributed to the homogenization of the surface structure, less clustering of SiC particles, a uniform distribution of secondary intermetallic phases such as $CuAl_2$ and minimization of micro-crevices (Zaki, 2000). The localized attack was confined to Al 6013/20 SiC (p) interface. A large number of secondary phase particles were observed. After studies showed the presence of Cu 3.55 %, Fe 1.77 %, Mg 1.71 %, and some Cl (0.32%) a high dislocation density was observed at the interface Figure 8 (Zaki, 2000). The formation of coherent films was made more difficult by the protrusion of the particles. This factor adds significantly the erosion –corrosion caused by polystyrene particles. The surface is subjected to a cycle of destruction and reformation of a protective film as a result of impact of polystyrene particles. The corrosion product which accumulates at the interface may act as cathode and increase the cathode / anode area ratio causing an overall increase in the rate of corrosion.

Alloy Al 6013 / 20 SiC (p) in temper T4 offered of temper T4 offered a good resistance to erosion–corrosion. It can be used in water containing Silica or other particulate matter without undertaking any major risk. Al 6013 reinforced with 20 Vol. % SiC (p) was designed to have improved mechanical properties over those of AAl11 6061/SiC (p). The corrosion resistance of al 6013 /20 Sic (p) was determined in fog testing cabinet (Zaki, 2000). A schematic of salt spray chamber is given in Figure 9. The cabinet comprised of a basic chamber level matic test reservoir (1.0 gal salt solution), reservoir (3.0 gal), bubble tank, twin optic fog assembly, and accessories such as a lower assembly bubble tank heater, control valves, and cabinet heaters. The cross section of the assembly is shown in Figure 10. The results obtained for O, F, and T4 Tempers of the alloy composite in the fog cabinet are shown in Table 9.

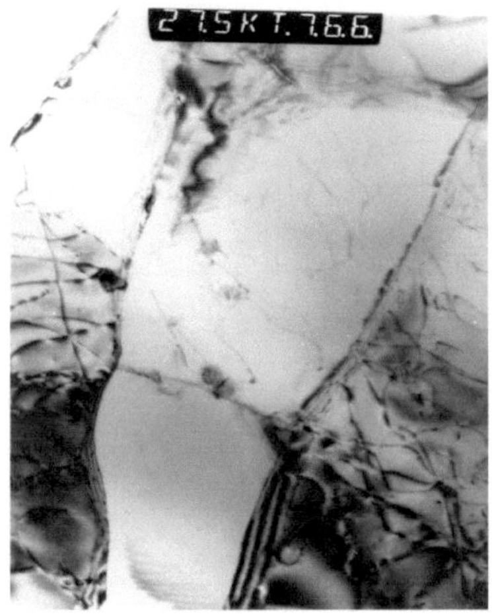

Fig. 8. TEM micrograph of Al 6013/SiC interface showing dislocation generations

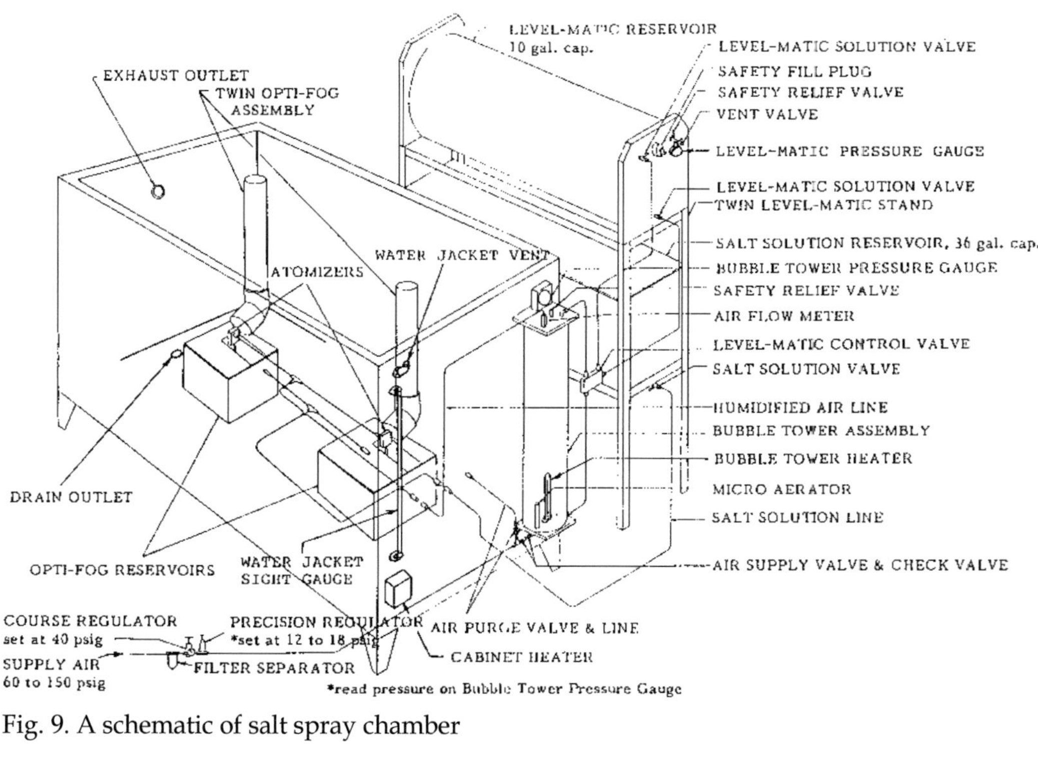

Fig. 9. A schematic of salt spray chamber

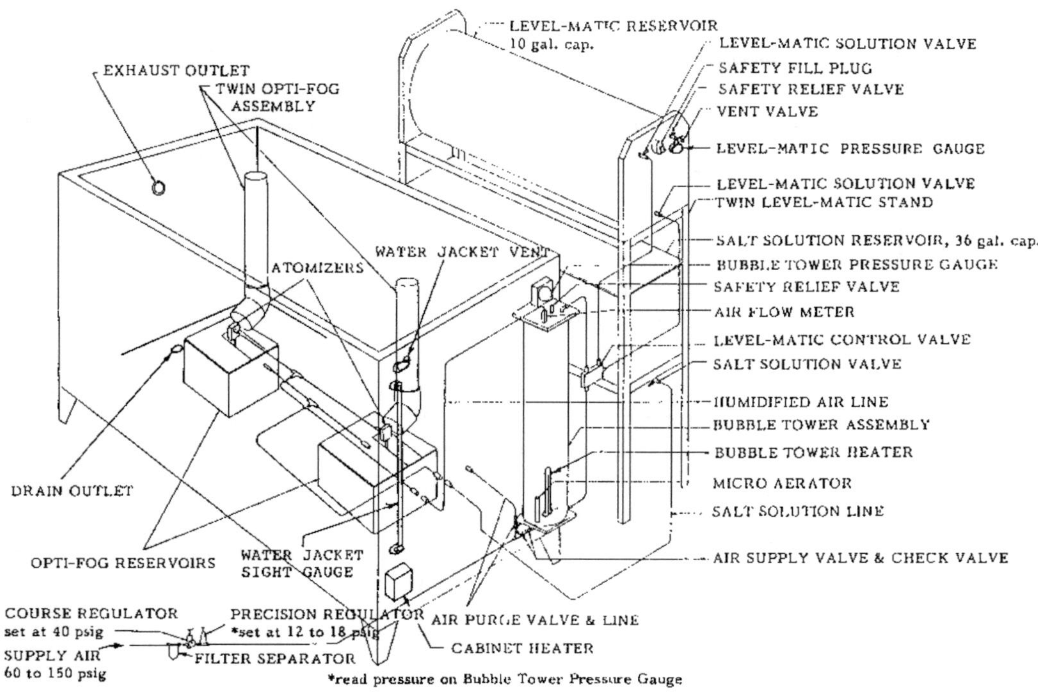

Fig. 10. Cross section of singleton salt fog corrosion test cabinet

Time	Temper-0	Temper-F	Temper-T4
200	10.23 V (19.8)	8.42 (15.78)	7.12 (13.35)
400	9.11 (17.8)	7.78 (14.58)	6.18 (11.09)
600	6.38 (11.96)	6.06 (9.49)	4.38 (8.21)
800	21.92 (9.23)	3.98 (7.46)	2.82 (5.28)
1000	4.66 (8.74)	3.76 (7.05)	2.63 (4.53)
1200	4.27 (8.01)	3.68 (6.90)	2.50 (4.83)

Table 9. Corrosion rates of Al6013/20SiC(p) in Salt Spray Chamber

A decrease in corrosion rate with increased exposure period was observed for all three tempers. The MMC temper T4 showed the highest resistance to pitting. The surface of the composite was often covered with a gelatinous product of aluminum hydroxide $Al(OH)_3$. The pit environment was acidic and bubbles of hydrogen rose from the surface forming corrosion chimneys. The hydrogen bubbles pump up $AlCl_3(OH)_3$ to the outside which reacts with water to form $Al(OH)_3$ (Burleigh et al., 1995). The pitting depth in temper T4 were lower than pitting depths in F and O tempers. It was reported that a high concentration of intermetallic compounds was observed at Al/SiC inter-phases which lead to localized corrosion (A). The corrosion rate of Al6013-20SiC (p) decreased for all tempers on increasing the temperature form 50 to 75°C and increased again on raising the temperature to 100°C. This change may be attributed to the changes brought about by the composition of the protective films from being bayerite ($AlO(OH)$) to boehmite (Al_2O_3, H_2) as shown by FTIR (Fourier transformation infra-red) spectroscopy).

The corrosion behavior of Al6013–20SiC (p) is a very strong function of Al (OH)3 and once the film formation is completed it becomes independent of oxygen (Beccario et al., 1994). The crystals of boehmite have been observed on the surface of the alloy. The data generated in highly aggressive environment shows promising applications potential of this alloy in salt water and humid environment typical of sea coastal environment in the Gulf Region.

3.4 Effect of Inhibitors

It has been shown in earlier sections that Al/SiC metal matrix composites such as Al 6013-20 SiC (p) exhibit improved mechanical and physical properties compared to wrought alloys. However, they are more susceptible to pitting than their monolithic counterparts (Beccario et al., 1994; Trazaskama, 1990). They also exhibit a higher corrosion rate at velocities greater than the 2.3 ms^{-1} (Zaki, 2000). A variety of surface modification techniques such as anodizing, chromate conversion coatings and organic finishing have been suggested for the protection of aluminum metal matrix composite from localized corrosion (Aylor & Moran, 1986; Lin et al., 1989; Mansfield et al., 1990). Cerium coatings have been the focus of attention in the last decade (Hinton & Arnold, 1986; Davenport et al., 1991).
Studies on to investigate the effect of inhibitors on Al 6013 – 20 SiC (p) included weight loss, Electrochemical and re-circulation loop studies (Zaki, 2009).
Following inhibits solutions were used

a. 1000 ppm $K_2Cr_2O_7$ + 1000 pm $NaHCO_3$ + 3.5 wt % NaCl
b. 1000 ppm Cerium chloride + 3.5 wt% NaCl
c. 1000 ppm sodium molybdate + 3.5 wt % NaCl

The results of inhibitive action of $K_2Cr_2O_7$ + 1000 pm NaHCO3 are summarized in Table 10.

Alloy Designation	Velocity (ms-1)	Corrosion rate in mpy(MDD) with no inhibitor	Corrosion rate in mpy(MDD)with inhibitors
Al 6013-20 SiC(p)-O	1.0	11.8(22.1)	3.07(5.76)
	1.9	11.6(21.7)	7.63(14.32)
	2.7	12.9(24.1)	8.4(15.77)
	3.8	13.6(25.5)	9.63(18.08)
Al 6013-20 SiC(p)-F	1.0	9.9(18.5)	3.61(5.68)
	1.9	10.4(19.5)	4.31(8.09)
	2.7	10.8(20.2)	5.53(10.38)
	3.8	11.3(21.2)	6.60(12.39)
Al 6013-20 SiC(p)-T4	1.0	9.6(18.5)	2.01(3.77)
	1.9	10.1(18.2)	2.70(5.07)
	2.7	10.8(20.2)	3.40(6.38)
	3.8	11.4(21.4)	3.80(7.13)

Note; All experiments were conducted in 3.5 wt% NaCl

Table 10. The results of inhibitor action of k2Cr2O7+1000 ppm NaHCO3 (Zaki, 2009)

The reduction in the corrosion rate with $K_2Cr_2O_7$ +$NaHCO_3$ has been attributed to the formation of protective layer of boehmite Al $(OH)_3$, $3H_2O$ and bayrite Al_2O_3, H_2O. The breakdown of the oxide layer leads to pitting. The reduction in the corrosion resistance at increased velocities is caused by continuous removal of protective layer by erodent particles. The protrusion of particulates also makes it difficult to achieve a passivating layer; hence the resistance to the impact of velocity is lowered.

The preferred site for localized corrosion is Al/SiC interface as this site is abundant in intermetallic compound (Zaki, 1998). The existence of thermal stresses and dislocation density at interface affects the kinetics of erosion corrosion and increases the sensitivity if Al/SiC interfaces to erosion-corrosion. Because of the encouraging results of inhibition treatment of Al7057, and Al1000, with cerium chloride and sodium molybdate, studies were further conducted on Al6013 -20 Vol. % SiC(p) MMC. The effect of inhibition treatment is shown in Table 11 below.

		Corrosion rate in 3.5% NaCl+1000 ppm		Corrosion rate in 3.5% NaCl+1000 ppm Namoo4, CeCl3
Sr. No	Temperature °C	Temper	Mpy(mdd)	Mpy(mdd)
1	50	0	4.72(8.86)	3.8(7.13)
		F	2.24(4.13)	1.8(3.38)
		T4	1.71(3.21)	0.9(1.69)
2	70	0	8.3(15.5)	5.06(9.5)
		F	6.53(12.26)	4.01(7.5)
		T4	2.54(4.77)	2.01(3.72)
3	100	0	12.90(24.2)	8.05(15.11)
		F	11.60(21.7)	8.26(15.41)
		T4	10.19(19.13)	5.41(10.15)

Table 11. Effect of Inhabition Treatment

As shown by table 11 cerium chloride is a more effective inhibitor than sodium molybdate as shown by a larger reduction in corrosion rate brought about by addition of cerium chloride compared to sodium molybdate. The corrosion rate of temper of the MMC is reduced from 19.13 mpy to 3.96 with Cerium Chloride at 100°C which is very significant. Electrochemical studies were also conducted at 50, 70 and 100°C to observe the effect of temperature on inhibition. The electrochemical data obtained by above studies is shown Table 12.

The results of studies summarized in Table 12 clearly established that cerium chloride is a more affective inhibitor than sodium molybdate. The large difference between the corrosion potential (E_{corr}) and the pitting potential (E_p) shows that the cerium chloride is a more affective inhibitor in 3.5 wt % NaCl. The corrosion potential (E_{corr}) shifts closer to E_p which shows the sensitivity of the MMCS to localized pitting in Sodium Chloride without inhibition. The cathodic polarization curve of temper T4 of the alloy in 3.5 wt% NaCl +1000 ppm $CeCl_3$ in dearated condition is shown Figure 11. The curves are overlaid on the main curve. A maximum reduction in current density (from 234 to 25.1uA/cm2) is exhibited by Temper T4 in cerium chloride (Zaki, 2009). The current densities recorded are summarized in the Table 13.

Solution	Temperature °C	Temper	R(K.ohms)	E_{corr}(mv)	I_{corr}(μA/cm^2	Corrosion rate mpy(mdd)
Cerium Chloride	50	O	9.004	-0.8	3.727	1.60(2.99)
	50	F	3.301	-0.783	0.57	2.80(5.28)
	50	T.4	2.43	-0.78	1.1	0.47(0.88)
Cerium Chloride	70	O	1.281	-0.909	4.757	2.04(3.81)
	70	F	9.14	-0.915	3.993	1.68(3.14)
	70	T.4	10.07	-0.993	2.155	0.92(1.71)
Sodium Molybdate	50	O	44.5	-0.8	150	1.24(3.22)
	50	F	11.91	-0.716	216.5	1.92(3.60)
	50	T.4	41.3	-0.791	72.8	0.47(0.88)
	70	O	58.77	-0.909	272.6	2.74(5.11)
	70	F	23.76	-0.916	137.0	2.12(4.06)
	70	T.4	34.87	-0.867	111.5	0.75(1.46)
Sodium Molybdate	100	O	100	-0.87	100	8.60(16.00)
	100	F	100	-0.868	100	6.50(12.16)
	100	T.4	58	-0.947	74	3.96(7.41)

Table 12.

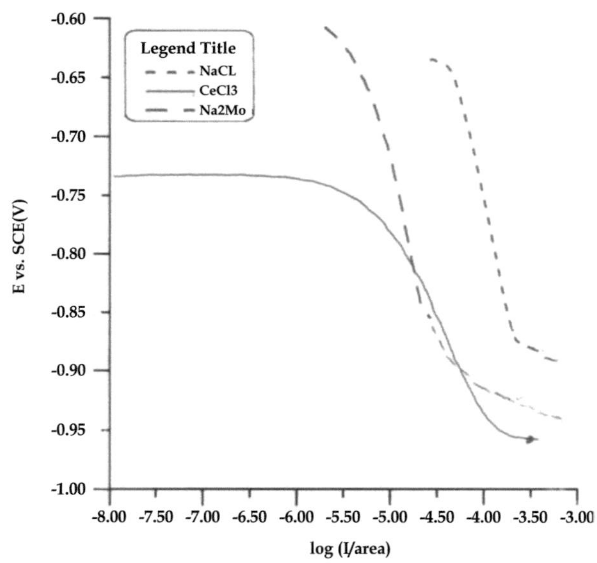

Fig. 11. A cathodic polarization curve of temper T4 of the alloy in 3.5 wt% NaCl + 1000 ppm $CeCl_3$ in deaerated condition (Zaki, 2009)

Sr. No.	Media	Icorr(μA | cm2)
1	3.5 wt % NaCl	234
2	3.5 wt% NaCl +1000 ppm Cecl3	25.1
3	3.5 wt% NaCl + 1000 ppm NaMoo4	178

Table 13. Current Densities of MMCS after Inhibition (Zaki, 2009)

Cerium chloride acts as a strong cathodic inhibitor for the alloy. Sodium molybdate on the other hand acts as an anodic inhibitor which acts by raising the pitting potential (Up) in the positive direction while maintaining E_{corr} negative to E_p. A typical cyclic polarization curve of the temper T4 of the alloy in 3.5 wt% NaCl + 1000 ppm $NaMoO_4$ is shown in Figure12. The corrosion potential tends to shifts to more positive values.

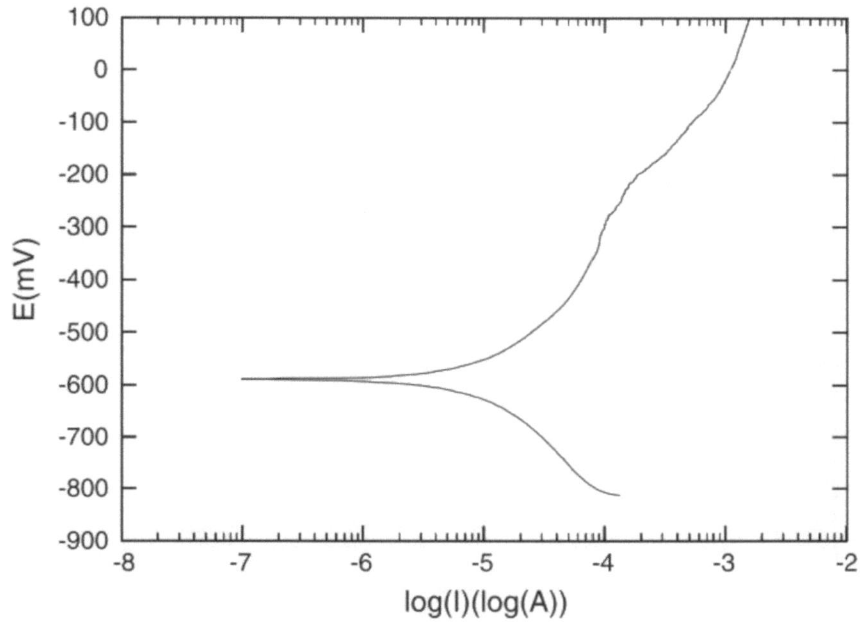

Fig. 12. A typical cyclic polarization curve of Al 6013-20 SiC (p)-T4 temper of the alloy in 3.5 wt.% NaCl + 1000 ppm sodium molybdate in deaerated conditions

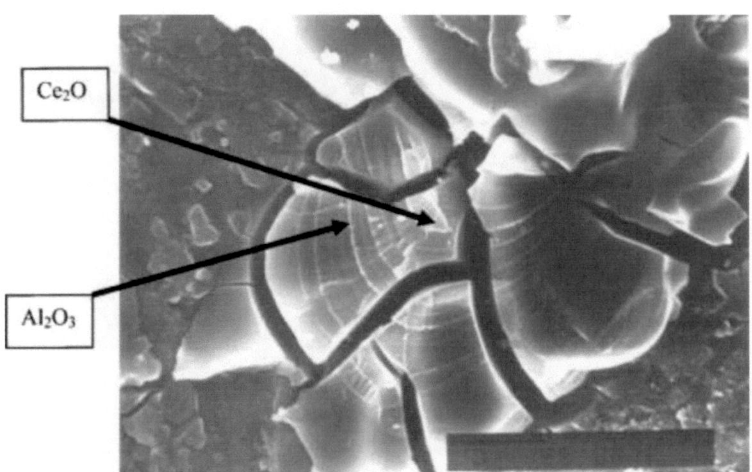

Fig. 13. Surface morphology of Al 6013-SiC (p) in 3.5 wt. % NaCl containing 1000 ppm CeCl3 (Zaki, 2009)

It is interesting to relate the surface morphology to localized corrosion. Typical features of surface morphology after inhibitor treatment are shown in figure 13. Deposition of two types of the particles in concentric rings is seen. These are particles of Ce_2O_3 and Al_2O_3. The square shaped particles of cerium oxide are shown in Figure 14. The oxide layer comprising of Ce_2O_3 and Al_2O_3 are very stable and protect the MMC from corrosion in 3.5 wt% NaCl. However, once the layer reaches a certain thickness, it flakes off. The broken oxide layer in

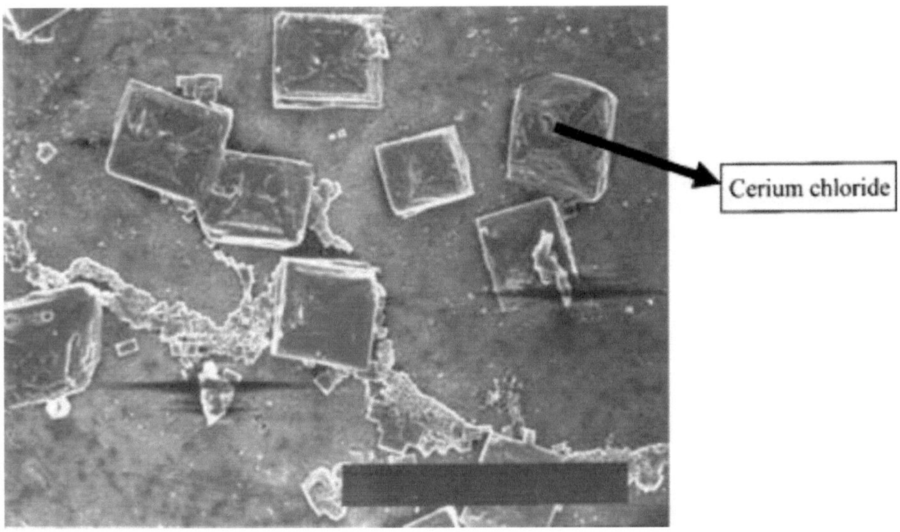

Fig. 14. Square-shaped particles containing predominantly cerium chloride formed on cathodic polarization (Zaki, 2009)

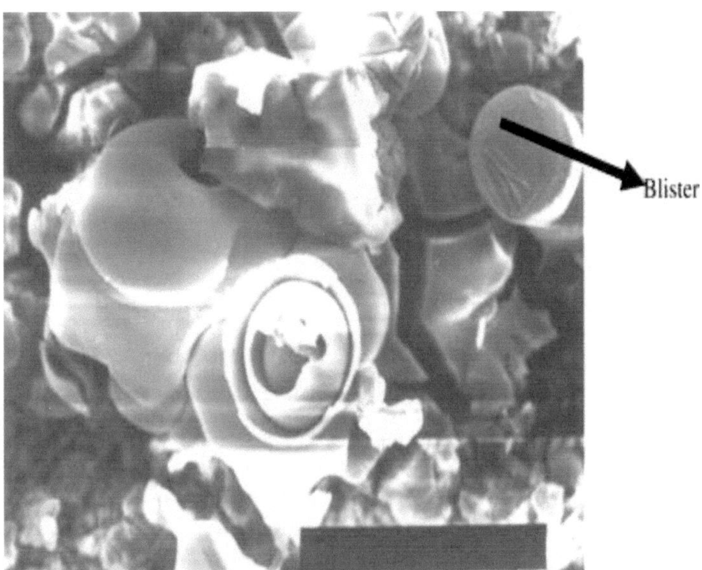

Fig. 15. Broken oxide layer forming blisters (mothballs)

the form of mothballs can be observed in Figure 15. It has been reported that cathodic reaction proceeds at the sites of intermetallic precipitates of copper and its solution becomes alkaline. The film of cerium oxide replaces the film of aluminum hydroxide with increased exposure time (Muhammad & Edwin, 2004; Misra et al., 2007). Whereas the studies on the inhibition of AlMMC are still lacking, there is sufficient evidence to show that cerium chloride is an effective inhibitor for corrosion protection of AlMMC Sodium molybdate is not as effective as cerium chloride shown by the studies reported above composite in chloride containing environment.

3.5 Corrosion mechanism

Despite decades of research no conclusive mechanism on the localized corrosion of Al/SiC(p) composites has been described - The role of intermetallic and dislocation generation at Al/Sic (p) interface has not been conclusively established. No attack a SiC particles has been reported in literature.

From several reliable studies it may be concluded that the pitting potential of monolithic alloys depends on the alloy composition and Ep which is more positive than that of reinforced material (Monticelli et al., 1997; Trazaskoma et al., 1990). The pitting resistance of several MMC investigated followed the order, Al2024 = Al6013 - 20Sic (p) > AL 6061>, Al 6013-20SiC (p) T4=Al5456 (Zaki 2000). In the studies conducted an abundant distribution of copper and secondary phase particles of Mg and Fe were observed.

Copper particles were also present in pit cavities. Analysis of corroded regions at the interface showed a greater concentration of copper compared to the surface away from the interface. The presence of $AlCl_3$ in the oxide film has been indicated by EDS studies (Trazaskoma et al., 1990). Results show a high concentration of copper (3.5%) and Fe (1.77%). There is therefore, a sufficient evidence to show that the increased reactivity at the interface is responsible for localized corrosion of composites. The intermetallic precipitates act as anodic or cathodic sites for initiation of localized corrosion. It is also observed that homogenization of the surface minimizes corrosion the reactivity at the interface is further minimized as shown by temper T4. The SiC particles do not provide any sites for initiation of pits. A higher concentration of copper in pit cavities may be attributed to higher velocities which transport copper ions. Dislocation generation at the interface further activates the interface.

Two more factors are reported to influence, the mechanism of corrosion; Na:YAG laser treatment and machining. Electrochemical studies undertaken showed that the corrosion potential (E_{corr}) increased by 79mv and the corrosion current density decreased by an order of magnitude for the laser treated specimens whereas the untreated surface showed extensive corrosion accompanied by abundant pits. The decrease of corrosion is reported to be due to reduction in the concentration of intermetallic precipitates.

The effect different machining conditions, WEDM, Carbide Turning and Diamond Turning on the electrochemical corrosion behaviour are shown in Figure 16. No significant difference in pitting corrosion potential between the three machining condition was observed. The magnitude of corrosion current for the three machining conditions however differed. Diamond turned specimens showed shallow pits at isolated sites accompanied by a high corrosion rate, whereas Carbide Turned specimens showed extensive pitting because of the hindrance of repassivation of pits due to micro and large crevices present on the surface, pits developed were deeper.

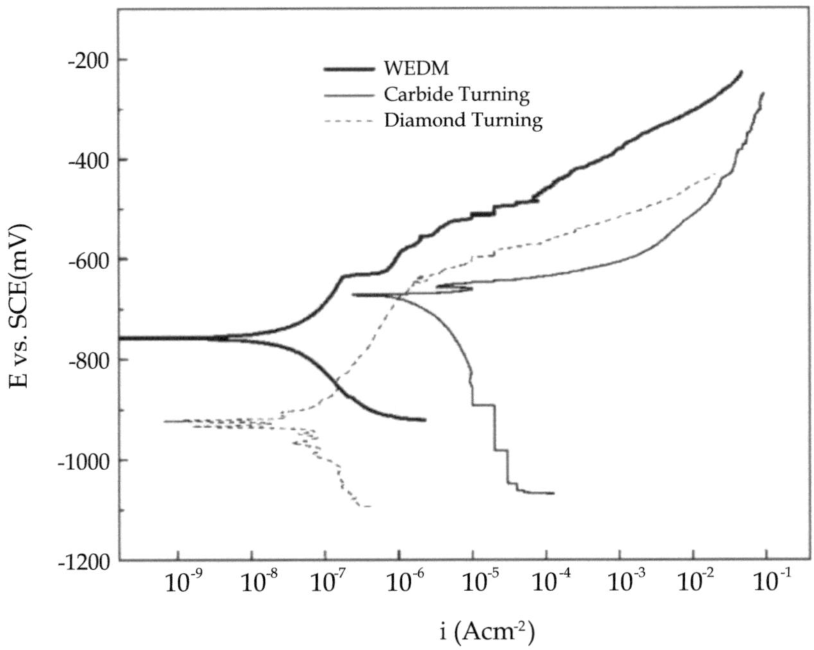

Fig. 16. Potentiodynamic polarization curves of the composite machined to different conditions (Yue et al., 2002)

In the eclectically discharged machined specimen, a resolidified layer of aluminium provided a blanketing effect on the substrate. A reasonable range of passivity was produces on the surface. The high resistance was provided by a layer of oxide on the resolidified aluminium layer created by machining. From the above evidence it is to be understood that surface morphology plays an important role at the Al /SiC interface. Scanning micro reference studies has been employed for direct mapping the active centers on surface of electrode with a low dimensional resolution in micrometers. Evidence of micro pitting has been observed at the open-circuit potentials which were more negative than the pitting potential. In their studies, the same conclusion was reached; i. e the SiC/Al interface is the active center for localized corrosion due to the precipitation of intermetallic compounds (Zaki, 2000).

The observation that heat treatment increases corrosion resistance is shown by temper T4 of Al6013 /20 SiC (p), it is further supported by studies on temper T6 of Al2009/SiC (w), which showed a higher resistance to pitting compared to as rolled specimens It has been already stated above their heat treatment induces homogenization which causes a reduction in the concentration of intermetallic compounds and hence, reduces localized Corrosion.

From the above discussion it may be concluded that the composites are more sensitive to pitting than their monolithic counterparts unless they are subjected to T4 or T6 heat treatment. These is sufficient evidence to show AL/SiC interface is the main target of localized corrosion due to the presence of intermetallic particulates and inclusion which may form micro-galvanic cells and induce localized corrosion. It is also observed that SiC particles are not attacked.

Surface treatment has a significant effect are localized corrosion as shown by the effect of laser treatment and effect of machining on the surface. The scanning micro effloresce electrode studies have shown that Sic/Al interface is the centre f localized corrosion.
Although no conclusion mechanism of localized corrosion of Al MMC exists, there is sufficient evidence to show that Al/SiC interface acts as a centre for localized corrosion and a reduction in the concentration of intermetallic compounds is accompanied by a reduction in localized corrosion as shown by the effect of tempers T4 and T6 on localized corrosion.

4. Conclusion

Based on the studies conducted in the last they decades, the following are the major conclusion on the mechanical and corrosion behaviour of Al MMCs

1. The mechanical properties of fiber, particulate, or whisker reinforced composites are strongly dependent upon the micro-structural parameters, size, shape, orientation and volume fraction of the reinforcement.
2. The tensile strength and Vickers micro-hardness increases significantly with increasing volume fraction of reinforcement as exemplified by Al6013, 6061, 2024 reinforced with particulate and whiskers. The strain to failure also decrease with increased volume fractions of reinforcement - Sliding velocity leads to lower wear rates and lower frication coefficient as shown by SiC and B_4C reinforcements.
3. Increasing cutting speed increased tool wear. The highest wear rate was shown by 15 wt% SiC (p) and the lowest by 5 wt% SiC.
4. Because of accumulation of stress concentration and high dislocation density Al MMC's are sensitive to stress corrosion cracking is 3.5 wt% NaCl - Al6061/20vol% SiC(p)-T6 shows a good resistance to stress corrosion cracking. The polarization curves shifted to higher current densities.
5. The corrosion rates of MMC's decreased with exposure time in long term immersion tests. Heat treatment lowered corrosion rates because of the homogeneous distribution of the precipitates an reduction in their concentration on the Al/SiC interphase electrochemical studies on MMC's showed that the pitting potential decreased with increasing volume fraction of SiC(p) in Al6013 and 6061 reinforce with SiC particulate. Shallow pits contain intermetallic $CuAl_2$.
6. The effect of machining conditions on corrosion showed that electrical discharge machining provided higher resistance to pitting than carbide turning or diamond turning machining.
7. The lowest rate of corrosion was shown by temper T4 if A6013 - 20 SiC(p) is 3.5 wt% NaCl containing silica and other particulate matter.
8. Studies in salt spray chamber showed a good resistance of MMC's to Corrosion and Heat treatment enhances corrosion resistance in corrosive environment.
9. MMC's exhibited a beneficial effect of inhibitor treatment with cerium chloride and sodium molybdate. Cerium Chloride has paved away more effective inhibition than sodium molybdate.
10. The mechanism of corrosion of MMC's is not conclusively understood. It has been, however, shown that the Al/SiC interface us highly reactive due to the presence of intermetallic compounds.

5. References

ASTM; Recommended Practice Designation G 31.72, Standard Practice for Laboratory immersion Testing, (2004) Vol 03, ASTM, Ohio, USA.

Aylor D. M, Moran P. I, Effect of Reinforcement on the pitting Behavior of Aluminum Based Metal Matrix Composites, Jr Electrochem Soc, (1986), Vol 30, p951.

Beccario A.M, Paggi G, Cingaud D, Castellor P, Silicon Carbide Alloy Metal Matrix Composites , Br Jr Corrosion, (1994), Vol 29(1), p65.

Beccario A. M, Poggi G. J, Ginguad D, Castello, Effect of Hydrostatic Pressure on Passivating Powder of Corrosion Layers Formed on 6061-T6 Aluminum Alloy in Sea, Br. Corros Jr, (1994), Vol 29, 1, pp 65-69.

Berkely, D.W, Sallam H.E, Nayeb. Hasemi, H,; The effect of PH on the mechanism of Corrosion and Stress Corrosion and Degradation of Mechanical Properties of AA 6061 and Nextel 440 Fiber-Reinforced AA 6061Composites, Corros Sc,(1998), Vol 40, 2/3 pp, 141-153.

Burleigh T.D, Ludwiczak and Petro R.A, Itergranular Corrosion of an Aluminum Magnesium Silicon Copper Alloy, Corrosion Science, (1995), Vol 5199(1), p50.

Chen C, and Mansfeild F, Corrosion Protection of Al 6092/SiC (p) Metal Matrix Composite, Corr Sc, (1997),Vol 6, PP 1075-108.

Colman S. L, Scott V. D, Enaney M.C, Corrosion Behavior of Aluminum Based Metals Matrix Composites, (2011), Vol 297, 11, Dol: 10.10078/BF001117589, Jr of Mat Sc.

Davenport A. J, Isacs H. S, Kendig M. W; Investigations on the Role of Cerium Compound on the Corrosion Inhibition of Aluminum, Corrs Sc,(1991),Vol 32 (516), p653.

Dehlan Al Khalid; Hafeez Syed, Tensile Failure Mechanism of Al 6061Reinforced with Submicron Al2O3, AJSE, (2006), Vol 31, No 2C.

Feng.Z; Lin, C; LinJ; Lin j; Luo, J; "Pitting Behavior of SiC/2024 Al MMC".

Griffen, A.J; Turnbull, A; 'An Investigation on the Electrochemical Polarization Behavior of Al6061 MMC' Corros Sc (1994), Vol 36,1, 21-35.

Hihara L. H, Corrosion of Metal Matrix Composites, Shriers Corrosion, (2010), Vol 3, pp 2250-2569.

Hinton B. R. W, Dr. Arnold Ryan N. E; Cerium Conversion Coating for Corrosion Protection of Aluminum, Mat. Forum, (1986), Vol 9(3), pp 162.

Hongho Ding, Hawthorn, G.A, Hihara, L.H, Inhibative Effect of Sea Water on the Corrosion of Particulate Reinforce Aluminium Matrix Composites and Monolithic Alloys, Jr. Electrochem Soc, (2009) 156, (100. (35).C159.

Isacs, H.S, Vyas, B; "Electrochemical corrosion Testing, ASTM, STP 727, 1981, p, 3.

Jamaludin, S.B, Yusoff Z, Ahmed R.R, Comparative Study of Corrosion Behavior of A.A. 2009/20 Vol% SiC(w), Porpugaliay Electrochimica Acta, (2008), Vol 26, pp 291-301.

Lin, S; Shih, H; Mansfield F; Corrosion Protection of Aluminium Alloys and Metal Matrix Composites by Polymer Coatings; Corros Sc (1992), Vol 23, 9, pp 1331-1349.

Lin S, Greene H. Shih and Mansfeld F, Corrosion Protection of Al Metal Matrix Composites, Corrosion, (1989), Vol 45(8), p615.

Mansfield S, Lin S, Kim H, Shih, Pitting and Passivation of Al Based Metal Matrix Composites, J Electrochem Soc, (1990), Vol. 137, pp 75-82.

Mansfield F. S.L. Jeanjagnet; The Evaluation of Corrosion Protection Measures For Metal Matrix Composites, Corros Sc (1984), Vol 26, pp 727-734.

Misra, Ajit Kumar, Balsubramanium, Corrosion Inhibition, Material Chemistry and Physics, (2007), Vol 103, 2, 3, pp 385-393.

Monticelli C, Zucchi C, Bruuonoro, Trabanelli C, Stress Corrosion Cracking Behaviour of some Aluminum Base Metal Matrix Composites, Corrs Sc, (1997), Vol 39,10, pp. 1949-1063.

Muhammad Ashraf, P. Leela Edwin, Evaluation of Corrosion Inhibition by Cerium on Aluminum under Marine and Laboratory Environment, 2nd Jr of Chem. Tech, (2004), Vol 11, pp 672-677.

Rohatgi, P.K; 'Aqueous corrosion of Metal Matrix Composites, Comprehensive Composite Materials, (2003), Chapter 3.18, , pp 481-500, Elsevier

Schwartz, M.M "Composite Material Processing Fabrication and Applications" Prentice Hall, USA (1997).

Shorowords,K.M; Haseeb, A.S.M.A; Celic,j.P; "Studies on the wear Friction and Tribochemistry of MMC Sliding against Phenolic Brake Pads, Wear" (2004), pp1176-1181

Turnbull, A.Br, Corros Jr 1992, Vol 27, p.27-35.

Trzskoma, P.P; in "Metal Matrix Composite Mechanism and Properties" Academic Press, (1991), p, 383.

Trazaskama P. P, Pit Morphology of Aluminum Alloy in Silicon Carbide Alloy Metal Matrix Composites, Jr Corrosion, (1990), Vol 46, p402.

Trazaskoma P.P, Maccefferty E, Crowe C.R, Corrosion Resistance of Al Based Metal Matrix Composites, Corrs Sc, (1990), 46, p402.

Vaeeresh Kumar, G.B., Rao, C.P; Selvararj.N; Bhagya Shekar, M.S.B; Studies on Al 6061-SiC and Al 7075-Al2O3 Metal Matrix Composite; Jr for Mater and Mater characterization and Eng, Nov (2010), Vol 9, pp, 43-55.

Vogelsang, M; Arsenault, R.J; Fisher, R.M; In SituHVEM study of Dislocation Generation of Al/SiC Interface in Metal Matrix Composites, Met.Trans A, (1986), Vol 17A, p139.

Yue, T.M; Yu, J.K; Maki, H.G; "Corrosion Behavior of Aluminium 2009/SiC Composite Machined to Different Conditions". Jr Mat Sc Letters, (2002), Vol 21, 14, pp1069-1072

Yue, T.M; Wu, Y.X; Man, H.c; 'On the Role of CuAl2 Precipitates in Pitting Corrosion of Aluminium (2009), SiC Metal Matrix Composites' Jr of Materials Sc (2000), Letters 9, pp 1003-1006.

Zaki Ahmad and Abdul Aleem B. J, The Effect of Inhibitors on the Susceptibility of AL 6013, SiC Interface to Localized Corrosion, Jr of Mat Eng, Perf,(2009),Vol 18,2, pp129-136.

Zaki Ahmad, Mechanical Beauvoir and the Fabrication Characteristics of Aluminum Metal Matrix Composites, Jr of Reinforced Composite Material, (2007),Vol 1, 4, pp 3027-3033.

Zaki Ahmad, Abdul Aleem. B.J; Degradation of Aluminium Metal Matrix Composites in Salt Water and its Control, Mater and Design, (2001), Vol 23, pp173-180.

Zaki Ahmad, Paulette, P.T; and Aleem B.J.A; "Mechanism of Localised Corrosion of Aluminium Silicon Carbide Composites in a Chloride Containing Environment' Jr. Mat Sc (2000), 35, pp 2573-2579.

Zaki Ahmad & Abdul Aleem B. J, Corrosion Resistance of a New Al 6013-20 SiC in Salt Spray Chamber, Jr of Mat SC and Eng, (2000), Vol 9, 3l, p338.

Zaki Ahmed & Abdul Aleem B.J, Corrosion Resistance of New Aluminium Al 6013/20SiC(p) in Salt Spray Chamber,Jr Mat Sc and Eng,(2000),Vol 9(3),p338.

Zaki Ahmed, Paulette, P, T, Aleem B.J.A, Mechanism of Localize Corrosion of Aluminum Silicon Carbide Composites in Chloride Containing Environment, Jr Mater Sc,(2000),Vol 3, 5, pp2573-2579.

Zaki Ahmad & Abdul Aleem B. J, The Erosion Corrosion of Al -SiC Composites in Sea Water, (1998), Final Report, KACST, 14-65, King Abdul Aziz City of Science and Technology, Riyadh, Saudi Arabia.

Lima, P.; Bonarini, A. & Mataric, M. (2004). *Application of Machine Learning,* InTech, ISBN 978-953-7619-34-3, Vienna, Austria

Li, B.; Xu, Y. & Choi, J. (1996). Applying Machine Learning Techniques, *Proceedings of ASME 2010 4th International Conference on Energy Sustainability,* pp. 14-17, ISBN 842-6508-23-3, Phoenix, Arizona, USA, May 17-22, 2010

Siegwart, R. (2001). Indirect Manipulation of a Sphere on a Flat Disk Using Force Information. *International Journal of Advanced Robotic Systems,* Vol.6, No.4, (December 2009), pp. 12-16, ISSN 1729-8806

17

Interrelation Between Failure and Damage Accumulation in the Pre-Fracture Zone Under Low-Cycle Loading

Vladimir Kornev, Evgeniy Karpov and Alexander Demeshkin
Lavrentyev Institute of Hydrodynamics SB RAS
Russia

1. Introduction

When structures are loaded in catastrophic mode of operation, localization of irreversible strains occurs in regions of stress concentration. This is caused by the geometry of a structure or by the presence of fetaures (dints, holes, cracks, inclusions with mechanical characteristics different from properties of original material). If loading is in progressing, repeated loads cause gradual degradation of the material in regions of the localization of inelastic strain. As a result, this leads to generation and extension of cracks and to loss of a load capacity of the structure. Study of regularities of material degradation in regions of strain localization will permit one to appreciate possible structure resources in a catastrophic situation or consequence of failure, which affects subsequent behaviour of a structure in a common regime.

The distinctive features of fracture surface microrelief of metallic components in fatigue are fatigue striations oriented normally to the crack extension direction. It is appropriate to relate formation of fatigue striations to stepwise crack tip advance, and to record residual deflection of a beam under three-point bending when the loading corresponds to the low-cycle fatigue. A current striation may be formed after several loading cycles due to arrest of a fatigue crack after each advance of its tip in the Laird- Smith model (Laird & Smith, 1962), the material being embrittled in a pre-fracture zone at each loading type. Damage accumulation in the pre-fracture zone is associated with accumulation of inelastic strains in this zone.

The aim of the present work was to study damage accumulation in the regions of inelastic strains near the notch tip having a finite width. Two cases are considered: i) symmetric three-point bending of a beam (the edge notch is made on the underside of a specimen in the transverse symmetry plane, damage accumulation is estimeted by the increment of a residual deflection); ii) tension of a plain specimen with a narrow edge notch (direct viewing of fatigue crack propogation was performed using digitized microscope with resolution of about 22500 pxel/mm^2). Mechanisms of deformation, damage accumulation and failure of material under fatigue conditions have been proposed.

In the first case, the choice of geometry of a specimen and the loading type are governed by the following considerations.

Under symmetric three-point bending of a beam with the transverse notch, main inelastic strains are concentrated ahead of the notch tip where stress concentration occurs. Change of mechanical characteristics of material in this region under repeated loading conditions causes the increase in residual deflection of the specimen. Thus, the amplitude of the residual deflection can be used as a measure of damage accumulation in the zone of localization of inelastic strains. This provides, from macroscopic phenomena, a possibility for qualitative and quantitative estimating the changes directly exhibited by the material due to processes of fatigue fracture. Moreover, in this case, the notch can be considered as a model of an edge crack with the blunted tip, and the region of localization of inelastic strains can be considered as a pre-fracture zone ahead of the tip of this crack.

In the second case, consider a plain specimen with one edge notch since two symmetrical notches lead to uncertainty in a choice of the point of crack initiation. Besides, after a crack initiates, the symmetry of a specimen is broken in one of paired notches and its initial symmetry loses significance.

2. Low-cycle symmetric three-point bending of a beam with edge notch

In the performed tests, the effect of various loading conditions, change in the geometry of a specimen, and preliminary plastic strain of material from which the specimens were made on the process of damage accumulation were studied. The attempt was made to determine parameters, which may be extended from the particular cases considered in the tests, to more general loading conditions. A possibility for description of regularities of damage accumulations with the aid of simple analytical functions involving constants just as determined from experiments, so specific for every material is considered.

The specimens were made from aluminum alloy D16T in the original state, so from preliminary stretched materials with the various degrees of plastic strain. The composition of the D16T alloy is as follows: Al was as a base metal, the alloying elements were Fe (0.3%), Si (0,19%), Mn (0.76%), Cu (4.0%), Mg (1.29%).

The experiments were conducted on electromechanical testing machine (the rated capacity load was 100 kN). The loading was repeated three-point bending with unloading and it was given by travelling of a moveable cross-head with a constant velocity. Loading diagrams were recorded at each loading cycle.

The minimum force of a cycle was $P_{\min} \approx 0$ for all the conducted tests. The maximum force of a cycle $P_{\max}$ was considered in three forms: *i*) $P_{\max} = const$ (stationary low-cycle loading), *ii*) $P_{\max} \equiv P_{\max}(N)$ was increasing step-function of N for which the number of cycles at one step was constant (non-stationary low-cycle loading with increasing load), and *iii*) $P_{\max} \equiv P_{\max}(N)$ was analogous decreasing step-function. The $P_{\max}$ value in all tests was chosen to provide plastic material deformation ahead of the notch tip. Tests under stationary cyclic loadings were conducted for different $P_{\max}$ values in the range from the limit of elasticity to the strength limit of a specimen. Besides, tests with different notch depth to beam height ratios were conducted.

Increment of the residual deflection $\delta w = \delta w(N)$ depends on the cycle number. The increment of the δw deflection arises mainly due to mechanical properties of material in a pre-fracture zone near the notch tip, and therefore, as it was said before, it is considered as a measure of damage increment in this zone. The increment is considered to achieve the limit w^* value if subsequent deflection of the beam proceeds without increase in P.

2.1 Stationary loading

Fig. 1 demonstrates, as an example, the experimental diagram represented by curves of beam deflection w versus applied force P in stationary low-cycle testing a specimen made from D16T. The P value is sufficiently large in order that fracture to happen after the limited number of cycles (in this case, $N^* = 390$). This allows one to visualize all distinctive features of such diagrams obtained also for test materials for different P_{max} values. Curve 1 corresponds to single loading of a specimen until fracture occurs; group of curves 2 corresponds to cyclic loading up to the instant when a crack starts to extend for $P < P_{max}$. Each curve of group 2 corresponds to loading branch of one cycle. All the curves of group 2, except the first one, have the initial horizontal section $P = 0$. The length of this section is equal to the value of residual deflection accumulated at previous cycles. In this figure, δw is the distance between adjacent curves of group 2.
Damage accumulation can be divided into two specific stages. The first stage (Fig. 1, subgroup A of curves 2) is a stage of cyclic strengthening at which decrease in δw is observed with increase of N. At this stage, δw achieves some minimum that is characteristic for the given value of P_{max} after which the δw value becomes constant within the limits of measurement accuracy. The second stage (Fig.1, sub-group B of curves 2) is characterized by the increase in the δw value as N increases, and this stage is accomplished by growth of a crack when $P < P_{max}$. Accumulation of micro-defects during the first stage is likely to lead to formation of macroscopic defect, which then progresses during the second stage. Therefore, we call the second stage as a stage of development of a macro-defect. The ratio between the number of cycles in sub-groups A and B and the law of δw variation depend on material characteristics (Karpov, 2009; Kornev et al., 2010).

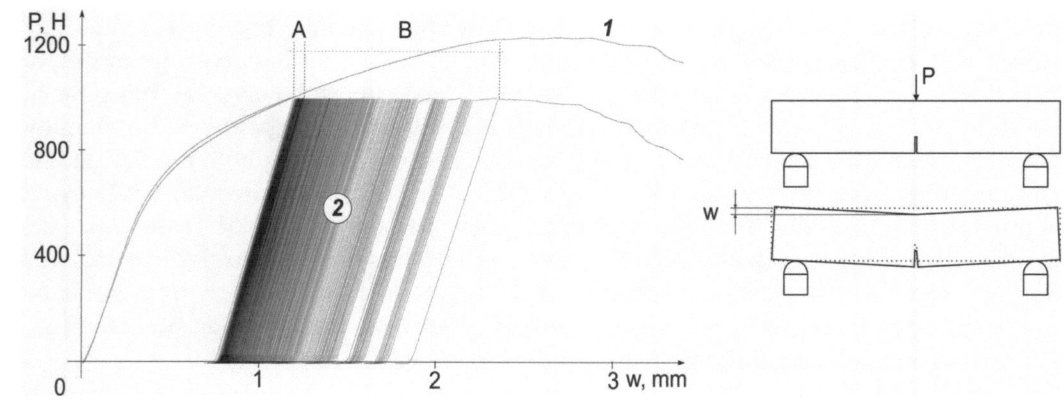

Fig. 1. Scheme of low-cycle test of specimen loaded in tree-point bending and plots of beam deflection as a function of force for every loading cycle for D16T alloy

Regularity in the residual deflection can be visualized as $\delta w(N)$ diagrams. An example of such diagrams is given in Fig. 2. Here pairs of curves are shown with numerals 1–5 for curves, one of the pairs being given by the analytical function and the second one being a saw-like profile. Here the saw-like profiles represent experimental $\delta w(N)$ curves for the D16T alloy. The analysis of experimental diagrams shows that curves can be approximated by plots of some power functions. These functions are to have asymptotes corresponding to the limits beyond which the process described by the diagrams can not take place. That is, the inverse power function with some scaling coefficient can be taken as approximating one.

Besides, in the general case, both descending and ascending branches of the experimental curves should be approximated by different curves.

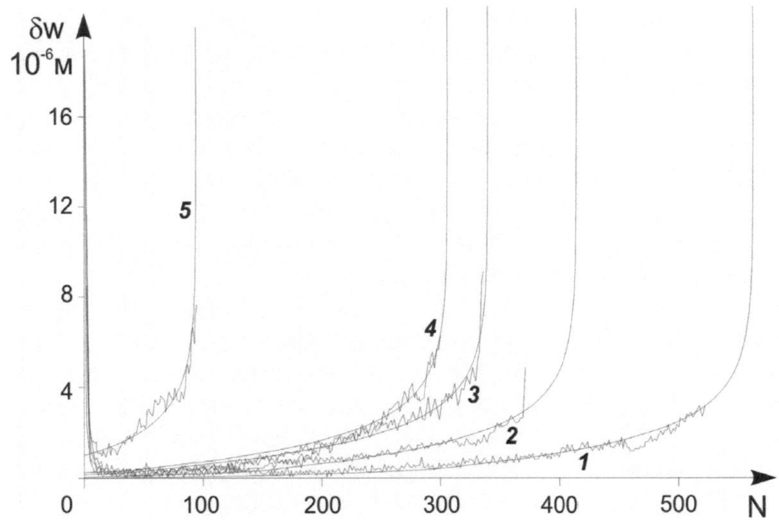

Fig. 2. Approximation of experimental curves of residual deflection increment for D16T alloy

The connection point of these curves corresponds to the minimum δw value. In this case, the form of curve, which approximately describes either descending or ascending branch, is unique for all values of $P_{\max}$. That is, the curves of approximating functions for all $P_{\max}$ values can be superposed by parallel shift. Fig. 2. shows such curves (branches of hyperbolas corresponding to saw-like profiles 1-5). These curves are given by functions of the form $f(N)=\lambda(\alpha+\beta N)^{-\gamma}$ normalized in such a way that the experimental minimum point δw would be a connection point of two curves approximating both the descending and ascending branches of the saw-like profile (α, β, γ, λ are experimental constants). If approximating functions have been defined, then the derivatives of these functions calculated for integers of the variable *N* can be used as a magnitude characterizing material damage. The stage of cyclic strengthening *A* of duralumin, which differs from steels by essentially larger grain sizes, is limited to several cycles even for low loads (Kornev et al., 2010). As compared with the *B* stage, the stage *A* can be neglected.

2.2 Non-stationary loading

Under non-stationary loading conditions, the process of cyclic strengthening begins again each time when $P_{\max}$ increases, and only the last stage explicitly includes both *A* and *B* stages of damage accumulation. However, if instead of change in δw, the analogous change in $\sum \delta w(N)$ (summation is performed over numbers of all cycles at one loading step), the stages *A* and *B* are also evident as in the case of the stationary loading. Fig. 3 demonstrates the experimental diagrams represented by curves of w versus P for non-stationary cyclic loading of alloy D16T.

The non-stationary loading with decreasing $P_{\max}$ shows that the initial overload of material provokes consequences, which may affect fracture process when loading with the low $P_{\max}$

value is applied again. Given in Fig. 4 are curves of w versus P for cyclic loading of D16T duralumin when decrease in P_{max} is step-like. Here, after some initial P_{max} value, monotonic nonlinear increase of $\delta w(N)$ is continued even for significant decrease in the P_{max} value. Beginning from the fourth step, $\sum \delta w(N)$ also increases. As a result, in spite of decreasing load, the specimen rapidly losses its load capacity. The tests also show that the increase in P_{max} after several steps of such loading leads to significant inelastic strain with strengthening for values of the force P, which earlier were appropriate to the linear dependence of the deflection on the applied force. This fact evidences that repeated loadings of a material preliminary experienced overloading lead to significant material damage even this loading is not followed by noticeable increase in the residual strain.

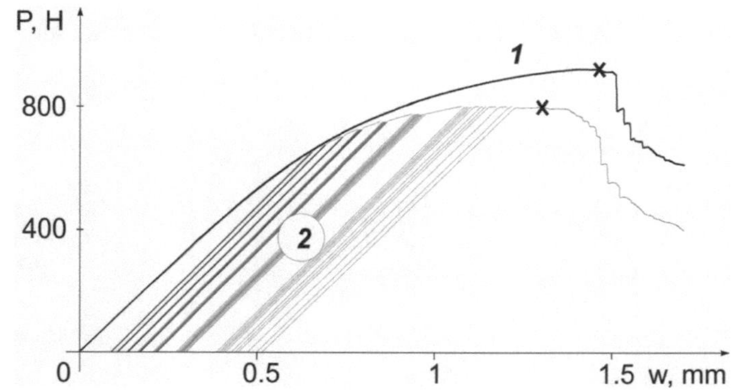

Fig. 3. Non-stationary cyclic loading with increasing maximal applied force for D16T alloy

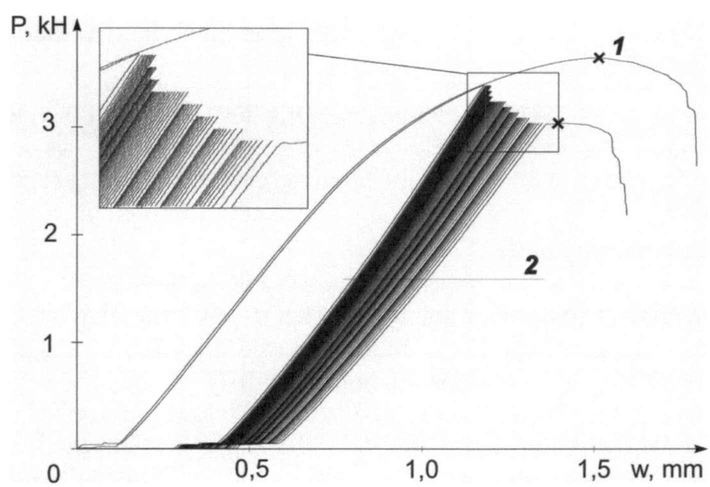

Fig. 4. Non-stationary cyclic loading with decreasing maximal applied force after initial overloading for D16T alloy

2.3 The typical ratio

The typical value for both stationary and non-stationary low-cycle loading is the ratio between limit deflections for single and repeated loading (w^* and w^{**}, respectively). The tests show that this ratio for material is constant when different schemes of stationary and non-stationary loading are applied (Kornev et al., 2010). This allows one to use this ratio for comparison of results obtained on specimens with various geometrical dimensions and for different loading regimes. For duralumin, we have $w^* \geq w^{**}$.

2.4 Preliminary inelastic strain

Preliminary inelastic strain of a material, from which the specimens have been made, essentially influences material resistance to cyclic fracture. As an example, Fig. 5 displays the experimental diagrams with curves of w versus P (Fig. 5 (a)) and the $\delta w(N)$ diagram (Fig. 5 (b)) for D16T duralumin with various degrees of preliminary stretching: diagram **1** for original materials, diagram **2** for materials stretched by 5%, and diagram **3** for materials stretched by 10%. All the specimens were loaded for the same P_{max} value, but in Fig. 5 (a), diagrams for three tests are displaced, for convenience, from each other along the horizontal axis.

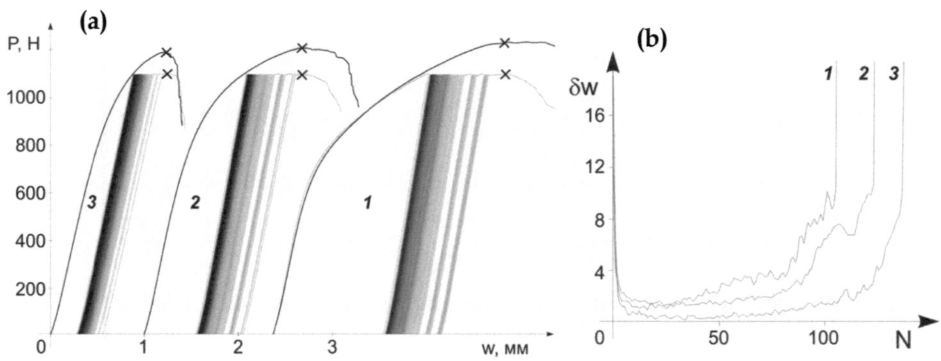

Fig. 5. (a), (b). Low-cycle loading of aluminum alloy after preliminary plastic deformation for D16T alloy

In Fig. 5(b), the area under the curve characterizes the limit deflection w^{**}, which decreases as preliminary stretching increases. However, the decrease in w^{**} is followed by the decrease in δw. This leads to increase of the limiting number of loading cycles.

2.5 Variation in the notch depth

Comparison of tests conducted on beams with notches of different depths has shown that if $\delta w(N)$ diagrams have been plotted for some notch depth l, the diagrams can be used to obtain analogous diagrams for other l values. Assume that $\delta w(N)$ diagrams have been plotted for the notch depth l_1, and for the notch depth l_2 there is the only diagram with curves of w versus P for single loading. The value of maximum applied force $P_{max} = P_1$ corresponds to some deflection $w = w_1$ for l_1 under single loading, the limit deflection of cyclic loading for this notch depth being $w^{**} = w_1^{**}$, and the limit deflection of the single loading being $w^* = w_1^*$. In this case, the curve of damage accumulation $f_1(N)$ for $P_{max} = P_1$, $l = l_1$ can be used to obtain the curve $f_2(N)$ for $P_{max} = P_2$, $l = l_2$. Here P_2 is the force for which a specimen with the notch depth l_2 has the deflection $w = w_2$ such that

$$\frac{w_1}{w_1^{**}} = \frac{w_2}{w_2^{**}}, \text{ where } \frac{w_2^{**}}{w_2^{*}} = \frac{w_1^{**}}{w_1^{*}} \Rightarrow w_2^{**} = \frac{w_1^{**} w_2^{*}}{w_1^{*}}.$$

3. Development of fatigue crack under low-cycle tension conditions of a plain specimen with edge notch

Tests on a plain specimen loaded in tension with the narrow edge notch were conducted. During the test, direct observation of fatigue crack propagation was performed using a digitized microscope with the resolution of 22500 pixel/mm². The testing scheme is shown in Fig. 6. The field of view is outlined by dashed line.

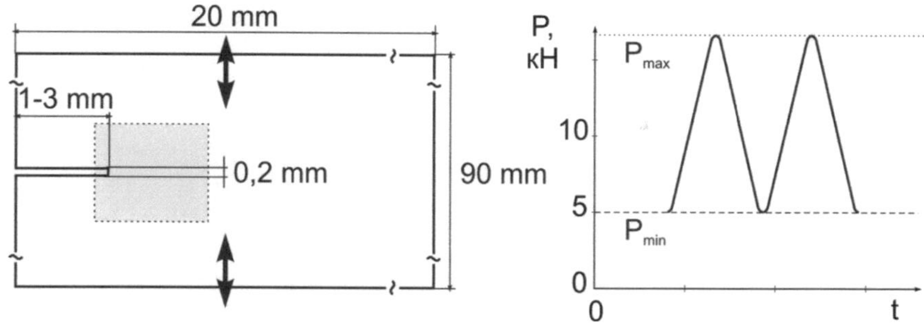

Fig. 6. Test scheme

The specimens were made from D16T alloy preliminary heat-treated at 500°C to give more plastic material. Plain specimens with notches of different lengths (1–3 mm) were used. The minimum load was the same in every cycle, the maximum load was chosen such that three different loading types were provided: *i*) near yield strength, *ii*) near the limit of load capacity, *iii*) the average value of them.

Photographs in Fig. 7 illustrate stages of crack propagation near crack-like defect in two cases *i*) continuous tension with constant rate, *ii*) low-cycle tension.

In the case of continuous tension, the following stages may be observed. First, intense plastic deformation ahead of the notch tip occurs and two zones of strain localization are formed with a delta-shaped area between these zones. The area we term as a pre-fracture zone since further it will define the crack extension direction. At this stage, several focuses of fatigue crack initiation are formed, which are located at notch angles as it is seen in the photograph (a.1). The pre-fracture zone is formed not by a prospective crack, but by the notch itself and its shape is unclear. Further (a.2), one of microcracks develops as a crack propagating within the zone of plastic strain localization irrespective of the pre-fracture zone specified by the notch. This zone becomes more structured and its tip is separated from the crack tip. Here the pre-fracture zone tip starts to shift towards the developing crack. At the next stage (a.3), crack branching takes place, the branches being formed just as near the crack, so at its faces. This evidences the significant extent of material embrittlement in the vicinity of crack extension. The angle at the pre-fracture zone tip starts to decrease. Then (a.4) the branch nearest to the pre-fracture zone tip has some advantages and defines the final direction of crack extension. When the crack tip joins the pre-fracture zone tip (a.5), the critical state is achieved after which the crack starts to extend very fast. The final failure of a specimen is preceded by a short stage (a.6), at which the angle at the pre-fracture zone crack becomes similar to the crack opening angle and one of pre-fracture zone edges defines a path of the subsequent crack extension unambiguously. The crack is very short before the critical state: its length is less than the notch width.

Under repeated low-cycle loading conditions, the pre-fracture zone created by a notch plays no noticeable role especially in the cases when cycle loading starts at insignificant plastic

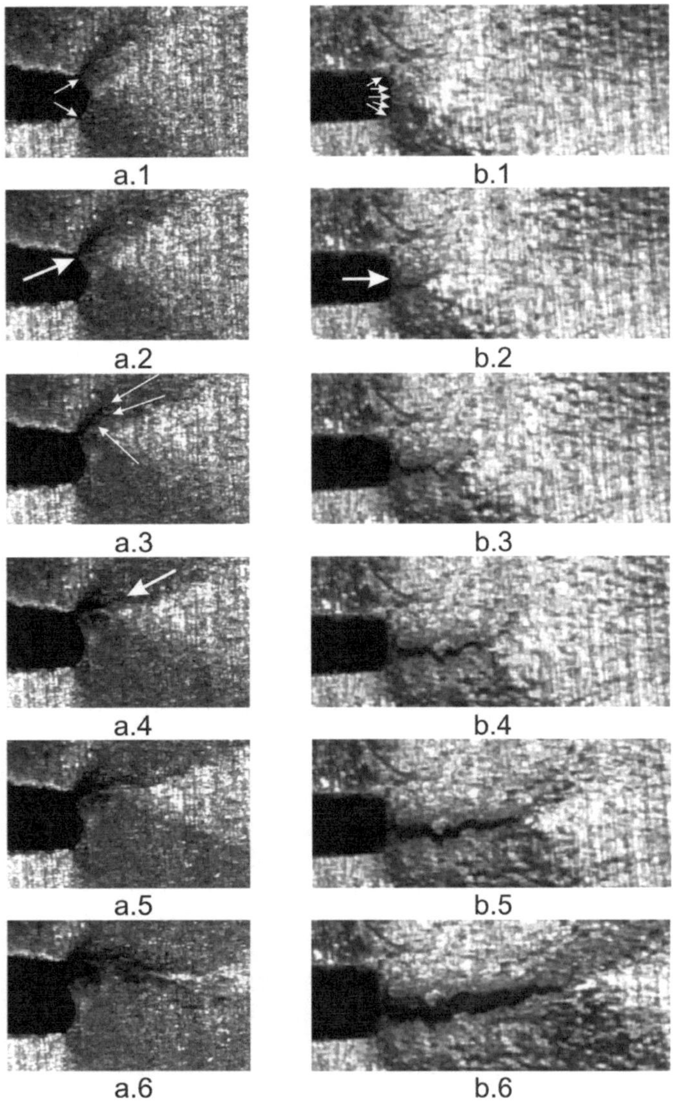

Fig. 7. Stages of crack propagation for D16T alloy; a.1. occurrence of incipient cracks under single loading conditions; b.1. the same under low-cycle loading conditions; a.2. growth of one of incipient cracks under single loading conditions; b.2. the same under low-cycle loading conditions; a.3. crack branching and moving of pre-fracture zone of a notch close together with the crack tip under single loading conditions; b.3. occurrence of "tooth" under low-cycle loading conditions; a.4. development of crack branch nearest to the pre-fracture zone of a notch under single loading conditions; b.4. growth of a crack beyond pre-fracture zone of a notch, formation of the proper pre-fracture zone for a crack under low-cycle loading conditions; a.5. merging pre-fracture zones of the crack and notch under single loading conditions; b.5. final stage of crack development under low-cycle loading conditions characterized by continuous growth of visible sizes of the pre-fracture zone and by decrease of the angle between whiskers; a.6. crack in critical state followed by fast final fracture under single loading conditions; b.6. the same under low-cycle loading conditions

deformations of a specimen. In this case, zones of localization of maximum plastic strains at notch angles are slightly structured and the crack can start to develop from any point of the front bound of the notch (b.1). When one of microcracks starts to extend, a delta-shaped pre-fracture zone is formed ahead of the microcrack (b.2). When the crack is short, the length of the pre-fracture zone is defined by that of the edge notch. The angle between whiskers in the pre-fracture zone tip is close to the right angle, and visible sizes of this zone are of the order of surface roughness occurring due to plastic flow of material. Then, at some distance from the front edge of the notch, temporal crack arrest takes place in all the considered cases. Keeping constant length, the crack begins to open at the expense of blunting and increasing the length of its front edge, and then it changes both the direction and rate of propagation. The path, which the crack takes, deviates from the mean direction of its propagation and the former coincides with the upper or lower edge of the pre-fracture zone. After this, the crack returns to the mean path of propagation. The "tooth" formed at a crack face is clearly seen (b.3). Then the rate of crack growth is steadily increased, crack opening being continued. The pre-fracture zone length increases (b.4). The next to last stage is characterized by continuous growth of visible sizes of the pre-fracture zone and by decrease of the angle between whiskers (b.5). At last, the critical state is achieved (b.6): the crack is blunted, significant crack opening takes place that is comparable with the original notch width, then the crack produces a path branch sharply deviated from the mean path of propagation and then final failure of material occurs after which loading must be ceased. As opposed to continuous loading, the crack length to the instant of the critical state is close to the notch length.

Given in Fig. 8 are plots characterizing crack propagation as a function of the number of cycles. At the left, some geometric characteristics of cracks versus the number of cycles are given. At the right, (P-ε) plots of specimens are given, where P is applied force, ε is the overall elongation of a specimen. Both minimum and maximum loads of loading cycles are shown on these plots, as well as the maximum strain on the first loading cycle. The value of maximum force in a cycle was chosen such that to provide areas of plastic strain ahead the notch tip.

Plots a.1 and a.2, b.1. and b.2, and c.1. and c.2 correspond to specimens with the notch of 1 mm, 2 mm and 3 mm in length, respectively. Curves 1, 2, 3 and 4 on the left plots correspond to the overall length of crack along its face, the distance of the crack tip from the front notch edge, the difference between these values, and the notch width increase of which shows opening of the crack mouth, respectively. All these values demonstrate nonlinear growth of cracks for which hyperbolic functions are applicable for description of both this growth and damage accumulation at the notch tip in the case of three-point bending of a beam. The general structure of material undoubtedly influences crack propagation as it can be seen on plots in Fig. 8 (a.1, b.1, and c.1), deviations of a crack due to structural heterogeneity being leveled out at its further propagation.

Fig. 9. illustrates separate parts of a crack corresponding to different stages of cyclic loading. Here a crack at the notch 3 mm in length is shown.

It seems likely that origination of a "tooth" is not a consequence of a heterogeneous structure of the material, but this is associated with sizes of defects from which a fatigue cracks start to propagate. In this case, a residual durability of the structure can be defined by its appearance.

Let us discuss interaction between plasticity zones formed in the vicinity of the notch tip and propagating cracks under single and cyclic loading conditions. Fig. 11 shows: a) schemes of zones of plasticity localization in the vicinity of the notch tip (region I) with features near angles (regions II and III), b) schemes of crack initiation under single loading conditions, c) scheme of crack initiation under cyclic loading conditions, d) scheme of fatigue crack propagation with origination of a "tooth" at the boundary of the zone of plasticity localization for a notch. Only in the case when load capacity is exhausted, initiation of cracks on the front notch edge at several points takes place: in original material under single loading conditions (Fig. 11.b), in material accumulated damages under fatigue loading (Fig. 11.c).

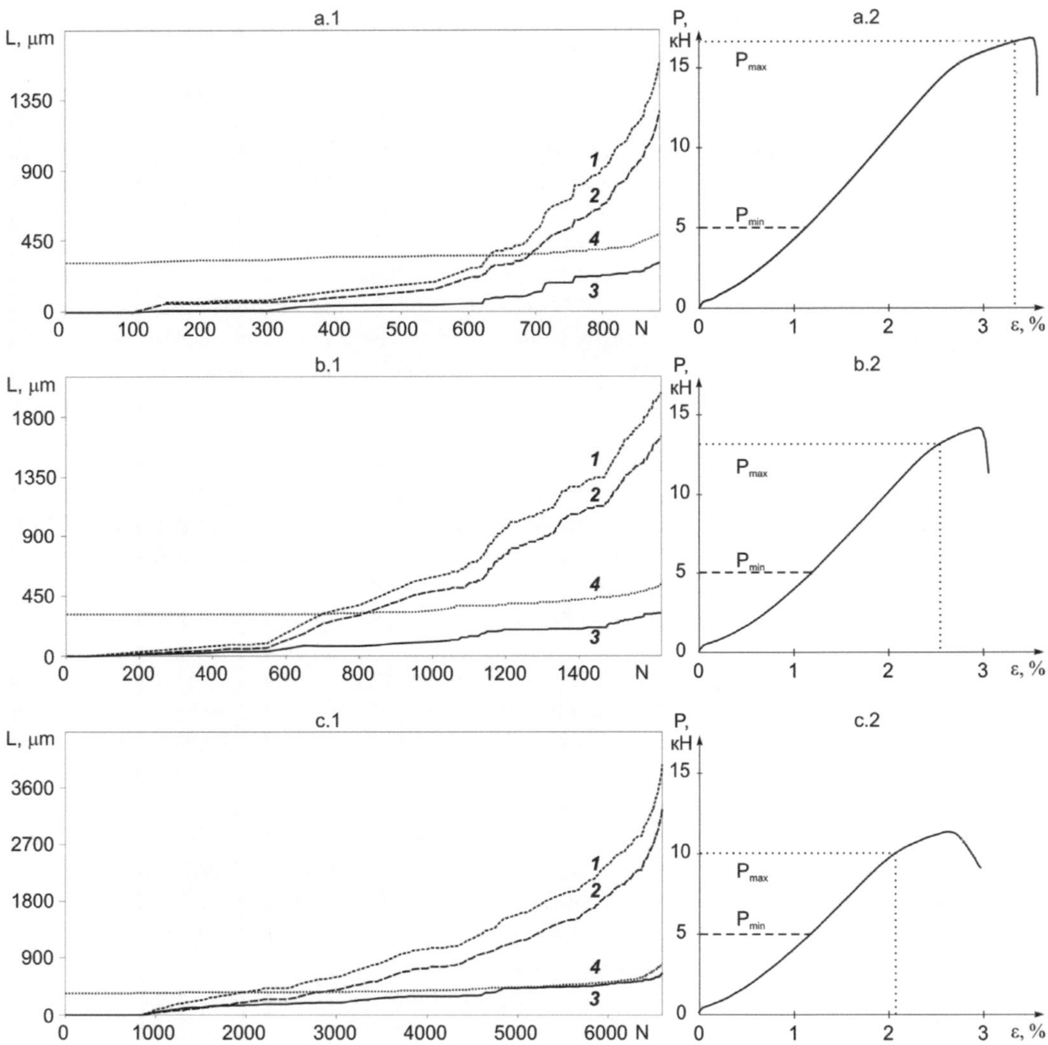

Fig. 8. Characteristics of fatigue crack propagation for D16T alloy

Given in Fig. 10 are forms of three cracks for notches of various lengths (notch lengths in mm are shown inside contours of final pre-fracture zones). Here it is seen that a "tooth" is inherent to all the cracks, the distances from the crack onsets to "teeth" are in direct proportion to lengths of notches. This "tooth" can be identified with pronounced fatigue striations originated on crack faces. In Fig. 10, all three "teeth" are marked with circles. The distance to a "tooth" is close to one third of the overall length of each crack. The segment on the horizontal axis in Fig. 8 is assigned to origination of a "tooth". This segment is nearly the same for all the cracks with respect to the total number of cycles. These segments correspond to about 600, 1100 and 4800 cycles, respectively, on a.1, b.1, c.1.

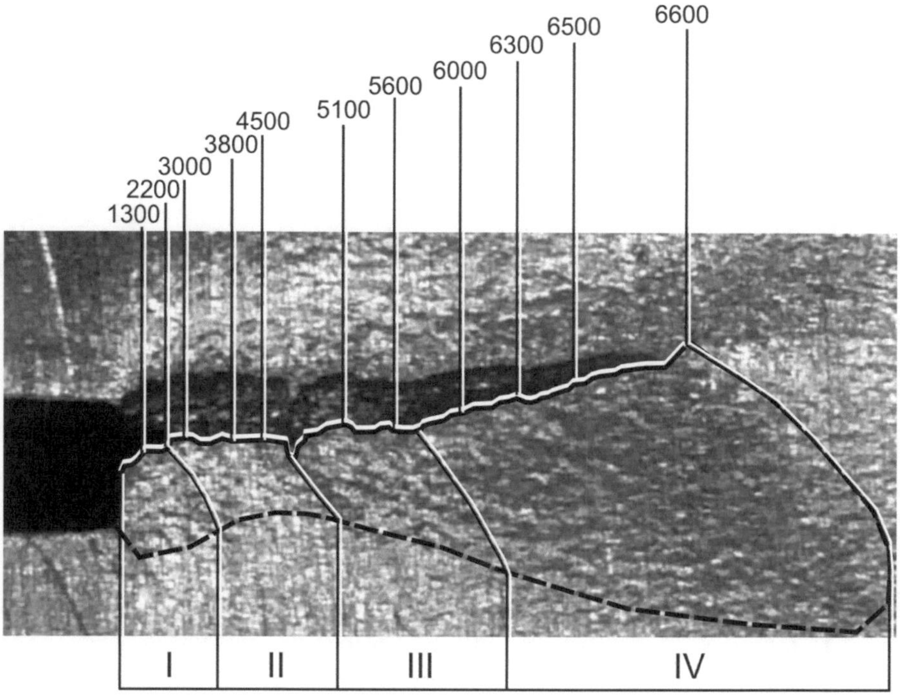

Fig. 9. Propagation of crack in pre-fracture zone for D16T alloy. Given at the top are the numbers of loading cycles, dashed line at the bottom outlines the visible area of maximum plastic strains. This area is divided into four regions: a crack initiates and propagates from the beginning in the closed state, and then it takes the mean propagation path (region I); the stage of stable crack propagation before a "tooth" arises (region II); stable crack propagation after "tooth" proceeds with increasing rate and continuous growth of the pre-fracture zone (region III); the final stage of crack propagations that is characterized on plots by clear nonlinearity (region IV)

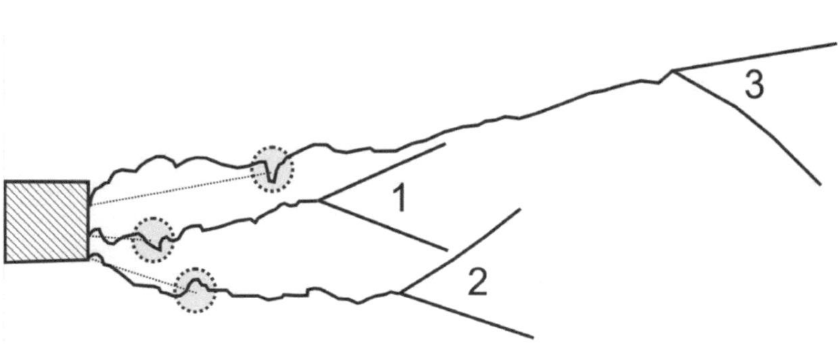

Fig. 10. Cracks and pre-fracture zones for various notch lengths

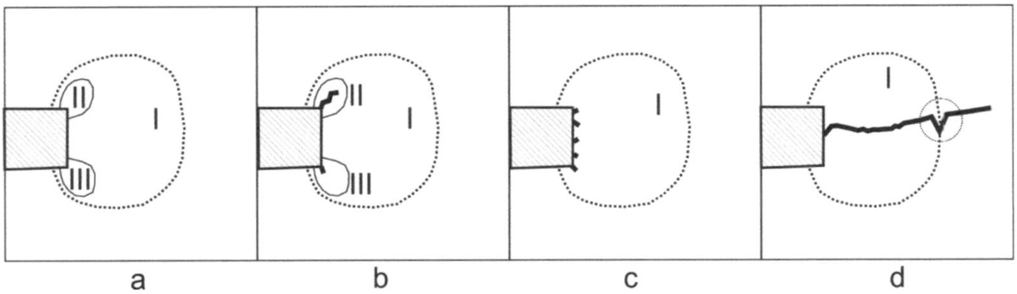

Fig. 11. Cracks and pre-fracture zones

At the initial stage, a fatigue crack passes through material located in the region of plasticity localization near the notch (region I in Fig 11.a, regions I and II in Fig. 9). Further, when the crack continues to propagate, the zone of plasticity localization ahead of the sharp crack tip serves as a pre-fracture zone. The former zone is sufficiently small. As a result of the fact that the crack tip passes from one region to another, a pronounced fatigue striation is originated in the form of a "tooth" (marked by circle in Fig 11.d).

4. Comparison between test results and theoretical notions

The model of low-cycle fatigue describing pulsing loading of a specimen with the inner macrocrack has been proposed in (Kornev, 2004, 2010). This model is appropriate to the scheme by Laird-Smith (Laird & Smith, 1962; Laird, 1967). Within the framework of the proposed model in (Kornev, 2004, 2010), information on material strain in the pre-fracture zone has been obtained in detail: processes of damage accumulation, step-wise crack tip advance, and failure of structures for pulsing loading are described. Accumulation of damages is associated with inelastic strain of materials in the pre-fracture zone. The simple relations for the critical fracture parameters and the fatigue life have been obtained. Attention should be paid to the following circumstances: in the foregoing model, the information of damage accumulation and the hypothesis concerning the crack arrest are essentially used. Emphasize that when damages are accumulated, just as linear, so nonlinear summation of damage in the pre-fracture zone material may occur in the context of the considered model. In deciding between one and another way for summation of damages, no experimental data on damage accumulation at every loading step were available. The experimental data described in the previous section make up this deficiency.

The model in (Kornev, 2004, 2010) describes occurrence of striations under fatigue fracture. Material is considered to consist of quasi-brittle fibers separated by thin layers, which possess quasi-ductile fracture type before strain, and after inelastic strain of layers, the fracture type changes to quasi-brittle. The fiber diameters coincide with diameters of grains of tested materials (the diameters of grains are $\approx 10^{-2}$ cm), and the fiber widths coincide with the thickness of layers separating subgrains. The width of the layers is $\approx 10^{-4}$ cm. Further we assign numeral 1 to the fiber material and numeral 2 to the thin layer material. Properties of the layer material allow description of occurrence of some marks (whiskers or ears) from fatigue striations. In essence, in work (Kornev, 2004, 2010) there is considered the behavior of the simplest composite medium, material of which changes its fracture type under inelastic strain.

For deriving sufficient fracture criteria (Kornev, 2004, 2010) for low-cycle fatigue, modification of the classical Leonov-Panasyk-Dugdale model (Kornev, 2004) is used, where the pre-fracture zone is a rectangle ahead of the crack tip. The modification of the classical Leonov-Panasyk-Dugdale model allowed one to describe not only the pre-fracture zone length $\Delta_{1\sigma+}$ at every loading cycle, but a magnitude of inelastic strain under tension $\varepsilon_{\sigma+}$ for material of the pre-fracture zone fiber nearest to the macrocrack center

$$\varepsilon_{\sigma+} = \frac{1 - \dfrac{\sigma_{m1}}{\sigma_a}\sqrt{\dfrac{r_1}{2l}}\dfrac{k_1}{\sqrt{n_1}}}{\dfrac{5}{\pi(\eta+1)}\dfrac{G_1}{\sigma_{m1}}},$$

$$\frac{\Delta_{1\sigma+}}{r_1} = \frac{k_1^2}{2n_1}\frac{\left(\dfrac{5\sqrt{2}}{4(\eta+1)}\dfrac{G_1}{\sigma_{m1}}\varepsilon_{\sigma+}\right)^2}{\left(1 - \dfrac{5}{\pi(\eta+1)}\dfrac{G_1}{\sigma_{m1}}\varepsilon_{\sigma+}\right)^2} \qquad (1)$$

Here σ_a is the amplitude of pulse loading; σ_{m1} is the limit of elasticity; n_1 and k_1 are integers ($n_1 \geq k_1$, k_1 is the number of damage-free material fibers); $n_1 r_1$ is the averaging interval for the first material; r_1 is the specific linear dimension of the first material structure; $2l_0$ and $2l = 2l_0 + 2\Delta_{1\sigma+}$ are lengths of initial and fictitious cracks, respectively; G_1 is the shear modulus of fibers; $\eta = 3 - 4\mu$ and $\eta = (3-\mu)/(1+\mu)$ are coefficients for plane strain and plane stress state, respectively, where μ is the Poisson ratio; for relations (1), the restriction $1 - 5G\varepsilon_{\sigma+}/(\pi\sigma_{m1}(\eta+1)) > 0$ holds.
Under cyclic pulse loading conditions, when the scheme of three-point bending is used, hysteresis loops take the form given in Fig. 12.
These hysteresis loops with translation differ from the standard statement in the model in (Kornev, 2004, 2010), in which the scheme of rigid loading under unloading is accepted. In Fig. 12, $\varepsilon_1^{(i)}$ is the limit elongation of original materials for $i = 0$; that after the first inelastic strain is $i = 1$; that after the second inelastic strain is $i = 2$, and etc. First three loops are depicted with lines widen from one loop to another, and the onset of the fourth loop is depicted with dots.

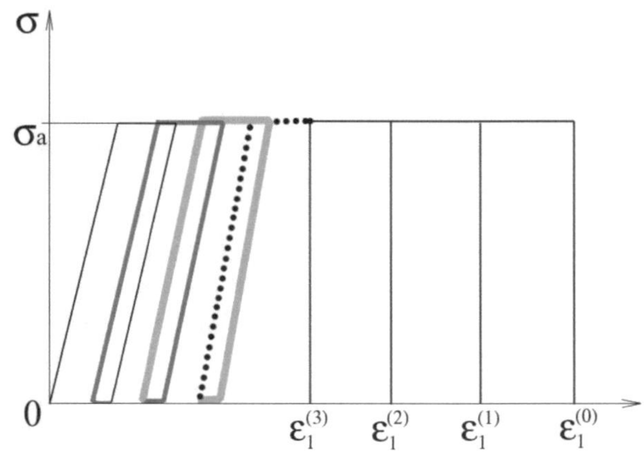

Fig. 12. Scheme of material damage

Consider a typical situation. Let failure occur at the fourth loading cycle. During the strain process, the material elongations at the real crack tip coincide with the limit material elongation $\varepsilon_1^{(3)}$ after the third inelastic strain of the fiber nearest to the real crack tip (Fig. 12), and the crack tip being advanced.

Recall that the crack length in the model in (Kornev, 2004, 2010) changes by the pre-fracture zone length $\Delta_{1\sigma+}$ after the step-wise real crack tip advance, and under the repeated loading, materials in the pre-fracture zone become brittle (Romaniv et al., 1990; Laird & Smith, 1962; Kornev, 2004, 2010). Each such advance is associated with the certain number of cycles when linear and non-linear damages are summed. The performed tests show that the initial state of material, which falls into the pre-fracture zone, influences the process of the step-wise crack tip advance. The magnitude of inelastic strain under stretching $\varepsilon_{\sigma+}$ and the pre-fracture zone length $\Delta_{1\sigma+}$ depend on: *i)* load amplitude σ_a ; *ii)* initial crack length *2l*; and *iii)* preliminary inelastic strain of material. If the basic parameter of inelastic strain $\varepsilon_{\sigma+}$ slightly depends on *2l*, then for the pre-fracture length $\Delta_{1\sigma+}$, the analogous dependence is pronounced. This corresponds to the passage of the second fracture process to the third one on the Paris curve (Romaniv et al., 1990; Shaniavski, 2003).

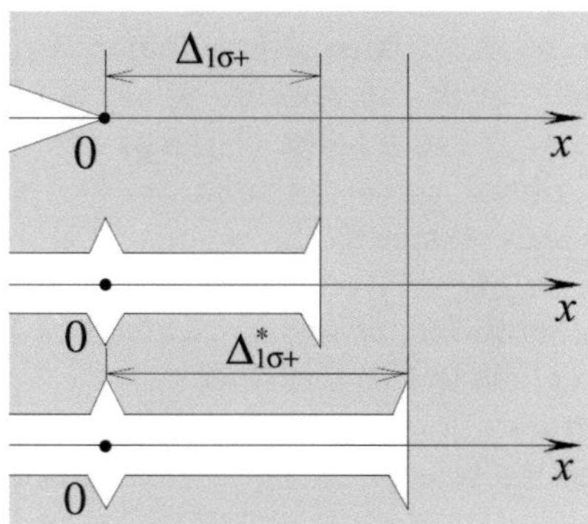

Fig. 13. Schematic drawing of material damage

As explained the step-wise advance of the real crack tip by the length $\Delta^*_{1\sigma+} = \chi\Delta_{1\sigma+}$, where χ is such a coefficient that for quasi-ductile layers $\chi = 1$, and for quasi-brittle layers $\chi > 1$.

The most important question is as follows: what is a distance from the crack onset where the crack will arrest? In the model in (Kornev, 2004, 2010), the case of original material loading is considered when the layers possess the quasi-ductile fracture type. Because of this, the ratio $\chi = 1$ holds. After preliminary inelastic strain of a material from which specimens have been made, embrittlement of layer materials occurs, and the ratio $\chi > 1$ can hold. The step-wise crack tip advance in composite materials with quasi-ductile and embrittled layers elucidates the schematic drawing in Fig. 13. At the scheme top, the right-hand tip of a blunted crack is shown before its start. In the middle part of the scheme, the crack tip advances by the segment $\Delta_{1\sigma+}$, then the crack is blunted at the mesoscale, see the ratio $\chi = 1$. At the scheme bottom, the crack tip advances by the segment $\chi\Delta_{1\sigma+}$, and then the crack is blunted at the mesoscale, see the ratio $\chi > 1$.

In essence, just as in the model in (Kornev, 2004, 2010), so in foregoing experimental results, the distinct influence of the single parameter on the process of damage accumulation under cyclic pulse loading is traced. In the model in (Kornev, 2004, 2010), it is the parameter that characterizes inelastic strain of material in the pre-fracture zone $\varepsilon_{\sigma+}$, and in experiments, it is the parameter, describing inelastic (residual) deflection of beams. In experiments, the residual deflection of beams characterizes both advance of the real crack tip and development of the pre-fracture zone after each loading cycle. The experimental results obtained agree with the diagram of fatigue failure with various kinds of relief elements formed as a result of fatigue fracture seen in Fig. 3.23 from (Shaniavski, 2003), just as the theory described in (Kornev, 2004, 2010), so the experimental data being correlated with the second and third stages of the Paris curve.

The model in (Kornev, 2004, 2010) gives no preference to linear or nonlinear summation of damages in metals (Romaniv et al., 1990; Coffin & Schenectady, 1954) under cyclic loading. The performed tests show that damage accumulation, in general, is always nonlinear, however, if the loading is performed for P_{max} near to the limit of the specimen plasticity, there exists, as a rule, the large interval on the axis of the number of cycles N, where the δw value is close to its minimum for given P_{max}value. Within this interval, the damage accumulation may be considered to be linear as it is under non-stationary loading with P_{max} values close to the limit of plasticity. In both cases, the model proposed in (Kornev, 2004, 2010) and the obtained experimental results agree well. In this case, the hypothesis concerning a crack arrest holds (Kornev, 2004, 2010): a crack arrests at the interface between a fiber and a layer and then it is blunted. This layer is the first one located beyond a pre-fracture zone $\Delta_{1\sigma+}$, and the zone itself is located in material not subjected to some plastic strain.

When a loading regime with increased P_{max} is considered (this regime models catastrophic overload of a structure with crack), the model like that in (Kornev, 2004, 2010) is proposed, and the experimental results obtained agree essentially worse. Such disagreement of theoretical notions (Kornev, 2004, 2010) with the experimental results is explained by the relations $\chi = 1$ or $\chi > 1$. After the first catastrophic series of loadings, the intensity of damage accumulation process in material significantly increases. In this loading regime, the refined hypothesis $\chi > 1$ $\chi > 1$ concerning crack arrest holds: a crack arrests at the fiber-layer interface at the distance $\Delta^*_{1\sigma+} > \Delta_{1\sigma+}$ and it is blunted, this interlayer is located beyond the pre-fracture zone $\Delta_{1\sigma+}$, and the last is in material that has already been subjected to some plastic strain and embrittlement (Romaniv et al., 1990; Laird & Smith, 1962; Kornev, 2004, 2010).

5. Conclusion

Recording the loading diagrams for beams with the edge crack allows description of both damage accumulation at the macroscale and failure of constructions at every loading cycle. The damage accumulation in the pre-fracture zone is associated with the residual deflection of beams after unloading.

Preliminary plastic deformation of aluminum alloy leads to decrease in its durability. However, here the intensity of damage accumulation also reduces under low-cycle three-point bending of beams. This results in increase in the number of loads, which structure may withstand.

In the chapter, loading regimes with gradual overloading and with increased loading have been considered. These regimes model a common situation and catastrophic overloading under cyclic loading. The results obtained may be useful for prediction of fatigue life of a structure with the crack and in analyzing situations after overloading.

Direct observation in the vicinity of the notch tip during the process of low-cycle tension of a plain specimen allows one to trace the behavior of a fatigue crack at various stages of its

propagation: *i*) crack initiation in the vicinity of stress concentration; *ii*) crack propagation within the plasticity zone; *iii*) onset of generation of a narrow pre-fracture zone formed by the crack itself; *iiii*) fast crack propagation beyond the plasticity zone.
Schemes allowing description of deformation, damage accumulation, and failure of material under fatigue with account of the preliminary inelastic deformation of the material and effect of stress concentration on crack initiation have been proposed.

6. Acknowledgment

The work was financially supported by Russian Foundation for Basic research (No 08-01-00220), in the context of the project No 23.16 included into the program of Presidium of Russian Academy of Sciences, and of *Integration Project of SB RAS, UB RAS, FEB RAS No* 119.

7. References

Romaniv, O.N.; Yarema, S.Ya.; Nikiforchin, G.N.; Makhutov, N.A.; Stadnik, M.M. (1990). *Fatigue and cyclic fracture toughness of structural materials, Vol. 4. Fracture mechanics and strength of materials, in four volumes*, Naukova Dumka, ISBN 5-12-000489-X, Kiev, Russia (in Russian)

Shaniavski, A.A. (2003). *Safety fatigue fracture of elements of aircraft constructions. Synergetic in engineering applications*. Monografiya, ISBN 5-94920-015-2, Ufa, Russia (in Russian)

Laird, C.; Smith, G.C. Crack propagation in high stress fatigue. *The Philosophical Magazine, A. Journal of Theor. Experim. and Aplied Physics*, Vol.7. No.77, (1962), pp. 847-857, ISSN 1478-6435

Laird, C. The influence of metallurgical structure on the mechanism of fatigure crack propagation. *Fatigue Crack Propagation, ASTM STP 415; Am. Soc. Testing Mats.* (1967), pp. 131-168, ISBN 0-8031-1250-5

Kornev, V.M. Two-scale model of low-cycle fatigue. Change from quasi-ductile to brittle fracture. *Strain and fracture of materials*. No.2, (2008), pp. 2-11 (in Russian), ISSN 1814-4632

Kornev, V.M. Distribution of stresses and crack opening displacement in the pre-fracture zone (Neuber-Novozhilov approach). *Physical Mesomechanics*. Vol.7, No.32, (2004), pp. 53-62 (in Russian), ISSN 1683-805X

Nikitenko, A.F. (1997). *Yield and long strength of metallurgical materials*. Institute of Hydrodynamics, Siberian Branch of Russian Academy of Sciences, Novosibirsk, Russia (in Russian), ISBN 5-7795-0024-X

Coffin, L.F.; Schenectady, N.Y. A Study of the effects of cyclic thermal stresses on a ductile metal. *Transactions of the ASME*. Vol.76., No.6., (1954), pp. 931-950, ISSN 0742-4795

Karpov, E.V. Deformation and fracture of a spheroplast under low-cycle loading at various temperatures. *Journal of Applied Mechanics and Technical Physics*, Vol.50, No.1, (2009), pp. 163–169, ISSN 0869-5032

Kornev, V.M. Two-scale model of low-cycle fatigue. Embrittlement of pre-fracture zone material. *Procedia Engineering*. Vol.2, No.1, (2010), pp. 453-463, ISSN 1877-7058

Kornev, V.; Karpov, E.; Demeshkin, A. Damage accumulation in the pre-fracture zone under lowcyclic loading of specimens with the edge crack. *Procedia Engineering*. Vol.2, No.1, (2010), pp. 465-474, ISSN 1877-7058

Part 6

Microstructures, Nanostructures and Image Analysis

18

Nanostructure, Texture Evolution and Mechanical Properties of Aluminum Alloys Processed by Severe Plastic Deformation

Abbas Akbarzadeh
Department of Materials Science and Engineering, Sharif University of Technology, Tehran, Iran

1. Introduction

Various research works have been conducted to replace heavy steel body constructions with lighter aluminum ones to achieve stronger energy consumption and environmental standards. The most important technical obstacle to this goal is the inferior ductility of most aluminum sheet alloys. It has been reported that control of the microstructure and the texture of materials is essential for improvement of their mechanical properties (Lee et al., 2002). Reducing the grain size of polycrystalline metallic materials to the nanosize (d < 100 nm, nanocrystalline) or submicron levels (100 nm< d <1 μm, ultra-fine grain) is an effective and relatively economic way of improving mechanical properties such as strength, toughness, or wear resistance in structural materials (Kim et al., 2006; Prangnell et al., 2001) which even can give rise to superplastic behavior under appropriate loading conditions (Pérez-Prado et al., 2004). Since it is practically difficult to reduce the grain size of many metallic materials such as aluminum alloys below 5 μm by a conventional cold working and recrystallization process, several new methods are developed to manufacture ultrafine grained (UFG) materials (Kim et al., 2006). These methods can be classified into two main groups namely bottom-up and top-down processes. In the bottom-up procedures, such as rapid solidification, vapor deposition and mechanical alloying, an ultra-fine microstructures is configured from the smallest possible constituents which are prohibited to grow into the micrometer domain (Pérez-Prado et al., 2004). In the top-down procedures, on the other hand, an existing microscale microstructure is refined to the submicrometer scale, e.g. by a process such as severe plastic deformation (SPD) (Pérez-Prado et al., 2004; Saito et al., 1999). The ancient Persian swords are the interesting examples of severe upset forging for development of fine microstructures (Sherby and Wadsworth, 2001).
By now, various SPD processes such as accumulative roll bonding (ARB) (Saito et al., 1999), cyclic extrusion compression (CEC) (Richert J. & Richert M., 1986), equal channel angular pressing (ECAP) (Valiev et al., 1991), and high pressure torsion (HPT) (Horita et al., 1996) have been proposed and successfully applied to various materials. The common feature of these techniques is that the net shape of the sample during processing is approximately constant, so that there is no geometric limitation on the applied strain (Prangnell et al., 2001). Among these processes, accumulative roll bonding has some unique features. Firstly,

unlike the ECAP, CEC and HPT processes which require forming machines with large capacity and expensive dies, the ARB process can be performed by a conventional rolling mill without any special die. Secondly, in comparison to the other methods, the productivity of the ARB process is relatively high because this process implies the potential of industrial up scaling to a continuous production of UFGed metallic sheets or plates (Saito et al., 1999). In principle at least, the use of heavy deformation in metal processing, with the objective of producing metal alloys with superior properties, has a long history which may be traced back to the early metal-working of ancient China [16], the blacksmith's production of high-quality Damascus steel in the Middle East [17] and the fabrication of the legendary
Wootz steel in ancient India [18].
The ARB process is a repetitive procedure of cutting, stacking and roll bonding of similar sheets for a desired number of cycles, Fig. 1, which is practically limited by technological constraints such as the occurrence of edge cracks. In this process the thickness of the sheet varies between fixed limits and by repeating the procedure very high strains can be accumulated in material, and as a result significant structural refinement can be achieved (Huang et al., 2003; Saito et al., 1999; Tsuji et al., 2003a).

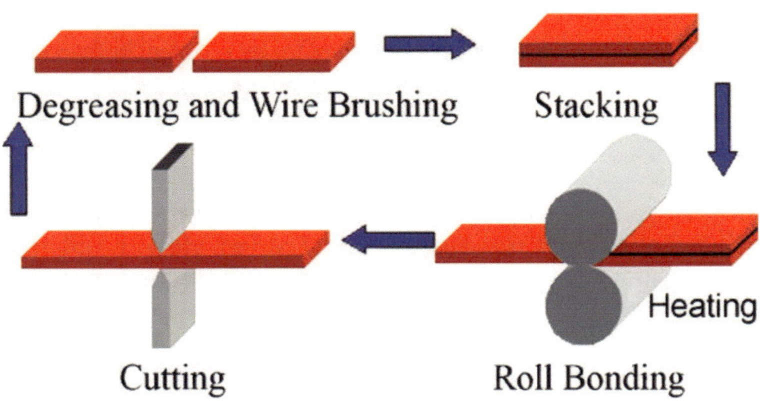

Fig. 1. Schematic illustration showing the procedure of the ARB process (Pirgazi et al., 2008a)

The first scientific paper on the ARB was published in 1998 and afterwards extensive studies have been conducted regarding the microstructural evolution and mechanical properties of various materials processed in different ARB cycles. Pure aluminum (Huang et al., 2003; Saito et al., 1998; Tsuji et al., 2003a), AA5083 (Saito et al., 1999), AA6061 (Lee et al., 2002; Park et al., 2001), AA8011 (Kim et al., 2002, 2005), AA3103 (Chowdhury et al., 2006a) and AA8090 (Chowdhury et al., 2006b) aluminum alloys, magnesium alloys (del Valle et al., 2005; Pérez-Prado et al., 2004), and Ti-IF steel sheets (Reis & Kestens, 2005, Reis et al., 2005; Tsuji et al., 2002a) are the most important materials which have already been successfully produced by the ARB process. The results of these investigations mainly indicate that during the first stages of ARB, ultra-fine grains with diameter less than 1 μm partially form in the sheets, and the volume fraction of these grains increases with increasing the number of cycles, so that after high levels of strain the sample is completely covered with ultra-fine lamellar grains which are not equi-axed and represent an aspect ratio bigger than one (Huang et al., 2003; Park et al., 2001; Saito et al., 1998; Tsuji et al., 2003a). The investigations on the mechanical properties by previous researchers show that the ARB is a promising process for improving this feature of metallic sheets. It has been reported that a significant increase in strength and hardness, more than two or three times of the values of starting materials can be achieved by ARB process in aluminum alloys and IF steel sheets (Lee et al., 2004; Park et al., 2001; Saito et al., 1998; Tsuji et al, 2002b).

Microstructural evolution during ARB has been studied by several researchers. TEM investigation and local crystallography were used (Huang et al., 2003) to study the microstructural evolution of AA1100 up to eight cycles of ARB. Their results show that at large strains, almost a homogenous lamellar structure is formed across the thickness. This microstructure is subdivided by high angle boundaries and low angle dislocation boundaries. The spacing of dislocation boundaries decrease and their misorientations increase by accumulating strain. EBSD analysis illustrates three mechanisms of UFGs development at different levels of strains (Pirgazi et al., 2008b). It was shown that development of subgrains is the major mechanism during the first two cycles of ARB. This mechanism is followed by strain induced transition of low angle grain boundaries to high angles and formation of a thin lamellar structure at medium levels of strain. Fragmentation of these thin lamellar structures into more equi-axed grains is considered as the dominant mechanism after sixth cycle of ARB. A similar microstructural evolution during ARB of AA6061 has been reported (Park et al., 2001). In addition to fabrication of UFGed and nanostructured materials for metallic sheets, some other unique features for ARB process are applied to different sheets to fabricate multi-layer composites (Min et al., 2006).

The texture analysis is a powerful tool to investigate the microstructural and substructural evolution of plastically deformed materials because it provides information on the fragmentation behavior of grains. When orientation contrast microscopy is used, the microstructural data can be linked with local textural information. It is commonly known that in materials produced by severe plastic deformation processes, the band contrast of the corresponding Kikuchi-lines decreases as a result of the presence of lattice defects, such as regions with high dislocation density, subgrains and grain boundaries. In the TSL® software, the contrast of the Kikuchi lines is quantified by the image quality (IQ) parameter. The IQ parameter is sensitive to a wide variety of additional material and instrumental factors which makes it almost impossible to deconvolute the IQ signal in order to convert it to univocal quantitative information on the local microstrain. Nevertheless, the EBSD technique is the only possible tool for nanoscale analysis of relatively large areas in severely deformed materials at present (Tsuji et al., 2002a).

It has been reported (Chowdhury et al., 2006a) that the texture development in an AA3103 alloy during accumulative roll bonding process shows symmetry at all stages and the major components can be characterized as the Dillamore {4411} <11 11 8> component along with the S component with a scatter around the brass component. It has been also reported that in AA8011 aluminum alloy sheets processed by ARB, the deformation texture is dominated by the Dillamore component and the shear texture was developed near the surface of the sheets. This surface shear texture disappears rapidly as the surface area of the material re-appears in the center of the composite sample in the next ARB pass (Kim et al, 2002, 2005). Similar results have been published by (Heason & Prangnell, 2002a). They have reported that in AA1100 alloy processed by the ARB, coarse unrefined bands can be retained even at very high strains. They have also proposed that the strong texture developed during ARB processing may lead to this inhomogeneity (Heason & Prangnell, 2002a). It has also been reported that the textures in Ti-IF steel produced by ARB display a conventional cross sectional gradient with typical shear component in the subsurface planes and plane strain compression components in the midsection (Reis & Kestens, 2005; Kolahi et al., 2009).

It has been reported that in AA8011 aluminium alloy sheets processed by ARB, the main deformation texture orientation is the Dillamore {4 4 11}<11 11 8> component and the shear texture is developed on the surface of the sheets (Kim et al., 2002). This surface shear texture disappears rapidly as the surface area of the material reappears in the centre of the composite sample during the next ARB cycle. The role of second phase particles on the grain size reduction of aluminium sheets during the accumulative roll bonding process has been investigated by comparing the microstructure and texture of a single phase (AA1100) and a particle containing aluminium alloy (AA3003) during various ARB cycles (Pirgazi & Akbarzadeh, 2009).
Circular shaped hollow sections like tubes and cylinders, as a category of engineering components, are also expected to achieve high strengths by nanostructure. Tube spinning is used as a common tube fabricating process (Wong et al., 2003), based on which a cold-bonding process titled "spin-bonding" with the advantages of ARB of sheets has been recently proposed (Mohebbi & Akbarzadeh, 2010a, 2010b) for manufacturing of high strength thin-walled tubes and cylinders. The SPD process proposed in that work is, in fact, repeatedly spin-bonding of layered tubes (accumulative spin-bonding, ASB) to induce large plastic strain on tubes similar to the ARB of sheets, Fig. 2 (Mohebbi & Akbarzadeh, 2010a).

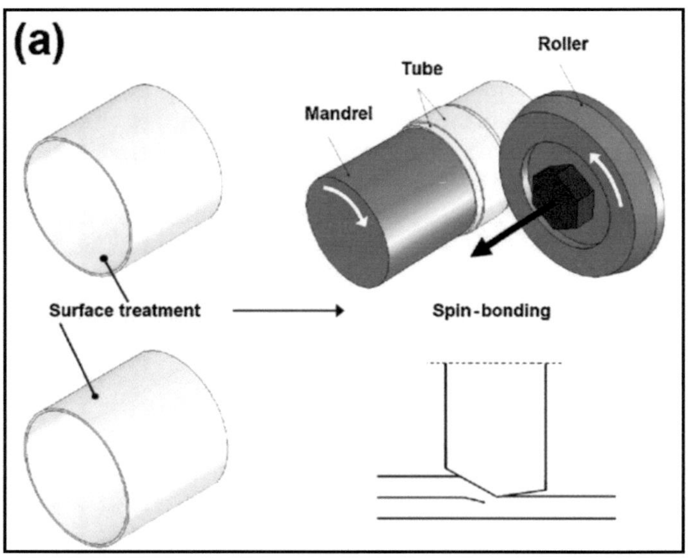

Fig. 2. Schematic illustration of the spin-bonding (Mohebbi & Akbarzadeh, 2010a)

In many cases, the inner and outer surfaces of hollow sections are exposed to different environments, and different characteristics are required inside and outside. In these cases, various bimetallic or clad tubes of stainless steels or high-alloy steels and super alloys clad onto carbon or low alloy steels, are utilized in boilers, heat exchangers, nuclear power plants and petroleum and chemical industries (Chen et al., 2003). So far, several methods have been used to produce composite tubes and cylinders. While centrifugal casting (Sponseller et al., 1998) and extrusion (Chen et al., 2003) can be used to fabricate the thick-walled bimetallic tubes, explosive bonding (Berski et al., 2006), ball attrition (Zhan et al., 2006), thermo-hydraulic fit method and hydraulic expansion method (Wang et al., 2005) are capable of manufacturing thin-walled cylinders. The most widely used cladding process is roll-

bonding of two or more sandwich sheets (Zhan et al., 2006). Although it is possible to fabricate seamed composite tubes by sandwich sheets, it is ideal to plan a method with advantages of roll-bonding to produce seamless thin-walled tubes and cylinders. The spin-bonding method in cold-bonding of cylinders was also proposed by utilizing the tube spinning process for manufacturing of clad tubes (Mohebbi, 2009a).
Tube spinning is an incremental and localized deformation in which material between the roller and the mandrel gradually deforms to the final thickness. Actually the deformation area is limited to a part of work piece which is in contact to the roller, so that the deformation is constrained strongly by surrounding metals. This is why the stress and strain have such a complicated distribution in this incremental process. There is always an inhomogeneous material flow due to the local deformation between the roller and the preform surface with a high strain rate (Mohebbi & Akbarzadeh, 2010b, 2010c, 2009b).

2. Experimental methods

The materials used to study the ARB process were fully annealed sheets of Al-Mn alloy (AA3003) and commercial purity aluminum (AA1100) with initial grain sizes of 40 and 34 µm. Thickness of the sheets was respectively 0.4 and 0.5 mm, and the chemical compositions are presented in Table 1. Two sheets of 150×50 mm2 were degreased (in acetone) and wire brushed (by a stainless steel brush with wires of 0.4 mm in diameter). After the surface treatment, the two sheets were stacked on top of each other and preheated to a temperature of 250 °C for 5 minutes. The plane strain rolling was performed along the longest dimension by 50% reduction in thickness at 200 °C without any lubrication and the mean strain rate was 51 sec^{-1}. Afterwards the roll bonded sample was cut into two sheets of approximately the initial dimension and the procedure was repeated up to a total of 8 cycles so that an accumulated equivalent strain of 6.4 was achieved.

Alloy	Fe	Cu	Si	Ti	Mg	Mn	Al
AA1100	0.57	0.12	0.13	0.03	0.02	0.013	Bal.
AA3003	0.436	0.216	0.18	0.009	0.005	1.076	Bal.

Table 1. Chemical composition of the aluminium alloys (Pirgazi & Akbarzadeh, 2009)

The microtexture and microstructure measurements were performed on the section which is perpendicular to the transverse direction of rolling and is located at the mid-thickness of the sheets. The analysis was carried out by employing an Orientation Image Microscopy (OIM) attached on a Philips XL30 ESEM microscope equipped with a LaB6 filament. The electron back-scattering diffraction (EBSD) mappings were carried out with step sizes in the range of 0.12 µm to 80 nm and the OIM software developed by TSL® was used for data acquisition and post-processing procedures.
Since the ARBed samples were severely deformed, acquiring good Kikuchi-patterns was not evident. By means of electro polishing a very smooth surface was produced and by adjusting the operating parameters of the SEM and the software (OIM Data Collection) a

satisfying result could be obtained. In this way, after using a grain confidence index (CI) standardization (5°, two points) and neighbor orientation correlation (CI = 0.1) cleanup procedures on the raw data, confidence indices of 0.9 and 0.79 were obtained for the samples processed by 4 and 10 cycles, respectively.

The Vickers micro hardness test was utilized to investigate the mechanical properties of initial and ARB processed samples. The values reported for HV represent the average of ten measurements taken at randomly selected points across the thickness of the sheets using loads of 50 and 100 g for 20 s. The mechanical properties of initial and ARBed samples were measured by tensile tests at room temperature executed with an Instron tensile testing machine. The test specimens were prepared with the tensile axis parallel to the rolling direction. The test conditions and the specimen size were chosen according to the ASTM-E8 standard.

Orientation distribution functions (ODFs) were calculated by harmonic series expansion with truncation at L = 16. The initial texture of the fully annealed AA3003 alloy sheet prior to the ARB process was also characterized using the EBSD analysis. In order to obtain a statistically acceptable result, this measurement was carried out with a step size of 2 μm over an area with dimension of 300 μm×300 μm located at mid thickness and as a result more than 22,000 orientations were analyzed.

The texture evolution during ARB process was predicted with the Alamel model (Van Houtte et al., 2005, 2006). In this model the texture is discretized in a set of N individual orientations which are considered in pairs of two. Each pair of orientations will accommodate the externally imposed strain but shear strains can be relaxed in equal and opposite measure for each grain belonging to a pair. In the ODF calculations the orthorhombic sample symmetry was imposed which is usually assumed for the conventional rolling process. Only the {111} <110> slip systems were taken into account for the texture calculations.

The tubes used for ASB process were prepared from commercially pure aluminum (AA1050). Thickness of the tubes was 0.8 mm. Work pieces were annealed at 350 °C for 2 hrs before the first cycle. Spin-bonding process was repeated up to four cycles. In the spin bonding process two surfaces to be bonded (inner surface of the external tube and outer surface of the internal one) were degreased in acetone and wire brushed as surface treatment. After scratch brushing, surface treated tubes were positioned against each other and fitted on the mandrel for tube spinning at room temperature with conditions of Table 2 (Mohebbi & Akbarzadeh, 2010c). At this stage, while the tube and mandrel rotate about their axes, a roller with a degree of freedom about its own axis moves along the direction of the tube axis to reduce its thickness to 50% leading to bonding of the tubes. No preheating was performed in this work. More details of the spin-bonding is available in (Mohebbi & Akbarzadeh, 2010a).

The longitudinal sections of the specimens were observed by an optical microscope after polishing and etching in the Tucker solution for 15 s. Transmission electron microscopy (TEM) micrographs and the corresponding selected area diffraction (SAD) patterns were also obtained from the specimens after 1, 2 and 4 cycles of ASB. To do so, thin foils were prepared by twin-jet polishing from the mid-thickness of the tubes normal to the tube surface. The microstructural evolution was also analyzed by electron backscattering diffraction (EBSD) pattern. This was performed on the longitudinal sections perpendicular to the peripheral direction of the tubes after various cycles of ASB. The measured area was 25 μm×30 μm located at the center of mid-thickness of the tubes.

Parameter	Value
Workpiece	
Inner diameter (mm)	35
Wall thickness (mm)	2.5
Initial length (mm)	50
Roller	
Diameter (mm)	54
Attack angle (°)	25
Smoothing angle (°)	5
Flow forming conditions	
Feed rate (mm/rev)	0.1
Reduction (%)	40
Speed of rotation (rpm)	30

Table 2. Dimensions of tools and work piece and the process conditions

EBSD patterns were obtained at 25 kV and analyzed by TSL software regarding the quantitative analysis of grain boundaries and misorientation distributions. The step size was 100 nm for the specimens after the first and second cycles and 60 nm for the specimens after the third and fourth cycles of ASB. Mechanical properties of the specimens were determined at ambient temperature with strain rate of 2.6×10^{-3} sec^{-1}. The samples were prepared from the circumferential direction of the tubes after straightening and machining according to the ASTM E8M standard. Vickers microhardness values were also measured through the thickness of the specimens by applying the load of 0.49 N.

The material used in ASB process was the commercial purity aluminum (AA 1050). Thickness of the tubes was 0.8 mm. The inner diameter of the internal tube and the external one were 51 and 52.8 mm, respectively and their length was 40 mm. Work pieces were annealed at 350 °C for 2 hrs before the bonding process. Since they were processed via tube spinning, work pieces were diametrically true after annealing. The bonding surfaces were degreased in acetone and scratch brushed. The time between surface preparation and bonding process was kept to less than 300 s to minimize the formation of contaminant film and a thick oxide layer on the bond surfaces of the tubes. Afterwards, the two tubes to be joined were positioned against each other. The process is schematically illustrated in Fig. 2.

The bond strength of the Al/Al bimetal layers was measured using the T-peel test. Samples were 8 mm wide and 50 mm long and were cut at the longitude of the tubes so that their width was in the circumferential direction. The breaking off force per width of the sample was measured as the bond strength (N/mm). Optical microscopy was employed to examine the bond interface of the longitudinal section of deformation zone. The surfaces to be examined were prepared by standard metallographic procedure to polished conditions without etching.

3. Results and discussions

3.1 Microstructure

3.1.1 ARB processing

Figure 3 shows the optical micrograph observed in the RD-ND section of the sample produced by 10 ARB cycles, indicating that a good bonding with no delamination has been achieved under the present ARB conditions. To investigate the microstructural evolution of aluminium sheets processed by ARB, the boundary misorientation maps were extracted from the EBSD data. These mappings were recorded on a section near the centre of the RD–ND plane of the samples. Because of the severe deformation of the ARB processed samples, acquiring good Kikuchi-lines was too difficult. However, by means of electropolishing in Barker's reagent (5 mL HBF4 in 200 mL H2O) and preparing a very smooth surface and adjusting the parameters of SEM (e.g. gun high voltage, working distance and spot size) and the OIM Data Collection software (e.g. exposure time, binning and step size), it was significantly improved.

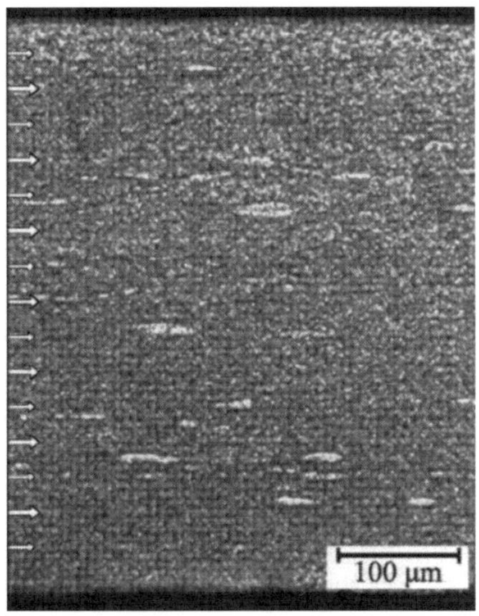

Fig. 3. Optical micrograph of longitudinal cross section of aluminum sheet (AA1100) after 10 ARB cycles

The orientation scans of the AA1100 and AA3003 aluminum sheets processed by various ARB cycles are depicted in Figs. 4 and 5. The boundary misorientation maps of these samples are also depicted in these figures. In these maps, the high angle grain boundaries (HAGB) with misorientations larger than 15° are drawn in bold black lines, while the low angle grain boundaries (LAGB) with misorientations between 2° and 15° are drawn in thin grey lines. In the case of AA1100 alloy, it is observed that after two cycles of ARB, the microstructure is covered with elongated grains surrounded by many low angle grain boundaries (Fig. 4a). The microstructures of the samples produced by four and six ARB cycles are generally similar to the sample produced by two ARB cycles. However, with the increasing number of cycles the grain sizes decrease gradually. After the sixth cycle, the structure consists of very fine elongated grains surrounded by high angle grain boundaries which are usually parallel to the rolling direction (Fig. 4c).

It is observed that the evolution of ultra-fine grains in the AA1100 alloy occurs by various mechanisms of grain refinement at different strains. At low strains ($\varepsilon<1.6$) grain subdivision is the dominant mechanism which leads to the formation of a fibrous microstructure of elongated grains and promotes the formation of subgrains surrounded by a high fraction of low angle grain boundaries (LAGB) (Fig. 4a) and some coarse unrefined bands which contain predominantly subgrains are seen in this specimen. At medium levels of strains

($1.6<\varepsilon<4.8$) the grain subdivision ratio decreases and strain induced transformation of low angle to high angle grain boundaries (HAGB) caused by the accumulation of dislocations in exiting subgrain boundaries is the controlling mechanism which results in a lamellar structure at the specimens processed by 4 and 6 ARB cycles (Figs. 4b and c). The fraction of HAGB presented in Figs. 4 and 5 confirm the increase of the HAGBs in expense of the LAGBs.

In comparison with the first sample, less pronounced unrefined regions of subgrains are also seen in these samples. The proposed mechanism of grain refinement at high levels of strain ($\varepsilon>4.8$) is the progressive fragmentation of the thin lamellar grains into more equi-axed grain structures which leads to development of a relatively homogeneous submicron grain structure without the unrefined regions after the eighth cycle (Fig. 4d).

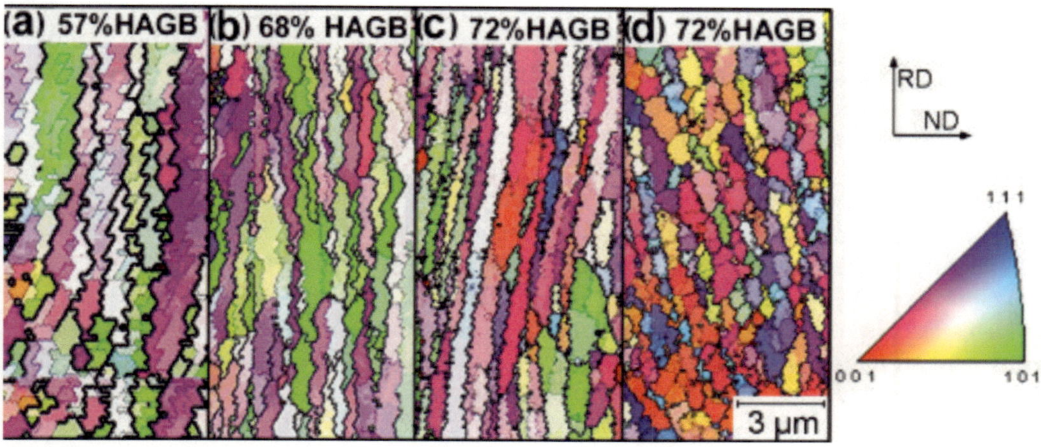

Fig. 4. Orientation scans obtained from the EBSD measurements of the AA1100 ARB processed by (a) 2, (b) 4, (c) 6 and (d) 8 cycles. Thin grey lines represent the misorientation (θ) of $2° \leq\theta <15°$ and bold black lines represent $15° \leq \theta$

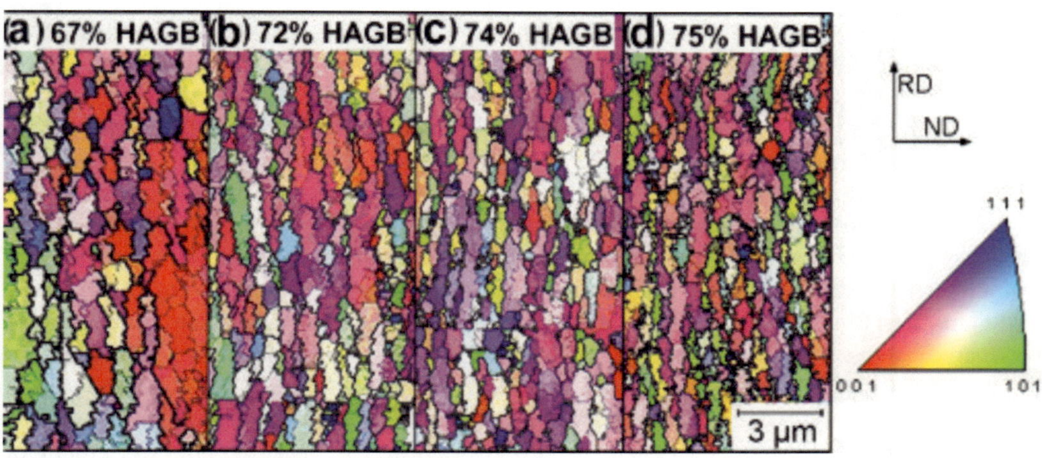

Fig. 5. Orientation scans of the AA3003 samples produced by (a) 2, (b) 4, (c) 6 and (d) 8 ARB cycles

Fig. 5 displays that in comparison with the AA1100 alloy, in the AA3003 alloy which contains a significant volume fraction of second phase particles, an ultra-fine grain structure develops at a much higher rate. It is reported that two types of precipitates evolve in the AA3003 alloy: orthorhombic (Fe,Mn)Al6 and cubic α-Al(Fe,Mn)Si. The total volume fraction of these precipitates is between 4 and 5% and the large majority of them are elongated and have an average length between 1-5 μm (Richert J. & Richert M., 1986). In a similar research on AA8097, it has been shown that these second phase particles cause the texture randomization and promotion of grain refinement (Heason and Prangnell, 2002b).
According to Fig. 5, the volume fraction of HAGBs after the second cycle in the AA3003 alloy is 67% compared to 57% in the AA1100 alloy. This rapid refinement of grains in the AA3003 alloy at relatively low strains results from extensive HAGBs generation in local deformation zones around the second-phase particles. During deformation, large lattice rotation around the second phase particles leads to the increase of local misorientation and production of new HAGBs. Dispersion of these particles also develops a random and more heterogeneous plastic flow behavior within the matrix and promotes the fragmentation of the lamellar structures seen in the AA1100 alloy (Pirgazi & Akbarzadeh, 2008c). This behavior as well as a gradual increase in the percentage of HAGBs and a reduction in the grain aspect ratio during the next cycles results in the formation of a more homogenous submicron grains structure in the AA3003 alloy (Fig. 5).
Although many low angle grain boundaries are also observed in the microstructure, they are usually perpendicular to the rolling direction. Figure 4d displays the microstructure developed in the AA1100 aluminium sheet processed by eight ARB cycles. Significant changes are observed at this level of strain and the grain morphology changes into more equi-axed structures. In comparison with the AA1100 alloy, the developed microstructures in AA3003 alloy during the ARB process are somewhat different. According to Fig. 5, it is clearly observed that in this alloy, the formation of ultrafine grains occurs at the earlier stages of the process. Figure 5a shows that even after the second cycle, the microstructure is completely covered with very fine grains whose morphology is more equi-axed than the same sample of AA1100 alloy (Fig. 4a).
With the number increasing of cycles, the grain size continuously decreases and after the eighth cycle the whole specimen shows a homogenous submicrometre grain size structure. Further analysis revealed that the microstructures of these samples differ in grain size and the fraction of HAGBs. Variations of microstructural parameters of the aluminium sheets during the ARB process are depicted in Fig. 6 with regard to the changes of the average thickness and length of grains. In the AA1100 alloy, the grain thickness (measured by the linear interception method on the EBSD maps) drastically decreases during the first four cycles. While with the increasing strain during the next cycles, the grain thickness reduces slightly and approached to a constant value of about 500 nm at large strains. Similar results are observed in the AA3003 alloy but the grain thickness in this case is smaller than the former case and reaches the minimum of 400 nm after the eighth cycle. Comparison of the grain length in the ARB processed samples revealed that there is a more dramatic difference between the evolved microstructures in the two alloys after a large number of ARB cycles. In the AA3003 alloy, the average length of grains decreases continuously while in the case of AA1100 alloy a drastic reduction is observed after the sixth cycle (Fig. 6b).
By analysing the EBSD data after each ARB cycle, quantitative measurements were made of the misorientation distributions and the results are summarised in Fig. 7. The Mackenzie distribution for boundaries in a random polycrystal is shown by the black bold line. In both

alloys, with ARB processing, the fraction of high angle grain boundaries increases and approaches to a near random misorientation distribution very much similar to the theoretical Mackenzie distribution. However, there are some differences between the misorientation profiles produced in the two alloys during the ARB process. It can be observed that in AA3003 alloy the fraction of HAGBs initially increases rapidly and reaches over 70% after the fourth cycle. During the next cycles, few further changes occur and the fraction of HAGBs after the eighth cycles reaches to 75%. In contrast, the fraction of HAGBs in AA1100 alloy increases more slowly with the increasing strain during the ARB process and reaches a saturation value of about 72% after the sixth cycle.

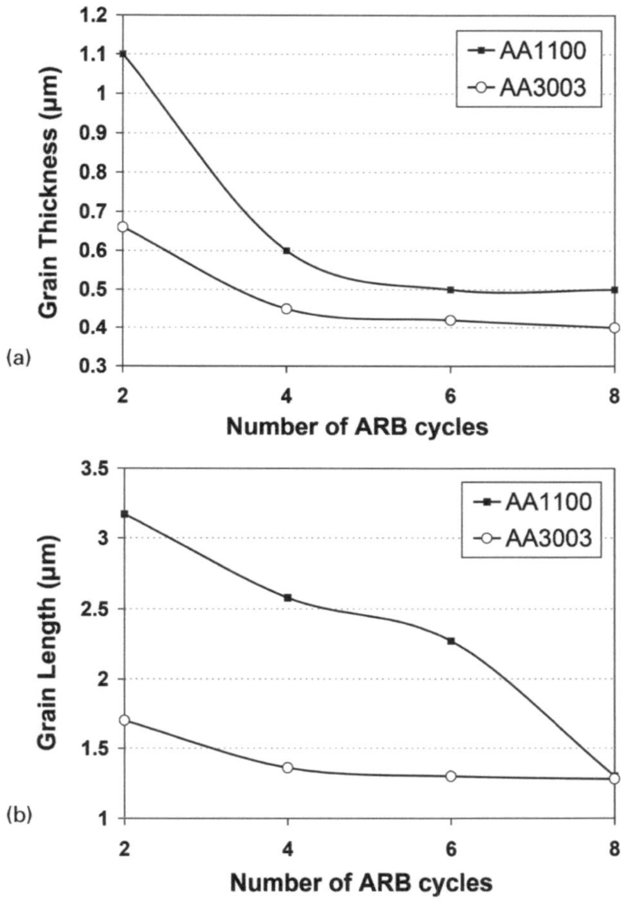

Fig. 6. Variation of microstructural parameters of aluminium sheets during ARB process, (a) grain thickness and (b) grain length

A general definition for the materials with a submicrometer grain structure proposed by (Prangnell et al., 2001) includes two main criteria: the average spacing of HAGBs must be less than 1 μm, and the fraction of HAGBs must be greater than 70%. The authors' observations proved that the development of such submicrometer or ultrafine grains in a single phase (AA1100) and a particle containing alloy (AA3003) occurs at different levels of strains during the ARB process. It was shown that (Heason & Prangnell, 2002a) the evolution of ultrafine grains in the AA1100 alloy occurs by various mechanisms of grain refinement at different strains.

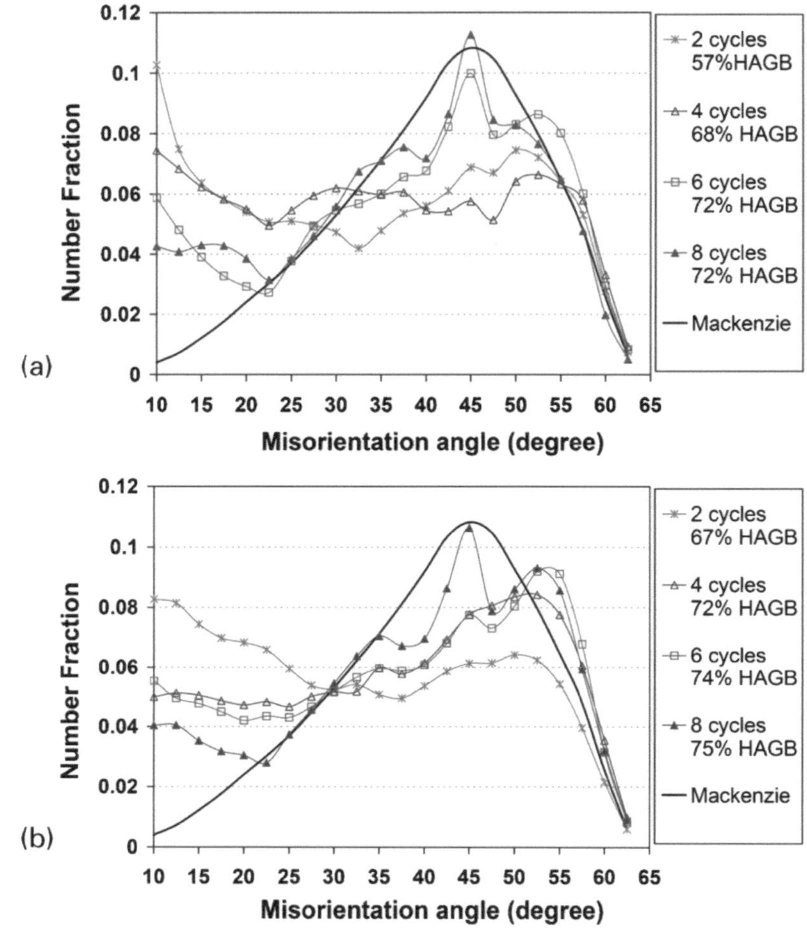

Fig. 7. Misorientation distributions in the samples of (a) AA100 and (b) AA3003 aluminium sheets processed by various ARB cycles

At low strains (ε< 1.6), grain subdivision is the dominant mechanism which leads to the formation of a fibrous microstructure of elongated grains. This mechanism also promotes the formation of subgrains surrounded by a high fraction of low angle grain boundaries (Fig. 4a) and some coarse unrefined bands which contain predominantly subgrains are seen in this specimen. At medium levels of strains (1.6 < ε < 4.8) the grain subdivision ratio decreases and the strain induced transformation of low angle to high angle grain boundaries caused by the accumulation of dislocations in exiting subgrain boundaries is the governing mechanism which results in a lamellar structure at the specimens processed by four and six ARB cycles (Figs. 4b and c).

The misorientation profiles plotted in Fig. 7 confirm the increase of HAGBs in expense of the LAGBs and the tendency of misorientation of the lamellar structure to a Mackenzie distribution. In comparison with the first sample, less pronounced unrefined regions of subgrains are also seen in these samples. The proposed mechanism for grain refinement at high levels of strain (ε> 4.8) is the progressive fragmentation of thin lamellar grains into more equi-axed grain structures (Fig. 4d). The last mentioned mechanism enables the ARB process to develop a relatively homogeneous submicrometer grain structure in aluminium sheets.

Figure 5 displays that in comparison with the AA1100 alloy, in the AA3003 alloy which contains a significant volume fraction of second phase particles, an ultrafine grain structure develops at a much higher rate. It has been reported that two types of precipitates evolve in the AA3003 alloy: orthorhombic (Fe,Mn)Al_6 and cubic α-Al(Fe,Mn)Si (Rios & Padilha, 2003). The total volume fraction of these precipitates is between 4 and 5% and the large majority of them are elongated and have an average length between 1 and 5 mm. According to Fig. 7, the volume fraction of HAGBs after the second cycle in the AA3003 alloy is 67% compared with 57% in the AA1100 alloy. At the same time, the transverse spacing of HAGBs in these specimens is 0.66 and 1.1 mm respectively (Fig. 6a). This rapid refinement of grains in the AA3003 alloy at relatively low strains results from extensive generation of HAGBs in local deformation zones around the second phase particles. During deformation, large lattice rotation around the second phase particles leads to the increase of the local misorientation and production of new HAGBs. Dispersion of these particles also develops a random and more heterogeneous plastic flow behaviour within the matrix and promotes fragmentation of the lamellar structures seen in the AA1100 alloy. This behaviour as well as the gradual increase in the percentage of HAGBs and reduction in the grain aspect ratio during the next cycles results in the formation of a more homogenous submicrometre grain structure in the AA3003 alloy (Fig. 6). TEM micrograph depicted in Fig. 8 confirms that the grain size of the sample processed by 10 ARB cycles reaches a nanoscale value. The microstructure mostly consists of grains with a size in the range of 200-300 nm which are surrounded by clear boundaries.

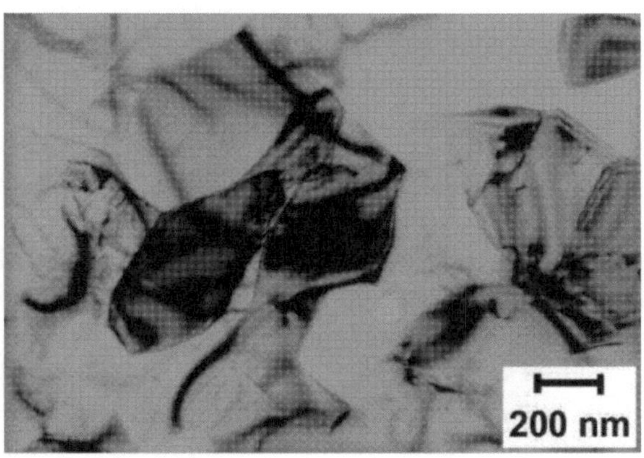

Fig. 8. TEM micrograph showing ultrafine grains in rolling plane of the AA1100 sample after 10 ARB cycles

3.1.2 ASB processing

Optical micrographs of the longitudinal sections of ASB samples are shown in Fig. 9. Previous study on the bond quality of spin-bonding process showed that the threshold thickness reduction for Al/Al bonding by conditions of this work is about 24% (Mohebbi & Akbarzadeh, 2010a). Therefore, it is expected that a good bond is obtained at thickness reduction of 50%. While no interface was observed on the sections after polishing, the bond interfaces are distinguished by etching (Fig. 9). Since subsequent ASB cycles improve the previous bonds, the last bond interface of each specimen is clearer than the previous ones. It is obvious in this figure that the external layer undertakes more thickness reduction than the internal one at each cycle, so that the ratio of external layer thickness to internal one is between 0.40 and 0.45. It should be noted that during the process more elongation were observed in the length of external tube than the internal one, confirming the mentioned thickness strain distribution.

Number of ASB cycles	1	2	3	4
Fraction of the HAGBs	0.54	0.76	0.83	0.84
Grain thickness (nm)	409	306	176	186
Grain length (nm)	885	717	465	419

Table 3. Results of the EBSD analysis

TEM micrographs and the corresponding selected area diffraction (SAD) patterns of the specimens ASBed by one, two and four cycles are shown in Fig. 10. The specimen after one cycle shows a grain size of about 1.5 µm. Large number of subgrain boundaries can be observed in this specimen. The specimen after two cycles illustrates ultra-fine grains with the size of approximately 600 nm. The grain size of about 150 nm is observed in the 4-cycle specimen. The SAD pattern illustrates a more complex shape comparing to the first and second cycles of ASB.

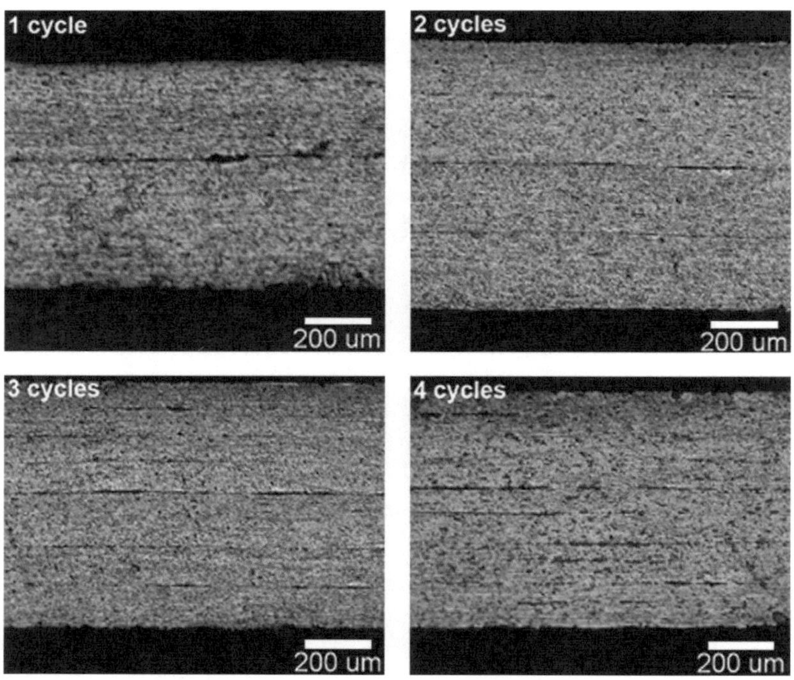

Fig. 9. Optical micrographs obtained from the longitudinal section of the AA1050 specimens after various cycles of ASB

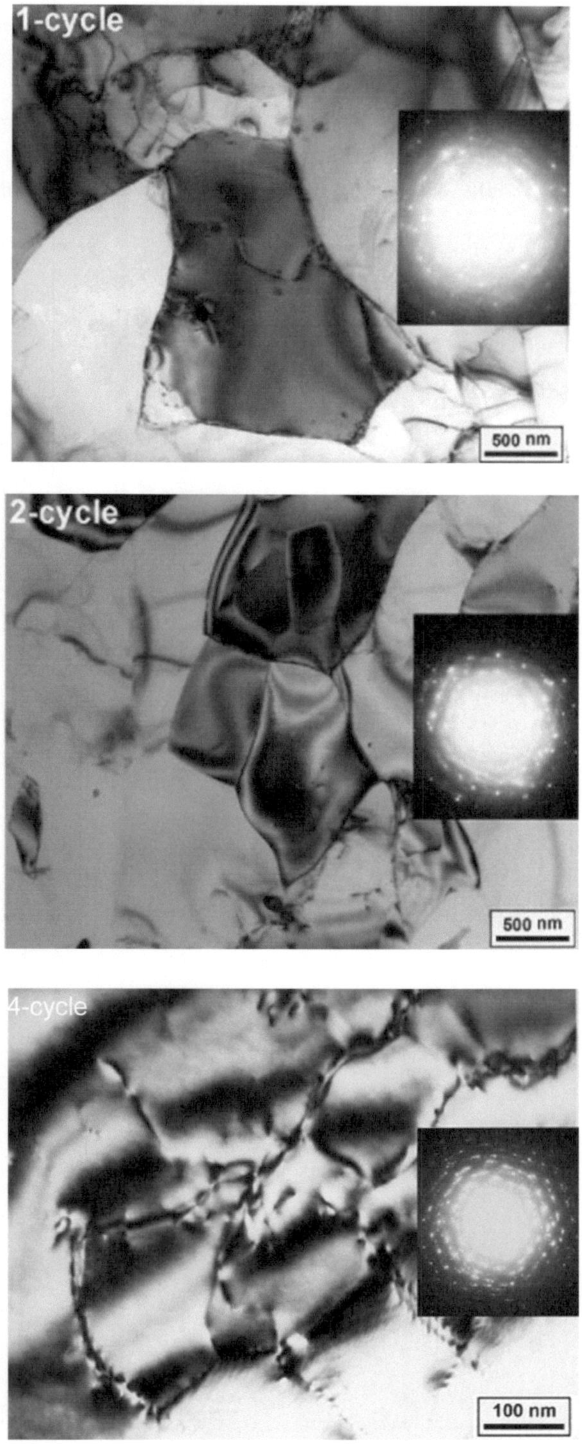

Fig. 10. TEM micrographs and SAD patterns of AA1050 specimens processed by one, two and 4 cycles of ASB

Fig. 11 illustrates the low angle and high angle grain boundary maps recorded on the radial (r)-longitudinal (z) plane of the ASBed specimens. In this figure, which is obtained from the EBSD analysis, low angle grain boundaries (LAGBs) are depicted by red lines where the misorientation between two points of a step is between 2° and 15° and high angle grain boundaries (HAGBs) with misorientations larger than 15° are drawn in the black lines. Comparing to the very locally analysis of TEM micrographs, this figure clearly shows the

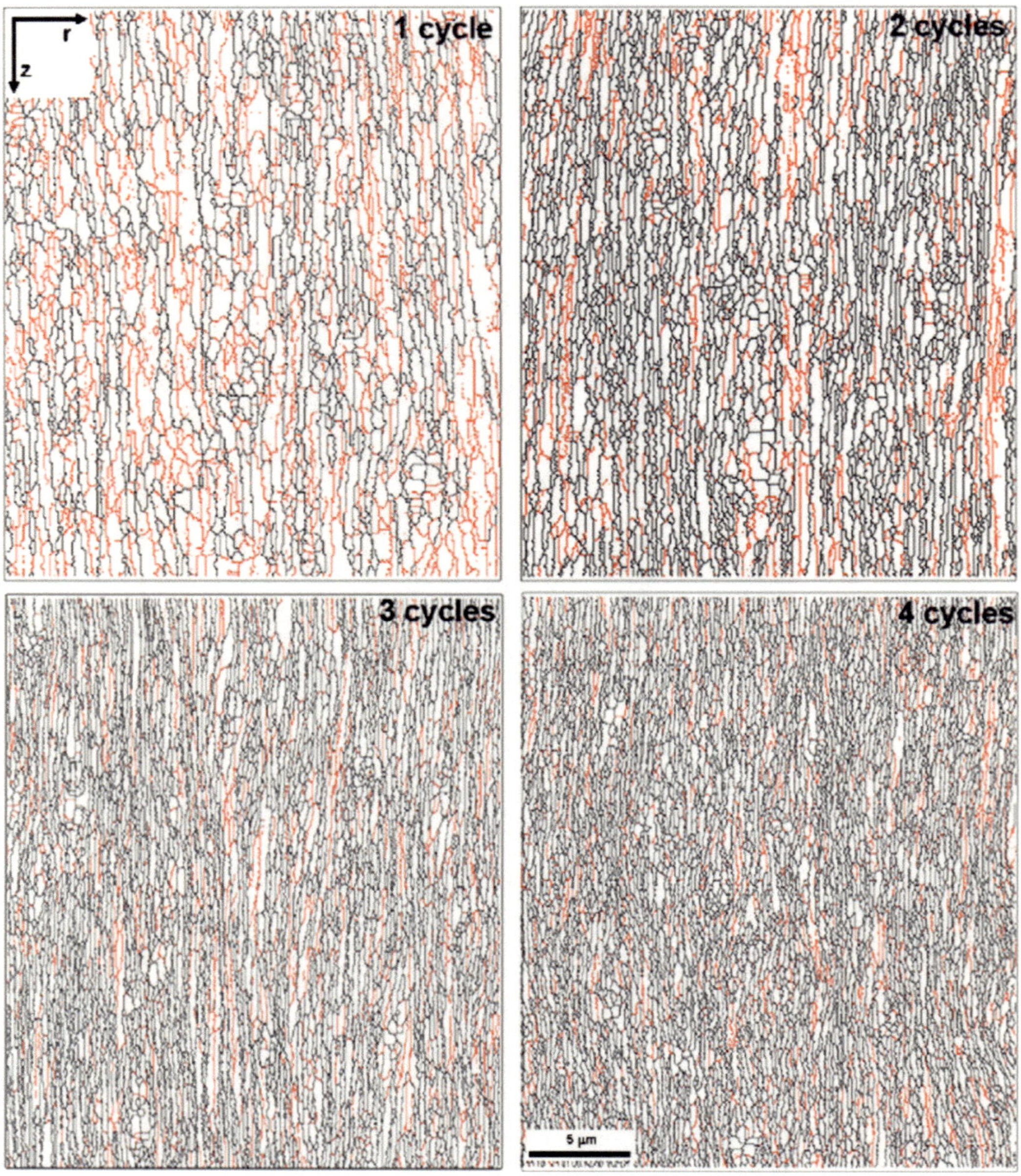

Fig. 11. High angle (black) and low angle (red) grain boundary maps of AA1050 specimens processed by various cycles of ASB

grain refinement and increment of HAGBs fraction by increasing the ASB cycles. Quantitative analysis of the EBSD patterns is presented in Table 3 and Figs. 12 and 13. Regarding the elongated grains, grain size is described by the grain thickness and grain length in radial and longitudinal directions, respectively. This is considered and applied for both subgrain and grain boundaries.

Fig. 12 demonstrates that both grain thicknesses and lengths decrease by increasing the ASB cycles, so that the grain thickness and length are, respectively, 186 and 419 nm after the fourth cycle. In addition to grain refinement, it can be seen in Figs. 11 and 12 that the HAGBs fraction is increased from 54% after the first cycle to 84% after the fourth cycle of ASB. The grain boundary misorientation increments through the ASB cycles are clearer in depiction of misorientation distributions, Fig. 13. The Mackenzie distribution for a random polycrystalline material is shown by black line. It is observed that while the misorientation distribution after the first cycle is considerably far from the random, it becomes much closer to a random distribution during the next ASB cycles.

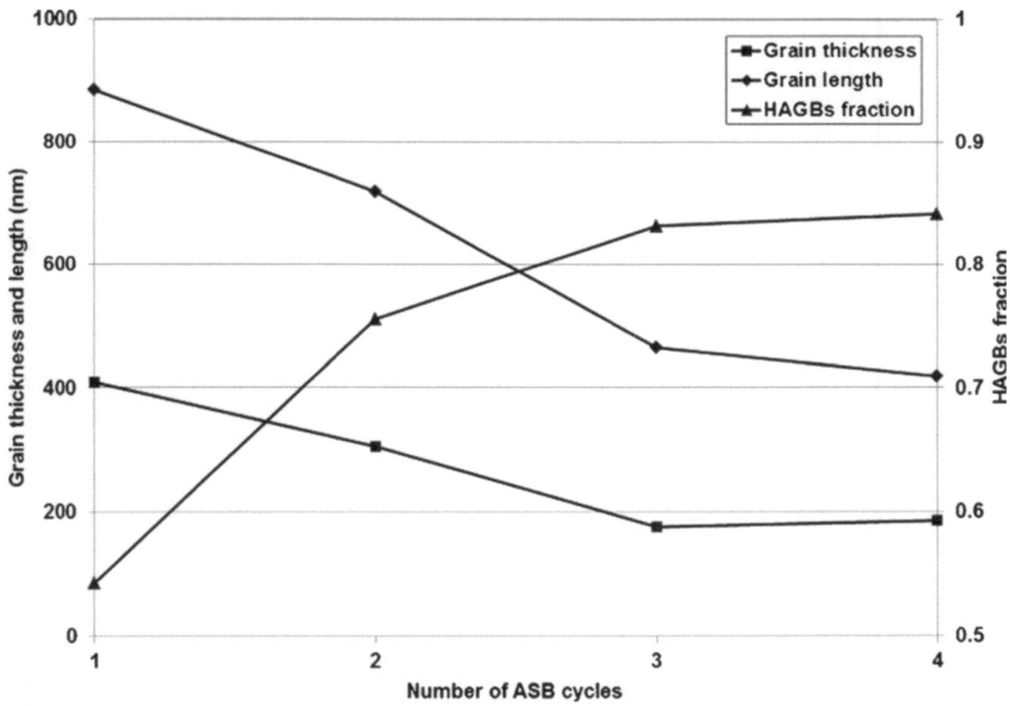

Fig. 12. Evolution of grain thickness and length and fraction of the HAGBs during the ASB cycles obtained from the EBSD analysis

An analytical model was recently developed (Mohebbi & Akbarzadeh, 2010a) to study the strain history of this process. In this model, the radial strain increment at each revolution is calculated. Total strain at each time is then determined by summation of these increments. The strain rate is calculated by dividing these strain increments to the deformation time of each one (Mohebbi & Akbarzadeh, 2010b). It was demonstrated that the material undergoes a small deformation with high strain rate at any exposure to roller, while at the time between the steps, the strain rate downfalls to zero. For thermally activated deformation and restoration processes, which occur even at room temperature by large strains (Estrin et al., 1998), the microstructural evolution and flow stress are dependent on the deformation temperature, strain and strain rate. The strain rate ($\dot{\varepsilon}$) and deformation temperature (T) are included into the Zener-Hollomon Parameter (Z), which is defined as:

$$Z = \dot{\varepsilon} \exp (Q/RT) \quad (1)$$

where Q is the activation energy for operative process and R is the gas constant (Zener & Hollomon, 1944).

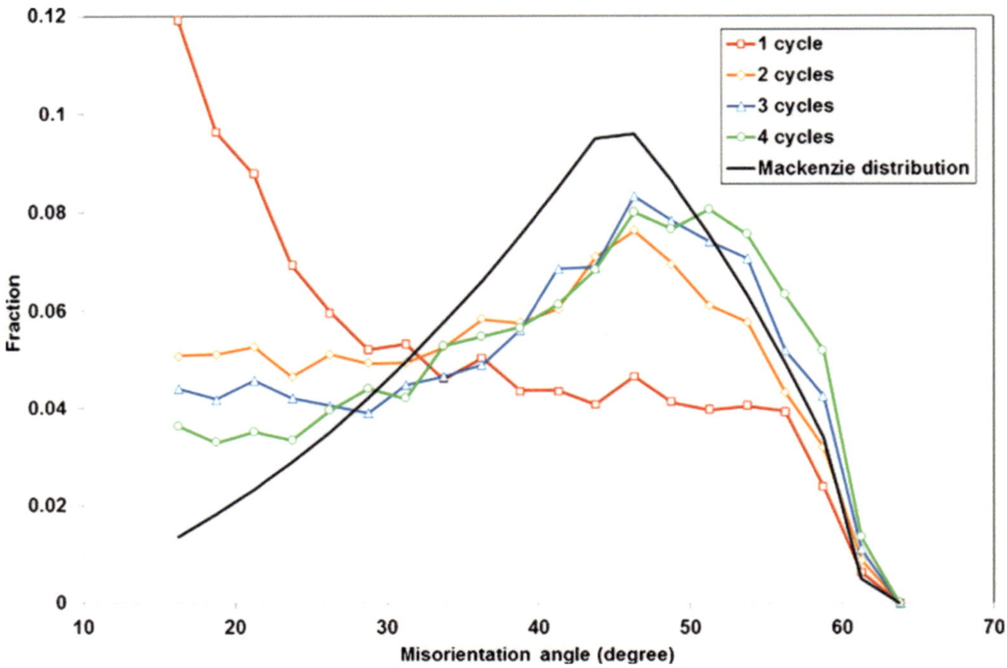

Fig. 13. Misorientation distribution of the specimens processed by various cycles of ASB comparing to the Mackenzie distribution

Generally, by increasing this parameter, subgrain size at dynamic recovery and steady state grain size at dynamic recrystallization are decreased and the flow stress is increased (Humphreys & Hatherly, 2004). Tsuji et al. (2003b) have studied the effect of strain rate on microstructural changes in deformation of the ultra-fine grained (UFG) aluminum produced by ARB. They had expected that higher speed of deformation can result in finer grain subdivision, because larger amount of dislocations would operate at higher strain rates. However, their results showed that grain size rather increases with increasing the strain rate, due to heat of deformation. Therefore, they concluded that higher strain rate and simultaneous cooling are favorable for producing finer grain size in SPD. However, as deformations with strain rates higher than 10 s^{-1} are adiabatic processes (Ryazanov et al., 2003), simultaneous cooling is not effective on deformation temperature and therefore, conclusion of Tsuji et al. (2003b) cannot be correct.
Controlled cooling of the specimen in tensile test maintains the normal temperatures of the tests conducted at 10^{-2} to 10^{-1} s^{-1}. For higher strain rates, the only way is to perform the test

by applying strain in increments small enough to determine the negligible temperature rises during the incremental cooling periods (Sevillano et al., 1981). Therefore, it can be said that increasing the strain rate in common SPD processes causes a high temperature rise due to continuous induction of a large strain, which prevents significant promotion of Z. In case of ASB, however, it is possible to increase the strain rate without temperature rise. Deformation with high strain rate and low temperature rise is a very interesting characteristic of ASB process for evaluation of the pure effects of strain rate on the microstructural evolutions and resulted mechanical properties without temperature rise.
This research group's work has shown (Mohebbi & Akbarzadeh, 2010b) that due to large amount of redundant strain and high Zener–Hollomon parameter, as the characteristics of ASB process, the grain refinement and the rate of microstructural evolution are expected to be higher in comparison to other SPD processes. This grain refinement occurred by common mechanism of grain refinement in SPDs of pure aluminum. Increase of dislocation density and its accumulation in cell structure by dynamic recovery leads to formation of subgrains at early stages. This is followed by increase of the misorientations by strain induced transition of low angle to high angle grain boundaries at next stages (Pirgazi et al., 2008b). This mechanism is confirmed by the TEM micrographs and SAD patterns as well as the EBSD analysis. As can be seen in Figs. 11–13, a large fraction of LAGBs is formed after the first cycle. After the fourth cycle, while the grains are more refined, the fraction of HAGBs is considerably increased and the boundary misorientation distribution is close to the random one (Mackenzie distribution).

3.2 Microtexture

Figure 14 shows φ_2 = 45, 65 and 90° sections of the initial orientation distribution functions (ODFs) of both aluminium sheets before the ARB process (fully annealed sheets). It is seen that the initial texture of these samples is mainly composed of a dominant rotated cube {001}<100> which is characteristic of the recrystallisation state. The ODFs corresponding to the samples of Figs. 3 and 4 are illustrated in Figs. 15 and 16 respectively. Only the φ_2 = 45 and 90° sections of ODF are represented because these sections contain the most important texture components for ARB processed aluminium sheets. It is seen that the main texture components in the mid layers of both materials are Copper {1 1 2}<1 1 1> located at φ_1, Φ, φ_2 ≡ 45, 65 and 90° and Dillamore {4 4 11}<11 11 8> at φ_1, Φ, φ_2 ≡ 90, 27 and 45° sections and there is no evidence of shear components. The ODFs of these ARB processed samples were generally similar and the overall texture intensity and the concentration of the contour lines increase continuously with increasing strain. However, it should be noted that in comparison with the AA3003 alloy, a more severe texture was observed in AA1100 alloy during the ARB process and the strength of the overall textures developed at the eighth cycle was 18.4 and 28.2 times the random levels, respectively.
The presence of second phase particles can also change the microtextural evolution in the ARB processed aluminium sheets. The comparison of the ODFs presented in Figs. 15 and 16 indicates that the developed texture in the AA1100 alloy is much stronger than the particle containing alloy. The evolution of a strong texture during the ARB process is attributed to the cyclic nature of this process. Heason and Prangnell (2002a) investigated the texture evolution during the ARB process of the AA1100 aluminium sheets by using a model based on the FC-Taylor approach. According to their model, the surface shear texture which deforms in plane strain compression (in the mid layer) rotates towards the copper orientation during the next cycle. In contrast, this model predicts that all the rolling

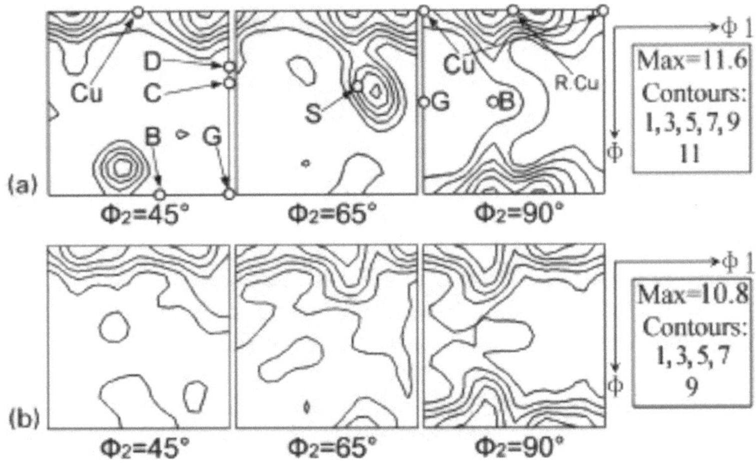

Fig. 14. φ_2=45, 65 and 90° sections of ODF of fully annealed (a) AA3003 and (b) AA1100 aluminium sheets before ARB process: standard texture components are indicated with symbols: Cu: cube; R.Cu: rotated cube; C: copper; B: brass; G: Goss; D: Dillamore

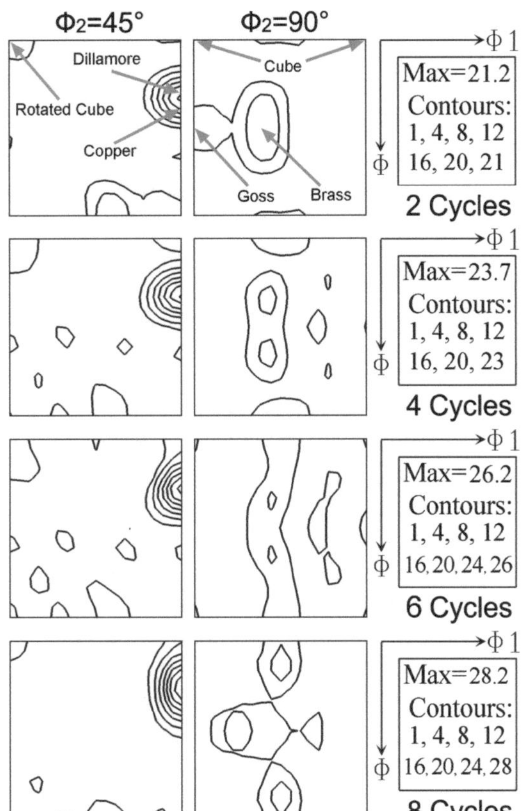

Fig. 15. φ_2=45 and 90° sections of ODFs corresponding to AA1100 samples processed by various ARB cycles: these ODFs were calculated assuming orthorhombic sample symmetry

components (copper, S and brass) rotate towards the shear orientation when they come to the surface of the sheet. This can also explain the increase in the intensity of copper and Dillamore components (which are close to each other) as the main texture components observed in this study. These authors have also reported that this strong texture leads to the development of coarse unrefined bands and prevents the full grain refinement to a submicrometre scale during the ARB process of a single phase aluminium alloy. Whereas, the presence of the second phase particles in the AA3003 alloy results in large local lattice rotation and can cause different textures compared with the rest of the matrix. Thus, the overall texture intensity will be much weaker than the AA1100 alloy.

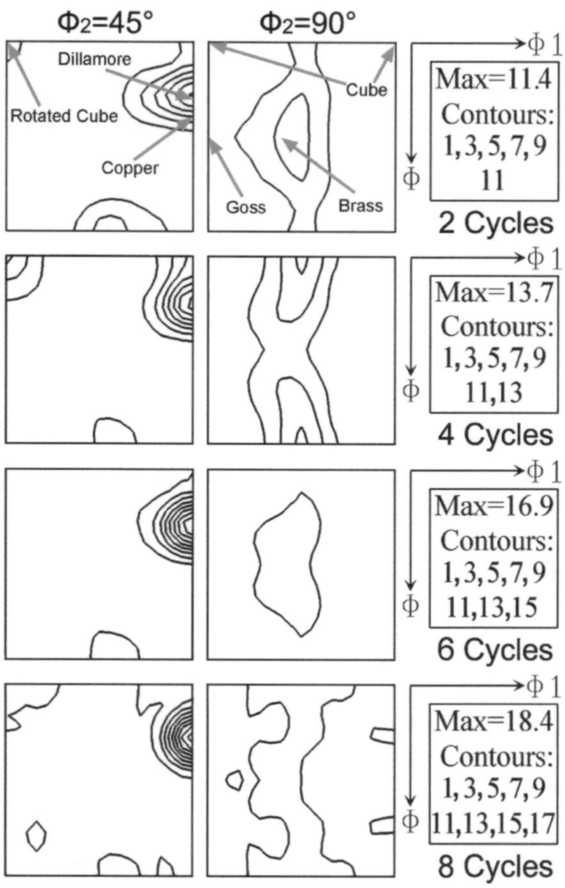

Fig. 16. φ_2=45 and 90° sections of ODFs corresponding to AA3003 samples processed by various ARB cycles

Kim et al. (2005) reported similar results in ARB processed AA8011 aluminium alloy. According to their study, the textural change from a shear texture to a rolling texture at the sheet centre during the ARB process contributed to an increase in the fraction of high angle boundaries. Also, a large number of second phase particles in the AA8011 alloy sheets weakened the texture intensity which is due to the inhomogeneous deformation around the second phase particles. A weak and more spread texture caused by the presence of the second phase particles contributes to the transformation from low angle subgrain boundaries to high angle grain boundaries by increasing the misorientation between two adjacent subgrains and removes the unrefined bands from the microstructure. For a better understanding of the texture evolution, the main FCC fibers were calculated and plotted in Fig. 17. Generally, with increasing the number of cycles, the intensity of the α fiber decreased while the intensities of the β and τ fibers increased. The α fiber, running from Goss to brass in Fig. 17a, indicates how the brass component decreased during the ARB process. It is seen that during all ARB cycles the intensity of Goss component is negligible. Fig. 17b shows the intensity of the β fibers.

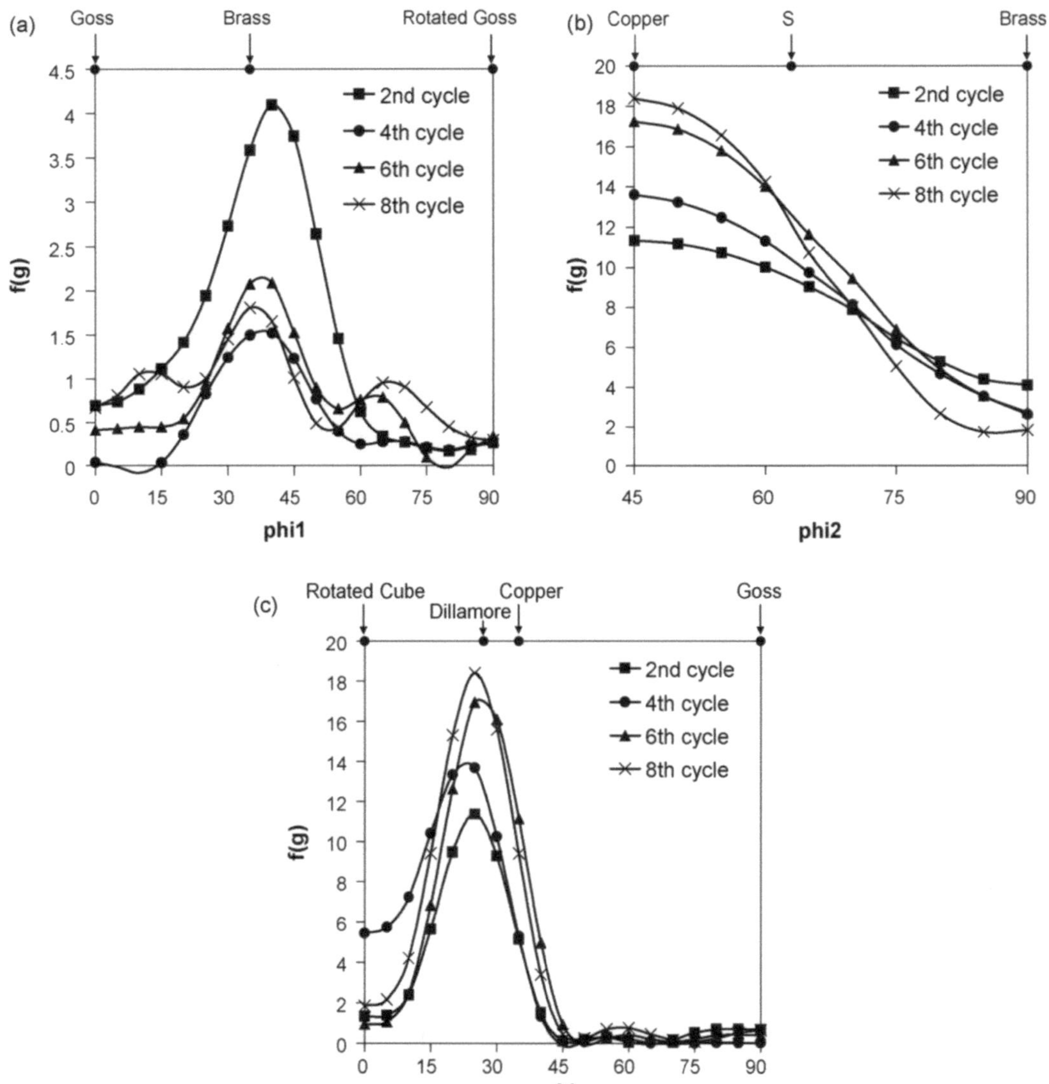

Fig. 17. Intensity of the FCC fibers in the ARB processed aluminum sheets: (a) β-fiber, (b) τ-fiber, and (c) α-fiber

It shows that in the early stages of the ARB process the intensity of the β fiber, running from copper over S to the brass orientation, is almost homogeneous, but with increasing the number of cycles the intensity of the copper component gradually grows at the expense of the brass and S components. The τ fiber is presented in Fig. 17c. It is observed that in this material the texture evolution is mainly centered around the copper and Dillamore components, and with an increasing number of cycles these components further intensify.

3.2.1 Modeling of texture evolution

In order to interpret the texture change involved in the ARB process, the texture development was simulated with the Alamel model (Van Houtte et al., 2005, 2006). The deformation during ARB processing is approximated by a two-dimensional velocity gradient tensor which is composed of a compressive strain (ε_{zz}, $\varepsilon_{yy} = 0$ and $\varepsilon_{zz} = -\varepsilon_{xx}$) added with a simple shear strain (ε_{xz}). The texture simulations have been performed on the initial texture of Fig. 14 for different strain modes with various ratios of the simple shear component (ε_{xz}) to the plane strain compression components. The imposed shear during the ARB process is characterized by the shear coefficient K ($K = \varepsilon_{xz}/\varepsilon_{zz}$).

The texture simulation for the subsurface region has been performed with a large value of the shear coefficient (K= 2.4) due to localization of shear deformation in this region. The texture prediction for the mid-thickness part has been performed under the deformation condition with K= 0.1. It is assumed here that even in the middle layers of the sheet there will still be a small amount of frictional shear as the ARB experiment was carried out in totally dry conditions (i.e. without lubrication). Due to the specific stack-and-roll geometry of the ARB process, the subsurface texture after the first ARB pass appears in the central layer of the composite sample in the second ARB pass. Hence, in order to simulate the texture evolution in the middle layer in the ARB pass n_i the Alamel model was applied on the subsurface from previous pass $n_{(i-1)}$ with a deformation mode that was characterized by K= 0.1. Fig. 18 presents results of the texture predictions for second, fourth, sixth and eighth ARB cycles. The calculated textures are in very good qualitative and quantitative agreement with the experimentally measured ODFs (Figs. 16 and 18).

The model calculation has produced the textures which exhibit a strong rolling β-fiber. It is important to notice, though, that the Dillamore and S components are far more important than the brass orientation, which corresponds very well to the experimentally observed texture, cf. Fig. 16. Furthermore, intensities of both S and brass orientations increase insignificantly whereas a considerable strengthening of the Dillamore orientation is observed during ARB process. Hence, it is shown here, on the basis of crystal plasticity modeling, that the sequence of rolling, cutting and stacking, which is characteristic for the ARB process eventually triggers a mechanism which leads to strengthening of the Dillamore component in the middle layers of the ARB sample.

Rotation of different texture components during ARB cycles, i.e. shear to copper and Dillamore components in the midsection under the plane strain compression mode and copper, S and brass to the shear component in the surface layer due to the shear deformation, is the unique feature of ARBed aluminum sheets (Heason & Prangnell, 2002a; Kim et al., 2005). This can be considered as the result of introduction of surface layer to the midsection during the next cycle and it is also associated with the increase in the number of interfaces. In fact, these interfaces induce additional strain during rolling and are responsible for the unique property.

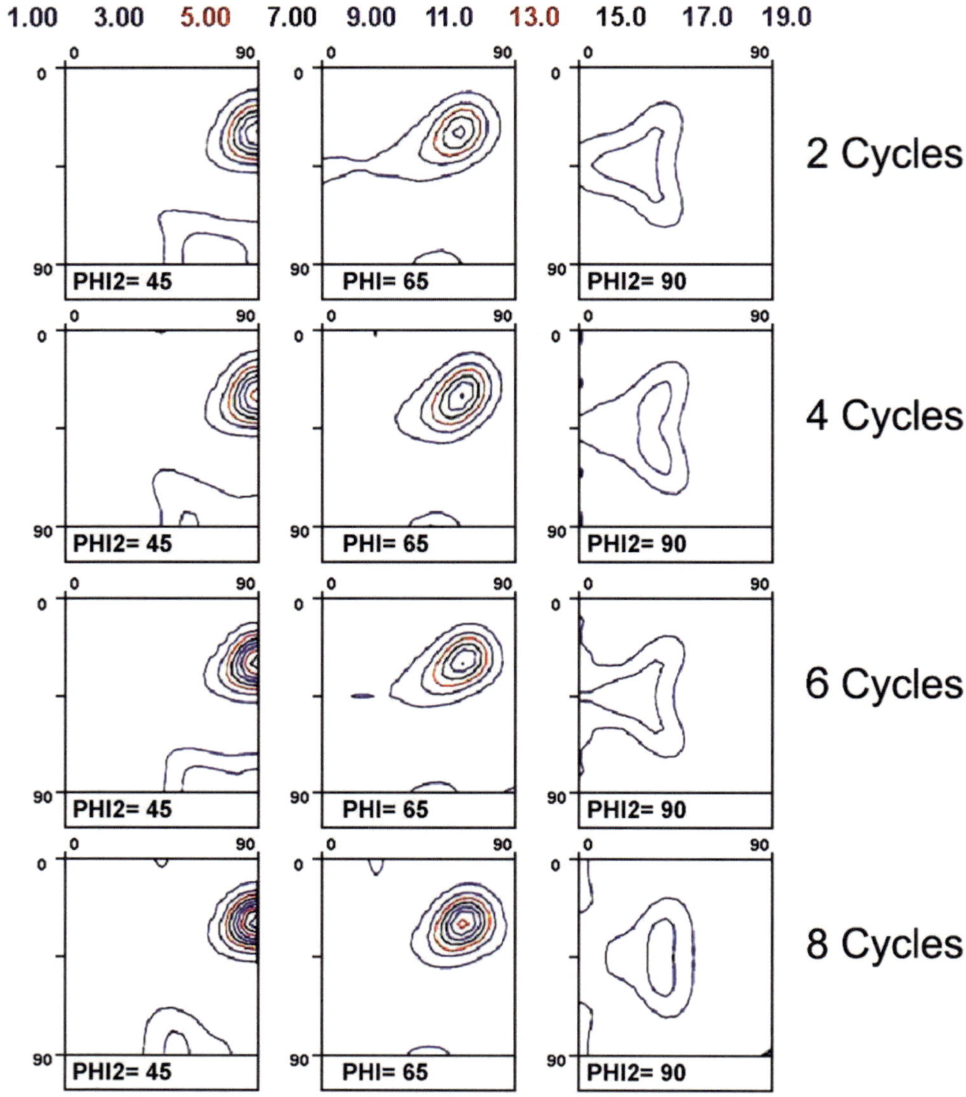

Fig. 18. Simulation of the texture development in the center region during different ARB cycles

3.3 Mechanical properties
3.3.1 ARB samples
Second phase particles in AA3003 alloy can also accelerate the rate of dynamic recovery which is very likely to occur in warm rolled aluminum sheets. Accumulation of dislocations in the vicinity of these particles leads to rapid transformation of low angle sub-grain boundaries to high angle grain boundaries and decreases the dislocation density inside the grains. Static recovery during pre-heating may also contribute to the reduction of dislocation density and promote the softening behavior. The variation of the microhardness with increasing strain during the ARB process confirms the occurrence of recovery in this

phenomenon, Fig. 19. The hardness curve in this figure corresponds well with the microstructural changes mentioned above. Initial work hardening caused by the reduction of grain size and the growth of dislocation density inside the crystalline lattice leads to a rapid increase in microhardness at the first ARB cycle (ε_{vM} = 0.8). During the following cycles, static or dynamic recovery caused by the interpass annealing treatment, the heat of deformation and the accumulation of dislocations in the vicinity of the second phase particles, prevents a further increase of the hardness and a plateau around 80 HV is established.

An interesting feature of this figure is the significant increase of the hardness value (close to two times of the initial value) after the eighth cycle, while the sample exhibits a fully recovered microstructure and probably still provides an acceptable ductility. According to the results presented here, different mechanisms of grain refinement can be attributed to the different levels of strain. During the first two cycles of ARB (ε < 1.6) grain subdivision is the dominant mechanism. This mechanism promotes the formation of a fibrous microstructure of elongated grains.

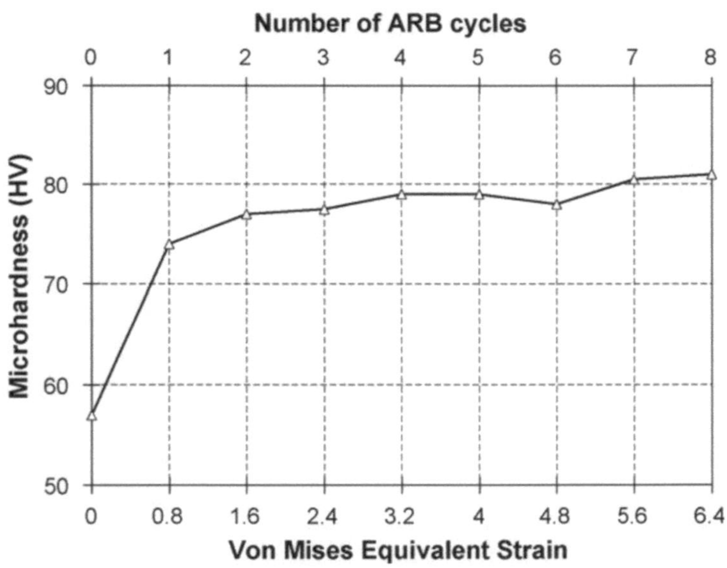

Fig. 19. Variation of Vickers microhardness of AA3003 alloy with increasing strain during ARB process

The results of the mechanical properties clarified that the strength of the ARBed aluminum sheets increases with increasing the number of cycles, Fig. 20. In order to study the relation between the mechanical properties and the microstructural changes, the tensile strength and the 0.2% yield strength were plotted as a function of minus square root of grain size in Fig. 21. The strength of the ARBed aluminum sheets is in a good conformity with Hall–Petch relationship, even for submicron grain structures. It was shown (Pirgazi et al., 2008b) shows that the amount of work hardening after yielding is rather small and necking occurs suddenly and causes the limited uniform elongation. It should be noted that the described microstructural evolution corresponds to the changes in mechanical properties very well. This suggests that the strength of the ARB samples is mainly attributed to grain boundary strengthening, though the details of the strengthening mechanism in the SPD materials should be discussed more. As has been already known, the ARB processed specimens have the dislocation substructures inside the UFGs as well as the elongated grain morphology. This means that the strength of these specimens might be affected not only by grain refinement strengthening but also by strain hardening. It has been proposed that when a deformation structure consists of a mixture of low and high angle grain boundaries, the strength can be considered as the sum of dislocation strengthening from LAGBs and grain size strengthening from HAGBs (Hansen, 2004). However, the detailed discussion about the Hall–Petch relationship and strengthening mechanisms should be done in the future works. In a similar study by Park et al. (2001) on AA6061 alloy processed by the ARB at 250 °C it has been reported that the rapid increase in strength at relatively low strains is mainly due to the work hardening caused by an increase in dislocation density and formation of the sub-grains.

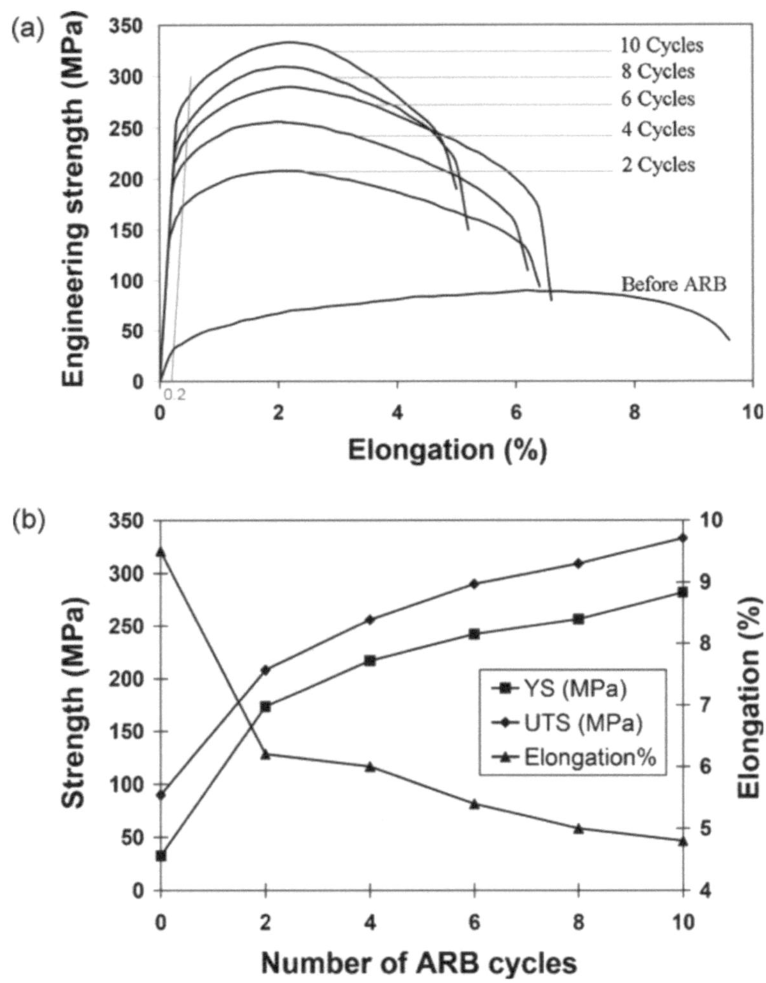

Fig. 20. (a) Engineering stress–strain curves and (b) the corresponding tensile properties of AA1100 aluminum sheets processed at different ARB cycles

During the following cycles, in which the incremental increase of the strength becomes smaller, the dislocation density is almost constant (Parh et al., 2001; Costa et al., 2005). The low density of dislocations at relatively high strains is due to the dynamic recovery (Costa et al., 2005), or absorption of dislocations into the grain boundaries (Prangnell et al., 2001). In any way, the increase of the volume fraction of ultra-fine elongated grains due to increase in the fraction of HAGBs can be contributed to the strengthening in the medium levels of strain ($1.6 < \varepsilon < 4.8$). Finally, the large strength at relatively high strains ($\varepsilon > 4.8$) can be explained by the development of ultra-fine grained structures.

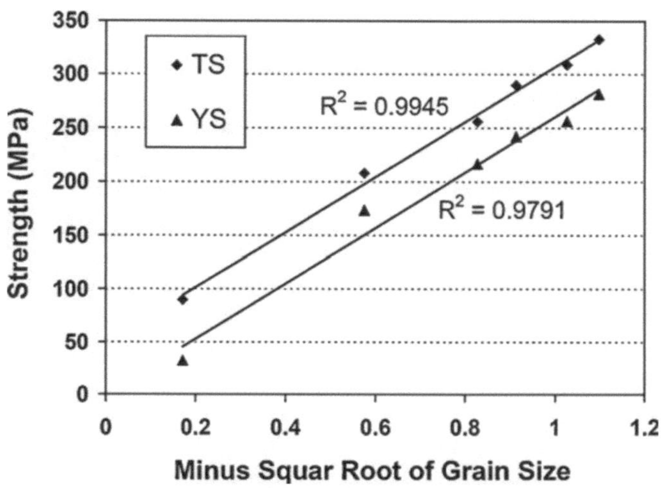

Fig. 21. Strength of ARBed AA1100 aluminum sheets versus the grain size. The grain sizes reported in this figure are the size of grains bounded by high angle grain boundaries

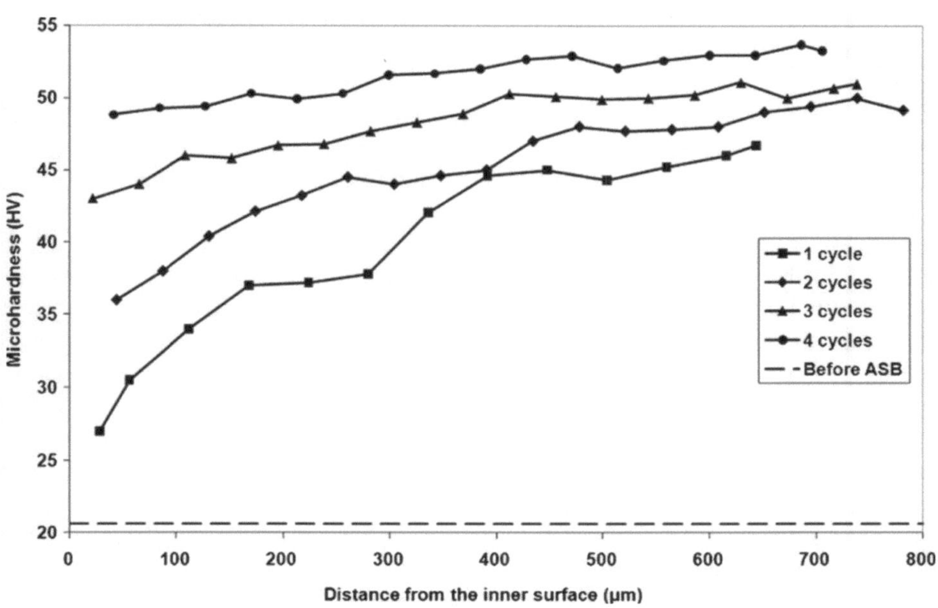

Fig. 22. Distribution of microhardness across the thickness of ASBed specimens

3.3.2 ASB samples

Distributions of microhardness across the thickness of specimens are compared in Fig. 22. The hardness of annealed specimen before ASB is shown by a dash line. It can be seen in this figure that the hardness is increased from the inner layers to the outer ones. Two trends in the hardness evolutions during the ASB cycles can be derived from this figure. First, the degree of hardness is increased, and second, the hardness distribution across the thickness of the tubes is homogenized. Fig. 23 shows the tensile stress-strain curves of the ASBed tubes.

Tensile properties derived from these curves are summarized in Fig. 24. The yield strength (at 0.2% offset strain) and the tensile strength are increased by the ASB cycles, although the growth rate is diminished. So that, the yield strength is increased up to 194 MPa after four cycles of ASB, which is about five times higher comparing to the initial material (40 MPa). Fig. 24 also shows that the difference between the yield and the tensile strengths is slightly decreased during the ASB cycles. Regarding the rupture elongation, a sharp drop is observed after the first cycle from 38% to 9%. Decrease of the elongation at subsequent cycles is quite low as compared to that of the first cycle. Hardness distribution across the thickness of the tubes confirms that materials below the outer surface undergo heavier plastic strain in comparison to the internal ones. This is due to not only higher thickness reduction but also higher redundant strains at the outer regions (Mohebbi & Akbarzadeh, 2010c).

Although this inhomogeneous hardness distribution is observed at all specimens, its intensity is decreased by increasing the number of ASB cycles. Two reasons can be mentioned for this scheme of hardness evolution. First, at each cycle of ASB, a tube is used as internal tube and so, its severely strained external layer is located within the interior regions. By this periodical entrance of the highly deformed outer materials within the thickness, the hardness distribution becomes more homogeneous by increasing the number of ASB cycles. Second, due to saturation of dislocation density and microstructural evolutions, the hardness and strength are always saturated at large strains (Valiev et al., 2000). Therefore, despite of inhomogeneous deformation, hardness becomes homogenized at large strains.

As it can be seen in Figs. 23 and 24, the yield strength and the tensile strength of the specimens are increased by increasing the ASB cycles, although its rate is gradually diminished. Grain refinement by mentioned mechanism, i.e. increase of dislocation density and formation of subgrains at early stages as well as increase of the misorientations at next stages are responsible for this strengthening (Pirgazi et al., 2008b; Sevillano et al., 1981). While based on this mechanism a steady state plateau is expected for strength of material, Fig. 24 shows that yield and tensile strengths are continuously increased. One may conclude that the strain induced by four cycles is not enough for strength saturation, especially at high value of Z in this work. However, it is shown that this continuous increase of the strength can be attributed to the through thickness inhomogeneity of hardness. Considering the mentioned hardness evolutions, it can be said that outer materials are severely work hardened after the two first cycles in comparison to inner ones. Therefore, in spite of strength saturation of the outer material, inner material has the possibility of work hardening at following cycles. This work hardening has two effects on the tensile properties. The first effect is continuous increase of the strength. By increasing the ASB cycles, inner materials are work hardened and therefore, the strength is increased. The second effect is related to the ratio of tensile to yield strength in tensile tests. As a result of work hardening of the inner materials, the overall flow strength of the specimen is increased during tension. That is why there is a high difference between the yield strength and the tensile strength in this work in comparison to that of ARB (Pirgazi et al., 2008b). By this explanation, more hardness inhomogeneity causes higher ratio of tensile to yield strength. Therefore, decrease of the difference between the yield and tensile strengths by the ASB cycles in Fig. 24, is mostly related to the hardness homogenization.

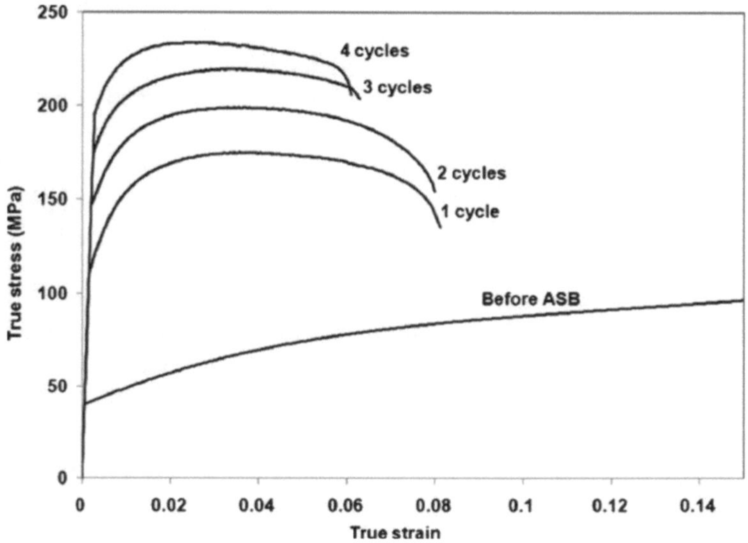

Fig. 23. Tensile stress–strain curves of the tubes processed by various ASB cycles

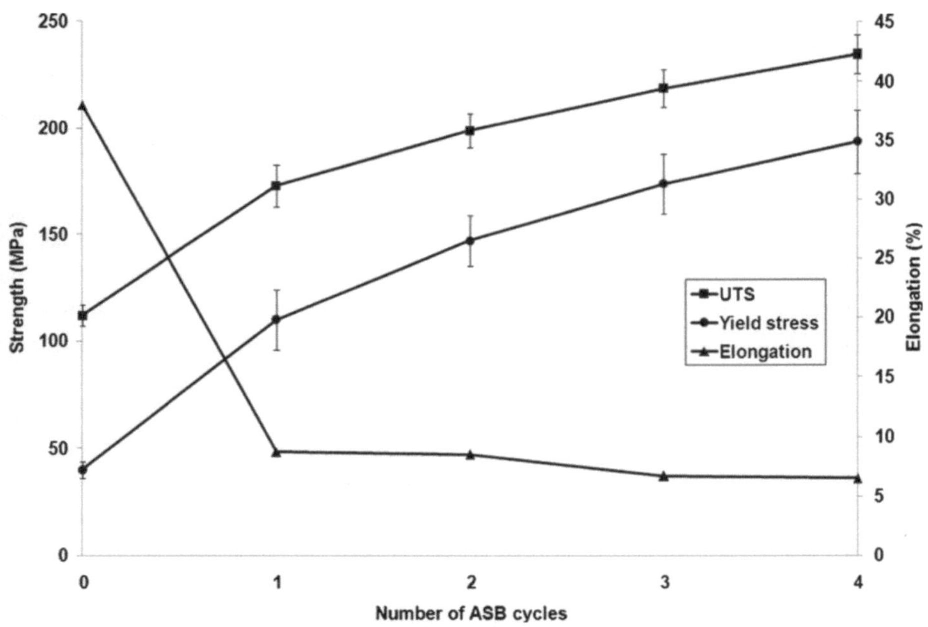

Fig. 24. Development of the tensile properties of tubes processed by various ASB cycles

Fig. 24 demonstrates a severe drop in elongation after the first cycle of ASB while its rate of reduction is very low at next cycles. This result is similar to the reported ones for ARB (Lee et al., 2002; Pirgazi et al., 2008b). Uniform elongation, which is reduced by increasing the ASB cycles, is related to onset of the plastic instability or necking. Plastic instability, on the other hand, is a function of work hardening and consequently, of the ratio of tensile to yield strength (Semiatin & Jonas, 1984). Therefore, it can be said that since this ratio is decreased by increasing the ASB cycles, the uniform elongation is decreased. In other words, inner materials of specimens after the early cycles of ASB, whose work hardening is not saturated, are work hardened during the tensile test, leading to delay in necking. At next cycles, due to saturation of work hardening across the entire thickness, plastic instability starts earlier and causes a less uniform elongation.
It is seen in Fig. 23 that post-uniform elongation includes a considerable portion of the total elongation. Previous studies on SPD have demonstrated that strain rate sensitivity of fcc metals increases by decreasing the grain size (Wang & Ma, 2004; Wei et al., 2004). High strain rate sensitivity, on the other hand, causes the increase of both uniform and post uniform elongations (Semiatin & Jonas, 1984). Takata et al. (2008) have illustrated that uniform elongation of commercially pure aluminum increases with increasing the strain rate in the specimens with a grain size larger than 1 μm, while post-uniform elongation increases with increasing the strain rate in the sub-micrometer grain size specimens. Therefore, it can be concluded that high strain rate sensitivity is effective on post-uniform elongation of aluminum with sub-micron grains. In fact, the high magnitude of post-uniform elongation in Fig. 23 can be mostly related to the high strain rate sensitivity.

4. Conclusions

EBSD analysis confirmed that the development of UFGs by ARB process is carried out via several mechanisms at different levels of strain. Grain subdivision as well as development of sub-grains is the major mechanism at the early stages of ARB. Strain induced transition of low angle to high angle grain boundaries and the formation of a thin lamellar structure occurs at the medium levels of strain. Finally, the progressive fragmentation of these thin lamellar structures into more equi-axed grains is the dominant mechanism at relatively high strains which leads to reduction of grain size to less than 500 nm. The presence of second phase particles in AA3003 aluminum sheets promotes the grain refinements and accelerates the occurrence of dynamic recovery. With an increasing number of cycles, the overall texture intensity increases and a very strong texture is developed which exhibits a limited number of sharp components. Large lattice rotation around these particles leads to the increase of local misorientation, evolution of a much weaker texture and development of a more homogeneous submicrometre grain structure in the AA3003 alloy. The Dillamore and the copper components are the main texture components of this material and they are sharpened with increasing the number of cycles. This texture evolution could be modeled with the Alamel crystal plasticity model and taking into account the specific geometry of the ARB process.
Generally, with increasing the number of cycles, the intensity of the α fiber decreased while the intensities of the β and τ fibers increased. It is shown that in the early stages of the ARB process the intensity of the β fiber, running from copper over S to the brass orientation, is

almost homogeneous, but with increasing the number of cycles the intensity of the copper component gradually grows at the expense of the brass and S components.

It is demonstrated that due to incremental deformation in ASB process, high value of strain rate without considerable temperature rise is applied, leading to a high degree of Zener-Hollomon parameter, as a characteristic of this SPD process. The grain structure of commercially pure aluminum is refined by this process and an average grain thickness and length of 186 and 419 nm are developed after the four cycles of ASB, respectively. TEM micrographs and SAD patterns as well as the EBSD analysis indicate that this grain refinement occurred by formation of subgrains at early stages of severe deformation followed by increase of the misorientations by transition of low angle to high angle grain boundaries at the next stages.

The characterization of mechanical properties revealed that the strength of the aluminum sheets considerably increased by the first two ARB cycles which is attributed to work hardening caused by increase in dislocation density and formation of subgrains. It was also included that the strength of ARBed aluminum sheets obeys the Hall–Petch relationship and corresponds well with microstructural variations. Microhardness distributions across the thickness of the tubes illustrate that because of high equivalent strain at outer regions, the hardness is increased from the inner regions to the outer ones. Due to periodical entrance of the external layer within the thickness and the consequent hardness saturation, the hardness and its homogeneity is increased with increase of the ASB cycles. The yield and tensile strengths of the material are significantly increased up to the values of 194 and 235 MPa, respectively. The scheme of hardness development leads to decrease of the ratio of tensile strength to yield strength and consequently to decrease of the uniform elongation.

5. Acknowledgments

Professors Leo Kestens and Roumen Petrov's permission for access to EBSD analysis and their help in the EBSD mapping and discussion about the texture analysis, at Ghent University in Belgium, are gratefully acknowledged. The author would like to express his sincere gratitude to Mr. Hadi Pirgazi and Mr. M.S. Mohebbi for their excellent research works at Sharif University led to this manuscript. The support of research office of Sharif University of technology is also acknowledged.

6. References

Berski, S., Dyja, H., Maranda, A., Nowaczewski, J. & Banaszek, G. (2006). *Journal of Materials Processing Technology*, Vol. 177, pp. 582–586

Chen, Z., Ikeda, K., Murakami, T, Takeda, T., Xie, J.X. (2003). *Journal of Materials Processing Technology*, Vol. 137, pp. 10–16

Chowdhury, S.G., Srivastava, V.C., Ravikumar, B. & Soren, S. (2006a). *Scripta Materialia*, Vol. 54, pp. 1691–1696

Chowdhury, S.G., Dutta, A., Ravikumar, B. & Kumar, A. (2006b). *Materials Science and Engineering A*, Vol. 428, pp. 351–357

Costa, A.L.M., Reis, A.C.C., Kestens, L. & Andrade, M.S. (2005). *Materials Science and Engineering A*, Vol. 406, pp. 279–285 del Valle, J.A., Pérez-Prado, M.T. & Ruano, O.A. (2005). *Materials Science and Engineering A*, Vols. 410/411, pp. 353–357

Estrin, Y., Toth, L.S., Molinari, A. & Bréchet, Y. (1998). *Acta Materialia*, Vol. 46, pp. 5509–5522

Hansen, N. (2004). *Scripta Materialia*, Vol. 51, pp. 801–806

Heason, C.P. & Prangnell, P.B. (2002a). *Materials Science Forum*, Vols. 408–412, pp. 733– 738

Heason, C.P. & Prangnell, P.B. (2002b). *Proc. of Conference on Nanomaterials by Severe Plastic Deformation-NANOSPD2,*, Dec. 9-13, Vienna, Austria

Horita, Z., Smith, D.J., Furukawa, M., Nemoto, M., Valiev, R.Z. & Langdon, T.G. (1996). *Journal of Material Research*, Vol. 11, pp. 1880–1890

Huang, X., Tsuji, N., Hansen, N. & Minamino, Y. (2003). *Materials Science and Engineering A*, Vol. 340, pp. 265–271

Humphreys, F.J. & Hatherly, M. (2004). *Recrystallization and Related Annealing Phenomena*, Elsevier, Oxford

Kim, H.W., Kang, S.B., Xing, Z.P., Tsuji, N. & Minamino, Y. (2002). *Materials Science Forum*, Vols. 408–412, pp. 727–732

Kim, H.W., Kang, S.B., Tsuji, N. & Minamino, Y. (2005). *Metallurgical and Materials Transaction A*, Vol. 36, pp. 3151–3163

Kim, Y.S., Kang, S.H., & Shin, D.H. (2006). *Materials Science Forum,* Vols. 503/504, pp. 681–686

Kolahi, A., Akbarzadeh A. & Barnett, M.R. (2009). *Journal of Materials Processing Technology*, Vol. 209, pp. 1436-1444

Lee, S.H., Saito, Y., Sakai, T. & Utsunomiya, H. (2002). *Materials Science and Engineering A*, Vol. 325, pp. 228–235

Lee, S.H., Lee, C.H. & Lim, C.Y. (2004). *Materials Science Forum*, Vols. 449–452, pp. 161–164

Min, G., Lee, J.M., Kang, S.B. & Kim, H.W. (2006). *Materials. Letters*, Vol. 60, pp. 3255–3259

Mohebbi, M.S. (2009a). M.Sc. Thesis, Sharif University of Technology, Tehran, I.R. Iran

Mohebbi, M.S. & Akbarzadeh, A. (2009b). *Proc. of the 2nd Int. Conference on Ultrafine Grained and Nanostructured Materials (UFGNS2009)*, Nov. 14-15, Tehran University, Tehran, I.R. Iran

Mohebbi, M.S. & Akbarzadeh, A. (2010a). *Journal of Materials Processing Technology*, Vol. 210, pp. 510–517

Mohebbi, M.S. & Akbarzadeh, A. (2010b). *Materials Science and Engineering A*, Vol. 528, pp. 180–188

Mohebbi, M.S. & Akbarzadeh, A. (2010c). *Journal of Materials Processing Technology*, Vol. 210, pp. 389–395

Park, K.T., Kwon, H.J., Kim, W.J., & Kim, Y.S. (2001). *Materials Science Engineering A*, Vol. 316, pp. 145–152

Pérez-Prado, M.T., del Valle, J.A. & Ruano, O.A. (2004). *Scripta Materialia,* Vol. 51, pp. 1093–1097

Pirgazi H., Akbarzadeh, A., Petrov, R., Sidor, J. & Kestens, L. (2008a). *Materials Science and Engineering A*, Vol. 492, pp. 110–117

Pirgazi, H., Akbarzadeh, A., Petrov, R. & Kestens, L. (2008b). *Materials Science and Engineering A*, Vol. 497, pp. 132–138

Pirgazi, H. & Akbarzadeh, A. (2008c). *Proc. of the 2nd Conference on Nanostructures (NS2008)*, March 11-14, Kish University, Kish Island, I.R. Iran

Pirgazi, H. & Akbarzadeh, A. (2009). *Materials Science and Technology*, ol. 25, No. 5, pp. 625-631

Prangnell, P.B., Bowen, J.R. & Gholinia, A. (2001). *Proceedings of the 22nd Risø´ International Symposium on Materials Science*, pp. 105–126

Reis, A.C.C. & Kestens, L. (2005). *Solid State Phenomena*, Vol. 105, pp. 233–238

Reis, A.C.C. , Kestens, L. & Houbaert, Y. (2005). *Materials Science Forum*, Vols. 495–497, pp. 351–356

Richert, J. & Richert, M. (1986), *Aluminum*, Vol. 62, pp. 604–607

Rios, P. R. & Padilha, A. F. (2003). *Materials Research*, Vol. 6, No. 4, pp. 605–613

Ryazanov, A.I., Pavlov, S.A. & Kiritani, M. (2003). *Materials Science and Engineering A*, Vol. 350, pp. 245–250

Saito, Y., Utsunomiya, H., Tsuji, N. & Sakai, T (1999). *Acta Materialia*, Vol. 47, No. 2, pp. 579–583

Saito, Y., Tsuji, N., Utsunomiya, H., Sakai, T. & Hong, R.G. (1998). *Scripta Materialia*, Vol. 39, No. 9, pp. 1221–1227

Semiatin, S.L. & Jonas, J.J. (1984). *Formability and Workability of Metals*, American Society for Metals, Ohio, USA

Sevillano, J.G., Van Houtte, P. & Aernoudt, E. (1981). *Progress in Materials Science*, Vol. 25, pp. 69–134

Sherby, O.D. and Wadsworth, J. (2001). *Journal of Materials Processing Technology*, Vol. 117, pp. 347–353

Sponseller, D.L., Timmons, G.A. & Bakker, W. T. (1998). *Journal of Materials Engineering and Performance*, Vol. 7, No. 2, pp. 227-238

Takata, N., Okitsu, Y. & Tsuji, N. (2008). *J. Materials Science*, Vol. 43, pp. 7385–7390

Tsuji, N., Ueji, R. & Minamino, Y. (2002a). *Scripta Materialia*, Vol 47, pp. 69–76

Tsuji, N., Ito, Y., Saito, Y. & Minamino, Y. (2002b). *Scripta Materialia*, Vol. 47, pp. 893–899

Tsuji, N., Saito, Y., Lee, S.H. & Minamino, Y. (2003a). *Advanced Engineering Materials*, Vol. 5, No. 5, pp. 338–344

Tsuji, N., Toyoda, T., Minamino, Y., Koizumi, Y., Yamane, T., Komatsu, M. & Kiritani, M. (2003b). *Materials Science and Engineering A*, Vol. 350, pp. 108–116

Valiev, R.Z., Krasilnikov, N.A. & Tsenev, N.K. (1991). *Materials Science Forum Engineering A*, Vol. 137, pp. 35-41

Valiev, R.Z., Islamgaliev, R.K. & Alexandrov, I.V. (2000). *Progress in Materials Science*, Vol. 45, pp. 103–189

Van Houtte, P., Li, S.Y., Seefeldt, M. & Delannay, L. (2005). *International Journal of Plasticity*, Vol. 21, pp. 589–624

Van Houtte, P., Kanjarla, A.K., Van Bael, A., Seefeldt, M. & Delannay, L. (2006). *European Journal of Mechanics A-Solids*, Vol. 25, pp. 634–648

Wang, X., Li, P. & Wang, R. (2005). *Int. J. of Machine Tools Manufacture*, Vol. 45, pp. 373–378

Wang, Y.M. & Ma, E. (2004). *Materials Science and Engineering A*, Vols. 375–377, pp. 46–52

Wei, Q., Cheng, S., Ramesh, K.T. & Ma, E. (2004). *Materials Science and Engineering A*, Vol. 381, pp. 71–79

Wong, C.C., Dean, T.A. & Lin, J. (2003). *Int. J. of Machine Tools Manufacture*, Vol. 43, pp. 1419–1435

Zener, C. & Hollomon, J.H. (1944). J. Applied Physics, Vol. 15, pp. 22–32

Zhan, Z., He, Y., Wang, D. & Gao, W. (2006). *Surface and Coatings Technology*, Vol. 201, pp. 2684–2689

19

Statistical Tests Based on the Geometry of Second Phase Particles

Viktor Beneš[1], Lev Klebanov[1], Radka Lechnerová[2] and Peter Sláma[3]
[1]*Charles University in Prague, Faculty of Mathematics and Physics, Department of Probability and Mathematical Statistics*
[2]*Private College of Economic Studies, Ltd., Prague*
[3]*COMTES FHT a.s., Metallography, Dobřany*
Czech Republic

1. Introduction

The actual trends in new material development impose the necessity of a thorough knowledge of the relationships between properties and microstructure. In order to meet the requirements on the performance of some materials in their applications, that are often very sophisticated, a very fine tuning of the manufacturing process and its parameters is needed. Therefore, it is indispensable to be able to distinguish between small differences in microstructures produced by the different variations of processing parameters. Another important field of material characterization is the description of microstructure heterogeneities. Such heterogeneities are often related to risks of premature damage nucleation and preferential defects (void, cracks, corrosion, etc.) occurrence and propagation.

One of the most important elements of the microstructure of metallic materials is the set of second phase particles. Particle size and shape distributions and the type of spatial dispersion (homogeneous, long-range or short range ordered, clustered, etc.) are often the major attributes of a particular microstructure (Humphreys & Hatherly, 2004; Polmear, 2006). Thin foils made from aluminium-manganese based alloys, such as AA3003, are the material most frequently used as fins in automotive heat exchangers (Hirsch, 2006). This application imposes very strict requirements on properties and related foil microstructures. The development of an appropriate production technology is contingent on the perfect knowledge of the impact of processing parameters on microstructure transformation, including the changes of the set of particles (Hirsch, 2006; Slámová et al., 2006).

In statistical setting, we deal with microstructures containing random objects in a space or plane, which may be second phase particles, pores, grains and their sections or projections. The question frequently asked is whether two microstructures come from a material with the same geometrical characteristics of microstructure. This statement forms a null hypothesis H_0 and the aim is to develop a statistical two-sample test of H_0 against an alternative hypothesis that the geometrical characteristics are different. In the literature, parametric models of microstructures as random sets are mostly used (Derr & Ji, 2000; Ohser & Mücklich, 2000) and the authors recommend Monte Carlo testing which is based on the possibility of simulating a random set under the null hypothesis. The evaluation of the test is based on a comparison of the test statistics (describing some characteristics (Tewari & Gokhale, 2006a;b) of the random set) obtained from simulated models with that obtained from the microstructure. In the present paper, we apply a nonparametric approach. We will consider some geometrical characteristics, which can be measured by means of image analysis or estimated from the observation in a window. Thus, we obtain a vector (typically of a large dimension) of data, which does not form a random sample since the vector components may be stochastically dependent. For statistical testing, such a dependency is a serious problem. For the data from a single window, such a problem should be considered (Kupczyk, 2006). Under the assumption that we can observe the microstructure in a few independent windows, we put all the data from the individual windows and observations within windows together. This global information for both sets can be compared using $\mathfrak{N}$-distances from probability theory, see (Klebanov, 2005). The same holds in the functional data approach, where data is first transformed to a function. When the number of independent windows is large, the test of H_0 can be transformed into a univariate distribution-free two-sample test. For a smaller number of measurements, the permutation test for H_0 based on $\mathfrak{N}$-distances can be used. In the paper, we consider microstructures with dispersed particles. A statistical method is developed independently of whether we deal with 2D or 3D data. Therefore, in the application presented from metallography, we will call the feature particles, which in fact are 2D particle sections. We distinguish two basic sets of particle characteristics, namely the individual particle geometry and the spatial distribution of particles. Even if these are rather different descriptive tools, we try to construct the test so that it can be applied to both of them in an analogous way. We describe a variety of methods in Section 2 including also a functional data analysis. In the applied part of the paper in Section 3, we present data from thin foils demonstrating the use of the test for a comparison of metallographic samples of aluminium alloys, observed by means of optical microscopy. The numerical results are in Section 4. Finally, in Section 5 we present a general discussion of the methods and interpretations of results. The theoretical statistical background is given in the Appendix.

2. Methods

Several statistical methods for the testing of differences between microstructures containing particles are suggested in this Section. In Subsections 2.1 and 2.2 we describe particle systems by a vector of parameters and the testing is reduced to a test of the null hypothesis that the vector of these random parameters has the same distribution for two random sets A and B. In Subsection 2.3 the functional data approach is used, we compare functions which fit the observed data. Finally in Subsection 2.4 a simulation study concerning the power of the test based on $\mathfrak{N}$-distances is performed in order to see its behavior with respect to different alternative hypotheses.

2.1 Vector approach – individual particle parameters

Here we describe method (I). Generally the individual particles observed in a window (metallographic sample) are not independent. We will assume that n windows of the same size and magnification are observed for two microstructures A and B. Let the windows be each sufficiently far from the other or taken from independent samples so that we can assume independence among windows. We measure the same number k of particles from each window. In our application, the image analyzer scans particles in a window in a meandering way so that we obtain a representative set of particles by taking the first k particles measured from each window. Assume that m geometrical parameters are measured for each particle (corresponding to microstructures A, B). Two independent samples $X_1, \ldots, X_n$ from A and

$Y_1, \ldots, Y_n$ from B with matrices of size $k \times m$ are thus obtained. Now we transform each matrix X_i, Y_i to a vector $\tilde{x}, \tilde{y}$, respectively, of size km:

$$\tilde{x} = (x_{1,1}, \ldots, x_{k,1}, x_{1,2}, \ldots, x_{k,2}, \ldots, x_{1,m}, \ldots, x_{k,m}), \tag{1}$$
$$\tilde{y} = (y_{1,1}, \ldots, y_{k,1}, y_{1,2}, \ldots, y_{k,2}, \ldots, y_{1,m}, \ldots, y_{k,m}),$$

and evaluate an empirical counterpart of the $\mathfrak{N}$-distance (11):

$$\widehat{\mathfrak{N}} = \frac{1}{n} \left[\sum_{i=1}^{n} \sum_{j=1}^{n} \left(2\mathcal{L}(\tilde{x}_i, \tilde{y}_j) - \mathcal{L}(\tilde{x}_i, \tilde{x}_j) - \mathcal{L}(\tilde{y}_i, \tilde{y}_j) \right) \right]^{\frac{1}{2}}, \tag{2}$$

where e.g.

$$\mathcal{L}(s,t) = \|s - t\| \tag{3}$$

is the Euclidean distance between vectors s, t, which is a strongly negative definite kernel, cf. Appendix.

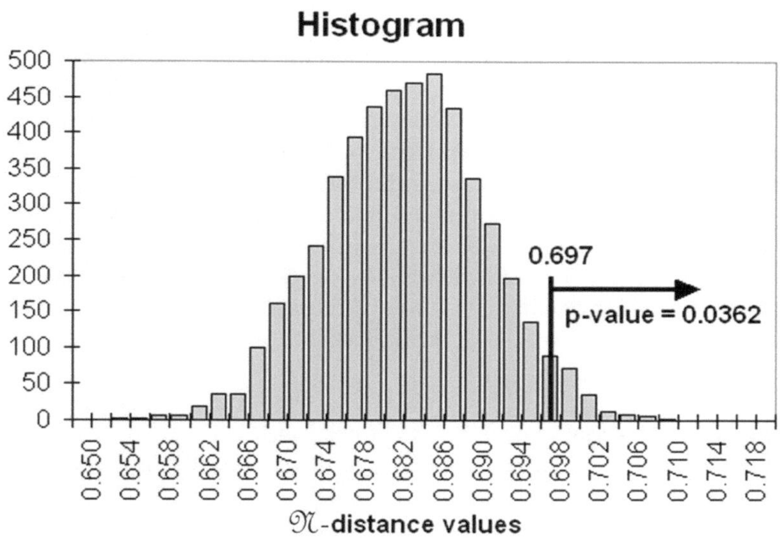

Fig. 1. Histogram of $\mathfrak{N}$-distances from 5000 random permutations ($m = 1, k = 500, n = 20$). The value 0.697 corresponds to the non-permuted case and the p-value is the probability (relative frequency) of $\mathfrak{N}$-distance being larger than this value.

We describe the permutation test (Lehmann & Romano, 2005) of the null hypothesis in more detail, which is used when n is small. Consider K random permutations of $1, \ldots, 2n$. Apply each permutation to long vector $(\tilde{x}_1, \ldots, \tilde{x}_n, \tilde{y}_1, \ldots, \tilde{y}_n)$, and then split the permuted set to the first n and last n vectors and evaluate (2) to obtain K empirical $\mathfrak{N}$-distances (K is recommended to be about 1000). Under the null hypothesis, permutations do not modify the distribution of the random variable $\mathfrak{N}$. From the histogram of these distances including the non-permuted case, we obtain the p-value for the test, which is the probability (under the validity of the null hypothesis) that the random $\mathfrak{N}$-distance is larger than its measured value. A typical example of the test is in Fig. 1, here we reject H_0 since the p-value is smaller than 0.05. If the p-value

were greater than 0.05, H_0 would not be rejected. This rule we use in all tests throughout the whole chapter.
We recommend also a simpler test in which the samples are split randomly into three sub samples $\tilde{x}, \tilde{x}', \tilde{x}''(\tilde{y}, \tilde{y}', \tilde{y}'')$, respectively, of size n/3 (assuming it is an integer). Then put (12)

$$U_i = \mathcal{L}(\tilde{x}_i, \tilde{y}_i) - \mathcal{L}(\tilde{x}_i, \tilde{x}_i^{'}) \qquad (4)$$
$$V_i = \mathcal{L}(\tilde{y}_i^{'}, \tilde{y}_i^{''}) - \mathcal{L}(\tilde{x}_i^{''}, \tilde{y}_i^{''})$$

$i = 1, \ldots, n/3$. The null hypothesis is now equivalent to the hypothesis that U_i, V_i come from the same distribution, which can be tested by an arbitrary univariate two-sample test, e.g. a Kolmogorov-Smirnov test (using STATS package in R language, (Ihaka & Gentleman, 1996)), whose statistic has the form

$$\max_x |H_n^U(x) - H_n^V(x)|,$$

where $H_n^U(x)$ and $H_n^V(x)$ are empirical distribution functions of $U_1, \ldots, U_n$ and $V_1, \ldots, V_n$ correspondingly.
For small values of n, i.e. $n < 120$, however, the loss of information when splitting the files to a size $n/3$ leads to the situation where the use of the asymptotic statistics for the Kolmogorov-Smirnov test is not recommended (Buening & Trenkler, 1978, p.135). The test based on splitting is distribution-free (independent of the underlying distribution of observations), but it has smaller power than the permutation test, see (Klebanov, 2005). Namely, based on simulated samples from multivariate normal distributions and location alternative, it was shown that splitting test has about the same power as permutation test, but based on three times smaller sample size. For the one-dimensional case and the samples from normal distribution and location alternatives, the permutation $\mathfrak{N}$-test has the power very closed to optimal t-test. However, for the samples from the mixture of normal distribution $\mathfrak{N}$-test may be more powerful than t-test. In all situations permutation $\mathfrak{N}$-test is more powerful than Kolmogorov-Smirnov test.

2.2 Vector approach – the spatial distribution of particles

Here we do not evaluate the measurement directly but first the measured information is transformed.
To test the difference in spatial distribution of particles we use m mutual characteristics of particle centroids, among them:

a) a distribution function of the nearest neighbour distance (G-function) (Tewari & Gokhale, 2006b),

b) a contact distribution function (F-function) (Tewari & Gokhale, 2006a),

c) a pair correlation function (pcf) (Ohser & Mücklich, 2000).

Concerning the spatial distribution, we distinguish complete independence (CI), attraction (clustering) and repulsion (regularity). Functions F and G coincide when CI takes place, for clustered patterns graph G is to the left of F while for regular patterns F is to the left from G. These are distance characteristics while pcf is a second-order characteristic, being identically equal to 1 under CI. Peaks of pcf correspond to typical distances between pairs of points.
The edge-corrected estimators of these functions (Kaplan-Meier estimators for F, G, Ripley's estimator for pcf, using SPATSTAT package in R language) are obtained. We do the estimation

in each of n windows for both microstructures A, B. From the estimated curves we construct vectors $G_j = G(j\triangle)$, $F_j = F(j\triangle)$, $pcf_j = pcf(j\triangle)$, $j = 1, \ldots, k$, where $\triangle > 0$ is a given step and $k\triangle$ the range considered. Choosing m of these three characteristics we construct $2n$ vectors of size km, cf. (2). If the number of independent windows n is smaller than 120, we evaluate $\mathfrak{N}$ in (2). Using K random permutations of these $2n$ vectors we perform the permutation test of

H_0 : microstructures A, B have the same distribution of characteristics involved,

exactly as in the method (I) above. If $n > 120$, we evaluate (5) and use the Kolmogorov-Smirnov test. The tests are not much dependent on $\triangle$ if it is small. All this is called method (II).

The characteristics of the spatial distribution are dependent of the intensity λ_A, λ_B of particle centroids of microstructures A, B, respectively. The intensity is the mean number of particle centroids per unit volume and it is estimated as

$$\lambda_A = \frac{n_A}{|W|}, \ \lambda_B = \frac{n_B}{|W|},$$

where $|W|$ is the size of the window and n_A, n_B are the corresponding numbers of particle centroids observed in the window. Clearly, microstructures with different intensities have different nearest neighbour distances, etc. Consider the problem of an investigation of the difference between A and B purely in the spatial distribution independently of the different intensities. For the case $n_A \neq n_B$, we can scale the image window B by $\sqrt{n_B/n_A}$. This transformation leads to the same estimated density of particles of A and transformed B (in windows of different size). Then we evaluate functions F, G, pcf and continue by testing based on $\mathfrak{N}$-distances as above, this is our method (III).

Consider finally the problem of testing the difference in the density of particles. Here we need a parametric model and we restrict it to a point process model with no interactions (Poisson process). Under the assumption that both microstructures can be modeled by a stationary Poisson process, there is a theoretical test of the hypothesis: $H_0 : \lambda_A = \lambda_B$. We reject H_0 at a confidence level α, if the statistics:

$$T = \frac{|n_A - n_B|}{\sqrt{n_A + n_B}} > u_{1-\frac{\alpha}{2}} \tag{5}$$

(see (Ng et al, 2007)), where for $0 < a < 1$, u_a denotes the a-quantile of the standard Gaussian distribution.

2.3 Functional data approach

In Subsection 2.2 we dealt in fact with functions (F, G, pcf) which describe the spatial distribution of particles. Since in the computer we have always discrete data, i.e. a finite number of the values of a function, typically at equidistant argument points, we used the vector analysis for testing the null hypothesis by means of $\mathfrak{N}$-distances. However, within this theory it is also possible to deal with functions, this approach belongs to the field of statistical analysis of functional data.

In the functional data approach we used the Bernstein polynomials (Korovkin, 2001) as a suitable approximation for corresponding functions F, G, pcf. Let us remind that the Bernstein polynomial of the degree n for the function $f(x)$ is defined as

$$B_n(x; f) = \sum_{j=0}^{n} f\left(\frac{j}{n}\right) \binom{n}{j} x^j (1-x)^{n-j}.$$

The test for the null hypothesis is constructed by means of $\mathfrak{N}$-distances, with the strongly negative definite kernel $\mathcal{L}$ for two functions $f(x)$ and $g(x)$ defined on an interval $[0,a]$ as

$$\mathcal{L}(f,g) = \Big(\int_0^a (f(x) - g(x))^2 dx\Big)^{1/2}.$$

In this case the empirical analog on $\mathfrak{N}$-distance is defined as

$$\mathfrak{N} = \Big(\frac{2}{M^2}\sum_{i=1}^{M}\sum_{j=1}^{M}\mathcal{L}(F_i^A, F_j^B) - \frac{1}{M^2}\sum_{i=1}^{M}\sum_{j=1}^{M}\mathcal{L}(F_i^A, F_j^A) - \frac{1}{M^2}\sum_{i=1}^{M}\sum_{j=1}^{M}\mathcal{L}(F_i^B, F_j^B)\Big)^{1/2}. \quad (6)$$

Here F_j^A and F_j^B are the Bernstein polynomials for the function F (correspondingly, G or pcf), constructed for the jth window of microstructures A and B.
To compare the microstructure A with microstructure B we use permutation test, that is we combine the functions F_j^A and F_j^B in one long vector of functions, make a random permutation, and after that we split the vector into two parts, calculating after that $\mathfrak{N}$-distance between corresponding parts. The described operation has to be repeated many times, which is possible thanks to fast computers. This ends our method (IV).
Further, using the functional data approach, we suggest a new comparison technique, qualitatively different from the previous ones. Suppose again we have two microstructures (A and B), observed in M windows. Denote by n_j the number of particles in j-th window from microstructure A, and by k_j the corresponding number from B. Corresponding coordinates of particle centroids are denoted by $(X_1^{(j)}, Y_1^{(j)}), \ldots, (X_{n_j}^{(j)}, Y_{n_j}^{(j)})$ for jth window from A, and by $(U_1^{(j)}, V_1^{(j)}), \ldots, (U_{k_j}^{(j)}, V_{k_j}^{(j)})$ for jth window from B ($j = 1, 2, \ldots, M$).
Method (V) is based on a smoothing procedure in each window by convolving a discrete two-dimensional distribution concentrated in particle centroids with two-dimensional Gaussian distribution with zero mean vector and standard deviations $\sigma_j(A) = 1/\sqrt[4]{n_j}$ and $\sigma_j(B) = 1/\sqrt[4]{k_j}$, i.e. we pass to the functions

$$\mu_j(x,y) = \frac{1}{\sigma_j^2(A)}\frac{1}{n_j}\sum_{s=1}^{n_j} K\Big(\frac{x - X_s^{(j)}}{\sigma_j(A)}, \frac{y - Y_s^{(j)}}{\sigma_j(A)}\Big) \quad (7)$$

for A, and

$$\nu_j(x,y) = \frac{1}{\sigma_j^2(B)}\frac{1}{k_j}\sum_{s=1}^{k_j} K\Big(\frac{x - U_s^{(j)}}{\sigma_j(B)}, \frac{y - V_s^{(j)}}{\sigma_j(B)}\Big) \quad (8)$$

for B. Without loss of generality, we may suppose that the window is a unit square: $Q = \{(x,y) : \ 0 < x < 1, \ 0 < y < 1\}$. Define strongly negative definite kernel $\mathcal{L}$ for two functions $f(x,y)$ and $g(x,y)$ given on Q as

$$\mathcal{L}(f,g) = \Big(\int_0^1\int_0^1 (f(x,y) - g(x,y))^2 dxdy\Big)^{1/2}.$$

In this case the empirical analog on $\mathfrak{N}$-distance (6) is used with μ_i, ν_j as arguments of $\mathcal{L}$.
To compare the microstructures A and B we use again the permutation test, that is we combine the functions μ_j and ν_j in one long vector, make a random permutation, and after that we split the vector into two parts, calculating after that $\mathfrak{N}$-distance between corresponding parts. The described operation has to be repeated many times.

It is possible to study a more general situation. Namely, we need not consider only coordinates (X,Y) of the particle centroids, but also individual characteristics of the particles. Then, we will have instead of two-dimensional vector (X,Y) the vectors of higher dimensionality, and instead of the functions μ and ν depending on two arguments we will have corresponding functions depending on three or more arguments. Theoretically, it is possible to consider an arbitrary number of characteristics of the particles, but the calculations for more than three arguments are very time consuming. Therefore, as the method (VI), we consider the case of three arguments, and as an example the case of three parameters: two coordinates of the particle centroid and the area of the particle (section).

Again, we have two microstructures (A and B), observed in M windows. Denote by n_j the number of particles in j-th window from microstructure A, and by k_j the corresponding number from the microstructure B. Corresponding coordinates of particle centroids and their areas are now denoted by $(X_1^{(j)},Y_1^{(j)},Z_1^{(j)}),\ldots,(X_{n_j}^{(j)},Y_{n_j}^{(j)},Z_{n_j}^{(j)})$ for j-th window from A, and by $(U_1^{(j)},V_1^{(j)},W_1^{j}),\ldots,(U_{k_j}^{(j)},V_{k_j}^{(j)},W_{k_j}^{(j)})$ for j-th window from B $(j=1,2,\ldots,M)$.

We make a smoothing procedure in each window by convolving a discrete three-dimensional distribution concentrated in particle centroids and their areas with three-dimensional Gaussian distribution with zero mean vector and standard deviations $\sigma_j(A) = 1/\sqrt[4]{n_j}$ and $\sigma_j(B) = 1/\sqrt[4]{k_j}$, i.e. we pass to the functions

$$\mu_j(x,y,z) = \frac{1}{\sigma_j^3(A)}\frac{1}{n_j}\sum_{s=1}^{n_j} K\Big(\frac{x-X_s^{(j)}}{\sigma_j(A)},\frac{y-Y_s^{(j)}}{\sigma_j(A)},\frac{z-Z_s^{(j)}}{\sigma_j(A)}\Big)$$

for A, and

$$\nu_j(x,y,z) = \frac{1}{\sigma_j^3(B)}\frac{1}{k_j}\sum_{s=1}^{k_j} K\Big(\frac{x-U_s^{(j)}}{\sigma_j(B)},\frac{y-V_s^{(j)}}{\sigma_j(B)},\frac{z-W_s^{(j)}}{\sigma_j(B)}\Big)$$

for B.

Without loss of generality, we may suppose that the window is a unit cube:
$Q = \{(x,y):\ 0<x<1,\ 0<y<1,\ 0<z<1\}$.

Define strongly negative definite kernel $\mathcal{L}$ for two functions $f(x,y,z)$ and $g(x,y,z)$ given on Q as

$$\mathcal{L}(f,g) = \Big(\int_0^1\int_0^1\int_0^1 (f(x,y,z)-g(x,y,z))^2 dxdydz\Big)^{1/2}.$$

Then again the empirical analog on $\mathfrak{N}$-distance (6) is used with μ_i,ν_j as arguments of $\mathcal{L}$.
Further, we apply the same testing procedure as for two-dimensional case, described above.

2.4 The power of the test

In order to understand and describe the properties of the suggested testing based on $\mathfrak{N}$-distances, it is necessary to study the power of the tests, which quantifies the probability of a correct rejection of H_0. It is possible to do this by means of simulations of special cases. For the use of Kolmogorov-Smirnov test for comparison of U,V in (12) a study of the power is presented in (Klebanov, 2005). We present here another study of the power of our test when using permutation testing in the vector approach.

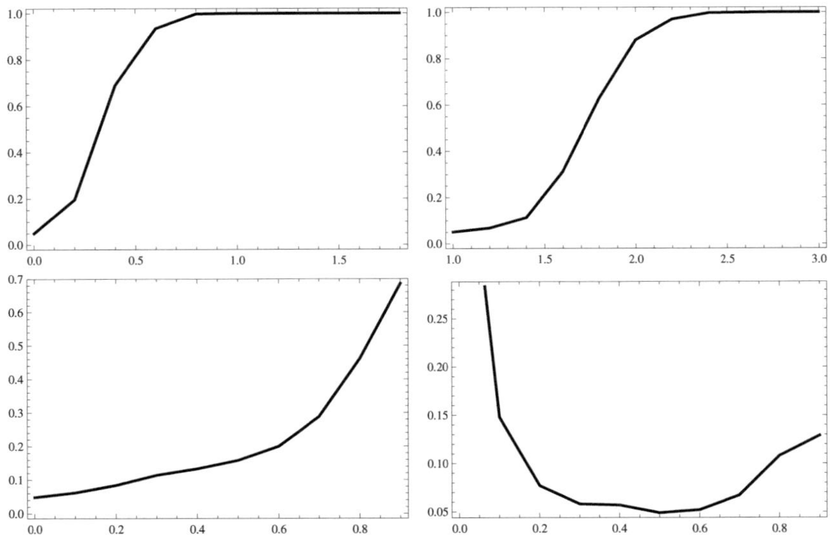

Fig. 2. Estimated probability, that the null hypothesis is rejected by the permutation test on 0.05 significance level, given the alternative of location (upper left), scale (upper right), correlation (lower left and right).

Consider a $k-$dimensional Gaussian distribution with mean $(\mu, \ldots, \mu)$, and variance matrix terms $\Sigma_{ii} = \sigma^2$, $\Sigma_{ij} = \rho\sigma^2$, $i \neq j$, $\rho \in [0,1]$. This distribution can be simulated as

$$X_j = \sigma\sqrt{\rho}Z + \sigma\sqrt{1-\rho}Y_j + \mu, \; j = 1, \ldots, k,$$

where $Z, Y_1, \ldots, Y_k$ are independent identically distributed standard Gaussian random variables. Two independent random samples of size n, with parameters $(\mu_1, \sigma_1, \rho_1)$,$(\mu_2, \sigma_2, \rho_2)$, respectively are compared with null hypothesis of equal distributions and alternatives of

(i) location: $\mu_1 \neq \mu_2$. (ii) scale: $\sigma_1 \neq \sigma_2$. (iii) correlation: $\rho_1 \neq \rho_2$.

Numerical results of the simulation and testing are in Fig. 2, where it is $\mu_1 = 0$, μ_2 horizontal axis (upper left); $\sigma_1 = 1$, σ_2 horizontal axis (upper right); $\rho_1 = 0$, ρ_2 horizontal axis (lower left); $\rho_1 = 0.5$, ρ_2 horizontal axis (lower right). The number of windows is n =20, k =100 grains, 100 permutations, averaged over 1000 simulations. In the lower right graph the number of windows is 40. The parameters not involved in the alternative are $\mu_1 = \mu_2 = 0$ in (ii), (iii), $\sigma_1 = \sigma_2 = 1$ in (i), (iii), $\rho_1 = \rho_2 = 0$ in (i), (ii). We can observe that for the location alternative the power function increases more rapidly than for the scale and correlation alternatives. For both location and scale alternatives the power function increases more rapidly than for the correlation alternatives.

3. Materials

Further we present an application of suggested statistical methods. A Czech company AL INVEST Břidličná, a.s. provided five Al-Mn alloys denoted A, C, L, P, Z, the composition of which is in Table 1. The alloys with high manganese contents are A, C, L, P, high silicon contents have A, C and L, they differ in the zinc contents present in C and not present in A, L. Considering the high solubility of Zn in Al, all Zn is dissolved in aluminium matrix. Therefore,

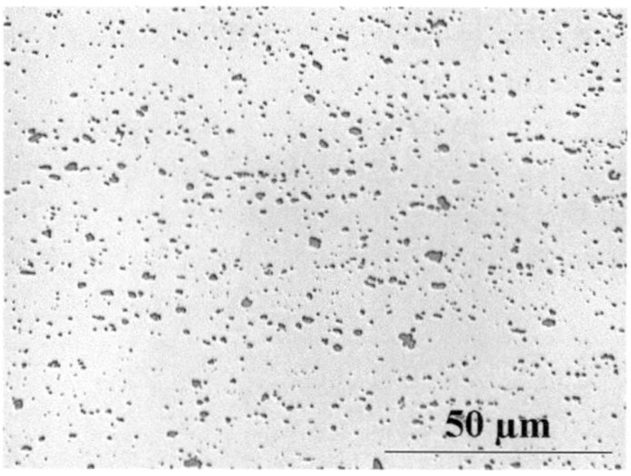

(a) Material A

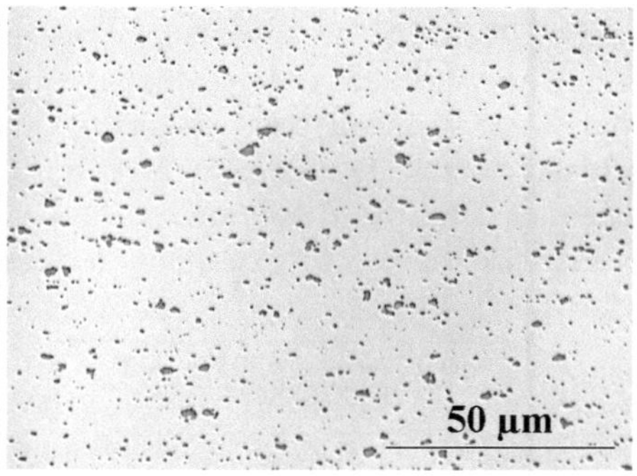

(b) Material C

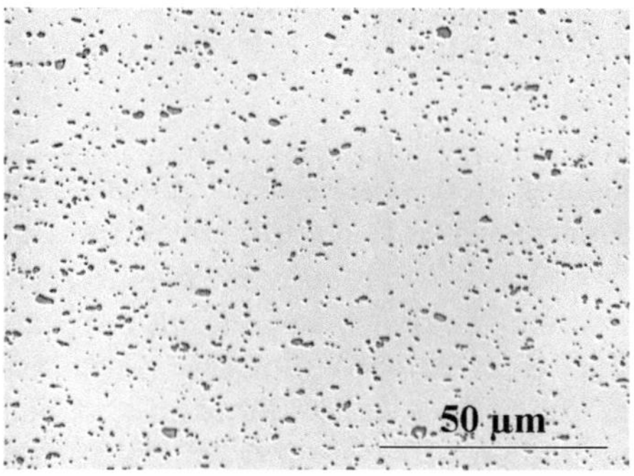

(c) Material L

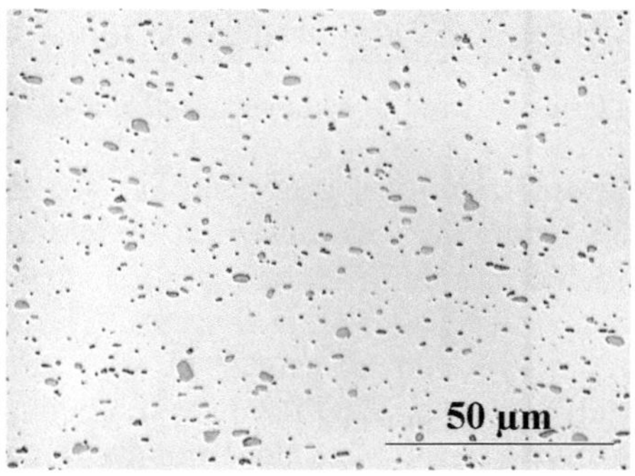

(d) Material P

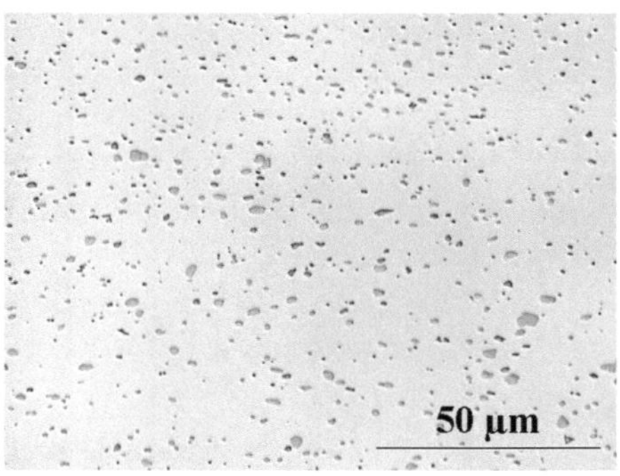

(e) Material Z

Fig. 3. A metallographic sample of material (a) A, (b) C, (c) L, (d) P, (e) Z, respectively, a transverse section of the foil. In (a), (b), (c), (alloys with higher contents of Si) particles α-$Al_{12}(Mn, Fe)_3Si$ prevail. In (d), (e), (alloys with lower contents of Si) particles $Al_6(Mn, Fe)$ prevail.

Zn does not participate in second-phase particles and its presence in alloy C does not affect its particle volume fraction and size distribution. Alloys *P* and *Z* have lower silicon contents, *P* has lower copper contents while *Z* has lower manganese contents. The most important factor influencing particles volume fraction and size distribution in the set of alloys considered is the content of silicon. Coarse particles are mostly primary, undissolved particles of α-Al_{12}(Mn,

$Fe)_3Si$ and $Al_6(Mn, Fe)$. Fine particles, present especially for alloys with a higher content of Si, are mainly precipitates α-$Al_{12}(Mn, Fe)_3Si$, cf. (Slámová et al., 2006).
All alloys were twin-roll cast in strips of 8.5 mm in thickness. All specimens were homogenized at high temperature after a 35% reduction in thickness and then cold rolled to thickness of 0.4 mm. The samples of 0.4 mm thickness were annealed again at 350 °C in order to increase the ductility of the material so as to facilitate the cold rolling up to the final foil thickness of 0.10 mm.

Alloy	Mn	Si	Fe	Cu	Zn	Mn+Si+Fe	(Mn+Si)/Fe	Mn/Si
A	1.09	0.54	0.47	0.16	0.004	2.1	3.5	2.02
C	1.02	0.56	0.53	0.15	1.02	2.1	3.1	1.8
L	1.02	0.59	0.48	0.11	0.01	2.1	3.3	1.7
P	1.01	0.20	0.61	0.04	-	1.8	2.0	5.1
Z	0.86	0.10	0.61	0.16	-	1.6	1.6	8.7

Table 1. Composition of experimental materials [wt. %].

Metallographic samples after grinding and polishing were observed by optical microscopy, see Fig. 3, where the vertical size of the window corresponds to the total thickness of the foil. Twenty windows along the foil, see Fig. 4, with distance 1 mm between neighbouring ones were measured by an image analyzer. The number of particles and individual parameters of each particle were measured in each window and for each material. The individual parameters are Area, EqDiam (i.e. diameter of the circle with the same area as parameter Area), Minimal Feret, Maximal Feret (extremes of the breadth, Ohser & Mücklich (2000), of a particle w.r.t. directions). The Shape Factor can be evaluated as a fraction of Minimal Feret and Maximal Feret. Parameters Area, EqDiam and Shape Factor were used to the input data to our test.

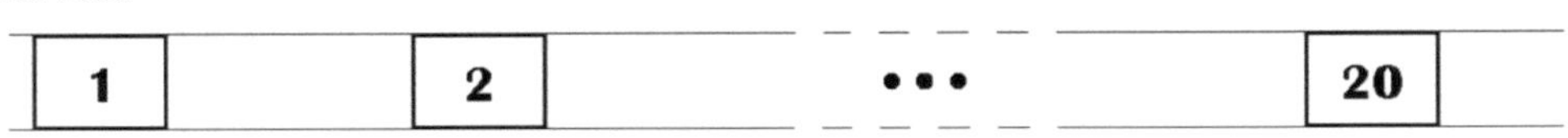

Fig. 4. Scheme of sampling windows along the foil of thickness 0.1 mm. The distance between neighbouring windows is 1 mm.

4. Numerical results

The preliminary analysis concerns the mean size of particle sections and their density, see Tables 2, 3, apparently A, C, L differ from P and Z, see also Fig.5. For the testing, since the number of windows $n = 20$ is small, we use the permutation test described in Section 2. The results of method (I) are in Table 4. The microstructures do not differ in the shape factor of particles. Concerning the particles size we observe again two separated groups $\{A, C, L\}$ and $\{P, Z\}$ (according to silicon contents). Between them there is a significant difference, while within the groups this is not the case. Nevertheless in Table 4 we observe a difference in pairs $A - L$, $P - Z$, while we cannot reject the null hypothesis in pairs $A - C$, $C - L$.
For the methods (II)-(IV) of the spatial distribution of particle section centroids first the F, G and pcf functions were estimated. The estimators of functions F, G and pcf for all windows of material P are presented in Fig. 6, i.e. in each figure a), b), c) there are 20 graphs. We observe a small variability of the estimators. Let us note that there is a similarly small variability of these estimators in all other materials. In Fig. 7 we compare the average estimators (from

Alloy	MeanArea	Standard Deviation	Density	Total number
A	0.696	0.864	0.073	19 773
C	0.683	0.873	0.070	18 834
L	0.689	0.796	0.080	21 611
P	1.080	1.287	0.047	12 605
Z	1.223	1.386	0.041	10 997

Table 2. The mean area of observed particle sections (in [μm^2]), their density (in [μm^{-2}]) and total number for materials A, C, L, P, Z, evaluated from all 20 windows.

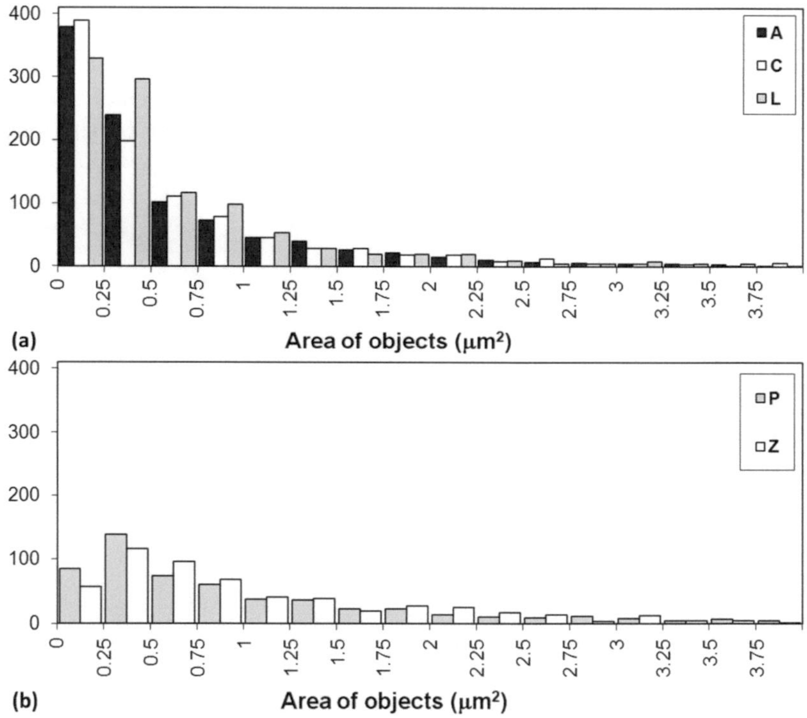

Fig. 5. Histograms of areas of particle sections in the first window of each material. In a) we have A, C, L and in b) there are P and Z.

Alloy	Mean Number	Standard Deviation
A	988,65	44.76
C	941,7	53.50
L	1080,55	41.36
P	630,25	46.55
Z	549,85	33.47

Table 3. The mean number of observed particle sections and their standard deviation in one window (evaluated from all 20 windows).

20 windows) of the functions F, G and pcf evaluated for all materials. We obtain the results that for the material L, A, C these are greatly different from the results for materials P and Z. The estimators of pcf of materials P and Z are practically the same and the differences of

	All	Area	Eqdiam	ShapeFactor
A-C	0.343	0.369	0.335	0.050
A-L	0.040	0.042	0.035	0.203
A-P	< 0.001	< 0.001	< 0.001	0.565
A-Z	< 0.001	< 0.001	< 0.001	0.529
C-L	0.302	0.355	0.288	0.046
C-P	< 0.001	< 0.001	< 0.001	0.283
C-Z	< 0.001	< 0.001	< 0.001	0.322
L-P	< 0.001	< 0.001	< 0.001	0.422
L-Z	< 0.001	< 0.001	< 0.001	0.488
P-Z	0.015	0.020	0.002	0.639

Table 4. The p-values for two-sample tests of individual particle parameters (method I) evaluated for all $m = 3$ parameters together (the column All) and then for single parameters ($m = 1$) with $n = 20, k = 500$ and with the number of permutations equal to 5000.

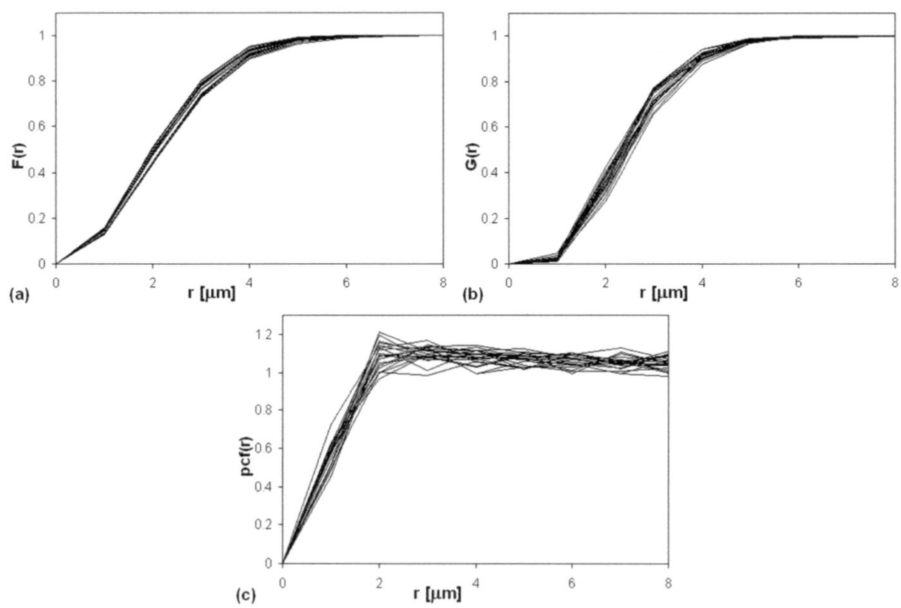

Fig. 6. Estimators of (a) F-function, (b) G-function and (c) pcf for material P. Graphs obtained from each of 20 windows are drawn in the same figures in order to observe a small variability of the estimators among the windows.

estimators of the functions F and G are small, while the corresponding functions of materials L, A, C are shifted to the left. This is caused mostly by the different particle density
The results for method (II) of the two-sample tests for the spatial distribution in Table 5 lead to the interpretation that there are significant differences in spatial distribution between any materials of different groups $\{A, C, L\}$ and $\{P, Z\}$. It is interesting to observe what happens within the groups. We can see that while distribution functions F, G still yield differences, the pair correlation function does not reveal any. Only the pair $C - A$ which has the most close contents of silicon, does not reveal any difference in any characteristics.

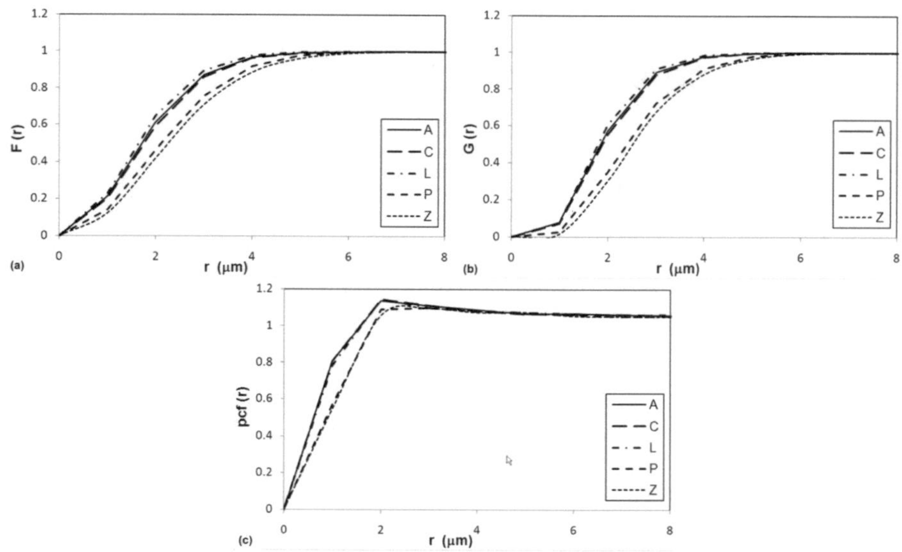

Fig. 7. Graphs of (a) F-function, (b) G-function and (c) pcf for all three materials obtained by averaging the estimators from 20 windows. We observe that for materials L, A, C they look differently from P, Z.

	All	F	G	pcf
A-C	0.6	0.166	0.034	0.999
A-L	0.004	< 0.001	< 0.001	0.784
A-P,A-Z	< 0.001	< 0.001	< 0.001	< 0.001
C-L	< 0.001	< 0.001	< 0.001	0.584
C-P,C-Z	< 0.001	< 0.001	< 0.001	< 0.001
L-P,L-Z	< 0.001	< 0.001	< 0.001	< 0.001
P-Z	< 0.001	< 0.001	< 0.001	0.545

Table 5. The p-values for two-sample tests of the spatial distribution (method II) evaluated for all three parameters together ($m = 3$, column All) and then for single parameters ($m = 1$) with $n = 20$, $k = 11$, $\triangle = 1\ \mu m$ and with the number of permutations equal to 5000.

Let us analyze the number of particles for the microstructures P and Z, we can test the difference between $n_P = 13866$, $n_Z = 11945$ from all 20 windows, thus in (5) we have $T = 12 > 1.96$ and we reject an H_0 of equal particle density at the significance level $\alpha = 0.05$. Here the Poisson process model assumption is violated, as suggested by the shape of pcf in Fig. 6 we have a type of a regular model, i.e. mild repulsion since there are nonoverlapping particles around the centroids. Clearly, if we reject the null hypothesis $\lambda_P = \lambda_Z$ for the Poisson process model using (5), we reject it for the regular model too, since it is less dispersed, i.e. the numbers of particles observed in windows vary more slowly.

Further a finer analysis of the spatial distribution of particles is applied using method (III). If we eliminate the effect of particle density on the spatial distribution by means of the scale change as suggested in Section 2, the results change as presented in Table 6. The pure effect of the spatial distribution of particle centroids is such that there is no significant difference between materials $P - Z$, $A - C$. But this is moreover the case also for individual functions F

	All	F	G	pcf
A-C	0.483	0.132	0.527	0.540
A-L	0.003	0.002	0.509	0.004
A-P	0.019	0.108	0.022	0.028
A-Z	0.013	0.178	0.122	0.009
C-L	< 0.001	< 0.001	0.502	< 0.001
C-P	0.001	0.060	0.057	0.001
C-Z	0.005	0.195	0.158	0.003
L-P	< 0.001	0.010	0.037	0.001
L-Z	< 0.001	0.005	0.256	< 0.001
P-Z	0.628	0.611	0.416	0.479

Table 6. The p-values for two-sample tests of the spatial distribution (with the effect of particle density eliminated, method III). Evaluation for all three parameters together (first column) and then for single parameters ($m = 1$) was obtained with $n = 20$, $k = 11$, $\triangle = 1\ \mu m$ and with the number of permutations equal to 5000.

(holds for $A - P,\ A - Z,\ C - Z$) and G ($A - L,\ A - Z,\ C - L,\ C - Z,\ L - Z$). That means some differences between two groups are removed.
Finally we present results of two-sample tests when using the functional data approach in Subsection 2.3. First for the comparison based on functions F, G, pcf we use the versions with scale change to eliminate the effect of particle density. Similar results as in Table 6 are expected using this method (IV), see Table 7. Even if individual $p-$values in both tables differ, the decisions about H_0 are almost completely the same.

	F	G	pcf
A–C	0.59	0.57	0.32
A–L	0.02	0.46	< 0.01
A–P	0.17	0.02	0.02
A–Z	0.42	0.13	< 0.01
C–L	< 0.01	0.42	< 0.01
C–P	0.07	0.04	< 0.01
C–Z	0.43	0.14	0.01
L–P	0.02	0.04	< 0.01
L–Z	0.01	0.26	< 0.01
P–Z	0.58	0.44	0.46

Table 7. The p-values for two-sample tests of the spatial distribution (with the effect of particle density eliminated), using the functional data approach, method (IV). Evaluation was obtained with the number of permutations equal to 100.

Further the tests based on multidimensional smoothing of particle characteristics are applied, which are qualitatively different methods. They are not sensitive to the particle density, on the other hand it may reveal local inhomogeneities and differences. First we give the results of comparison by method (V), that is corresponding $p-$values of the test based on $\mathfrak{N}$-distances of functions (7), (8), for different pairs of microstructures in terms of the particle centroid coordinates only, see Table 8. In many cases, but not in all, the results are similar to those in Tables 6, 7 (spatial distribution only is investigated in both methods). Different results are obtained especially for pairs $A - C,\ P - L$.

A-C	A-L	A-P	A-Z	C-L	C-P	C-Z	L-P	L-Z	P-Z
< 0.01	0.0297	< 0.01	0.475	< 0.01	< 0.01	< 0.01	0.257	< 0.01	0.059
< 0.01	0.07	< 0.01	0.19	< 0.01	< 0.01	< 0.01	0.11	0.02	0.04

Table 8. The p-values of $\mathfrak{N}$-test for corresponding coordinates comparisons in the functional data approach, method (V). The top row corresponds to the choice $\sigma_j(A) = 1/\sqrt[4]{n_j}$ and $\sigma_j(B) = 1/\sqrt[4]{k_j}$, the bottom row to the choice $\sigma_j(A) = \sigma_j(B) = \frac{1}{3}$.

A-L	A-P	P-Z
< 0.01	< 0.01	< 0.01

Table 9. The p-values of $\mathfrak{N}$-test for the comparisons based on coordinates and areas of the particles in the functional data approach, method (VI).

Finally a simultaneous analysis of spatial distribution and an individual particle parameter (area of the section) was performed using the functional data approach, method (VI). The results of the comparison are given in Table 9. We do not consider all pairs of microstructures since for those of different groups (different silicon contents) the particle areas surely cause the rejection of null-hypothesis. As we can see from Table 9, this is the case also in other pairs.

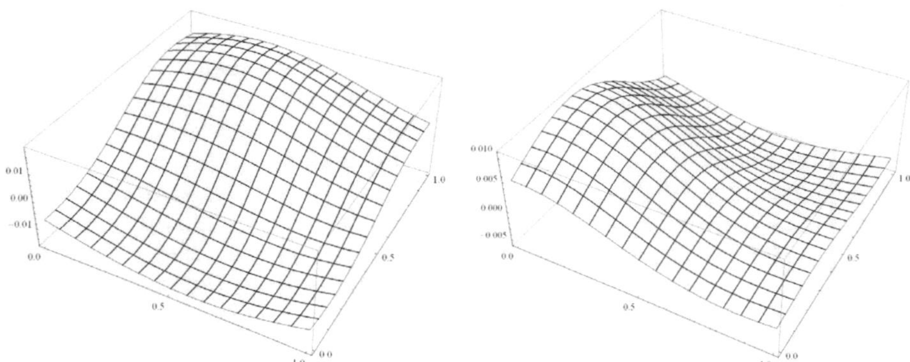

(a) A–C (b) A–Z

Fig. 8. Average value of the difference $\mu_j - \nu_j$ of functions (7),(8) taken from all 20 windows, microstructure A–C, A–Z.

5. Concluding remarks

This chapter brings an extension and continuation of research started in (Benes et al., 2009). New statistical methods are developed for the comparison of microstructural images of random objects in metallography and other applications. They are based on an appropriate interaction of approaches from mathematical statistics, image analysis and stochastic geometry. A proper two-sample test derived from $\mathfrak{N}$-distances enables one to evaluate a large amount of information from a few observed windows. The tests presented are easy to apply to metallographic images observed by light microscopy and image analysis. In comparison with the above mentioned paper here we suggest further methods based on functional data analysis and we analyze a broader set of foils from aluminium-manganese based alloys.

The first group of methods based on vectors of characteristics obtained from image analysis measurements and further transformation of data is well-established and the results easily understandable. From the practical point-of-view, it should be mentioned that $\mathfrak{N}$-distances are scale dependent, so that when evaluating qualitatively different information simultaneously ($m > 1$) one has to choose the scale carefully to be comparable for all parameters. This is guaranteed in the analysis of spatial distribution since all three functions used fall within a similar range. In the analysis of particle characteristics the size scale has to be modified comparably to the range of the shape factor. We conclude that mostly recommended methods are (I) and (IV) (or (III)).

On the other hand the newly proposed methods (V) and (VI) based on multidimensional functional data analysis need further investigations to be able to claim their usefulness. In comparison with method (I) they are able to evaluate different numbers of particles in windows. Theoretically, they are independent on the dimensionality, and therefore they are of great potential use in multidimensional statistical analysis. At first sight corresponding tests seem to be strict and sensitive to both local differences in spatial distribution and in particle characteristics. One must be very careful when interpreting the results of a functional data analysis. It should be also added that functional data analysis combined with permutation testing is more time-consuming (especially in the multivariate case), but feasible when using fast computers, and, especially, clusters.

Some of results obtained by classical and functional data analysis may seem to be contradictory, but this is not the case, the reason is that various methods for comparison of spatial distribution are different in their nature. We may consider pairs of microstructures A-C and A-Z. The functional data method using functions μ, ν in (7),(8) rejects the null hypothesis for A-C and does not reject it for A-Z. This conclusion is related to Fig. 8, where we can see the average values of the difference $\mu_j - \nu_j$ taken from all 20 windows, where the range is two times smaller for A-Z than for A-C. The same observation holds for individual windows, too. We investigated also the sensitivity of method (V) with respect to the choice of bandwidth σ_j. In the top row of $p-$values in Table 8 there is the asymptotically optimal bandwidth by theory. One can observe how the $p-$value slightly changes with a broader bandwidth in bottom row, but there is no evidence of a systematic change.

It is possible to study statistical properties of the tests by simulations. Concerning the resolution of the test, the power of the variants of the test based on $\mathfrak{N}$-distances was compared in (Klebanov, 2005). Besides our study in Subsection 2.4, in paper (Bakshaev, 2008) a large comparative study of the power of several two-sample tests (Kolmogorov-Smirnov, Cramer-von Mises, Anderson-Darling, Wilcoxon, Mann-Whitney, $\mathfrak{N}$-distances) was made. It appears that for a multidimensional case the test based on $\mathfrak{N}$-distances has the highest power.

The results of the application of the two-sample test in metallography can be transformed from conclusions about the geometry of the microstructure to conclusions relevant materials research. Since the production of all three materials was based on the same processing, the only difference is in the chemical composition of the alloys. Therefore, we can conclude that a differentiation for high and low silicon contents is apparent, while small differences in composition within different groups $\{A, C, L\}$ and $\{P, Z\}$ do not have a clearly apparent impact on the microstructure. We can observe that while P and Z have different particle densities, they do not differ in particle size and shape, nor in the pure spatial distribution (interactions).

6. Acknowledgement

The research was supported by the Czech Science Foundation, project GAČR P201/10/0472, and by the Czech Ministery of Education, project MSM 0021620839. Our memory belongs to Margarita Slámová, who was the coauthor of the paper (Benes et al., 2009). She died prematurely in 2009 and we missed her during the preparation of this chapter a lot.

7. Appendix

Here we give a mathematical background of $\mathfrak{N}$-distances and related statistical testing. This background comes from (Klebanov, 2005). Let $\{\mathfrak{X}, \mathfrak{A}\}$ be a measurable space, $\mathcal{L} : \mathfrak{X}^2 \to \mathbb{R}^1$ is a negative definite kernel on $\mathfrak{X}$ ($\mathcal{L}(x,x) = 0$ and $\mathcal{L}(x,y) = \mathcal{L}(y,x)$) if and only if

$$\int_{\mathfrak{X}} \int_{\mathfrak{X}} \mathcal{L}(x,y) h(x) h(y) dQ(x) dQ(y) \leq 0 \tag{9}$$

for an arbitrary probability measure Q on $\{\mathfrak{X}, \mathfrak{A}\}$ and a measurable function h on $\mathfrak{X}$ such that $\int_{\mathfrak{X}} h(x) dQ(x) = 0$. We say that $\mathcal{L}$ is strongly negative definite if the equality in (9) implies $h = 0$ almost surely with respect to the measure Q.

Let $\mathcal{L}$ be a strongly negative definite kernel on $\mathfrak{X}$, $\mathcal{B}_{\mathcal{L}}$ the set of all probabilities μ on $\{\mathfrak{X}, \mathfrak{A}\}$ for which there exists the integral $\int_{\mathfrak{X}} \int_{\mathfrak{X}} \mathcal{L}(x,y) d\mu(x) d\mu(y) < \infty$. For $\mu, \nu \in \mathcal{B}_{\mathcal{L}}$ put $\mathcal{N}(\mu, \nu) =$

$$= 2 \int_{\mathfrak{X}} \int_{\mathfrak{X}} \mathcal{L}(x,y) d\mu(x) d\nu(y) - \int_{\mathfrak{X}} \int_{\mathfrak{X}} \mathcal{L}(x,y) d\mu(x) d\mu(y) - \int_{\mathfrak{X}} \int_{\mathfrak{X}} \mathcal{L}(x,y) d\nu(x) d\nu(y). \tag{10}$$

Then

$$\mathfrak{N}(\mu, \nu) = \Big(\mathcal{N}(\mu, \nu) \Big)^{1/2} \tag{11}$$

is a distance on $\mathcal{B}_{\mathcal{L}}$, it is called $\mathfrak{N}$-distance. We will use $\mathfrak{N}$-distances in the following using two approaches. First in classical data analysis when $\mathfrak{X} = \mathbb{R}^k$ is the Euclidean space of $k-$dimensional vectors. Secondly in functional data analysis where $\mathfrak{X} = L_2$ is the space of square integrable functions.

Let $\mathcal{L}(x,y)$ be a strongly negative definite kernel on $\mathbb{R}^k$, X, Y are two independent random vectors in $\mathbb{R}^k$, define one-dimensional independent random variables U, V by

$$U = \mathcal{L}(X,Y) - \mathcal{L}(X,X'), \qquad V = \mathcal{L}(Y',Y'') - \mathcal{L}(X'',Y''). \tag{12}$$

Here all vectors X, X', X'', Y, Y', Y'' are mutually independent, equalities of distributions $X \stackrel{d}{=} X' \stackrel{d}{=} X''$, $Y \stackrel{d}{=} Y' \stackrel{d}{=} Y''$ hold. We have $X \stackrel{d}{=} Y \iff U \stackrel{d}{=} V \iff \mathfrak{N}(X,Y) = 0$. Consider testing of the hypothesis $H_0 : X \stackrel{d}{=} Y$ for multivariate random vectors X, Y. This hypothesis is equivalent to $H_0' : U \stackrel{d}{=} V$, where U, V are random variables taking values in $\mathbb{R}^1$. Consider two independent samples $X_1, \ldots, X_n;\ Y_1, \ldots, Y_n$ from general multivariate populations X and Y, respectively. A one-dimensional test to U and V can proceed in the following ways:

a) split each sample randomly in three equal parts X, X', X'', Y, Y', Y'' and use (12); this leads to a loss of information,

b) simulate the samples from X' and X'' (as well as from Y' and Y'') by independent choices from observations $X_1, \ldots, X_n$ (and from $Y_1, \ldots, Y_n$, correspondingly); thus we do not test the hypothesis $X \stackrel{d}{=} Y$, but the one of the corresponding empirical distributions,

c) permutation test using Monte Carlo approximation.

8. References

Bakshaev A. (2008). Nonparametric tests based on N-distances, *Lithuanian Math J*, Vol. 48, No. 4, 357–376.

Beneš V, Lechnerová R, Klebanov L, Slámová M, Sláma P. (2009). Statistical comparison of the geometry of second-phase particles. *Mater Charact*, Vol. 60, 1076-1081.

Buening H, Trenkler G. (1978). *Nonparametric statistical methods*, Walter de Gruyter, Berlin (in German).

Derr R, Ji C. (2000). *Fitting microstructural models in materials science.* http://www.stat-or.unc.edu/webspace/miscellaneous/cji/scan05.pdf.

Hirsch J. (2006). *Virtual fabrication of aluminium products*, John Wiley & Sons, New York.

Humphreys FJ, Hatherly M. (2004). *Recrystallization and related annealing phenomena*, Elsevier, London.

Ihaka R, Gentleman R. (1996). R: A language for data analysis and graphics, *J Comput Graph Stat*, Vol. 5, No. 3, 299–314.

Klebanov LB. (2005). *$\mathfrak{N}$-distances and their applications*, The Karolinum Press, Prague.

Korovkin PP. (2001). *Bernstein Polynomials*, In: Hazewinkel M, Encyclopaedia of Mathematics, Springer, New York.

Kupczyk J. (2006). Application of significance tests in quantitative metallographic analysis of a C-Mn-B steel. *Mater Charact*, Vol. 57, 171–175.

Lehmann E, Romano P. (2005). *Testing statistical hypotheses*, Springer, New York.

Ng HKT, Gu K, Tang ML (2007). A comparative study of tests for the difference of two Poisson means. *Comp Stat & Data Anal*, Vol. 51, 3085–3099.

Ohser J, Mücklich F. (2000). *Statistical analysis of microstructures in materials science*, Wiley.

Polmear IJ. (2006). *Light alloys - from traditional alloys to nanocrystals*, The Fourth edition, Elsevier, London.

Slámová M, Sláma P, Cieslar M. (2006). *The influence of alloy composition on phase transformations and recrystallization in twin-roll cast AlMnFe alloys*, In: Poole WJ, Wells MA, Lloyd DJ, editors. Aluminium alloys. Mat. Sci. Forum, Vol. 519-521, 365–370.

Tewari A, Gokhale AM. (2006). Computations of contact distributions for representation of microstructural spatial clustering. *Comput Mater Sci*, Vol. 38, 75–82.

Tewari A, Gokhale AM. (2006). Nearest neighbor distributions in thin films, sheets, and plates. *Acta Mater*, Vol. 54, No. 7, 1957–63.

20

Microstructural Evolution During the Homogenization of Al-Zn-Mg Aluminum Alloys

Ali Reza Eivani[1,2], Jie Zhou[2] and Jurek Duszczyk[2]
[1]*Materials Innovation Institute (M2i), Mekelweg 2, 2628 CD Delft,*
[2]*Department of Materials Science and Engineering, Delft University of Technology, Mekelweg 2, 2628 CD Delft, The Netherlands*

1. Introduction

1.1 Background

Aluminum and aluminum alloys are probably the most ideal materials for extrusion, and they are the most commonly extruded. Most of commercially available aluminum alloys can be extruded. Principal applications include parts for the aircraft and aerospace industries, pipes, wires, rods, bars, tubes, hollow shapes, cable sheathing, for the building, automotive and electrical industries. Sections can be extruded from heat-treatable or non-heat treatable low-, medium- and high-strength aluminum alloys [1].

In the last 30 years, the development of aluminum extrusion technology has, in the main, been focused on the billet metallurgy, die design and process control for low- and medium-strength aluminum alloys in the 6xxx series for architectural applications, in order to maximize extrusion speed and at the same time fulfill the requirements in product specifications in terms of dimensions, shape, surface and mechanical properties. As a result, there is a wealth of information available on the relationship between alloy chemistry, microstructure and extrudability of these alloys [2]. In comparison, the fundamental knowledge and extrusion technology, especially those for medium- and high-strength aluminum alloys in 7xxx series, are rather scarce in the open literature [2].

7xxx series aluminum alloys, almost exclusively for air transport applications in the past but now increasingly used in the rail and road vehicles, must comply with much more stringent performance specifications than 6xxx series aluminum alloys for architectural applications. Although many investigations on the behavior of medium- and high-strength aluminum alloys at individual processing steps have been performed, systematic research linking all these processing steps is lacking, while the extrusion behavior is associated with alloy composition and a series of microstructural evolutions throughout the whole chain of material processing from casting through homogenization to extrusion. Such research is particularly needed for the aluminum extrusion companies that are currently shifting the application fields of extrusions from architecture to ground transport where medium-strength alloys (7003, 7005, 7010, 7020, 2011, 2017 and 2618) and high-strength alloys (7049, 7050, 7075 and 2024) are increasingly used. This chapter concerns one of the mostly used medium-strength alloys, AA7020, as a representative of Cu-free 7xxx series aluminum alloys. Table 1 shows the nominal chemical composition of the AA7020 aluminum alloy.

Element	Si	Fe	Cu	Mn	Mg	Zn	Ti	Cr	Zr	Al
Wt. %	<0.35	<0.35	<0.2	0.05 – 0.5	1.0 – 1.4	4.0 - 5.0	Zr + Ti = 0.08-0.25	0.1 – 0.4	0.08 – 0.20	Bal.

Table 1. Nominal chemical composition of the AA7020 aluminum alloy

It should be noted that despite a broad range of applications, the AA7xxx series alloys have a number of characteristics that are not favorable for material processing, for example, low extrudability, high extrusion pressure required together with low solidus temperature which can cause incipient melting, makes their production at low throughputs. In addition, in the 7xxx series aluminum alloys including the AA7020 alloy, due to the long solidification interval (the temperature gap between the liquidus and solidus), microsegregation tends to be significant and homogenization needs lots of time, which causes the formation of second phase particles, some of which may be detrimental to the hot workability of the alloy as well as its final mechanical properties. Furthermore, the formation of a peripheral coarse grain structure is quite common in these alloys which can significantly degrade the mechanical properties. These issues will be discussed as limiting factors in the extrusion of the AA7020 aluminum alloy.

1.2 Limiting factors in the extrusion of AA7020

Numerous limiting factors, e.g., the formation of defects, low extrudability and the complications in the final microstructure of the product in 7xxx series aluminum alloys can impair the productivity of the extrusion process [2]. Most of the limiting factors are directly or indirectly related to the chemical composition, metallurgical features formed DC casting and evolving during the homogenization treatment, and extrusion conditions [2]. Using an optimum chemical composition within the allowance range of a specific alloy, in combination with optimum homogenization and extrusion conditions can result in a significant increase of the extrudability of the material and improvement of the mechanical properties of the final product. This requires the knowledge of the factors presented below, i.e., hot workability and peripheral recrystallization, both of which are strongly dependent on chemical composition of the alloy and the homogenization treatment.

1.2.1 Hot workability

For wrought aluminum alloys, hot workability is an important index of manufacturability. It refers to the capacity of an alloy to withstand hot deformation at a maximum rate without inducing flow non-uniformity or structural defects. If translated into extrusion (termed extrudability), it is defined as the maximum speed for a sound extrudate with sufficient dimension and shape accuracy. Hot workability is in fact affected by all parameters affecting the fracture of the material under processing [1]. It is strongly dependent on the size and density of second-phase particles which are in turn dependent on the chemical composition and homogenization treatment. On the other hand, compressive stresses superimposed on shear stresses during the deformation process can have a significant influence on closing small cavities or limiting their growth and thus enhancing workability. Because of the important role of the stress state, it is not possible to express workability in absolute terms. Workability depends not only on material characteristics but also on process variables, such as strain, strain rate, temperature, and stress state [1]. In other words, extrudability can be deteriorated by two factors: (i) unfavorable processing conditions and (ii) the presence of large second-phase particles.

Investigation of the effect of processing conditions on the hot workability of AA7xxx series aluminum alloys is out of the scope of this chapter. However, the effect of particles which would deteriorate the hot workability of the alloy is the main subject of this chapter.

High-strength aluminum alloys (7xxx series) are known for having rather poor hot workability due to the presence of dissolved and later precipitated elements in the form of large particles that raise flow stress and cause the actual temperature of the extrudate to increase above the solidus temperature, leading to hot tearing as shown in Fig. 1. Hot tearing represents the separation or failure of the product as a result of a sequence of phenomena consisting of local melting, crack formation and final fracture of the product. Hot tearing may occur as a result of the presence of large intermetallic particles or even the multiplication of the two mentioned factors (local melting and large particles). Therefore, in hot extrusion, applicable extrusion speed depends on the alloy composition and the microstructure formed during direct-chill casting and developed during homogenization, cooling and reheating to the initial billet temperature.

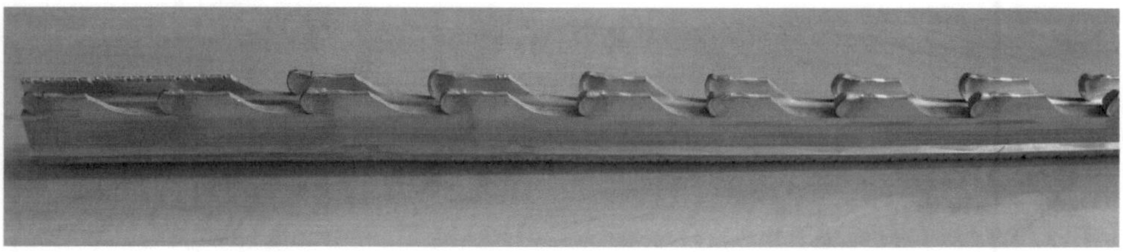

Fig. 1. Extreme case of hot tearing in AA7075 aluminum alloy [3]

In the case of the 7xxx series alloys, the two phases that deteriorate the extrudability are (i) compounds of Al-Fe-Mn-Si elements, which are especially important if they are located at the grain boundaries and (ii) Al-Mg-Zn-Cu eutectic phases which are mostly located at the dendrite boundaries.

1.2.1.1 Particles composed of Al, Fe, Mn and Si

It has been proven that in the case of the AA7020 aluminum alloy Al, Fe, Mn and Si-containing particles have the $Al_{17}(Fe_{3.2},Mn_{0.8})Si_2$ stoichiometric composition and are mostly located at the grain boundaries and therefore are called GB particles [4]. GB particles deteriorate the hot workability of the alloy in addition to mechanical properties since they are hard and brittle especially when located at the grain boundary regions. In order to avoid the detrimental effects of these particles, they should be dissolved during the homogenization treatment. If the particles are not dissolvable, they should be spheroidized.

1.2.1.2 Particles composed of Al, Mg, Zn and Cu

During the solidification of the 7xxx series aluminum alloys containing Mg, Zn and Cu, some intermetallic particles such as Al_6CuMg_4, $Al_2Mg_3Zn_3$, AlCuMg, $MgZn_2$, Al_2Cu and $MgZn_2$ phases form [5-9]. In addition, mutual solutions of different phases can result in the formation of new particles, for example, mutual solid solution of Al_6CuMg_4 and $Al_2Mg_3Zn_3$ compounds (T phase), solid solution between AlCuMg and $MgZn_2$ compounds (M phase), solid solution formed by $Al_5Cu_6Mg_2$ and Mg_2Zn compounds (Z phase) and between Al_2CuMg and Al_2Cu compounds (S phase). The investigation of these particles is interesting for hot working since they mostly have low melting points, which may result in incipient melting during hot deformation.

1.2.2 Peripheral recrystallization

In addition to high flow stress and low solidus temperature, local recrystallization is another factor limiting the extrusion window of medium- and high-strength aluminum alloys. For ground and air transport applications, a qualified extrudate should not contain any undesirable microstructural features, most notably local recrystallized, excessively large grains.

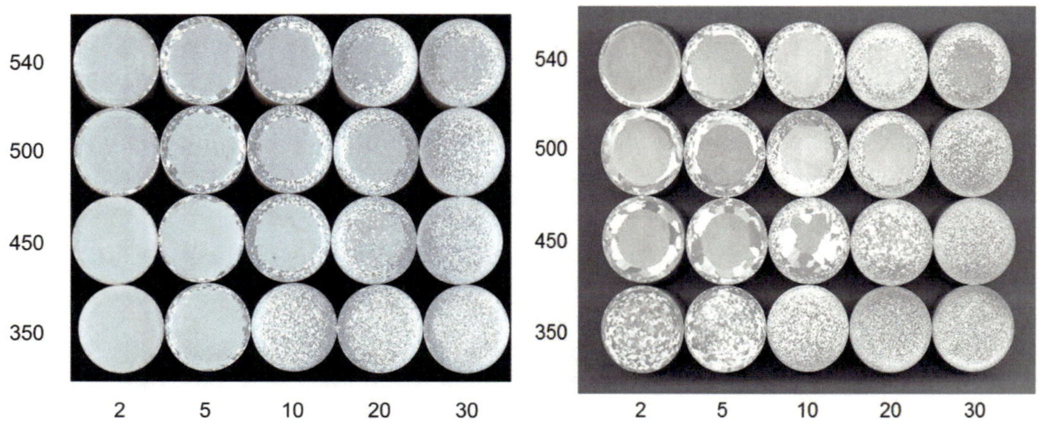

(a) (b)

Fig. 2. Macrostructures of AA6005 25 mm bar varying with ram speed (mm/s) and billet temperature (a) as extruded and (b) after solution heat treatment [10, 11]

(a) (b)

Fig. 3. Grain structures of (a) the core and (b) periphery of AA7020 extruded at 450 °C, 3 m/min and 15:1 [12]

The microstructure of an extruded aluminum product varies from an unrecrystallized fibrous structure to a thoroughly recrystallized fine or coarse grain structure (Fig. 2), depending on the chemical composition, homogenization treatment and extrusion conditions, i.e., speed and temperature, and cooling procedure [10, 11]. The most undesirable microstructure is the one with a peripheral coarse grain (PCG) structure [2] which is a well known defect in hot extruded aluminum alloys. In this case, the peripheral surfaces of the structure are fully recrystallized, having large grains, while the core is composed of unrecrystallized elongated grains as shown in Fig. 3.

PCG degrades the properties of the extruded product such as strength, fracture toughness and stress corrosion resistance [2]. It is actually a perpetual problem that extruders encounter in meeting the specifications of aircraft alloys that base their strength requirements on typical longitudinal properties of the unrecrystallized core and assume implicitly that no recrystallized outer band structure is present. It is generally understood that the peripheral recrystallization is a complex interplay of billet composition (grain growth inhibitors, i.e., Mn, Cr, or Zr), microstructure, deformation conditions and critical temperatures (solvus, solidus and recrystallization) [2, 3, 10-12].

1.3 Application of homogenization treatment

A homogenization treatment after DC casting for the 7xxx alloys is meant to serve the purposes of dissolving second-phase particles and generating disposoids that are able to inhibit recrystallization and PCG zone formation. The metallurgical features that occur during DC casting and should be studied during the homogenization treatment are presented below.

1. The mechanical properties of extruded products are largely dependent on alloying elements present in solid solution. These elements increase the strength mainly through solid solution or precipitation hardening [13]. During casting of aluminum alloys, a large fraction of alloying elements segregate to the liquid and result in an inhomogeneous distribution of alloying elements. Therefore, removal of the inhomogeneous distribution of alloying elements on a microscale is of prime importance during the homogenization treatment.
2. Segregation can also result in the formation of eutectic constitutive particles with low melting points in the grain boundary regions or inside the grains [14, 15]. As mentioned earlier, the presence of low melting point (LMP) phases which may cause incipient melting during hot deformation can deteriorate the hot workability of aluminum alloys. Therefore, one of the aims of the homogenization treatment is to dissolve LMP phases.
3. In addition, the formation of some hard particles with sharp edges mostly from impurities, e.g., Fe, in combination with some alloying elements such as Mn and Si is expected during DC casting. These particles also decrease the hot workability and limit the range of process parameters applicable during extrusion [2, 14, 16-19]. Therefore, it is necessary to dissolve these particles as well, in order to obtain high mechanical properties and extrudability. If the dissolution of these particles is exhibitively energy and time consuming, these particles should be spheroidized.
4. Eliminating the PCG structure or decreasing its extent is of great interest to the aluminum extrusion industry. It is generally known that the formation of small dispersoid particles can pin the low and high angle grain boundaries and therefore, inhibit recrystallization and grain growth. Therefore, an optimum homogenization treatment should take the formation of fine, well-distributed dispersoid particles into account.

1.4 Previous works on homogenization treatment of aluminum alloys

Although there have been a number of investigations on the homogenization treatment of the 7xxx series aluminum alloys in recent years [5-8, 20-26], most of the efforts have been focused on the nature and evolution of the $Al_2Mg_3Zn_3$ (T), Al_2CuMg (S), $(CuZnAl)_2Mg$ and $MgZn_2$ (η) phases [5-9] and the formation and distribution of dispersoids during homogenization [20-26]. In addition, in comparison with other aluminum alloys, the information on the 7xxx series aluminum alloys is rather scarce in the literature. Although some researchers have studied the microstructural changes and the evolution of the eutectic and low melting point phases during homogenization in the case of the 7xxx series aluminum alloys, there is still a lack of information in the case of the AA7020 aluminum alloy. Moreover, there has been no comprehensive quantitative study on the dependence of the particles on homogenization treatment parameters. Most of the investigations are concerned with the microstructural evolution and phase transformations during homogenization, describing the phenomena in a qualitative manner. For example, Lim et. al. [6] investigated the effects of compositional changes and preheating conditions on the evolution of constitutive particles, the M, T, S phases and dispersoids in AA7175 and AA7050 alloys. Senkov et. al. [27] studied the effect of homogenization treatment on the microstructural evolution of four newly developed 7xxx series aluminum alloys to obtain optimized conditions. Jackson and Sheppard [14] studied the effect of homogenization treatment on the microstructural changes of AA7075, 7150 and 7049 alloys. They focused on the evolution of the microstructure, low melting point phases and the M, T, S, Al_3Mg_2 (b), β, $Al_{18}Cr_2Mg$ (E) and $CrAl_7$ phases. Fan et al. [28] studied the evolution of microstructure in an Al-Zn-Mg-Cu alloy during homogenization. Ciach et. al. [29] conducted theoretical and experimental studies on the dendritic structure and its dissolution in aluminum-zinc alloys. However, the research on the commercially important AA7020 alloy is scarce. No quantitative investigation on the microstructural evolution in AA7020 regarding the grain boundary particles and low melting point phases during homogenization has been carried out.

As mentioned earlier, in addition to the dissolution of particles, dispersoid formation is another important aim of the homogenization treatment. Dispersoids exert a retarding force or pressure on low angle and high angle grain boundaries, which has a profound effect on the behavior of aluminum alloys in terms of recovery, recrystallization and grain growth [30]. A large amount of experimental research in addition to modeling efforts [15, 20, 24-26, 31-42] has been carried out to understand the conditions under which the dispersoids form and the effect of homogenization parameters on the evolution of dispersoids, with the aim of maximizing the effect of pinning the grain boundaries to retard recrystallization and grain growth. Most of the previous research [15, 20, 25, 42] focused on Al_3Zr as the only dispersoids present in the microstructures of the AA7xxx series aluminum alloys. However, the formation of the other types of dispersoids in other series of aluminum alloys has been a subject of extensive research. For example, in the AA3xxx series aluminum alloys, the formation of Mn-containing dispersoids and their effect on the recrystallization behavior were investigated [31-33]. In the AA6xxx series aluminum alloys, different kinds of dispersoids, i.e., Zr-, Mn- and Cr-containing ones were found to play individual roles in recrystallization inhibition [34-38]. In the case of the AA7xxx series aluminum alloys, the formation of Zr- and Sc-containing dispersoids has been investigated [15, 20, 25, 42]. For example, Robson et al. [15, 20, 25, 42], investigated the effect of Zr addition on the dispersoid formation and recrystallization fraction after hot deformation. It was concluded that using an optimum two-step homogenization treatment, a smaller fraction of recrystallization could be obtained. Robson [25] further studied the effect of Sc on the formation of dispersoids, as Sc was expected to eliminate the dispersoid free zones, as observed in the scandium free AA7050 alloy, thus greatly increasing the recrystallization resistance.

Comprehensive investigations to characterize Cr- and Mn-containing dispersoids formed in the AA7xxx series aluminum alloys during homogenization are scarce, although Cr-, Zr-, and Mn-containing dispersoids commonly co-exist in these alloys. Since analyzing all the individual dispersoids during the investigations to optimize the homogenization treatments is practically impossible, clear characterization of different types of dispersoids with a combination of analytical methods is of prime importance.

The objectives of this research were to determine the effect of homogenization treatment on the evolution of the particles, especially the grain boundary and low melting point ones, and to establish the correlations of the process parameters such as time and temperature with the fractions of these particles in the structure. Thorough investigations were performed on the effect of homogenization treatment on the evolution of these particles using optical microscopy (OM), X-ray diffraction (XRD) analysis, field emission gun-scanning electron microscopy (FEG-SEM), electron probe microanalysis (EPMA) and differential scanning calorimetry (DSC) and the results were quantitatively analyzed. The dependence of the fractions of the particles in the structure on homogenization parameters was also investigated using various quantitative methods. In addition, a comprehensive investigation on the formation of Cr-, Zr-, and Mn-containing dispersoids which commonly co-exist in the AA7020 aluminum alloy was performed.

2. Experimental procedure and data processing

Cubic samples of 2 cm in size were cut from the centre of a direct-chill (DC) cast AA7020 ingot. The chemical compositions of the variants of the AA7020 alloy used in this study are shown in Table 2. Isothermal homogenization treatments were performed in a salt bath at temperatures of 390-550 °C for 2-48 h, as shown in Table 3. Following the heat treatments, the samples were quenched in water.

Element (Wt %)	Si	Fe	Cu	Mn	Mg	Zn	Ti	Cr	Zr	Al
N1	0.31	0.28	0.2	0.34	1.24	4.36	0.001	0.10	0.08	Bal.
N2	0.30	0.30	0.19	0.35	1.20	4.37	0.002	0.10	0.13	Bal.
N3	0.29	0.31	0.2	0.36	1.22	4.37	0.001	0.10	0.20	Bal.
Nominal composition	<0.35	<0.35	<0.2	0.05-0.5	1.0-1.4	4.0-5.0	Zr+Ti= 0.08-0.25	0.1-0.4	0.08-0.20	Bal.

Table 2. Chemical compositions of the AA7020 alloy variants used in this study

Temperature (°C)	Time (hours)
	2
390	4
430	8
470	16
510	24
550	32
	48

Table 3. Homogenization treatment conditions used in this study

Optical microscopy (OM) was performed using an OLYMPUS BX60M light microscope on the samples etched using Barker's etchant. Images were analyzed using the Soft Imaging Software (SIS) image processor. Three samples in each homogenization condition were prepared and the analysis was performed on two images with approximately 6.2 mega pixel image quality and the average values are reported. The differences between the measured data from different samples and different images are represented by error bars.
The samples were examined using field emission gun-scanning electron microscope (FEG-SEM). The optimum operating voltage and current were 10 kV and 1 nA, respectively. With these settings, dispersoids as small as 10 nm in diameter could be detected.
The SEM images of the GB particles after different homogenization treatments were quantitatively analyzed to investigate their dissolution during homogenization. 20 GB particles were analyzed in each case, and the width was measured and the average value calculated. During homogenization at high temperatures, i.e., 510 and 550 °C, some of the GB particles were completely dissolved in the structure. The dissolved GB particles were also considered in the calculation with a null width. The average initial number density of the GB particles in 20 micrographs of the structure was counted to be 2×109 μm^{-2}. The average number density of the GB particles after homogenization was also counted employing the same method and, if it was less than the average initial number density, indicating the full dissolution of some of the GB particles, a zero width was put into the calculations.
Discs having a diameter of 3 mm were punched from the samples and ground down to a thickness of less than 60 μm, followed by electro-polishing in a solution of 30% nitric acid and 70% methanol cooled to -25 °C in a double-jet polishing unit at 20V.
Energy dispersive X-ray (EDX) analysis was performed with an analyzer attached to the FEG-SEM to determine the chemical compositions of the particles in the as-homogenized microstructures. In the case of small particles (< 500 nm), in order to keep the analysis volume in the EDX measurements as small as possible, the analysis was performed on TEM samples with an average thickness of 100 nm or less.
Electron Probe Microanalysis (EPMA) was performed using an electron beam with energy of 15 keV and beam current of 50 nA employing Wavelength Dispersive Spectrometry (WDS). The composition at each analysis location of the sample was determined using the X-ray intensities of the constituent elements after background correction relative to the corresponding intensities of reference materials. The thus obtained intensity ratios were processed with a matrix correction program CITZAF [43]. The points of analysis were located on lines with increments of 2 μm and involved the elements of Cr, Mn, Cu and Zr. Al was measured by difference.
A BRUKER-AXS D5005 diffractometer with Cu Kα1 wavelength was used to identify the phases present in the as-cast and as-homogenized conditions. Quantitative XRD (QXRD) analysis was performed using the direct comparison method [44] to estimate the weight percent of the phases in the structure. Application of this method requires the weight percent of the phase of interest (i.e., GB particles) in the as-cast structure, as the baseline. To calculate the weight percent of the GB particles in the as-cast structure, the surface fraction of the GB particles was calculated using FEG-SEM together with EDX analysis. The analysis was performed on 20 images at a magnification of 1000 and all the particles present in each image were analyzed. Assuming a uniform distribution of the GB particles in the structure, the surface fraction can be approximated to be equal to the volume fraction. The volume fraction of the GB particles was converted to weight percent using the density of the GB

particles (3709 kg/m³ [45, 46]) and the density of AA7020 aluminum alloy (2780 kg/m³ [47]). The only assumption made was the density of the other particles (a mixture of various phases) other than the GB particles being equal to the density of the AA7020 aluminum alloy.

To determine the volume fraction of particles from the data obtained by optical and SEM microscopy, a simple rule was used. It was assumed that the average surface fraction measured in a large number of images from different positions in the substrate was representative of the volume fraction [48]. It has been mathematically proven that the average surface fraction is equal to the volume fraction, provided that an enough large number of sections are investigated [48]. In this research, the investigation was performed on such a number of images that a constant average value was obtained, being not significantly changed by adding another image to the measurements.

The number density and radii of dispersoids obtained from SEM micrographs are in the form of the number of particles on many cross sections in the observation area in 2-D. 2-D cross section observations of the volume generally do not directly correspond to the coherent values in 3-D. In other words, the average particle diameter and number density of particles in each size group of the size distribution are not correct representatives of the real numbers in 3-D. The reason is that the crossing plane may not cut the particle in the middle and therefore, an observed specific cross section with a constant size may be a cross section of a particle which is cut through the middle or a cross section of a larger particle which is not cut through the middle. This point is schematically illustrated in Fig. 4 [48]. It can be seen that a mono-dispersed system of diameter D_j in 3-D can result in different circular sections in 2-D. It is shown in Fig. 5 that particles of large sizes can contribute to increasing the 2-D observed number density of particles with smaller sizes depending on the geometry of the cutting plane.

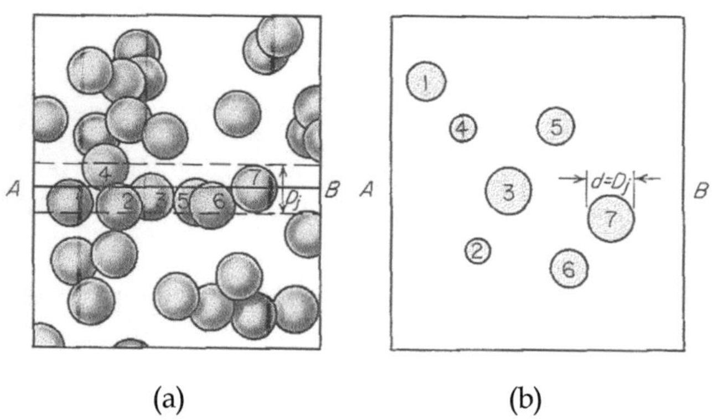

Fig. 4. (a) Distribution of mono-sized particles of D_j in 3-D and (b) observed A-B cross section of the particle distribution in 3-D [48]

The solution to this problem is to subtract the contribution of large particles to the 2-D measured sizes of smaller particles. For this purpose, different methods such as Scheil's method, Schwartz's method, Schwartz-Saltykov method have been proposed and used [48]. These methods can be used to find the distribution of particle sizes from a distribution of section diameters. The three methods differ in the details of how the numbers of sections contributed by larger spheres are determined.

In addition to the methods mentioned above, there are other methods which work with the distribution of section areas to determine 3-D particle sizes. Among these methods Johnson's and Johnson-Saltikov methods are well known. Johnson's derivation is applicable only to single-phase structures. However, Saltykov's improvement of Johnson's method applies to a distribution of particles as well as grain sizes [48]. Since the method is applicable for the prediction of grain and subgrain sizes in addition to particles, Johnson-Saltykov method was used in this research.

According to Saltykov's method [48], the most rational scale for the classification of particles (or grain sizes) is a linear logarithmic scale of diameter. Using the Johnson-Saltykov method, the analysis and calculations in the logarithmic scale can be simplified and facilitated. An advantage of this method is that a size distribution of particles can be obtained directly [48]. However, it must be noted that the resulting size distribution graphs will be presented by logarithmic size categories.

Thermal analysis of the as-cast and homogenized materials was carried out by means of a DSC analyzer at a heating rate of 20 °C/min over a temperature range of 35 to 700 °C. Samples were cubes weighing 12 mg each and Al_2O_3 powder was used as the reference. The analysis was performed under the protection of Ar gas. To ascertain the effect of homogenization treatment on the dissolution of the LMP phases the DSC profiles were quantitatively processed. For this purpose, the area underneath the peak was correlated to the fraction of the LMP phases in the structure [3].

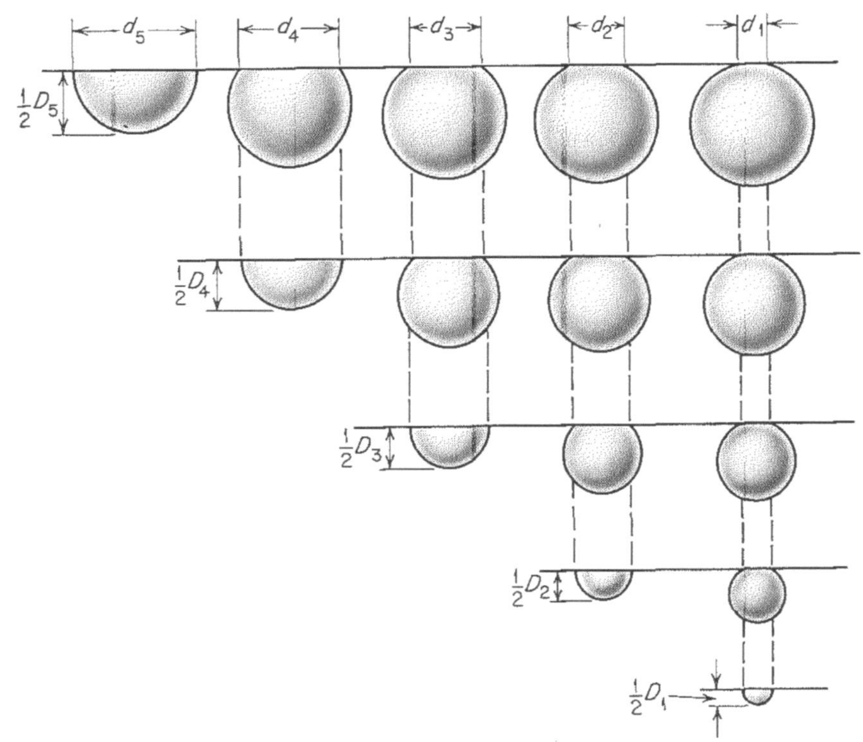

Fig. 5. Contribution of single size particles, i.e., D_5 in different 2-D size groups depending on the geometry of cutting plane [48]

3. The as-cast microstructure

3.1 Grain boundary (GB) particles

Low and higher magnification secondary electron FEG-SEM images of the as-cast microstructure of an AA7020 aluminum alloy variant (N2) are shown in Fig. 6. The constitutive particles elongated along the grain boundaries can be clearly seen. The average width of these grain boundary (GB) particles is 640 nm. The perturbations on the surfaces of the GB particles are illustrated by arrows in Fig. 6 (b).

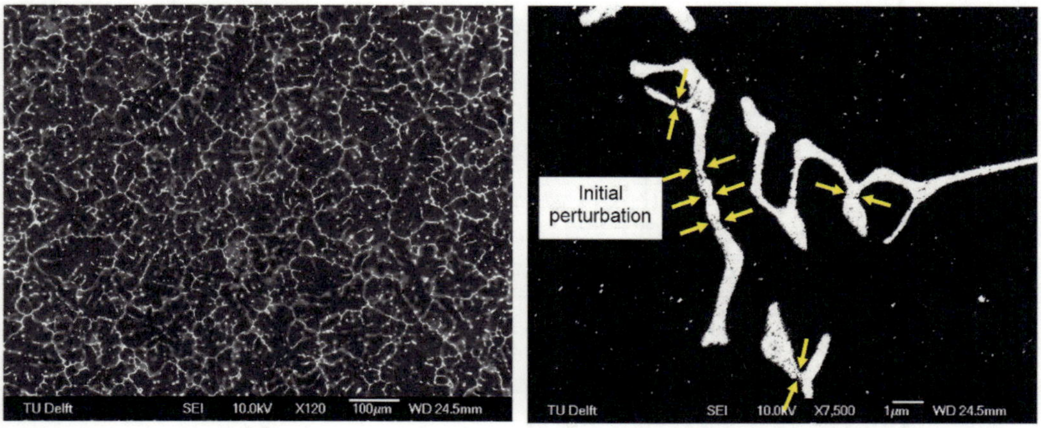

(a) (b)

Fig. 6. (a) Low and (b) higher magnification SEM micrographs showing the GB particles in the as-cast microstructure (alloy variant N2) [49]

To determine the compositions of the GB particles, EDX analysis on more than 20 GB particles having the same morphology was performed. The results showed that the majority of the GB particles had similar compositions, as given in Table 4. By using an image analyzing software together with EDX analysis on a large number of different secondary phases in the as-cast structure, the initial fraction of the GB particles with respect to all of the secondary phases was calculated to be 74±3 wt.%.

With XRD analysis, a phase in a mixture can be identified if its volume fraction is higher than 5% [44]. The results of the image analysis indicated that the volume fraction of the GB particles was close to 7%. Therefore, it was possible to determine the identity of these particles using XRD analysis [44]. The results, shown in Fig. 7, illustrate that only one secondary phase could be detected, which was $Al_{17}(Fe_{3.2},Mn_{0.8})Si_2$ (PDF No. 01-071-4015 [45]). Comparison of the XRD results with the EDX analysis, as given in Table 4, shows a good agreement.

The chemical composition of thermodynamically stable Al-Fe-Mn-Si compounds may be presented by $Al_{16}(Fe,Mn)_4Si_3$ or $Al_{15}(Fe,Mn)_3Si_2$ [4]. The crystallography of the intermetallic phases containing aluminum, silicon, iron and manganese implies that they should be considered as phases with multiple sublattices [50]. Therefore, these compounds may be simply considered as a solution of the Al-Fe-Si particles and Mn or vice versa. In this case, their formation and stability at different conditions obey the thermodynamics of solutions. Since Fe and Mn can reside on the same sublattices [50], the Al-Fe-Mn-Si particles can be considered $Al_{16}(Fe_{(1-y)},Mn_y)_4Si_3$. From the role of the $RT((1-y)\ln(1-y)+y\ln y)$ term in the Gibbs free energy of solutions [51], it can be stated that a compound with equal values of Fe and Mn has the lowest total free energy.

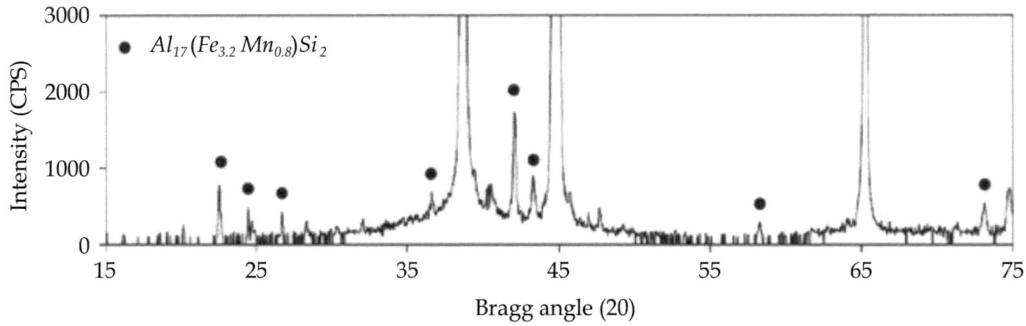

Fig. 7. X-ray diffraction pattern of the as-cast material showing the presence of the GB particles in the alloy variant N2 [49]

Element	Al	Fe	Mg	Si	Zn	Cu
EDX	72.1	16.1	2.8	4.3	2.7	2.0
XRD	62.2	24.2	6.0	7.6		

Table 4. Measured mean compositions (wt. %) of the grain boundary constitutive particles in the as-cast material (alloy variant N2) together with the calculated chemical compositions of the suggested phase identity based on the XRD analysis

In the DC-cast AA7020 aluminum alloy, the amount of Fe is larger than Mn in the grain boundary regions. The larger amount of Fe may be attributed to the partitioning coefficients of Fe and Mn, which result in severer microsegregation of Fe toward the grain boundaries and therefore, a higher concentration of Fe in these regions [52]. Fe has a small solid solubility in aluminum [52]. Therefore, the excess Fe rather than what is consumed in Al-Fe-Mn-Si particles must form other intermetallic compounds. In this case, if y=0.5, in addition to the thermodynamically stable $Al_{16}(Fe_{(1-y)},Mn_y)_4Si_3$ phase, some separate Al-Fe-Si and Al-Fe particles are expected to form to consume the remaining insoluble Fe at the grain boundaries. However, as mentioned above, the solution formation results in a decrease in the Gibbs free energy of the system determined by the $-RT((1-y)\ln(1-y)+y\ln y)$ term. Therefore, in this system, the stable Al-Fe-Mn-Si particles dissolve some of the excess Fe and form the meta-stable $Al_{17}(Fe_{3.2},Mn_{0.8})Si_2$ particles and the remaining Fe incorporates in other intermetallic compounds. The amount of the Fe dissolved in the stoichiometric $Al_{16}(Fe_{(1-y)},Mn_y)_4Si_3$ particles should be so much that the total energy of the system is minimized by the formation of $Al_{17}(Fe_{3.2},Mn_{0.8})Si_2$, Al-Fe-Si and Al-Fe particles. The same may be valid for the replacement of Si atom with exceeding Al in the compound from the stoichiometric values.

3.2 Low melting point (LMP) phases

The presence of the low melting point (LMP) phases in the as-cast structure was determined using DSC. The DSC profile of the as-cast structure is shown in Fig. 8. It is clear that there is an endothermic reaction at 576 °C and the melting of the bulk sample occurs at 661 °C. In order to ensure that the endothermic peak is associated with the melting of the LMP phases rather than their dissolution, two samples were heated at 10 °C higher and lower than the reaction temperature, i.e., 566 and 586 °C, for 1 min. DSC analysis of these samples showed that, after these treatments, the endothermic reaction still occurred, which indicated that the 1 min treatment did not result in the dissolution of the corresponding phases.

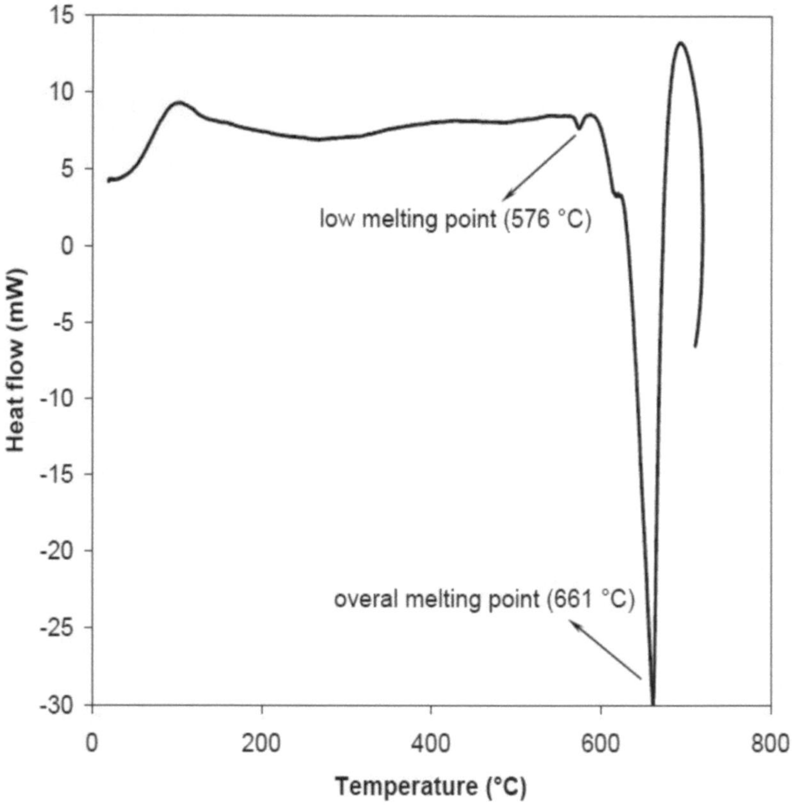

Fig. 8. DCS profile of the as-cast sample (alloy variant N2) at a heating rate of 20 °C/min [53]

The microstructures of the samples were investigated using field emission gun-scanning electron microscope (FEG-SEM). During the analysis, the phases in the as-cast structure, for example, $Al_{17}(Fe_{3.2},Mn_{0.8})Si_2$ and Al-Fe-Si, were detected whose compositions and morphologies were the same as those present in the as-cast structure. The only difference observed in the structures was that for the sample treated at 586 °C for 1 min, the morphology of the Al-Cu-Mg-Zn particles changed from a round shape in the as-cast structure, shown in Fig. 9 (a), to a sponge-like one with perturbations as shown in Fig. 9 (b). These morphological changes must be due to the melting of Al-Cu-Mg-Zn particles during heating up to 586 °C and re-solidification of the particles upon water quenching. This suggests that the endothermic reaction observed in the DSC profile is indeed due to the melting of Al-Cu-Mg-Zn particles. The primary elements present in the LMP phases and their concentrations are shown in Table 5, based on the EDX analysis.

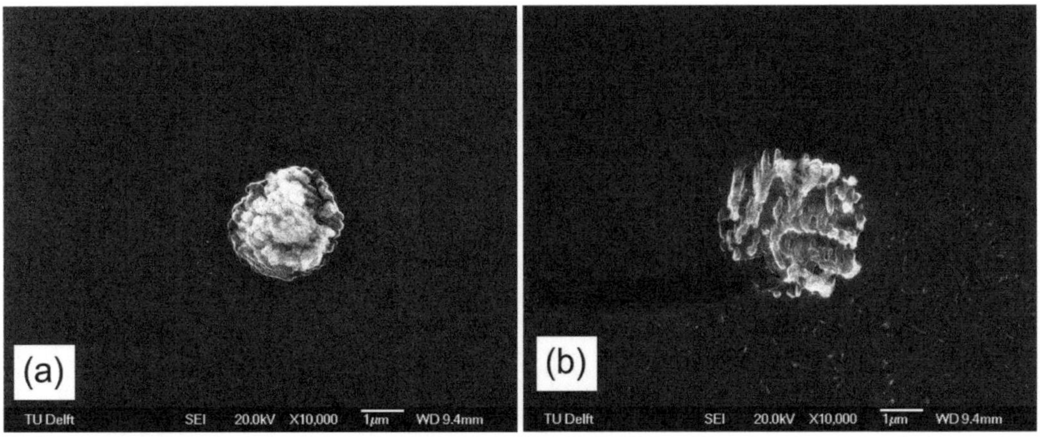

Fig. 9. (a) An Al-Cu-Mg-Zn particle with a round shape in the as-cast structure and (b) a sponge-like Al-Cu-Mg-Zn particle with perturbations in sample N2 after heating to 586°C for 1 min and water quenching [53]

Element	Al	Cu	Mg	Zn
wt%	57±7	17±3	8±2	6±2

Table 5. Measured mean composition of the Al-Cu-Mg-Zn particles in the as-cast alloy variant N2

4. Effects of homogenization

4.1 Microstructural evolution during homogenization

Fig. 10 shows the optical microstructures of the material (alloy variant N2) after 2 h homogenization at different temperatures. Homogenization at 390 and 430 °C led to an increase in the volume fraction of particles. At 470 °C, the volume fraction appeared to be unchanged, while at 510 and 550 °C, it decreased.

Fig. 11 shows low and higher magnification secondary electron FEG-SEM images of the dominant particles formed during homogenization at 390 °C. The grain boundaries are still delineated by the GB particles while the initial continuity of the GB particles shown in Fig. 6(a), is deteriorated by spheroidization. Moreover, large needle-shaped and round precipitates appear in the structure. Examples of these precipitates together with large Al-Fe-Si particles are illustrated in Fig. 11(b). These particles, as pointed at in Fig. 11(a), are dispersed inside the grains. EDX analysis on more than 20 particles with similar morphologies determined the chemical compositions of these precipitates and the results are shown in Table 6.

It was possible to identify these compounds formed during homogenization at low temperatures using XRD analysis. The results given in Fig. 12 (a) show that in addition to the previously present GB particles (Fig. 7), new particle are present in the homogenized microstructure, i.e., $MgZn_2$ (η) and Mg_2Si (β) particles. However, the XRD pattern of the sample homogenized at 550 °C, presented in Fig. 12 (b), shows that no new particles have been formed during homogenization at such a high temperature, which is consistent with the results from the optical microscopy analysis, shown in Fig. 10.

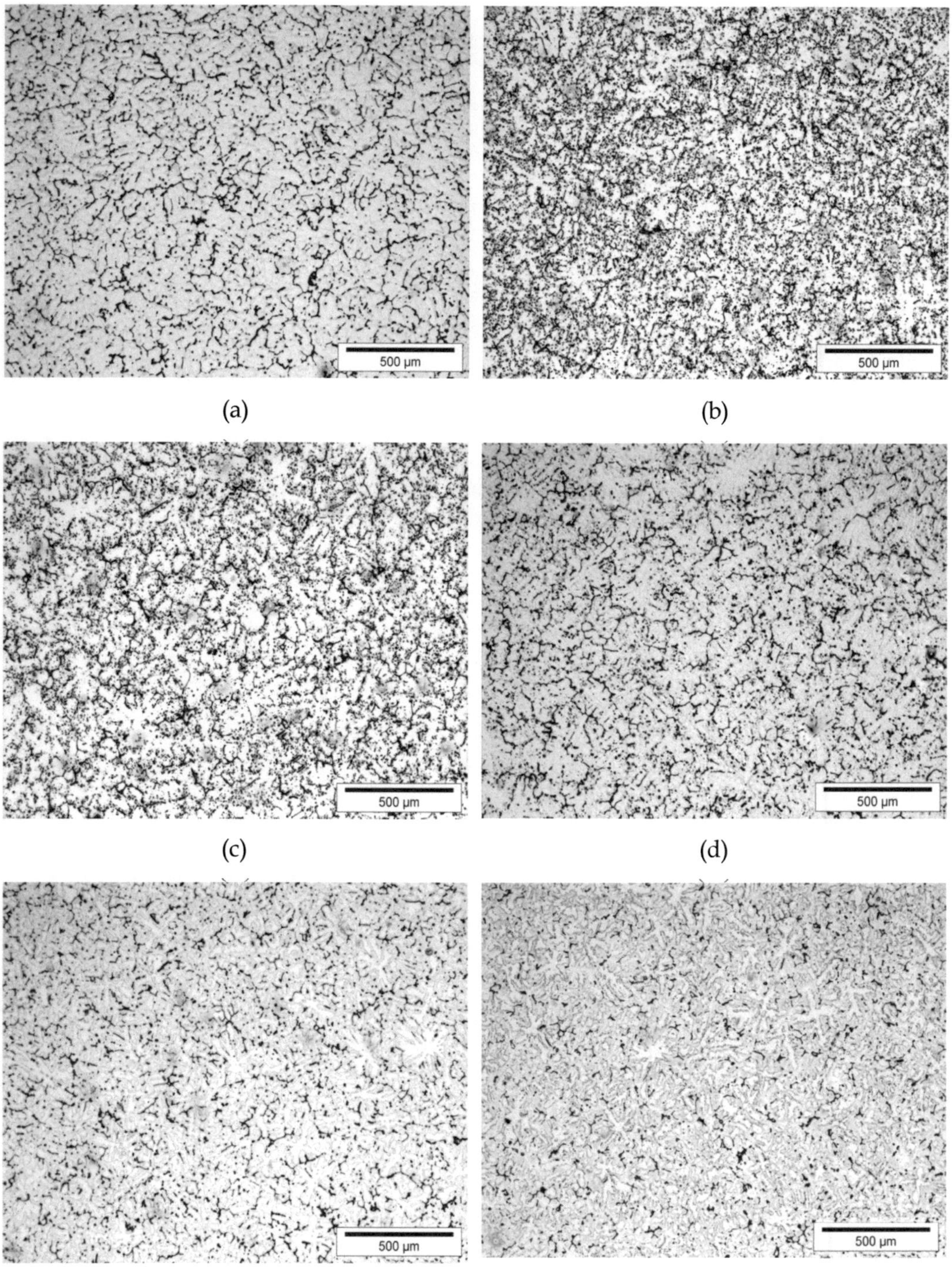

Fig. 10. Effect of the temperature of homogenization for 2 h on the evolution of particles in alloy variant N2, (a) the initial structure, (b) 390, (c) 430, (d) 470, (e) 510 and (f) 550 °C [49]

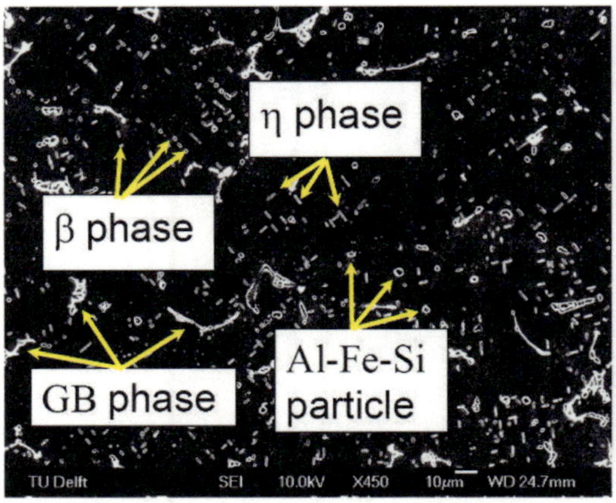

(a)

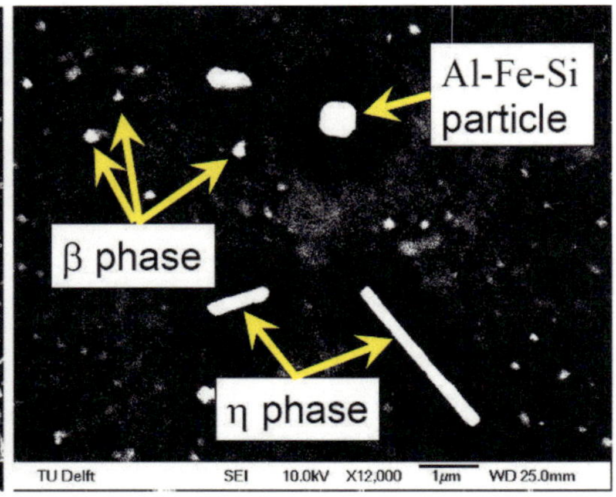

(b)

Fig. 11. (a) Low magnification FEG-SEM image of the alloy variant N2 homogenized at 390 °C, showing the GB particles and (b) the needle-shaped and round $MgZn_2$ (η) and Mg_2Si (β) particles together with large Al-Fe-Si particles [49]

Element (Wt%)	Al	Mg	Zn	Si	Fe
η phase (EDX)	63±4	4±2	28±1	3±1	2±1
η phase (XRD)	...	15.7	84.3	...	...
β phase (EDX)	56±4	22±3	5±2	15±3	2±2
β phase (XRD)	...	63.38	...	36.62	...

Table 6. Measured mean compositions (wt. %) of the needle-shaped and round precipitates in the as-homogenized microstructure of the alloy variant N2 together with the stoichiometric chemical compositions based on the XRD results

It was also found that even after homogenization at a high temperature, i.e., 550 °C, some of the particles were not dissolved in the structure. These retained particles are mostly the GB particles and other particles which together with their EDX spectrums are shown in Fig. 13 (a) and (b). EDX suggested that the particles shown in Fig. 13 (a) and (b) were $Al_{13}Fe_4$ and Al_8Fe_2Si, respectively.

The investigations carried out using the FEG-SEM of the samples homogenized at 390 and 430 °C indicated the presence of needle-shaped and round precipitates, as shown in Fig. 11. The morphologies of these particles and their chemical compositions indicated these particles to be $MgZn_2$ and Mg_2Si precipitates, which is in agreement with [54-56]. The formation of these precipitates may be attributed to the super-saturation of the structure with alloying elements occurring during solidification at high cooling rates applied during DC casting. When the as-cast alloy is exposed to a homogenization treatment at a low temperature (< 470 °C), there is a tendency for the alloying elements to precipitate out. As the temperature increases (> 470 °C), the solubilities of these elements in the α-Al matrix increase [4, 57] and the formation of new particles is not expected. Thus, it can be concluded that the formation of new particles or the dissolution of old ones depend primarily on the homogenization temperature.

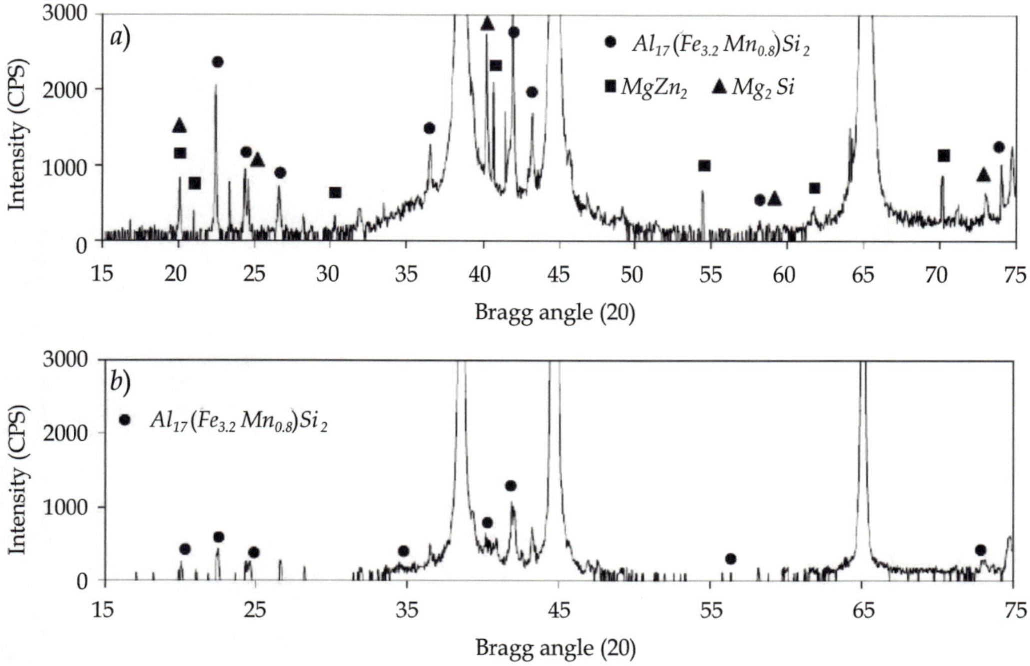

Fig. 12. X-ray diffraction patterns of the alloy variant N2 homogenized at (a) 430 and (b) 550 °C showing the presence of the GB particles, $MgZn_2$ (η) and Mg_2Si (β) particles [49]

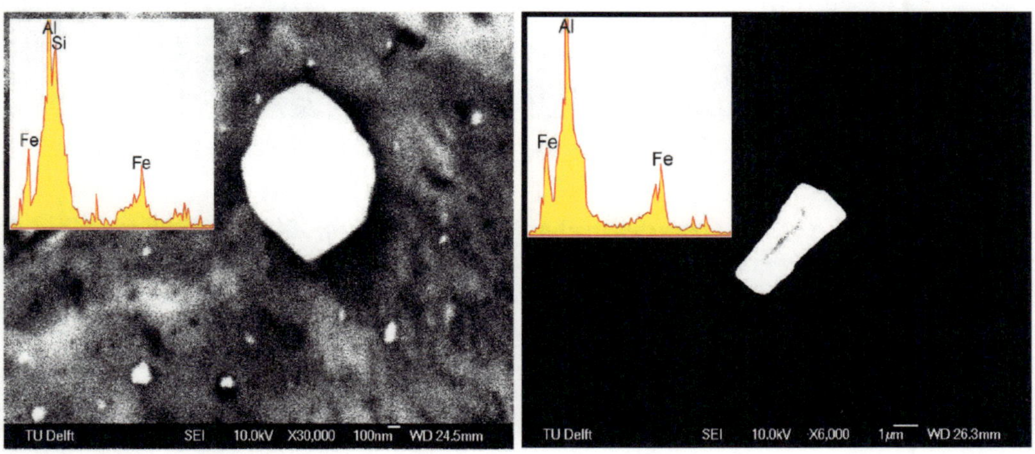

(a) (b)

Fig. 13. Particles remaining in the microstructure of the alloy variant N2 after homogenization at 550 °C for 48 h, (a) Al_8Fe_2Si and (b) $Al_{13}Fe_4$ particle [49]

4.2 Evolution of the GB particles during homogenization

The evolution of a typical GB particle during homogenization at 390 and 550 °C is shown in Figs. 14 and 15, respectively. It is clear that the dominant process at lower temperatures is the spheroidization of the GB particle, while at higher temperatures the decrease in the width of the GB particle is the main evolution process.

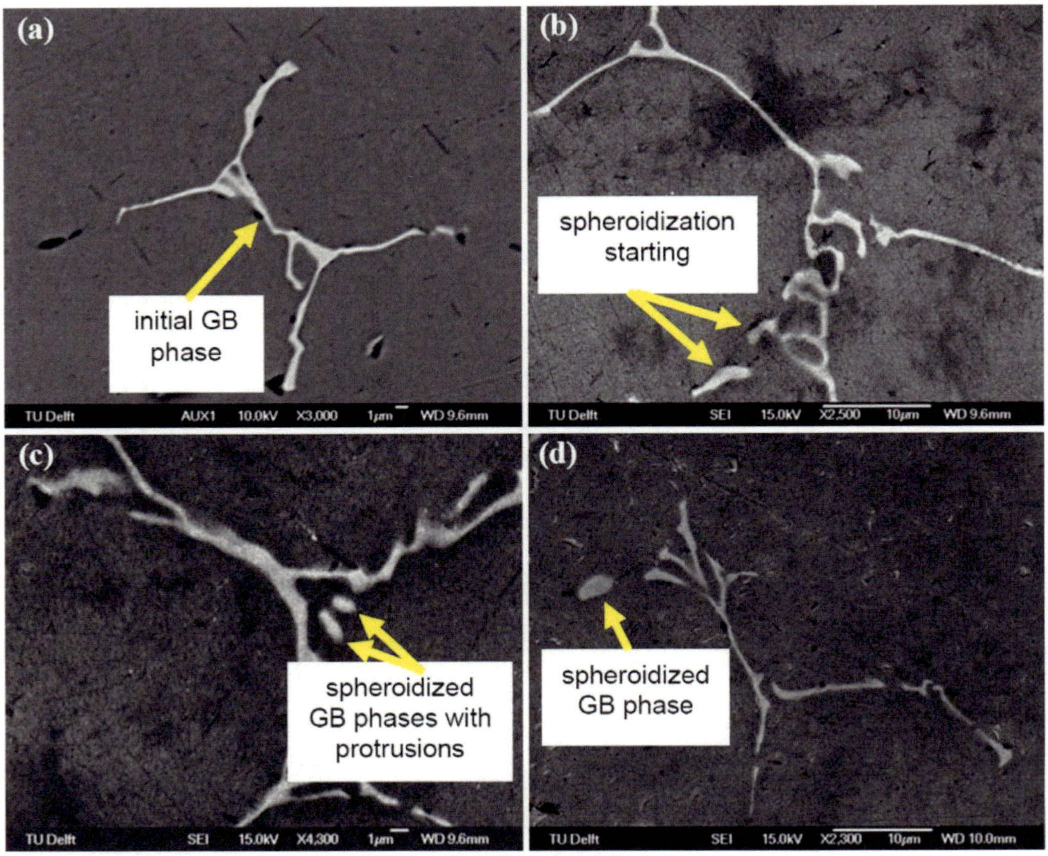

Fig. 14. Typical shapes of the GB particle after homogenization of the alloy variant N2 at 390 °C, (a) initial, (b) 2, (c) 8 and (d) 24 h [49]

Unlike the evolution of other particles in aluminum alloys [2, 14, 15], the evolution of the GB particles during homogenization, depending on the process parameters and the nature of the particles, may occur in the form of spheroidization or dissolution. The spheroidization mechanism of these particles is quite interesting. However, more interestingly, the dissolution of the GB particles obeys a specific dissolution mechanism introduced hereafter as the thinning, discontinuation and full dissolution (TDFD) mechanism.

4.2.1 Spheroidization during homogenization at low temperatures

The analysis of the SEM images indicates that although the fraction of the GB particles does not decrease during homogenization at 390 °C, the morphological changes towards spheroidization take place, as can be seen in Fig. 14. The analysis of 20 pictures from the as-cast structure and the one homogenized at 390 °C for 48 h indicates that after homogenization the fraction of spheroidized particles increases by two times compared with the as-cast structure.

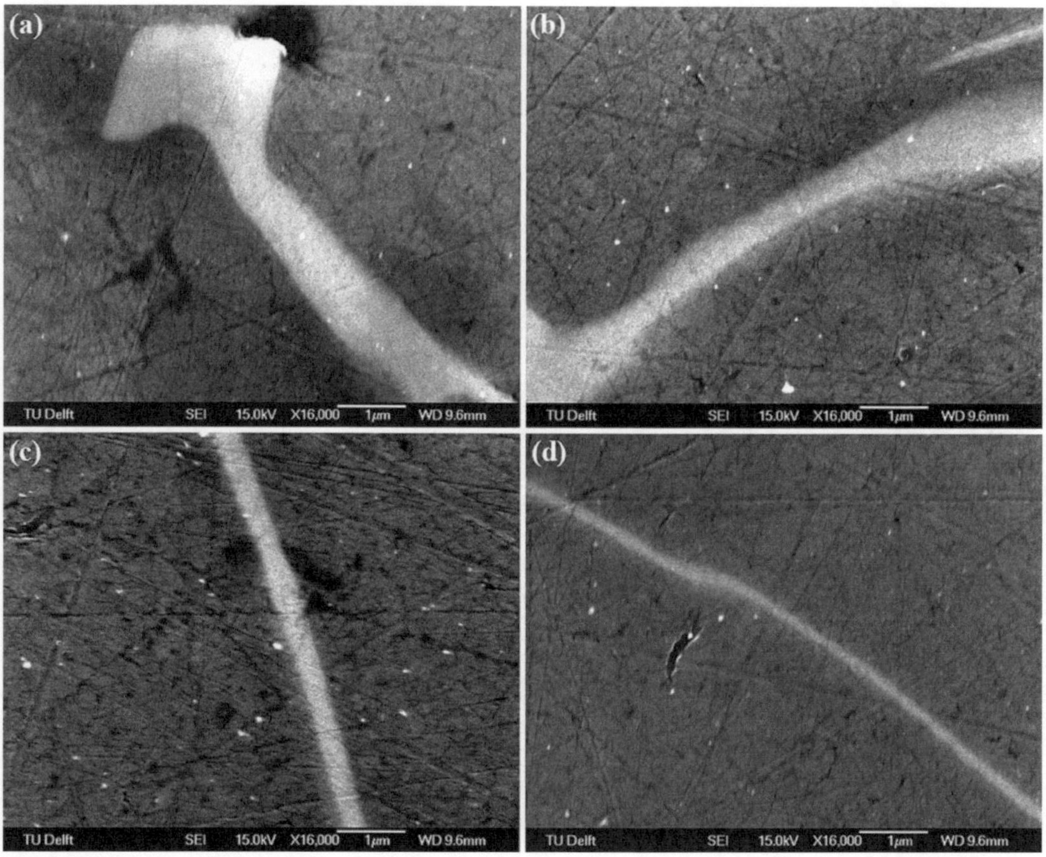

Fig. 15. Decrease in the width of a GB particle after homogenization of the alloy variant N2 at 550 °C, (a) initial, (b) 2, (c) 8 and (d) 24 h [49]

The proposed mechanism of the spheroidization of the GB particles is illustrated in Fig. 16, based on the experimental observations from the FEG-SEM images (a typical one is shown in Fig. 14). Fig. 16(a) shows a GB particle with initial protrusions on its surface. Afterwards, spheroidization occurs and the GB particle takes an ellipse shape, Fig. 16(b). The spheroidization continues till the GB particle takes a spherical shape with protrusions on its surface, Fig 16(c), and the process ends with removing the protrusions till the GB particle resembles a sphere, Fig. 16(d). The driving force for spheroidization is the decrease in the surface energy of the GB particle with decreasing interfacial length between the GB particle and the aluminum matrix [15, 58].

As mentioned earlier, one of the main aims of homogenization treatment prior to hot deformation is to dissolve detrimental particles, especially those located at the grain boundary regions. Although this goal would not be achieved if the particles are not dissolved but spheroidized, spheroidization of particles can be beneficial in the sense that these particles cause less stress concentrations at sharp tips and edges and therefore, inhibit crack initiation which can lead to an improvement in the hot workability of the material.

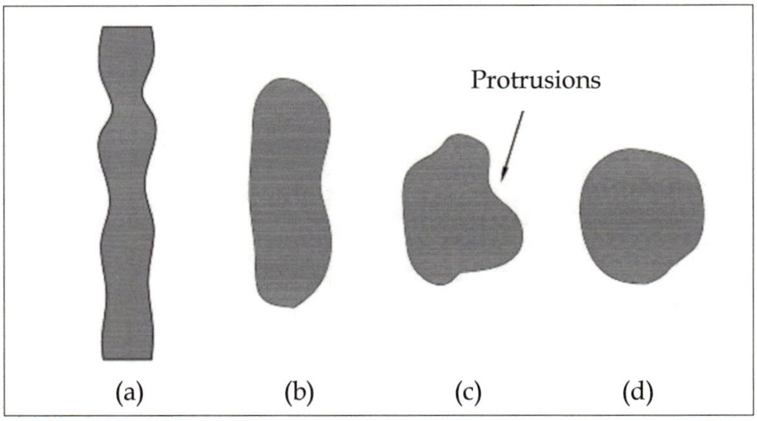

Fig. 16. Schematic view of the spheroidization mechanism describing the evolution of a GB particle during homogenization [49]

4.2.2 Thinning, discontinuation and full dissolution (TDFD) during homogenization at high temperatures

In order to understand the dissolution sequence of the GB particles at high homogenization temperatures, the evolution of a GB particle was investigated at different time intervals during homogenization at 550 °C. It was found that the dissolution process started with the thinning of the GB particle without primary spheroidization. Fig. 15 shows that the average width of the GB particle decreases from 640 nm to 130 nm by a homogenization treatment at 550 °C for 24 h. The thinning process continues until the GB particle become discontinuous in some regions (Fig. 17) and finally the full dissolution of the GB particles occurs. The occurrence of discontinuities during spheroidization of an eutectic particle has been reported elsewhere, for example in [59].

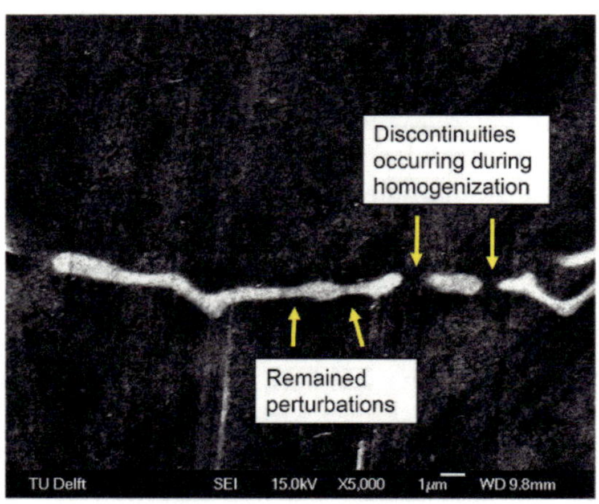

Fig. 17. A GB particle after homogenization at 550 °C for 8 h, showing the thinning and discontinuation (alloy variant N2)

The dissolution mechanism of a GB particle at different stages is schematically illustrated in Fig. 18. The driving force is the increases in the solubility limits in the matrix at high temperatures and therefore the presence of concentration gradients of Mn, Fe and Si in the structure. Fig. 18(a-c) schematically illustrates the overall thinning process of a GB particle. Assuming that during homogenization an overall decrease in the width of a GB particle occurs at a constant rate in different regions regardless of the widths, Fig. 18(a) through (c), the parts having smaller widths meet each other sooner than other parts, as shown by arrows in Fig. 18(d). Therefore, the discontinuities, Fig. 18(e), occur as a result of the inherent perturbations, Fig. 6(b), of the surfaces of the GB particle, shown by arrows in Fig. 18(a). Afterwards, the dissolution continues with the same mechanism as occurring to the small parts till the GB particle disappears. The remaining perturbations which help the continuation of dissolution of the GB particle with a similar mechanism are shown by arrows.

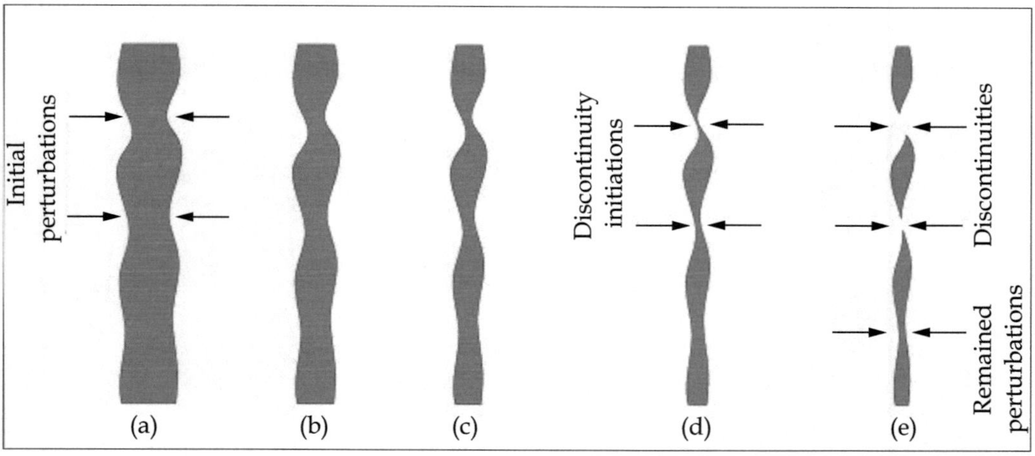

Fig. 18. Schematic description of the thinning, discontinuation and full dissolution (TDFD) mechanism responsible for the evolution of a GB particle during homogenization at high temperatures [49]

The experimental observations of the shapes of the particles, presented in Fig. 19, show that the tips of the particles may initially have rectangular, ellipsoidal, needle or circular cross sections, Fig. 19(a) through (c). However, as shown in Fig. 19(d) through (f), during the dissolution, the tips get a circular cross section. Assume the initial shape of the tips to be rectangular having two steep tips at the edges. According to the Gibbs-Thompson equation [15], a large concentration of the solute at the interface is resulted in, which indeed increases the dissolution rate significantly. The steep edges dissolve sooner and therefore a circular tip will be produced. The same is valid for an ellipsoidal cross section.

5. Quantitative analysis of particle dissolution

5.1 All the particles

5.1.1 Results of quantitative optical microscopy (QOM)

Fig. 20 shows the effect of homogenization time on the volume fraction of all the particles present in the structure. The calculation was based on the changes in the volume fraction of the particles in the structure, as shown in Fig. 10. It is clear that during homogenization at

390 and 430 °C the fraction of particles increases, while during homogenization at 510 and 550 °C the fraction of the particles decreases, indicating dissolution.

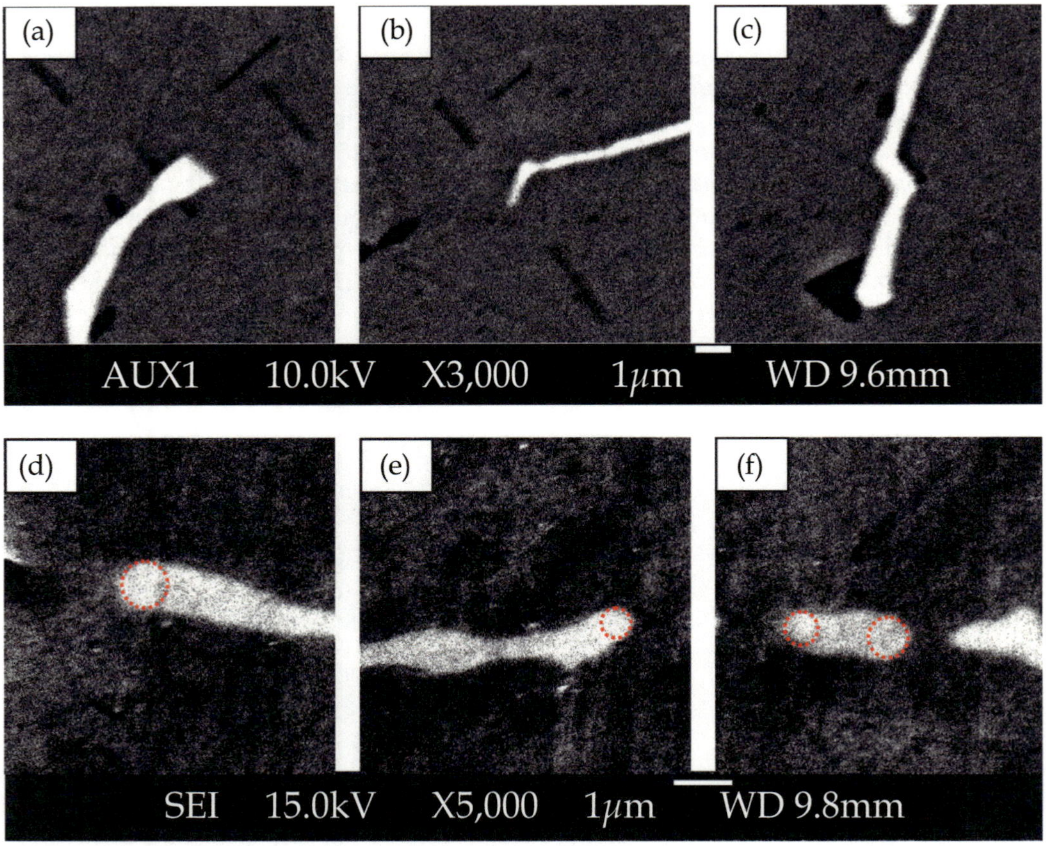

Fig. 19. Shapes of the tips of an $Al_{17}(Fe_{3.2},Mn_{0.8})Si_2$ particle in the alloy variant N2, (a) through (c) in the initial structures and (d) through (f) after homogenization at 550 °C for 8 h. In (d) through (f), circles have been drawn on the tips of the particles, showing the perfectness of the circular cross section of the tips [60]

The most noticeable in Fig. 20 is the increase in the fraction of all the particles at the initial stage of homogenization at low temperatures, which acts in contrary to the aim of the homogenization process [2, 14, 15]. The increases in the fraction of particles during the first 2 h of homogenization at 390 and 430 °C, as shown in Fig. 20, are due to the formation of a large number of new particles, namely $MgZn_2$ (η) and Mg_2Si (β). As stated above, it indicates that part of the elements precipitate out by forming precipitates during the first 2 h of homogenization.

5.2 GB particles

5.2.1 Results of quantitative XRD analysis (QXRD)

Fig. 21 compares the strongest XRD peaks and other ones related to the GB particles in the AA7020 alloy variant N2 samples homogenized at different temperatures for 8 h. This figure was obtained by focusing on the 41 to 44 ° Bragg's angle (2θ) range in the XRD patterns, i.e.,

Figs. 7 and 12. Compared with the as-cast structure, it is obvious that the intensity of the peaks related to the GB particles in the sample homogenized at 390 °C is almost unchanged, while at 430 and 470 °C it is decreased slightly. Homogenization treatment at higher temperatures, namely 510 and 550 °C, however, resulted in marked decreases in the intensity of the peaks related to the GB particles in the XRD pattern in comparison with that of the as-cast structure.

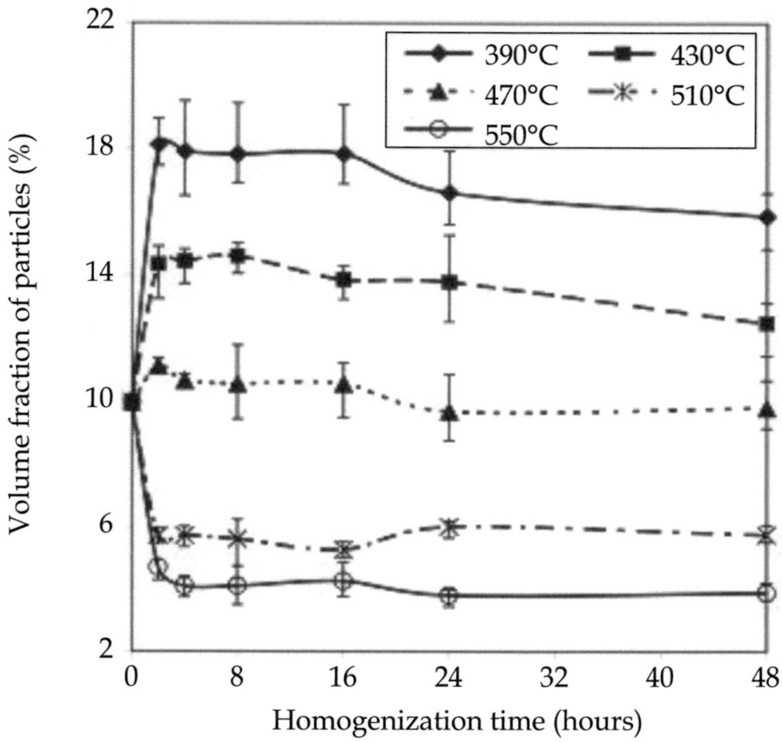

Fig. 20. Effect of homogenization time on the volume fraction of particles (alloy variant N2) [49]

To quantify the results, the fraction of the GB particles in the structure after homogenization was calculated according to the results of the XRD analysis using the direct comparison method [44]. As it was mentioned earlier, the initial fraction of the GB particles with respect to all of the secondary phases in the as-cast structure, necessary to form a baseline for the direct comparison method, was calculated to be 74±3 wt.%. Fig. 22 illustrates the weight percents of the GB particles as a function of homogenization time at different temperatures. It can be seen that the fraction of the GB particles after homogenization at 390 °C remains unchanged and at 430 °C it is decreased slightly, which is not in agreement with the results of the quantitative optical microscopy (QOM) shown in Fig. 20. Therefore, the increase in the fraction of particles during homogenization shown in Fig. 20 is due to the formation of new precipitates (η and β), but not due to the increase in the GB particles. However, at higher temperatures, the fraction of the GB particles decreases, which is in line with the behavior shown in the optical microscopy measurements.

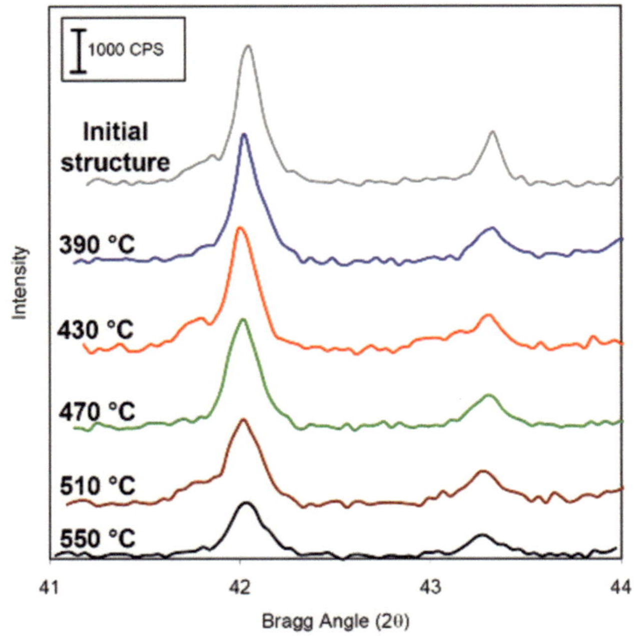

Fig. 21. The strongest (left) and another (right) XRD peaks related to the GB particles in samples homogenized at different temperatures for 8 h (alloy variant N2) [49]

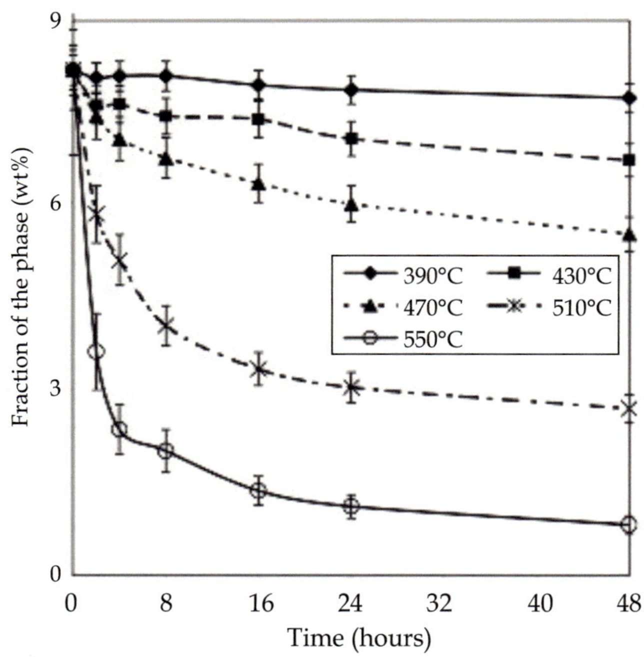

Fig. 22. Fraction of the GB particles (wt.%) in the structure of the alloy variant N2 after homogenization at various temperatures, based on the QXRD analysis [49]

5.2.2 Results of quantitative analysis using FEG-SEM (QSEM)

The measured average widths of the GB particles as a function of homogenization time are shown in Fig. 23. It can be seen that, at 390 and 430 °C, the average width of the GB particles is almost unchanged from the initial value. The minor variations in the width of the GB particles at these temperatures are within the margin of error. However, at higher temperatures, dissolution occurs, evidenced by the decreases in the width of the GB particles. It can also be seen in Fig. 23 that the most rapid dissolution occurs at the first few hours of homogenization regardless of homogenization temperature. Moreover, the dissolution rate is much higher at 550 °C.

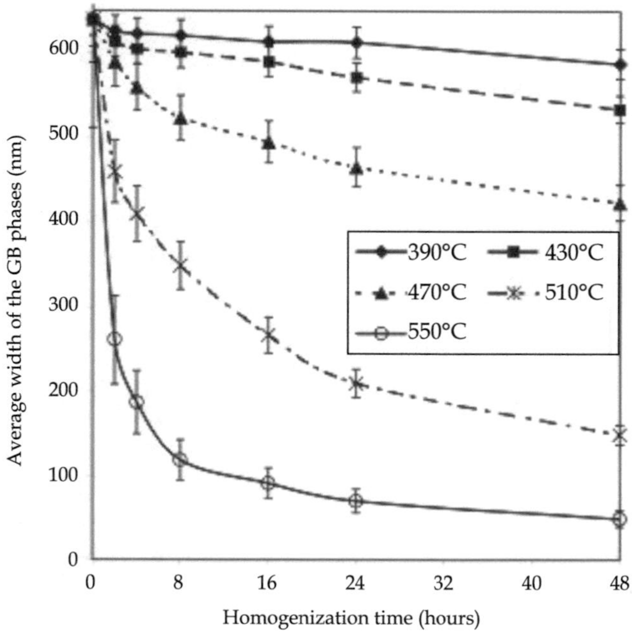

Fig. 23. Average GB widths in the alloy variant N2 as a function of homogenization time, based on the QSEM analysis [49]

To account for the dissolution of the GB particles, a study on the solubility limits of the elements composing the main GB particles, namely Fe, Mn and Si, is essential. The data on the solubility limits of the elements in Al-Mn-Fe-Si are scarce in the literature [4]. In this case, the solubilities of the elements in Al in the four-component Al-Mn-Fe-Si regions adjacent to the ternary systems may be estimated, based on three-component regions of the Al-Fe-Mn, Al-Fe-Si and Al-Mn-Si ternary systems [4]. Table 7 shows the solid solubility limits of iron, manganese and silicon in the four component Al-Fe-Mn-Si system at different temperatures [4]. It is clear that at low temperatures, i.e., lower than 470 °C, the solubility limits of the elements forming the GB particles are small in the α-Al matrix. Therefore, considerable dissolution of the GB particles is not expected at these temperatures. This is in agreement with the results of the quantitative image analysis of the OM images, presented in Figs. 20, 21 and 22. At higher temperatures, however, the solubility limits increase, as shown in Table 7, which indicates that the most of the alloying elements in the GB particles are dissolved in the α-Al matrix. It is also clear that even at such high temperatures (e.g., 550

°C), the solubility limits of the elements in the Al matrix are less than the weight percentages of those elements in the composition of the alloy (N2). Therefore, complete dissolution of all the GB particles is not expected even after holding for a long time. This explains the observations in Fig. 10 (f) and 15 (d) that there are still some GB particles remaining in the structure after homogenization at 550 °C. EDX analysis confirmed the existence of these particles even after homogenization at 550 °C for 48 h.

	Solubility limit Wt%		
T (°C)	**Fe**	**Mn**	**Si**
390	0.002	0.026	0.03
430	0.004	0.05	0.06
470	0.009	0.15	0.08
510	0.016	0.25	0.11
550	0.044	0.44	0.2

Table 7. Solid solubility limits of iron, manganese and silicon in the four component Al-Fe-Mn-Si system at different temperatures [4]

5.2.3 Comparison between the QXRD and QSEM results

The benefit of using QXRD analysis is that the results obtained are primarily related to the GB particles while the QOM analysis includes the GB particles, the later formed η and β precipitates and other particles. In addition, QOM gives useful quantitative information on the quantities of the particles formed during homogenization. In Fig. 20, as discussed earlier, the increase in the volume fraction of particles during homogenization at low temperatures, is due to the formation of new precipitates (η and β) rather than the increase in the GB particles. On the other hand, the volume fraction of the GB particles is almost unchanged, as shown in Fig. 22. Therefore, the combination of the results from QOM and QXRD quantitative analyses is essential to investigate the evolution of the microstructure during homogenization treatment.

The slight decrease in the fraction of the GB particles during homogenization at low temperatures means that the kinetics of the dissolution of these particles at low temperatures (< 430°C) is relatively slow. Since the fraction of the GB particles decreases significantly during homogenization at 510 °C and higher, it is concluded that in order to dissolve the GB particles, applying a homogenization treatment at 510 °C or higher is necessary.

In the case of QSEM analysis, an unchanged width means that no dissolution has occurred and any decrease in the width of the GB particles is the result of dissolution. The general trend of the change in the width of the GB particles during homogenization using QSEM analysis (Fig. 23) agrees with that of the fractions of the particles from QXRD analysis.

5.3 LMP phases

5.3.1 Results of quantitative analysis using FEG-SEM (QSEM)

Fig. 24 shows the radii of the particles obtained from FEG-SEM after homogenization at various conditions. For these experimental data, EDX analysis was first performed on the particles which morphologically resembled the LMP phases. The radii of more than 10 particles in each homogenization condition, after ensuring that these were the LMP phases, were measured and the average is reported. The particle radius of zero in this figure indicates that no particles with the composition of the LMP phases were found in the structure, having the same methodology for the prediction of these particles with FEG-SEM at the same magnification. Therefore, the radius of the particles in this case was considered to be zero. It can be observed that the average radius of the LMP phase particles decreases gradually with time during homogenization at 430 and 470 °C.

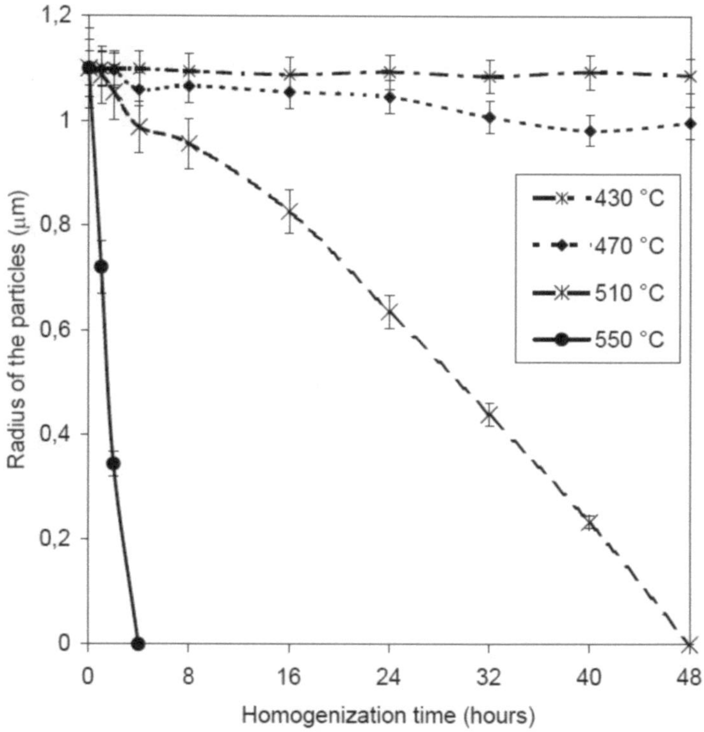

Fig. 24. Average radii of LMP phase at different homogenization conditions in the alloy variant N2 [53]

5.3.2 Results of quantitative differential scanning calorimetry (QDSC)

Fig. 25 shows that after homogenization at a moderate temperature, e.g., 470 °C, the LMP phases are still present in the structure, even after 48 h. At 510 °C, however, the LMP phases are fully dissolved within 48 h and at 550 °C within 2 h, as shown in Fig. 25. This is due to the large increases in the diffusion rates of the elements (i.e., Mg, Cu and Zn) at a high homogenization temperature. Therefore, it is necessary to employ a homogenization treatment at 510 °C for 48 h or 550 °C for 2 h to dissolve the LMP phases.

Fig. 26 gives a close-up view of the peaks in the DSC profiles of the samples homogenized at 470 °C for different hold times. It is clear that the peak intensity decreases with increasing homogenization time up to 48 h, which indicates the decreasing volume fraction of the LMP phases during homogenization at 470 °C.

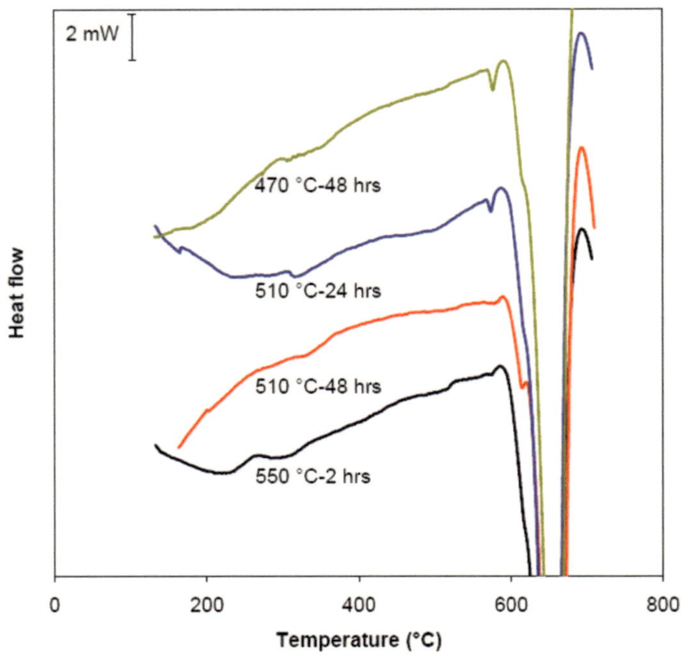

Fig. 25. DCS profiles of the samples homogenized at 470, 510 and 550 °C for different times [53]

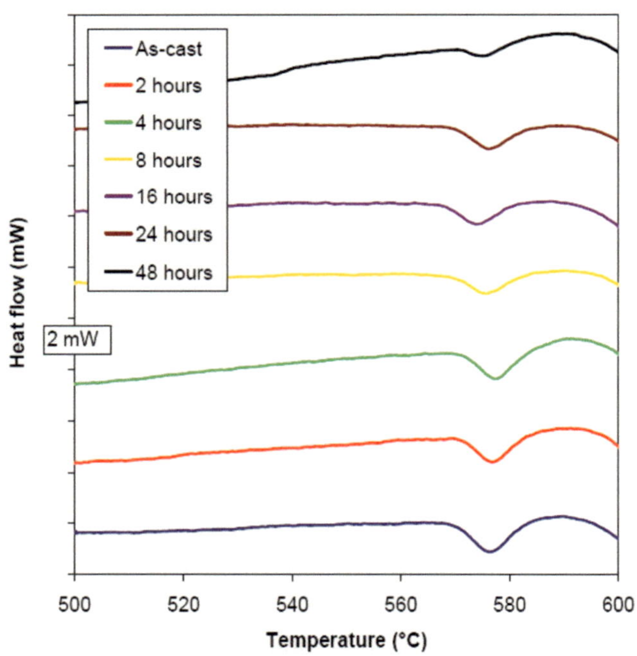

Fig. 26. Regions close to the peaks of the DSC profiles of the samples homogenized at 470 °C for different times (alloy variant N2) [53]

Fig. 27 shows the calculated volume fractions of the LMP phases at different homogenization conditions from the intensities of the DSC peaks, using Eqs. (4) and (5). It is clear that the volume fraction of the dissolved LMP phases increases with time during homogenization. Moreover, the volume fraction of the dissolved LMP phases is larger at a higher temperature and full dissolution only occurs at 510 or 550 °C at reasonable times (less than 48 h).

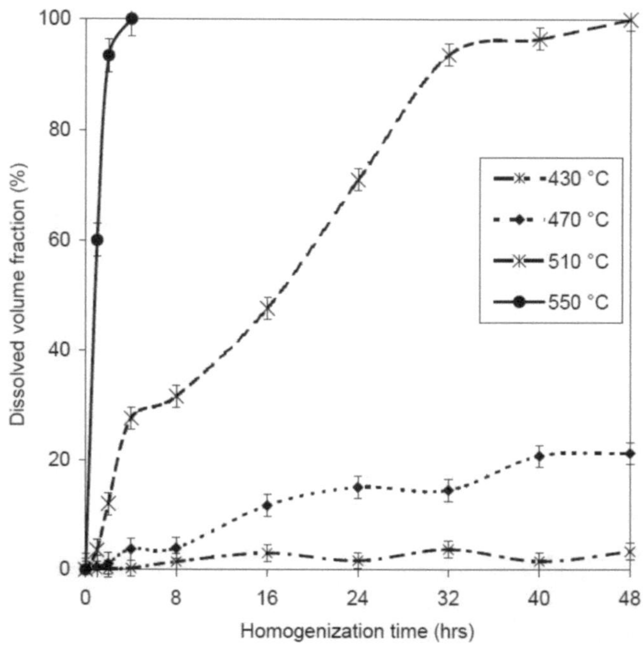

Fig. 27. Volume fractions of the dissolved LMP phases at different homogenization conditions (alloy variant N2) [53]

6. Formation of dispersoids

6.1 Detection of different dispersoid types

Table 8 shows the results of EDX analysis on more than 450 dispersed particles in a size range of less than 100 nm, with different shapes and at different locations with respect to the grains, i.e., grain interior and grain boundary regions in the alloy variant N3 (Table 2) homogenized under various conditions. It was possible to differentiate between 4 types of dispersoids, Zr- (Type 1), Cr- (Type 2), Mn-containing (Type 3) dispersoids and the ones containing a mixture of various elements (Type 4). The number fraction of each dispersoid type, out of 450 dispersoids counted, is also presented in Table 8. It can be seen that 62% of the dispersoids are Zr-containing ones, which indeed account for the majority of the dispersed particles. The number fractions of the other types (2, 3 and 4) are 23, 14 and 1%, respectively. Also shown in Table 8, at the grain boundary regions, i.e., 5 μm from both sides of the grain boundary particles ($Al_{17}(Fe_{3.2},Mn_{0.8})Si_2$ particles), the number fraction of the Mn-containing dispersoids (Type 3) reaches 93 %. However, this type of dispersoids was only observed after homogenization at 510 °C and higher, and especially for a holding time of 4 h or longer. It can be seen that the number fraction of Zr- and Cr-containing dispersoids at the grain boundary regions are only 2 and 1%, respectively.

Dispersoid type	Mg	Fe	Zn	Zr	Cr	Mn	Al	Number fraction (%) out of total 450 dispersoids analyzed	Number fraction (%) out of 75 dispersoids anlyzed at the GB at T≥510°C, time≥4 hrs
Type 1 (Zr-containing)	0.08	0.67	0.1	11.9	0.3	0.09	86.86	62	2
Type 2 (Cr-containing)	0.23	11	0.82	0.6	34.1	0.0	53.22	23	1
Type 3 (Mn-containing)	0.09	7.3	0.26	0.2	0.02	12.8	79.33	14	93
Type 4	4.59	3.2	10.24	0.07	0.12	0.07	81.71	1	4

Table 8. Chemical compositions and number fractions of different types of dispersoids detected in the microstructure of the alloy variant N3

Fig. 28 (a) gives a close-up view of the different types of disperoids. It is clear that the semi-spherical Zr-containing ones (Type 1) are the smallest, while the Cr-containing ones (Type 2) which are fully spherical are the largest. The small sizes of the Zr-containing dispersoids may be attributed to the very low diffusion rate of this element in the aluminum matrix and high nucleation rate [61]. Type 3, also semi-spherical in shape, is not present in the grain interior, thus not shown in Fig. 28 (a). It is presented in Fig. 29. Type 4 which has an elliptic or rod shape morphology is also shown in Fig. 28 (a). To assure the trustfulness of the particles determined with FEG-SEM, TEM analysis was also performed on the same sample and the results are presented in Fig. 28 (b). It can be seen that the same three different types of dispersoids are present in the TEM image with an approximately the same size range.

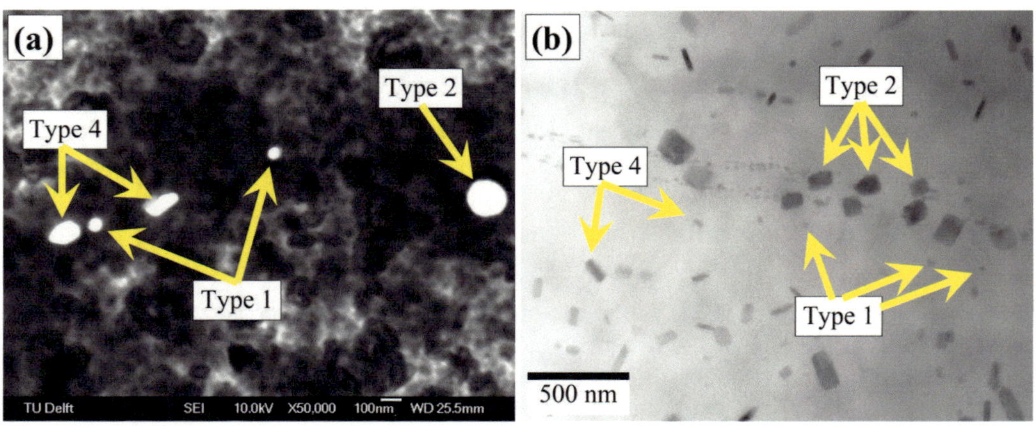

Fig. 28. (a) Close-up view of the different types of dispersoids in a sample homogenized at 510 °C for 8 h and (b) TEM image of the dispersoids in the same sample (alloy variant N3) [62]

Fig. 29 shows the distributions of the different types of dispersoids after homogenization. Fig. 29 (a) and (b) which illustrates the distributions of Zr- and Cr-containing dispersoids (Type 1 and 2) were captured in the grain interior. The number densities of these types decrease significantly with decreasing distance toward the grain boundaries. Fig. 29 (c) illustrates the distribution of Mn-containing dispersoids in the vicinity of an $Al_{17}(Fe_{3.2},Mn_{0.8})Si_2$ particle. It is clear that most of the particles in this region are Mn-containing dispersoids. Their number density inside the grain interior is however almost zero. During the analysis, it was not possible to capture a region with a sufficiently large number of Type 4 dispersoids. Therefore, no figure showing the distribution of this type can be presented in Fig. 29.

Quantitative Zr, Cr and Mn measurements from a line-scan across a grain in the as-cast microstructure of the AA7020 alloy are shown in Fig. 30. It can be seen that the concentrations of Zr and Cr in the grain interior are higher than what would be expected from the peritectic Al-Zr and Al-Cr phase diagrams [65, 66]. The lowest concentrations of Zr and Cr are found at the grain boundaries, which is in agreement with the finding of other researchers [15, 20, 25, 39]. The fluctuations in the Zr concentration across the analyzed region reflect the underlying dendritic structure within each grain. The measurements made close to the centre of the dendrite arms show that the Zr and Cr concentrations exceed their nominal values of 0.2 and 0.1 wt.% in the alloy variant N3, respectively. These regions solidified first during DC-casting and were thus enriched in Zr and Cr. However, a large fraction of grains contained Cr below its nominal value. In particular, low Cr levels were

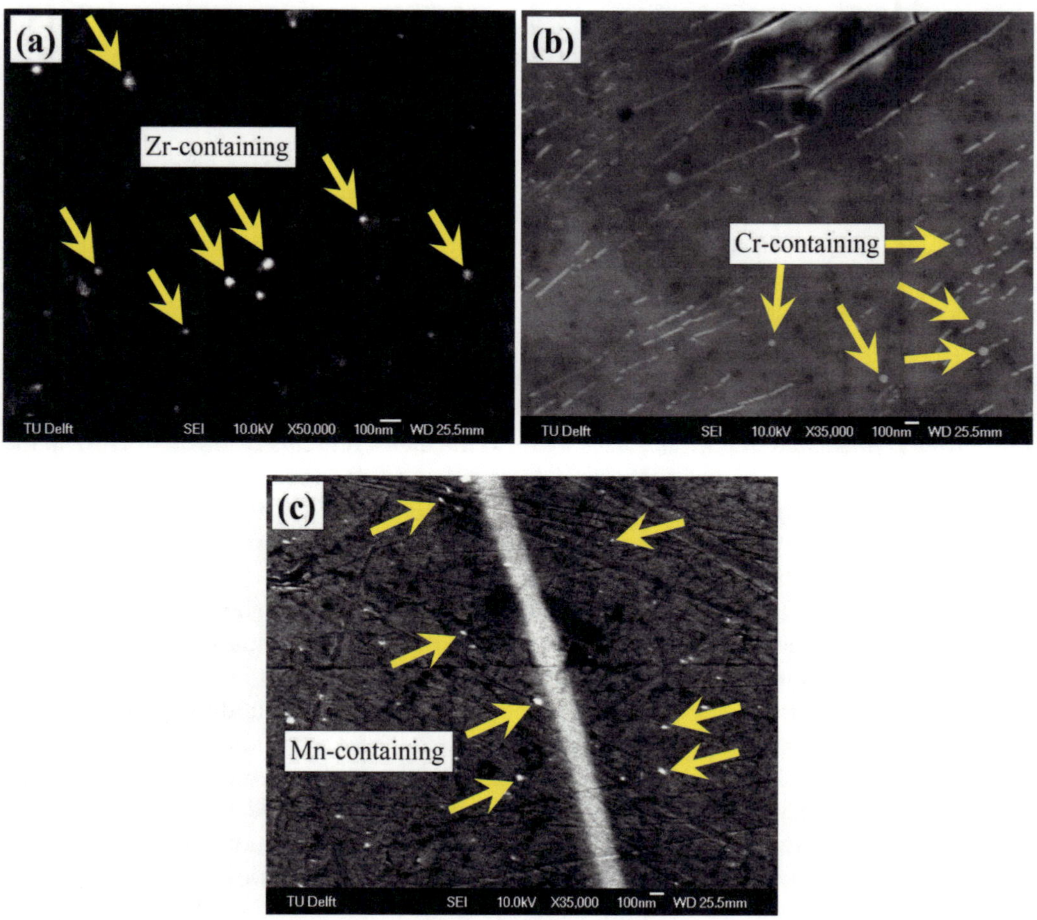

Fig. 29. Distributions of the different types of dispersoids after homogenization at 510 °C for 8 h, (a) Type 1 (Zr-containing), (b) Type 2 (Cr- containing) and (c) Type 3 (Mn-containing) (alloy variant N3) [62-64]

found near the grain boundaries and interdendritic regions. Therefore, it can be concluded that the reason for the small numbers of Cr-containing dispersoids in the grain boundary regions is the negligible concentrations of this element in these regions due to microsegregation during DC-casting. However, the fluctuations in the Zr level are not significant. The lowest Zr level is close to 0.13 wt.%.

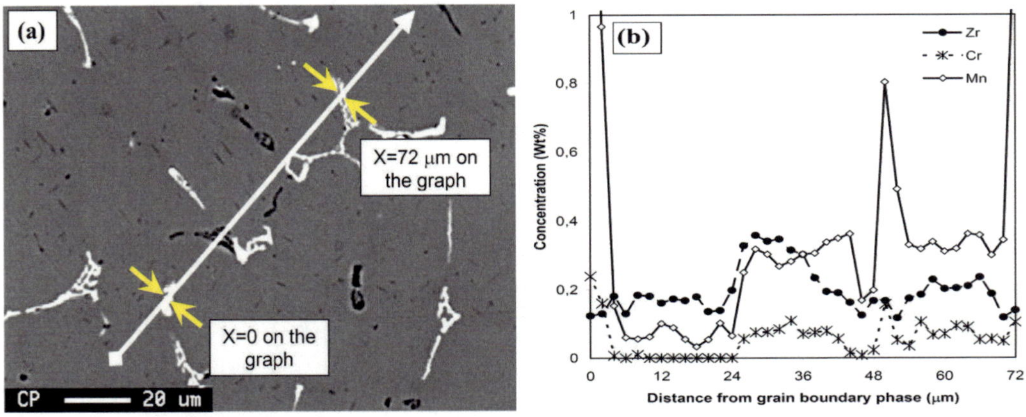

Fig. 30. (a) An EPMA scan and (b) analysis showing segregated Zr, Cr, Mn and Al at the cell boundaries of the as-cast microstructure in the AA7020 aluminum alloy variant N3 [62-64]

As indicated earlier, most of the dispersoids in the grain boundary regions are Mn-containing dispersoids. They are almost absent in the grain interior. The results of the EPMA analysis confirmed that the Mn concentration in the as-cast microstructure in the grain interior is less than the solid solubility limit of Mn at the mentioned homogenization temperatures [48]. Therefore, the reason for no Mn-containing dispersoids formed in the grain interior is a very small concentration of Mn in this region. The peak in the Mn concentration presented in Fig. 30, corresponds to the measurements made on or close to the intermetallic $Al_{17}(Fe_{3.2},Mn_{0.8})Si_2$ particles (GB particles) at the grain boundaries. The small concentrations of Mn in the grain interior and the peaks in Fig. 30 indicate that, during solidification, most of the Mn element localized in the grain boundary regions formed the $Al_{17}(Fe_{3.2},Mn_{0.8})Si_2$ particles located at the grain boundaries.

As mentioned earlier, the Mn-containing dispersoids form only at high temperatures (> 510 °C) and holding times longer than 4 h and in the grain boundary regions. During homogenization at high temperatures, $Al_{17}(Fe_{3.2},Mn_{0.8})Si_2$ particles may dissolve in the microstructure, thus increasing the Mn concentration in the grain boundary regions. Fig. 31 shows the EPMA measurements of the Mn concentration from a line scan across a grain in a sample homogenized at 550 °C for 8 h. It is clear that the Mn concentration in the regions close to the $Al_{17}(Fe_{3.2},Mn_{0.8})Si_2$ particles increases, which is attributed to the dissolution of $Al_{17}(Fe_{3.2},Mn_{0.8})Si_2$ particles during homogenization under this condition. This results in the formation of Mn-containing dispersoids close to the grain boundaries. However, the Mn concentration in the grain interior only changes slightly as confirmed by the present EPMA analysis (Fig. 31), mainly because of the low diffusion rate of Mn in the aluminum matrix [67]. As a result, the Mn concentration is too small in the grain interior to form the Mn-containing dispersoids.

From Fig. 29, it is clear that the Cr-containing dispersoids (Type 2) have a localized distribution in the grain interior. The localized distribution of the Cr-containing dispersoids together with the fact that they are larger than the other types of particles may indicate that this type of dispersoids form heterogeneously, which is in agreement with other investigations [34]. The heterogeneous nucleation of the Cr-containing dispersoids on the u-phases in the case of Al-Mg-Si alloys has been documented [34]. It was observed that during the heating of the as-cast Al-Mg-Si alloys to 580 °C an intermediate phase, referred to as the 'u-phase' nucleated on the Mg_2Si needles. The phase was rich in Mn or Cr. Upon continued heating, the dispersoids containing Mn and Cr nucleated heterogeneously on the 'u-phase' precipitates before these precipitates were dissolved.

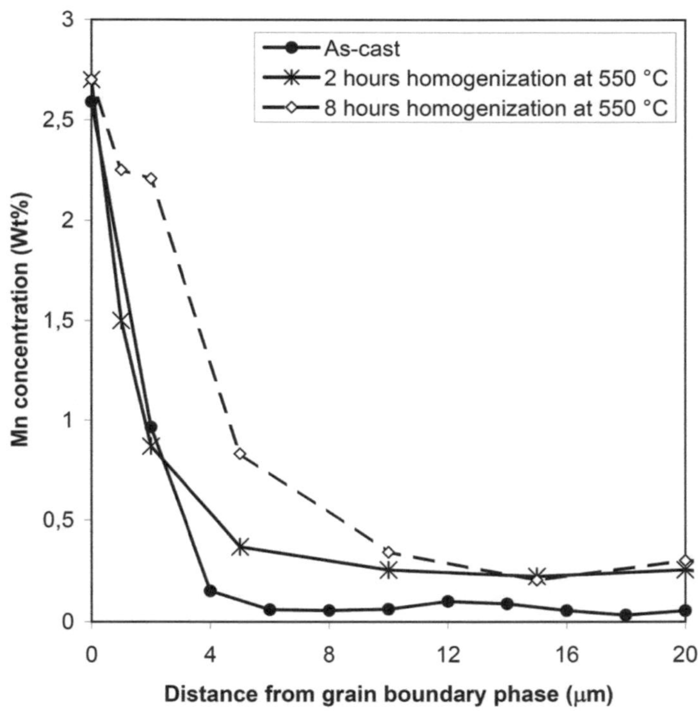

Fig. 31. EPMA scan analysis showing the Mn concentration across a grain of the AA7020 aluminum alloy (alloy variant N3) homogenized at 510 °C for 8 h [62, 64]

To make a solid conclusion on various dispersoids formed during homogenization, they were characterized with respect to formation temperature, size, location and morphology. Table 9 shows the characteristics of the four types of dispersoids formed in the homogenized AA7020 alloy. The deviation expresses the ratio of the number of the particles (with the mentioned characteristics) that showed different chemical compositions to the total number of the particles (with the mentioned characteristic) that were analyzed. The Zr-, Cr- and Mn-containing dispersoids have deviations of only 3, 5 and 4.5 %, respectively. It indicates that the characteristics presented in Table 9 are reasonably accurate to differentiate between these types of dispersoids. It should be noted that the deviation of Type 4 is not reported since the dispersed particles of this type are rare in the microstructure.

Dispersoid type	Major elements	Formation temperature (°C)	Location	Radius (nm)	Morphology	Distribution	Deviation (%) *
Type 1	Zr	390 – 550	Center of the grains	<25	Semi-spherical	Uniform	3
Type 2	Cr, Fe	390 – 550	Center of the grains	25-100	Fully spherical	Localized	5
Type 3	Mn, Fe	T ≥ 510 °C	Close to the GB phases	<50	Semi-spherical	Uniform	4.5
Type 4	Mg, Zn, Fe	390 – 550	All	25-100	Elliptic	Uniform	---

* The deviation expresses the ratio of the number of the particles with the mentioned characteristics that showed different chemical compositions to the total number of analysed particles with the same characteristics.

Table 9. Characteristics of different types of dispersoids in AA7020 after 4 h homogenization at different temperatures

6.2 Evolution of Zr-containing dispersoids during homogenization

As mentioned in Table 8, Zr-containing dispersoids constitute about 62 % of all dispersoids present in the microstructure of the alloy (N3) after homogenization. Therefore, they can be considered the most important ones for recrystallization inhibition. In addition, due to their higher number density, it is easier to quantify them and their evolution during homogenization. Fig. 32 presents typical FEG-SEM images of the Zr-containing dispersoids and related size distribution graphs in the central region of a grain. The size distribution graphs were obtained using the Johnson-Saltykov method as mentioned in the experimental procedure [48] and therefore, the x axis has logarithmic size distribution categories, which is finer at lower values and vice versa.

It should be mentioned that in order to evaluate the efficiency of a homogenization treatment on the inhibition of recrystallization, all relevant parameters including size, size distribution and volume fraction of particles should be taken into consideration, which have been incorporated into an equation of Zener drag pressure [30]. This equation has recently been developed to include the effect of size distribution of dispersoids [3, 68]. The intention of preparing this chapter is to investigate the microstructural evolution during homogenization and therefore, discussion on the effect of homogenization treatment on recrystallization inhibition has been avoided. For more details, the reader is referred to other references [3, 63, 64, 68, 69].

The number density of the Zr-containing dispersoids decreases significantly with increasing distance towards the grain boundaries. This can be attributed to the segregation of zirconium during solidification. It is clear that the sizes and number densities of the Zr-containing dispersoids change with homogenization condition.

Comparison between Figs. 32 (a) and (b) shows that at a given temperature, the sizes increase slightly with time while the number densities remain almost unchanged. In addition, the number of particles of larger sizes, i.e., larger than 15-20 nm, increased with increasing homogenization time at a certain temperature. Comparison between Figs. 32 (b) and (c) for a given composition of the alloy variant N3 and different homogenization temperatures demonstrates that at a higher homogenization temperature e.g., 550 °C, Fig. 32 (c), a larger fraction of particles are of large sizes and the number densities are relatively small, which may be considered as a sign of coarsening. In addition, paying attention to the effect of chemical composition in Fig. 32 (b) (for alloy variant N3) and Fig. 32 (d) (for alloy variant N1), one can find reasonably similar size distributions with different number densities. The sizes are almost constant while the number densities increase significantly with increasing Zr content in the alloy.

Fig. 33 (a) shows the effect of homogenization time on the diameters of the Zr-containing dispersoids at different homogenization temperatures. It is clear that at each homogenization temperature, the average dispersoid diameter increases with increasing holding time and then tends to reach a constant value. In addition, the average diameter is also a function of homogenization temperature. At a higher temperature, the dispersoid diameter is larger.

The effect of homogenization time on the number density of the dispersoids at different homogenization temperatures is illustrated in Fig. 33 (b). It can be seen that with increasing holding time, the number density increases and then stays at a certain level. In addition, the number density of the dispersoids formed at 470 °C is significantly larger than that formed at 390 or 550 °C, while the difference between 390 and 550 °C in the dispersoid number density is considerably smaller.

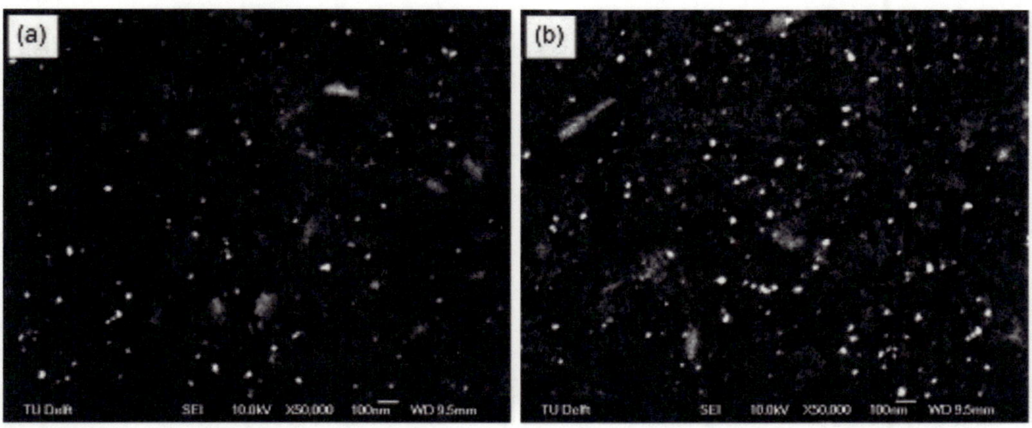

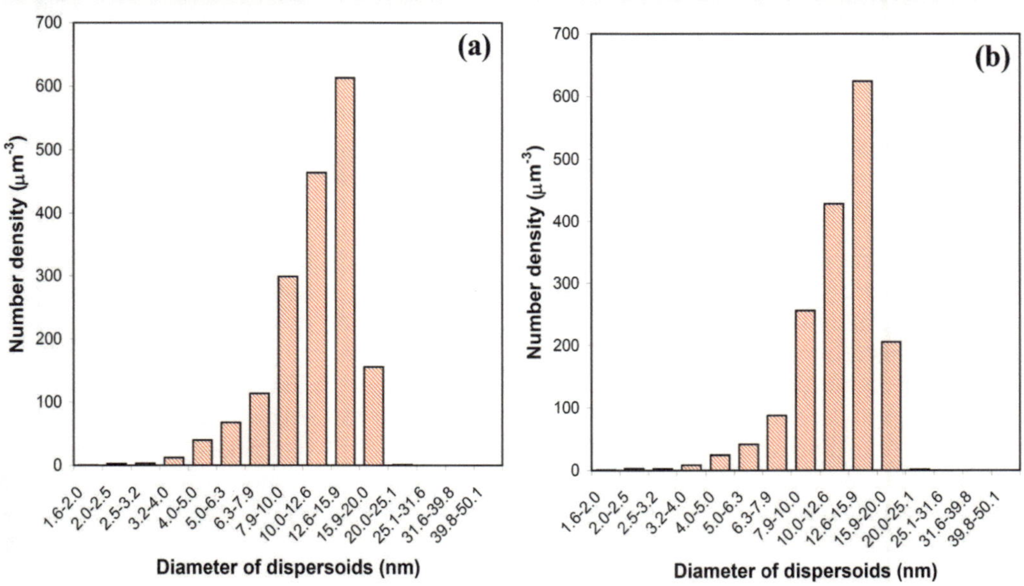

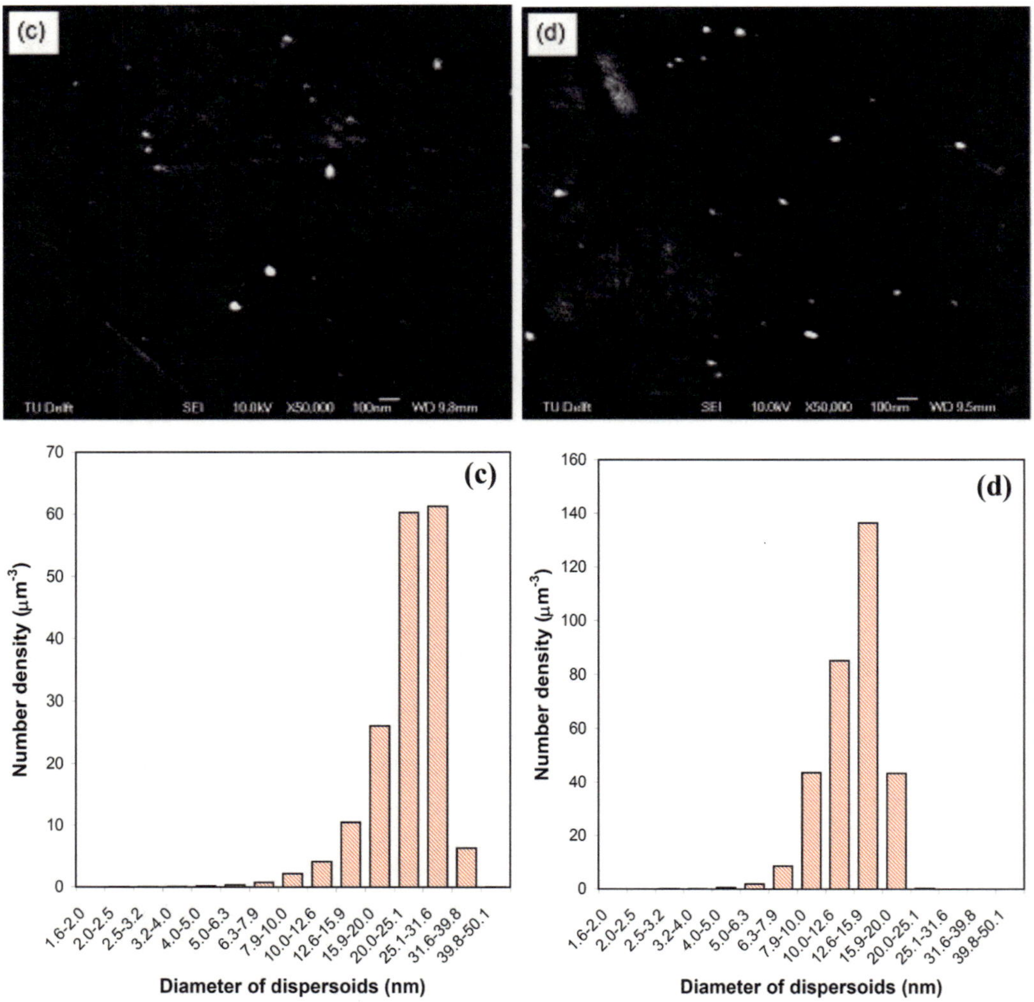

Fig. 32. Typical SEM micrographs showing the effects of homogenization parameters and Zr content on the sizes and size distributions of the Zr-containing dispersoids (a) alloy variant N3, T=470 °C for 8 h, (b) N3, T= 470 °C for 24 h, (c) alloy variant N3, T=550 °C for 24 h and (d) alloy variant N1, T= 470 °C for 24 h [63, 69]

Fig. 34 (a) shows the effect of Zr content on the average diameter of dispersoids formed at 470 °C as a function of time. It can be seen that the average diameter of the dispersoids formed in the alloy with the highest Zr content are larger than those in the other two alloys. However, this effect is not very strong, as the average dispersoid diameters in the N1 and N2 variants do not differ much from each other. An increase in the average diameter of the dispersoid particles with increasing homogenization time towards a constant value is also observed for the alloy variants with different Zr contents. Fig. 34 (b) presents the effect of Zr content on the number density of the Zr-containing dispersoids homogenized at 470 °C. It is clear that the Zr-content has a strong effect on the number density of the dispersoids. The number density of the dispersoids for the alloy with a Zr content of 0.2 wt.% is almost two times as much as that in the alloy with a Zr-content of 0.13 wt.%.

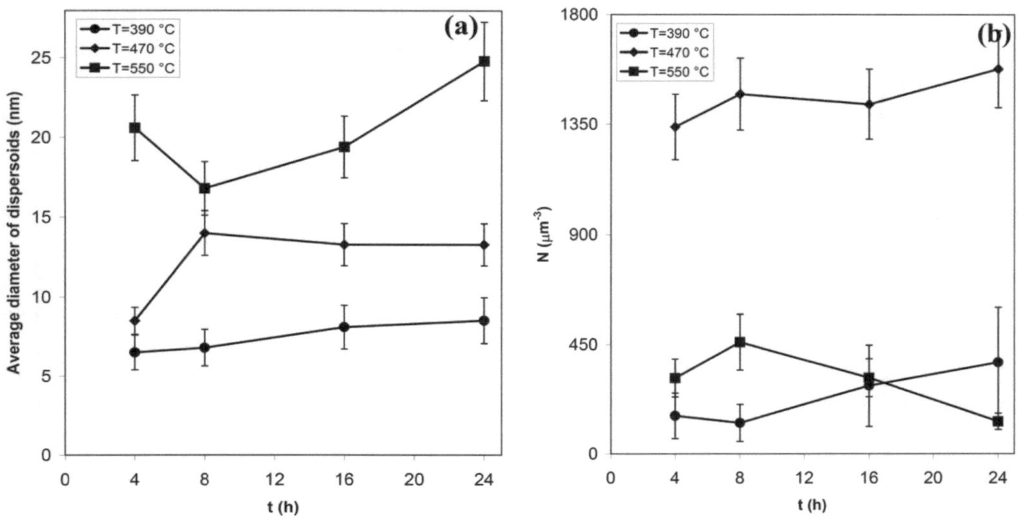

Fig. 33. Effect of homogenization time on the (a) average diameter and (b) number density of the Zr-containing dispersoids in the alloy variant N2 homogenized at different temperatures [62, 69]

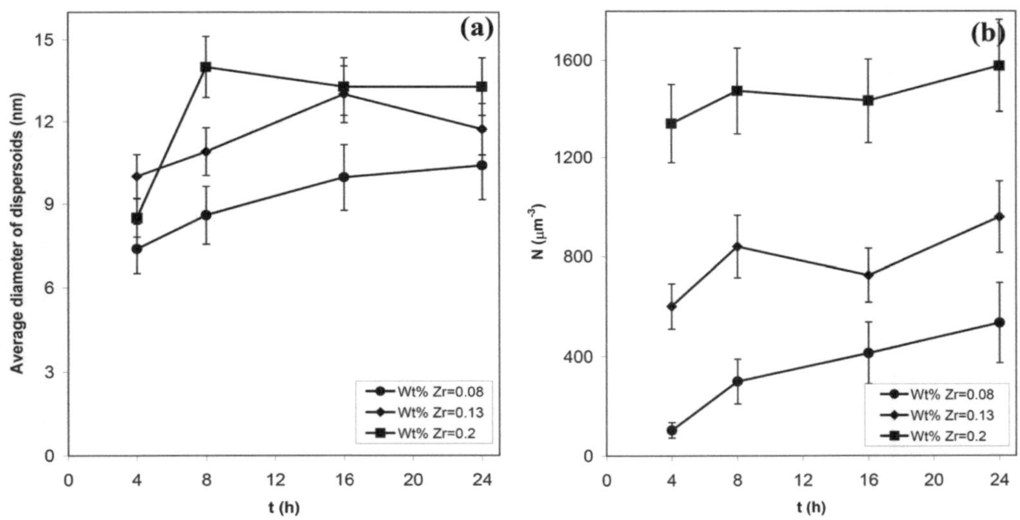

Fig. 34. Effect of Zr content on the (a) average diameter and (b) number density of dispersoids formed at 470 °C as a function of homogenization time [62, 69]

7. Conclusions

Main particles detected in the as-cast microstructure of AA7020 aluminum alloys were categorized to be grain boundary ones, low melting point particles and dispersoids. The evolution of these particles during the homogenization treatment of the AA7020 aluminum alloy was quantitatively analyzed and the following conclusions have been drawn.

1) The particles distributed along the grain boundaries which constitute more than 70% of the secondary phases present in the as-cast structure of the AA7020 aluminum alloy are $Al_{17}(Fe_{3.2},Mn_{0.8})Si_2$ particles. The low melting point phases are indeed present in the as-cast microstructure of the AA7020 aluminum alloy, which may cause incipient melting at 576 °C. These phases contain Al-Cu-Mg-Zn and dissolve during homogenization at 550 °C for 2 h.
2) The width of the grain boundary particles remains unchanged during homogenization at low temperatures. It however decreases at higher temperatures. The extent of the dissolution is more dependent on homogenization temperature than on time. The evolution mechanisms of the GB particles during homogenization consist of spheroidization during homogenization at low temperatures and thinning, discontinuation and full dissolution (TDFD) at high temperatures.
3) Four different types of dispersoids are formed in the AA7020 aluminum alloy variants during homogenization. In addition to the well-known Al_3Zr dispersoids, three other types of dispersoids are also present in the homogenized microstructure of the AA7020 aluminum alloy. The number densities of Zr- and Cr-containing dispersoids are large in the grain interior and very small in the grain boundary regions. These two types of dispersoids appear to be fully spherical and are formed at all the homogenization conditions. The Mn-containing dispersoids form only when the homogenization temperature is equal to or higher than 510 °C and holding time longer than 4 h. The number density of these dispersoids is close to zero in the grain interior but becomes high in the grain boundary regions. The number density and sizes of the Zr-containing dispersoids increase with increasing Zr content of the alloy and homogenization time.

8. Acknowledgments

This research was carried out under the project number MC 4.04203 in the framework of the Research Program of the Materials innovation institute M2i (www.m2i.nl). The authors acknowledge Mr. K. Kwakernaak and Mr. E.R. Peekstok for assistance in microstructure examination.

9. References

[1] ASM Handbook, 10th ed., vol. 14, Metal Forming, ASM International, Metals Park, OH (1992).
[2] T. Sheppard, *Extrusion of Aluminum Alloys*, Kluwer Academic Publishers, Dordrecht (1999).
[3] A.R. Eivani, Modeling of Microstructural Evolution during Homogenization and Simulation of Transient State Recrystallization leading to Peripheral Coarse Grain Structure in Extruded Al-4.5Zn-1Mg Alloy, PhD Thesis, June 2010, Delft, The Netherlands.
[4] N.A. Belov, D.G. Eskin and A.A. Aksenov, *Multicomponent Phase Diagrams: Applications for Commercial Aluminum Alloys*, Elsevier Science, New York (2005).
[5] L.L. Rokhlin, T.V. Dobatkina, N.R. Bochvar and E.V. Lysova, *J. Alloys Compd.* 367, 10 (2004).
[6] S.T. Lim, Y.Y. Lee and I.S. Eun, *Mater. Sci. Forum* 519-521, 549 (2006).
[7] C. Mondal and A.K. Mukhopadhyay, *Mater. Sci. Eng. A* 391, 367 (2005).
[8] R.K. Gupta, N. Nayan and B.R. Ghosh, *Cana. Metall. Q.* 45, 347 (2006).

[9] H. Ahmed, A.R. Eivani, J. Zhou and J. Duszczyk, Proc. Symp. *Aluminum Alloys: Fabrication, Characterization and Application*, TMS Annual Meeting, New Orleans (2008).
[10] N. Parson, and C. Jowett, Proc. 3rd Australasian Pacific Aluminium Extrusion Conference, Sydney, Australlia, (2005).
[11] N. Parson, S. Barker, A. Shalanski and C. Jowett, Proc. 8th Int. Aluminum Extrusion Technology Seminar, vol. 1 (2004).
[12] A.R. Eivani, H. Ahmed, J. Zhou and J. Duszczyk, Proc. Symp. *Aluminum Alloys: Fabrication, Characterization and Application*, TMS Annual Meeting, New Orleans (2008).
[13] J.W. Martin, *Precipitation Hardening*, 2nd ed., Butterworth-Heinemann, Oxford (1998).
[14] A. Jackson and T. Sheppard, Proc. 6th Int. Aluminum Extrusion Technology Conf., Chicago, Aluminum association, Washington DC, vol. 1 (1996).
[15] J.D. Robson and P.B. Prangnell, *Acta Mater.*, 49, 599 (2001).
[16] G.T. Hahn and A.R. Rosenfield, *Metall. Trans. A*, 6, 653 (1975).
[17] D.S. Thompson, *Metall. Trans. A*, 6, 671 (1975).
[18] G.G. Garrett and J.F. Knott, *Metall. Trans. A*, 9, 1187 (1978).
[19] F.Y. Xie, T. Kraft, Y. Zuo, C.H. Moon and Y.A. Chang, *Acta Mater.*, 47, 489 (1999).
[20] B. Morere, C. Maurice and R. Shahani, J. Driver, *Metall. Mater. Trans. A* 32, 625 (2001).
[21] E. Clouet, A. Barbu, L. Lae and G. Martin, *Acta Mater.* 53, 2313 (2005).
[22] Z. Jia, G. Hua, B. Forbord and J.K. Solberg, *Mater. Sci. Eng. A* 444, 284 (2007).
[23] B.L. Ou, J.G. Yang and M.Y. Wei, *Metall. Mater. Trans. A* 38, 1760 (2007).
[24] J.D. Robson, M.J. Jones and P.B. Prangnell, *Acta Mater.* 51, 1453 (2003).
[25] J.D. Robson, *Acta Mater.* 52, 1409 (2004).
[26] J.D. Robson, *Acta Mater.* 52, 4669 (2004).
[27] O.N. Senkov, R.B. Bhat, S.V. Senkova and J. Tatalovich, *Mater. Forum* 28 (2004).
[28] X. Fan, D. Jiang, Q. Meng and L. Zhong, *Mater. Lett.* 60, 1475 (2006).
[29] R. Ciach and B. Dukiet-Zawadzka, *J. Mater. Sci.* 13, 2676 (1978).
[30] F.J. Humphreys and M. Hatherly, *Recrystallization and Related Annealing Phenomena*, 3rd ed., Elsevier Science Inc., Oxford (1995).
[31] M. Peters, J. Eschweiler and K. Welpmann, *Scripta Metall.* 20, 259 (1986).
[32] O. Daaland and E. Nes, *Acta Mater.* 44, 1413 (1996).
[33] Y.J. Li and L. Arnberg, *Acta Mater.* 51, 3415 (2003).
[34] L. Lodgaard and N. Ryum, *Mater. Sci. Eng. A* 283, 144 (2000).
[35] M. Cabibbo, E. Evangelista, C. Scalabroni and E. Bonetti, *Mater. Sci. Forum* 503-504, 841 (2006).
[36] L. Lodgaard and N. Ryum., *Mater. Sci. Tech.* 16, 599 (2000).
[37] D.H. Lee, J.H. Park and S.W. Nam, *Mater. Sci. Tech.* 15, 450 (1999).
[38] R.A. Jeniski, B. Thanaboonsombut and T.H. Sanders, *Metal. Mater. Trans. A* 27, 19 (1996).
[39] J.D. Robson, *Mater. Sci. Eng. A* 338, 219 (2002).
[40] S.V. Senkova, O.N. Senkov and D.B. Miracle, *Metal. Mater. Trans. A* 37, 3569 (2006).
[41] O.N. Senkov, R.B. Bhat, S.V. Senkova and J.D. Schloz, *Metal. Mater. Trans. A* 36, 2115 (2005).
[42] A. Deschamps and Y. Bréchet, *Mater. Sci. Eng. A* 251, 200 (1998).
[43] J.T. Armstrong, in K.F.J. Heinrich, D.E. Newbury (Eds), *Electron Probe Quantitation*, Plenum Press, New York (1991).
[44] B.D. Cullity, S.R. Stock and S. Stock, *Elements of X-Ray Diffraction*, 3rd ed., Prentice Hall, New Jersi (2001).

[45] Alphabetical indexes for experimental patterns, Sets 1-52, International Center for Diffraction Data (ICDD), 2005, Powder Diffraction File Number (PDF no.) 01-071-4015.
[46] M. Cooper, *Acta Crystallogr.* 23, 1106 (1967).
[47] ASM Handbook, 10th ed., vol. 2, *Non-Ferrous Alloys*, ASM International, Metals Park, OH (1992).
[48] R.T. DeHoff and F.N. Rhines, *Quantitative Microscopy*, McGraw-Hill, New York (1968).
[49] A.R. Eivani, H. Ahmed, J. Zhou, J. Duszczyk, Metal. Mater. Trans. A. 40, 717 (2009).
[50] R.W. Cahn and P. Haasen, *Physical Metallurgy*, North Holland, Amsterdam (1996).
[51] D. Gaskell, *Introduction to the Thermodynamics of Materials*, Taylor & Francis Co., New York (2003).
[52] R. Nadella, D.G. Eskin, Q. Du and L. Katgerman, *Prog. Mater. Sci.* 53, 421 (2008).
[53] A.R. Eivani, H. Ahmed, J. Zhou, J. Duszczyk, Mater. Sci. Tech. 26, 215 (2010).
[54] M. Dumont, W. Lefebvre, B. Doisneau-Cottignies and A. Deschamps, *Acta Mater.* 53, 2881 (2005).
[55] G. Sha and A. Cerezo, *Acta Mater.* 52, 4503 (2004).
[56] A. Melander and P.A. Persson, *Acta Metall.* 26, 267 (1978).
[57] ASM Handbook, 10th ed., vol. 3, *Alloy Phase Diagrams*, ASM International, Metals Park, OH (1992).
[58] M. Conserva, E. Di Russo and A. Giarda, *J. Metall.* 6, 367 (1973).
[59] E. Ho and G.C. Weatherly, *Acta Metall.* 23, 1451 (1975).
[60] A.R. Eivani, H. Ahmed, J. Zhou, J. Duszczyk, Phil. Mag. 42, 1109 (2010).
[61] M. S. Zedalis and M. E. Fine, *Metall. Trans. A.* 17, 2187 (1986).
[62] A.R. Eivani, H. Ahmed, J. Zhou, J. Duszczyk, C. Kwakernaak, Mater. Sci. Tech. DOI 10.1179/026708310X12635619988267 (2010).
[63] A.R. Eivani, H. Ahmed, J. Zhou, J. Duszczyk, Mater. Sci. Eng. A. 527, 2418 (2010).
[64] A.R. Eivani, S. Valipour, H. Ahmed, J. Zhou, J. Duszczyk, Metal. Mater. Trans. A. 40, 2435 (2009).
[65] J. Murray, A. Peruzzi and J.P. Abriata, *J. Phase Equilib.* 13, 227 (1992).
[66] N. Saunders, *Z. Metallkd* 80, 894 (1989).
[67] Y. Du, Y.A. Chang, B. Huang, W. Gong, J. Jin, H. Xu, Z. Yuan, Y. Liu, Y. He and F.Y. Xie, *Mater. Sci. Eng. A* 363, 140 (2003).
[68] A.R. Eivani, S. Valipour, H. Ahmed, J. Zhou, J. Duszczyk, Metal. Mater. Trans. A. 42, 1109 (2011).
[69] A.R. Eivani, H. Ahmed, J. Zhou, J. Duszczyk, Adv. Mater. Res. 89-91, 177 (2010).